S0-BKS-934

ADVANCES IN LASER SCIENCE–III

AMERICAN INSTITUTE OF PHYSICS
CONFERENCE PROCEEDINGS NO. **172**
NEW YORK 1988

OPTICAL SCIENCE AND ENGINEERING SERIES 9

SERIES EDITOR: RITA G. LERNER

ADVANCES IN LASER SCIENCE–III

PROCEEDINGS OF THE THIRD INTERNATIONAL LASER SCIENCE CONFERENCE

ATLANTIC CITY, NJ 1987

EDITORS:

ANDREW C. TAM
IBM ALMADEN RESEARCH CENTER

JAMES L. GOLE
GEORGIA INSTITUTE OF TECHNOLOGY

WILLIAM C. STWALLEY
UNIVERSITY OF IOWA

L.C. Catalog Card No. 88-71879
ISBN 0-88318-372-2
DOE CONF-871147

Printed in the United States of America.

Contents

I. C. New Short Wavelength Sources

I. D. Novel Lasers and Studies

II. Nonlinear Optical Phenomena and Applications

II. A. Nonlinear Optics in Atomic and Molecular Gases

II. B. Nonlinear Optics of Semiconductors

II. C. Nonlinear Optics in Thin Films and Solids

II. D. Nonlinear Optical Phase Conjugation

II. E. Dephasing-Induced Processes in Nonlinear Optics

III. Atomic, Molecular, and Ionic Spectroscopy

III. A. Laser Cooling and Trapping of Particles

III. B. Multiple Excitation or Ionization of Atoms and Molecules

III. C. Atomic and Molecular Spectroscopy

IV. Condensed Matter and Surface Spectroscopy

IV. A. LASER SPECTROSCOPY OF SOLIDS

IV. B. NOVEL OPTICAL STUDIES OF SURFACES

IV. C. SOLITONS, LASER PROPAGATION, AND PARTICULATE INTERACTION

V. Photochemistry, Photophysics, and Photobiology

V. A. Photochemistry, Photofragmentation, and Photophysics

V. B. Energy Transfer and State-Resolved Processes

V. C. Laser Photobiology

V. D. Ultrafast Laser Spectroscopy and Dynamics

VI. Diagnostic and Analytical Applications

VI. A. Photothermal Spectroscopy

VI. B. High-Temperature Diagnostics

PREFACE

This book is the third volume of the Proceedings of papers presented at the International Laser Science (ILS) Conferences. The annual ILS Conferences have been established to survey the core laser science areas including lasers and their properties, spectroscopy, laser-induced processes, and a selection of laser applications. This meeting is a Topical Conference of the American Physical Society (APS) and is also the annual meeting of the APS Topical Group on Laser Science. Earlier ILS Conferences were held in Dallas in 1985 (ILS–I) and in Seattle in 1986 (ILS–II). ILS–III, from which the current volume is collected, was held in Atlantic City, New Jersey, November 1–4, 1987, with the co-sponsorship of the Optical Society of America. The next Conference, ILS–IV, will take place in Atlanta in October 1988, in cooperation with the 1988 American Vacuum Society (AVS) Annual Meeting. Future ILS Conferences (e.g., ILS–V to meet at Stanford University in August 1989) are planned in coordination with those of related scientific societies.

ILS–III was characterized by an outstanding technical content, which at the same time maintained an open APS style which encouraged active participation from a broad range of laser scientists. The ILS–III Conference began auspiciously with three opening keynote lectures by the 1986 Nobel Laureates in Chemistry: Dudley R. Herschbach (Harvard University), Yuan T. Lee (University of California at Berkeley), and John C. Polanyi (University of Toronto). These scientists were honored by the Royal Swedish Academy of Sciences for their research in chemical physics, specifically reaction dynamics, which is basic to a detailed understanding of chemical reactions. The Nobel citation acknowledges the development of two important techniques for elucidating what happens when molecules collide and atoms rearrange themselves to form new molecules. These techniques, crossed molecular beams and infrared chemiluminescence, provide a basis for the determination of the potential energy surfaces associated with specific reactions. Furthermore, these techniques are now in extensive use coupled strongly with recent advances in laser technology as a means of probing the chemical reactions of ground and (laser-prepared) excited state molecules, interrogating product states by laser spectroscopy or ionization, and generating new coherent light sources. Hence, the Program Committee of ILS–III felt that it was very appropriate to have the Conference begin with keynote lectures by these Nobel Laureates. The abstracts of these keynote lectures, as well as brief biographies of the Laureates, are recorded in the beginning of this book.

There were many other significant contributions, plenary papers, invited papers and contributed papers (both oral and poster) that formed the framework of the ILS–III Conference. We also note the following special events:

- An evening Panel Discussion Meeting entitled, "Dye Lasers: The State-of-the-Art," organized by Michael G. Littman, Princeton University;
- An enlightening post-Banquet speech by John N. Howard, Editor of Applied Optics, on "The Scientific Contributions of Lord Rayleigh."

Plenary lectures were delivered at the beginning of each half-day session (typically with five parallel sessions). The plenary lecturers who spoke at ILS–III were R. N. Zare, J. P. Taran, R. L. Byer, B. P. Stoicheff, G. W. Flynn, and R. C. Powell. These plenary lectures and many of the invited and contributed papers presented in parallel sessions and in the poster and postdeadline sessions, are covered in this volume.

This Proceedings volume, as with earlier volumes, is not intended to replace the publication of new and complete research results in those technical journals that cover the field. Rather, the brief accounts of research summarized here are intended to provide summarized descriptions of the research topic chosen, with sufficient description (aided by a listing of the key references) to enable the reader to understand the basis of what has been accomplished and where to find more detailed information. In some instances historical perspectives and thoughts on future directions are also presented.

The content of this volume is organized in a fashion designed for subsequent use rather than as the format of a technical digest provided at many conferences; the organization follows that of the general areas covered in the Conference rather than the order of presentation at the meeting. Also, to ensure quality and readability, each paper in this volume has been reviewed by at least one referee.

Many people have contributed extensively to ILS–III and its Proceedings in addition to the undersigned editors. Marshall Lapp, Conference Co-Chair, has provided vitality and leadership throughout the series of ILS Conferences. The many program organizers, who formulated the areas of laser science to be covered and the actual sessions in each area, contributed extensive efforts to arrange the key invited speakers from all over the world; their names are acknowledged in the ILS–III Conference program printed in Bull. Am. Phys. Soc. **32**(No. 8) (1987) and in Optics News **13**(No. 10) (1987). International invitations are frequently possible only because of the efforts of Rolf Gross, the ILS–III International Vice-Chair. Special thanks are due to Lynn Borders, ILS–III Administrative Assistant, who did an amazing job of handling the administrative details of the Conference and the Proceedings; her efforts were essential for the success of the Conference. Financial support for ILS–III was provided by the following organizations: the Air Force Office of Scientific Research, the Office of Naval Research, the National Science Foundation, the Petroleum Research Fund of the American Chemical Society, the University of Iowa, and IBM Research Division (Almaden Research Center).

The Editors:
Andrew C. Tam, Program Chair
IBM Almaden Research Center
James L. Gole, Program Vice-Chair
Georgia Institute of Technology
William C. Stwalley, Conference Chair
University of Iowa

ABSTRACTS OF KEYNOTE ADDRESSES BY DUDLEY R. HERSCHBACH, YUAN T. LEE, AND JOHN C. POLANYI

Dudley Herschbach received his BS in mathematics in 1954 and his MS in chemistry in 1955, both from Stanford. He received a second master's degree in physics in 1956 and his PhD in chemical physics in 1958, both from Harvard. Since then, Herschbach has held positions at Harvard, except for the period from 1959 to 1963, when he was an assistant and an associate professor at Berkeley. He has been Baird Professor of Science at Harvard since 1976. Currently he is doing theoretical calculations of electron configurations as well as experiments using coincidence measurements of velocity and rotational angular momentum vectors to undo averaging over initial impact parameters and molecular orientations.

VIBRATIONAL DYNAMICS AND REACTIVITY OF CROWDED MOLECULES

Dudley R. Herschbach
Harvard University
Department of Chemistry
Cambridge, Massachusetts 02138

Laser spectroscopy greatly facilitates probing condensed-phase structure and reactivity, particularly in the nanoliter volume of diamond-anvil high pressure cells. Vibrational frequency shifts and line broadening induced by pressure reveal distinctive contributions from repulsive and attractive solute-solvent forces. Illustrative data and analysis are presented for iodine, pyridine, and other solutes in benign solvents. In the liquid range, typically extending up to 10-12 kbar, bond length changes are usually only of the order 0.02%; thus molecules are only slightly distorted and a perturbation treatment is appropriate. For quasidiatomic modes, several observed features are consistent with model calculations employing approximations familiar for gas-phase vibration-to-translation energy transfer. At higher pressures, typically in the solid phase above 30 kbar, chemical bonds begin to weaken due to repulsions caused by overlap of electron clouds of neighboring molecules. High pressure thus acts like a catalyst in promoting chemical reactions at low temperatures.

Yuan Lee received his BS in 1959 from National Taiwan University and his MS in 1961 from National Tsing Hua University (also in Taiwan). He received his PhD in chemistry from Berkeley in 1965. He carried out postdoctoral research beginning in 1965 at both Lawrence Berkeley Laboratory and Harvard prior to his joining the chemistry faculty at the University of Chicago. Since 1974 he has been a professor of chemistry at Berkeley and a principal investigator with the Materials and Molecular Research Division at Lawrence Berkeley. In recent years, Lee and his group, using seven molecular beam machines in the laboratory, have been leaders in studying the chemical reactions of large organic molecules such as those significant for combustion chemistry and atmospheric chemistry.

MOLECULAR BEAM STUDIES ON THE DYNAMICS OF PRIMARY PHOTODISSOCIATION PROCESSES

Yuan T. Lee
University of California
Department of Chemistry
Berkeley, California 94720

Application of the method of molecular beam photofragmentation translational spectroscopy to the investigation of primary photochemical processes of polyatomic molecules will be the main theme of this lecture. Concerted unimolecular dissociation processes, especially those that dissociate simultaneously into three products will be given as examples to illustrate how information concerning the dynamics and mechanism of dissociation processes can be obtained from precise measurements of angular and velocity distributions of products in an experiment in which a well defined beam of molecules is crossed with a laser. Recent studies on the primary and secondary decompositions of RDX is shown to be dominated by concerted dissociation processes.

* This work was supported by the Office of Naval Research under Contract Number N00014-83-K-0069.

John Polanyi received his BSc (1949), MSc (1950) and PhD (1952) in chemistry from the University of Manchester. After positions at the National Research Council (Ottawa) and at Princeton, Polanyi moved in 1956 to the University of Toronto, where since 1974 he has held the position of University Professor of Chemistry. He says of his group that "our current major interest these days is to induce reactions at sub-monolayer coverages on surfaces - single crystal surfaces. We are trying to move our reaction dynamics from the three-dimensional world of gas to the two-dimensional world of the adsorbed state".

PHOTO-DISSOCIATION, -REACTION, -DESORPTION AND -EJECTION OF ADSORBED SPECIES

John C. Polanyi
University of Toronto
Department of Chemistry
Toronto, Ontario, Canada M5S 1A7

Recent work, performed in this laboratory, on the effects of mid-uv radiation in giving rise to single-photon photo-excitation of adsorbates will be reviewed. Adsorbates included CH_3Br[1], H_2S[1,2] and OCS, at sub-monolayer coverage on clean LiF(001) crystals at 115K in UHV. The surface was irradiated by pulsed excimer radiation, principally at 193, 222 and 308 nm, photo-fragments and parent molecules leaving the surface energy distribution by time-of-flight mass spectrometry. The effects identified include *photo-fragmentation* of molecules held at the surface with resultant modified molecular dynamics and cross section, *surface aligned photo-reaction (SAP)* between co-adsorbed species, as well as, *photo-desorption* (detachment of slow-moving parent molecules) and *photo-ejection* (detachment of fast-moving parent molecules). Mechanisms and implications of some of these novel processes will be discussed.

1. E. B. D. Bourdon, J. P. Cowin, I. Harrison, J. C. Polanyi, J. Segner, C. D. Stanners and P. A. Young, J. Phys. Chem. 88, 6100 (1984).

2. E. B. D. Bourdon, P. Das, I. Harrison, J. C. Polanyi, J. Segner, C. D. Stanners, R. J. Williams and P. A. Young, Faraday Discuss. Chem. Soc. 82 (1986).

I. LASER SOURCES AND NOVEL LASER STUDIES

SOLID STATE LASERS - THE NEXT 10 YEARS

Robert L. Byer
Stanford University
Applied Physics Laboratory
Stanford, California 94305

Major advances in solid state laser technology historically have been preceded by advances in pumping technology. The helical lamp used to pump the early ruby lasers was superseded by the linear flashlamp now used to pump Nd:YAG lasers. The next advance in pumping technology is the diode laser array. The improvements in power and efficiency of the diode laser coupled with the fortuitous spectral overlap of the diode laser emission wavelength with the Nd ion absorption bands near 805 nm have led to a revolution in solid state laser capability. Progress has been rapid with new ions and wavelengths reported in the near infrared from 946 nm to 2010 nm. Frequency extension via nonlinear interactions has led to green and blue sources of coherent radiation. Linewidths of less than 10 kHz have been demonstrated. Overall electrical efficiencies of greater than 10% have been achieved. As diode laser sources decrease in cost, high average power diode laser pumped solid state laser sources will become available. Power levels exceeding 1 kW appear possible. Potential applications of these compact all solid state laser sources to spectroscopy, quantum noise limited sensors, astronomy, and materials processing will be discussed.

LASER GENERATION OF VUV RADIATION AND SPECTRA OF RARE GAS EXCIMERS

Boris P. Stoicheff
University of Toronto
Department of Physics
Toronto, Ontario, Canada M5S 1A7

Tunable and coherent vacuum ultraviolet radiation is now available over broad regions from 200 to ~60 nm. It is generated by third harmonic and four-wave frequency mixing in gases and metal vapors, and is of sufficient intensity for absorption and fluorescence spectroscopy. We have recently applied these sources to studies of Xe_2, Kr_2, and Ar_2 produced by supersonic jet expansion. Vibronic spectra of Xe_2 and rovibronic spectra of Kr_2 and Ar_2 arising from transitions between the ground and three lowest (stable) excited states have been recorded at resolving powers of $1 - 5 \times 10^5$. Spectroscopic constants have been derived, potential energy curves computed, and radiative lifetimes measured and compared with theory.

GREEN INFRARED-PUMPED ERBIUM UPCONVERSION LASERS

W. Lenth, A. J. Silversmith, and R. M. Macfarlane
IBM Research Division, Almaden Research Center, San Jose, CA 95120

ABSTRACT

Upconversion pumping with near-infrared cw dye lasers was used to achieve laser action at 550 nm on the $^4S_{3/2} \rightarrow {}^4I_{15/2}$ transition of Er^{3+}. In $YAlO_3:Er^{3+}$, upconversion excitation occurs via sequential two-photon absorption. In $YLiF_4:Er^{3+}$, energy transfer upconversion involving two Er^{3+} ions in the intermediate $^4I_{11/2}$ state causes excitation of the upper laser level. These upconversion lasers operate at temperatures up to 90 K. With pump powers of several hundred milliwatt near 800 nm we have obtained 550-nm output powers in the milliwatt range.

INTRODUCTION

There is currently much interest in the development of visible solid-state laser sources that are pumped by infrared semiconductor diode lasers. One approach that has been demonstrated recently uses the nonlinear optical susceptibility of noncentro-symmetric crystals for intra-cavity second harmonic generation[1] or sum-frequency mixing[2] in Nd:YAG lasers pumped by GaAlAs diode laser arrays. We have investigated another approach in which nonlinear pumping processes based on two-photon absorption are used to excite green emission with near-infrared pump lasers. Laser operation at 550 nm was obtained in $YAlO_3:Er^{3+}$ and $YLiF_4:Er^{3+}$ crystals at temperatures up to 77 K and 90 K, respectively.[3,4]

Upconversion pumping is possible in laser materials that exhibit metastable intermediate energy levels which can be populated efficiently. Subsequent excitation of higher lying states that emit visible light can occur via energy transfer processes involving two excited ions in the intermediate state or by absorption of a second photon from the intermediate level. Several trivalent rare earth ions, such as Er^{3+}, Ho^{3+}, Tm^{3+}, and Nd^{3+}, have energy level systems that offer the potential for upconversion laser operation at red, green and blue wavelengths. The generation of incoherent visible light based on upconversion excitation of rare earth materials has been studied extensively in the past.[5] Stimulated emission by upconversion excitation using filtered flashlamp radiation at 77 K was reported in 1971 by Johnson and Guggenheim at wavelengths of 670.9 nm and 551.5 nm in BaY_2F_8 doubly doped with Yb^{3+} and Er^{3+}, and Yb^{3+} and Ho^{3+}, respectively.[6] Two successive energy transfer processes from the optically pumped Yb^{3+} ions to the Er^{3+} or Ho^{3+} laser ions were used as the dominant excitation mechanism for the visible emission. In our work, the use of pump lasers resonantly tuned to the spectrally narrow transitions typical of rare earth ions, as well as the use of longitudinal pumping arrangements permitted efficient cw excitation of upconversion lasers.

UPCONVERSION LASER OPERATION

We have used longitudinal pumping with cw dye lasers at wavelengths around 800 nm for upconversion excitation of the $^4S_{3/2}$ level of Er^{3+} (see Fig. 1). The $YAlO_3:Er^{3+}$ and $YLiF_4:Er^{3+}$ laser crystals were 0.3 cm long and mirror coatings were directly deposited onto the spherically polished faces (r = 2 cm). The crystals were placed inside a helium exchange gas cryostat, which permitted continuous variation of the temperature between 10 and 300 K. The laser transitions terminate at a Stark component of the erbium $^4I_{15/2}$ manifold at 218 cm^{-1} ($YAlO_3$) and 252 cm^{-1}

($YLiF_4$) above the ground state. Cooling of the laser crystals reduces the population of the lower laser level so that reabsorption losses become insignificant.

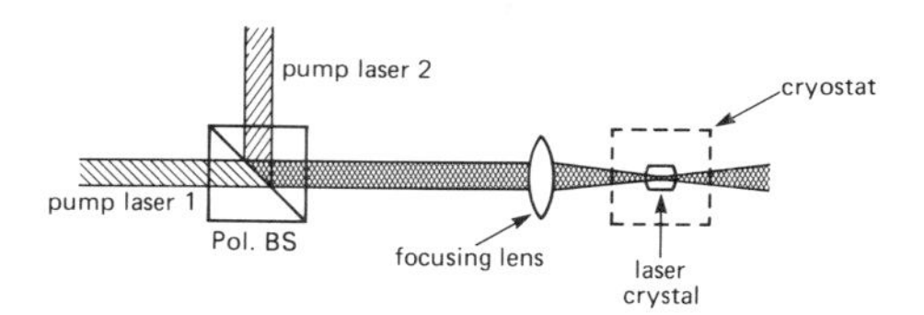

Figure 2. Experimental arrangement for the two-photon pumped upconversion laser. The two pump lasers are orthogonally polarized and combined using a polarizing beam splitter.

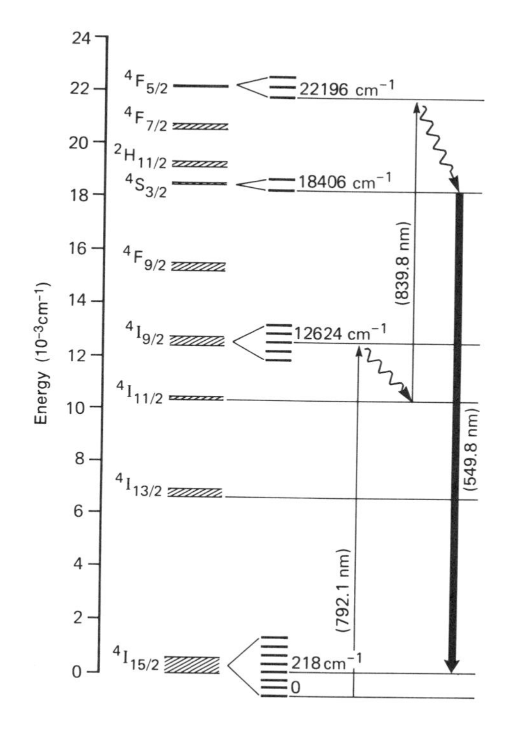

Figure 1. Energy level diagram of $YAlO_3:Er^{3+}$ showing the pumping and lasing transitions.

In the case of $YAlO_3:Er^{3+}$, sequential two-photon absorption was used for pumping and the relevant parts of the Er^{3+} energy level diagram and the excitation scheme are illustrated in Fig. 1. The first dye laser at 792.1 nm is tuned to the most effective absorption transition between the $^4I_{15/2}$ and $^4I_{9/2}$ multiplets. Subsequent non-radiative relaxation populates the $^4I_{11/2}$ (10,290 cm^{-1}) and $^4I_{13/2}$ (6,602 cm^{-1}) levels, which have long lifetimes of 1.2 ms and 5.3 ms, respectively.[7] The second pump laser is tuned to one of several possible absorption transitions from these intermediate metastable states to higher excited states above the $^4S_{3/2}$ upper laser level, which has a lifetime of 160 μs at temperatures <77 K.[7] The most efficient upconversion excitation was achieved when the second pump laser was tuned to 11,906 cm^{-1}, corresponding to the $^4I_{11/2}$ (10,290 cm^{-1}) $\rightarrow$ $^4F_{5/2}$ (22,196 cm^{-1}) transition. The light from the two pump lasers was combined using polarization coupling and focused onto the crystal (see Fig. 2). In addition to this sequential two-photon pumping scheme, phonon-assisted energy transfer involving two ions in the $^4I_{11/2}$ state results in upconversion excitation of green emission using only the first pump laser. However, the lasing threshold could not be reached by pumping with one laser alone. The upconversion laser operates at 549.6 nm and the laser transition terminates on the 218 cm^{-1} Stark component of the $^4I_{15/2}$ multiplet. This $^4S_{3/2} \rightarrow {}^4I_{15/2}$ transition has an emission cross section of 2×10^{-19} cm^2, which is 5-10 times larger than those for any of the other green emission lines in $YAlO_3:Er^{3+}$ with the exception of two transitions that have comparable cross section but terminate on the lowest Stark component and that at 51 cm^{-1}. At least at temperatures above 30 K, the laser transition at 549.6 nm seems to be homogeneously broadened due to a fast non-radiative relaxation time for the lower level at 218 cm^{-1}. The dependence of the 549.6 nm output power on pump power at 30 K is shown in Fig. 3. The absorption of the first laser at 792.1 nm was 30% and approximately 15% of the second laser was absorbed when the first laser provided 200 mW. With pump powers of approximately 200 mW from each laser, 1 mW of green cw output power was obtained using an output coupling of only 1%.

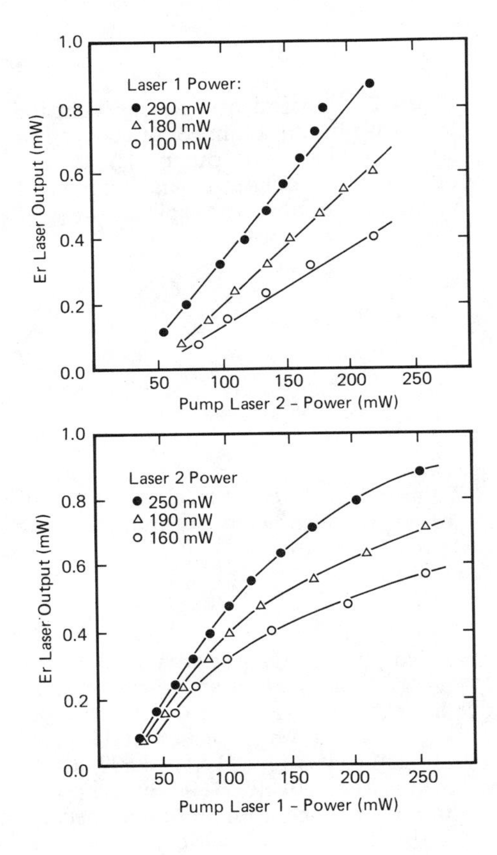

Figure 3. Output power of the $YAlO_3:Er^{3+}$ laser at 549.6 nm *versus* input power of one pump laser at 792.1 nm while the power of the other at 839.8 nm is held fixed. Pump laser power levels are measured before the focussing lens; and the power reaching the crystal is 70% of these values. Similarly, the output level is measured outside the cryostat and not corrected for the 80% transmission of the windows.

The laser operated in the fundamental TEM_{00} mode. Since the cavity mode spacing of the 0.3-cm long crystal was comparable to the spectral width of the emission line, only one longitudinal mode oscillated in spite of the spatial hole burning typical of solid-state lasers. The laser linewidth was measured to be 30 MHz using a scanning Fabry-Perot etalon with a free spectral range of 2 GHz. Lasing was obtained at temperatures up to 77 K. The laser performance was found to depend on temperature and the lowest pump threshold and highest output power were obtained at ~30 K. Due to the temperature dependence of the effective cavity length, the laser wavelength tuned with temperature over the 0.03-nm wide Er^{3+} emission line.[4] At ~30 K, the cavity mode frequency was matched to the spectral maximum of the laser gain.

The main reason for operating the 549.6 nm $YAlO_3:Er^{3+}$ laser at low temperatures is the need for depopulation of the lower laser level so that reabsorption losses are greatly reduced. In addition, spectral narrowing occurs, which leads to an increase of the peak emission and absorption cross sections. Another advantage of low temperatures is that the excited state populations in the intermediate metastable multiplet and the upper laser state reside predominantly in the lowest Stark component. At the pump power levels used for laser excitation, weak fluorescence from the $^2H_{9/2}$ (24,479 cm^{-1}) and $^2P_{3/2}$ (32,773 cm^{-1}) states was observed.[4] These levels can only be excited by cooperative energy transfer processes that involve Er^{3+} ions in the $^4S_{3/2}$ state. These processes constitute a depletion mechanism for the upper laser level. This observation illustrates that upconversion laser operation that involves the participation of many energy levels can be very complex.

In the case of $YLiF_4:Er^{3+}$ (1%), upconversion laser operation at 550.0 nm was readily obtained by pumping with only one dye laser near 800 nm. Several strong absorption transitions between the $^4I_{15/2}$ and $^4I_{9/2}$ manifolds could be used for excitation. Pumping on these transitions results in strong population of the metastable $^4I_{11/2}$ level, which has a lifetime of 2.9 ms. A very efficient energy transfer process leads to excitation of $^4F_{7/2}$ and subsequent nonradiative relaxation populates the $^4S_{3/2}$ upper laser level, whose lifetime is 0.4 ms (see Fig. 4). When the pump laser is turned off rapidly, the upconversion laser continues to oscillate for several hundred microseconds until the population in the intermediate $^4I_{11/2}$ state decays and excitation of the $^4S_{3/2}$ state via energy transfer stops.

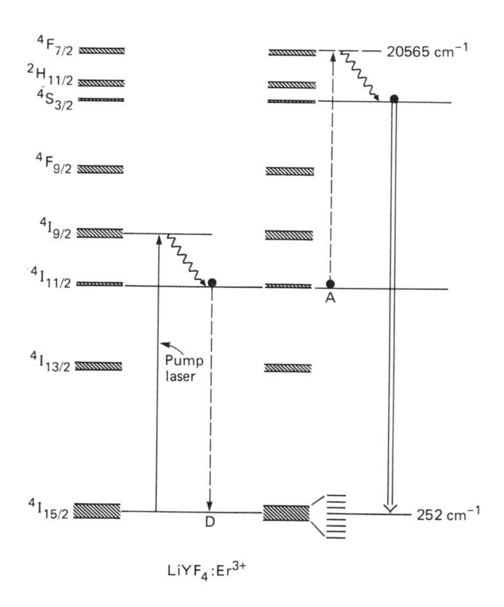

Figure 4. Energy level diagram of $YLiF_4:Er^{3+}$ and illustration of upconversion excitation by energy transfer.

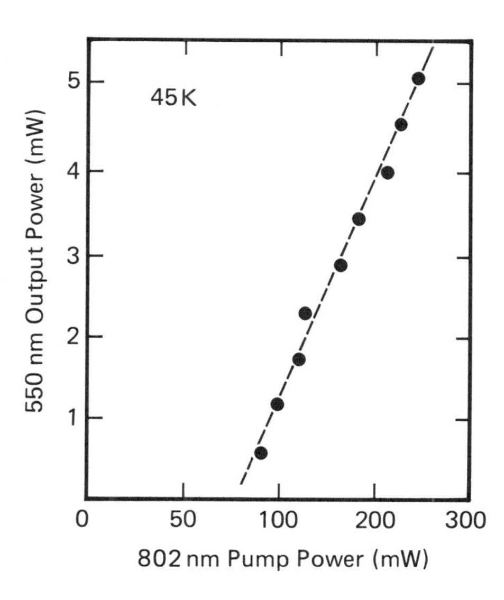

Figure 5. Output power of the $YLiF_4:Er^{3+}$ laser at 50 K as a function of cw dye laser pump power. The 550.0 nm output power is averaged over the semi-random train of laser pulses.

The dependence of the 550.0 nm output power on pump power is shown in Fig. 5. At 50 K, the laser threshold is reached with 80 mW pump power. With 240 mW of 802 nm power incident on the cryostat window a green output power of 5 mW was observed polarized perpendicular to the c-axis of the crystal. At 77 K, the output power decreased to approximately 1.5 mW. The output coupling of the mirror coating on the rear crystal facet was only 0.5%. Upconversion laser operation was obtained at temperatures up to 90 K. The laser performance was dependent on temperature and the optimum operation temperature was near 50 K. In spite of the nonlinear upconversion pumping mechanism, the laser power increased linearly with pump power. This is thought to arise from saturation effects that result from the strong population build-up in the intermediate metastable states and the associated depopulation of the $^4I_{15/2}$ ground state. Measurements of the laser light with a high-speed photodetector showed that the laser output consisted of a series of semi-random narrow pulses of 150-80 ns duration and an average repetition rate of 40-250 kHz for average laser output powers of 0.5-5 mW. Further investigations of the laser dynamics are needed to fully understand the origin of this spiking of the cw-pumped laser.

CONCLUSIONS

In conclusion, green laser emission has been obtained by two-photon upconversion excitation of two Er^{3+} doped crystals using cw pump lasers near 800 nm. In $YAlO_3:Er^{3+}$ sequential two-photon absorption was used for obtaining cw laser operation at 550 nm. In $YLiF_4:Er^{3+}$ energy transfer processes following excitation of a metastable intermediate level with a near-infrared cw dye laser resulted in efficient upconversion pumping of green laser emission. The photon energy of the upconversion lasers corresponds to approximately 70% of the sum of the energy of the pump photons. The upconversion pumping schemes depend critically on the lifetimes of the intermediate energy levels, the Er^{3+} concentration and the host material. Further investigations of these dependencies will permit optimization of upconversion laser materials and performance. The results obtained suggest that longitudinal pumping with infrared laser sources can be a practical approach for operating visible upconversion lasers.

REFERENCES

1. T. Baer and M. S. Keirstead, in Digest of Conference on Lasers and Electro-Optics (Optical Society of America, Washington, D.C., 1985), paper THZZ1.
2. W. P. Risk, J.-C. Baumert, G. C. Bjorklund, F. M. Schellenberg, and W. Lenth, Appl. Phys. Lett. 52, 85 (1988).
3. A. J. Silversmith, W. Lenth and R. M. Macfarlane, J. Opt. Soc. Am. 3, 128 (1986); Proc. of 1986 Annual Meeting of OSA, paper PDP12.
4. A. J. Silversmith, W. Lenth, R. M. Macfarlane, Appl. Phys. Lett. 51, 1977 (1987);
5. F. E. Auzel, Proc. IEEE 61, 758 (1973).
6. L. F. Johnson and H. J. Guggenheim, Appl. Phys. Lett. 19, 44 (1971).
7. M. J. Weber, Phys. Rev. B 8, 54 (1973).

MODELING THE DYNAMICS OF TITANIUM SAPPHIRE LASERS

A.M.Buoncristiani † and C.H.Bair
NASA Langley Research Center, Hampton, VA 23665

John Swetits † and Lila F.Roberts ††
Old Dominion University, Norfolk, VA 23508

ABSTRACT

We have developed a model of an end-pumped, injection seeded Titanium doped Sapphire ring laser. The model of the active region has two primary ingredients-continuity equations describing the temporal development of the right and left traveling waves through the medium and a set of rate equations describing the evolution of the electron population in a four level model of the vibronically broadened transitions between levels of the Titanium ion. The model describes the spatial-temporal behavior of the photon flux within the cavity. We report here on the effects of a reverse wave suppressor mirror on the onset of lasing.

INTRODUCTION

NASA's plans to develop laser systems for the active remote sensing of the atmosphere from space will require lasers with efficient and stable output at a number of specific wavelengths in the visible and infra-red regions. The performance requirements imposed on such lasers are severe; for example, the Ti:Sapphire laser proposed for remote sensing should operate with an energy of 1 Joule per pulse, at a repetition rate of 10 Hertz and with a laser linewidth of 1 picometer. In order to develope a stable laser capable of operating to such strict requirements it is necessary to have a clear understanding of the laser physics of the material. Furthermore, special techniques such as injection seeding may be needed to control and shape the output pulse. In order to gain some understanding of the physical characteristics of lasers using these new materials and to aid in the search for better materials we have developed a model for the operation of a tunable solid state laser. The laser modeled is a simple ring cavity, allowing end-pumping of the active medium and injection seeding through the output coupler. The model describes the evolution of events in time and along one spatial dimension. When it is necessary to take into account the transverse characteristics of the laser beam or the distribution of excitation in the laser medium an average over the transverse dimensions is taken to represent the value of that quantity at the axial point in question. We assume that no elements in the laser cause a shift in the phase of the signal apart from that due to propagation; we treat the amplitude of a single mode of the field (including polarization states).

MODEL

The Photon Equations. We consider a flux distribution of photons traveling in one dimension through a medium assumed to be homogeneous with refractive index n. The concentration of optically active ions within the medium is N_T and at any one time

---

† National Research Council Resident Research Asssociate
†† Supported by NASA under Training Grant NGT-47-003-804

t the population inversion at point x on the axis is given by n(x,t). The photon flux distribution at a point x in the medium and at a time t is described by the two functions: $F_{\pm}(x,t\,;\lambda)$. An accounting for the change in each of the photon fluxes is given by the continuity equations,

$$\frac{1}{v}\frac{\partial F_{\pm}}{\partial t} \pm \frac{\partial F_{\pm}}{\partial x} = (\sigma(\lambda)\, n - \alpha(\lambda))\, F_{\pm} + S_{\pm}(x,t\,;\lambda) \qquad (1)$$

The left hand side represents the net change in the corresponding flux as it moves through the material with speed $v = c/n$. The right hand side represent the individual contributions to the change in flux from its interaction with the medium. The terms in parentheses give the small signal gain; $\sigma(\lambda)$ is the emission cross-section and $\alpha(\lambda)$ is the loss coefficient in the medium (in cm^{-1}). The functions $S_{\pm}(x,t;\lambda)$ represent the contribution of spontaneous emission to the corresponding flux; they depend upon the emission characteristics of Titanium in Sapphire [1,2] and on the geometrical configuration in the cavity. The equations for photon flux propagation in the remainder of the cavity are the same as those above with the speed of light in the medium v replaced by the speed of light in vacuuo c and with the right hand sides set equal to zero.

The Rate Equations. In a four-level model of the lasing action of Titanium in Sapphire the concentration of electrons in the i th energy level at position x and time t is denoted by n_i (x,t). A narrow band pump excites electrons from the ground state to a level in the pump band; the electrons undergo a radiationless transition to the upper laser level.. The population densities of the lasing levels n_1 and n_2 are driven by stimulated emission and absorption. Each of the levels can be depleted by radiationless transitions. The rate equations giving the time dependence of the population densities for the upper and lower electron levels for this system are given below.

$$\frac{dn_2}{dt} = W_P - (n_2 - \frac{g_1}{g_2} n_1) \int d\lambda\, \sigma(\lambda)\, F(\lambda) - \frac{n_2}{\tau_2}$$

$$\frac{dn_1}{dt} = \frac{n_2}{\tau_{fl}} - (n_2 - \frac{g_1}{g_2} n_2) \int d\lambda\, \sigma(\lambda)\, F(\lambda) - \frac{n_1}{\tau_1} \qquad (2)$$

$F(\lambda) = \{ F_+ (x,t\,;\lambda) + F_- (x,t\,;\lambda)\}$ is the total photon flux at x and t. In these equations τ_i is the lifetime for the i th level, while τ_{ij}^{-1} is the rate for the transition from level i to level j. The total decay from level 2 has both radiative and non-radiative contributions so that the decay rates are related by

$$\tau_2^{-1} = \tau_{fl}^{-1} + \tau_{NR}^{-1}$$

where τ_{fl} is the fluorescent lifetime of the material and τ_{NR}^{-1} is the transition rate for non-radiative transitions. $W_P(x,t)$ is the rate at which the upper laser level is populated;

it depends on the character of the pump beam (strength and transverse profile), upon the absorption characteristics of the active medium and upon the branching ratio between the pump level and the upper lasing level. The population densities, n_i, for each of the laser levels are constrained by the relation

$$N_T = n_0 + n_1 + n_2$$

where N_T is the dopant concentration. The degeneracy of the ith electron level is g_i.

COMPUTATION

Equations (1) and (2) comprise the basis for the model. To these equations initial conditions and boundary conditions appropriate for the laser system under consideration are appended. Evaluation of the parameters involved are determined from measurements on Ti:Sapphire [1,2] and from the laser system under study. Equations (1) form a quasi-linear first order system of partial differential equations for the photon flux; these can be reduced, by standard techniques, to a pair of ordinary differential equations along the characteristic curves for the system. These characteristic curves turn out to be worldlines ($x \pm vt$) for photons circulating in the cavity. This fact provides the basis for an algorithm for the computation; we solve the system of equations by treating the photon flux as propagating along their respective worldlines. In order to study the effects of the strength of the injection seed we have taken an average over the spatial dimensions of the cavity to produce a set of rate equations. We have studied the quantitative behavior of this system in detail and the results of this study will appear elsewhere. A computer program calculates the time dependence of the spectral distribution for the photon flux intensity traveling in both directions in the cavity and the population of electronic levels within the crystal. Figure 1 shows the computed spectral distribution for one of the circulating waves as it appears at the output coupler.

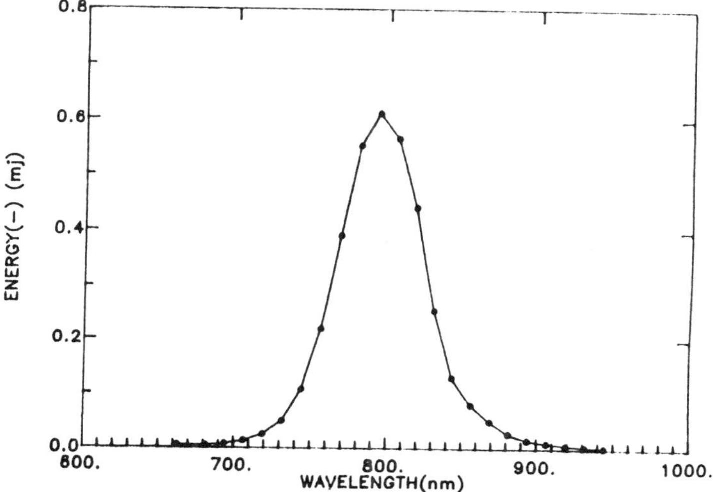

Figure 1. Computed spectral distribution of the ring laser output.

Part of our experimental study of injection seeding in Titanium doped Sapphire lasers included an examination of the effects of a reverse wave suppressor mirror (RWSM) external to a ring laser which is injection seeded by a laser diode [3]. As the reflectivity of the external mirror is increased the isolation (ratio of energy in the reverse wave to that in the forward wave) decreases. The isolation of the RWSM was calculated for various values of pump energy iabsorbed as a function of reverse wave mirror reflectivity and these calculations were compared to measured values of the isolation ; The results are displayed in Figure 2.

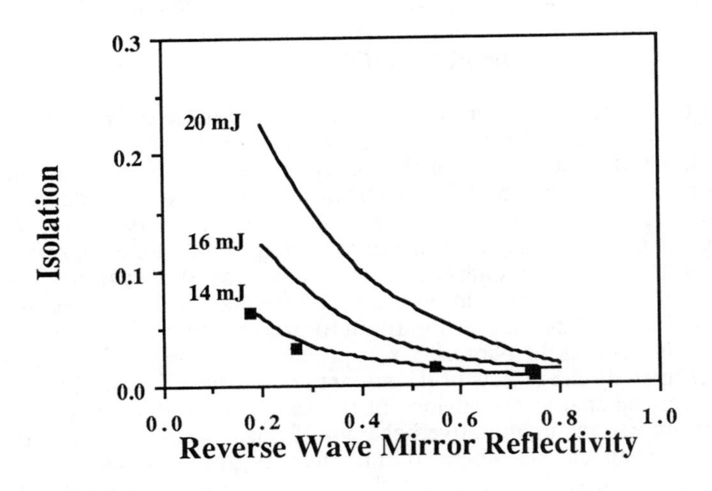

Figure 2. Computed values of the isolation in a ring laser with an external reverse wave suppressor mirror are shown as solid lines; measured values of the isolation are shown as squares.

REFERENCES

1. P.F.Moulton, Jour. Opt. Soc Am., Vol. 3, No. 1, p 125 (1986).

2. C.E.Byvik and A.M.Buoncristian, IEEE J. Quantum Electronics, Vol. QE-21, p 1619 (1985).

3. P.Brockman, C.H.Bair et.al., Optics Letters,Vol. 11, No. 11,p 712 (1986).

4. C.H.Bair, P.Brockman et.al., in Technical Digest of the Topical Meeting on Tunable Solid State Lasers Oct. 26-28, Williamsburg, VA, Series Vol. 20, p 67 (1987).

LASER DIODE PUMPED 1.07μ AND 535 nm Nd:BEL

Richard Scheps, Joseph Myers, and E.J. Schimitschek
Naval Ocean Systems Center, Code 843, San Diego, CA 92152

D.F. Heller
Allied Corporation, Corporate Technology Center, Morristown NJ 07960

ABSTRACT

Performance data for a laser diode pumped cw Nd:BEL laser is presented. Two phased laser diode arrays are used as the pump source, each emitting 500 mW. The heat sink for the arrays is temperature controlled to allow for wavelength tunability. A Nd:YAG rod was pumped under similar conditions and the results are compared. Although the absorption bandwidth for Nd:BEL is substantially broader than for Nd:YAG the Nd:BEL was found to have a higher threshold for lasing. Both rods had slope efficiencies of 42%. The dependence of the output power on output mirror reflectivity was measured, with Nd:BEL showing a greater sensitivity to reflectivity than Nd:YAG. The optimum reflectivities were found to be .98 and .97 for Nd:BEL and Nd:YAG respectively. The maximum TEM00 cw power achieved for each rod at these reflectivities was 250 mW for Nd:BEL and 283 mW for Nd:YAG. We conclude that under the conditions used in this work, both BEL and YAG hosts perform comparably.

INTRODUCTION

Of the various hosts for diode pumped Nd based lasers, BEL offers several unique advantages. These include a relatively broad spectral absorption bandwidth and the ability to be made athermal. In addition, the crystal growth and polishing technology of this material is relatively mature. Unlike some other hosts, the absorption bandwidth is centered near the Nd:YAG absorption band, so that in many cases the same diode array pumps can be used for both BEL and YAG hosts. In this work we present a comparative study of laser diode pumped cw Nd:BEL and Nd:YAG. We show that both hosts perform well, and specify under what conditions BEL might be the better host.

EXPERIMENT

The outputs of two 500 mW phased arrays were collimated and combined using a polarizing beamsplitter cube. The laser diodes are mounted on active heat sinks for wavelength control. The output spectrum of the laser diodes as measured with an OMA show 5 to 6 longitudinal modes operating simultaneously with a total bandwidth of about 1.5 nm. With both arrays operating at 500 mW, approximately 700 mW could be focussed onto one face of a 1 cm long Nd:host rod which is coated HR at 1.06μ and is greater than 85% transmissive at 808 nm. The focussed spot size consists of two unequal intensity lobes separated by 50μ; 99% of the pump energy was contained in a rectangular area 75μ wide by 10μ high. These are "low brightness" arrays with an emitting length of 400μ. The resonator geometry was nearly hemispherical.

RESULTS

The output power in the IR as a function of laser diode pump power is shown for both hosts in Figure 1. The comparison was performed using a 95% output coupler. It can be seen that both hosts have a 42% slope efficiency, with BEL having a higher threshold.

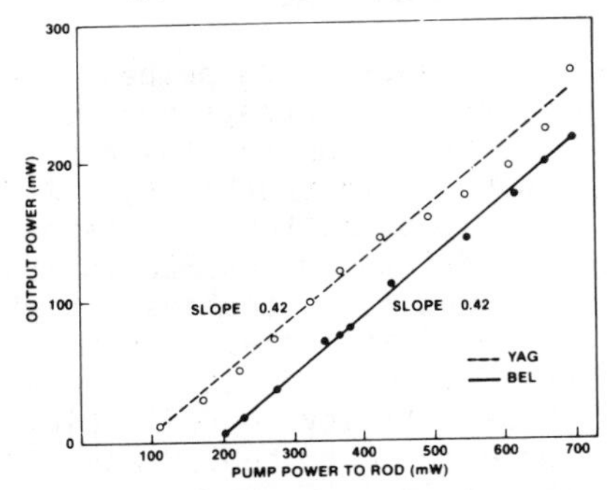

Figure 1. Output power dependence on pump power.

The dependence of the output power on the mirror reflectivity was measured for reflectivities between 95% and 99.9%. The results indicate that BEL is more sensitive to output reflectivity than YAG. A peak output of 250 mW was measured for BEL at 98% reflectivity while the peak for YAG was 283 mW at 97% reflectivity. The threshold pump power as a function of output coupling was also measured. At 98% the pump threshold for BEL was 92 mW, while at 97% the threshold for YAG was 61 mW. From this data one can derive the single pass loss for BEL and YAG (.002 and .012, respectively) and the small signal gain dependence on pump power. For BEL this number was one-third that for YAG, which is approximately the same ratio as the stimulated emission cross sections for Nd in the two hosts.

From an empirical point of view one can factor the overall efficiency into the product of three measurable quantities, as shown in Table I for BEL and YAG. The first factor is the electrical to optical pump light conversion efficiency for the laser diode. Although these efficiencies are a factor of 2 lower than presently available, the laser diodes were adequate for this work. The lower efficiency for the diodes used to pump BEL reflects the higher junction temperature needed to wavelength tune to the absorption peak. The power consumption of the coolers is not included in the overall efficiency. The collection efficiency of the optics is a function of the NA of the lenses and the reflectivity of the AR coating. The higher number for BEL is a result of using higher quality coatings. The third factor is the pump light to IR conversion efficiency of the laser rod, and is given for the optimum output coupling for each host. The lower number for BEL is due to the higher pump threshold.

Table I. Measured efficiency factors for Nd:host lasers

Host	Electrical to 808/810 nm	Collection optics	808 nm to 1.06μ or 810 nm to 1.07μ @R	Overall efficiency (%)
YAG	.170	.63	.40 @ 97%	4.3
BEL	.135	.75	.35 @ 98%	3.5

The measured efficiencies compare well with predicted efficiencies for Nd:YAG[1]. Using the measured resonator single pass loss of 1.2%, the Stokes efficiency of .76 and the estimated[2] quantum efficiency of .80, one can calculate for Nd:YAG an optical conversion efficiency of .24. This agrees well with the measured optical conversion efficiency of .25 (the product of the third and fourth columns in Table I).

Using an intracavity lens and a KTP crystal we obtained several milliwatts of green light from both YAG and BEL. The conversion efficiencies were extremely low due to the lens losses and alinement sensitivity of the resonator. As a result of the poor conversion efficiency, longitudinal mode coupling[3] was not observed and good amplitude stability of the second harmonic output was obtained.

CONCLUSIONS

Under the conditions used in the present work, both BEL and YAG perform comparably well. The broader absorption band in BEL did not appear to enhance this host's operation because of the long absorption path length and the relatively narrow spectral bandwidth of the laser diode arrays. However, for side pumping or for pumping with broader spectral bandwidth arrays, the broader absorption in BEL will become a factor. In addition, it is important to recognize that a minimum of 24% of the pump light is converted to heat (the quantum defect) so that at high pump fluence athermal BEL might prove a superior host.

ACKNOWLEDGEMENTS

The authors wish to acknowledge the financial support of the NOSC IED program administered by Dr. John Silva and Mr. Kenneth Campbell, and the technical support of J. Krasinski, T.C. Chin, and R.C. Morris from Allied Corporation.

REFERENCES

1. W. Koechner, SAIC Report 168-352-040, SAIC, McClean VA 22102, January, 1987.
2. L. DeShazer, private communication.
3. T. Baer J. Opt. Soc. Am. B, 3, 1175 (1986).

750 mW, 1.06 um, CW TEM_{00} OUTPUT FROM A $Nd:YVO_4$ LASER END PUMPED BY A SINGLE 20 STRIPE DIODE ARRAY

R.A. Fields, M. Birnbaum, and C.L. Fincher
Aerospace Corp., P.O. Box 92957, Los Angeles,Calif. 90009

J. Berger, D.F. Welch, D.R. Scifres, and W. Streifer, Spectra Diode Labs, 80 Rose Orchard Way, San Jose, Calif. 95134

ABSTRACT

We report the highest electrical to optical conversion efficiency (12.5%), and the highest cw output power (750 mW), from any solid state laser crystal end pumped by a single multiple stripe diode laser array. The optical slope efficiency, which refers to the 1.06 µm output power divided by the optical pump power incident on the end face of our laser crystal, was measured to be 64% in the best case and corresponds to nearly a 95% Nd quantum conversion rate, the highest yet reported.

The diode laser end pumped laser (DLEPL) is currently the most efficient solid state laser. The first demonstration that a high power 10 stripe electrically efficient diode laser array could be effectively imaged for optically end pumping within the mode of a small Nd:YAG hemispherical laser resonator resulted in an 8% electrically efficient device[1]. More recently, improvements in the power per stripe of laser diodes has resulted in a Nd:YAG DLEPL having a total efficiency of 10.8% at 415 mW TEMoo output and a maximum output of 465 mW[2]. Here we report still higher DLEPL efficiency and output power by using the same high power diodes to pump $Nd:YVO_4$, a material recently demonstrated[3] to have a lower lasing threshold and better absorption of diode light than Nd:YAG in a comparable DLEPL device.

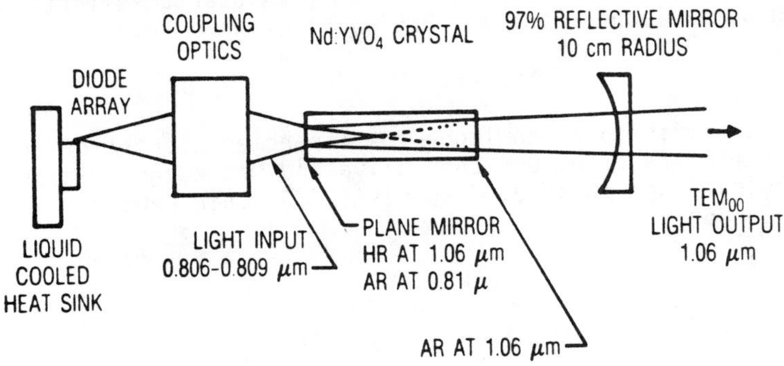

Figure 1. Schematic illustration of the experimental setup.

A schematic of the DLEPL device reported here is depicted in figure 1. A 4.5 mm long $Nd:YVO_4$ laser crystal was pumped by either a 100 µm (SDL 2430) or 200 µm (SDL 2460) -wide laser diode array with their respective coupling optics. The coupling optics for the 100 µm diode laser (scheme A) consisted of a f= 6.5 mm, 0.615 NA compound lens for collimation followed by a 10 mm achromat focusing lens. The coupling for the 200 µm-wide diode (scheme B) has been well described elsewhere[2] and consists of the same collimating lens followed by a 4x anamorphic prism pair with a f= 25.6 mm focusing lens. Scheme A focuses the image to approximately two times the object size with no measurable transmission loss. Scheme B delivers a more effective beam distribution in the crystal due to the combination of one-to-one imaging, an approximately 80% decrease in large axis beam divergence and a near circular beam near the focus. The beam quality of scheme B is compromised by nearly a 20% transmission loss because of reflective losses of a large fraction of the highly diverging diode rays.

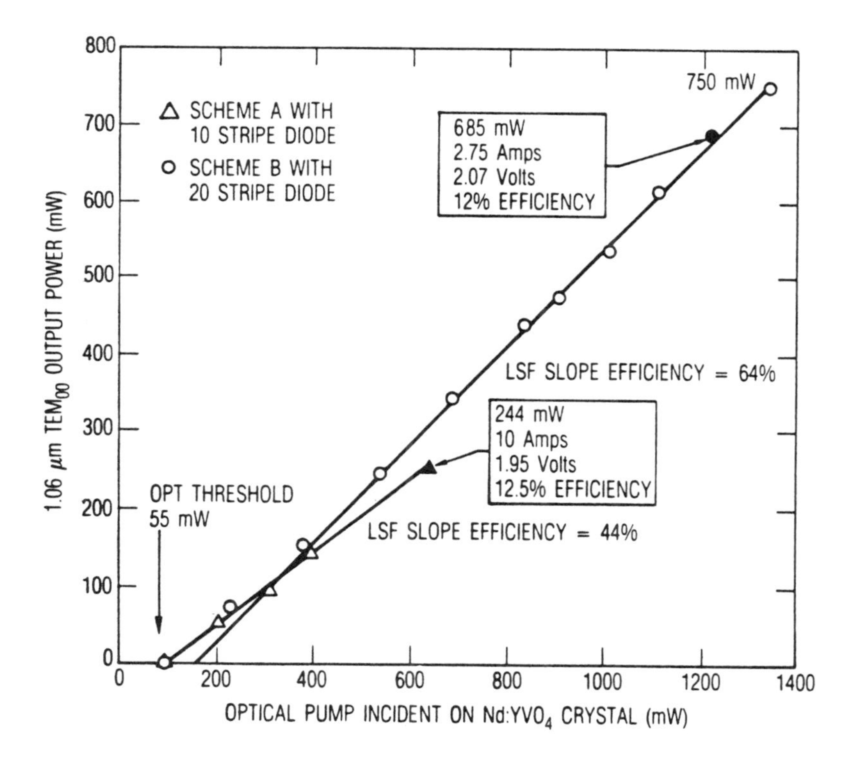

Figure 2. Experimental plot of optical power incident on the crystal vs. TEM_{00} 1.06 µm output for schemes A and B.

The results from scheme A and scheme B diode pumping of $Nd:YVO_4$ are presented in figure 2. The 244 mW output from scheme A displayed an overall electrical to laser light output efficiency of 12.5%, the highest yet observed for a DLEPL. The more powerful diode in scheme B yielded over 750 mW output at an efficiency of 11.9% (peak eff. 12.0 % at 685 mW). The optical slope efficiency of 64% (optical power incident on the crystal vs. TEMoo 1.06 um ouput) from scheme B is the highest yet observed for any Nd doped crystal. The theoretical limit for optical conversion in this case was 67% and can be obtained by computing the product of the measured coating transmission (96%), diode light absorption by the crystal (92%), and the energy difference between the pump and output photons (76%). This results in a cw laser quantum efficiency of 95%.

The high overall efficiency of scheme A despite the lower output power can be traced to the lower lasing threshold of the 10 stripe diode (half that of the 20 stripe diode) and the better transmission of the coupling optics. The larger image of scheme A optics did not allow focusing of the 200 um diode within the 260 um TEMoo beam waist diameter. The lower slope efficiency observed in the scheme A data is attributed to poor alignment. Early diode degradation did not permit further optimization.

Our maximum output power of 750 mW and electrical efficiency of 12.5 % are temporarily the highest observed for a DLEPL and should be improved upon with higher power, lower divergence diodes currently under development. The device's output power near the maximum was increasing linearly as a function of optical input power. This indicates that considerably more power can be extracted before the system is affected by saturation or thermal side effects. The lack of thermal side effects is strongly linked to the high quantum efficiency, since for every watt of pump power over threshold we are loading the crystal with less than 300 mW. Our data also indicates that quantum efficiency is strongly affected by the power density. The lower slope efficiency near threshold is in agreement with our earlier low power DLEPL slope of 50%[3]. This suggests that a power density of approximately 1.7 kW cm^{3} for Nd:YVO_4 is required to achieve near unity quantum conversion. This relationship, in addition to the effect of cross section on DLEPL performance is currently being investigated.

REFERENCES

1. D. L. Sipes, Appl. Phys. Lett. 47, 74 (1985)
2. J. Berger, D. F. Welch, D. R. Scifres, W. Streifer, and P. S. Cross, Appl. Phys. Lett. 51, 1212, (1987)
3. R. A. Fields, M. Birnbaum, C. L. Fincher, Appl. Phys. Lett. 51,1885,(1987)

A Tunable, Nd-doped Lithium Niobate Laser at 1084 nm

L.D. Schearer
Laboratory for Atomic and Molecuar Physics
University of Missouri-Rolla, Rolla, MO 65401

M Leduc
Laboratoire de Spectroscopie Hertzienne
L'Ecole Normale Superieure
Paris, France

Abstract

Over 250 mW of Cw laser emission at 1084 nm is obtained from Nd:$LiNbO_3$ when the rod is end-pumped along the crystalline 'y' axis by 1 W from a Kr^+ laser at 752 nm. The laser can be tuned over 3 nm at the 1084 nm peak with a thin, uncoated etalon in the cavity. Thresholds of 30 mW of absorbed pump power were obtained with a weak output coupler, rising to 220 mW with a 35% transmitting output mirror. No pump-induced photorefractive effects were observed.

Introduction

Early reports of laser emission in Nd-doped $LiNbO_3$ crystals were confined to pulsed operation due to poor crystal quality and large photorefractive effects. Recently crystals of better quality have become available and with the addition of 5 mole percent of MgO into the $LiNbO_3$, photorefractive damage is much reduced. Thus, recent work has resulted in cW operation of the device[1,2]. Here we describe the cW laser properties of this material when pumped by 1 W of Kr^+ laser light at 752 nm. This device is characterized by good efficiency, high gain, and low thresholds, and with the addition of a thin, solid etalon within the cavity, is easily tunable in a region of the spectrum which contains the helium triplet resonance transition (2^3S-2^3P). Optical pumping of the helium metastable atom at 1083 nm has led to a many interesting applications[3]; tunable laser emission at this wavelength will extend the range of applications[4].

Experimental Results

Figure 1 shows schematically the cavity used in the earlier experimental work. The meniscus lens shown serves both as a focussing element for the pump radiation and as a curved mirror forming one end of the laser cavity. The focal length of the lens and the radius of curvature of the mirror are about 4.2 cm. The $LiNbO_3$ crystal is 4x4x10 mm and contains about 1% Nd. The long dimension coincides with the crystalline b-axis. The laser output is linearly polarized along one of short edges coinciding with the c-axis. The ends of the crystal are AR coated at 1084 nm. A lens of focal length 2.75 cm is positioned so that the laser emission from the output mirror is nominally plane parallel. The total cavity length is about 25 cm. The output coupler is a plane reflector with transmissions varying from 1 to 35%. The gain and threshold measurements were made with this cavity. To obtain the tuning characteristics a thin, uncoated etalon and/or a Lyot filter is inserted in the cavity. For later work this crystal was coated on one end with a high reflection coating at 1084 nm and high transmission below 850 nm. This surface then became one end of the laser.

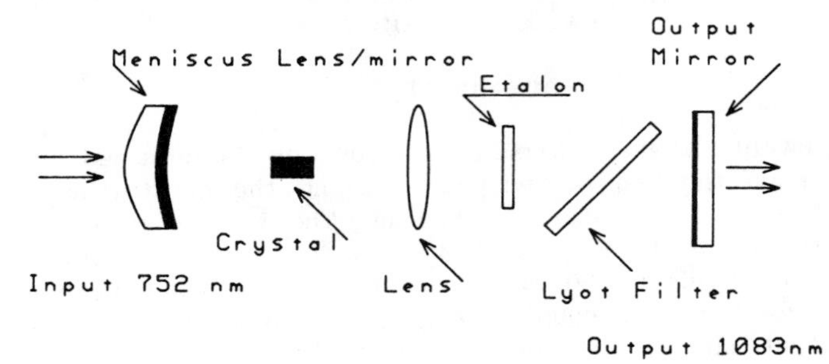

Figure 1

A. Laser Performance

Table 1 shows the threshold powers and power output levels when the crystal is pumped with a Kr^+ laser at 752 nm. At 752 nm the 1 cm long crystal absorbs 78% of the incident pump radiation. The laser output is independent of the polarization of the pump beam.

Table 1

Transmission(%) Output Mirror	Threshold(mW) Absorbed Power	Output Power(mW) at 1 W Input	Slope Eff.(%)
35%	172 mW	260 mW	43%
16	62	250	35
4	39	175	24
1	23	85	11

We used several other pump sources with the second of the cavity configurations described above. These results are summarized in Table 2.

Table 2

Pump Wavelength	Incident Pump Power	Power Output
752 nm	1 W	260 mW
598 nm	300 mW	30 mW
805 nm	250 mW	2 mw
598 nm(pulsed)[1]	300 mW(ave)	4 mw(ave)

[1] 5 kW peak at 6 kHz rep rate; 250 ns width.

B. Tuning Characteristics

The fluorescence width of the Nd-doped $LiNbO_3$ is significantly greater than in Nd:YAG crystals. This increased width suggested the possibilty of tuning the laser output frequency. The tuning characteristics obtained with a 16% transmitting output coupler and a 0.2 mm thick solid etalon in the cavity are shown in fig. 2. As the laser reached the limit of the tuning range, it lased simultaneously at two wavelengths. A 6 mm thick Lyot filter added to the cavity suppressed the second wavelength.

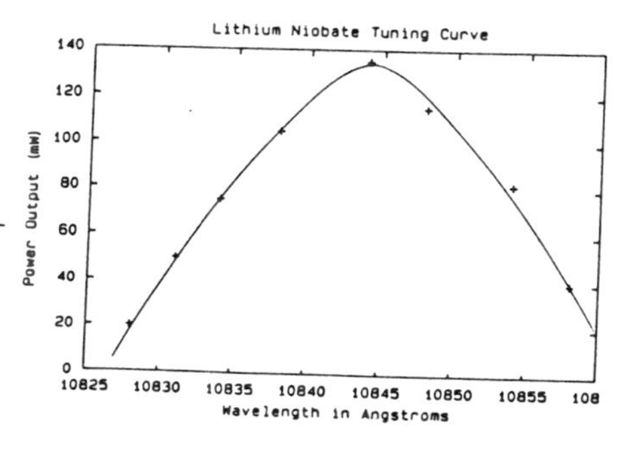

Figure 2

Conclusions

The Nd-doped $LiNbO_3$ material appears to have thresholds, gain, and efficiencies comparable to Nd:YAG and may be a promising contender as a solid-state laser device when one considers its electrooptic properties. For our purposes, the ease with which it can be tuned to the helium metastable resonance transition, offers substantial advantages for the development of efficient and compact, optically-pumped helium magnetometers.

References

1. T.Y. Fan, A. Cordova-Plaza, M.J.F. Digonnet, R.L. Byer, and H.J. Shaw, J.Opt.Soc.Amer. 3, 140 (1986).
2. A. Cordova-Plaza, M.J.F. Digonnet, and H.J. Shaw, IEEE J. Q.Electr. QE-23, 262 (1987).
3. L.D. Schearer Ann. Phys. Fr.10, 845 (1985).
4. D.S. Betts and M.Leduc, Ann. Phys. Fr.11, 267 (1986).

STUDY OF REPEATED Q-SWITCHED TRAIN PULSE Nd: YAG LASER

Sun De-cai, Wu Xing, YaoJian-quang
Dept. of Precision Instrument Engineering, Tianjin Univ., PRC

ABSTRCT

A new type of repeated Q-switched train pulse Nd: YAG laser is presented in this paper. The relations of pulse width, peak pulse power and pulse energy with Q-switch modulation period T and effective pumping rate R are derived both in theoretical results and numerical calculation results. The stable operation of the laser system is achieved and the experimental results are presented. The application of the laser system as pumping source is also discussed.

INTRODUCTION

Q-switched train pulse lasers have been reported in some papers before. The train pulse lasers reported could not operate at repeatitive frequency. The laser system developed by us is a new type of repeated Q-switched Nd: YAG laser which can put out a pulse train through the periodic modulation of Q-switch during each pumoping period and can operate at repeatitive frequency. The pulse number in a pulse train can be changed from 1 to 16. The interval between two adjacent pulses can be changed from tens of microseconds to hundreds of microseconds. The average energy of one pulse is tens of mJ. The pulse width is less than 30 ns.

THEORETICAL ANALYSIS

The known four-level rate equations are:

$$dN/dt = R - \lambda N - \beta \Phi N$$
$$d\Phi/dt = \beta \Phi N - \Phi (T_1 + \varepsilon) \qquad (1)$$

N—the difference in stored energy between the upper and lower laser states.
R—the effective pumping rate
λ—the sum of pumping rate and spontaneous decay rate
T_1-the fractional output mirror transmission
ε—the round-trip cavity loss
Φ—the energy of the coherent electromagnetic field inside the cavity
β—the stimulated emission rate

Through the given conditions and analytic calculations from the rate equations, we get the expressions of pulse energy E, peak pulse power P_m and pulse width Δt of one pulse in a pulse train when the Q-switch is operated at repeatitive frequency 1/T.

The expressions are:

$$P_m = T_1 \frac{h\nu}{2L} C\left[\frac{R}{\lambda} - \left(\frac{R}{\lambda} - N_e\right)e^{-\lambda T} - Nt - Nt\ln\frac{\frac{R}{\lambda} - \left(\frac{R}{\lambda} - N_e\right)e^{-\lambda T}}{Nt}\right]$$

$$E = \frac{ch\nu}{2L} \frac{T_1}{T_1+\varepsilon} Nt \ln \frac{\frac{R}{\lambda} - \left(\frac{R}{\lambda} - N_e\right)e^{-\lambda T}}{Nt}$$

$$\Delta t = \frac{E}{P_m}$$

Here Ne—the value of N at the time when a pulse is formed
$Nt = (\varepsilon + T_1)/\beta$
T— the time interval between the Q-switch consecutive operations

Through numerical calculations on the expressions, three curves expressing the above three formulars are derived. We find that P_m increases as Q-switch modulation period T, that is, the interval between two adjacent pulses, increases. When T increases to some point, P_m reaches a constant. The principle is that the initial population inversion increases as T increases, then it will reach a constant when T is bigger. The curve E-T shows the same principle as the above one. The curve Δt--T shows that Δt decreases as T increases, that is, the bigger the initial population inversion, the smaller the pulse width Δt.When the initial population inversion reaches a constant, Δt also becomes a constant.

EXPERIMENTAL RESULTS AND CONCLUSIONS

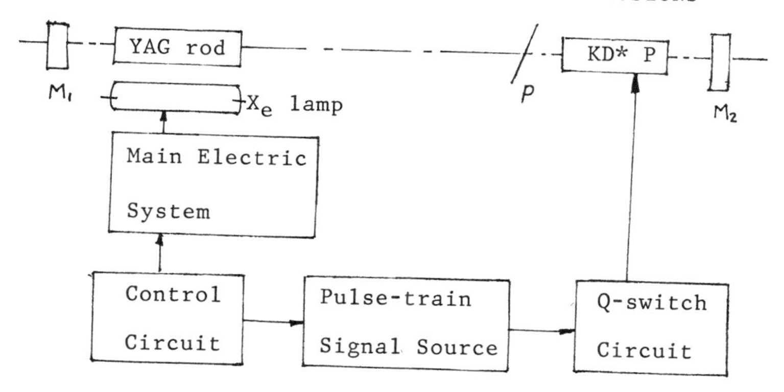

Fig.1 Principle diagram of the laser system.

We developed a repeated Q-switched train pulse Nd:YAG laser system. The principle diagram of the laser system is shown in Figure 1. In the main electric system, the charge voltage is 1100V, the capacity bis 150 μF, the repeatitive frequency is 10 Hz.

Experiment 1. The laser system put out the pulse train containing 10 pulses, with pulse interval 20 μs. The pictures of the pulse and the pulse train are shown in Figure 2 and Figure 3.

The pulse width is Δt=30 ns. The pulse average energy is 15 mJ.

Experiment 2. The laser system put out the pulse train containing 3 pulses with pulse interval 60μs. The pictures of one pulse and the pulse train are shown in Figure 4 and Figure 5. The pulse width Δt is 2ons. The pulse average energy is 40 mJ.

From the above experiments, we find that the experiment results conform to the theoretical conlusions. When the pulse interval T increases, the pulse energy increases and the pulse width decreases.

The laser system can be used as a new pumping source to pump other laser systems. The operation of $LiF(F_2^-)$ colour center laser is studied by using the repeated Q-switched train pulse Nd:YAG laser as pumping source. The satisfactory results have been achieved. The output of 0.53 μm pulse train can also be realized by using second harmonic generation technique in this laser system.

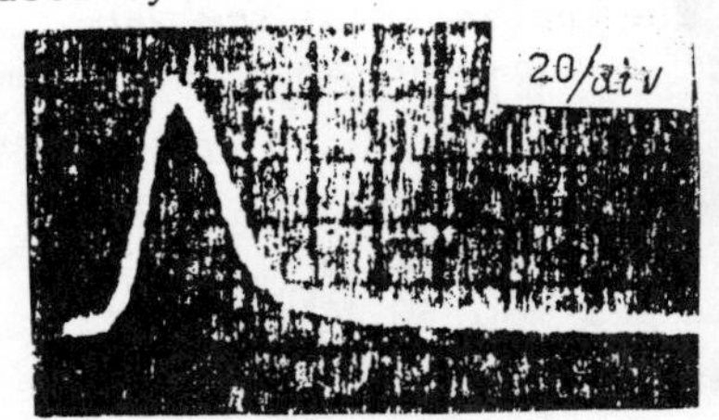

Fig.2 A single pulse of the pulse train

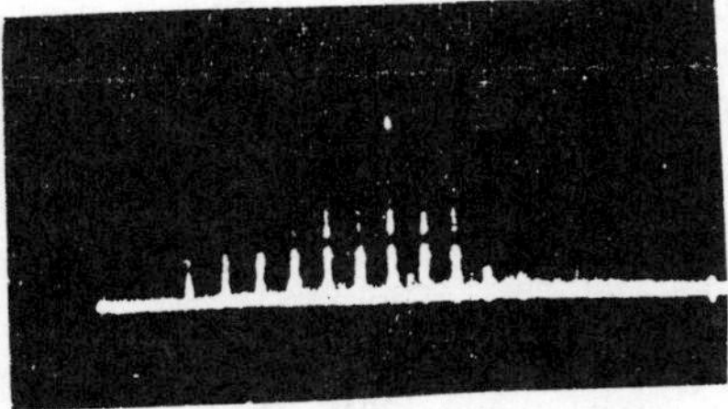

Fig. 3 A pulse train of 10 pulses

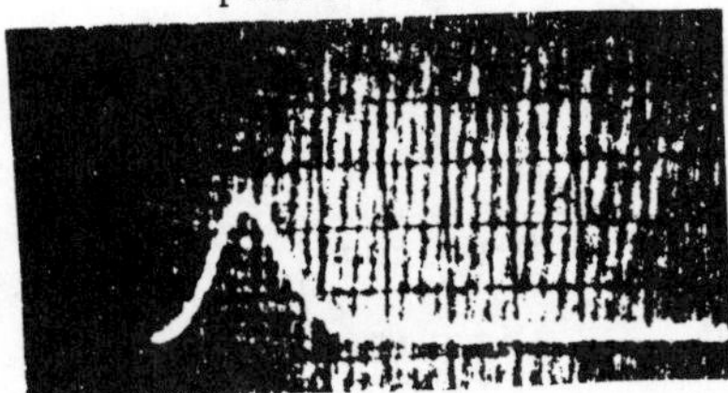

Fig.4 A single pulse of the pulse train

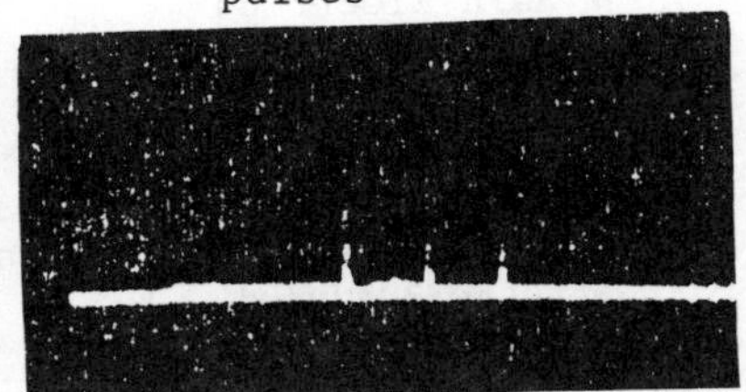

Fig. 5 A pulse train of 3 pulses

References

[1] R.B. CHESIER, Proceedings of the IEEE, Vol. 58, No. 12, P1899.

[2] W.G. Wagner, J. Appl. Phys. Vol.34 P2040-2046, 1963.

[3] Li Shi Chen, Tianjin University Journal No. 2 (1978).

RESONATOR OPTIMIZATION OF A CW MODE-LOCKED Nd:YLF LASER

Herman Vanherzeele
E. I. DuPont de Nemours & Company, Experimental Station, Wilmington, Delaware 19898

ABSTRACT

Thermal lensing effects in a cw pumped Nd:YLF rod have been characterized, permitting the optimization of the laser resonator. In mode-locking regime, a well-designed system generates an average output power similar to that of a typical Nd:YAG laser, however, with an increased stability and with pulse durations that are significantly shorter. As a result, Nd:YLF may become the preferred choice for many applications.

INTRODUCTION

Investigation of the spectroscopic and physical properties of Nd:YLF have suggested that this uniaxial material may be a better candidate than Nd:YAG for generating short pulses with high peak power[1-4]. The relatively large bandwidth (1.35 nm) of Nd:YLF compared to Nd:YAG (0.45 nm) is particularly atractive in this respect [4]. However, cw gain in Nd:YLF is an issue, because both (competing) transitions at 1053 nm and 1047 nm, respectively σ ($E \perp c$) and π ($E \parallel c$) polarized, have a lower cross section than the 1064 nm line in Nd:YAG [2]. As a result, one expects lower average power from a Nd:YLF rod than from a similar Nd:YAG rod [4], despite the lower threshold for Nd:YLF. It is the purpose of this paper to demonstrate that this important drawback can be overcome in an optimized resonator.

THERMAL LENSING EFFECTS IN Nd:YLF

Optimization of a resonator design requires a good knowledge of the thermal lensing of the gain medium. Thermal lensing measurements in Nd:YLF have been reported [3] for pulsed operation (at a fixed input energy). However, data are given only for π polarization ; for σ polarization, the thermal lens was reported to be too weak to be measurable. For this work more detailed and accurate data are needed. A Quantronix 116 laser head with one krypton lamp and a 4 x 79 mm Nd:YLF rod (1% doping) are used. To investigate the effect of the pump power dissipated in the rod on the transmitted wavefront, a collimated and polarized 633 nm HeNe laser illuminates the full aperture of the rod. From the measurement of the power transmitted through a slit (rather than a circular aperture as commonly used for these measurements in isotropic materials), the thermal lens f of the rod is inferred for both σ and π polarized light, both in a principal plane (containing the wave vector and the c-axis of the rod) and in a plane perpendicular to it. The experimental results are shown in Fig.1. The data are accurate to within ±25%. The focal length f of the thermal lens clearly is different for both polarizations, and, moreover, for a given polarization, a different lens effect is measured in both planes previously mentioned. For σ polarized beams, for which the lens is weaker (see Fig.1, curve 1), $f<0$ in a principal plane, and $f>0$ in a plane perpendicular to it ; the magnitude of the lens in both planes is the same within the experimental errors. On the other hand, for π polarized beams (Fig.1, curves 2 and 3), $f<0$ in both planes, with the weaker action in the principal plane. From these data it will be obvious that the thermal lensing effects inside a Nd:YLF resonator will be anything but negligable, despite the fact that the lens is much weaker than for Nd:YAG. The anisotropy of the lens effect, for a given polarization, particularly merits attention. This effect, if not compensated for, causes astigmatism. It gives rise toelliptically distorted beam profiles inside and outside

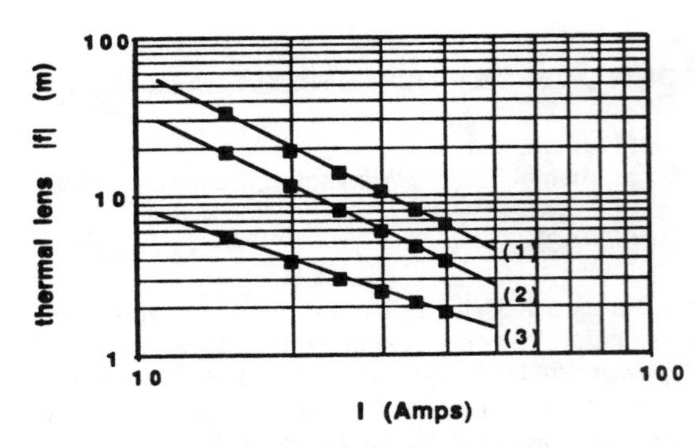

Fig.1 : Thermal lens f (m) at 633 nm for a 4x79 mm Nd:YLF rod as a function of lamp current (Amps) ; (1) : σ polarized light (f<0 in the principal plane, and f>0 in the plane perpendicular to it), (2) : π polarized light in the principal plane (f<0), (3): π polarized light in a plane perpendicular to the principal plane (f<0). The dots represent the experimental data.

the resonator, thereby reducing the maximum extractable TEM_{00} power from the rod. On the other hand, thermally induced birefringence has also been investigated. This effect, although measurable, turns out to be too weak to cause any noticable loss inside a resonator.

RESONATOR DESIGN AND SYSTEM PERFORMANCE

Several resonator candidates with a cavity length matching a 100 MHz mode-locker, were theoretically analyzed to ensure a combination of all of the following desirable properties : (i) optimum rod filling for TEM_{00} operation (ideally, the rod should be the limiting aperture), (ii) good astigmatism compensation of the thermal lensing at both lasing transitions, (iii) good dynamical stability over a wide range of lamp currents, (iv) good pointing stability and low mirror misalignment sensitivity, and (v) small beam divergence. A resonator design that adequately combines all of the above properties, while requiring only one set of optics for both 1047 nm and 1053 nm operation, is schematically represented in Fig.2. A spherical lens (f=38 cm) ensures an optimum filling of the rod (beam radius, $HW1/e^2M$, inside the rod is 1mm). A cylindrical lens (f=470 cm) corrects the thermally induced astigmatism to better than a few percent over the entire range of usefull electrical input powers (lamp current 30-40 Amps). The mode volume in the rod does not change over this range of lamp currents. The far field beam divergence ($FW1/e^2M$) is 1.6 mrad at both wavelengths. Notice that without the cylindrical lens, the resonator is unstable. The resonator offers a good dynamical stability, and has a very low mirror misalignment sensitivity (because of the spherical lens). As a result, ab initio alignment of the resonator can be done easily, despite the presence of the intracavity lenses.

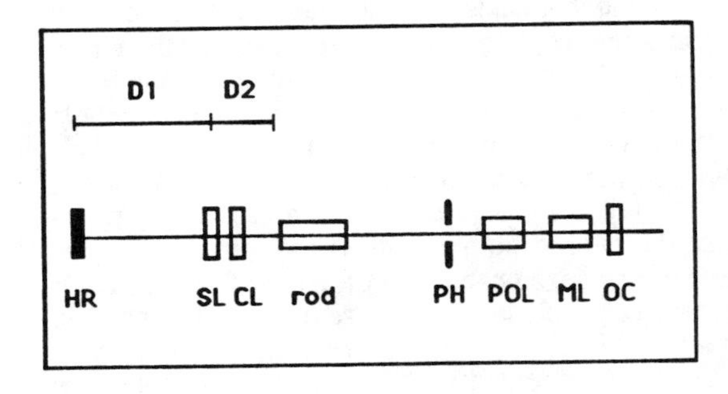

Fig.2 : Schematical representation of the Nd:YLF resonator discussed in the text ; HR = high reflector (R=100 cm), OC = 12% output coupler (R=-120 cm), CL = cylindrical lens (f=470 cm), SL = spherical lens (f=38 cm), ML = mode-locker, PH = pinhole, POL = polarizer. (For 1047 nm operation, D1 = 55 cm, and for 1053 nm operation D1 = 48 cm ; D2 = 9 cm in both resonators)

The resonator was built, starting out from a commercial unit (Quantronix 416). It is harmonically mode-locked with 10 W of rf power at 100 MHz. Lasing is restricted to 1053 nm by proper alignment of the rod with respect to a Brewster plate polarizer inside the cavity. With a 12% output coupler, the average mode-locked output power is 10 W for a lamp current of 37 Amps. The output power is stable to within 1.2 % (peak to peak value) without using any stabilizing feed back mechanisms. This is better than what one usually observes in Nd:YAG. The beam profile, measured at several locations, has an excellent Gaussian shape, without noticeable ellipticity. Experimentally obtained data for the spot sizes agreed to within a few percent with the ones calculated from the ABCD matrix for the resonator, thus confirming the thermal lens data. The pulse duration, measured with a real time autocorrelator, is 40 psec (FWHM) assuming a Gaussian pulse shape. Spectral measurements obtained with a scanning Fabry-Perot, furthermore showed that the pulses are nearly bandwidth limited. The excellent short and long term temporal stability was further evidenced by monitoring the timing fluctuations (only a few psec) of the optical pulses in the resonator. Detuning the cavity length by approximately 0.5 μm from its ideal length (either side) results in a 200 MHz mode-locked operation of the laser with no change in average output power. However, a 25% pulse broadening is observed in this regime, and the timing fluctuations are greatly enhanced.

Frequency doubling the Nd:YLF output in a 5 mm long KTP crystal, cut for type II second harmonic generation at 1053 nm (phase matching angle 35 deg with respect to the X-axis in the X,Y plane), produces nearly 2 W of green light. This high average output power combined with the relatively short pulse durations and the enhanced stability suggest the use of this system as a pump source for synchronously mode-locked dye laser oscillators operating in tandem. With an average power at 526 nm as small as 600 mW, a synchronously pumped dye laser (Coherent single jet model 700 with a one plate birefringent tuning element) currently produces 130 mW average power at the peak of the gain of rhodamine 6G. The dye laser pulses typically have a duration of 1.2 psec (FWHM), assuming a sech^2 pulse shape.

CONCLUSIONS

In summary, the thermal lensing of a cw pumped Nd:YLF rod has been characterized, and the thermally induced astigmatism has been pointed out for the first time. An optimized resonator has been designed both for 1047 nm and 1053 nm operation, using only one set of optics to correct the thermally induced astigmatism, while still providing an optimum mode volume in the rod. The resonator combines good TEM_{00} mode quality with high average output power and excellent short and long term stability. In addition, the pulse width in mode-locking regime is significantly shorter than for similar Nd:YAG lasers. We therefore believe that this optimized Nd:YLF system may become a prefered choice for many applications, including pumping dye laser oscillators and amplifiers. Further system improvements, including an electronic cavity length control to eliminate thermal drift problems, as well as a monolithic design of rod and intracavity optics, are currently under investigation.

REFERENCES

1. E. J. Sharp, D. J. Horowitz, and J. E. Miller, J. Appl. Phys. 44, 5399 (1973)
2. T. M. Pollak, W. F. Wing, R. J. Grasso, E. P. Chicklis, and H. P. Jenssen, IEEE J. Quant. Electron. QE-18, 159 (1982)
3. J. E. Murray, IEEE J. Quant. Electron. QE-19, 488 (1983)
4. P. Bado, M. Bouvier, and J. Scott Coe, Opt. Lett. 12, 319 (1987)

RECENT PROGRESS IN FREQUENCY STABILIZATION OF DIODE LASER

T. Yabuzaki
Department of Physics, Kyoto University, Kyoto 606, Japan

M. Kitano
Radio Atmospheric Science Center, Kyoto University,
Uji, Kyoto 611, Japan

ABSTRACT

The frequencies of diode lasers have been highly stabilized by using Doppler-free saturated absorption spectra of alkali-metal atoms, and long term stability in terms of Allan variance σ better than 10^{-13} has been achieved. In this method, the diode laser frequency has been directly modulated to obtained the derivative of resonance, which is used as the error signal to control the diode current. New methods are studied to obtain nearly derivative of Doppler-free resonances, without frequency modulation. The frequency stability of a diode laser locked at the center of a resonance is similar to above case of the saturated absorption.

1. INTRODUCTION

In recent years, diode lasers have been used in wide variety of fields from fundamental researches such as laser spectroscopy to practical applications such as optical communications. However, the frequency stability is rather poor under the free-running condition compared with gas lasers, because it is very sensitive the the diode temperature (frequency shift being typically $\sim$10 MHz/mK) and driving current (typically $\sim$10 MHz/μA). The frequency stabilization is now becoming more and more important in many applications. In order to stabilize the laser frequency, it is necessary to control both of the diode temperature and current very precisely. In most of high frequency-stabilization done so far, the temperature is regulated highly and then the current is controlled so that the laser frequency is locked at an external frequency reference. A conventional way is to use a Fabry-Perot cavity, but the resonance frequency is affected by the change of temperature and by the vibrations, so it is much better to use atomic and molecular absorption lines, which provide absolute reference frequencies.

This paper describes about the recent our works on the very high frequency-stabilization of GaAlAs lasers, in which Doppler-free spectra of the D-lines of cesium (894 nm/852 nm) and rubidium atoms (795 nm/780 nm) have been used. At first the stabilization using the saturated absorption spectra[1-3] is described, by which stability in terms of square root of Allan variance σ (which we will call simply Allan variance hereafter) better than 10^{-13} has been achieved for the averaging time 1-500 s. However, in order to obtain the derivative shape of resonance, used as an error signal to control the diode current, the laser frequency is directly modulated, the modulation depth being several MHz. Such direct modulation is undesirable in many applications. Some Doppler-free spectroscopy has been found to provide the error signal, or derivative shape of resonance, by switching the polarization of light or by applying an oscillating longitudinal magnetic field to atoms in stead of frequency modulation of laser light.

2. FREQUENCY STABILIZATION BY USING SATURATED ABSORPTION SPECTRA

The GaAlAs laser used is a Fabry-Perot type with output power of about 3 mW. The temperature of the diode laser was controlled by a system with double control loops containing two thermoelectric coolers,[3] by which the fluctuation of the temperature was reduced down to 3×10^{-5} ℃. This temperature fluctuation corresponds to the frequency fluctuation less than 1 MHz. Under the control of the temperature, we observed the Doppler-free spectra of D-lines of Cs and Rb by using the technique of saturated absorption spectroscopy and stabilized the laser frequency at the centers of resolved hyperfine components of these absorption lines. Here we describe the case of the Rb-D_1 line at 794.8 nm. In the experiments we have used natural abundant Rb, i.e. 73% ^{85}Rb and 27% ^{87}Rb. Figure 1(a) shows the diagrams of associated energy levels together with hyperfine transitions for these two isotopes, and (b) shows the relative positions and intensities of hyperfine components. In addition to the Lamb dips occurring at these components, crossover resonances is observed at the center of adjoining components for each isotope, if the separation is near or less than Doppler-broadening $\sim$400 MHz.

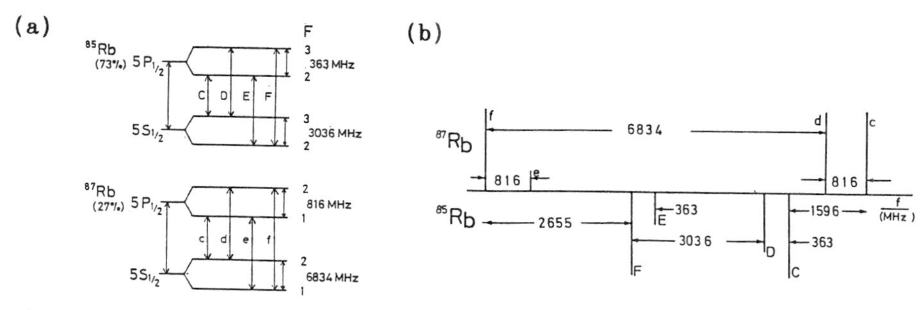

Fig.1. (a) Energy diagram of Rb isotopes, (b) relative positions and intensities of eight transitions.

Figure 2 shows the optical system to observe the saturated absorption and also to stabilize the laser frequency at the center of the Doppler-free spectrum. The output of the temperature-controlled diode laser is splitted into two beams and applied to the Rb cell from opposite directions as a saturating beam ($\sim$500 mW) and a probe beam ($\sim$30 mW). A third beam,

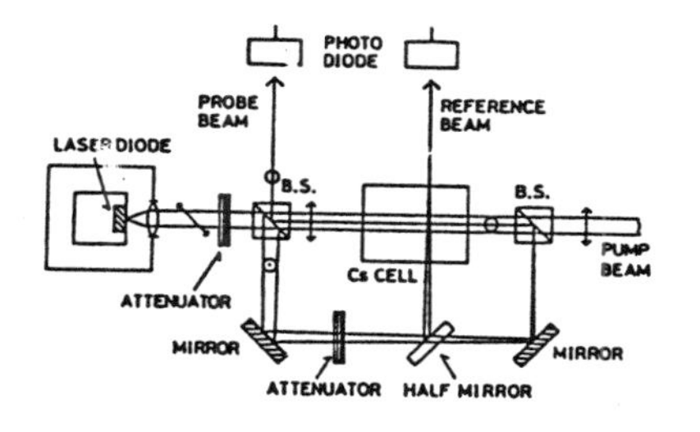

Fig. 2. Optical system.

which we call reference beam, is also applied perpendicularly to above beams in order to monitor and to subtract the Doppler-broadened background from the absorption signal of the probe beam. The cell was cylindrical with 2 cm diameter and 2 cm length, which was used at room temperature.

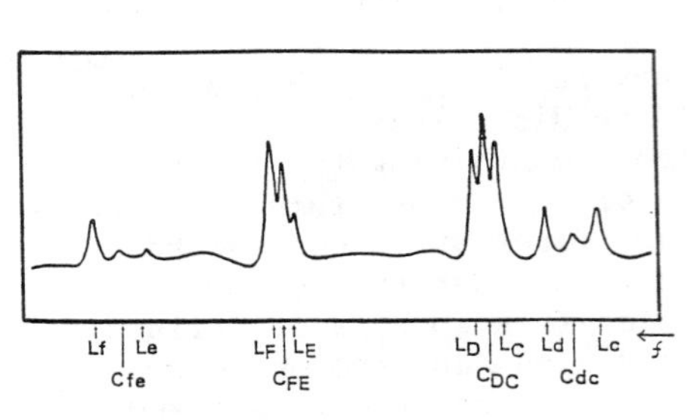

Fig. 3. Saturated absorption spectra of the Rb-D_1 line.

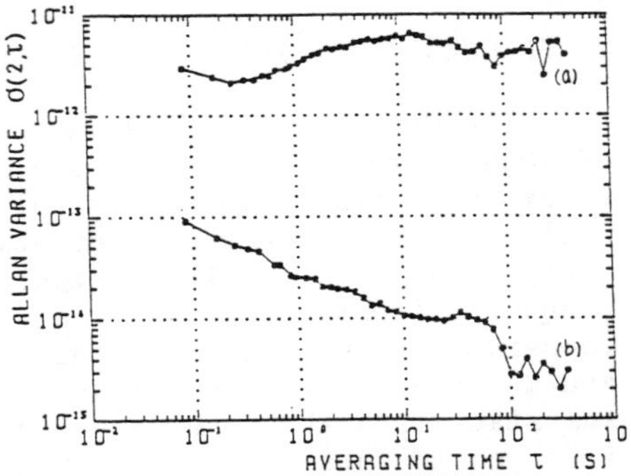

Fig. 4. Frequency stability in terms of Allan variance.

Figure 3 shows the purely Doppler-free saturated absorption signal for the Rb-D_1 line. The line width in the present case is mainly determined by the spectral width of the diode laser. The derivative shape of resonance was obtained by modulating the laser frequency modulated at audio-frequency with modulation depth of few MHz, and the probe signal, after subtracting the broad background, was lock-in detected. This was used as a error signal fed back to the driving current of the diode laser. For this feedback system we adopted PID (proportional, integral, and differential) control. Figure 4 shows the Allan variance σ for the frequency fluctuation, estimated from the fluctuation of the error signal, as a function of averaging time τ for the cases of P control and PID control with optimized parameters. We see that long-term stability has been considerably improved compared with the works done so far. This very high stability is, of cause, because of the use of a very narrow atomic spectrum, but it is also due to the improvements of the feedback system and well isolation from the vibrational and acoustic noise.

3. POLARIZATION SWITCHING AND ZEEMAN MODULATION SPECTROSCOPIES - FREQUENCY-STABILIZATION UNDER NO FREQUENCY MODULATION

As mentioned above, we could get considerably high frequency-stability using saturated absorption spectra of alkali-metal vapors. In this case, the frequency of diode laser was directly modulated to get the derivative shape of Doppler-free signals. However, such frequency modulation is not desirable in many applications. We have studied about the Doppler-free spectroscopies, which offer the derivative shape, or nearly derivative shape, of resonances under no frequency modulation of the laser light. Figure 5 shows the two optical configurations. In both cases, the laser beams are applied to the cell from opposite directions as in the saturated absorption spectroscopy, but the light beams are circularly polarized and a magnetic field is applied to the cell. In the case (a), sense of circular polarization is switched alternatively, and in the case of (b), an oscillating magnetic field is applied to the cell along the direction of the laser beams. In both cases, the detected absorption signal of the probe beam is applied to the lock-in amplifier tuned at the frequency of polarization switching or of the oscillating magnetic field.

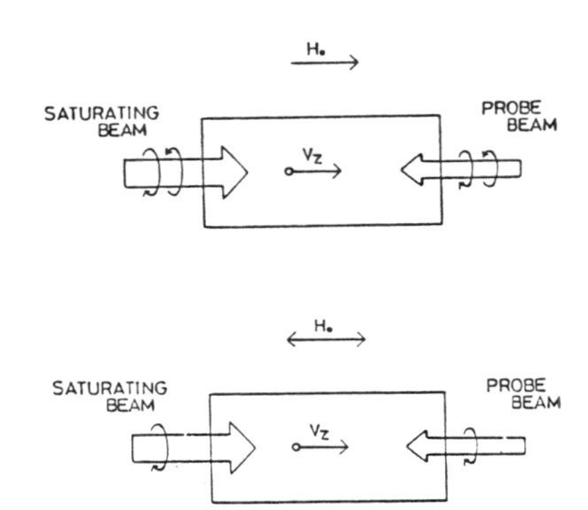

Fig. 5. Optical configurations studied.

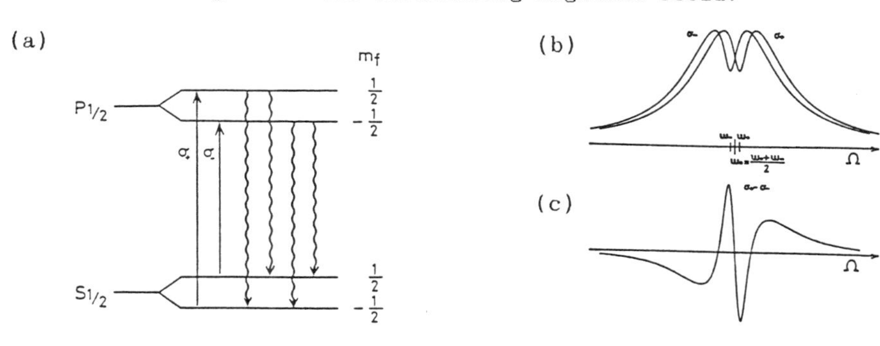

Fig. 6 (a). Simplified energy diagram, (b) probe signals, and (b) difference of probe signals for $\sigma_{\pm}$ polarizations.

The principle is very simple. Let us consider a simplified energy diagram shown in Fig. 6 (a), where both of the ground and the excited states consist of two Zeeman sublevels. The transition frequencies for $\sigma_{\pm}$ are shifted to opposite directions with the same amount in the presence of magnetic field. So, the absorption signals of the probe beam for $\sigma_{\pm}$ light become as shown in Fig. 6 (b), so that the difference of these signal, which is large in the vicinity of the sharp Lamb dips, is obtained by lock-in detection as shown in Fig. 6 (c). Namely the signal obtained in this way crosses zero at the center of resonance in the absence of magnetic field, so that the resonance frequency is not affected by the change of magnetic field intensity. When the longitudinal oscillating magnetic field is applied, the direction of the field is switched alternatively, so it is basically equivalent to the switching of circular polarizations. The difference is that the magnetic field intensity changes through zero, which gives additional effect like Hanle effect for the transitions causes optical pumping.

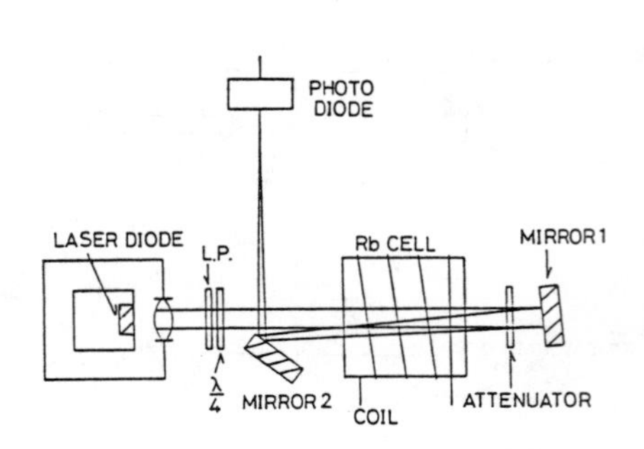

Fig. 7. Optical system for Zeeman-modulation spectroscopy.

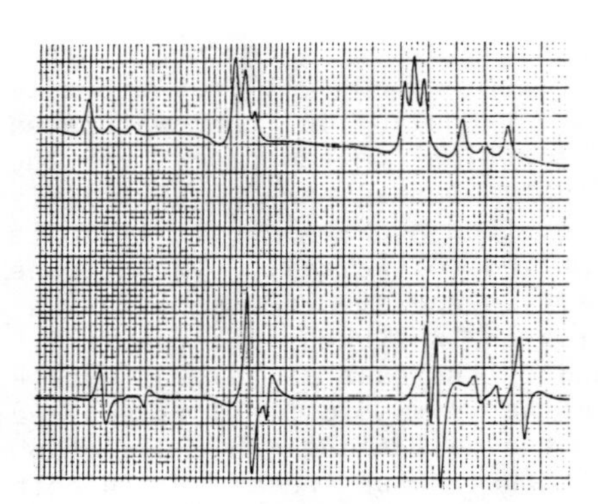

Fig. 8. Lock-in detected signal (lower trace) with saturate absorption spectra (upper trace).

As an example, we show in Fig. 7 the optical system used for the Zeeman modulation experiment, corresponding to Fig. 6 (b). The output of the temperature-controlled diode laser is circularly polarized and applied to the cell as a saturating beam, and reflected backward light is used as a probe beam. An oscillating magnetic field with amplitude about 10 G and frequency of 5 kHz is applied to the cell. Figure 8 shows the lock-in detected signal (lower recorder trace), together with saturated absorption spectra taken simultaneously (upper recorder trace). We did not subtract the Doppler-broadened background in the present case. We see in Fig. 8 that the shape of each signal is approximately given by the derivative of the saturated absorption spectrum, although the relative intensities are not so.

We have tried to lock the laser frequency at the center of one of the Doppler-free spectra, by feeding back the signal obtained in this way to the driving current of the diode laser. Figure 9 shows the frequency stability in terms of Allan variance σ as a function of averaging time τ , for the Zeeman modulation case. Upper and lower curves shows the cases when the P and PID controls are adopted in the feedback system. From Fig. 9, we see that the stability obtained in this system is a little bit lower than the case of saturated absorption shown in Fig. 4. However, the most important point is that the laser frequency is unmodulated to get the error signal. The similar results have been obtained in the case of polarization switching.

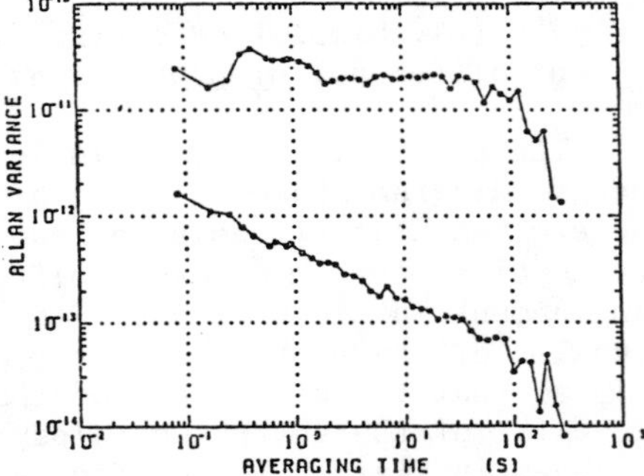

Fig. 9. Frequency stability under no frequency modulation

5. DISCUSSIONS AND CONCLUSIONS

In this paper, we have mentioned, at first, about the frequency stabilization of a diode laser using a saturated absorption spectra of alkali-metal vapors, taking rubidium as an example. The stability obtained is very high, which is better than 10^{-13} in terms of Allan variance for the averaging time 1-500 s, which was estimated from the variation of the error signal. The problem in this frequency stabilization is that we have to modulate the laser frequency to get the error signal to control the diode current. Then, we have developed several spectroscopical methods to get the error signal (derivative shape of Doppler-free resonances) without frequency modulation of laser light. In this paper, we have shown the methods where circular polarizations of light applied to the atoms are alternative switched or an oscillating magnetic field is applied, in stead of frequency modulation. The frequency stability using these signals has been performed, and long-term stability compatible or a little bit lower than that of the case of saturated absorption has been obtained. Details of these works are to be reported elsewhere. Since the optical system is rather simpler than that of saturated absorption spectroscopy, the techniques mentioned here are expected useful in wide variety of applications of diode lasers. We have not discussed about the improvements of short-term stability, i.e. spectral narrowing of diode lasers, which has been studied extensively by using an external mirror or by using a fast feedback system. Since the width of Doppler-free spectra obtained in this work is determined by the spectral width of the diode laser used, so if one use a laser with narrow spectrum, the long-term stability is expected to be improved further. The works in this direction is underway in our laboratory.

ACKNOWLEDGMENT

We should like to thank Professors T. Ogawa and T. Hashi for their useful discussions, and to our former and present students, H. Hori, A. Ibaragi, Y. Kitayama, T. Kawamura, M. Shibutani, and K. Kawamura for their experimental assistance. This work is supported in part by the Ministry of Education, Science and Culture, Japan, under the Grant-in-Aid for Scientific Research.

1) T. Yabuzaki, A. Ibaragi, H. Hori,M. Kitano, and T. Ogawa, Jpn. J. Appl. Phys. 20, L451 (1981).
2) H. Hori, Y. Kitayama, M. Kitano, T. Yabuzaki, and T. Ogawa, IEEE J. Quantum Electron. QE-19, 169 (1983).
3) T. Yabuzaki, H. Hori, M. Kitano, and T. Ogawa, Proc. Int. Conf. on Laser's 83, ed. by Powell (STS Press, McLean, VA, 1983).

IR DIODE LASERS MADE BY MOLECULAR BEAM EPITAXY OF PbEuSe

M. Tacke, P. Norton[*)], H. Böttner, A. Lambrecht, H.M. Preier[**)]
Fraunhofer-Institut fuer Physikalische Messtechnik,
D-7800 Freiburg, West Germany

ABSTRACT

Double Hetero (DH) Structures and Graded Index (GRIN) Structures were made from the ternary material $Pb_{1-x}Eu_xSe$ by variation of the Europium content. The lattice mismatch of these structures does not influence the diode operation notably. Lasers were made with emission wavelength between λ = 8 µm and 3.2 µm. DH lasers with active layers from PbSe show the highest cw operation temperatures reported for DH-Lasers in the mid IR: up to 154 K with currents below 0.5 A. Such lasers cover a spectral range from λ = 8.2 µm to 5.8 µm.

INTRODUCTION

One of the major goals of diode laser development is high temperature operation. Most of the diode lasers used to date in the mid infrared have cw operation temperatures well below 80 K. This hinders wide spread use of diode laser spectrometers, e. g. for gas analysis. DH lasers generally are superior to diffused homostructure lasers with regard to their accessible operation temperatures. There has been considerable effort to develop DH lasers by molecular beam epitaxy (MBE) of IV-VI compounds for the mid IR. D.L. Partin used this approach with MBE of materials based on PbTe. He reported on a DH laser with a maximum cw operation temperature of 147 K, where the active layer consisted of PbTe and the confinement layers of PbEuSeTe[1]. The selenium content of the quaternary material was chosen for optimum lattice match. Our approach is the use of PbSe as base material. The ternary PbEuSe was expected to yield low lattice mismatch[2]. In fact, this ternary has a lattice mismatch to PbSe of some 0.1 % for a composition with band gap energy of 0.4 eV as compared to 0.15 eV of PbSe[3]. For this reason, lasers from the technologically simpler ternary compound were studied.

EXPERIMENTAL RESULTS

Lasers were grown on PbSe substrates. The lower confinement layers had a thickness of some 3 µm. The active layer of DH-lasers had a width around 0.5 µm, while in GRIN lasers the active layers with constant composition had some 0.25 µm width and were bounded

*) Present address: Varian, Stuttgart, West Germany
**) Present address: Spectra Physics, Bedford, USA

with graded layers of comparable thickness. The lasers were laterably defined by contact stripes of 20 µm width; chips of some 300 µm length were cleaved from the wafers and mounted with conventional In pressure contacting.

Figure 1 shows the threshold current of lasers with PbSe active layer as a function of temperature. 150 K operation temperatures were exeeded with total laser currents below 500 mA. Since such lasers were operated at least for short times up to 1 A, 0.5 A is regarded as acceptable for long term operation. In pulsed mode, these lasers worked up to more than 200 K with 5 µs pulses below 3 A current. The lower threshold currents were found for a laser with 4 % Eu in its confinement layers as compared to a 2 % Eu laser. Hence we conclude that lattice mismatch, which should have larger detrimental effect in the 4 % Eu laser, does not noticeably affect the DH laser characteristics.

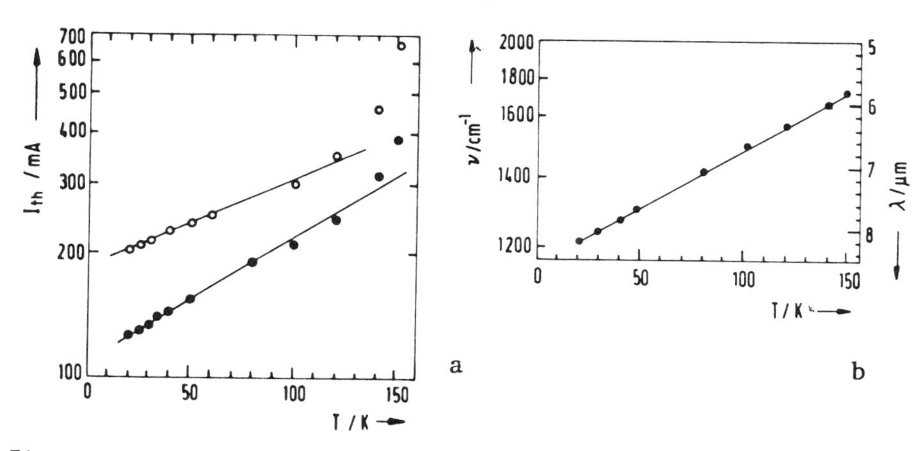

Figure 1. a: The threshold current of lasers with PbSe active layer and ~2 % Eu (o) and ~4 % Eu (•) content in the confinement layers. b: The tuning range that is accessible with these lasers between 20 and 150 K.

Figure 2 shows the emission wavelengths of lasers with PbEuSe as active material. Every dot corresponds to the threshold wavelength of a single laser. There obviously is some spread within the wavelengths for a given wafer. This is well known for MBE wafers, and pronounced in our data, since our MBE system does not provide for a substrate rotation during growth. Above the Eu scale, error bars give our experimental concentration settability. Four runs have been made for 0.8 % Eu within three months, the data of these wafers are shown separated sideways. Within the range of scatter for an individual wafer, and the concentration error, the lasers are comparable.

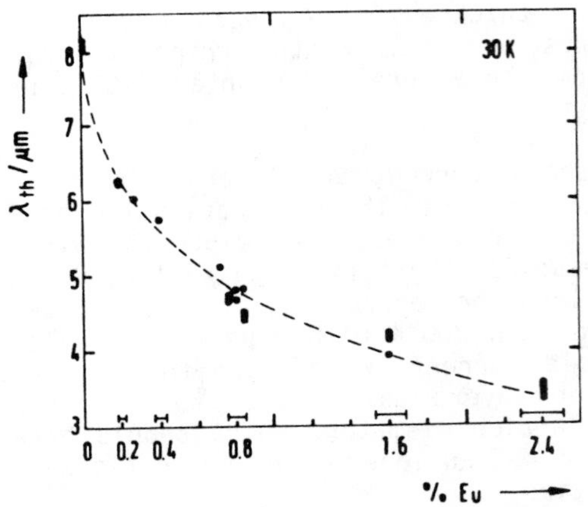

Figure 2. The wavelength at threshold of lasers with different Europium contact in the active layer (T = 30 K).

Probably due to reduced carrier lifetimes in the ternary compounds, the maximum operation temperatures are reduced upon the Europium admixture. The lasers with 2.4 % Eu worked cw up to 100 K and pulsed up to 140 K.

CONCLUSIONS

PbEuSe has proven to be a valuable material for diode lasers in the mid IR. DH lasers from PbEuSe with PbSe active layer operated to the highest temperatures reported for mid IR DH diode lasers. Optimisation of these structures may result in still higher operation temperatures. Due to a reduction in operation temperatures with decreasing emission wavelength, the practical wavelength limit for lasers from this material will be near 3 µm.

REFERENCES

1. Dale L. Partin, Appl. Phys. Lett. 43, 996 (1983).
2. P. Norton, M. Tacke, Journ. of Cryst. Growth 81, 405 (1987).
3. P. Norton, M. Tacke, A. Lambrecht, reported at the IV. European Workshop on MBE, Les Diablerets, Switzerland, April 1987.

HIGH-SPEED MODULATION OF SEMICONDUCTOR DIODE LASERS: EFFECTS OF DEEP-LEVEL TRAPS

E. Bourkoff and X. Y. Liu
The Johns Hopkins University, Dept. of Electrical & Computer Engineering
Baltimore, MD 21218

T. L. Worchesky
Martin Marietta Laboratories, 1450 S. Rolling Rd.
Baltimore, MD 21227

ABSTRACT

We have investigated the effects of deep-level traps on the high-speed modulation behavior of semiconductor diode lasers. A set of three coupled, nonlinear rate equations which model the dynamic behavior of the diode laser has been numerically solved. Our model includes not only the effects of deep-level traps[1], but also of gain saturation[2], spontaneous emission into the lasing modes, varying carrier and photon lifetimes, and varying injection current densities. We show how the laser modulation frequency decreases with increasing trap density and how this behavior depends upon various other laser characteristics.

INTRODUCTION

High-speed modulation of semiconductor diode lasers for the purpose of high-data rate communications has been of interest in recent years. The modulation bandwidth is limited by the resonance frequency of the modulation response or the 3db bandwidth (if such a resonance is absent). The resonance frequency corresponds to the frequency of self-sustained pulsations or to that of relaxation oscillations. We can, therefore, predict the modulation behavior of diode lasers. In this paper, we will discuss the effects of various parameters, including deep-level traps, on both self-sustained pulsations and relaxation oscillations.

RATE EQUATIONS

There exist several models for both self-sustained pulsations and relaxation oscillations, and the model of optically saturable absorbing centers is a common one. In a paper by Chik, Dyment, and Richardson[3], the authors point out that their experimental results favor the saturable absorber model and did not agree with the deep-level trap model. But some experiments do show the presence of deep-level traps in the active layers. The actual mechanism causing self-sustained pulsations is far from clear. It is highly desirable to further investigate the effects of deep-level traps in order to get a better understanding of this mechanism.

We have adopted a set of three coupled nonlinear rate equations. In our model, we combined Copeland's and Channin's models[1,2], taking into account deep-level traps, gain saturation, and spontaneous emission into the lasing mode. The rate equations follow:

$$dS/dt = g(N-N_e)S - S/\tau_p + \gamma N/\tau_c - \sigma_o cS(T_o-T) \quad (1)$$

$$dN/dt = J/ew - g(N-N_e)S - N/\tau_c + dT/dt \quad (2)$$

$$dT/dt = \sigma_o cS(T_o-T) - \sigma_e v_t NT \quad (3)$$

where

$$g(S) = g_o/(1+S/S^*) \tag{4}$$

S, N, and T are the photon, carrier, and trap densities, respectively. J is the injection current density. g is the gain parameter, for which we adopt the gain saturation relation in a two-level system. S^* is the gain saturation parameter, γ the probability of spontaneous emission into the lasing mode, N_e the minimal carrier density to achieve transparency. τ_c and τ_p are the carrier and photon lifetimes; T_o is the total trap density; σ_o and σ_e are the optical and electron capture cross-sections; c is the light speed in the active layer; v_t the carrier thermal velocity; and w the thickness of the active layer. Unless otherwise specified, we will use the values listed below throughout the paper:

$g_o = 4.0\times10^{-6}$ cm^3/sec, $N_e = 1.667\times10^{18}$ cm^{-3}, $\gamma = 10^{-5}$, $S^* = 1.75\times10^{15}$cm^{-3}
$\tau_c = 3\times10^{-9}$ sec, $\tau_p = 3\times10^{-12}$ sec, $T_o = 2\times10^{17}$ cm^{-3}, $\sigma_o = 3\times10^{-16}$ cm^2
$\sigma_e = 1.5\times10^{-17}$ cm^2, $c = 8.33\times10^9$ cm/sec, $v_t = 4.42\times10^7$ cm/sec, $w = 0.2$ μm

When traps are absent, the threshold current is given by

$$J_{th}=(ew/g_o\tau_c\tau_p)(1+g_o\tau_pN_e) \tag{5}$$

But with traps present, the threshold current will be slightly increased. Based on the assumption that $\gamma \approx 0$,and $S\approx0$ at threshold, we get, from the steady-state rate equations, the following threshold current:

$$J_{th}=(ew/g_o\tau_c\tau_p)(1+g_o\tau_pN_e+\sigma_oc\tau_pT_o) \tag{6}$$

The increase in the threshold current due to the presence of traps is about 7% based on the values listed above. The definition for the threshold current given in Eq. 6 is consistent with the calculated DC response from the steady-state rate equations.

RESULTS

The dynamic rate equations were numerically solved, and we obtained curves of oscillation frequency vs. both trap and injection current densities, as in Ref. 1. We notice that the curve of frequency vs. trap density is very similar to the curve of frequency vs. density of saturable absorbers given by Ref. 3.

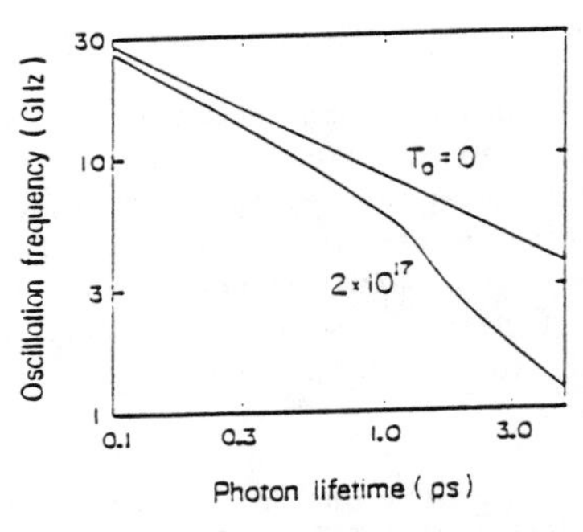

Fig. 1. Oscillation frequency vs. photon lifetime.

Fig. 1 shows the numerically-obtained dependence of oscillation frequency on the photon lifetime. With T_o=0, $\gamma \approx 0$, and $S^*\approx\infty$, the oscillation frequency given by Lau and Yariv[4] reduces to

$$f_T\approx(1/2\pi)\sqrt{(J/J_{th}-1)/\tau_c\tau_p} \tag{7}$$

It is obvious that log f_T depends linearly on log τ_p, in agreement with Fig. 1. However, when $T_o\neq0$, we no longer obtain a linear dependence. Similar behavior of the oscillation frequency is observed when the carrier lifetime is varied.

Gain saturation and spontaneous emission have little effect on oscillation frequency. On the other hand, they have profound effects on the damping rate of relaxation oscillations and the onset of self-sustained pulsations. The larger γ or $1/S^*$, the larger the damping rate, causing a higher trap density for the onset of self-sustained pulsations.

We have also numerically investigated the large-signal pulse modulation with 50% duty cycle, the results shown in Fig. 2. For low bias level, we observe self-sustained

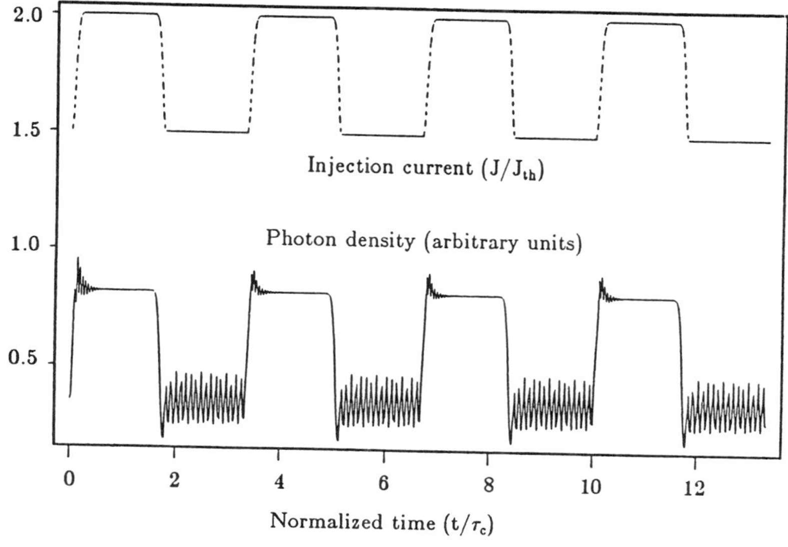

Fig. 2. Pulse modulation.

pulsations when the pulse is off, and overshooting and ringing when the pulse is turned on. For higher bias level (not shown), the damping rate is much larger and self-sustained pulsations gradually become relaxation oscillations. With higher pulse amplitude, the overshooting and ringing are almost completely suppressed when the pulse is turned on. When the pulse is turned off, we observe relaxation oscillations instead of self-sustained pulsations. As expected, higher modulation frequency can be achieved with higher levels of bias.

SUMMARY

The presence of deep-level traps on the order of 10^{16} cm^{-3} will cause a slight increase in the threshold current of about a few percent, as well as an appreciable decrease in the oscillation frequency. The modulation behavior of semiconductor diode lasers clearly depends upon the trap density and upon the carrier and photon lifetimes. The gain parameter and spontaneous emission have little effect on the oscillation frequency, but they have profound effects on the damping rates of relaxation oscillations or on the on-set of self-sustained pulsations. If traps are present, the oscillation frequency is lower for a given normalized injection-current level. Incorporation of both the deep-level trap model and saturable absorbing center model is presently under investigation and comparisons with experiments are being made.

REFERENCES

1. J. A. Copeland, Electron. Lett. 14 , 809 (1978).
2. D. J. Channin, J. Appl. Phys. 50 , 3858 (1979).
3. K. D. Chik, J. C. Dyment, and B. A. Richardson, J. Appl. Phys. 51 , 4029 (1980)
4. K. Y. Lau and A. Yariv, IEEE J. Quantum Electron. QE-21 , 121 (1985)

ULTRASHORT PULSE HIGH INTENSITY LASER ILLUMINATION OF A SIMPLE METAL

H. M. Milchberg, R. R. Freeman and S. C. Davey
AT&T Bell Laboratories
Murray Hill, NJ 07974

ABSTRACT

We have observed the self-reflection of intense, sub-picosecond 308 nm light pulse incident on a planar Al target and have inferred the electrical conductivity of solid density Al. The pulse lengths were sufficiently short that no significant expansion of the target occurred during the measurement.

We describe the experimental study of the electrical conductivity of a solid density material, in this case a simple Drude metal, over an extended range of elevated temperature with *little or no change in its density*.

The self reflection of a laser pulse, focused onto a smooth target at fixed pulsewidth and spot size, was monitored over four orders of magnitude in energy. The frequency shifts of the reflected light were also recorded as a function of the laser energy for a given pulsewidth. As discussed below, these novel frequency shift measurements determined the velocity of the interface expansion and the average electron temperature during the pulse. For pulsewidths of $\leq 400 fsec$, we estimate that the interface expansions did not exceed $\approx \lambda/15$.

Pulses of light from a synchronously mode-locked oscillator operating at 616 nm and compressed to 400 fsec were amplified to $\approx$ 0.6 mJ at 10 Hz, doubled in KDP to 308 nm and amplified by two XeCl excimer amplifiers up to 7 mJ/pulse. The overall ASE component of the total laser energy output was measured to be less than 1 part in 10^3. Elaborate precautions to eliminate ASE were necessary, for we found that as little as 1% fractional ASE on the leading edge of the pulse was sufficient to decrease greatly the measured reflectivity, presumably due to the production of long scale length vapour or plasma in front of the target.

The linearly polarized pulses were focused with a 200 mm lens (f/10) to a spot area of $10^{-5} cm^2$ in vacuum. The targets were mounted at 45 degrees with respect to the beam, and the position of the tightest focus was adjusted and monitored using the time integrated x-ray energy yield from the plasma as a guide. A computer controlled translation device moved the target such that each pulse at 10 Hz encountered fresh material.

Figure 1 shows the reflectivity of S and P polarized light pulses of 400 fsec duration from a 400Å thick Al film for intensities between 5×10^{11} and 10^{15} W/cm^2. The reflectivity shows three distinct regions as a function of intensity: an initial drop for intensities $\leq 10^{14}$ W/cm^2, a minimum value extending over a factor of 3 in intensity, and an increase in the reflectivity beyond $\approx 3\times 10^{14}$ W/cm^2.

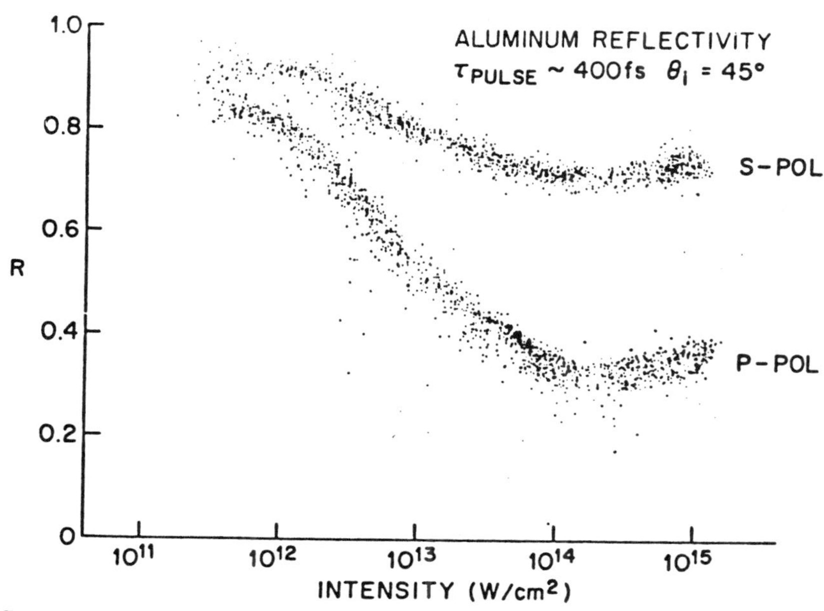

Figure 1: Self-reflection data for S and P polarized light. Each point represents a laser pulse.

We have analyzed the reflectivity data on the basis of a Drude model of AC conductivity, where the complex index of refraction, n, is written: $n^2 = 1 + i4\pi\sigma/\omega$, where σ is the complex AC conductivity. This complex index of refraction is incorporated in the Fresnel equations[1], which describe the reflectivity of a lossy interface. The assumption of a Drude model for the conductivity is based on the nearly free electron Fermi surface of Al[2] and its lack of interband transitions at the laser wavelength. The parameter of interest, the DC conductivity, σ_{dc}, is included in σ.

In this experiment we determine the dependence of the DC conductivity, σ_{dc} $(=\dfrac{N_e e^2}{\nu m_e})$ upon T_e, the electron temperature.

In analyzing the data it is necessary to relate T_e to the laser intensity since the relevant material properties are functions of T_e. In a separate series of measurements[3], we recorded the spectrum of the reflected light on a shot-to-shot basis using an optical multichannel analyzer. As a function of increasing intensity the spectrum broadened and *shifted* to higher frequencies. We interpret this effect in terms of instantaneous mass motion normal to the target, where the light is reflected by electrons moving in hot material with an average directed velocity given[4] by $v_o^2 = (2/(\gamma - 1))^2 \times (ZkT_e/m_i)$ for unsteady flow into vacuum. Here, m_i is the ion mass, Z is the effective ion charge (determined in a self-consistent manner from a local thermal equilibrium (LTE)[5] model), and $\gamma = C_p/C_v$, where $C_{p(v)}$ is the electron heat capacity at constant pressure (volume). A complete range of spectral shift vs intensity data was taken to correlate the intensity of the laser with the electron temperature in the target. Figure 2 is a plot of v_o^2 vs laser input energy for p polarized light. Note the linear behavior for laser energies below ≈ 1mJ.

The next step in determining σ_{dc} is to establish how N_e depends upon T_e as a function of temperature. N_e can be readily modeled within the framework of LTE provided all the microscopic ionization and recombination rates are sufficiently high that nearly equilibrium conditions for sufficiently low ion stages are obtained during the heating/reflecting pulse. Because the Fermi temperature of cold Al is nearly 12 eV, with all 3 of aluminum's valence electrons itinerant, LTE shows N_e to be a *weak* function of temperature for temperatures less than 50 eV.

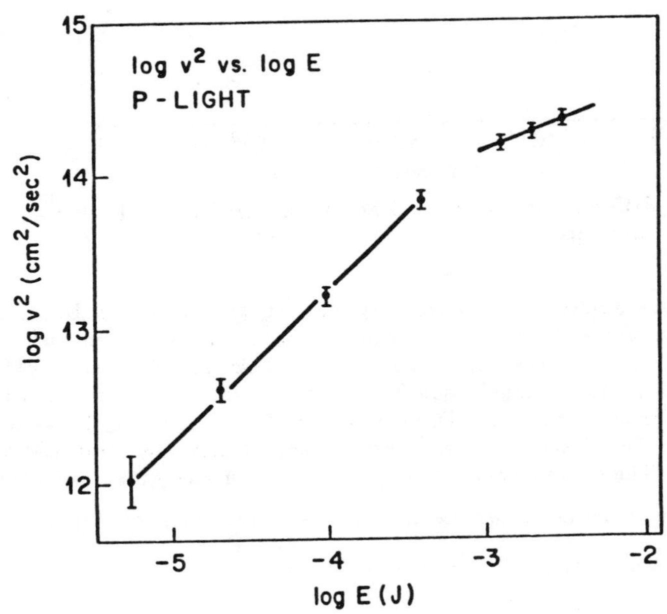

Figure 2: Log Vo^2 vs. log E, where Vo is the material expansion velocity obtained from Doppler shifts and E is total *incident* energy. For P polarization.

With the determination of T_e and the calculation of $N_e(T_e)$, the sum of the electron thermal energy, the energy of the directed mass motion from the target and the energy stored in the various ion stages is equal, to within experimental uncertainties, to the measured absorbed energy for both S and P light. Thus our analysis passes an important check for internal consistency: energy balance. This result, along with the result of Fig. 2, suggests that the material is "intensity integrating", i.e., its temperature depends primarily on the incident energy at these pulsewidths for energies below $\approx$ 1mJ. This has implications for the production of short pulse x-rays from these plasmas and suggests that, at least in this regime, x-ray pulses are likely to be of longer duration than the driving laser pulses.

We have used the experimental results and the results of the LTE model to numerically invert the Fresnel equations to determine σ_{dc} vs T_e. On Fig. 3 the resistivity, $\rho,(=1/\sigma_{dc})$ vs I and T_e is plotted for both S and P polarized light. The similarity of the S and P results supports the validity of our Drude/Fresnel model. These curves show clearly the three regions of transport, including a well defined "saturation" in the resistivity near 200 $\mu\Omega-cm$ at $T_e > 40$ eV.

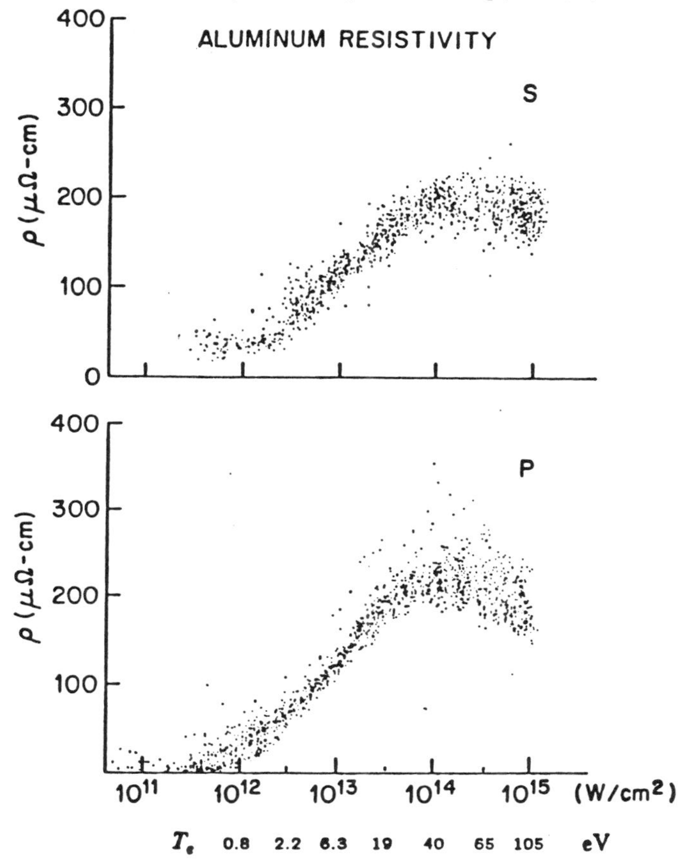

Figure 3: Resistivity vs. electron temperature for S and P polarizations.

At low T_e, the resistivity increases (conductivity decreases) approximately as T_e, a result which is expected: Since the laser couples energy to the electrons, for the short pulse lengths and high energies employed here[6], $T_e >> T_{phonon}$; under these circumstances the electron scattering is dominated by phonon creation, a process whose rate scales as the density of electron final states, or as T_e, for T_e not too much larger than T_{Fermi}.

At the highest temperatures and intensities, our data is consistent with the expectations of a high temperature plasma; that is, a $T_e^{-3/2} \times Z$ dependence[7] of the resistivity. In this limit, the Fermi-Dirac statistics become Maxwellian, all final states are accessible to the scattered electrons, and the Coulomb scattering cross section averaged over a Maxwellian velocity distribution directly yields a $T_e^{-3/2}$ temperature dependence.

The maximum in the resistivity near $200\mu\Omega-cm$ over nearly an order of magnitude in laser intensity is interpreted[7] as "resistivity saturation", that is, there exists a minimum mean free path of the electrons, regardless of the degree of disorder of the material or the temperature. We can estimate this length by computing $l_{\min} = (\sigma_{dc} m_e / N_e e^2) \cdot v_e$, where $v_e = \sqrt{2kT_e/m_e}$ is the electron velocity at the resistivity maximum. For $T_e = 50$ eV, $l_{\min} \approx 3\text{Å}$. This value is comparable to the interatomic spacing in Al, and suggests the picture that the now localized electron cannot travel less than an interatomic dimension before suffering another scattering event.

This rather remarkable record of the change in the controlling mechanisms of the electrical transport of a simple metal is possible only because the material could be heated to electron temperatures up to and far in excess of its Fermi temperature, without significantly changing the density.

The authors would like to acknowledge discussions with R. C. Dynes and R. M. More.

REFERENCES

1. M. Born and E. Wolf, *Principles of Optics,* (Pergamon Press, N.Y., 1980).
2. N. W. Ashcroft and N. D. Mermin, *Solid State Physics,* (Holt, Rhinehart and Winston, N.Y., 1976).
3. R. R. Freeman and H. M. Milchberg, to be published.
4. Ya. B. Zeldovich and Yu. P. Raizer, *Physics of Shock Waves and High Temperature Hydrodynamic Phenomena,* (Academic Press, N.Y., 1966).
5. H. R. Griem, *Plasma Spectroscopy,* (McGraw-Hill, N.Y. 1964).
6. J. G. Fujimoto, J. M. Lui, E. P. Ippen, and N. Bloembergen, Phys. Rev. Lett. *53,* 1837 (1984).
7. N. F. Mott, *Metal-Insulator Transitions,* (Taylor and Francis, London, 1974).

PRODUCTION OF INTENSE, COHERENT, TUNABLE, NARROW-BAND LYMAN-ALPHA RADIATION*

R. S. Turley, R. A. McFarlane
Hughes Research Laboratories, Malibu, CA 90265

D. G. Steel, J. Remillard
University of Michigan, Ann Arbor, MI 48109

ABSTRACT

Nearly transform limited pulses of 1216 Å radiation have been generated by sum frequency generation in 0.1 to 10 torr of mercury vapor. The summed input beams, consisting of photons at 3127 Å and 5454 Å originate in 1 MHz band-width ring-dye laser oscillators. The beams are amplified in pulsed-dye amplifiers pumped by the frequency doubled output of a Nd:YAG laser. The 3127 Å photons are tuned to be resonant with the two-photon 6^1S to 7^1S mercury transition. The VUV radiation can be tuned by varying the frequency of the third non-resonant photon. We have also observed difference frequency generation at 2193 Å and intense fluorescence from the 6^1P state at 1849 Å. We have studied the intensity and linewidth dependance of the 1849 Å fluorescence and 1216 Å sum frequency signals on input beam intensity, mercury density, and buffer gas pressure and composition.

INTRODUCTION

We are currently involved in a number of investigations which require a coherent radiation source near the hydrogen Lyman-α wavelength at 1216 Å. These include a measurement of the two-photon ionization cross section of atomic hydrogen at Lyman-α, technology for cooling and trapping of atomic hydrogen, and measurements of the heights and widths of the (n=2) Feshbach and Shape resonances in H^-. These experiments require a source that is narrow-band (preferably transform limited), broadly tunable, bright, and coherent. We have constructed such a source and measured the effects of mercury density, phase matching conditions, and strong input fields on the output intensity and linewidth.

VUV PRODUCTION

The Lyman-α radiation is produced by summing in mercury vapor two photons at 3127 Å and a third photon at 5454 Å. The input beams originate in 1 MHz band-width ring-dye lasers. They are amplified in pulsed-dye amplifiers which are pumped by the doubled output of a Nd:YAG laser. The resulting beams have transform-limited linewidths of 300 MHz. The beams were focussed with a 25 cm focal length achromatic lens to a 100 μm spot in a cell containing 0.1 torr of mercury. The intensities of the beams at the focal point were 2×10^9 W/cm^2 at 3127 Å and 1×10^{10} W/cm^2 at 5454 Å. The mercury cell construction was similar to the cell described in reference 1. In addition to the sum frequency generation, we also observed difference frequency generation at 2193 Å and fluorescence from the 6^1P state at 1849 Å.

For most of the measurements reported here, the mercury cell was operated at a temperature of 180°C. A few of the measurements were taken at 150°C. Absorption measurements with a low pressure mercury lamp determined that this corresponds to a mercury density of 10^{15} cm^{-3}.

* The work reported here was partially supported by the Air Force Office of Scientific Research under Contracts F49620-82-C-0004 and F49620-87-C-0083.

INTENSITY AND LINEWIDTH OPTIMIZATION

The peak Lyman-α intensity occurred when the 3127 Å beam was tuned 0.5 cm^{-1} to the blue of the low intensity resonance wavelength of 31964.1 cm^{-1}. The 1849 Å and 2193 Å signals were maximum when the 3127 Å signals remained fixed at 31964.1 cm^{-1}. The full width at half maximum of the 1849 Å fluorescence signal as a function of the wavelength of the 3127 Å beam was 2.6 cm^{-1}, which is indicative of power broadening. Our results can be compared to the strontium measurements of Scheingraber and Vidal[2] who reported a similar blue Stark shift of the resonant frequency. Mahon and Tomkins[1] have reported an observation of the Stark broadened absorption linewidth in mercury at these intensities, but did not report the blue shift in the resonance frequency.

To further investigate the saturation of the 6^1S to 7^1S transition, we measured the fluorescence signal at 1849 Å as we varied the mercury cell temperature, input beam intensity, and buffer gas pressure and composition. The 1849 Å signal is proportional to the population of the 7^1S state since it results from a cascade from the 7^1S state to the 6^1S ground state through the 6^1P state.

The intensity of the fluorescence signal decreased by a factor of 5 when the 5454 Å beam was allowed to pass through the cell. This indicates that the branching ratios for sum and difference frequency generation are large enough to significantly reduce the population of the 7^1S state. Hence, although Smith et al.[3] noted a problem with stimulated emission at 1.0 μm resulting from 7^1S to 6^1P transitions in a plane wave geometry, this will not be a problem for Lyman-α output in a tight focussing geometry like ours. In order to maximize the 1849 Å signal, the remaining measurements were made without the 5454 Å beam being present.

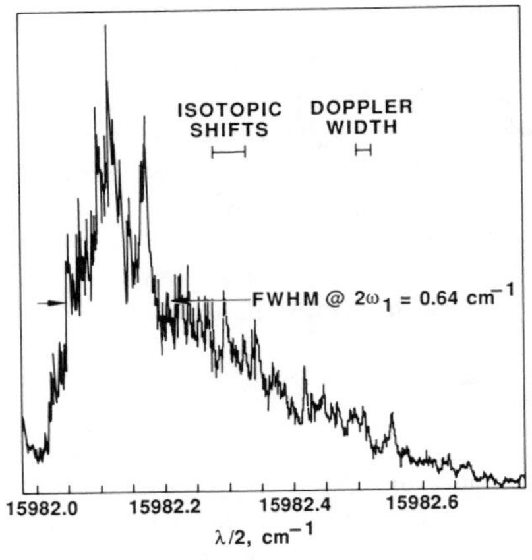

Figure 1. Fluorescence as a function of detuning of the 3127 Å input beam.

The fluorescence intensity as a function of the frequency of the 3127 Å input beam is shown in Figure 1. The measured 0.64 cm^{-1} AC Stark broadened linewidth is significantly larger than either the linewidth of the input beam, the Doppler broadening, or the isotopic frequency shifts.

Figure 2 is a plot of the intensity of the 1849 Å fluorescence as a function of the temperature of the mercury cell. Increasing the mercury cell temperature beyond 145°C reduces the fluorescence output. Absorption measurements with a low pressure mercury lamp show a monotonic decrease in cell transmission of 2537 Å light up to 220°C. This suggests that the 7^1S population decrease is caused by an increase in an excited state collisional process such as excimer formation.[4]

We further investigated the saturation of the two photon absorption by monitoring the 1849 Å fluorescence as a function of input beam intensity. Figure 3 is a plot of this signal as a function of the energy of the 3127 Å pulses. The smooth curve in the figure is a fit to the functional form $I = \alpha E^x$, where I is the florescence signal and E the pulse energy in mJ. The best fit was for $\alpha = 0.99$ and $x = 1.8$. The fact that the functional dependance does not deviate from the expected low intensity quadratic form implies that the occupation probability of the 7^1S state is much less than 1. In other words, the cascade from the 7^1S state is sufficiently fast to inhibit fluorescence reduction even though the input intensities are large enough to produce

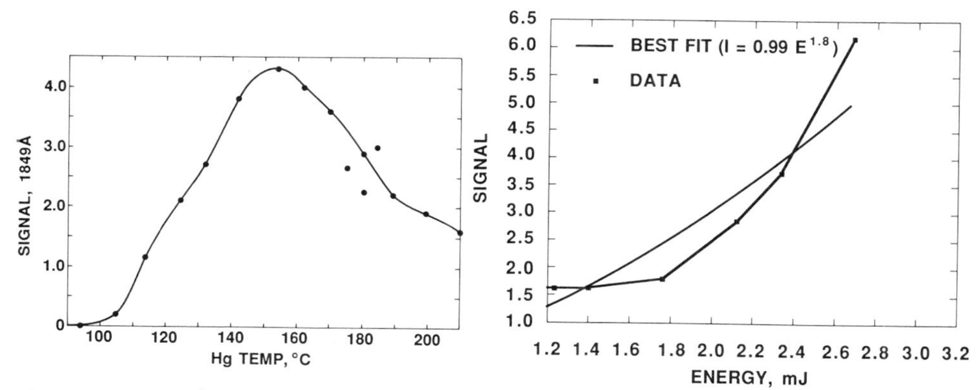

Figure 2. 1849 Å fluorescence as a function of mercury cell temperature.

Figure 3. Fluorescence signal as a function of 3127 Å pulse energy.

significant power broadening and AC Stark shifts. This is different from the observations of Scheingraber and Vidal[2] in strontium, where a dip in third-harmonic generation is attributed to significant population in the upper two-photon resonant state.

Switching buffer gas from helium to krypton resulted in a slightly less intense 1849 Å signal at a given input power and cell temperature. However, the fluorescence had the same dependance on the input intensity for both gasses. Varying the krypton or helium buffer gas pressure resulted in no detectable change in the florescence. However, we do expect the partial pressure of krypton to have a strong effect on the sum frequency generation. Krypton is negatively dispersive in this wavelength regime, affecting the phase matching conditions of the sum frequency process.

CONCLUSIONS

In summary, sum frequency generation in mercury with transform limited input beams provides a powerful spectroscopic source for production of tunable, bright, narrow-band, coherent radiation near the hydrogen Lyman- α transition. Although the VUV output increases rapidly with mercury temperature, apparent excimer formation limits the maximum temperature to 145°C.

With 3 mJ 300 MHz linewidth beams at 3127 Å and tight focussing, the power broadening is larger than the isotopic frequency shifts, the Doppler width, or the input beam linewidth. This ensures that the beams interact with each atom in the mercury column. In spite of the high intensity of the input beams, there is evidence that the population of the 7^1S state is not significant enough to reduce the intensity of sum frequency generation.

REFERENCES

1. R. Mahon and F. Tomkins, IEEE JQE QE-18, 5 (1982).
2. H. Scheingraber and C. R. Vidal, Opt. Commun. 38, 75 (1981).
3. A. V. Smith, G. R. Hadley, P. Esherick, and W. J. Alford, Opt. Lett. 12, 708 (1987).
4. S. M. Skippon and T. A. King, Appl. Phys. B37, 223 (1985).

STUDY OF SOFT-X-RAY GENERATION BY LASER-HEATING SOLID AND GASEOUS Ta PLASMAS WITH SUBPICOSECOND PULSES

H. W. K. Tom and O. R. Wood, II
AT&T Bell Laboratories, Holmdel, New Jersey 07733

ABSTRACT

Using 100 fsec optical pulses, we are able to couple laser energy into solid Ta before appreciable vaporization takes place and thus have been able to compare the efficiency of x-ray emission in the 17 to 30 nm region from laser-heated solid and gaseous density Ta plasmas. We find that a 0.4% x-ray conversion efficiency from a solid density Ta plasma requires a laser fluence of 200 J/cm^2 whereas a gaseous Ta plasma with optimal density profile requires a fluence of only 3 J/cm^2. This fact is explained by the short attenuation length and shorter pulse duration of x-rays emitted from the heated solid density plasma.

INTRODUCTION

Recent interest in optical excitation of solid targets with ultrashort pulses has been motivated by the possibility of generating short duration pulses of x-rays useful for pumping short wavelength lasers[1] and for time-resolved studies of materials. Short durations are expected because the leading edge of the x-ray pulse should follow the laser heating pulse while the falling edge may follow the rapid diffusive cooling of the solid density plasma.[2] To help understand the physical processes involved in short pulse excitation of solid targets, we have performed a comparative study of the efficiency of x-ray emission in the 17 to 30 nm region from solid and gaseous Ta plasmas.

LASER HEATING GASEOUS PLASMA

In Fig. 1 we show the x-ray yield in 4π steradians in the 17 - 30 nm region[3] as a function of delay-time between two 2.6 J/cm^2, 100 fsec duration pulses.[4] The first pulse that arrives heats the solid and starts the evolution of a gaseous plasma. The second pulse heats the gaseous plasma. In contrast to experiments using longer duration pulses, here the gas evolves in the dark, and the gas hydrodynamics during the heating pulse are negligible. We see that the x-ray yield increases dramatically with delay time up to a saturation value. At times greater than a few psec the data may be fit (solid curve in Fig. 1) by a saturating exponential of the form $1 - \exp(-\Delta t/\tau)$ where Δt is delay time and τ = 22 psec. The efficiency of heating the optimally evolved gas is 19 times greater than the efficiency of heating the unprepared solid with a single pulse of 5.2 J/cm^2 fluence. This same behavior is observed for 600 fsec pulses except τ = 66 psec. Similar results to those shown in Fig. 1 have been obtained recently by Kuhlke et al.[5] using 300 fsec pulses at 310 nm

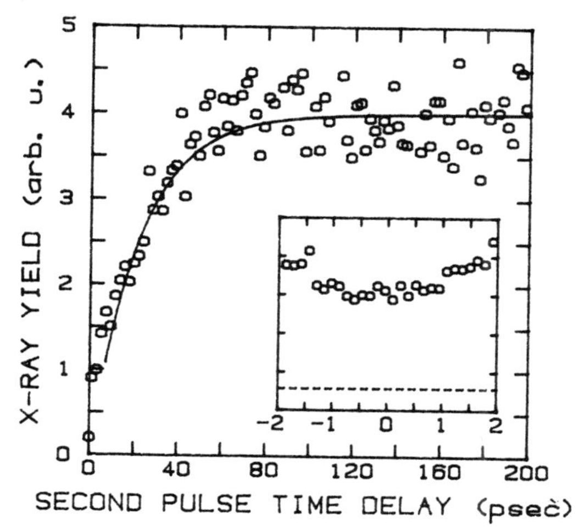

Figure 1. X-ray yield in a laser-heated Ta plasma versus delay time between prepulse and heating pulse.

in spite of the fact that 13% of their laser energy was in the form of amplified spontaneous emission (which could preheat the target).

The saturation shape of the curve is easily explained by the laser energy being most efficiently absorbed by inverse Bremsstrahlung and resonance absorption in the region of the gaseous plasma that has an electron density around the critical density. As the thickness of this region grows to exceed the laser absorption length, the laser energy absorbed and, therefore, the x-ray conversion approaches saturation. The approach can be shown to be exactly exponential for the case of isothermal expansion.[6]

The inset in Fig. 1 shows the yield at the earliest time delays. At zero delay, the two pulses act as a single pulse of 5.2 J/cm^2 fluence and produce 4 times the x-ray yield of two infinitely-time-separated 2.6 J/cm^2 pulses (dashed line in Fig. 1). We note that no fast drop is observed during the first psec. Surprisingly, the yield is the same whether the laser energy is deposited in a single 100 fsec pulse, a single 600 fsec pulse, or two 100 fsec pulses delayed by 1 psec. This indicates that the material that is emitting x-rays remains sufficiently hot for at least one psec. Since the solid density plasma should cool by electron thermal diffusion on a 100 - 200 fsec timescale, and there is considerable blow off in 1 psec, this indicates that x-rays are largely emitted from ablated material and not from the hot solid density material. Because the material in the blow-off cools mostly by expansion, the duration of the x-ray emission is probably of order 10 psec. We have attempted to measure the x-ray emission time with an x-ray streak camera and found it to be less than 20 psec[7] (the time resolution of our instrument).

LASER-HEATING SOLID DENSITY PLASMA

The x-ray conversion efficiencies for laser-heating the solid with single short pulses are shown in Fig. 2. The conversion efficiencies of 100 fsec (x's) and 600 fsec (O's) duration pulses approach 0.4% at the highest input intensities (> 5 X 10^{14} W/cm^2). Consistent with the lack of sensitivity to how the laser energy is coupled into the system during the first psec (inset to Fig. 1), we find that 100 fsec and 600 fsec duration pulses have the same intensity dependence, namely $(\text{Intensity})^2$, and pulses of equal fluence have the same conversion efficiency. We also plot the conversion efficiency for laser heating the optimally evolved gas density profile with 600 fsec pulses (I's). Our 'optimal' profile was created by prepulsing with a 9 J/cm^2, 600 fsec duration pulse 200 psec before the heating pulse. The conversion efficiency has an (Intensity)$^{1.5}$ dependence and reaches 0.4% with laser intensities of only 10^{13} W/cm^2.

DISCUSSION AND THEORY

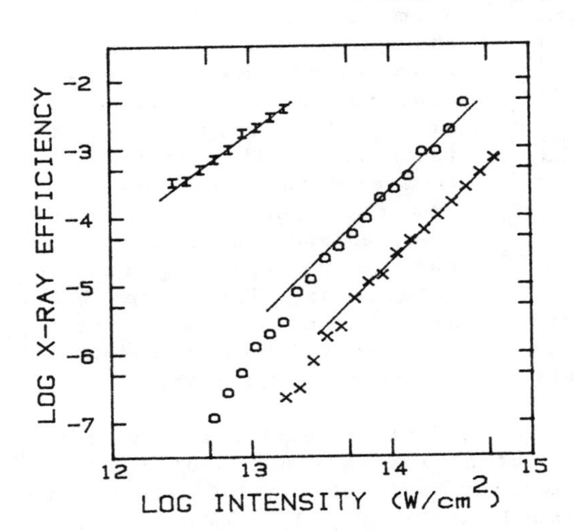

Figure 2. Efficiency of converting incident laser energy to radiation in the 17 - 30nm region versus intensity

Why the x-ray yield is so much greater when heating gaseous rather than solid density plasmas can also be explained with relatively simple physical arguments. For the gas we'd expect some incident laser energy to be 'wasted' on heating overdense and underdense regions of the plasma and for only ~1/3 of the laser energy to be absorbed at the critical density. The thickness of this region would be about the inverse Bremsstrahlung length, e.g. 5000 Å. Detailed numerical simulations including laser absorption and electron thermal diffusion,[8] ionization kinetics[9] and the Ta ionization potentials[10] support this picture. For our 2.6 J/cm^2, 100 fsec duration pulse heating a gaseous profile prepared with a 2.6 J/cm^2 prepule, the simulation gives an average ionization stage of 3.0 and an electron temperature of 13 eV, 1 psec after the heating pulse. The peak intensity of the recombination radiation[14] in the 17 - 30 nm region emitted into 4π is 130 MW per cm^2 of heating laser beam cross section. This corresponds to an x-ray conversion efficiency for the total (prepulse and heating pulse) laser energy, of 0.26%, assuming a pulse duration of 100 psec.

For the solid, the laser energy is coupled into the top 1000 Å by electron diffusion during the 100 fsec timescale of the pulse,[8] yet only atoms in the top 100 Å can radiate due to the strong absorption in transition metals[11] and overdense plasmas in our detection bandwidth. Taking reflective loss (~2/3) into account, our single 5.2 J/cm^2 100 fsec duration pulse will provide 18.2 eV/atom which will divide between about 7 eV in ion potential energy (Z = .8) and 9.5 eV electron temperature. The peak power emitted from the top 100 Å is 24 MW per cm^2 of heating beam cross section: assuming a 10 psec duration, the x-ray conversion efficiency is 4.6×10^{-5}. This value is somewhat larger than the experimental value given in Fig. 2

(7×10^{-6} at 5×10^{13} W/cm^2). The most intense 100 fsec pulse shown in Fig. 2 provides 55 J/cm^2 of incident fluence. Using the same model this pulse would supply 182 eV/atom over a 1000 Å slab and in equilibrium would divide with 55 eV in ion potential energy ($Z = 2.9$) and 30 eV electron temperature. The x-ray power would be 13 GW per cm^2 of heating beam cross section. If this radiates for 10 psec, then the x-ray conversion efficiency would be 0.15% in close agreement with the experimentally measured value of 0.1%.

CONCLUSIONS

The most important result from the point of view of using short laser pulses to create ultrashort x-ray pulses is that the material that radiates most efficiently does not cool rapidly at solid density diffusive rates and thus the x-ray emission pulse may have a falling edge determined by expansion cooling making the pulse duration as long as 5 to 10 psec. At still higher fluences, increased reflection,[12] increased diffusive cooling of the highly excited and ionized solid and nonlinear effects such as multiphoton ionization and stimulated light scattering may become significant. In this region, laser-heating of solids may become more efficient. Experiments with more energetic 100 fsec duration pulses are needed to resolve this question.

REFERENCES

1. R. G. Caro et al. in Short Wavelength Coherent Radiation: Generation and Applications, D. T. Attwood and J. Bokor, eds. (AIP, N. Y., 1986), p. 145
2. R. W. Falcone and M. M. Murname, in Short Wavelength Coherent Radiation: Generation and Applications, D. T. Attwood and J. Bokor, eds. (AIP, N. Y., 1986), p. 81.
3. Absolute calibration of radiation in 4π is referenced to Cd^+ photoionization yield, see, for example O. R. Wood, II et al., Opt. Lett. 11, 198 (1986).
4. R. L. Fork et al., Appl. Phys. Lett. 41, 223 (1982).
5. D. Kuhlke et al., Appl. Phys. Lett. 50, 1785 (1987).
6. T. P. Hughes, Plasmas and Laser Light, (Wiley, N. Y., 1975), p. 285.
7. O. R. Wood, II et al., in XV International Conference on Quantum Electronics Technical Digest Series 1987, Vol. 21, (Optical Society of America, Washington, D. C. 1987), p. 2.
8. S. D. Brorson et al., Phys. Rev. Lett. 59, 1962 (1987).
9. N. Nakano and H. Kuroda, Phys. Rev. A 27, 2168 (1983).
10. T. A. Carlson, et al., Atomic Data 2, 63 (1970).
11. Absorption depth in the 20 - 40 eV region for Cu from H.-J. Hagemann, W. Gudat and C. Kunz, J. Opt. Soc. Amer. 65, 742 (1975).
12. H. Milchberg et al., Bull. Amer. Phys. Soc. 32, 1628 (1987).

SHORT-WAVELENGTH LASING AND SUPERFLUORESCENCE FROM CHANNELED BEAMS

G. Kurizki, Chemical Physics Department, Weizmann Institute of Science, Rehovot, Israel 76100

One of the main trends in the area of stimulated X-ray emission has been the development of sources based on beams of relativistic or subrelativistic emitters. The emission from such sources is continuously tunable, owing to the Doppler shift towards progressively shorter wavelengths as the beam velocity increases. The better known sources in this category are various versions of free-electron lasers (FELs), encompassing structures wherein a spatially periodic perturbation acts either on the electron beam or on the emitted radiation. This article is intended to draw attention to recent work[1,2] concerning a different type of high-velocity emitters, namely, relativistic electrons or positrons channeled in crystals or in artificial periodic structures. The appeal of such emitters is the possibility to use them as a source of fully coherent, tunable, stimulated X-radiation.

The possibility to achieve coherence by stimulating channeling radiation (CR) stems from its analogy to the radiation of fast-moving two-level atoms. This is an essential feature of channeling[3], a type of motion occurring whenever a charged particle enters a crystal at a sufficiently small angle relative to a set of crystal planes or axes, so that the potential exerted on it by these planes (axes) is larger than the kinetic energy associated with its motion perpendicular to them (Fig. 1a). This perpendicular (transverse) motion is therefore trapped in the potential while the longitudinal motion is nearly free, because of its high kinetic energy. Channeled positrons are trapped and oscillate about potential minima that are located midway between crystal planes, occupying (quantum-mechanically) nearly-discrete levels of transverse energy (Fig. 1b). For electrons the channeling potential is obtained by turning the one in Fig. 1b upside down, so that they oscillate about minima on the crystal planes.

CR is emitted[3] as the channeled particle makes transitions between transverse energy levels e and g separated by ΔE_{eg} in the limit of nonrelativistic velocities $\beta=v/c\ll1$. As $\beta\to1$, the Doppler-shifted emission frequency becomes $\omega=\omega_{eg}/(1-\vec{\beta}\cdot\hat{q})$, where $\hat{q}$ is the unit vector of emission, $\omega_{eg}=(\Delta E_{eg}/\hbar)\gamma^{-\alpha}$, $\alpha\simeq1/2$ for planar channeled positrons, $\alpha\simeq1/3$ for channeled electrons and $\gamma=(1-\beta^2)^{-1/2}$. The $\gamma^{-\alpha}$ dependence accounts for changes in the frequencies of transverse oscillations due to relativisitc mass increase. For nearly forward emission $\omega\simeq(\Delta E_{eg}/\hbar)2\gamma^{2-\alpha}$. Hence, keV photon energies are attained at electron or positron energies above 50-100 MeV ($\gamma\gtrsim10^2$) for ΔE_{eg} of few eV or less. In addition to its tunability, CR has other advantages: good spectral brightness and narrow width of the peaks, as well as collimation within a cone of width γ^{-1} about the mean (longitudinal) direction of motion $\hat{z}$.

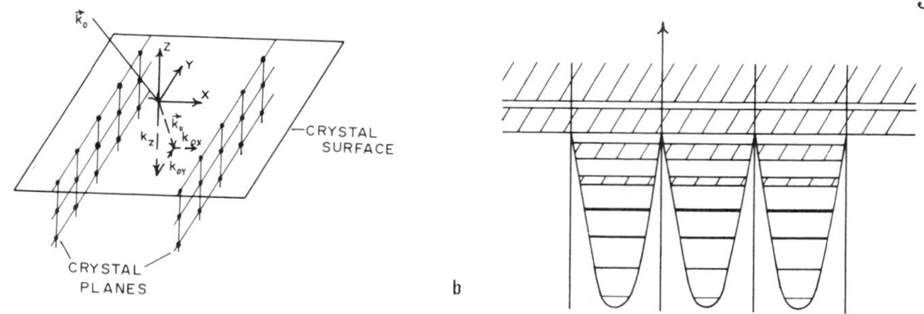

Figure 1 - (a) Geometry of positive particle channeling between crystal planes, and (b) the corresponding band structure of allowed transverse energies (shaded areas and lines).

In the dipolar regime of CR ($\gamma \lesssim 10^3$ for channeled electrons) stimulated emission can be described as an interaction of the radiated field with the channeled beam polarization $\vec{P}(\vec{r},t)=\sum_i \vec{\mu}_i \delta(\vec{r}-\vec{z}_i(t))$. Here $\vec{\mu}_i$ is the dipole operator for the e-g transition of the i'th particle, which is labelled only by its longitudinal position $z_i(t)=\beta_z c(t-t_o)$, t_o being its time of entry into the structure. The ability to label each particle by $z_i(t)$ only is due to the periodic arrangement of channels in the structure, so that each particle is represented by Bloch waves oscillating in phase in all channels. Hence, as long as the transverse spread of the channeled beam is smaller than the transverse variation of the emitted wave, the beam emits as a linear chain of moving dipoles centered about the minimum of the channeling potential. The effective interaction Hamiltonian between the electromagnetic wave and $\vec{P}(\vec{r},t)$ is then given by[1]

$$H_{int} = -\int d\vec{r}[\vec{P}(\vec{r})\cdot\vec{E}(\vec{r})+(\vec{P}(\vec{r})\times\vec{\beta})\cdot\vec{B}(\vec{r})] \tag{1}$$

where $\vec{E}$ and $\vec{B}$ are the electric and magnetic fields of the wave and $\vec{P}\times\vec{\beta}$ is the motional (Röntgen) magnetization . Upon Fourier-expanding $\vec{E}$ and $\vec{B}$ one finds that this magnetization reduces the matrix element for emission in a mode with wavevector $\vec{q}$ by a factor of $(1-\vec{\beta}\cdot\hat{q})$, assuming the dipole moments $\vec{\mu}_{eg}$ are polarized along $\vec{E}_{\vec{q}}$.[1] The field-polarization coupling in the Maxwell-Bloch equations (MBE) derivable from (1) is reduced by the same factor. These equations are written for the slow envelopes of $\vec{P}=\vec{\mathcal{P}}\exp[i(\omega t-\vec{q}\cdot\vec{r})]$ and $\vec{E}=\varepsilon \exp[-i(\omega t-\vec{q}\cdot\vec{r})]$ and the population-inversion density W, allowing for the transition linewidth T_2^{-1}, inversion relaxation time T_W, cavity lifetime $\delta/2$, pumping rate Λ and beam density ρ. Two emission regimes describable by the MBE will be considered:

a) Small-signal lasing regime[2]: Consider a beam channeled through a periodic structure enclosed in a Bragg resonator. If the beam is pumped by a perpendicular ($\vec{\beta}\cdot\hat{q}=0$) optical or infrared laser beam at a rate $\Lambda\rho=T_W^{-1}$ required to maintain uniform inversion throughout the structure, then the small-signal steady-state solution of the MBE is

$$|\varepsilon(z)|^2=|\varepsilon(0)|^2\exp\{\{[-L_2^{-1}+(L_2^{-2}+4L_a^{-2})^{1/2}]-\delta/c\}z\} \tag{2}$$

where the gain G (in square brackets) is determined by the dephasing length $L_2=c\beta_z T_2$ and amplification length

$L_a=[\hbar c^2\beta_z/2\pi\omega_{eg}(1-\vec{\beta}\cdot\hat{q})\Lambda\ T_W\rho\mu_{eg}^2]^{1/2}$. In the limit $L_a>>L_2$

(strong dephasing) $G\simeq L_2/L_a^2$, as predicted on using the Golden Rule for the rate of stimulated transitions[4]. The coherence length, over which energy losses are still small enough to maintain resonance with the field, is rather large, easily exceeding 1mm. The lasing threshold for emission at $\lambda=100Å$ required by (2) amounts to very high current densities, 10^8A cm^{-2} for electrons channeled in crystals. It is hoped that the distributed-feedback configuration (Bragg reflection of $\vec{E}$ by the crystal planes[5]) can reduce this threshold to 10^6A/cm^2. In artificial structures composed of cylindrical channels with diameters ~100Å the predicted threshold[3] is much lower: 10^5-10^6 A/cm^2, which may be reduced to 10^3-10^4 A/cm^2 by distributed feedback.

b) Superfluorescence[1]: For current densities below the lasing threshold it may still be possible to enhance CR emission, by creating a superfluorescent (SF) pulse. If the requirements discussed below are met, then, keeping in mind that the channeled beam emits as a linear chain of dipoles, we find that SF enhances the emission rate per particle by a factor of $\lambda/2a=\rho A\lambda/2$, where a is the mean longitudinal separation between dipoles and A is the cross section of the structure. The conditions for SF are: (1) Complete population inversion of the e-g transition must be achieved, on pumping the channeled beam by a laser pulse at 90° (see above). (2) The pumping pulse duration T_p should satisfy $T_p<<T_c=T_1(2/\rho A\lambda)$, where T_c is the SF emission time, shortened relative to the single-particle transition lifetime T_1. For T_1~nsec and enhancement factors $\gtrsim10$, we need $T_p\lesssim10$ psec. (3) In order to have irreversible decay $T_c\gtrsim L_S/c$, where L_S/c is the escape time of a photon from the pumped region of length L_S. (4) To ensure above-threshold operation, $T_2>>T_c$. For $\lambda\sim10Å$ and enhancement $\gtrsim10$, conditions (3) and (4) are satisfied simultaneously at current densities above 100A/cm^2, either by positrons with GeV energies channeled in crystals for $L_S\lesssim1$ mm, or by electrons in artificial structures of the type discussed above, for $L_S\lesssim1$ cm. Thus, SF at wavelengths as low as 10Å seems to be within the present state-of-the art, whereas lasing at such wavelengths is contingent on the availability of channeled beams with current densities above MA/cm^2.

References
1. G. Kurizki, in Relativistic Channeling, ed. J. Ellison and R. Carrigan, (Plenum, New York, 1987), p. 505.
2. G. Kurizki, M. Strauss, J. Oreg and N. Rostoker, Phys.Rev. A35, 3424 (1987). G. Kurizki and A. Friedman (to be published).
3. For an updated collection of reviews - see book in Ref. 1.
4. V.V. Beloshitskii and M.A. Kumakhov, Phys.Lett. A69, 247 (1978).
5. M. Strauss, A. Ron and N. Rostoker (to be published).

High-Repetition Studies of the XeCl Laser Comparing Inductively Stabilized and Normal Shaped Electrodes

M. Sentis, R.C. Sze, F. Hommeau,
B. Forestier, B. Fontaine

Institut de Mecanique des Fluides de Marseille
C.N.R.S.
Universite D'Aix-Marseille II
Marseille, France

Introduction

Previously, it has been shown that a segmented electrode structure with a small inductor connected in series with each segment is capable of stabilizing excimer discharges so that relatively long laser times is sustained (110ns FWHM and total lasing times greater than 220ns)[1,2]. The present work places such an electrode structure in a high repetition rate discharge test bed using x-ray preionization. The performance regarding pulse to pulse repeatability and pulse to pulse energy degradation are compared with the performance of a normal shaped electrode.

Comparison of Electrodes in Circuit with Peaking Capacitors

A typical CLC circuit is used used for laser excitation in XeCl on the high repetition rate test bed called LUX[3]. X-ray preionization is provided by a WIP electron beam gun capable of kilohertz operation. The peaking capacitor is set up as a very short line with small inductors separating the array of capacitor components. This allowed lasing with x-ray preionization to be as long as 80ns in normally unsegmented solid electrodes. The laser presently is operated in the burst mode with a large capacitance connected to the charging power supply. Figure 1 gives a typical histogram of the time evolution of the power of 250 shots at various pulse repetition rates from 100Hz to 1.5 KHz. Some of the drop in energy as a function of the number of pulses is due to the drop in charging voltage, but most of this drop is due to degradation of the discharge. The time evolution of each pulse can be interogated and the time evolution of different pulses can be compared. Comparisons (Fig.2.) of energy drop at the end of the 250 pulses are compared as a function of pulse repetition rate for unstabilized shaped electrodes and for the inductively stabilized electrodes before and after segmentation. We see that the worse performance is the stabilized electrode before segmentation. This is understandable since the electrode is unprofiled. However, note that the segmented stabilized electrode performs significantly better than the regular solid shaped electrode. In fact 1.5KHz operation is only possible with the stabilized electrode. The highest pulse repetition rate of 1.5KHz.is limited by the operation of the x-ray WIP gun. In Fig. 3 energy fluctuation is compared between the stabilized and unstabilized electrodes as a function of the pulse repetition rate. Again we see substantial improvement in the case of the stabilized electrode.

Operation in Circuit without Peaking Capacitors

Such a circuit has slow current risetimes due to the large series inductance of the thyratron. This circuit can only be operated effectively with stabilized electrodes. The voltage, current and lasing waveforms are shown in Fig.4. Note the steady state voltage at 3kv is sustained for over 150ns. The current risetime is over 100ns and is very slow. Lasing is for over 80ns. Since the power deposition is low due to the large thyratron inductance, energy of the lasing pulse is substantially lower due to poor energy extraction efficiency. This is reflected in the long delay before lasing begins as shown in Fig.4 and the fairly large fluctuations in the pulse to pulse reproducibility observed in its histogram data at both 100 Hz and 1 KHz operatons.

References:

1. R.C. Sze, "Inductively Stabilized rare-gas halide minilaser for long-pulsed operation", J. Appl. Phys., 54, 3, pp.1224-1227 (March 1983)

2. R.C. Sze, "Inductively Stabilized Excimer Lasers", Proceedings of the International Conference on Lasers '83, pp.512-517, STS Press (1984)

3. M. Sentis, B. Forrestier, B. Fontaine, R.C. Sze, M. Vannini, "Gas Flow and High Power Studies of a High prf X-ray Preionized Discharge Pumped XeCl Laser", Conference on Lasers and Electro-optics 1987, Baltimore, MD. (May, 1987)

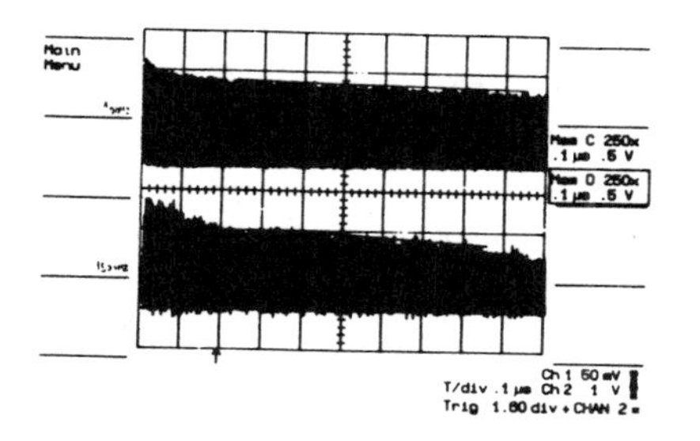

Figure 1. Pulse to pulse time resolved power histogram for the segmented inductively stabilized electrode and using the CLC circuit at 100 and 1500 Hz. operation.

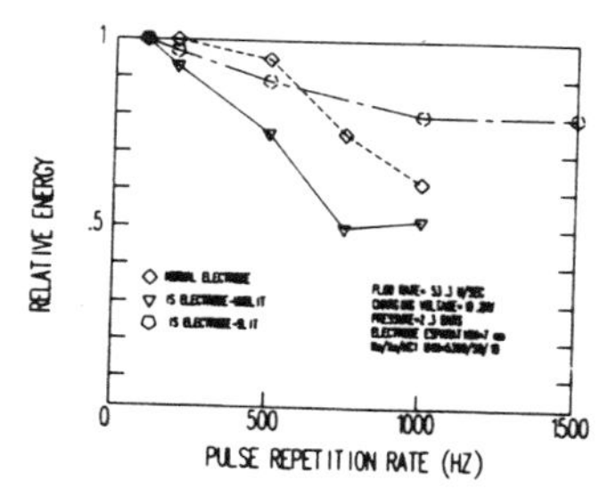

Figure 2. Relative energy degradation at end of 250 pulses for normal shaped electrode, unsegmented inductively stabilized electrode and segmented stabilized electrode.

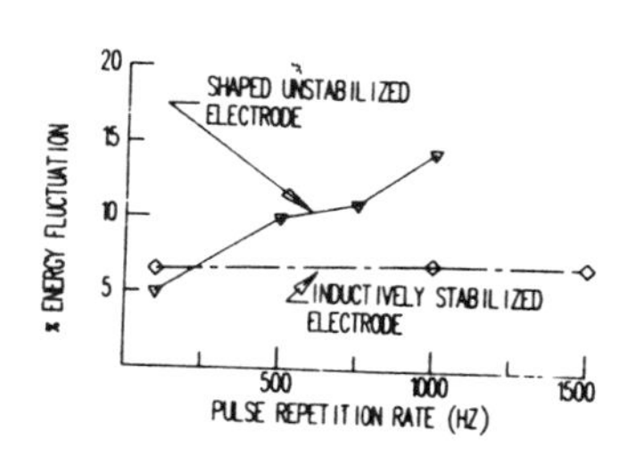

Figure 3. Energy fluctuations as a function of pulse repetition rate for unstabilized and stabilized electrodes.

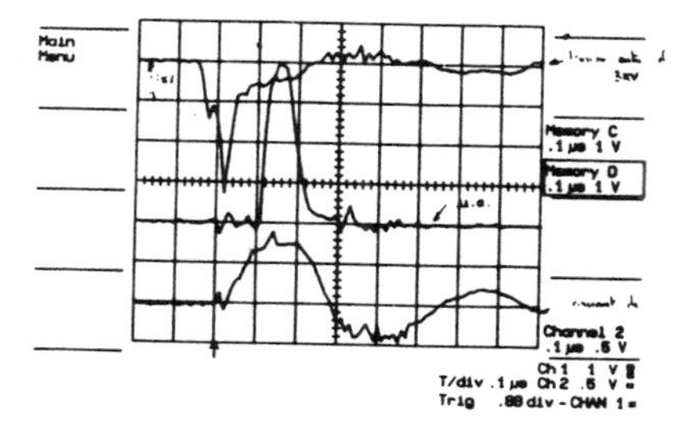

Figure 4. Voltage, current and lasing characteristics as a function of time for the stabilized electrode operating with no peaking capacitors.

PROGRESS IN THE GAMMA-RAY LASER PROGRAM AT TEXAS 1: FLASH X-RAY TECHNIQUES FOR PUMPING NUCLEAR MATERIALS

F. Davanloo, T. S. Bowen, J. J. Coogan,
R. K. Krause, and C. B. Collins

Center for Quantum Electronics,
University of Texas at Dallas
P. O. Box 830688, Richardson, TX 75083-0688

ABSTRACT

Progress is reported in the characterization of a repetitively pulsed, flash x-ray device producing intense nanosecond pulses. The most recent advances consist of reliable collections of spectral distributions of the fluxes emitted by different anodes. Such knowledge of the x-ray photon spectra facilitates use of these flash x-ray sources to excite nuclear fluorescence for the evaluation of candidate materials for a gamma-ray laser. For other applications, as well, these devices can offer a table-top laboratory alternative to laser plasma x-rays and to synchrotron radiation.

INTRODUCTION

Recently we described[1-4] a flash x-ray source producing intense nanosecond pulses at high repetition rates. The extremely low profile of our x-ray diode had resulted in an effective impedance of about 1-4 Ω at the frequencies characteristic of those in the wave-forms of the driving voltages. Thus, the diodes were reasonably well-matched to the Blumlein during the most important part of the discharge current. As a result, the output pulses were found to have durations comparable to the transit times of the lines, an aspect not seen in previous devices.[5]

Higher average x-ray powers were obtained both by scaling these devices in size, and by operating them at higher charging voltages.[6] Preliminary attenuation measurements with a combination of K edge filters and aluminum foils of known thicknesses indicated that at least 25% of the total x-ray energy lay in the K lines in agreement with previous observations.[3-4] Here we present a further step in the characterization of these devices by recording the spectra of the x-ray energies emitted from anode materials into both K lines and continuum. The continuum will be important to excite nuclear candidate materials for a gamma-ray laser when transition energies are poorly known.

CONTINUUM SPECTRUM

Knowledge of the spectra of the energies of the primary photons emitted from flash x-ray devices is important in a variety of applications. Unfortunately, the calibration of x-ray spectrometers for such measurements is very difficult if dispersive elements are used. For this reason Si-crystal detectors offer considerable appeal for the characterization of repetitively pulsed sources. Provided the detector geometry is arranged so that two photons are not detected from the same shot, the energy of each photon collected can be determined from the usual pulse height analyses of nuclear physics. Of course, such a method of x-ray spectroscopy is inapplicable to single shot systems such as those powered from Marx-like generators.

Operating times of several hours per day at repetition rates of several hertz are routine with the x-ray devices we have developed. Such consistent performance facilitated the accumulation of spectral output using a Si(Li) crystal as a detector connected to an EG&G system 5000 spectrometer. This particular system is used mainly for measuring x-ray energies below 100 keV.

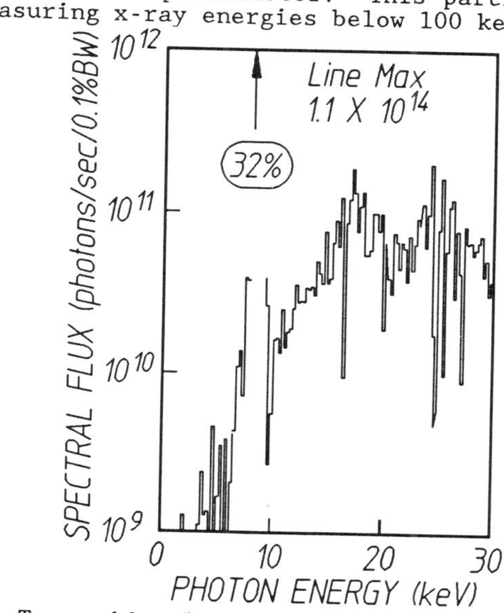

Figure 1: Typical spectral distributions of flux emitted from the scale-0.33 device with a copper anode. Charging voltage was 25 kV.

To avoid pulse pileups, the detector was placed a long distance from the x-ray source; and lead collimators were placed in front of the x-ray device and the detector. A helium column was placed in a line between the source and detector to reduce absorption by the air. Shielding around the detector itself prevented x-rays scattered from the surrounding objects from entering the detector. Several tests were performed in order to determine whether photon multiples were correctly eliminated. The detector aperture was reduced by a known factor, while the accumulation time was increased by the same factor and the spectra were compared. Collimators were adjusted, and the procedure was repeated until no appreciable differences in collected spectra were observed. Once the counting and detector conditions were properly adjusted, data was accumulated for two different scale flash x-ray systems, the 0.33-scale and EXRAD II.[6] Figure 1 shows the spectral distribution of flux emitted from the scale-0.33 device with a charging voltage of 25 kV and an anode material of copper. A third of the energy was found to be in the K lines. The remainder was distributed over a fairly broad band of true continua. The spectrum of Fig. 2 was collected by operating the EXRAD II system with a tungsten anode at a charging voltage of 60 kV. This spectrum includes a correction for the detector efficiency and is scaled in terms of relative keV/keV as a function of energy. As seen in Fig. 2 the continuum is distributed over a fairly broad band within a range of 15 - 70 keV.

CONCLUSIONS

Our application concerns the use of x-ray devices developed in this work to excite nuclear transitions. The ultimate signal-to-noise ratio observed for the resulting fluorescence will depend only upon the total radiation that can be delivered to an extended absorber in a working period. Since energies of potential pump transitions in most systems are very poorly known, the availability of broad continuua from our devices make it a more viable source for mounting such double resonance experiments at the nuclear level, than would be a synchrotron which would have to be tuned over the working range, line at a time.

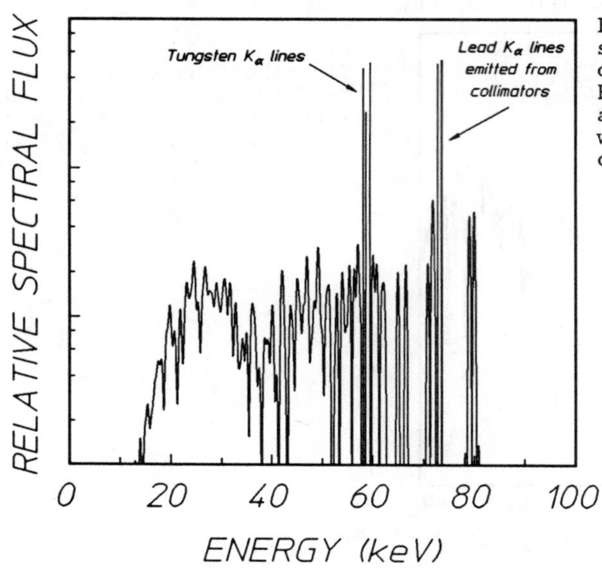

Figure 2: Relative spectral distribution of flux emitted from EXRAD II operating with a tungsten anode and with a charging voltage of 60 kV.

ACKNOWLEDGEMENT

This work was supported by the Innovative Science and Technology Directorate of the Strategic Defense Initiative Organization and directed by the Naval Research Laboratory.

REFERENCES

1. C. B. Collins, F. Davanloo, and T. S. Bowen, Rev. Sci. Instrum. 57, 863 (1986).

2. F. Davanloo, T. S. Bowen, and C. B. Collins, Rev. Sci. Instrum. 58, 2103 (1987).

3. F. Davanloo, T. S. Bowen, and C. B. Collins, in Advances in Laser Science-I, edited by W. C. Stwalley and M. Lapp (AIP Conference Proceedings No. 146, New York, 1986) pp. 60-61.

4. T. S. Bowen, J. J. Coogan, F. Davanloo and C. B. Collins, in Advances in Laser Science-II, edited by M. Lapp, W. Stwalley and G. A. Kenney-Wallace (AIP Conference Proceedings No. 160, New York, 1987) pp. 98-100.

5. P. Krehl, SPIE Review 689, 26 (1986).

6. F. Davanloo, T. S. Bowen, J. J. Coogan, and C. B. Collins, in Center for Quantum Electronics Report No. GRL/8602, University of Texas at Dallas, 1987, pp. 47-70.

PROGRESS IN THE GAMMA-RAY LASER PROGRAM AT TEXAS 2: COHERENT TECHNIQUES FOR PUMPING A GAMMA-RAY LASER

S.S. Wagal, P.W. Reittinger, T.W. Sinor and C.B. Collins
Center for Quantum Electronics, University of Texas at Dallas
P.O. Box 830688, Richardson, TX 75083-0688

ABSTRACT

We report the refinement of a technique of high resolution nuclear spectroscopy which we call Nuclear Frequency Modulation Spectroscopy, and we report our latest experimental results for developing "rf-sidebands" in the ^{119}Sn nucleus in foils of Sn and FeSn.

INTRODUCTION

In our search for coherent techniques for pumping a gamma-ray laser we are interested in the possibility of using virtual states of nuclear excitation to mediate non-linear processes in the nucleus. Currently we are investigating apparent two photon Mossbauer transitions. In theory these could make possible the frequency upconversion of optical laser photons to gamma-ray energies.[1-5] As early as 1968 it had been reported that applying a radiofrequency alternating magnetic field to a ferromagnetic foil containing ^{57}Fe induced frequency dependent nuclear hyperfine structure in the material.[6] Then in 1976 it was shown that the rf induced hyperfine structure could be transferred to a non-ferromagnetic foil containing ^{57}Fe through the use of ferromagnetic "driving" foils.[7] In our attempt to understand the rf-sideband phenomenon and its potential application towards the development of a gamma-ray laser, our research has followed two paths: developing a technique of nuclear spectroscopy which utilizes the frequency dependence of the rf-induced hyperfine structure, and studying the rf-sideband phenomenon in Mossbauer isotopes other than ^{57}Fe.

EXPERIMENTAL

We have developed a technique of nuclear spectroscopy which we call Nuclear Frequency Modulation Spectroscopy (NFMS).[8-10] This technique obtains a spectrum by using rf-sidebands induced in a material to modulate the gamma-ray absorption cross section of a Mossbauer absorber in contrast to a conventional Mossbauer spectrometer which uses controlled doppler shifts to scan through the energies of the gamma-ray source. In Fig. 1 we show a collection of typical NFM spectra of an iron foil enriched with 98% ^{57}Fe using a ^{57}Co source in a Pd matrix. The spectra show gamma-ray absorption as a function of the driving frequency of an induction coil in which the iron foil absorber was mounted. The three spectra were obtained at different energies by giving the source a different constant velocity relative to the absorber in each case. If one knows the energies of gamma-ray absorption by an unperturbed

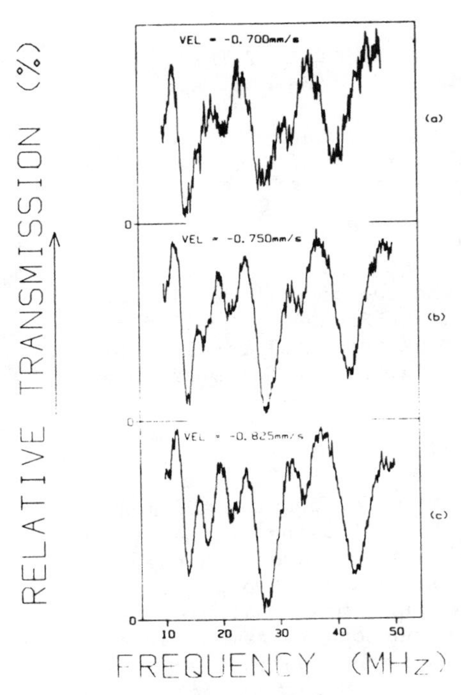

Fig.1 NFMS data

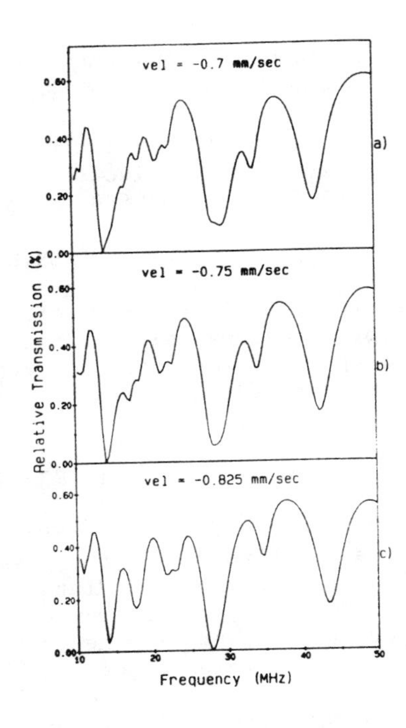

Fig.2 NFMS simulation

absorber, information obtainable from a conventional Mossbauer spectrum, then one can easily simulate an NFM spectrum simply by assuming that rf-sideband absorption occurs at energies displaced from the unperturbed absorption energies by integral multiples of the frequency of the rf field in which the absorber is immersed (Fig. 2). It is our intention to use NFMS to reveal inherent properties of the rf-sidebands which are obscured in a conventional Mossbauer spectrum. Simultaneous to the development of NFMS we have been investigating the rf-sideband phenomenon in the ^{119}Sn isotope. Since a Sn foil is non-ferromagnetic it

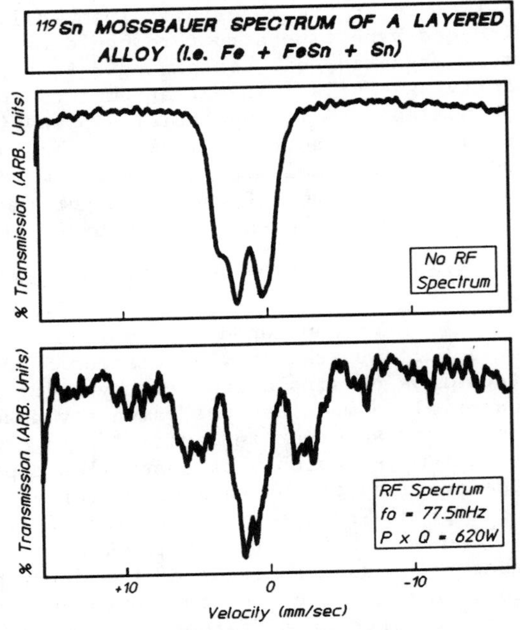

Fig. 3 Mossbauer spectra of FeSn with and without rf field.

was hoped that rf-sidebands could be induced in a Sn foil through the use of ferromagnetic driving foils similar to the way in which Chien and Walker induced sidebands in a stainless steel absorber.[7] Unfortunately, this technique has not been successful in producing rf-sidebands in ^{119}Sn. On the other hand, when Sn was alloyed with Fe, it proved to be extremely simple to induce sidebands in the resulting ferromagnetic absorber (Fig.3). The spectra in Fig. 3 were obtained with a ^{119}Sn source in a $CaSnO_3$ matrix. The Fe-Sn alloy was mounted in the induction coil of an L-C tank circuit. The first spectrum shows the FeSn spectrum when the rf is off, emphasizing the characteristic doublet structure of FeSn due to the magnetic dipole interaction. Once the rf is applied we clearly get sidebands displaced from the unperturbed absorption peaks by integral multiples of the frequency of the applied field.

SUMMARY

NFMS has proven to be a useful technique for observing rf-sidebands unencumbered by the unperturbed "parent" absorption peaks which plague a conventional Mossbauer spectrum. Therefore, NFMS is useful for searching for resonances and quantum structure in the rf-sidebands. The work with ^{119}Sn and Sn alloys has revealed that the relationship of the rf-sideband phenomenon to the hyperfine interactions in a material is not clearly understood.

REFERENCES

1. C.B. Collins, S. Olariu, M. Petrascu and I. Popescu, Phys. Rev. Lett. 42, 1397 (1979).
2. C.B. Collins, S. Olariu, M. Petrascu and I. Popescu, Phys. Rev. C 20, 1942 (1979).
3. S. Olariu, I. Popescu and C.B. Collins, Phys. Rev. C 23, 50 (1981).
4. S. Olariu, I. Popescu and C.B. Collins, Phys. Rev. C 23, 1007 (1981).
5. C.B. Collins, F.W. Lee, D.M. Shemwell, and B.D. DePaolo, J. Appl. Phys. 53, 4645 (1982).
6. N.D. Heiman, L. Pfeiffer and J.C. Walker, Phys. Rev. Lett. 21, 93 (1968).
7. C.L. Chien and J.C. Walker, Phys. Rev. B 13, 1876 (1976).
8. B.D. DePaolo, S.S. Wagal and C.B. Collins, JOSA B 2, 541 (1985).
9. P. Reittinger, S. Wagal, C.B. Collins in Advances in Laser Science-2, edited by M. Lapp, W.C. Stwalley and G.A. Kenney-Wallace (AIP Conference Proceedings no. 160 Seattle, 1986) p. 101.
10. P.W. Reittinger, S.S. Wagal, T.W. Sinor and C.B. Collins, Rev. Sci. Instr. (pending).

* Supported by SDIO (IST) through NRL and in part by ONR.

PROGRESS IN THE GAMMA-RAY LASER PROGRAM AT TEXAS 3: OBSERVATION OF NUCLEAR FLUORESCENCE

J. Anderson, Y. Paiss, C. D. Eberhard, and C. B. Collins

Center for Quantum Electronics,
University of Texas at Dallas,
P. O. Box 830688, Richardson, TX 75083-0688

ABSTRACT

The nuclear analog of the ruby laser is shown in Fig. 1. It embodies the simplest concepts for a gamma-ray laser. Not surprisingly, the greatest rate of progress has occurred in this direction. Studies aimed at upgrading the nuclear database have yielded a singularly useful technique for the calibration of large pulsed bremsstrahlung sources used for pumping in the region from 150 keV to 1.3 MeV. In addition, experiments utilizing bremsstrahlung from a 6 MeV linear accelerator have indicated the possibility of accessing the broad octupole states above 2 MeV in ^{111}Cd, ^{113}In and ^{115}In. Effective integrated cross sections $(\pi b_a b_o \sigma_o \Gamma/2)$ on the order of 10^4 cm^2 keV were observed. Finally the upconversion pumping, shown schematically in Fig. 2, of the first of 29 potential candidate materials for a gamma-ray laser was reported.

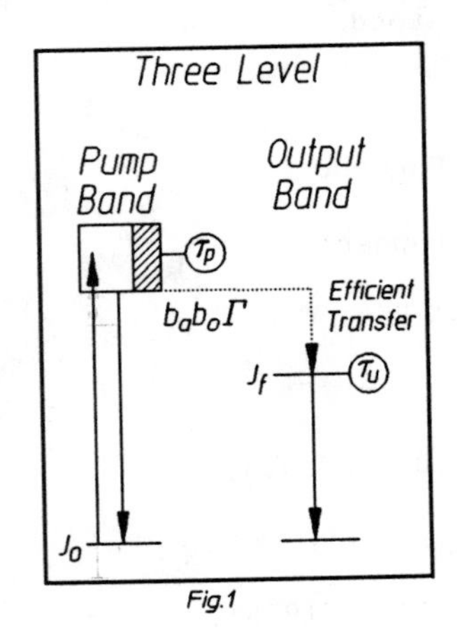

Fig.1

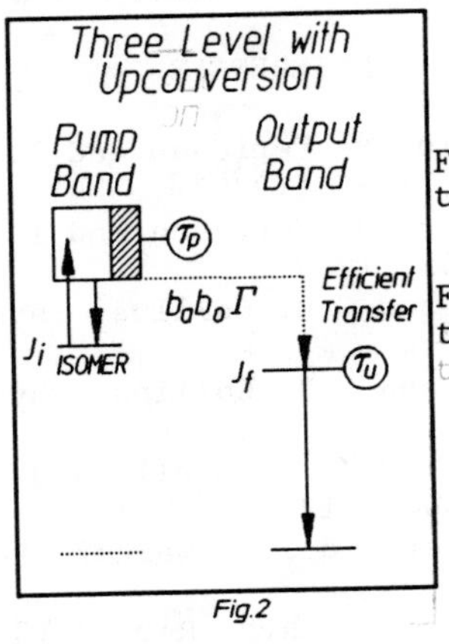

Fig.2

Figure 1: Nuclear analog to Cr^{3+} in ruby.

Figure 2: Nuclear analog to Group VI lasers.

NUCLEAR FLUORESCENCE WITH PULSED SOURCES

In a sample exposed to a pulse of intense bremsstrahlung, the number of isomers formed through nuclear excitation to a pump level followed by decay to the metastable state is directly proportional to $[\Sigma_i\ (\pi b_a b_o \sigma_o \Gamma/2)_i\ \phi(E_i)]$, where the summation is over the pump states below the end point energy of the beam, $(\pi b_a b_o \sigma_o \Gamma/2)_i$ is the effective integrated cross section of the i-th pump level at energy E_i in units of cm^2 keV, and $\phi(E_i)$ is the time-integrated spectral intensity of the photon flux at E_i in cm^{-2} keV^{-1}. In the energy region from 150 keV to 1.3 MeV, the known gateways[1,2] of ^{77}Se and ^{79}Br were used to determine location and effective cross section[3] of the 1078 keV state in ^{115}In. Now, by exposing samples of ^{77}Se, ^{79}Br, and ^{115}In to an intense bremsstrahlung pulse and by observing the decay of the isomers formed, a calibration of the flux in the pulse can be ob-

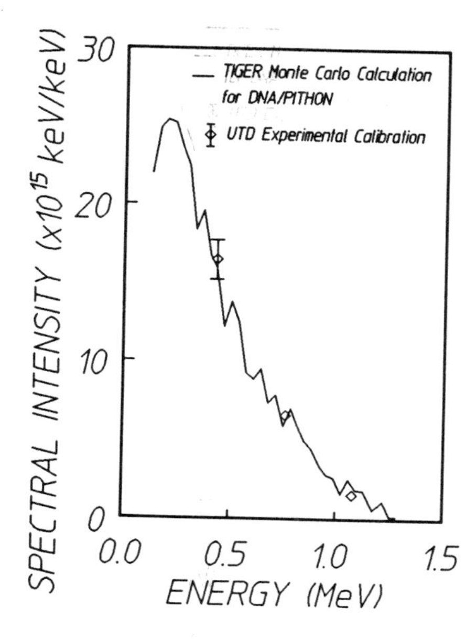

Figure 3: Experimental calibration of DNA/PITHON device.

tained at several discrete photon energies. Figure 3 shows the results of applying this calibration procedure to the output of the DNA/PITHON device. Our UTD/APEX device has been calibrated by this method and then used to examine new photoactivation reactions. Some effective cross sections have been found to be several orders of magnitude larger than expected on the basis of the existing nuclear data base.

NUCLEAR FLUORESCENCE FROM CW SOURCES

Samples of ^{111}Cd, ^{113}In, and ^{115}In were irradiated with a Varian Clinac 1800 linear accelerator with end point energy of 6 MeV, and the activation was measured. The contributions due to the known gateway states below 1.3 MeV are negligible compared to the total activation. If the activation is produced by a broad state above 1.3 MeV, then the effective integrated cross sections are on the order of 10^4 cm^2 keV. These values are much higher than any previously observed, suggesting that the excitation may be accessing the collective octupole states just above 2 MeV in the isotopes ^{111}Cd, ^{113}In, and ^{115}In.

UPCONVERSION IN THE ISOMER ^{180}Ta^m

There is one naturally occurring isomer, ^{180}Ta^m, ranked by astrophysicists as nature's rarest stable nucleus, Nevertheless, it is the most abundant of 29 candidates for a gamma-ray laser. Figure 4 shows the energy level diagram. An enriched sample containing 1.4 mg of ^{180}Ta^m and 30 mg of ^{181}Ta diluent was exposed to the bremsstrahlung of the linear accelerator. Figure 5 shows the spectra taken from the Ta sample both before and after the irradiation. The signature of the deexcitation of the metastable through the mechanism indicated in Figs. 2 and 4 is the appearance of the ^{180}Hf K_α x-ray, whose intensity decayed with a half-life of 8.1 h corresponding to the accepted value for the ^{180}Ta ground state.

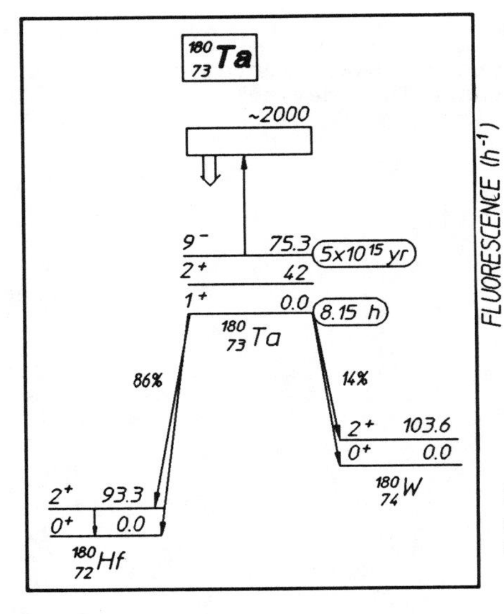

Figure 4: Energy levels and decay schemes of ^{180}Ta.

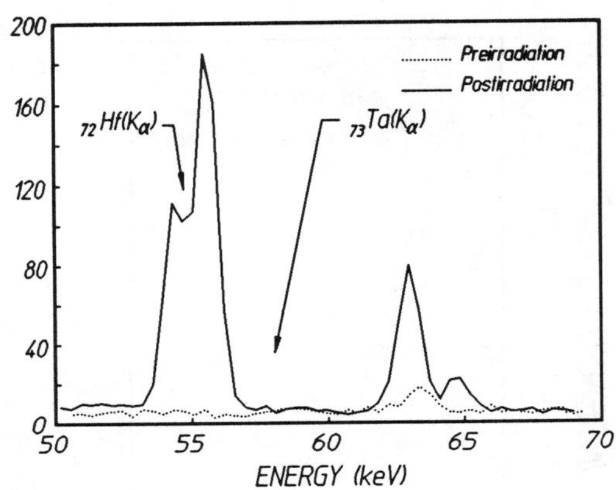

Figure 5: Spectra of enriched Ta sample showing the final products of the reaction $^{180}Ta^{m}(\gamma,\gamma')^{180}Ta$.

The significance of this observation is twofold:

1. The first real isomer to be tested for a gamma-ray laser was successfully pumped down with an astonishingly large cross section of 40,000 cm^2 keV on a scale where 10 cm^2 keV describes a fully allowed process.

2. The nuclear analog to the ruby laser is a fully viable scheme for a gamma-ray laser, and $^{180}Ta^{m}$ narrowly misses being an acceptable candidate. It performed about 10^4 times better than would have been expected theoretically.

ACKNOWLEDGEMENT

This work was supported by the Innovative Science and Technology Directorate of the Strategic Defense Initiative Organization and directed by the Naval Research Laboratory.

REFERENCES

1. J. A. Anderson and C. B. Collins, Rev. Sci. Instrum. 58, 2157 (1987).

2. J. A. Anderson and C. B. Collins, Rev. Sci. Instrum. (pending).

3. C. B. Collins, J. A. Anderson, Y. Paiss, C. D. Eberhard, R. J. Peterson, and W. L. Hodge, Phys. Rev. C (pending).

STUDIES OF THE AR/XE LASER USING DIFFERENT PUMPING TECHNIQUES

B. L. Wexler, A. Suda*, B. J. Feldman, K. Riley*
Laser Physics Branch, Naval Research Laboratory, Washington,D.C. 20375

ABSTRACT

We have begun studies of the Ar/Xe laser, operating on the 5d-6p Xe transitions, using electron-beam, e-beam sustained, and x-ray preionized discharges to investigate the kinetics and potential of the laser with these different pumping techniques. Preliminary results include a specific output energy of 0.75 J/l using the e-beam alone, enhancements of up to 10 using the e-beam sustained discharge, and the first reported lasing from an x-ray preionized discharge, using a device with a 5 cm electrode gap. Line competition has also been observed.

INTRODUCTION

The Ar/Xe laser, operating on the 5d-6p Xe transitions, has been demonstrated to have high efficiency (3-5%) and specific output energy (to 6 J/l) when pumped in an e-beam sustained discharge[1]. When operated in an electron-beam pre-ionized mode, however, a lower efficiency (1.3%) and specific energy (.67 J/l) are reported[2]. With UV pre-ionization and a fast transverse discharge, an efficiency similar to the latter (1.4%), but at much lower energy loading and, therefore, specific output energy (0.1 J/l) has been reported[3]. All of these experiments used similar power loadings on the gas, but the last work had very short discharge pulses compared to the electron beam results, which limited the specific energy,while the UV preionization also limited the pressure to 1.4 atmospheres. These results, however, indicate the significant potential of this laser, which can be compared to the rare gas halide lasers in its high pressure operation, potential efficiency and specific energy, and through similarities in the kinetics processes involved. There are two obvious major differences. First, the lack of the halogen gas results in the significant advantages of long pulse operation, as there is no fuel "burn-up", and long gas lifetime and simple high repetition rate operation. Second, the laser operates on atomic, not molecular transitions, so that the linewidth can be narrow, the gain can be high and the lower laser levels exist and participate in the kinetics. The laser is also, apparently, limited to a much lower pump intensity than the excimers, but the long pulse lengths which have been achieved with the electron beam sustained or pre-ionized systems allow similarly high specific energies.

ENERGY LEVELS AND KINETICS

While the laser has been observed to operate on five transitions in the 5d-6p manifold, the dominant line is the $5d[3/2]^o_1$-- $6p[5/2]_2$ transition, at 1.73 μm. The most important kinetics processes populating the upper levels can be listed as:

$$Ar^+ + 2\ Ar \rightarrow Ar_2^+ + Ar \qquad Ar_2^+ + Xe \rightarrow Xe^+ + 2\ Ar \quad (1)$$

$Ar_2^+ + Xe \rightarrow ArXe^+ + Ar$ $Xe^+ + 2\,Ar \rightarrow ArXe^+ + Ar$ (2)

$ArXe^+ + e \rightarrow ArXe^{**} \rightarrow Xe^{**}(5d) + Ar$ (3)

The high efficiency observed in Ref. 1 was accounted for by postulating "recirculation pumping" from the Xe 6s metastable level, followed by ionization and subsequent recombination through equation (3). This channel is certainly desirable because of the much higher quantum efficiency possible due to ionization from the high lying metastable level rather than direct pumping of the upper level or ionization from the ground state, and may be operable in the e-beam or uv preionized discharge lasers, since the metastable lifetime is estimated to be about 0.5 μs. An examination of the rates indicates that high E/P is necessary for populating the metastable levels, while the recombination process operates at low E/P. In the e-beam sustained work, the metastable population was said to be created by the high energy e-beam itself, while the low energy electrons of the discharge were responsible for the recirculation pumping. Ringing of the preionized discharge, so that the E/P alternated between high and low values, was discussed in Ref.2.

EXPERIMENT

We have begun experiments with electron-beam pumping, electron-beam sustained discharge pumping, and x-ray pre-ionized discharge pumping in order to investigate the mechanisms and potential of this laser system under these different conditions. The electron beam apparatus used is a small device, operable at 175-250 kV gun voltage, with a current through the 2 x 40 cm foil window of 1 to 5 A $/cm^2$ and pulse lengths of 0.6 to 1.2 μs. The laser chamber has an electrode spacing of 3 cm, and an extractable volume of 125 cm^2. When operated with electron-beam pumping alone, a maximum output of 90 mJ was observed, at 250 kV gun potential with a 1 mil Ti foil window, using a mixture of 1 % xenon in 4 atmospheres of argon. The output energy observed as a function of gun voltage and pressure reflects optimization of energy deposition into the laser volume. Based on our calculation of energy deposition, we estimate the efficiency to be 1 %. The results are comparable to those achieved in the literature when the relatively short pulse length of our device is considered.

For operation in the e-beam sustained mode, a simple switch and capacitor circuit were used. A thicker, 1.5 mil foil and cathode grid used to prolong foil lifetime resulted in much lower laser output from the e-beam alone. At the maximum gun voltage, 12 mJ was observed from a 1.3 % mixture at 3 atmospheres pressure and application of the discharge resulted in a maximum total output of 48 mJ, an enhancement factor of 4. An enhancement factor of 11 was observed when the e-beam contribution was reduced to 3 mJ and the discharge circuit improved. In order to investigate the high specific energy, high efficiency regime of operation used in Ref. 1, modification of the e-beam to produce longer pulses and higher currents is necessary.

Interesting line competition effects have been observed. The laser begins on either the 2.63 μm or 2.65 μm line (our filter cannot distinguish between them). Output then switches to the 1.73 µm line, which dominates the output. That line also terminates, to be followed by lasing again at 2.6 µm. While the initial behavior can be explained on the basis of gain and degeneracy effects, the later behavior can not, implying interesting upper and lower level population dynamics.

Use of an x-ray preionized discharge to pump the laser has the potential of efficient, high repetition rate operation and reduced size and complexity. In comparison with UV preionization, x-ray preionization can uniformly preionize high pressure, large volume devices. We have been able to solve the problem of surface flashover in high pressure argon discharges by careful attention to the laser head design, which both eliminates electric field enhancements and utilizes a long flashover path length. The laser head is an acrylic cylinder with 22.5 cm inside diameter. The electrodes are 50 cm long and 12.5 cm wide, separated by 5 cm. The lower electrode was hollowed out underneath to allow passage of the x-rays through a thin region of the electrode, 7.5 cm wide and 40 cm long.

The discharge was driven using a "spiker-sustainer" type circuit now often used in excimer lasers. A short, high voltage pulse was applied to the laser head from a two stage Marx Bank, with capacitance 10 nF and maximum voltage 120 kV. A rail gap separated the laser head and spiker circuit from a pulse forming network of total capacitance 160 nF and pulse length 400 ns. The object was to create a high metastable population with the short high voltage pulse and then pump the laser using the lower voltage, long pulse from the PFN. However, initial results indicated that the lasing began with the spiker current but rapidly terminated when the PFN was switched on. Best results occurred using only the spiker circuit. In experiments to date, a maximum output of 38 mJ has been observed from the estimated 0.4 liter active volume at a total pressure of 2 atmospheres using a 1.7 % mixture. These results are expected to improve significantly with improvements in the system.

We have written a computer code to model this laser. Interestingly, the critical recombination pump rate (3) is not known and must be estimated. However, even with this approximation we have seen reasonable agreement in discharge voltage, current and laser pulse shape.

The results reported above are far from optimal. Significant increases in laser output and efficiency are expected to result from modifications to both the electron beam and preionized systems. In order to investigate the kinetics of the laser, diagnostics to measure the gain and important populations will be implemented. In conjunction with our model, these will enable us to better understand the physics and potential of this interesting laser under these different pumping conditions.

REFERENCES

* Geo-Centers, Inc., Ft. Washington, MD

1. N.G. Basov et al, IEEE J. Quantum Electron., **QE-21**, 1756, (1985).
2. S.A. Lawton et al, J. Appl. Phys. **50**, 3888, (1979).
3. F.S. Collier et al, IEEE J. Quantum Electron., **QE-19**, 1129, (1983).

FREE ELECTRON LASER AMPLIFIERS IN THE HIGH GAIN COMPTON REGIME WITH VARIABLE WIGGLERS

Kenneth R. Hartzell, Jr.
Department of Physics, Villanova University, Villanova, PA 19085

ABSTRACT

We present a theoretical model of a free electron laser (FEL) in the high gain Compton regime with a variable wiggler patterned after the work of Bonifacio, Pellegrini, and Narducci for the constant wiggler configuration[1]. This model incorporates a variable wiggler subject to the constraint $\lambda_w B_w$ = constant where $\lambda_w B_w$ are the spatial periodicity and field strength of the wiggler, respectively.[2] Treating the system as a single pass amplifier, the equations of motion are numerically integrated for various spatial dependencies of the wiggler wavelength. The subsequent behavior of the radiation field relative to position and wiggler geometry is then presented, demonstrating significant improvement over the constant wiggler configuration.

THEORETICAL DEVELOPMENT

The description of the FEL operation in the high gain Compton regime will be given completely in terms of scaled dimensionless variables. As functions of position η and time τ, the equations for the electron phase $\phi_j(\eta,\tau)$, energy $\Gamma_j(\eta,\tau)$, and complex field amplitude $A(\eta,\tau)$, subject to the constraint $\lambda_w B_w$ = constant, are[2]

$$\left[\frac{\partial}{\partial\eta} + \frac{\partial}{\partial\tau}\right]\phi_j(\eta,\tau) = \frac{1}{2\rho}\left[1 - \frac{1}{\rho^2\Gamma_j^2(\eta,\tau)}\right] \tag{1}$$

$$\left[\frac{\partial}{\partial\eta} + \frac{\partial}{\partial\tau}\right]\Gamma_j(\eta,\tau) = -\frac{1}{\rho}\left[\frac{A(\eta,\tau)}{\Gamma_j(\eta,\tau)}e^{i\phi_j(\eta,\tau)} + c.c.\right] \tag{2}$$

$$\left[\frac{\partial}{\partial\eta} + \frac{\partial}{\partial\tau}\right]A(\eta,\tau) = i\delta(\eta,\tau)A(\eta,\tau) + \frac{1}{\rho}\left\langle\frac{e^{-i\phi(\eta,\tau)}}{\Gamma(\eta,\tau)}\right\rangle \tag{3}$$

where $\delta(\eta,\tau)$ is the position and time dependent detuning parameter given by

$$\delta(\eta,\tau) = \left[1 - \frac{1}{\lambda_w(\eta)}\frac{\partial\lambda_w}{\partial\eta}\tau\right]\left[\delta_0 + \frac{1}{2\rho}\right]\frac{\lambda_w(0)}{\lambda_w(\eta)} - \frac{1}{2\rho}\,. \tag{4}$$

The averaging in the field equation is taken over all electrons in one bunch, while the index j references a particular electron within that bunch.

The only other term present is the Pierce parameter given by $\rho = [K(z)\gamma_0^2\Omega_p/4\gamma_R^2(z)\omega_w(z)]^{2/3}$ where $K(z) = eB_w(z)\lambda_w(z)/2\pi m_0c^2$ is the wiggler parameter, γ_0 is the electron injection energy, Ω_p is the relativistic plasma frequency, $\gamma_R{}^2(z) = [1 + K^2(z)]\lambda_w(z)/2\lambda$ is the square of the resonant energy, and $\omega_w(z) = 2\pi c/\lambda_w(z)$. The scaling of η and τ in terms of the position z and time t (measured

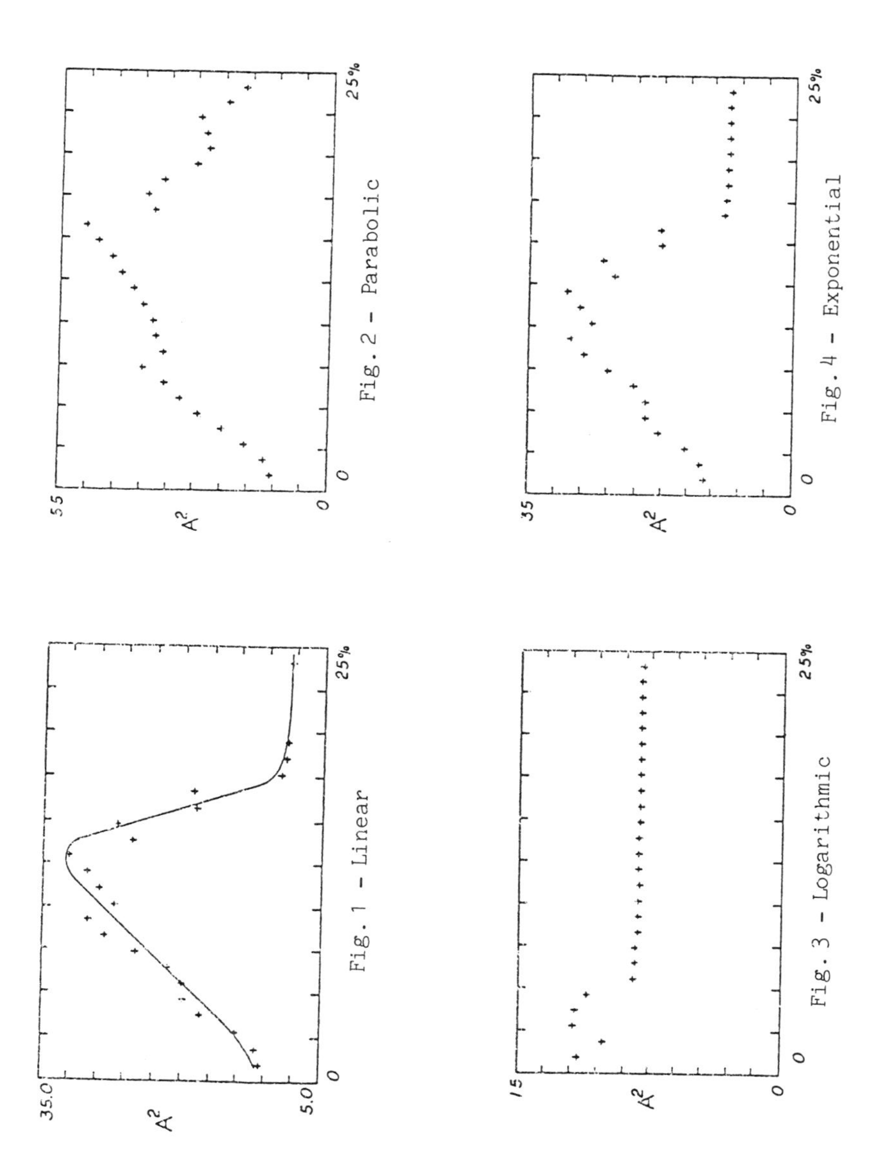

Fig. 1 - Linear

Fig. 2 - Parabolic

Fig. 3 - Logarithmic

Fig. 4 - Exponential

Figures (1) - (4): Graphs of output intensity of laser field vs. percentage decrease in wiggler wavelength over the total length of the device for different wiggler geometries. The smooth curve in Figure 1 is a qualitative average of the points and is not the result of any numerical curve fitting technique in the laboratory frame) along with the scaling of the dynamical variables in this formalism necessitates that both the wiggler parameter and the Pierce parameter be constant. This can be accomplished by requiring that $\lambda_w(z) B_w(z)$ = constant.

ANALYSIS AND RESULTS

The FEL system is treated as a single pass amplifier of an injected signal using one bunch of 8 electrons. These electrons are initially unbunched (spaced evenly to within quantum fluctuations over the spatial extent of the bunch) and all having the same injection energy γ_0. For direct comparison with the results of Ref. 1 the initial numerical values are chosen as ρ = .0021, δ_0 = 1.86, A^2 = 8.00, and $\lambda_w(0)$ = 3.2 cm. Equations (1) - (3) are integrated numerically using the fourth-order Runge-Kutta method for four different tapering geometries: linear, parabolic, logarithmic, and exponential.

Figure 1, a graph of output intensity vs. percentage decrease in wiggler wavelength over the total length of the FEL, reviews the results for the linear case[2]. The peak in output intensity occurs for a 13% decrease in λ_w yielding an amplification on the order of 400%, or an 8-fold improvement over the constant wiggler amplification of approximately 50%. Figure 2 shows the parabolic case, displaying the peak value of output intensity at 16% decrease in λ_w with a 650% amplification, or a 13-fold improvement over the constant wiggler. The logarithmic case is seen in Figure 3 to provide no improvement over the constant wiggler, while the exponential tapering of Figure 4 is seen to very closely resemble the linear case both qualitatively and quantitatively.

CONCLUSION

This theoretical model of a FEL amplifier does predict significant improvement in amplification of the input laser intensity over the constant wiggler configuration for several different geometries, indicating that gain enhancement via this technique may be attractive.

REFERENCES

1. R. Bonifacio, C. Pellegrini, and L. M. Narducci, Op. Commun., 50, 6, p. 373 (1984).
2. K. R. Hartzell Jr., Phys. Lett. A, 122, 9, p. 476 (1987).

FIBER-OPTIC RING LASER

M. V. Iverson and G. L. Vick
Rockwell International Corporation, Cedar Rapids, IA 52498

ABSTRACT

We have built a ring laser by placing an antireflection-coated GaAlAs laser diode into a loop of fiber. Its output is primarily a single spectral line; sidebands are greater than 35 dB down from that line. We also treat the laser theoretically, by calculating the total electric field, the gain condition, and the linewidth.

INTRODUCTION

A fiber-optic ring laser[1-3] contains a semiconductor optical amplifier[4-5] in an optical fiber ring. This novel laser uses an optical fiber ring instead of mirrors and a semiconductor optical amplifier instead of helium neon gas. These components form a resonant cavity. Internal modes of the optical amplifier are suppressed by antireflection coatings deposited on both end faces of the chip. The resonant modes of the ring are contained within a fiber-optic directional coupler. Its length is less than one meter. The external feedback of light reduces the electrical bias current required for threshold of laser action. A coupling ratio of 99/1 increases the external feedback of light beyond that for standard 50/50 directional couplers. Tai et.al. attempted a measurement of the linewidth[3] for a similar structure and found 45 kHz as an instrument-limited upper bound. In the theory by Tai et.al., the natural linewidth ($\Delta\nu_0$) of the amplifier without feedback was reduced by 10^{-4} times with good fiber to diode coupling efficiencies.[3] Tai et.al. assumed the natural linewidth ($\Delta\nu_0$) for their diode. Since fiber-optic ring lasers are applicable to coherent communication systems and to optical gyroscopes, a straight-forward model is desired without assuming the natural linewidth ($\Delta\nu_0$). Also a single-mode spectrum is desired to prevent spurious interference.

THEORY

We consider the electric field for a fiber-optic ring laser in Figure 1. An optical amplifier is placed on the bottom between opposite legs of a fiber-optic directional coupler (right). Six critical locations are denoted with capital letters. Amplifier, fiber and coupler parameters are defined as follows:

g	- amplifier gain coefficient	$l = \ell$	- length of the diode
t_1, t_2	- amplifier to fiber coupling transmissibilities	k_1, k_2	- propagation constants
		n_ℓ, n_f	- refractive indices
r	- coupling ratio (coupler)	α, α_c, β	- losses of the diode, coupler, and fiber
L	- length of the fiber ring		

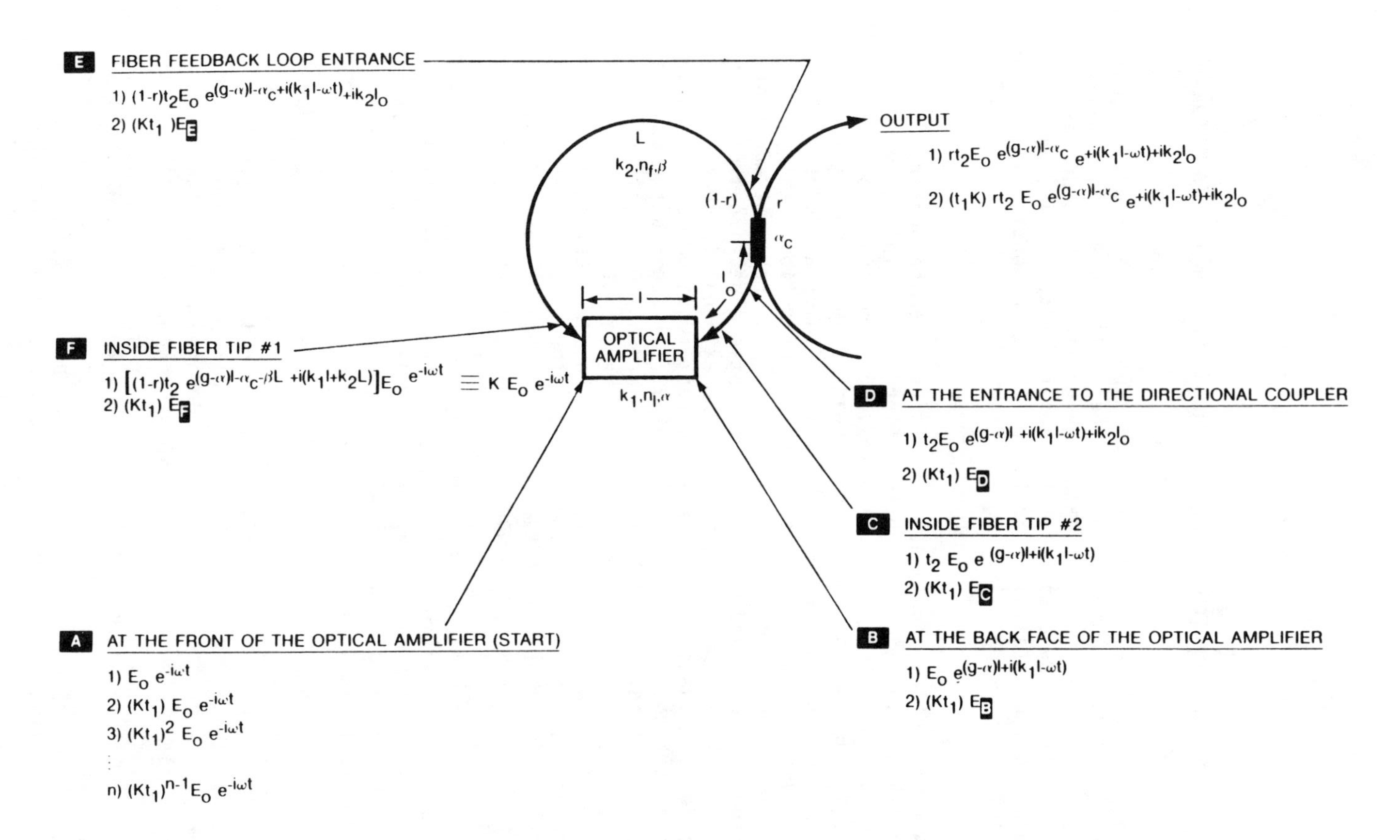

FIGURE 1. ELECTRIC FIELD ANALYSIS OF A FIBER-OPTIC RING LASER
START WITH THE OPTICAL AMPLIFIER AND WORK COUNTERCLOCKWISE.

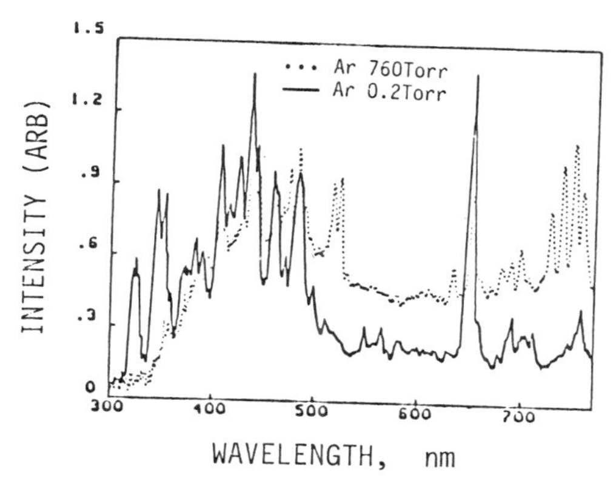

Fig. 2 The spectrum of the HCF emission monitored by an Optical Multichannel Analyzer (EG&G). The electrical input energy to the HCF was 200J.

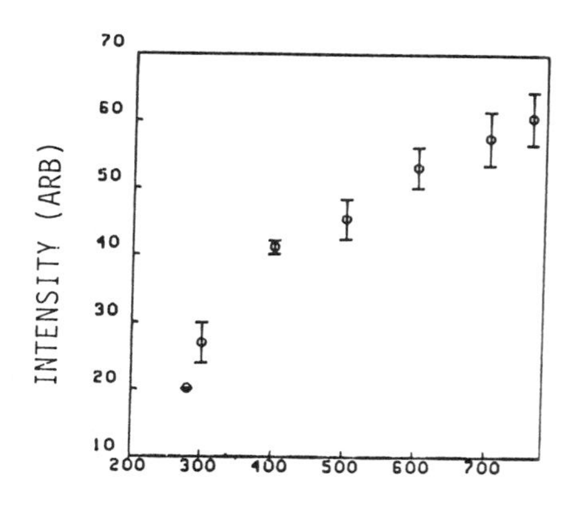

Fig. 3 The radiatd output of the HCF as a function of the argon fill gas pressure. The output was monitored by a PM tube (UVP1200) and an interference filter (400nm).

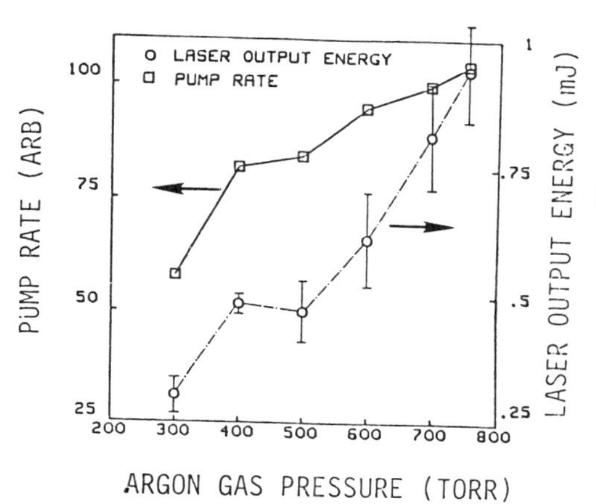

Fig. 4 The pump rate and laser output energy of LD490 as a function of argon fill gas pressure.

NONRESONANT OPERATION OF TWO-TRANSITION LASERS IN AN INHOMOGENEOUSLY BROADENED MEDIUM

D. L. Lin
State University of New York at Buffalo, Amherst, N. Y. 14260

S. G. Sun
Institute of Radio Engineering, Nanjing, P.R. China

ABSTRACT

The master equation for the nonresonant operation of inhomogeneously broadened two-transition laser is derived from the density matrix equation of motion. Operation characteristics are calculated from the steady state solution of the master equation and are compared with those of a one-transition laser. It is shown that the two-transition laser operation has a number of advantages.

INTRODUCTION

Recently, Li, Gong and Lin[1] formulated the general problem of nonresonant interaction between a three-level atom and cavity fields with arbitrary detunings. A number of novel properties were found in their discussions of the atomic level occupation probabilities. In a subsequent series of papers, Liu, Li and Lin[2] and others[3] then investigated photon distribution and other field properties. It is found in these studies that many of these new phenomena can be understood on the basis of two-photon processes.

On the other hand, the laser operation of a three-level Ξ-type atomic system was discussed by Zhu, Li and Liu[4], and the detuning effects on the laser action of a three-level Λ-type atom was discussed by Sun, Li and Bei[5]. It is interesting to note that the presence of two-photon processes tend to enhance the atom-field coupling strength. The influence of detunings on the operation characteristics is shown to be remarkable, especially when the two detunings have opposite signs. Since the active atom is assumed to be homogeneously broadened in these discussions. We consider the nonresonant laser operation of Λ-type three-level atoms in an inhomogeneously broadened medium. Our treatment is a straightforward generalization of Riska and Stenholm[6] who studied two-level laser operation in an inhomogeneously broadened medium.

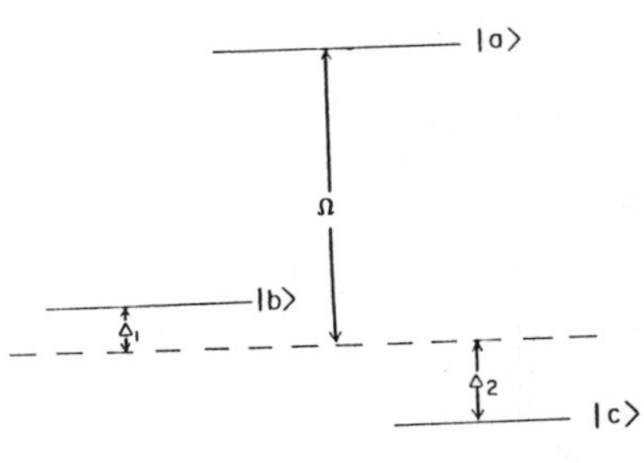

Fig.1. Energy level configuration of a Λ-type atom.

THEORY

As shown in Fig. 1. the three-level atom has a common upper level $|a\rangle$ and two lower levels $|b\rangle$ and $|c\rangle$ which may be resulting from a splitting of degenerate states in external magnetic field. The detuning parameters Δ_1 and Δ_2 are generally different from each other. The two lower levels are assumed to have the same decay rate γ. The Hamiltonian for the system of an atom interacting with a cavity mode of frequency Ω is given by

$$H = \Omega a^{\dagger}a + \sum_{i=a,b,c} \omega_i A_i^{\dagger}A_i + V_1(t) [a A_a^{\dagger}A_b + h.c.] + V_2(t) [aA_a^{\dagger}A_c + h.c.] \qquad (1)$$

$$V_i(t) = g_i \sqrt{n+1} \sin[Kz(t)] \qquad (2)$$

where $a^{\dagger}$ creates a photon and $A_i^{\dagger}$ creates an atom in the state i, g_i is the atom-field coupling constant and $K = \Omega/c$. The function $z(t)$ is the position of the active atom at the time t in the cavity of length L.

Following the method of Refs. 5 and 6, the master equation is derived from the equation of motion of the denstiy matrix in the Doppler limit $Ku \gg \gamma$. The result is

$$\frac{d}{dt} \rho_{n,n} = -\sum_{i=1}^{2} A_i \exp\left(-\frac{\Delta_i^2}{K^2\mu^2}\right)\left[1-\frac{n+1}{4}\sum_{m=1}^{2}\frac{B_m}{A_m}\left(1+\frac{\bar{\gamma}^2}{\bar{\gamma}^2+\Delta_i^2}\right)\right]\rho_{n,n}$$

$$+\sum_{i=1}^{2} A_i \exp\left(-\frac{\Delta_i^2}{K^2\mu^2}\right)\left[1-\frac{n}{4}\sum_{m=1}^{2}\frac{B_m}{A_m}\left(1+\frac{\bar{\gamma}^2}{\bar{\gamma}^2+\Delta_i^2}\right)\right]\rho_{n+1,n+1}$$

$$+ C(n+1)\rho_{n+1,n+1} - C n \rho_{n,n} \qquad (3)$$

where A_i, B_i, and C stand for the gain, saturation and loss coefficients respectively, and we have defined the mean decay rate of upper and lower levels $\bar{\gamma} = (\gamma_a + \gamma)/2$. The threshold condition, mean photon number and width of intensity distribution can all be calculated from the steady state solution of (3). A detailed account of this calculation will be published elsewhere, and here we just summarize some of the results in the following.

RESULTS AND DISCUSSION

The threshold condition follows directly from the steady state solution. For a two-transition laser, the threshold is in general lower than that for the one-transition laser. Therefore the three-level laser is easier to start oscillation than the two-level laser provided that all the other parameters are the same. Furthermore, it is found that the threshold for three-level lasers is easier to be kept unchanged than that of two-level lasers.

To study the stability of the operation, we plot the photon number probabilities for various δ_2 when δ_1 is fixed. Compared to the corresponding plot for the one-transition laser in Ref. 6, we find that the intensity of the two-transition laser is generally stronger. It is also observed in Fig. 2. that the most probable photon number does not change as much when δ_2 changes. Thus the non-resonant operation is more stable than the one-transition laser. Furthermore, it is of interest to point out that the half-width of the photon number distribution is a slowly varying function of detuning parameters in contrast to the one-transition case in which the half-width increases quickly with increasing detuning. This means that the laser field does not deviate very far from the coherent state.

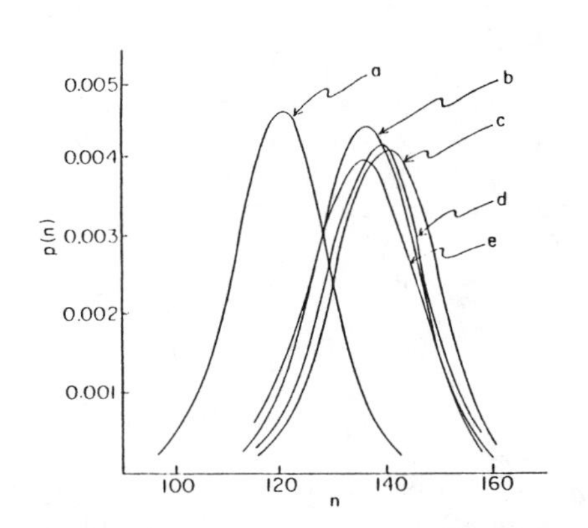

Fig. 2. Photon probability distribution for δ_1= 2.0 and δ_2 is (a) 1.0 (b) 2.0 (c) 2.5 (d) 2.8 (e) 5.0. Other parameters are A_1/C = 1.2, A_2/C = 1.5, B_1/A_1=0.005 and B_2/A_2 = 0.001.

The authors would like to thank X. S. Li for many helpful discussions throughout this work

REFERENCES

1. X. S. Li, C. D. Gong and D.L. Lin, Phys. Rev. A 36, (in press).
2. Z. D. Liu, X. S. Li and D. L. Lin, Phys. Rev. A 36, (in press).
3. X. S. Li and D. L. Lin, (to be published).
4. S. Y. Zhu, X. S. Li and Z. D. Liu, Phys. Rev. A (accepted for publication).
5. S. G. Sun, X. S. Li and N. Y. Bei, Phys. Rev. A (accepted for publication).
6. D. O. Riska and S. Stenholm, J. Phys. A 3, 189 (1970); ibid 536 (1970).

DYNAMICS OF COUPLED ELECTRON - NUCLEON MOTION IN A LASER FIELD

F. X. Hartmann and K. K. Garcia
Institute for Defense Analyses, Alexandria, Virginia

J. K. Munro, Jr. and D. W. Noid
Oak Ridge National Laboratory, Oak Ridge, Tennessee

ABSTRACT

Energy transfer processes in single particle coupled nucleon-electron models interacting with an intense laser field are studied using semi-classical quantization of the coupled classical Hamiltonian.

INTRODUCTION

With the development of intense laser sources operating at higher powers, shorter pulse widths or shorter wavelengths, the study of intense laser-matter interactions is a growing area of science -- some current work centers around the use of laser produced plasmas as x-ray or x-ray laser sources. In addition, new and interesting areas are the study of above threshold ionization, ultraviolet multiphoton picosecond processes, multiphoton ionization and the production of harmonic radiations from atoms subjected to strong laser fields. In these areas of research the intense laser field interacts with the electronic charge distribution of the atom. In addition to transfering energy to the electronic cloud of the atom, the possibility also exists for energy transfer between near-lying excited states of the nucleus. In general, the energy transfer could proceed directly from the laser to the nucleus, or indirectly via non-radiative energy transfer from the electronic motions to nuclear excitations.

Experiments show that the possibility for nuclear excitations exists due to transitions of atomic electrons to inner-shell vacancies. Additionally, laser driven excitation of the first excited state of ^{235}U has been reported (Ref. 1) and a second experiment is ongoing (Ref. 2). These experiments show the possibility of exciting ultra-low energy nuclear transitions from a nuclear ground state of possible interest in isotope separation or generating controlled gamma emissions from excited isomeric levels, by transition from long- to shorter-lived states.

In 1985, Baldwin, Biedenharn, Rinker, and Solem (Los Alamos, Ref. 3), addressed the energy transfer problem by focusing on the laser-electron interaction with a perturbative approach (Ref. 4) to estimate nuclear excitation probabilities. Alternately, Noid, Hartmann and Koszykowski (Oak Ridge, Ref. 5) proposed a study of the coupled dynamics of the system in a semiclassical approach. This work initially looked at the electron-nucleus system and, in continuation at Oak Ridge National Laboratory, has now looked at a simple model containing electron-nucleus-laser terms. Recently, Berger, Gogny, and Weiss (Livermore, Ref. 6) examined the physics of laser-driven classical electron motion with a perturbative approach to the nuclear matrix element.

In this brief report we describe our method for calculating energy transfer between two coupled systems in the presence of a laser field. This semiclassical approach (Ref. 7) is one which solves Hamilton's dynamical equations of motion for the electron and nucleon from initial quantum conditions, and treats the laser field classically as an explicit function of time. This approach is one which provides a route to models which include both nuclear as well as electronic collective degrees of freedom.

MODEL

A simple model of coupled electronic and nucleonic independent particle motion is used as a first step to study energy transfer processes in the presense of a laser field. The time dependent Hamiltonian, H(p,r,t) for the coupled electronic-nucleonic system is given by:

$$H(p,r,t) = H_n(p_n,r_n) + H_e(p_e,r_e) + H_c(r_e,r_n) + H_{laser}(r_e,r_n,t)$$

where H_n is the nuclear Hamiltonian, H_e is the electronic Hamiltonian, H_c is the electron-nuclear coupling term, and H_{laser} is the time dependent laser Hamiltonian. The nuclear Hamiltonian describes the motion of an odd proton in a Woods-Saxon well:

$$H_n(p_n, r_n) = m_n c^2 \left[\left(\frac{p_n^2}{m_n^2 c^2} + 1 \right)^{1/2} - 1 \right] + V_0 \left\{ 1 + \exp\left[\frac{(r_n - R_0)}{a_n} \right] \right\}^{-1}$$

Here m_n is the nucleon rest mass, V_o is the Woods-Saxon well depth, a_n is the well diffusivity and R_0 is the nuclear radius. The electron Hamiltonian is:

$$H_e(p_e, r_e) = m_e c^2 \left[\left(\frac{p_e^2}{m_e^2 c^2} + 1 \right)^{1/2} - 1 \right] - \frac{e^2 \eta (Z - k)}{r_e}$$

and the coupling term is:

$$H_{coup} = - \frac{\eta k e^2}{| r_e - r_n |}$$

where m_e is the electron rest mass, Z is the nuclear charge, η is a screening parameter, and k is an artificial coupling parameter. With k equal to zero, the system is "uncoupled", with k is equal to one, the system is "coupled." The laser Hamiltonian is given by:

$$H_{laser}(r, t) = \mu(r) \, E \cos\omega t \; .$$

This term contains the time-dependent contribution for the laser, characterized by electric field strength E and frequency ω. The quantity $\mu(r)$ is the dipole moment given by the electron-nucleon distance; the nuclear core is fixed at the origin of the coordinate system.

RESULTS

Laser terms and relativistic corrections, of particular importance to the electron's motion, have been added to the model reported previously (Ref. 5). This model still has dynamics characteristic of a generic nucleus (Blatt-Weisskopf transition rates) which provides a standard starting point. The nuclear frequencies are, however, high for nuclei of ultimate interest by two to three orders of magnitude; since the nuclear frequency sets the overall energy scale of the problem; the example which we depict here serves to llustrate the basic steps of the calculations and the qualitative conclusions.

For the case discussed in this paper $V_o = -30$ MeV, $R_o = 1.25$ Fm $A^{1/3} =$ 8.66 Fm for A=208, $a_n = 0.65$ Fm , Z = 82 , $m_e = 0.511$ MeV and $m_n = 938$ MeV. These are reasonable parameters for an odd proton outside a doubly magic core. The proton is initially in an l=3 state (with a nearby l=1 state) and, given the frequency scale set by the nucleon transition, the electron is started in the K state for purposes of illustration. The frequency is in units of inverse time units, where one time unit is the time for light to travel one Fermi. The frequency is converted to energy units of MeV by multiplying by 197. The electric field has units of $MeV^{1/2}$ $Fm^{3/2}$. In these convenient units $e^2 = 1.44$ MeV · Fm and $e^2/2\pi hc$ is 1/137.

Classical trajectories (Fig. 1) in momentum and coordinate space are generated from the Hamiltonian using Hamilton's equations: $\dot{p} = -\partial H/\partial r$, $\dot{x} = \partial H/\partial p$. The initial conditions for the trajectories are chosen from states of the separable Hamiltonian, that is, with the coupling term neglected. For each such

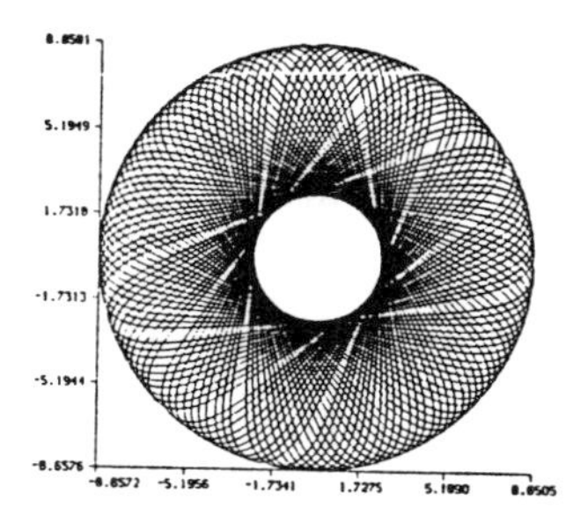

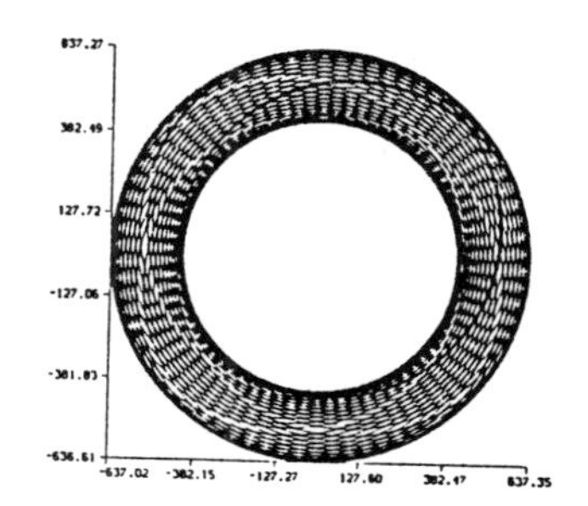

Fig. 1. Coupled relativistic nucleonic (radius = 8.66 Fm) and electronic (radius = 637 Fm) trajectories.

separated system, quantization of the action angle variables can, in general, include the fractional term described by Keller (Ref. 8). With the relativistic corrections though, the K orbit trajectories for Z greater than half the fine structure constant generated from initial conditions using the Langer correction (Ref. 9) are unstable. In a classical interpretation, the electrons orbit decays into the nuclear center. This behavior is associated with the singularity occurring in Sommerfeld's relativistic treatment of the Bohr atom (Ref. 10). The initial condition for the relativistic electron is thus taken to be that of the "old quantum theory" without the Langer correction. In light of the recent discussion by Jackson (Ref. 11), this is an interesting aspect to pursue.

The trajectories are used to calculate autocorrelation functions of the total system dipole moment. The absorption band shape (power spectrum) is given by the Fourier transform of the appropriate autocorrelation function. With the laser field off, the power spectrum is used to understand the basic dynamical frequencies. We find four basic frequencies in addition to the many harmonics and lower intensity frequencies. The lowest frequency of the four is associated with the orbital motion of the electron. The three higher frequencies are associated with the nucleon motion. In a crude sense, the nucleon moves in a "narrow" elliptic orbit close to the appearance of a linear vibration; this orbit precesses. The two highest frequencies arise from this "vibrational" frequency of the nucleon plus and minus half the "rotational" frequency. The rotational frequency appears between the electronic orbital and the nucleonic vibrational frequencies. When the total dipole moment autocorrelation function in the uncoupled case is transformed, components of the electronic frequency appear superimposed on the nucleonic frequencies. For the relativistic Hamiltonian (in the old quantum theory) the total dipole moment shows a stronger admixture of the electron and nucleon motion than the non-relativistic case.

The trajectory calculations with the laser term are carried out on a Cray computer at Oak Ridge National Laboratory. We integrate Hamilton's equations as a function of time for an ensemble of 64 trajectories each with a different initial phase for the laser; these phases are evenly distributed over two pi radians. We choose 64 trajectories for the phase ensemble since the vector capability of the Cray allows us to run up to 64 trajectories in parallel. The explicit time dependence of the laser pulse easily allows us to vary frequency and intensity and even, if desired, the envelope.

The laser is first scanned through a range of frequencies at a fixed high intensity. For each fixed frequency, the ensemble averaged nucleon energy, electron energy and total energy are computed as a function of time (Fig. 2). Plots of the maximum energy transfer reveal resonances at which the transfer is enhanced (one such resonance is crossed in the scan shown in Fig. 2). Fourier transforms of these time dependent ensemble averages are useful in extracting frequency information on the location of significant resonances for laser absorption in the coupled system. An investigation of the intensity dependence shows that the energy transfer, largest when on a resonance, falls with decreasing electric field strength. For the case in Fig. 2, the energy transfer roughly drops by one order of magnitude for each order of magnitude decrease in the electric field strength over the range 1x10(-6) to 1x10(-8); (the intensity is proportional to the square of the electric field strength).

The energy transfer values are related to the frequency spread of the particle's motion. Typically, the ensemble energy transfer, which is related to that frequency spread, reaches a constant after some time. We have run extended time durations to verify that this ensemble frequency spread is roughly constant for an extended time duration, thus, the laser absorption is essentially a quasi-periodic function of time.

Ensemble-averaged energy as a function of time in the laser field.

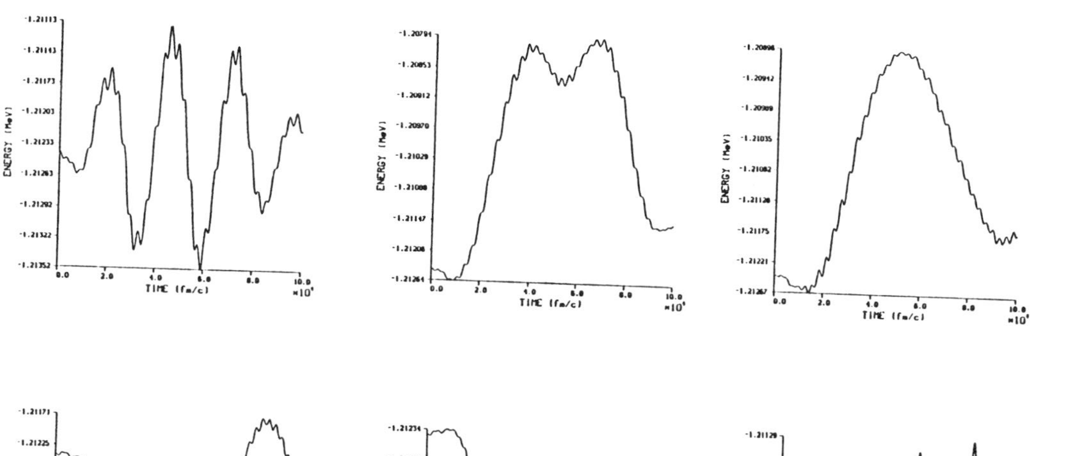

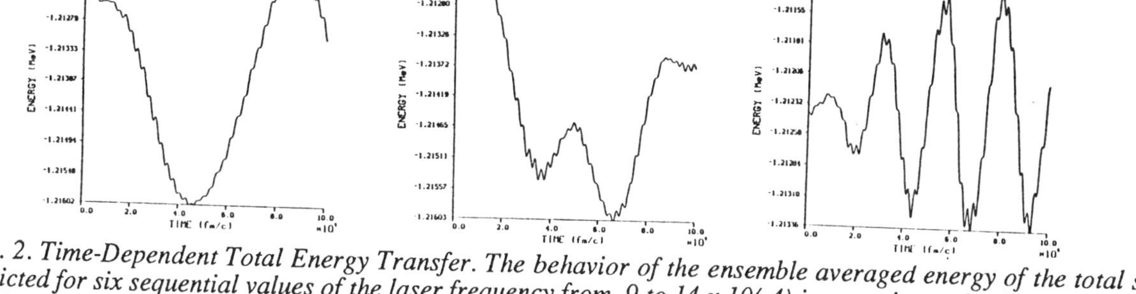

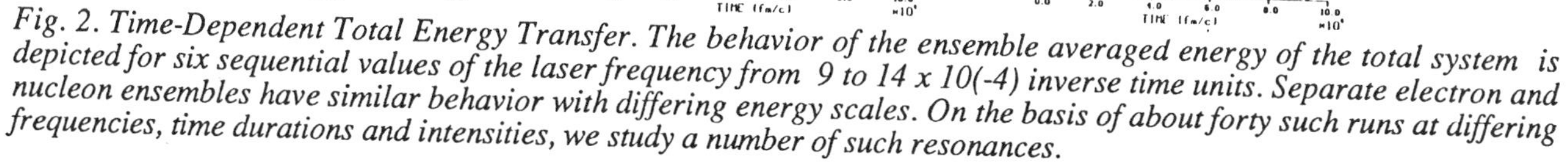

Fig. 2. Time-Dependent Total Energy Transfer. The behavior of the ensemble averaged energy of the total system is depicted for six sequential values of the laser frequency from 9 to 14 x 10(-4) inverse time units. Separate electron and nucleon ensembles have similar behavior with differing energy scales. On the basis of about forty such runs at differing frequencies, time durations and intensities, we study a number of such resonances.

CONCLUSION

An approach to exploring non-radiative energy transfer in coupled systems and intense laser fields has been applied to the study of energy transfer in a single-particle electron and single-particle nucleon model.

It is demonstrated that energy transfer to the nuclear motion occurs via coupling to the electronic motion in a laser field for a simplistic single-particle model of a valence nucleon and an inner shell electron at a laser resonance. Energy transfer in this system not only depends on the laser frequency but also on intensity. Electrons closest to the nucleus have greater coupling; electrons further from the nucleus though, are expected to be most affected by the laser.

The nucleon model sets the energy scale for the frequencies of the coupled system--currently this is too high for practical applications since the single-particle nucleon transitions are too high in energy. The specific features of the nucleon model can be scaled to examine lower frequency transitions and collective motions included in this approach. The electron part of the model basically describes a tightly bound electron and with suitable changes can be extended to treat outer shell electrons or collection motions (in the laser field) as appropriate.

We acknowledge the support of the Dept. of Energy (Basic Energy Sciences); DARPA (Directed Energy Office) and at IDA, the support of Dr. O. Kosovych. Continuing stimulating discussions with Prof. Y. Y. Sharon are also acknowledged.

REFERENCES

1. Y. Izawa and C. Yamanaka, Phys. Lett. 88B, 59 (1979).

2. J. A. Bounds, P. Dyer and R. C. Haight in Advances in Laser Science II, eds. M. Lapp, W. C. Stwalley and G. A. Kenney-Wallace (AIP Conference Proceedings 160, New York, 1987) p. 87.

3. G. A. Rinker, J. C. Solem and L. C. Biedenharn, ibid., p. 75.

4. M. Morita, Prog. of Theor. Phys. 49, 1574 (1973).

5. D. W. Noid, F. X. Hartmann and M. L. Koszykowski in Advances in Laser Science II , eds. M.Lapp, W. C. Stwalley and G. A. Kenney-Wallace (AIP Conference Proceedings 160, New York, 1987) p. 69.

6. J. F. Berger, D. Gogny and M. S. Weiss, submitted to J. Quant. Spect. and Rad. Transfer, 1987.

7. D. W. Noid, M. L. Koszykowski and R. A. Marcus, J. Chem. Phys. 67, 7 (1977) and Ann. Rev. Phys. Chem. 32, 267 (1981).

8. J. B. Keller, Ann. Phys. (NY) 4, 180 (1984).

9. R. Langer, Phys. Rev. 51, 699 (1937).

10. A. Sommerfeld, Atomic Structure and Spectral Lines, transl. by H. L. Brose, (Methuen, London: 1923) p. 251.

11. J. David Jackson, Physics Today 40, 34 (1987).

SYNTHESIS OF A VOLATILE COMPOUND OF THE (2-) Ho-160 NUCLEAR ISOMER

F. X. Hartmann+ and R. A. Naumann
Departments of Chemistry and Physics
Princeton University, Princeton, New Jersey

ABSTRACT

The synthesis of volatile compounds of radioactive holmium atoms are described and the gamma-ray spectra following the decay of the longest-lived isomer (Ho-160) presented. This latter isomer has an ultra-low energy transition of potential interest in laser driven nuclear inter-level transfer studies.

INTRODUCTION

The odd-odd nucleus Ho-160 has four identified long-lived nuclear isomeric levels (Ref. 1). The longest lived (5.02 hr) 2- state at 0.060 MeV excitation lies within 1.5 keV of a shorter lived 7 minute (1+) state, by an (E1) transition, whose exact position is unknown and is of potential interest in nuclear inter-level transfer studies (Ref. 2). We report on the synthesis of a volatile form of this isomer.

PROCEDURE

We use 15 MeV protons from the Princeton Cyclotron Facility to bombard a target of approximately 0.07 grams of dysprosium oxide (Dy_2O_3). The bombardments typically last for a total charge collection of 3.6 milli-Coulombs. The resultant material is converted to the chloride by heating to dryness in concentrated HCl (3 ml) and then dissolved in approximately 1.5 ml of slightly acidic water. A few drops of the ionic solution are passed through a 3x100 mm ion exchange column at warm temperature (70 C) and the holmium faction is eluted at elevated pressure (100 psi) using a 0.17 M alpha-hydroxyisobutyric acid solution. Approximately 0.52 ml of $HoCl_3$ stock solution (82 mg $HoCl_3 \cdot H_2O$ in a 2.5 ml H_2O/2.5 ml ethanol solution) are added along with an equimolar solution of Hfod (structure described below) in 95% ethanol. The Hfod is previously neutralized by a sodium hydroxide solution in 50% ethanol and 50% water. The product is extracted and dried using methylene chloride; purification by sublimation can be carried out at about 170 C for thirty minutes under vacuum. This procedure is developed from the approaches used in Refs. 3 and 4.

+ Work performed at Princeton University, current address: Institute for Defense Analyses, Alexandria, Virginia.

DISCUSSION AND RESULTS

Of the possible isotopes which can be produced from the stable dysprosium precursors in their natural abundances (Dy-156, 158, 160, 162, 163, and 164) the 5.02 hr Ho-160m isomer (which decays to excited states of Dy-160 as well as the unstable Ho-160 ground state) gives the most significant detectable activity. It is produced primarily from the Dy-160 (p,n) Ho-160 reaction. The remaining Ho isotopes and isomers which could contribute at a much lesser or negligible extent to the activity are: Ho-158 (27, 21 and 11.5 m), Ho-159 (33 m), Ho-160 (7m, 1h), Ho-161 (2.48h), Ho-162 (15m, 68m), Ho-163 (4570 y) and Ho-164 (37m, 29m).

The labelled holmium β-diketonates are also interesting because of their volatility. This volatility can be understood in the following picture. Removal of the $6s^2 5d^1$ valence electron results in the usual +3 chemical oxidation state of a typical lanthanide atom. These +3 ions display a smoothly decreasing ionic radius with increasing atomic number. For complexes with octahedral coordination this shrinkage in the lanthanide series results in increased steric crowding of the ligands. This crowding protects the metal center from reactions which could result in decomposition. For holmium compounds, some common β-diketonates are stable at temperatures in excess of 200 C (Ref. 5).

In light of a number of considerations concerning coordination number, fluorine content, effective shielding, metal ion radius, substitutional group size, aromaticity and methylene carbon substitutuion; two compounds are particularly attractive for use as gaseous sources: 1,1,1,2,2,3,3-heptafluoro-7,7-dimethyl-4,6-octanedione (fluoro-octanedione "Hfod") and 2,2,6,6-tetramethylheptane-3,5-dione (dipivaloylmethane, "Hdpm" or "Hthd"). The structure of the Ho(fod)3 compound is depicted in Fig. 1.

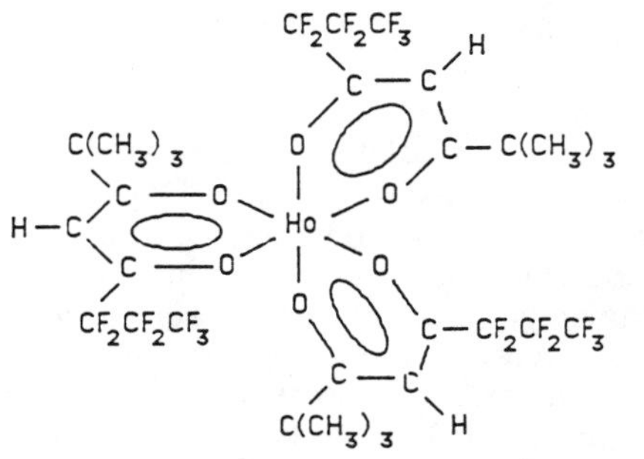

Fig. 1. Structure of Ho(fod)3.

We report on infrared spectra of the compound as observed in a specially constructed, all glass, infrared cell capable of being heated to elevated temperatures (200 C). These spectra show the removal of water from the outer hydration sphere of the compound. A GeLi gamma-ray spectrum of the labelled Ho(fod)3 is shown in Fig. 2. Gamma-rays in this spectrum can be assigned to de-excitations of the Dy-160 nucleus. We additionally report success in maneuvering the sources in closed glass tubes.

The Ho-160m nucleus is of potential interest in studies of the laser driven excitation of isomers to short lived states in the investigation of laser-electron-nucleon coupling. The volatile compounds produced here are also of importance for use as sources in an experiment to measure the mass of the electron neutrino (Ref. 6); in particular for the detection of the ultra-low decays of Ho-163 introduced internally to the counting gas of a heated gas proportional detector. The physical properties of the volatile β-diketonate compounds, in general, could make them useful in other radiochemical preparations or measurements.

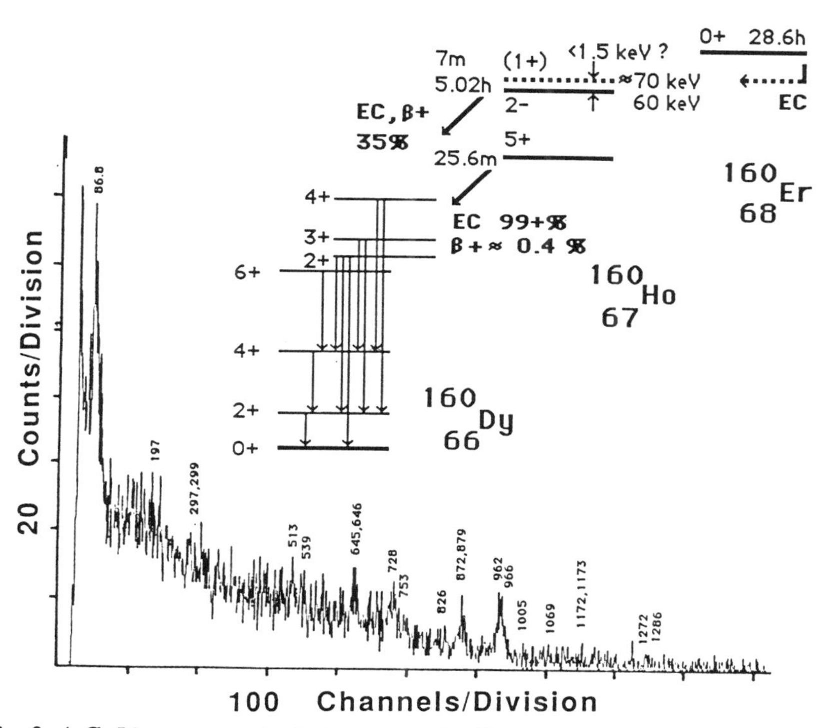

Fig. 2. A GeLi spectrum depicting some significant gamma-rays (MeV) associated with the decay of the labelled Ho(fod)3. The decay scheme is also illustrated.

This work was supported by the National Science Foundation. The authors acknowledge the supporting work of Dr. Timos Altzitzoglou.

REFERENCES

1. C. M. Lederer and V. S. Shirley, eds. Table of the Isotopes, 7th Ed. (Wiley, N.Y.,1978) and A. Artna-Cohen (compilations to be published).

2. J.A.Bounds, P.Dyer and R.C.Haight in Advances in Laser Science II, eds. M. Lapp, W.C.Stwalley and G.A.Kenney-Wallace (AIP Conf. Proceedings 160, New York, 1987) p.87 and G.A. Rinker, J.C. Solem and L.C. Biedenharn, ibid., p. 75.

3. K. J. Eisentrout and R. E. Sievers, Inorganic Synthesis 11, 94 (1967).

4. C.S.Springer, Jr., D.W. Meek and R.E.Sievers, Inorg. Chem. 6, 1105 (1967).

5. Mehrota, Bohra and Gaur, Metal β-Diketonates and Allied Derivatives, (Academic Press, N. Y.,1978).

6. F. X. Hartmann and R. A. Naumann, Phys. Rev. C 31, 1594 (1985).

RESONANT GAIN OF OPTICALLY PUMPED FAR INFRARED RAMAN LASERS

Jagdish Rai and Jerald R. Izatt
University of Alabama, Tuscaloosa, Alabama 35487

ABSTRACT

In this paper, we present theoretical and experimental results for the gain of optically pumped far-infrared (FIR) Raman lasers over a broad tuning range. Our results show laser emission at line-center and off-resonant frequencies, depending upon the Rabi oscillation frequency for the pump field. Theoretical calculations are compared to the emissions from $^{12}CH_3F$ and $^{13}CH_3F$.

INTRODUCTION

The process of stimulated Raman scattering provides a useful source of tunable coherent FIR radiation. The pump laser for this process is usually a high power, continuously tunable, multiatmosphere CO_2 laser which can be tuned over broad portions of the 9-11 μm spectrum; e.g. ~450 GHz in the 9R branch. Several molecules have been pumped, yielding FIR emission from 67 to 1130 μm. These molecules in general can be classified into two types. The first type, of which NH_3 is an example, has three pertinent levels, while the second type has a double Raman process which can be accurately described on the basis of a four level scheme. $^{12}CH_3F$ and $^{13}CH_3F$ are good examples of the four-level medium. The gain for both types of lasers can be calculated using a density matrix analysis [2,3]. However, most calculations are limited to the simple case of large pump offsets. In this paper, we present a density matrix analysis of the four level molecules for general offsets. In particular, we obtain the very important case of resonant emission. We have obtained, for the first time, the expression for the gain at and close to resonance. The gain explicitly depends on the Rabi frequency for the pump field. Our calculations can easily be applied to the three level case by making all terms corresponding to the missing level, in the density matrix equal to zero.

DENSITY MATRIX ANALYSIS

The density matrix ρ for the atomic system obeys the equation

$$i\hbar \frac{\partial \rho}{\partial t} = [H,\rho] \tag{1}$$

where H is the Hamiltonian of the system which consists of an atomic and interaction part. We treat the radiation field classically as a sum of the pump and FIR fields. The induced polarization is obtained from

$$P = \mathrm{Tr}(\mu\rho) \quad , \tag{2}$$

where μ is the dipole moment matrix. The gain of the laser can be obtained from the imaginary part of $\chi(P = \chi E,\ \chi = \chi' - i\chi'')$. The general expression for χ'' [U] is evaluated numerically as a function of the Rabi frequency $\beta = \mu E/2\hbar$. This is basically the small signal gain of the FIR lasers. The gain depends upon the population of the various levels very strongly. We have also studied various models for the population evolution numerically, considering two, three and four level atomic media.

The gain expression evaluated on the basis of the density matrix shows line center or resonant emission for small values of $\beta\tau$, where τ is the lifetime of the corresponding transition. This behavior for $\beta\tau \approx 1$ to 5, is shown in Fig. 1(a). As $\beta\tau$ increases, which may be either due to increase in pump intensity or increase in the life- time, the gain at resonance starts decreasing, and the major peaks of emission are for off-resonant frequencies. For $\beta\tau \approx 6$ to 10, this is shown in Fig. 1(b). Physically, the drop in resonant gain can occur due to saturation of the gain. Another mechanism for the drop in gain is due to Rabi oscillations leading to radiation trapping for the resonant atoms. However, the Raman emission in the large off-resonant case is a two photon process, and the gain increases as the pump detunings and intensities increase.

EXPERIMENTAL RESULTS AND CONCLUSIONS

The resonant and off-resonant emissions are found to occur for $^{13}CH_3F$ and $^{12}CH_3F$ molecules respectively. The experimentally observed emissions are shown in Figs. 2(a) and 2(b). For $^{12}CH_3F$, we see strong off-resonant emission for J-number from 12 to 24. However, for $^{13}CH_3F$, we have seen resonant emission for J-number, from 34 to 42. This is very well explained on the basis of our theoretical model as $\beta\tau$ is low due to a decrease in the lifetime and the dipole moments (both these quantities depend very strongly on J-numbers of the transition). For low J-number for $^{12}CH_3F$, as $\beta\tau$ is large, we see a strong off-resonant emission. We have also observed a very persistent asymmetry in the emission between lines which may be due to the K-structure for these lines. These effects are currently under investigation.

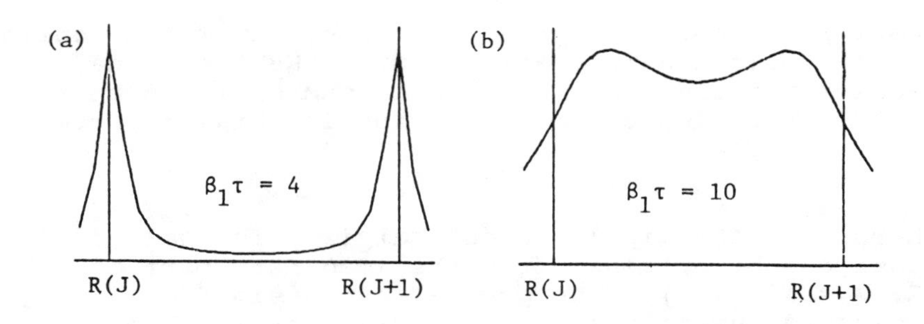

Figure 1. Theoretical gain curves for emission from a four-level system (a) resonant (b) off-resonant.

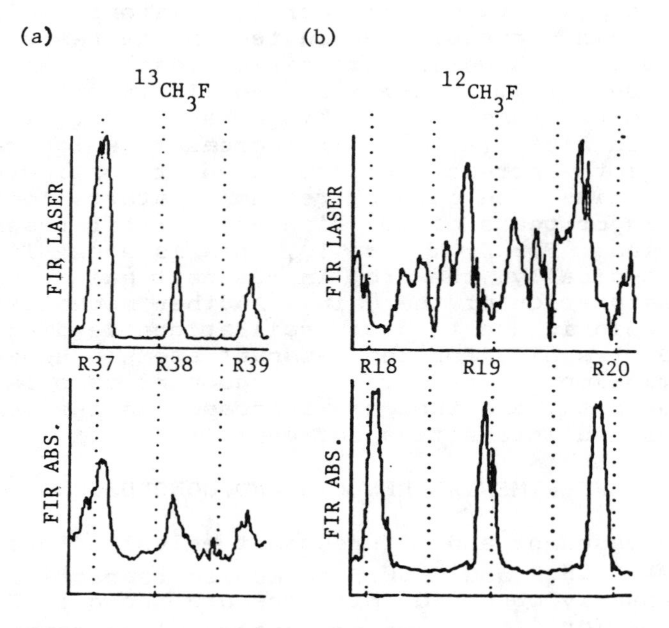

Figure 2. Sample emission spectra for ν_3 R-branch (a) resonant emission from $^{13}CH_3F$ (b) off-resonant emission from $^{12}CH_3F$.

REFERENCES

1. P. Mathieu and J.R. Izatt, Optics Lett. 6, 369 (1981).
2. Richard L. Panock and R.J. Temkin, IEEE J. Quant. Electron. QE-13, 425 (1977).
3. D.G. Biron, R.J. Temkin, B.G. Danly and B. Lax, IEEE J. Quant. Electron. QE-17, 2146 (1981).
4. Jagdish Rai and J.R. Izatt, to be published in Phys. Rev. A.

A KINETIC MODEL FOR A SOLAR-PUMPED IODINE LASER

L. V. Stock
Hampton University, Hampton, VA 23668

J. W. Wilson
NASA, Langley Research Center, Hampton, VA 23665

ABSTRACT

Good agreement between the theoretical predictions and the experimental data is found for a flashlamp-pumped iodine laser for different fill pressures and different lasants (i-C_3F_7I, n-C_4F_9I, and t-C_4F_9I). The following loss mechanisms of the laser output power have been identified: a relatively large amount of initial molecular iodine in the fill gas and laser light scattering as a function of pressure.

A kinetic model has been constructed[1,2] to describe the kinetics of a solar-pumped iodine laser. Present kinetic coefficients are given in table I. The kinetic model as it develops will be used to establish scalability of a solar-pumped iodine laser used in space for the generation and transmission of power. Recent experiments[3] are examined to further refine and develop the kinetic model.

TABLE I - Reaction Rate Coefficients Found Within Published Values.

[Along with the value found for α_i (See Text)]

Reactions		Reaction Rate Coefficients, $(cm^3)^n/sec$			
		Symbol	R = i-C_3F_7	R = n-C_4F_9	R = t-C_4F_9
I^*+R →	RI	K1	$.163 \times 10^{-11}$	$.749 \times 10^{-12}$	$.956 \times 10^{-12}$
$I+R$ →	RI	K2	$.133 \times 10^{-10}$	$.988 \times 10^{-11}$	$.705 \times 10^{-11}$
$R+R$ →	R_2	K3	$.438 \times 10^{-11}$	$.606 \times 10^{-11}$	$.222 \times 10^{-13}$
$R+RI$ →	R_2+I	K4	$.766 \times 10^{-16}$	$.776 \times 10^{-16}$	$.475 \times 10^{-16}$
I_2+R →	$RI+I$	K5	$.178 \times 10^{-11}$	$.224 \times 10^{-10}$	$.111 \times 10^{-11}$
I^*+RI →	$I+RI$	Q1	$.112 \times 10^{-15}$	$.660 \times 10^{-15}$	$.923 \times 10^{-15}$
I^*+I_2 →	$I+I_2$	Q2	$.339 \times 10^{-10}$	$.339 \times 10^{-10}$	$.339 \times 10^{-10}$
I^*+I+RI →	I_2+RI	C1	$.109 \times 10^{-32}$	$.868 \times 10^{-33}$	$.363 \times 10^{-32}$
$I+I+RI$ →	I_2+RI	C2	$.597 \times 10^{-31}$	$.218 \times 10^{-31}$	$.420 \times 10^{-31}$
I^*+I+I_2 →	$2I_2$	C3	$.800 \times 10^{-31}$	$.800 \times 10^{-31}$	$.800 \times 10^{-31}$
I^*I+I_2 →	$2I_2$	C4	$.393 \times 10^{-29}$	$.393 \times 10^{-29}$	$.393 \times 10^{-29}$
		α_i	$.339 \times 10^{-1}$	$.262 \times 10^{-1}$	$.77 \times 10^{-2}$

The iodine laser experiments used an elliptical collector with the flashlamp and the laser tube at the foci. The flashlamp and the laser tube radii are .35 cm, the pumping length is 30.5 cm, and the distance between the mirrors is 60 cm with a two percent output mirror transmission. The pumping pulse had a gaussian shape with FWHM of one-half a millisecond. The peak power output of the flashlamp approximates that of a 7000 K blackbody radiator. The laser tube is first evacuated to about 10^{-6} torr and then the lasant is allowed to evaporate into the laser cavity until the desired pressure is reached. The static gas is then immediately pumped by the flashlamp. Care was taken to ensure a repeatable laser output.

The coupling of the flashlamp to the laser tube has been accurately modeled. The power output is calculated yielding the time of laser threshold, the energy output, and the lasing time. These output parameters are compared to the experimental data for the three gases in figures 1-3. In order to understand the experimental data, an initial condition (table II) of a relatively large amount of molecular iodine $[I_2]_0$ is added in the kinetic model as a function of fill pressure. I_2 is observed as a discoloration of the lasant used to fill the laser tube. In addition to the $[I_2]_0$ the optical time constant[1] in the kinetic model is modified to incorporate a constant α_i as $\tau_c / (1 + \alpha_i P_0)$ becoming a linear function of the fill pressure P_0. This constant α_i is added to simulate the scattering of laser light by deposits on the Brewster windows and aerosols produced in the lasant. Varying these two parameters and the rate coefficients within known experimentally defined bounds [2,4] by using gradient search methods to minimize the differences

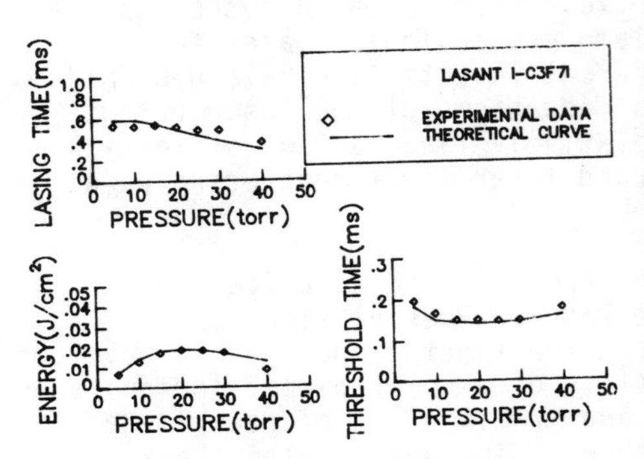

Figure 1.-- Experimental data and theoretical predictions for the lasant i-C_3F_7I.

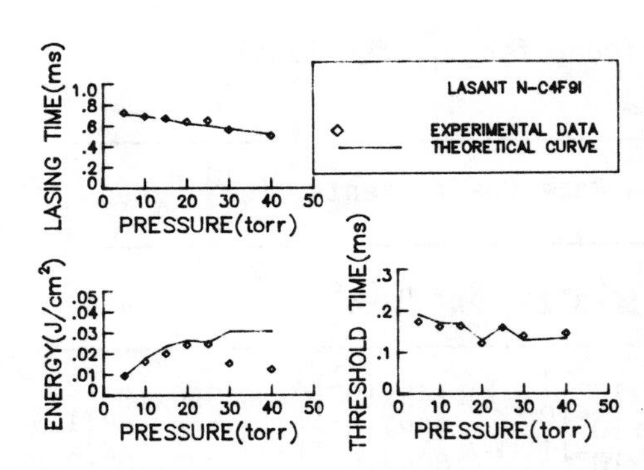

Figure 2.-- Experimental data and theoretical predictions for the lasant n-C_4F_9I.

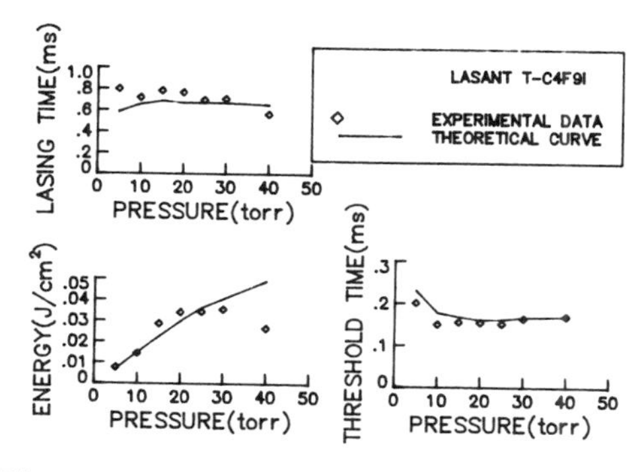

Figure 3.-- Experimental data and theoretical predictions for the lasant t-C_4F_9I.

between experimental data and theoretical predictions, there is reasonable agreement. For instance, at 25 torr less than a 6 percent difference between the theoretical calculations and the experimental data for the energy output of the laser for the three gases is found using the parameters given in tables I and II. This is well within the experimental reproducability given as ± 1 mJ/cm² for pressures less than 10 torr and ± 5 mJ/cm² for pressures greater than 15 torr.

TABLE II - Initial I_2 Density $[I_2]_0$ and I_2 Density After Equilibrium is Reached in the Laser Tube $[I_2]_f$. For Different Pressures and Lasants.

LASANTS	i - C_3F_7I ($\times 10^{15}$)		n - C_4F_9I ($\times 10^{15}$)		t - C_4F_9I ($\times 10^{15}$)	
Pressure (torr)	$[I_2]_0$	$[I_2]_f$ (molec/cm³)	$[I_2]_0$	$[I_2]_f$ (molec/cm³)	$[I_2]_0$	$[I_2]_f$ (molec/cm³)
5.	.42	2.23	.41	2.43	.41	.31
10.	.41	3.80	.40	4.30	.40	.48
15.	.42	5.17	.41	5.91	.41	.63
20.	.43	6.34	.42	7.18	.42	.79
25.	.46	7.39	.45	7.81	.45	.95
30.	.49	8.31	.48	9.86	.48	1.12
40.	.59	9.60	.57	12.06	.58	1.49

In table II $[I_2]_0$ is compared to $[I_2]$ after equilibrium is reached $[I_2]_f$. For the lasants i-C_3F_7I and n-C_4F_9I the $[I_2]_f$ is significantly higher than $[I_2]_0$ (table II). Since I_2 is a strong quencher it can be expected, therefore, that the energy output will also drop significantly if the lasant is repeatedly pumped without refilling; this is verified experimentally[3]. On the other hand, $[I_2]_f$ for t-C_4F_9I is not significantly greater than $[I_2]_0$ implying a repeatability of the output energy from multiple pulses of the same fill gas, and this is found experimentally. Furthermore, theoretically it is predicted that I_2 quenches the laser action in the gases $i\text{-}C_3F_7I$ and $n\text{-}C_4F_9I$ which is indicated by shorter lasing times for different pressures (figures 1, 2, and 3), as opposed to the longer lasing times of $t\text{-}C_4F_9I$. Longer lasing times and subsequent higher energy outputs of $t\text{-}C_4F_9I$ are primarily due to the smaller recombination rate K_3 which forms the dimer R_2 (table I) since it allows the free radical to recombine with iodine creating the parent gas rather than I_2. These properties - slow I_2 and R_2 formation - make the lasant $t\text{-}C_4F_9I$ a good candidate for a space-based solar-pumped laser.

Good agreement between the theoretical model and the experimental data for this system has been reached. The results are generally within experimental reproducibility. The reaction rate coefficients given here represent the best values obtainable within the context of this experimental system and are within the range of known uncertainties. The development of the kinetic model represents a continuing effort to establish the scalability of a solar-pumped space-based iodine laser to be used for the transmission and generation of power.

REFERENCES

1. L. V. Stock, J. W. Wilson, and R. J. De Young,"A Model for the Kinetics of a Solar-Pumped Long Path Laser Experiment," NASA TM 87668, May 1986.

2. J. W. Wilson, S. Raju, and Y. J. Shiu, "Solar-Simulator-Pumped Atomic Iodine Laser Kinetics," NASA TP 2182, August 1983.

3. Ja H. Lee, W. R. Weaver, and B. M. Tabibi, "Characteristics of t - Perfluorobutyl Iodide as a Solar-Pumped Laser Material." Conference on Lasers and Electro-optics, May 1985.

4. J. W. Wilson et al., "Threshold Kinetics of a Solar-Simulator-Pumped Iodine Laser," NASA TP 2201, 1984.

LASER INSTRUMENTATION FOR MULTI-MODE GRAVITATIONAL RADIATION DETECTORS

Jean-Paul Richard
University of Maryland, College Park, Maryland 20742

ABSTRACT

In multi-mode detectors, the antenna displacements can be amplified 400 times at the last resonator. Sensing these displacements with laser illuminated Fabry-Perot interferometers is discussed here. Laser shot noise and back action on the detector would be at the level of one phonon with cavity power of 10 watts and cavity finesse of ~ 30000. This would correspond to δh values of the order of 3.5 $10^{-22}/\sqrt{Hz}$ over a bandwidth of 700 Hz for a 1200 kg five-mode detector operating at 1600 Hz.

INTRODUCTION

In the present search for gravitational radiation, the strongest pulses are expected to be in the form of short bursts associated with catastrophic astrophysical events lasting a few milliseconds or less. Strain amplitudes δh of the order of 10^{-20} to 10^{-22} may correspond to frequent signals from far away sources[1]. The multi-mode resonant bar detector has been proposed[2,3] for the search and study of such events since bandwidths of the order of 50% or more of the frequency of the fundamental longitudinal mode of the antenna can be achieved and appear essential in the study of the structure and time of arrival of short pulses.

MULTI-MODE DETECTOR

The multi-mode detector is shown schematically in Fig. 1. It consists of an arbitrary number n of coupled harmonic oscillators. The largest one is the Weber resonant cylindrical antenna of effective mass M_1 which couples to the gravitational field. The frequency of its fundamental longitudinal mode is ν_o. The masses of the coupled oscillators are related by $M_{i+1}/M_i = \mu$ and their natural frequency when uncoupled is also ν_o. As sketched in fig. 2, the motion of the last resonator is sensed by two Fabry-Perot interferometers whose output are summed. Each face of the last resonator carries one of the end mirrors of one of the two interferometers. The specific case we will consider here consists of the following: a 1200 kg five-mode system with fundamental longitudinal frequency $\nu_o = \omega_o/2\pi = 1600$ Hz, $\mu = 0.05$, mode mechanical quality factors Q's equal to 9 10^6 and operating temperature of 0.05 K. The dynamic mass of the last resonator including a Fabry-Perot double-sided mirror is 4 g.

The analysis of a multi-mode detector response to a "short" pulse shows that the usable double-sided bandwidth of such a detector is approximately given by[2]:

$$\delta\nu \cong 2\nu_o \sqrt{\mu} \cong 1/\tau \qquad (1)$$

which corresponds to 720 Hz and an averaging time $\tau \cong 1.4$ ms.

The spectral density of the thermal noise force acting on M_5 and its resulting rms displacement if initially at rest are:

$$f_n^2(\omega) = 8k_B T_a M_5/\tau_a \quad \text{and} \quad \delta x_n^2 = 2k_B T_a \tau / M_5 \omega_o^2 \tau_a \qquad (2)$$

where k_B is the Boltzmann constant, T_a and τ_a, the temperature and the damping time of the antenna.

NOISE IN LASER INSTRUMENTATION

The quantum shot noise in Fabry-Perot instrumentation leads to an error in the position measurement of M_5 with an rms value:

$$\delta x_{sn} = \frac{1}{8\pi\, n_t} \sqrt{\frac{2\ hc\lambda}{\eta P\tau}} \qquad (3)$$

where n_t is the effective number of light-round-trips in each Fabry-Perot cavity, h is Plank's constant, c is the speed of light, λ is the laser wave length, η is the photo-detector quantum efficiency and P is the laser power at the photo-detector. The laser measurement also exhibits a quantum back-action effect due to fluctuations in the arrival rate of photons at the surface of the Fabry-Perot mirrors. For a laser which displays Poisson noise, the spectral density of that force is:

$$f_{qp}^2(\omega) = 8\ n_t^2\ Ph/\lambda c \qquad (4)$$

The laser shot noise and back action force should be of the order of their thermal counterparts if the detector design is optimized. In our numerical case, the effective value of n_t is 10^4, λ = 632nm, η = 0.85 and P = 1mW. Then, $\delta x_{sn} \cong 2.2\ 10^{-18}$m and $f_{qp}(\omega) \cong 1.3\ 10^{-15}\,N/\sqrt{Hz}$ and are of the order of the corresponding effects produced by the thermal noise force.

The rms fluctuations in the measured position of the last resonator is the sum of the contributions from f_n, f_{qp} and δx_{sn}:

$$(\delta x_T)^2 = (\delta x_n)^2 \left[1 + \frac{f_{qp}^2(\omega)}{f_n^2(\omega)} \right] + (\delta x_{sn})^2 \qquad (5)$$

where the first term is the thermal noise component. The second and third terms are the laser back action and shot noise.

DETECTOR SENSITIVITY

Filling in eq. (5) with the calculated values from eqs.2-4 and multiplying by $M_n \omega_o^2/k_B$, an expression is obtained for the equivalent detector noise temperature T_p:

$$T_p = T_a \frac{2\tau}{\tau_a} + \frac{\hbar\omega_o}{2\sqrt{\eta}\, k_B}\left[\frac{\sqrt{\eta}\,\beta_L\tau\omega_o}{2} + \frac{2}{\sqrt{\eta}\,\beta_L\tau\omega_o}\right] \quad (6)$$

where $\beta_L = K_L/K_n$, $K_n = M_n\omega_o^2$ and $K_L = 16\pi n_t P/\lambda c$. (7)

This expression exhibits the quantum sensitivity which can be achieved with laser instrumentation after thermal noise and classical sources of errors in the laser instrumentation are reduced to appropriate levels. This pulse noise temperature can be converted into a sensitivity in h through:

$$(\delta h)^2 = \frac{4k_B T_P}{M_1\omega_o^2 L^2} \quad (8)$$

where L is the length of the bar antenna. The numerical value obtained here for T_p is 190 nK. For L = 1.5 m, The resolution in h over the 720 Hz bandwidth is 9.3 X 10^{-21} or 3.5 X $10^{-22}/\sqrt{\mathrm{Hz}}$.

Noise contributions associated with fluctuations in the laser frequency and power output and with inbalances in the lenght and power of the Fabry-Perot cavities have been evaluated and can be reduced to acceptable level.

REFERENCES

1. L.L. Smarr, *Sources of Gravitational Radiation* (Cambridge University, Cambridge, 1979).
2. J.-P.Richard, in *Proceedings of the Second Marcel Grossmann Meeting on General Relativity, Trieste,* 1979, edited by R. Ruffini (North Holland, Amsterdam, 1982), p. 1239.
3. J.-P. Richard, Phys. Rev. Lett. 52, 165 (1984).

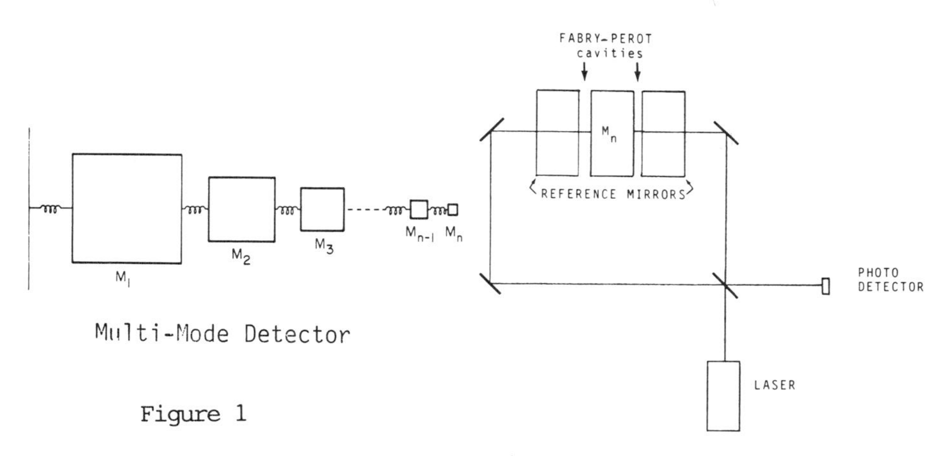

Multi-Mode Detector

Figure 1

Fabry-Perot Instrumentation

Figure 2

LANGMUIR PROBING STUDIES OF UV-PHOTOIONZED CO_2 LASER MIXTURES

Y.Z.WANG
Shanghai Institute of Optics and Fine Mechanics,Academia,Sinica.

ABSTRACT

A Langmuir probe incorporated with a boxcar averager is used to measure the temporally and spatially resolved electron density and electron temperature in a UV photoionization device for several gases and gas mixtures. The data provides information to optimize the effectiveness of the UV preionization operation in high pressure lasers.

INTRODUCTION

UV photoionization technique has been widly used in many TEA CO_2 lasers. In this way to improve the discharge uniform and to increase the output power of laser.

In the UV photoionization device preionization electron are produced by UV ionization of some gases molecules in the discharge volume. But they are also produced by the photo-electron emission of the cathode material and by the direct injection of the electron in UV discharge.

In this report,we used a Langmuir probe and a Boxcar to measure the preionization electron density and the electron temperrature in UV photoionization discharge.The time variable charcteristic of the electron density and electron temperature are studied.

EXPERIMENT

The schematic diagram of the experimental setup is shown in Fig1.

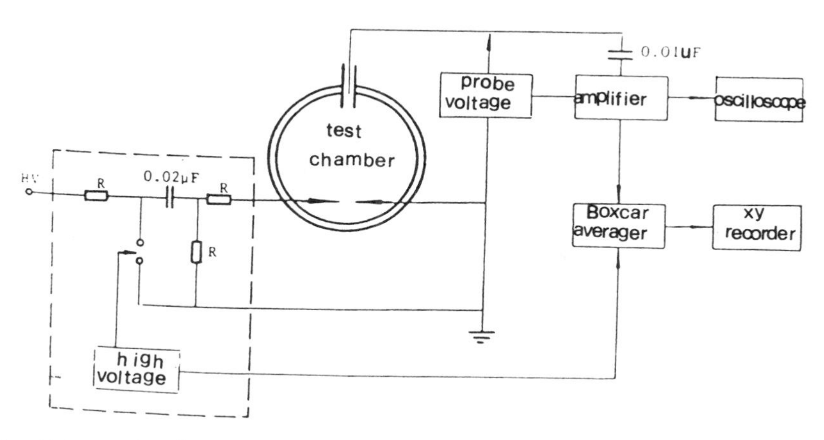

Fig.1. Schematic diagram of experimental setup.

The UV radiation is produced by a discharge column having six pairs of 1-mm diam pointed tungsten transverse electrodes.The distance between the discharge electrodes and the probe is vaeied by using a sliding cylinder mechanism.The energy storage capacitance is 0.02 uf,the voltage of the pulse power suppy,35kv;the pulse repetition rate 1-10 per second;the preionization chamber 9.6 cm in diameter;the diameter of the Langmuir probe,1mm.The probe voltage is varied from -75v to +75v.

RESULTS

The typical results of the measurements of the electron density and electron temperature are shown in Fig.2.

The electron density and electron temperature reach their peak values about 1.5 us after initiation discharge.

The charactristic decay time of the electron density is approximately 5us and that of the electron temperature is 15us under the present experimental conditions.There is a noticeable decrease of electron density when CO_2 gas is introduced into the mixture.

The cause of the reduction of the electron density is due to the absorption of the UV radiation by CO_2 gas.

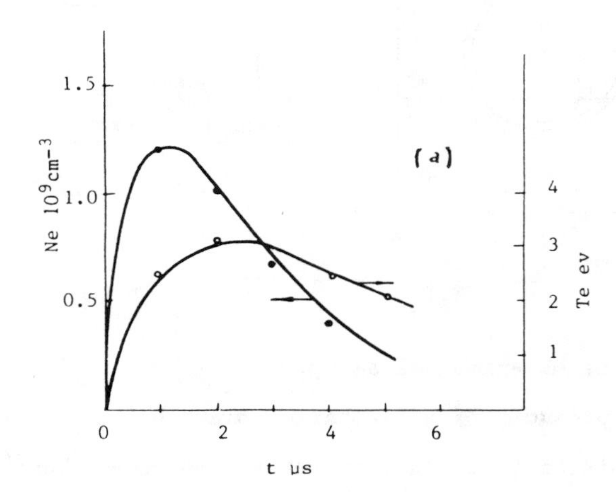

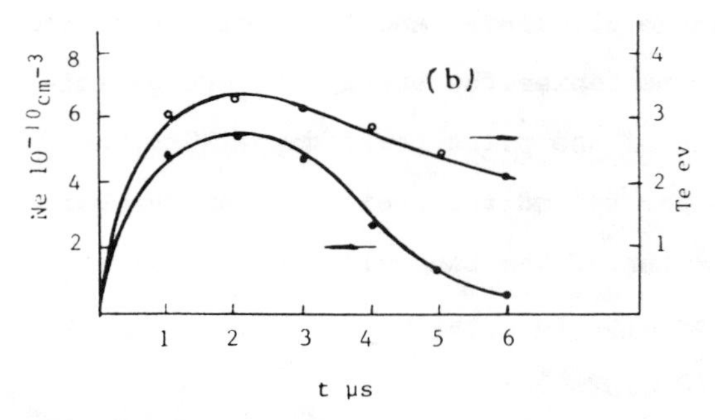

Fig.2. Electron density and electron temperature as function of time.
(a) $CO_2:N_2:H_e$ = 1:1:8,1 atm;
(b) $H_e:N_2$ = 1:1, 1 atm.

CONCLUSION

In this paper the temporally and spatially resolved electron density and electron temperature measurements in a preionization chamber are presented. The mechanism of the ionization is due to the single photon process. The knowledge of the time variation of the local electron density and electron temperature provide the important information to maximize the performance of a high-pressure laser operation.

HIGH-POWER, CONTINUOUSLY SOLAR PUMPED AND Q-SWITCHED IODINE LASER

J. H. Lee, D. H. Humes, and W. R. Weaver
NASA Langley Research Center, Hampton, Virginia 23665

B. M. Tabibi
Hampton University, Hampton, Virginia 23668

ABSTRACT

A train of high power (up to 160 W) Q-switched pulses at up to 1 kHz were continuously produced from a CW solar-simulator pumped iodine laser system. The Q-switching was accomplished by a fast mechanical chopper in the laser cavity and is the first reported Q-switching of a CW iodine laser. The dependence of CW and Q-switched laser powers on parameters such as the iodide vapor pressure, the flow rate, and the chopper speed are discussed.

INTRODUCTION

A direct solar-pumped iodine laser system is being developed for in-space laser power transmission.[1-3] A continuously solar-pumped 1 MW iodine laser system was studied and found to have potential advantages over other systems for space applications.[3] A CW high-power iodine laser was developed that operated longer than one hour and had a CW output power of up to 10 W.[4] However, the generation of high peak-power, short pulses is desirable for some space applications such as laser propulsion[5] and material processing. The generation of intense, short bursts of laser power at high repetition rates can be achieved by Q-switching a CW laser. This presentation reports results obtained from Q-switching a solar-pumped CW iodine laser.

EXPERIMENT

Figure 1 shows a diagram of the solar-simulator pumped iodine laser system of Reference 2 with the addition of a Q-switching chopper for this experiment. For the CW output the mechanical chopper is removed from the laser cavity. A Vortek arc lamp with a 0.15-m long arc is used as the solar simulator. The arc lamp and laser tube are optically coupled in an elliptical cylindrical reflector. The laser tube had a 20 mm ID and a 0.45 m length. The gain length was 0.15 m and the remainder was external to the reflector. The laser cavity consisted of an 85% reflective output and a 100% reflective rear mirror. The cavity length was 0.8 m. The solar simulator provided an irradiance up to 1250 SC (1 SC = 1.35 kW/m^2) at the surface of the laser tube.[3] The irradiance is varied with the arc current. The laser emission is at 1.315 μm.

The repetition rate of the Q-switching depended on the rotational speed of the chopper blade and the number of blade openings. Tests were made to determine the effect of two methods of varying the Q-switching repetition rate on laser power. One set

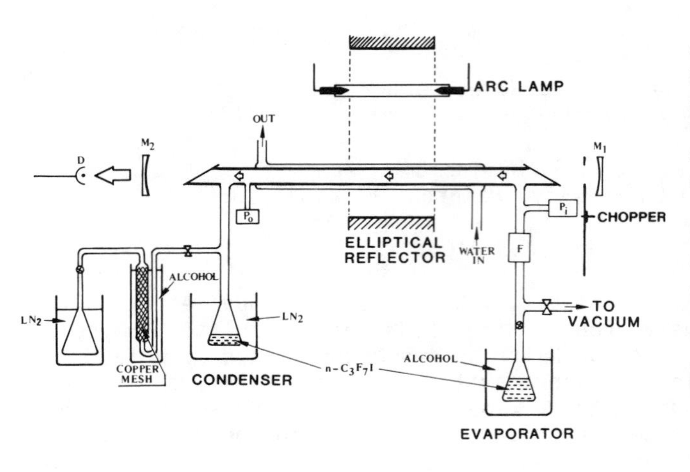

Figure 1. Solar simulator pumped iodine laser experiment equipped with a Q-switching chopper.

of tests used a constant number of blade openings and a varying rotational speed. The other tests used a constant rotational speed and a varying number of blade openings. In all cases the width of the openings was 20 mm, and the axis of the chopper blade was 153 mm from the center line of the laser tube. The maximum rotational speed of the chopper was 5500 rpm.

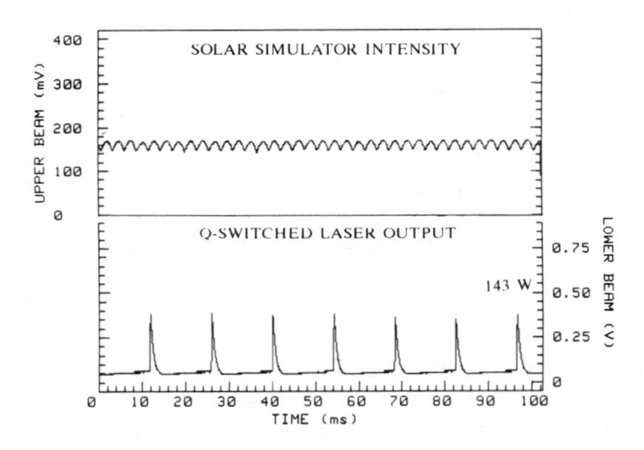

Figure 2. Q-switching lasr output (lower trace) and solar simulator signal (upper trace)

The flow of $n\text{-}C_3F_7I$ vapor was longitudinal and maintained by the 400 torr pressure differential between an evaporation and a condensation reservoir. A Ge photodiode, a pyroelectric power meter, and a photoelectric energy meter were used to monitor the laser output. The integrated power in the absorption band (250 - 290 nm) of the $n\text{-}C_3F_7I$ for the solar simulator spectrum was approximately two times that of the solar spectral irradiance for 500 solar constants.[2] An arc current of 300 A gives a lamp output of 34 kW. The total radiative power reaching the laser tube was 9.7 kW which gives an irradiance of 763 SC.

RESULTS

Figure 2 shows the output of detectors monitoring the Q-switched laser and the solar-simulator. The output power of the CW laser was fixed at 6 W in this test by keeping constant the flow rate

(3000 SCCM), the arc current (300 A), and the iodide pressure (25 torr). The chopper could provide pulses with durations from 140 μs (FWHM) down to 55 μs (FWHM) by increasing its speed from 27 m/s to 88 m/s. The maximum power was obtained at the maximum speed of 88 m/s, reaching 143 W for a 300 A simulator current.

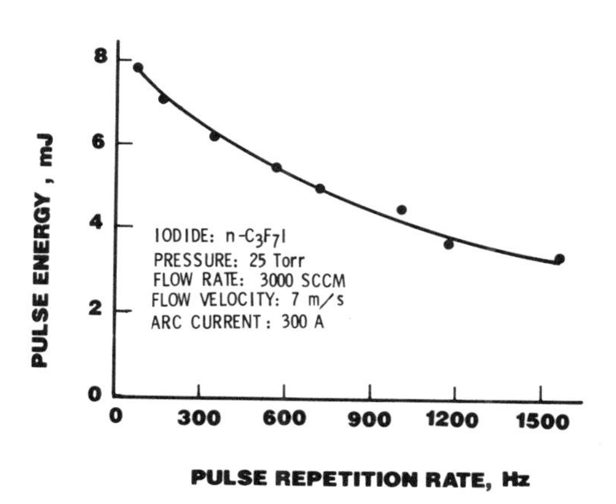

Figure 3. Energy of Q-switched pulse vs the pulse repetition rate.

The measured energy of the Q-switched pulses was found to depend on the pulse repetition rate and is shown in figure 3. The pulse energy is a maximum at about 60 Hz which corresponds to a period of ~17 ms. This time is comparable to the 20 ms dwell time for n-C_3F_7I in the pump length of the laser tube.

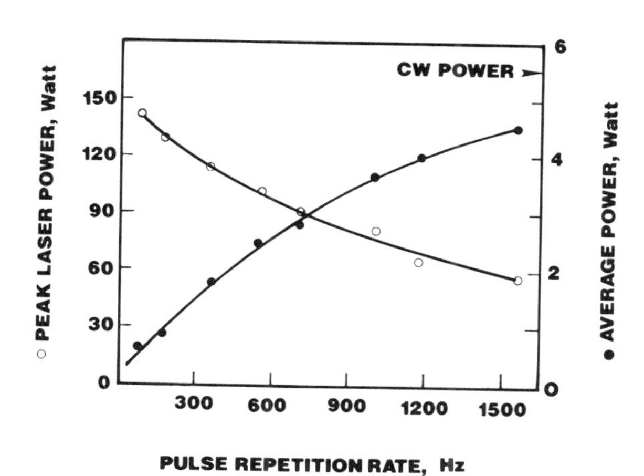

Figure 4. Peak laser power and average power output vs the pulse repetition rate. The CW power is indicated by an arrowhead.

Figure 4 shows the experimental results for the average power and the peak power versus the pulse repetition rate. These data points were obtained by varying the number of openings in the chopper blade while fixing the rotation at 5500 rpm. This rotation corresponds to a speed of 88 m/s which is the maximum obtainable with the chopper. The dependence of the peak power is similar to that of the pulse energy on the repetition rate (see figure 3). However, the average power increases with higher repetition rate and reached 81% of the CW power at 1.6 kHz.

CONCLUSION

The output of the CW iodine laser can be increased by Q-switching with a mechanical chopper. An increase of about 25 times was achieved. However, as the output increased the efficiency decreased. When the increase was only 10 times the CW laser power, the average power was 81% of the CW power. When the increase was 25 times the CW power, the average power was reduced to 12% of the CW power. These results give positive support to the potential of the CW solar-pumped iodine laser system for pulsed, high peak-power applications such as laser propulsion and materials processing.

REFERENCES

1. J.H. Lee and W. R. Weaver: Appl. Phys. Lett. 39, 137 (1981).
2. J. H. Lee; M. H. Lee; and W. R. Weaver: Proceedings of the International Conference on Lasers, '86, 1986, p. 150.
3. J. H. Lee; W. R. Weaver; D. H. Humes; M. D. Williams; and M. H. Lee: Paper TLS 23, International Laser Science Conf., Dallas TX, November 18-22, 1985, (AIP Conf. Proc. 146, Advances in Laser Science I, 1986, p. 179).
4. R. J. De Young; G. H. Walker; M. D. Williams; G. L. Schuster; and E. J. Conway: Paper 879038, IECEC, Philadelphia, PA., August 10-14, 1987.
5. G. A. Simons and A. N. Pirri: The Fluid Mechanics of Pulsed Laser Propulsion. AIAA J. 15, 835 (1977).

II. NONLINEAR OPTICAL PHENOMENA AND APPLICATIONS

RECENT ACTIVITIES IN CARS AT ONERA

J.P. Taran
Office National d'Etudes et de Recherches Aérospatiales
BP 72, 92322 Châtillon Cedex, France

ABSTRACT

The biasing caused by spatial averaging and the measurement errors resulting from vibrational saturation and Stark shifts have been studied theoretically and experimentally for Coherent anti-Stokes Raman Spectroscopy. Both phenomena are significant and limit the applicability of the method.

INTRODUCTION

Coherent anti-Stokes Raman Scattering (CARS) now is frequently used for temperature and concentration measurements in reactive media. However, precise quantitative measurements with CARS require experimental caution. This communication reviews two delicate aspects of the CARS[1] interaction which were analyzed at ONERA recently:[2,3]

- the phenomenon of spatial averaging in the focal volume, which is a potential source of biasing when strong gradients are present; its impact on the temperature measurements was studied;

- the vibrational saturation which is seen when scanning CARS is used to probe low density gases.

Both these studies were undertaken in order to resolve recurrent experimental difficulties which were leading to systematic measurement errors. Specific measurements thus were undertaken and models developed in order to analyse the difficulties and find remedies.

SPATIAL AVERAGING

All diagnostic methods have a finite spatial resolution. CARS and other light scattering techniques like Rayleigh, Raman or fluorescence scattering have spatial resolutions which are all in the millimeter range. This is generally comparable to or better than what solid probes can achieve. However, this resolution is frequently inadequate, particularly when stiff gradients are present as often happens in flames at atmospheric pressure and above.

In unsteady flows, the probability of simultaneously probing adjacent gas samples with different composition and temperature is significant. The phenomenon was modeled in order to evaluate the temperature measurement bias.[3] One assumes that two homogeneous nitrogen samples with temperatures T_1 and T_2 (with $T_1 > T_2$) are simultaneously present in the probe volume, and that they are separated by a sharp boundary. The probe volume is taken as a cylinder with length L and the hot sample as a cylinder segment of length ℓ_1, where ℓ_1 occupies a fraction α of the length L ($\alpha = \ell_1/L$). The CARS electric fields generated in the two samples are added, yielding a composite signal with a spectral shape which differs from normal CARS spectra and depends on α. Although the spectral shape is distorted, the data processing code will find a "best fit" spectrum and will output its associated temperature value, T_{CARS}, as the result of the data analysis.

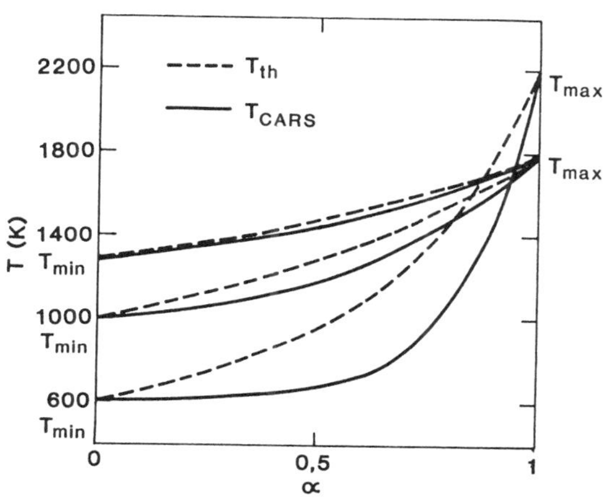

Fig. 1 - Thermodynamic and CARS temperatures vs fraction of probe volume in hot sample, for 3 pairs of high and low temperature values.

This result must be compared to some reference temperature of the sample contained in the probe volume. We take for this temperature the thermodynamic temperature, T_{th}, which is the temperature which the sample would have after being mixed homogeneously. The comparisons are shown in Fig. 1 for three temperature pairs (T_1, T_2). The agreement is excellent for the couple (T_1 = 1800 K, T_2 = 1300 K). This result is interesting as these temperature are typically seen in the exhaust of a kerosene-fueled simulated jet engine reactor operated at atmospheric pressure.[4] The measurement bias, which is here shown to be negligible, had been invoked as a possible reason for the discrepancy found between the CARS results and those of a sampling probe in that study.[4] The agreement is also fair for the intermediate set (1800 K, 1000 K). However, one observes that the discrepancy becomes very large when T_2 is lowered to 600 K. The CARS measurement then returns very low values up to α = 0.6, where the temperature defect is about 300 K below the thermodynamic temperature. Situations like this may be found, e.g., in stoichiometric premixed turbulent or laminar flames at atmospheric pressure, which present large gradients. The resulting errors are not acceptable.

An experimental demonstration of the effect can be given using a Bunsen burner, with multiplex CARS and a BOXCARS beam configuration. The spatial resolution, which we define as the length of probe volume outside of which no signal is generated, was 3 mm. Previously, spatial resolution was defined as the length over which 75% of the signal is created in a homogeneous sample;[1] it would be 1.5 mm in this case, i.e. a very optimistic value.

Two perpendicular radial temperature profiles were measured 9 mm above the burner, by moving the flame horizontally along orthogonal axes X and Y (Fig. 2). The Y axis was taken aligned with the volume element. The first radial profile was taken by moving the flame along the Y axis, thus yielding the least spatial resolution of about 3 mm on the flame front (Fig. 2a). The other profile along the X axis (Fig. 2b), for which the volume element came tangent to the flame front, afforded the finest resolution on this front. Also shown is the circular flame cross section together with CARS probe volume size and alignment in the two cases.

The flame front never is properly resolved, even in Figure 2b , since the length of the CARS probe volume is larger than the flame front "diameter". However, the hot gas contributions from the tips of the probe volume are overwhelmed by the cold gas ones located near the center of the volume, until the latter is almost entirely immersed in the combustion products. Consequently, the position of the flame front is fairly accurately determined, but the gradient appears less steep than it really is.

It is possible to reconstruct the front of Figure 2a from that of Figure 2b using our analysis and assuming that the resolution in Figure 2b is perfect. The result is shown as a solid line in Figure 2a. The calculation was performed assuming a spatial resolution of 3.5 mm, which is slightly larger than measured directly, but gives a finer agreement. Note that the solid line cannot be calculated at $r \geq 3.25$ mm because data is missing in Figure 2b beyond $r = 5$ mm.

The large difference between the two experimental profiles of Figure 2 clearly demonstrates that strong biasing may occur in certain flames in favor of the cold gases, even if these gases occupy only a small fraction of the probe volume. In turbulent flames, errors can also be expected. These will depend on the temperature fluctuations and the probability of probing inhomogeneous samples compared with the homogeneous ones.

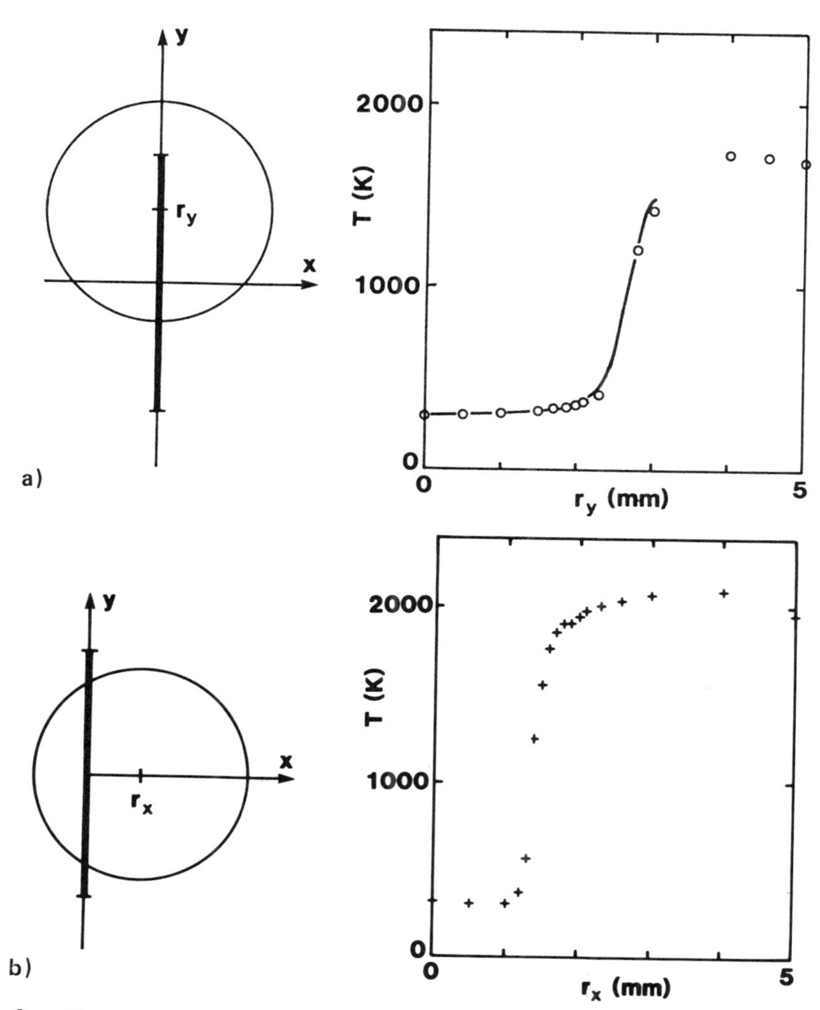

Fig. 2 - Temperature profiles in premixed flame: a) traverse for which the volume element remains perpendicular to flame front; b) orthogonal profile. The solid line in Fig. 2a shows the effect of averaging on the curve of Fig. 2b. Diagrams (left) also depict the relative positions of the CARS probe volume (heavy segment) and the flame front. r is the distance between the center of the probe volume and the flame axis.

From the preceding study, one may conclude that the best possible spatial resolution must be achieved at the outset. This entails use of steep focussing. We shall now see that this is not necessarily desirable, since other phenomena alter the CARS response, even before the dielectric breakdown limit is reached.

SATURATION

Saturation in CARS, which had been predicted very early,[5] has now been observed by several groups.[6-8] Recently, we conducted a quantitative study on a low pressure N_2 glow discharge.[2] The model is based on Bloch equations, using the Placzek approximation.

A detailed presentation of this work is not in order here. The main conclusions are the following:

1 the CARS spectra are broadened;

2 the Rabi frequency increases with the square root of the final state vibrational quantum number, potentially leading to large errors in relative vibrational population measurement;

3 the Stark shift is practically independent of this quantity;

4 the result depends only weakly on rotational quantum number, so that rotational temperature measurements are not appreciably affected by the use of high fields.

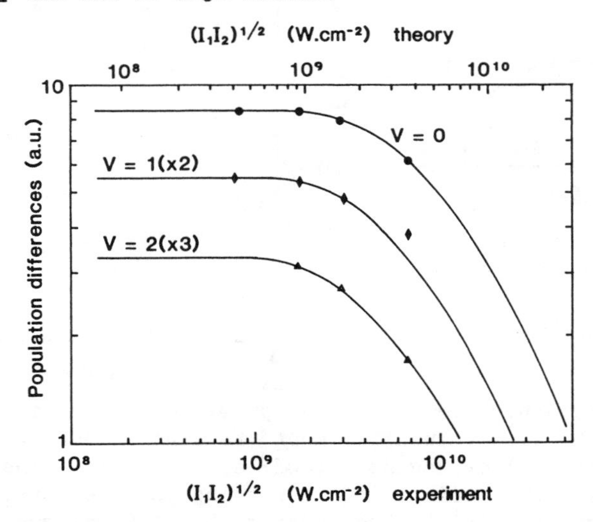

Fig. 3 - Ratioed BOXCARS signals from a 2.6 mbar discharge for $\vartheta = 0, 1, 2$ initial states vs product of pump beam intensities, and numerical predictions for the same transitions (solid lines). Simulations are for a 0.06 cm^{-1} bandwidth multimode dye laser and a single mode pump. Data and theoretical results have been multiplied by 2 for the $\vartheta = 2 \Leftarrow \vartheta = 1$ transition and by 3 for $\vartheta = 3 \Leftarrow \vartheta = 2$.

An example of the vibrational population measurement error made is shown in Figure 3. The population differences densities measured by CARS for the first three vibrational states in the discharge have been plotted vs pump power density product $I_1 I_2$. Saturation is seen to cause appreciable errors above 1 GW/cm^2, and the error is larger for the higher ϑ's. The experimental and the theoretical results are in good agreement, although the intensity scales require some adjustment. This is because the model makes numerous simplifying assumptions, e.g. plane waves are taken to simulate Gaussian beams.

The model also predicts original new results, like the spectral broadening of the anti-Stokes pulse. Typical spectral contents are shown in Figure 4 for pump pulses of duration t_p = 1 ns, with slight detuning (0.001 cm^{-1}) from resonance. Ω_R is the Rabi frequency of the transition. This finding confirms, and expands on, work done by Agarwal.[9] Moreover, it can be shown that because of the saturation, the CARS beam divergence may be changed; under certain conditions additional beams are generated.

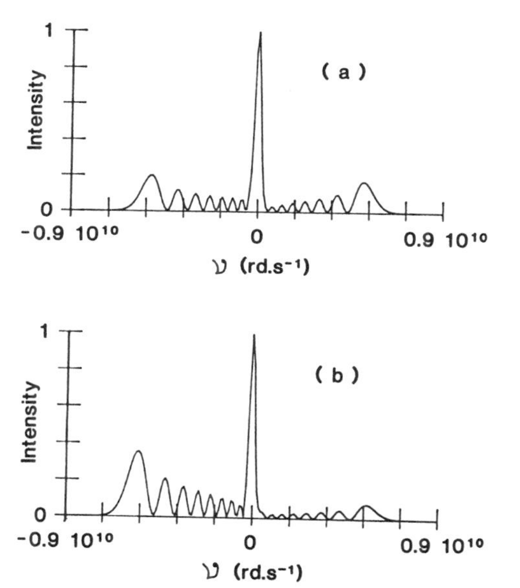

Fig. 4 - Anti-Stokes spectrum for a Gaussian pulse when $\Delta\omega t_p \simeq 2$ and $\Omega_R t_p = 30$. ($\Delta\omega$ = 0.001 cm^{-1}; $I_1 = 6.10^{10}$ W/cm^2; $I_2 = I_1/3$).
a) Stark effect neglected; b) Stark effect included.

CONCLUSION

Saturation and spatial resolution are intimately coupled problems for CARS. In order to avoid saturation, which affects spectral resolution, reduces signal intensity and causes a vibrational state dependent number density measurement error, one has to reduce pump beam power density. This in turn is done at the expense of spatial resolution, potentially causing additional biases in inhomogeneous media.

The two limitations will not be disposed of easily. Each CARS experiment must be designed so as to find the best compromise between the restrictions they impose.

REFERENCES

1. S.A.J. Druet and J.P.E. Taran, Progr. Quantum electron. 7, 1 (1981).

2. M. Péalat, M. Lefebvre, J.P.E. Taran and P.L. Kelley - "Sensitivity of vibrational CARS spectroscopy to saturation and Stark shifts", to be published

3. J.P. Boquillon, M. Péalat, P. Bouchardy, G. Collin, P. Magre and J.P. Taran - "Spatial averaging and Multiplex CARS Temperature Measurement Error", to be published.

4. R. Bédué, R. Gastebois, R. Bailly, M. Péalat and J.P.E. Taran, Comb. and Flame 57, 141 (1984).

5. P. Régnier, F. Moya and J.P.E. Taran, AIAA Journal 12, 826 (1974).

6. M. Duncan, P. Oesterlin, F. Konig and R.L. Byer, Chem. Phys. Letters 80, 253 (1981).

7. R.P. Lucht and R.L. Farrow, to be published.

8. A. Gierulski, M. Noda, T. Yamamoto, G. Marowski and A. Slenczka, Optics Letters 12, 608 (1987).

9. G.S. Agarwal and Surendra Singh, Phys. Rev. 25, 3196 (1982).

EXTRARESONANT STIMULATED RAYLEIGH SCATTERING FROM A TWO-LEVEL SYSTEM INTERACTING WITH TWO INTENSE FIELDS

H. Friedmann and A.D. Wilson-Gordon
Department of Chemistry, Bar-Ilan University, Ramat-Gan 52100, Israel

ABSTRACT

Pressure- and radiation-induced stimulated Rayleigh scattering in the Rayleigh wing are discussed quantitatively for the case where both the pump laser at fixed frequency ω_1 and the probe laser with variable frequency ω_2 are arbitrarily strong. The absorption spectra of the pump and probe fields, both expressed as a function of ω_2, exhibit a narrow dispersive lineshape centered at the fixed frequency ω_1 but with opposite tendencies: the field whose frequency is closest to the resonance frequency, ω_{ba}, of the two-level system always undergoes stimulated emission, the other field being absorbed. The intensity of these processes increases, up to a limit, with the intensity of both the pump and probe fields. A qualitative explanation for the origin of the dispersive lineshape is suggested.

In a previous publication,[1] we have demonstrated the close connection between Bloembergen's pressure-induced extraresonances[2] in four-wave mixing and the pressure-induced extraresonant probe-absorption spectrum. We have also noted that both processes can be obtained as a result of radiative relaxation.[1,3] Pressure-induced and radiative-relaxation-induced extraresonant absorption spectra have dispersive profiles with probe absorption at probe frequency detunings $|\Delta_2|\equiv|\omega_{ba}-\omega_2|$ larger than the pump frequency detuning $|\Delta_1|\equiv|\omega_{ba}-\omega_1|$ and stimulated probe emission for $|\Delta_2|<|\Delta_1|$. Amplification of probe radiation has been ascribed to stimulated Rayleigh scattering[1,4] or to two-wave mixing.[4]

In the present paper, we extend our calculations to the case where both the pump laser at fixed detuning Δ_1 and the probe laser with variable detuning are arbitrarily strong. The pump and probe absorption spectra are calculated as a function of $\delta = \Delta_1-\Delta_2 = \omega_2-\omega_1$. General formulae for these spectra have been derived in a previous publication.[5] We are interested here in the absorption spectrum of the pump and probe beams proportional to Im $\rho_{ba}(\omega_1)$ and Im $\rho_{ba}(\omega_2)$. These are shown in Figs. 1a and 1b for the collisional and collisionless cases, respectively, for various values of the pump and probe Rabi frequencies, V_1 and V_2.

On expanding Im $\rho_{ba}(\omega_1^2)$ and Im $\rho_{ba}(\omega_2)$ in powers of V_1 and V_2, we find an extraresonance at δ=0 in the presence of collisions in the terms of order V^3. The terms of order V^5 exhibit an extraresonance at δ=0 even in the absence of collisions. In these terms, the collision-dependent extraresonance competes with the purely intensity-dependent one. This explains the different trends in the absorption, at high intensity of both the pump and probe, in the collisional (Fig. 1a) and collisionless regimes (Fig. 1b). The most striking result is that at high V, the extraresonance is the dominant feature of the spectrum.

The dispersive lineshape of the probe spectrum has been ascribed to population grating.[4] Since the grating involves <u>many</u> atoms and the extraresonant dispersive feature results from the Bloch equations of

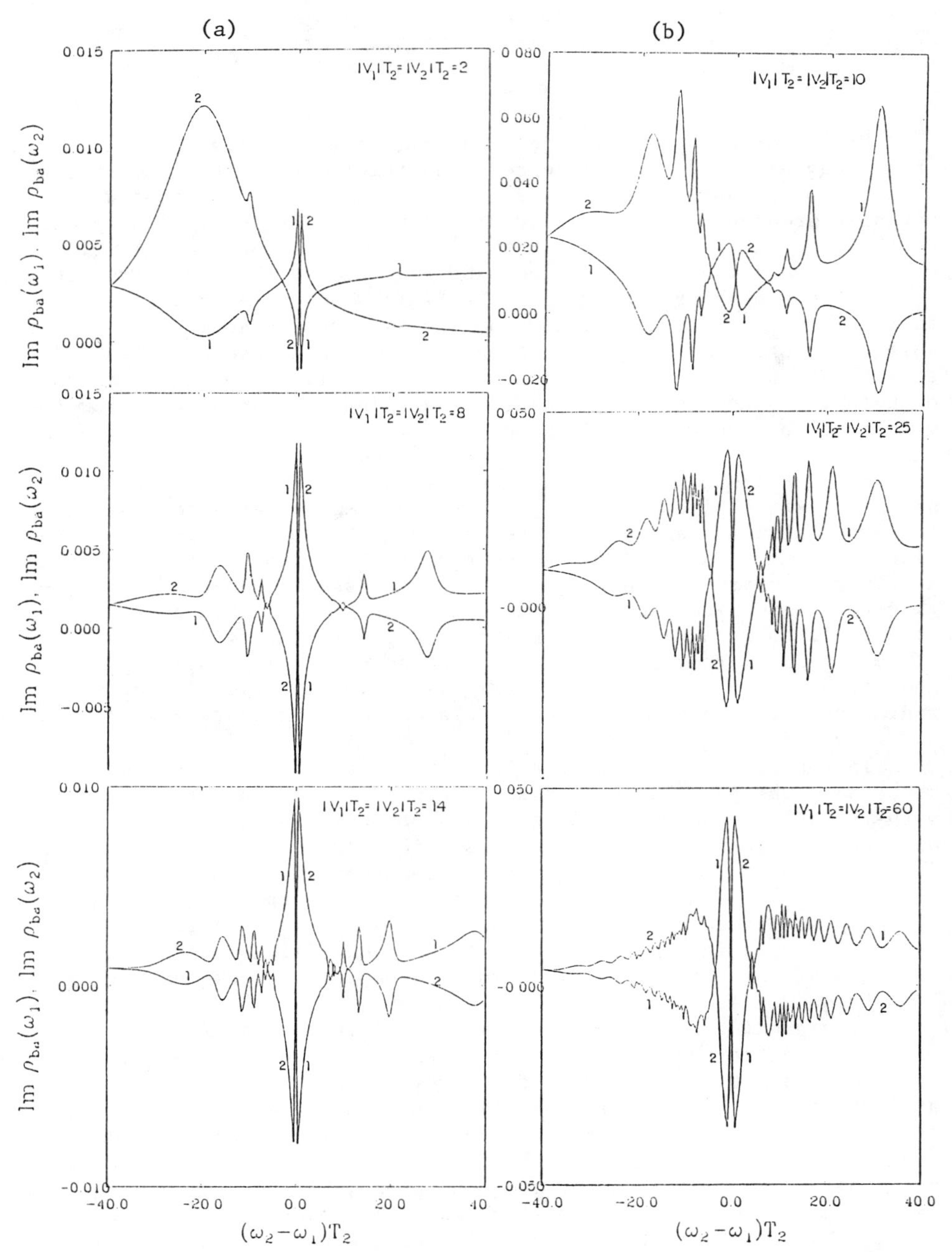

Fig. 1. Pump (1) and probe (2) absorption spectrum for (a) collisional and (b) collisionless cases: $\Delta_1 T_2 = -20$ and in (a) $(1/T_1) = 0.1(1/T_2)$.

a single two-level atom, we suggest that one of the lasers creates a virtual (dressed-atom) state which together with the upper state of the two-level atoms leads to a new resonance or extraresonance. We have shown[1] that the Rayleigh-like two-photon transitions are responsible for the extraresonance. In Fig. 2, we compare two two-photon processes for the case where $\omega_2>\omega_1$, that is, $|\Delta_2|<|\Delta_1|$. In Fig. 2a, a pump photon is absorbed and a probe photon is emitted. In Fig. 2b, the reverse process is depicted. The process of Fig. 2a will have a higher transition probability than that of Fig. 2b, since the coefficient of $|a\rangle$ in the expression for the dressed state $|a'\rangle$ as a linear combination of $|a\rangle$ and $|b\rangle$ will be larger than in the corresponding expression for $|a''\rangle$ (see Fig. 2b). This is due to the fact that $|\Delta_2|<|\Delta_1|$.

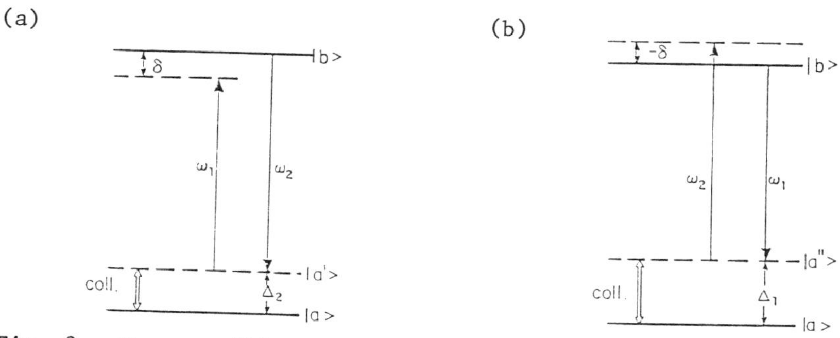

Fig. 2. Two possible two-photon processes for the case $|\Delta_2|<|\Delta_1|$.

The two-photon processes of Fig. 2 will take place provided δ is of the order of $(1/T_1)$, the width of the upper state. Obviously, if $\delta=0$, the two processes will become equally likely giving no net absorption or emission. We have taken the virtual state $|a'\rangle$ or $|a''\rangle$ as the origin of the two-photon Rayleigh process since transition to $|b\rangle$ can take place from there. The finite width of $|b\rangle$ allows absorption of ω_1 when $\omega_1 \approx \omega_2$. This is not possible starting from $|a\rangle$ rather than $|a'\rangle$ since the zeroth width of $|a\rangle$ only allows absorption at $\omega_1 = \omega_2$ and therefore does not contribute to net absorption or stimulated emission via a two-photon process. However if $|a\rangle$ also has a finite lifetime, due for instance to inelastic collisions, additional extraresonant behavior is expected.[6] In Fig. 2, occupation of state $|a'\rangle$ or $|a''\rangle$ is achieved by collisions. When the fields are sufficiently intense, this occupation can also be obtained by purely radiative processes.[3]

1. H. Friedmann and A.D. Wilson-Gordon, Methods of Laser Spectroscopy, eds. Y. Prior, et al (Plenum, NY, 1986), p.307.
2. A.R. Bogdan, Y. Prior, and N. Bloembergen, Opt. Lett. 6, 348 (1981).
3. H. Friedmann and A.D. Wilson-Gordon, Phys. Rev. A 26, 2768 (1982).
4. G. Grynberg, E. Le Bihan and M. Pinard, J. Physique 47, 1321 (1986).
5. H. Friedmann and A.D. Wilson-Gordon, Phys. Rev. A 36, 1333 (1987); see also, G.S. Agarwal and N. Nayak, J. Opt. Soc. Am. B 1, 264 (1984); Phys. Rev. A 33, 391 (1986).
6. A.D. Wilson-Gordon and H. Friedmann, Opt. Lett. 8, 617 (1983).

COMPETING NON-LINEAR AND COOPERATIVE LIGHT EMISSIONS *

K.-J. Boller, H.-J. Lau, J. Sparbier, and P.E. Toschek
Universität Hamburg, D - 2000 Hamburg 36, F.R. Germany

ABSTRACT

Two-photon excitation (λ_1,λ_2) of lithium vapor generates light emission by four processes involving the 3d,3s levels: 1.Two-photon emission, 3d-2s, 2.superfluorescence, 3d-2p, 3.parametric mixing,and 4.two-photon scattering. Upon stepwise tuning λ_2, emission spectra are recorded close to 610 and 812 nm, and thus the sum frequency $\omega_1 + \omega_2$ is scanned across the 4f resonance. Light observed at 610 nm is superfluorescence. The light at 609.2 nm varies in wavelength upon a shift of λ_1, but *not* of λ_2: It is two-photon emission. Two more emission peaks shift in step with λ_2: Parametric emission by five-wave mixing, and hyper-Raman-Stokes scattering of *two* photons. Interference of these emissions has been observed. - The pulse fluctuations of parametric emission are caused by the pump light, while quantum fluctuations dominate the other processes.

INTRODUCTION

The excitation of lithium vapor by pulsed dye lasers at $\lambda_1 \cong 670$ nm and $\lambda_2 \cong 460$ nm (0.6 and 0.4 MW, respectively) has enabled us to observe *simultaneously* light emission (close to 610 nm wavelength) by *four* different interaction processes which involve the Li 3d and 3s levels (Fig. 1), 1. two-photon emission, 3d-2s, 2. superfluorescence[1], 3d-2p, 3. parametric five and six-wave mixing and 4. two-photon excited scattering ("Hyper-Raman Hyper-Stokes emission"). The appearance of two-photon emission has been reported in the past[2].

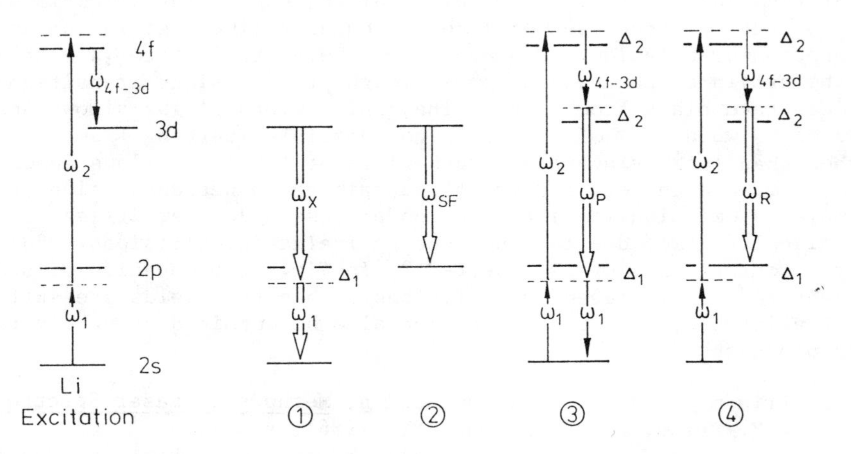

Fig.1 Two-photon excitation of Li 4f,3d (left).Competing emission involving the 3d level: two-photon emission(1), superfluorescence(2), five-wave-mixing(3), hyper-Raman hyper-Stokes emission(4)

COMPETING NON-LINEAR AND COOPERATIVE LIGHT EMISSIONS

The present experiment is based on two grating-tuned dye lasers (0.3 cm^{-1} bandwidth) which are excited by a 30-MW XeCl excimer laser. The light beams at λ_1 and λ_2 are coaxially focussed in a heat pipe. The generated light which emerges from the oven almost collinearly with the light at λ_2, is separated by a dichroic mirror and analysed by a 1-m-Czerny-Turner spectrograph, with photo-multiplier, and box-car integrator, or with a diode array and storage oscilloscope(Fig.2)

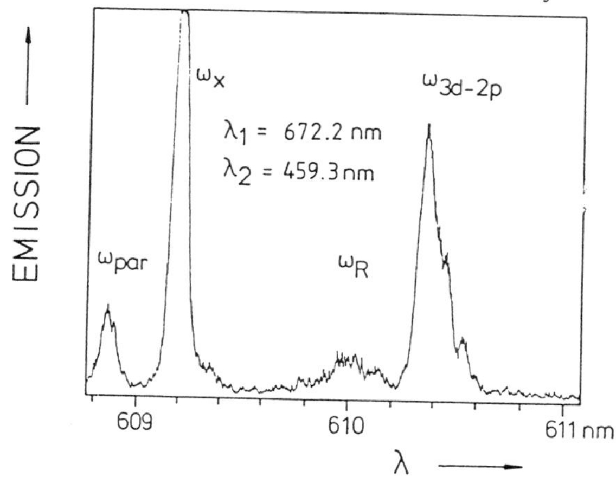

Fig.2 Emission spectrum close to superfluorescence at the 3d-2p transition (610.4 nm

Upon stepwise tuning λ_2, a series of spectra like the one in Fig. 2 has been recorded at about 610 nm (close to the 3d-2p transition)and 812 nm (close to 3s-2p). The sum frequency $\omega_1 + \omega_2$ is nearly resonant with but scanned across the 4f resonance. Generated by the powerful pump light, four emission lines occur within the tuning range $\Delta\lambda_2 \cong 1.5$ nm. The light emitted at 610.4 nm (corresponding to the 3d-2p resonance) is shown to consist of superfluorescence. Analogous emission from the 3s level has been observed at 812.6 nm. The light emitted at 609.2 nm (and at 811.9 nm) varies in wavelength only upon a shift of λ_1, but *not* λ_2. This light is (non-degenerate) two-photon emission, stimulated by the pump frequency ω_1 (Fig. 3).

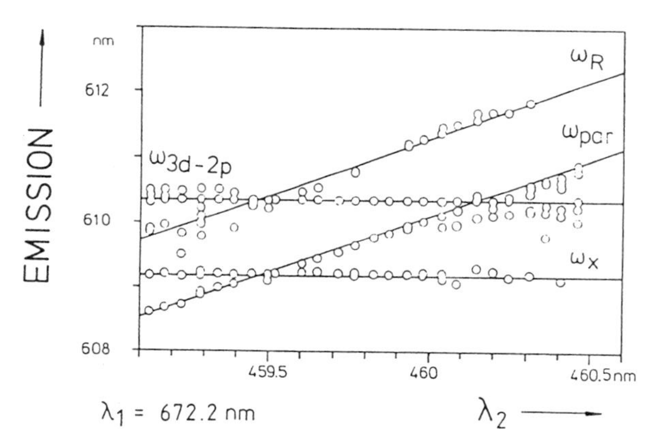

Fig.3 Wavelength of peak emission of the four emission processes *vs* pump laser wavelength λ_1

There is light emitted on two more frequencies, which shift in step with ω_2: The light at ω_p is linked to the pump light via the *virtual* level imposed by ω_1: It is parametrically generated by five-wave mixing[3]. The signal peaks when the direction of observation and the pump beams subtend a small angle, due to the requirement of phase matching. The light at ω_R involves the real 2p level (see Fig.1): It is a Raman-like emission which requires two pump photons and two Stokes photons. At particular tuning values, the ω_2-dependent resonances coincide with the ω_2-independent ones. Two (out of three) of these crossings, at λ_2 = 459.5 nm and at 460.1 nm, have been observed. The resulting emission exceeds linear superposition of the spectrally separated resonances. This fact indicates interference of the respective emission processes.

FLUCTUATIONS OF THE LIGHT EMISSION

We have also studied the fluctuations of the peak power and of the time delay of the light pulses generated by some of the above processes using a transient digitizer. Both variables display characteristic distributions. All peak fluctuations can be fitted by 2-parameter Γ-distributions. Their variance is largest for superfluorescence, smaller for two-photon emission (see Fig. 4), and smallest for parametric emission. The data are compatible with only the fluctuations of the pump light to characterize parametric emission, and quantum fluctuations to dominate the other processes.

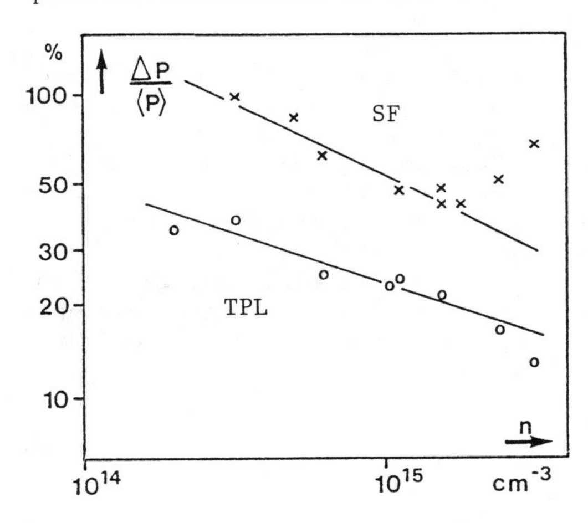

Fig.4 Peak power fluctuations of a two-photon laser (TPL) and superfluorescence (SF) *vs* Lithium vapor density

REFERENCES

1. M.F.H.Schuurmans, Q.H.F. Vrehen, and D.Polder, *Superfluorescence*, Advances in Atomic and Molecular Physics (Academic Press Inc. New York 1981) Vol *17*, p. 167
2. B. Nikolaus, D.Z. Zhang, and P. E. Toschek, Phys.Rev.Lett. *47*, 171 (1981)
3. A.H.Kung, J.F. Young, C.G.Bjorklund, and S.W.Harris, Phys.Rev. Lett. *29*, 985 (1972)

* Supported by the Deutsche Forschungsgemeinschaft

SECOND HARMONIC GENERATION IN MERCURY VAPOR

Yen-Chu Hsu, Chin-Mei Yang, and Bor-Chen Chang
Institute for Atomic and Molecular Sciences, Academia Sinica, P.O. Box 23-166, Taipei 10764, Taiwan, Republic of China

ABSTRACT

Second harmonic generations (SHG, $2\omega_2$) have been studied in the oriented mercury atoms ($6p^3P_1$), which were prepared by a linearly polarized laser light (ω_1) tuned to the vicinity of $6s^1S_0 \rightarrow 6p^3P_1$. Two different SHG are seen, as ω_2 is near the two-photon transition, $6p^3P_1 \rightarrow 8p^3P_1$. The predominant SH signal seen at λ_2= 5395.9Å is due to the resonance enhancement, while, the other is wavelength-tunable. Its tuning range spans about 5 cm^{-1}. This tuning SHG, referred as "extra-SHG" afterwards, shows much different polarization and time dependences than the major one. It can be used to examine the atomic orientation effect on SHG in the vapor phase.

INTRODUCTION

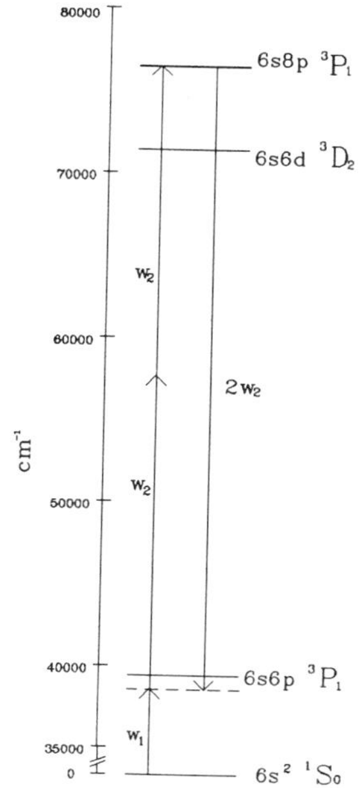

Fig. 1. Mercury energy levels.

Despite the fact that in an isotropic medium, $\chi^{(2)}$ vanishes under the electric dipole approximation, evidences of second harmonic generations[1] and sum frequency mixings[2] in atomic vapors have been reported over the past decade .Since then, several mechanisms[1,3] have been proposed to interpret these symmetry breakdown phenomena.

It is known that atoms ,while undergoing a transition, will oscillate between the ground and the excited states with a given Rabi frequency. The atoms can thus be oriented or aligned, provided that they are irradiated by a polarized light. In other words , the medium is no longer isotropic at that instant. This may account for the observed symmetry breakdown to some extent. Nevertheless, such effect has never been investigated previously. Here, the Hg, $6p^3P_1$ atoms were chosen for studying the atomic orientation effect on SHG in the vapor, since their orientations can be well defined by an optical excitation.

EXPERIMENTAL

Fig. 1 shows the excitation scheme used in this study. The doubled output (ω_1) of Lambda Physik excimer pumped dye laser, FL3002, was tuned to the transition frequency of $6s^1S_0 \rightarrow 6p^3P_1$ to prepare the $6p^3P_1$ mercury atoms. After 0 – 75ns time delay, the second dye laser beam was collinearly introduced and focussed at the center of the mercury heat pipe, which was typically maintained at 100°C and buffered with 10-torr helium. The produced SHG ($2\omega_2$) was filtered and then imaged onto a 0.5-m monochrometer (McPherson Model 219). Subsequent signal averaging was processed by a SRS 250 Gate Integrator.

While polarization measurements were taken, an additional polarizer and a pair of Fresnel Rhombs were used to define the polarization of the second dye laser. To minimize any possible polarization discrimination, a depolarizer and a narrow band filter (2700 ± 50Å) were used to substitute for a monochrometer.

RESULTS AND DISCUSSIONS

Intense SHG ($2\omega_2$) due to the $6p^3P_1$ Hg atoms was

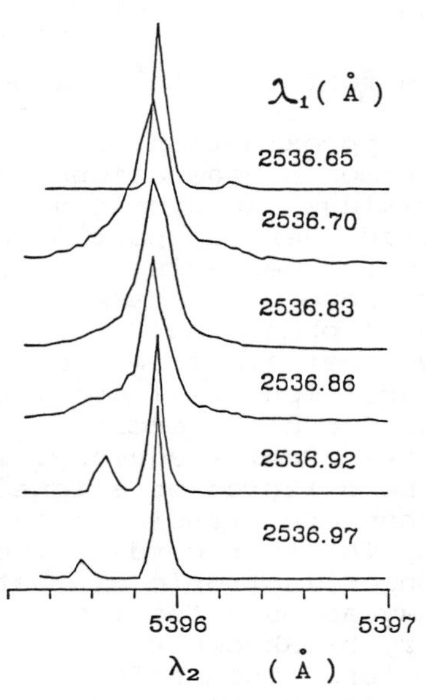

Fig. 2. The SHG excitation spectra obtained with six different λ_1 excitations. The resonant wavelength of Hg $6s^1S_0 \rightarrow 6p^3P_1$ is 2536.8Å.

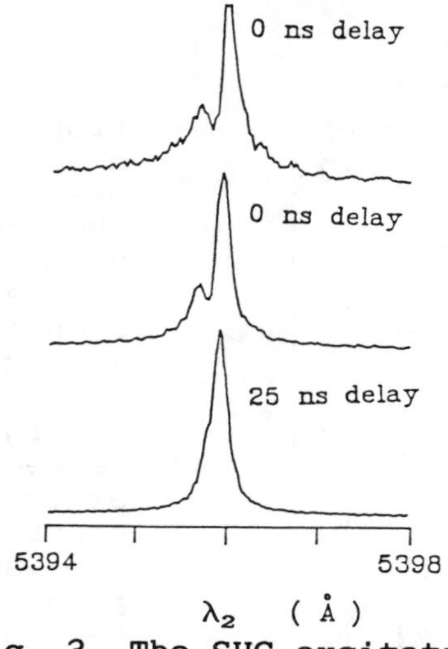

Fig. 3. The SHG excitation spectra obtained at λ_1 = 2536.9Å . Different λ_2 intensities (8 and 6 mJ, separately) were used at the top and the middle traces. The middle and bottom spectra are different in the time delay.

observed at λ_2=5395.9Å. Fig. 2 shows the SHG excitation spectra obtained at various ω_1, indicating that the 5395.9Å-feature is fixed regardless of ω_1. Since this SHG featrue is resonant with the two-photon process of $6p^3P_1 \longrightarrow 8p^3P_1$, it can be easily identified as a resonance enhanced SHG.

In addition, a less intense and frequency tunable SHG is seen in Fig. 2. By calculating the observed frequency shifts, $2\Delta\omega_2$ of these two SHG and the detuning frequency of ω_1, a relation $-\Delta\omega_1 + 2\Delta\omega_2 = 0$ was determined from the best fit. This leads us to rule out the possibility that the extra-SHG is given rise from the different hyperfine component of Hg.

It has also been found that the intensity of the extra-SHG highly depends upon the time delay of the two lasers. The experimental results are shown in Fig. 3, where the extra-SHG is obviously invisible with a 25ns-delay. On the contrary, the intensity of the 5395.9Å feature increases. With the additional evidence obtained from the dispersed emission, it is deduced that these two SHG were derived from two different processes. Since the tuning range of the extra-SHG is only 5 cm^{-1} and there is no state nearby $6p^3P_1$, we interpret that it is a non-resonant emission from the three-photon resonant state, $8p^3P_1$, as shown in Fig. 1.

According to the electric-quadrupole selection rule, the resonant SHG due to $6p^3P_1 \longrightarrow 8p^3P_1$ is allowed. In constrast to this, the extra-SHG cannot be accounted for with any higher-order moment. It is worthwhile to mention that the conversion efficiency of this non-resonant SHG is about 3×10^{-7}, which is comparable to the resonant SHG observed in the alkaline atomic vapors. Futhermore, as the relative polarization of the two lasers switches from parallel to perpendicular, the intensity ratio of the extra-SHG to the resonant SHG increases by an order of magnitude. We believe that the detailed studies on the extra-SHG can enlighten the atomic orientation effect on SHG in the gas phase. Analysis on the measured polarization ratio of SHG are in progress.

REFERENCES

1. (a) T. Mossberg, A. Flusberg, and S. R. Hartmann, Opt. Commun. 25, 121(1978);(b) Kenao Miyazaki, Takuzo Sato, and Hiroshi Kashiwagi, Phy. Rev. Lett. 43, 1154(1979); Phy. Rev. A 23, 1358(1981);(c) Jumpei Okada, Yukio Fukuda, and Masahiro Matsuoka, J. Phys. Soc. JPN. 50, 1301(1981).
2. D. S. Bethune, R. W. Smith, and Y. R. Shen, Phys. Rev. Lett. 37, 431(1976); Phys. Rev. A 17, 277(1978).
3. D. S. Bethune, Phys. Rev. A 23, 3139(1981).

ABSOLUTE RATE MEASUREMENTS OF TWO-PHOTON PROCESS OF GASES, LIQUIDS, AND SOLIDS

C. H. Chen, M. P. McCann, and M. G. Payne
Chemical Physics Section, Oak Ridge National Laboratory
Oak Ridge, Tennessee 37831-6378

Due to rapid improvements in high-power laser performance, two-photon absorption processes have become a very useful tool for studying the molecular structures of various gases, liquids and solids. However, measurements of absolute two-photon absorption cross sections were more or less ignored previously because of their small size. In this work, we obtained not only the two-photon absorption spectra, but also measurements of their absolute cross sections for various gases, liquids, and solids.

Resonance ionization spectra with the two-photon absorption process to various excited states for Ar, Kr,[1,2] H_2, Xe,[3] NO, CO,[4] and ArNO were obtained. By combining a coherent vacuum ultraviolet (VUV) beam; which was produced by a third harmonic generation process in a Xe cell, two-photon absorption and three-photon ionization spectra for Ar, Kr, and H_2 were obtained. Another process combining either two ultraviolet (UV) or one UV photon plus one visible photon was used for exciting various states of Xe, CO, NO, and ArNO. Vibronic state-selected molecules and van der Waals molecules were produced by a rare gas seeded nozzle beam. A mass spectrometer was used to detect the ions by resonance ionization spectroscopy (RIS).

The experimental schematic for two-photon resonance excitation of CO is shown in Fig. 1. The two-photon absorption process promotes $\mathrm{CO}(X^1\Sigma^+)_{v=0}$ to $\mathrm{CO}(A^1\pi)_{v=1}$, a blue laser beam excites $\mathrm{CO}(A^1\pi)_{v=1}$ to $\mathrm{CO}(B^1\Sigma^+)_{v=0}$. The $\mathrm{CO}(B^1\Sigma^+)$ molecules can be ionized by the UV beam which proceed the two-photon excitation. The two-photon absorption cross section of $\mathrm{CO(X)}_{v=0} \rightarrow CO(A)_{v=1}$ obtained was 2.2×10^{-47} cm^4 sec mol^{-1} and the ionization cross sections of $\mathrm{CO}(B^1\Sigma^+)_{v=0}$ was determined to be 1.4×10^{-20} cm^2. Molecular term energies for these two transitions were determined as follows:

$$\Delta E(J, J') = 66227.9 + 1.577J'(J'+1) - 0.346J(J+1)$$

for

$$CO(X)_{v=0} \xrightarrow{2h\nu} CO(A)_{v=1},$$

$$\Delta E(J', J'') = 20688 + 1.948J''(J''+1) - 1.577J'(J'+1)$$

for

$$CO(A)_{v=1} \xrightarrow{h\nu} CO(B)_{v=0}$$

Similar types of experiments were also performed in other gases.

Two-photon induced fluorescence spectra were obtained to study the excitation and decay mechanism for dyes,[5] neat benzene, toluene, xylene isomers, and polysubstituted methyl benzenes. By measuring the absolute number of fluorescent photons, two-photon absorption cross sections and two-photon induced fluorescence cross sections for aromatic compounds were obtained. The results are shown in Table I.

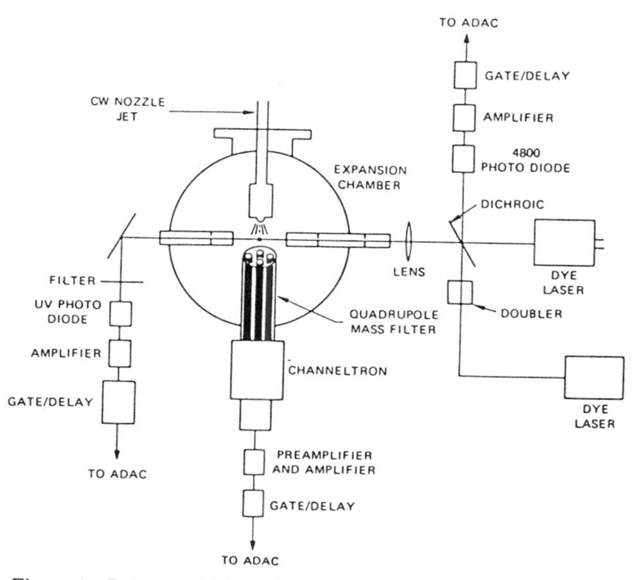

Figure 1. Data acquisition scheme. The two dye lasers are pumped by a Lumonics Nd:YAG laser. The "doubler" is the KDP crystal with appropriate red filter.

Table I. Two-photon absorption cross sections (δ) and two-photon induced fluorescence cross sections (δ_F) for benzene, toulene, and xylene. The unit of δ and δ_F is cm^4 sec mol^{-1} $photon^{-1}$.

Two-Photon Energy (nm)	Benzene		Toluene		o-xylene		m-xylene	
	δ	δ_F	δ	δ_F	δ	δ_F	δ	δ_F
200	<1.0 (-51)							
238	3.4 (-52)*	1.4 (-54)						
248			5.4 (-53)**	6.0 (-55)				
250	8.2 (-52)*	3.3 (-54)						
252	2.0 (-51)							
254			1.2 (-52)**	1.3 (-54)				
254	8.7 (-50)							
255	1.0 (-51)	1.7 (-51)						
255	6.0 (-52)*	2.5 (-54)						
259			1.1 (-52)**	1.2 (-54)	2.1 (-51)***	2.1 (-53)	1.2 (-51)***	1.6 (-53)
260	2.5 (-52)*	1.0 (-54)						
262					2.1 (-51)***	2.1 (-53)		
264			3.6 (-52)**	4.0 (-54)			1.2 (-51)***	1.6 (-53)
276								
277					5.0 (-52)***	5.0 (-54)	3.1 (-52)***	4.0 (-54)
177	1.24(-49)*	5.0 (-51)	1.0 (-49)**	1.1 (-51)			3.0 (-49)***	4.0 (-51)

* - represents δ obtained from the assumption of η_F of benzene is 0.004
** - represents δ obtained assuming η_F is 0.011 for toluene
*** - represents δ obtained by the assumption of η_F equal to 0.01, 0.013, and 0.01 for o-xylene, and m-xylene, respectively.

In addition, two-photon induced fluorescence spectra of various alkali halide crystals were obtained,[6,7] and the mechanism for formation and decay of V_K centers as well as F-centers were studied. The two-photon absorption cross sections and the fluorescence yields due to two-photon absorption were measured. The formation of F-centers for certain alkali halide crystals can be conveniently used to determine the band gaps of these crystals.[8]

Research sponsored by the Office of Health and Environmental Research, U.S. Department of Energy under contract DE-AC05-84OR21400 with Martin Marietta Energy Systems, Inc. M. P. McCann is a graduate student, University of Tennessee, Knoxville, Tennessee.

REFERENCES

1. M. P. McCann, C. H. Chen, and M. G. Payne, Appl. Spectrosc. 41, 399 (1987).
2. M. P. McCann, C. H. Chen, and M. G. Payne, Chem. Phys. Lett. 138, 250 (1987).
3. C. H. Chen, G. S. Hurst, and M. G. Payne, Chem. Phys. Lett. 75, 473 (1980).
4. W. R. Ferrell, C. H. Chen, M. G. Payne, and R. D. Willis, Chem. Phys. Lett. 97, 460 (1983).
5. C. H. Chen and M. P. McCann, Opt. Commun. 63, 338 (1987).
6. C. H. Chen, M. P. McCann, and J. C. Wang, Solid State Commun. 61, 559 (1987).
7. C. H. Chen and M. P. McCann, Chem. Phys. Lett. 126, 54 (1986).
8. C. H. Chen and M. P. McCann, Opt. Commun. 60, 196 (1986).

Measurement of Transition Moments Between Molecular Excited Electronic States Using the Autler-Townes Effect

A. M. F. Lau and D. W. Chandler
Combustion Research Facility, Sandia National Laboratories, Livermore, CA 94550

M. A. Quesada and D. H. Parker
Department of Chemistry, University of California, Santa Cruz, CA 95064

ABSTRACT

This paper reports the observation of the Autler-Townes splitting in the two-color four-photon ionization of hydrogen molecules, and the good agreement of theory with experiment. From the theoretical fits to the spectral profiles,the value of the electric dipole transition moment between the excited vibronic states E,F $^1\Sigma_g^+$, v=6 and D $^1\Pi_u$, v'=2 is determined to be 2.0 ± 0.5 a.u.

Analogous to the splitting of two degenerate molecular levels by a dc Stark field, the Autler-Townes effect (also known as ac, optical, or dynamical Stark splitting) is the splitting of two (nearly) resonant levels by a strong high-frequency field. When this energy level structure is probed by a second weaker laser, the spectrum reveals a doublet, from whose peak separation one may deduce the transition moment between the two levels coupled by the strong laser. We shall discuss the advantages of this method of measureing molecular transition moments over the traditional methods.

Figure 1 shows the multiphoton ionization scheme of the hydrogen molecule. The ions were collected as a function of detuning Δ_a of the uv (λ_a=193 nm) laser around the two-photon X-E,F resonance, while the detuning Δ_b of the infrared (λ_b=726 nm) laser was kept fixed.

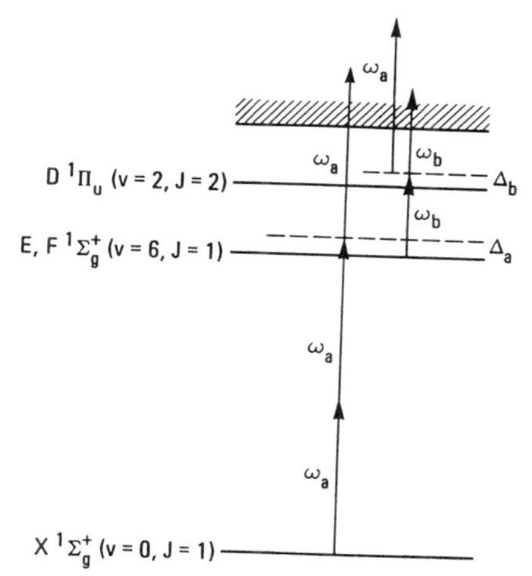

Figure 1. Schematic H_2 energy level diagram showing the multiphoton ionization with a UV (ω_a) and an IR (ω_b) lasers.

In the experiments,[1] two counterpropagating lasers, coincident in time, intersect in the ionization region of a time-of-flight mass spectrometer. One laser, acting as the probe laser, operates near 193 nm. The light is generated by Raman shifting the output of a frequency doubled dye laser in a high pressure gas cell of hydrogen. In this manner, we are able to generate 80 microjoules of light (5 nsec pulse at 10 Hz). This laser is used to probe the Stark splitting induced in the molecule by the second laser. The second laser, tuned to be in near-resonance between the E,F v=6, J=1 state and the D v=2 state, is a dye laser (10 nsec pulse width at 10 Hz) operating near 726.8 nm (LDS 750 laser dye). As the frequency of the UV laser is scanned through the Stark-split levels, H_2^+ ions are formed in the sample (3 microbar pressure), and extracted into the time-of-flight mass spectrometer with fields of around 200 volts per cm. The infrared beam is expanded and the center section is passed through a diaphragm of 4.0±10% mm diameter. Corning 754 filters were used in order to attenuate the beam. They were found to have a transmission of 0.44 at 726 nm. The overlap of the beams was optimized by maximizing the signal at the wavelength of the small peak. This gave reproducible results over the time period of a few weeks that the data was taken. The ionization spectra at different infrared intensities I_b for a given detuning Δ_b are shown on the left column of Figure 2.

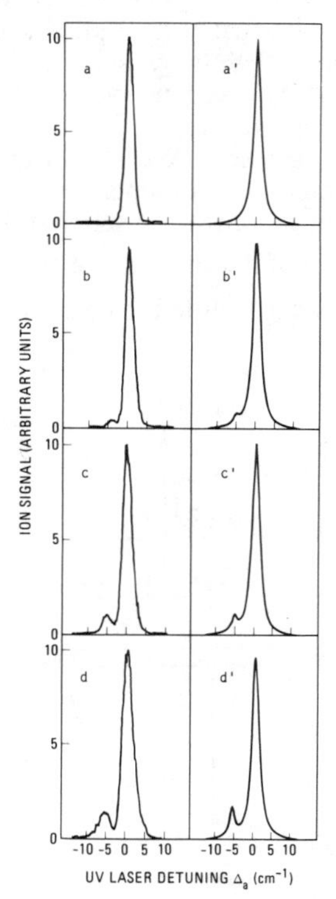

Figure 2. Comparison of the calculated multiphoton H_2^+-ion spectra (a'-d') with the corresponding observed spectra (a-d) at IR laser pulse energies: (a) 0 mJ; (b) 1.9 mJ; (c) 4.4 mJ; and (d) 10 mJ. The IR laser detuning Δ_b is 4.8 cm^{-1}.

To calculate the corresponding spectral profiles, we solve the density matrix equations describing the three-level molecular system interaction with the laser fields.[2] The longitudinal and transverse relaxations of the molecule due to radiative decays, photoionization, and the laser bandwidths are taken into account phenomenologically in the theoretical model. The solutions we obtain numerically are in the non-steady-state regime since multiphoton ionization continuously depletes the molecular population in the bound levels. The theoretical fits to the observed spectra are shown on the right-hand column of Figure 2. They yield a value for the vibronic transition moment between E,F, v=6 and D, v'=2 of 2.0±0.5 a.u., where the estimated errors are largely due to the uncertainty in the measured average IR intensity. This is in good agreement with the *ab initio* value 1.97 a.u. calculated by Huo and Jaffe.[3]

We note that in the above method, the determination of the peak separation of the Autler-Townes doublet is a frequency measurement, which can be done usually with high accuracy. Another advantage of the method is that in the strong saturation regime, the accuracy of the determined transition moment does not depend critically on precise knowledge of the molecular relaxation rates, the photoionization rates, or the laser bandwidths. In order to obtain absolute values of the transition moments, conventional methods generally require absolute measurements of populations, fluorescence intensities, or lifetimes of radiative decay to numerous lower states, whereas the method here is pairwise state-selective, is independent of absolute population, and requires only the measurement of laser intensity and frequency detuning. These features makes it particularly advantageous for measuring transition moments between excited electronic states.

This work is supported by the U. S. Department of Energy, Office of Basic Energy Sciences, Division of Chemical Sciences.

REFERENCES:

1. M. A. Quesada, A. M. F. Lau, D. H. Parker and D. W. Chandler, Phys. Rev. A 35, 4107 (1987).

2. A. M. F. Lau, Sandia Report *SAND85-8697* (Sandia Laboratories, Livermore, CA, 1985).

3. W. M. Huo and R. L. Jaffe, *Chem. Phys. Lett. 101*, 463 (1983), and private communication.

STIMULATED BRILLOUIN SCATTERING REFLECTIVITIES WITH A DUAL SPECTRAL-LINE PUMP

Ruth Ann Mullen, R.C. Lind, and G.C. Valley
Hughes Research Laboratories, Malibu, CA 90265

ABSTRACT

Reflectivity measurements using a dual spectral-line (DSL) pump having variable spectral-line separations and two different geometrically-constrained SBS medium lengths are summarized. Good empirical agreement between the experimental data and a transient solution for SBS gain is obtained provided that the interaction length for SBS is chosen to be the shorter of either the dual spectral-line beat length or the geometrically-constrained effective medium length. Such an analysis of the data appears to remain valid even for dual spectral-line separations large enough to violate the slowly-varying envelope approximation. The correctness of choosing the beat length of the dual spectral-line source as an effective pump coherence length for this analysis is discussed.

INTRODUCTION

Stimulated Brillouin scattering (SBS) of lasers with bandwidths larger than the inverse Brillouin lifetime is of considerable importance to the application of phase-conjugation to broadband, multi-line, or short-pulse laser systems. Reference (1) describes an experiment designed to investigate the effect of the relative sizes of the pump coherence length l_c and the geometrically-constrained SBS medium length l_m on phase-conjugate reflectivities. Earlier experimental and theoretical work in this area is described in references 2-4 and a closely analogous experiment using a modulated cw laser and SBS in fibers is described in reference 5.

EXPERIMENT

The experimental apparatus is shown in Figure 1. The two high pressure gas cells SBSA and SBSB were separately filled with different medium in order to obtain three different spectral line separations. Replacing one cell with a beaker of water and filling the other cell with 700 psi of CH_4 resulted in the largest spectral-line separation of 4.4 GHz. A 1.2 GHz spectral-line separation was obtained when one cell was filled with 700 psi of CH_4 and the other cell was filled with 300 psi of SF_6. When both cells were filled with 700 psi of methane, the DSL separation was zero and the effective pump bandwidth was equal to the transform-limited frequency width fw 1/e of the single-longitudinal mode YAG laser, 83 MHz. A third high pressure gas cell, labelled SBSC, was filled with 2000 psi of CH_4 and used to measure the reflectivity of the DSL pump. By focussing the DSL pump into this third cell, a geometrically-constrained effective medium length l_m roughly equal to the Gaussian beam confocal parameter of 2.7 cm was obtained. A much longer effective medium length of 106.7 cm was obtained by allowing the DSL beam to pass unfocussed through a cell of that length.

The reflectivities are plotted as a function of the ratio l_c/l_m in Figure 2. The much higher reflectivities associated with the large l_c/l_m data points are largely attributable to the much higher peak fluences (and consequently higher gains) available in the focussed geometry. The transient gains for each situation are given approximately by the pulsed gain formula[6] $G \sim \exp[2(g_B zF/\tau_B)^{1/2}]$, where F is the fluence of the beam ($J\ cm^{-2}$), g_B is the steady-state gain[1], τ_B is the Brillouin acoustic lifetime, and z is the SBS interaction length. When z is chosen to be the shorter of either l_c or l_m, reflectivity as a function of transient gain for all the DSL experimental data is in good agreement with the results of reflectivity measurements performed with a single-line, single-longitudinal mode pump under the same experimental conditions, as displayed in Figure 3.

DISCUSSION

In Reference 1, the effective laser coherence length l_c is chosen to be equal to the beat length of the DSL source without any explanation of why the occurence of multiple coherent beats within the long (collimated) interaction geometry does not act to lengthen the SBS interaction length z to $l_m/2$. One possible reason for this may be that, since the conditions for the collimated geometry were so near threshold in our experiment, the scattered light from one of the coherent beat regions may not have been an exact phase-conjugate. After propagating backwards over a region of DSL incoherence, the phases of the back-scatterred light from one region would then not match up correctly with the phases cf the pump in the next region, necessitating the re-starting of the stimulated scattering from noise. In the focussed geometry, the geometrically-constrained SBS medium length was always shorter than the DSL beat length.

REFERENCES

1. R.A. Mullen, R.C. Lind, and G.C. Valley, Opt. Commun. 63, 123 (1987).

2. G.C. Valley, IEEE J. Quantum Electron., QE-22, 704 (1986).

3. M.V. Vasilev, P.M. Semenov, and V.G. Sidorovich, Opt. Spektrosk. 56b, 193 (1984) [Opt. Spectrosc. (USSR) 56, 1221 (1984)].

4. M.D. Skeldon , P.Narum and R. Boyd, SPIE Proc. 613, 93 (1986).

5. E. Lichtman, A.A. Friesem, R.G. Waarts, and H.H. Yaffe, J. Opt. Soc. Am. B 4, 1397 (1987).

6. R.L. Carman, F. Shimizu, C.S. Wang, and N. Bloembergen, Phys. Rev. A 2, 60 (1970).

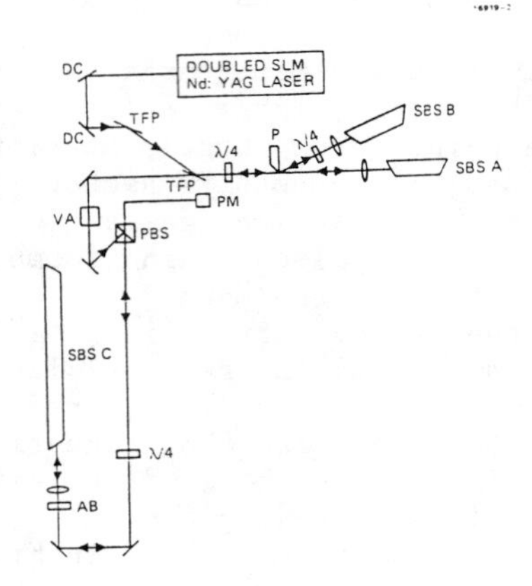

Figure 1. Experimental schematic, DC=dichroic mirror. TFP=thin film polarizer. $\lambda/4$=quarter-wave plate. P=prism. VA=variable attenuator (consisting of a rotable wave-plate and a fixed polarizer). PBS=polarizing beam splitter cube. AB=aberrator. PM=power meter.

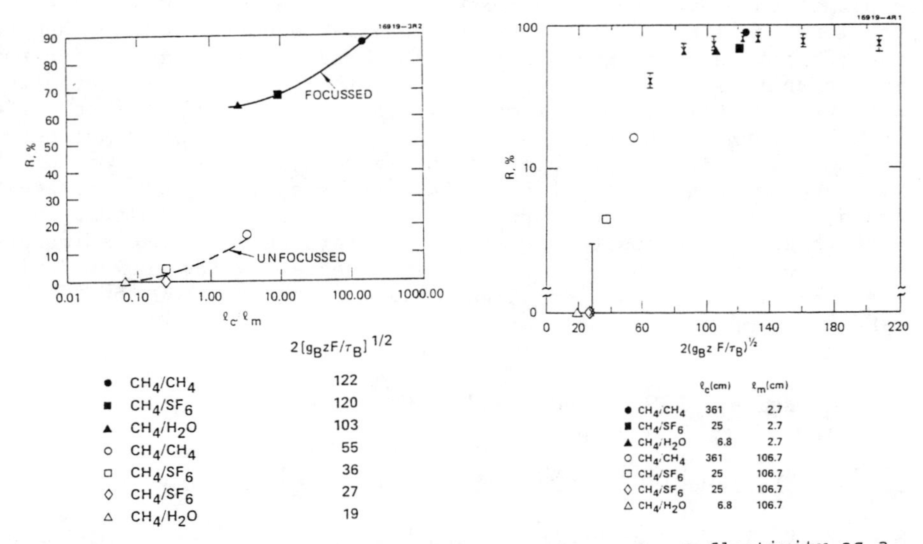

Figure 2. Reflectivity in 2000 psi of CH_4, as a function of the ratio ℓ_c/ℓ_m. Note that, because of the very different gains (see table 1), the data cannot be expected to lie on a single curve.

Figure 3. Reflectivity as a function of transient gain $2(g_B zF/\tau_B)^{1/2}$, where $z=\ell_m$ for $\ell_c>\ell_m$ and $z=\ell_c$ for $\ell_c<\ell_m$.

THE SUPPRESSION OF STIMULATED HYPER-RAMAN EMISSION IN Na DUE TO PARAMETRIC FOUR-WAVE MIXING

M. A. Moore,* W. R. Garrett, and M. G. Payne
Chemical Physics Section, Oak Ridge National Laboratory
Oak Ridge, Tennessee 37831-6378

ABSTRACT

We show that an interference effect involving four-wave mixing can almost totally suppress predicted examples of stimulated hyper-Raman emission in the forward direction of a laser pump beam. The effect is demonstrated in Na vapor.

Both experimental[1-3] and theoretical[4-8] studies over the past several years have shown that an optical interference effect may be seen in a nonlinear gaseous medium between states that have an odd-n number of photons connecting those states and also are coupled by a one-photon allowed transition. In the case of Xe it has been demonstrated by Miller *et al.*[2] that there is a suppression of ionization out of a three-photon resonance which has been theoretically proven to be caused by generation within the medium of a three-photon sum frequency which drives the atomic transition 180° out of phase with the direct three-photon pumping by the laser photons. The effect only occurs with unidirectional pump beams where four-wave mixing (FWM) can occur. Counterpropagating geometries do not allow the proper phase matching to occur, and this interference effect is not possible.

In a stimulated hyper-Raman (SHR) process (see Fig. 1), two pump laser photons are absorbed to produce a third photon at a frequency $\omega_{SHR} = 2\omega_L - E_1/\hbar$ and an excited atom at energy E_1 where in our case E_1 is the energy of either of the sodium 3p states. (The laser is tuned near two-photon resonance with either Na 3d or 4d.) Additionally, a field at $\omega_4 = 2\omega_L - \omega_{SHR}$ is generated that completes the transition back to the ground state. Since SHR emission results in transitions exactly to either of the $3p_{3/2}$ or $3p_{1/2}$ resonances, it represents the three-photon excitation of these states and should be suppressed by the field generated at ω_4 which is exactly 180° out of phase with the SHR signal. The ω_4 photon is resonant between the 3s ground state and the 3p state and is therefore trapped in the medium and cannot be seen.

The SHR emission is a stimulated process with gain along the laser beam in either direction. Since FWM can only occur in the forward direction the suppression effect may be demonstrated by comparing the SHR emission in both forward and backward directions. Other emissions in the forward direction may be discriminated against by putting a small aperture in the forward beam that allows only axial propagating SHR emission into the spectrometer.

A heat pipe is operated at a pressure of about 1 Torr with an argon buffer gas. Laser photons excite the sodium vapor and cause SHR photons to be produced. These are reflected by matching mirrors into a Jarrell Ash spectrometer, the choice of forward or backward beam determined by the angle of a rotatable mirror. Data scans were alternated between forward and backward directions at fixed laser detuning to compare forward to backward SHR intensities at exactly the same laser frequency.

---

* Graduate student, University of Tennessee, Knoxville, Tennessee

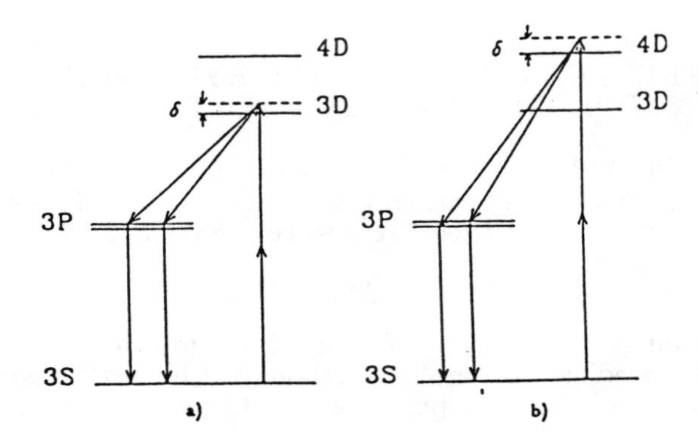

Figure 1. Schematic of Na energy levels and observed emissions, a) two-photon to near 3d resonance to 3p state, b) two-photon near 4d resonance to 3p state.

Results are shown in Figs. 2 and 3. When the laser is tuned exactly to the two-photon 3d resonance the emission seen is amplified spontaneous emission (ASE) rather than SHR. As is readily observable, there is a ASE signal going in both the forward and backward direction. As the laser is detuned by +0.1Å to the high energy side of the resonance, the SHR emission (no longer ASE) in the forward direction is decreased substantially while the backward SHR emission is basically unaffected. When the laser is detuned by 0.2Å as in Fig. 2c, the forward SHR emission is completely suppressed while the backward signal is weaker, but still in evidence.

The same result can be seen in Fig. 3 when the laser is tuned near the two-photon 4d resonance. As the laser is detuned from resonance so that the ASE becomes SHR emission, the forward directed SHR emission is strongly suppressed while the backward emission again is weaker, but does not entirely disappear even at a laser detuning of 0.3Å as seen in Fig. 3c.

We have also looked for the suppression effect for $\omega_{SHR} = 2\omega_L - E_{4p}/\hbar$ where E_{4p} represents the energy level for either of the sodium 4p states. These states are also connected to the ground state by an allowed one-photon transition. However, there are many alternate routes out of the 4p states to the ground state, and under these circumstances we see an axial phase matched parametric four-wave mixing (PFWM) process which "masks" the forward SHR suppression by the present measurement technique. The suppression is established in another way in an upcoming publication.

We note that an interference effect has also been predicted[9-11] and observed in two-photon resonant excitations that are also coupled by a phase matched PFWM process. In the present instance, this effect becomes operative only at higher pressures, to limit the PFWM output in a predictable fashion.

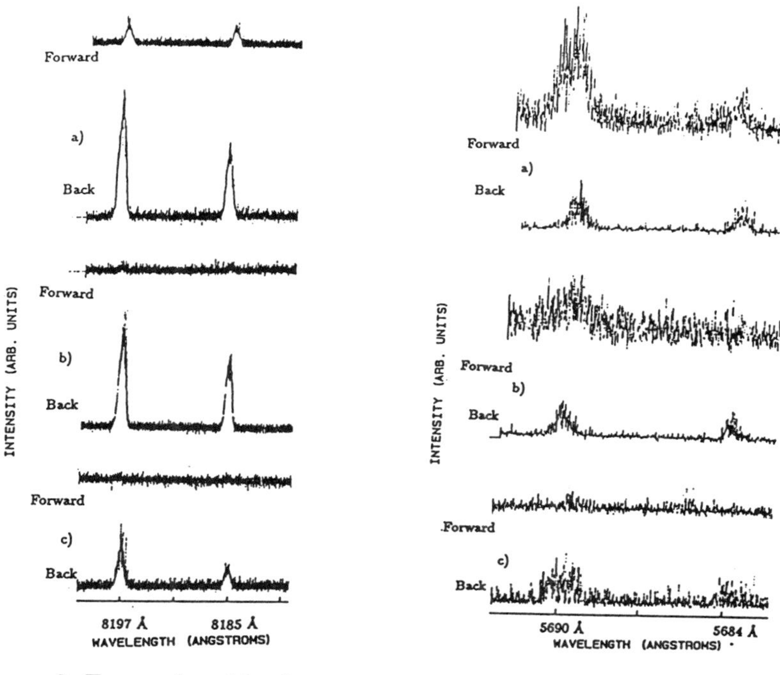

Figure 2. Forward and backward SHR emission from near 3d to 3p Na states at various detunings a) $\delta = 0$, b) $\delta = +1\mathring{A}$, c) $\delta = +0.2\mathring{A}$

Figure 3. Forward and backward SHR emission from near 4d to 3p Na states at various detunings a) $\delta = 0$, b) $\delta = +0.1\mathring{A}$, c) $\delta = +0.3\mathring{A}$

Research sponsored by the Office of Health and Environmental Research, U.S. Department of Energy under contract DE-AC05-84OR21400 with Martin Marietta Energy Systems, Inc.

REFERENCES

1. R. N. Compton, J. C. Miller, A. E. Carter, and P. Kruit, Chem. Phys. Lett. 71, 87 (1980).
2. J. C. Miller, R. N. Compton, M. G. Payne, and W. R. Garrett, Phys. Rev. Lett. 45, 114 (1980); J. C. Miller and R. N. Compton, Phys. Rev. A 25, 2056 (1982).
3. W. R. Garrett, W. R. Ferrell, M. G. Payne, and J. C. Miller, Phys. Rev. A 34, 1165 (1986).
4. M. G. Payne, W. R. Garrett, and H. C. Baker, Chem. Phys. Lett. 75, 468 (1980).
5. M. G. Payne and W. R. Garrett, Phys. Rev. A 26, 356 (1982).
6. M. G. Payne and W. R. Garrett, Phys. Rev. A 28, 3409 (1983).
7. J. J. Wynne, Phys. Rev. Lett. 52, 751 (1984).
8. G. S. Agarwal and S. P. Tewari, Phys. Rev. A 29, 1922 (1984).
9. E. A. Manykin and A. M. Afanas'ev, JETP 48, 931 (1965).
10. M. S. Malcuit, D. J. Gauthier, and R. W. Boyd, Phys. Rev. Lett. 55, 1086 (1985).
11. G. S. Agarwal, Phys. Rev. Lett. 57, 827 (1986).

RESONANCE HYPER-RAMAN SCATTERING IN THE GAS PHASE

Y. C. Chung and L. D. Ziegler
Northeastern University, Boston MA 02115

ABSTRACT

Spontaneous resonance nonlinear (hyper) Raman (RHR) spectra of NH_3, CH_3I and CS_2 are reported. This three-photon process is described at fifth order in the expansion of the electrical susceptibility (χ^5). In agreement with order of magnitude estimates, RHR scattering is about as strong as off-resonant Raman scattering. The observed RHR spectra are also nearly as intense as linear resonance scattering spectra due to the non-attenuation of the incident beam and greater incident powers in the blue than the uv. The analysis of RHR scattering is a probe of the dynamics, structure and vibronic coupling schemes of the resonant electronic state as illustrated by the RHR spectra of NH_3, CH_3I and CS_2.

INTRODUCTION

When intense pulsed radiation is scattered by a sample not only is inelastic scattering at $\nu_o - \Delta$ observed (linear Raman scattering), but very weak emission at $2\nu_o - \Delta'$ has also been seen. This nonlinear spontaneous scattering is known as hyper-Raman (HR) scattering. HR cross-section appears at the fifth order in the expansion of the electrical susceptibility. Due to the three-photon nature of HR scattering only g → u transitions appear in these spectra of centrosymmetric molecules. Analogous to the linear Raman effect, electronic resonance enhancement of HR scattering can be realized when twice the excitation frequency is coincident with a region of optical absorption. HR scattering cross-section (per molecule) is on the order of 10^{-5} to 10^{-6} that of linear Raman scattering. Electronic resonance enhancement increases the HR transition probability by ~ 10^4 to 10^6. Thus, resonance hyper-Raman scattering (RHR) is anticipated to be as intense as off-resonant linear Raman scattering. RHR spectra of NH_3, CH_3I and CS_2 are reported here and confirm these qualitative cross-section arguments. The leading terms describing RHR vibrational intensities are given by vibronic theory (A, B and C terms) and show how RHR vibrational intensities are related to one and two-photon electronic dipole allowed character, Franck-Condon factors and vibronic coupling strengths.

EXPERIMENTAL

The observed RHR spectra are excited by radiation in the 400 nm region produced by mixing Nd:YAG fundamental with dye laser output. Typically ~6 nsec pulses of ~ 5 mj/pulse at 10 Hz focused by a 15 cm focal length lens are used to excite the RHR spectra. Sample pressures of 200 torr, 150 torr and 2 atm were used to obtain RHR spectra of CS_2, CH_3I and NH_3 respectively. The reported spectra are all quadratic in laser intensity and linear in pressure dependence.

RESULTS AND DISCUSSION

The lowest lying electronic transition of ammonia, $X(^1A_1') \rightarrow A(^1A_2'')$, is a one-photon allowed Rydberg excitation which extends from 217 to 185 nm. When blue pulsed radiation at ~400 nm, which is two-photon resonant with the A state, is focused in a few atm of ammonia an inelastic scattering spectrum in the uv is observed. The RHR spectra of ammonia shows the same ν_2 and $\nu_1 + \nu_2$ vibrational progressions as found in the corresponding linear resonance Raman spectrum. The resonance enhanced rotational transitions have a very different pattern of relative intensities in the linear and non-linear spectra and can be analyzed to yield subpicosecond lifetimes of rovibronic levels of the resonant excited state.

The lowest lying Rydberg transition of methyl iodide is in the 200 - 185 nm region. Owing to the low symmetry of this molecule (C_{3v}), this transition is both one and two-photon dipole allowed. The linear and nonlinear resonance scattering spectra are shown below.

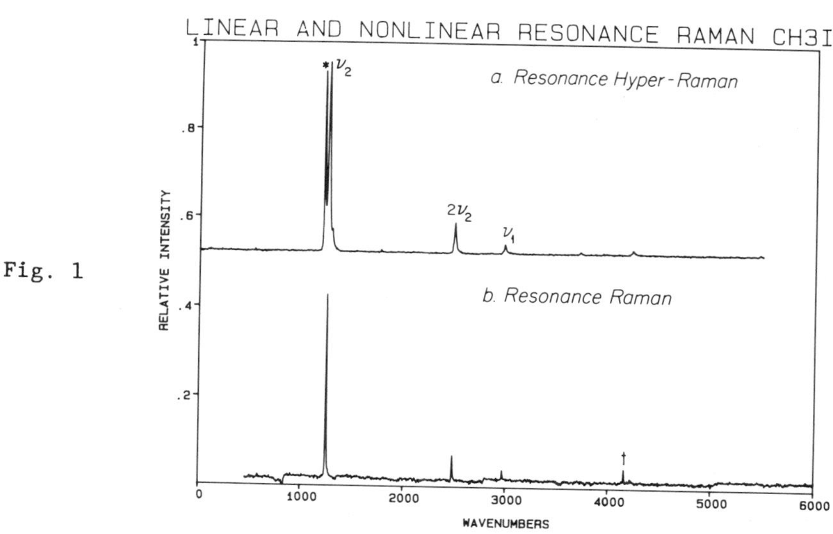

Fig. 1

The RHR spectrum of methyl iodide is extremely strong in accordance with the expectations of vibronic theory since molecules lacking a center of inversion may have electronic states which are simultaneously one and two-photon dipole allowed.

The lowest-lying one-photon dipole allowed transition of carbon disulfide is a $\Sigma_g^+ \rightarrow \Sigma_u^+(B_2)$ valence excitation (linear → bent configuration) in the region from 230 to 185 nm. The linear RR spectrum shows combinations and overtones of totally symmetric stretch and even bending overtones. RHR spectra excited at 406.2 nm and 393.6 nm are shown in fig. 2. Due to the parity selection rules odd overtones of the bending mode and antisymmetric stretch quanta are observed as indicated in the figure. This example illustrates that the hyper-Raman effect allows the observation of modes not active in linear Raman scattering.

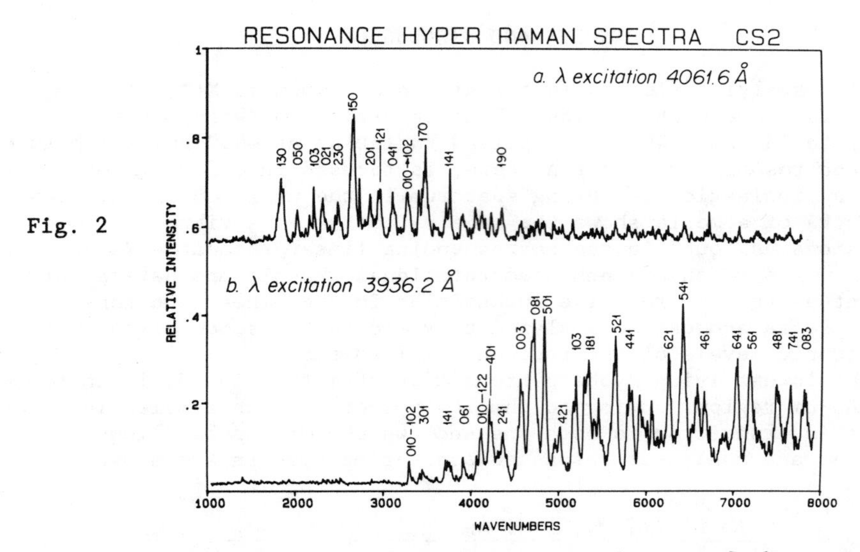

Fig. 2

The experimentally observed RHR signals are of the same order as the signals due to resonance linear scattering. This observation at first seems to contradict the order of magnitude estimates of relative scattering intensities of nonlinear to linear cross-section ($\sim 10^{-6}$). However, due to the negligible absorption of the incident RHR radiation much higher sample concentrations may be used in the nonlinear as compared to the linear scattering experiment. In addition, the lack of attenuation allows a much larger scattering volume to be observed in the RHR experiment. Finally, higher incident intensities are available in the blue than in the uv.

CONCLUSION

RHR spectra of many species should be observable as order of magnitude estimates reveal and as demonstrated by these nonlinear resonance Raman spectra. RHR scattering offers the following spectroscopic opportunities: (1) Electronic resonances in the vuv may be probed using readily available tunable near uv radiation (2) RHR scattering may be a useful technique for the study of non-fluorescent overlapping one and two photon allowed electronic states (3) RHR spectra may determine vibrational levels of the ground state which are inaccessible by other spectroscopic techniques.

Support of the National Science Foundation and the Petroleum Research Fund are gratefully acknowledged.

REFERENCES

1. L.D. Ziegler and J.L.Roebber, Chem. Phys. Lett. 136, 377 (1987).
2. L. D. Ziegler, Y. C. Chung and Y. P. Zhang, J. Chem. Phys. 87, 4498 (1987).
3. Y. C. Chung and L. D. Ziegler, J. Chem. Phys. (in press).
4. R. A. Desiderio, D. P. Gerrity and B. S. Hudson, Chem. Phys. Lett. 115, 29 (1985).

CONTROL OF RAMAN-LIKE COHERENT ON-RESONANT SOLITONS†

Farrès P. Mattar* and D. Kaup**
Department of Physics, New York University*, New York, NY 10003
and George R. Harrison Spectroscopy Laboratory
Massachusetts Institute of Technology*, Cambridge, MA 02139; and
Department of Physics, Clarkson University**, Potsdam, NY 13676

ABSTRACT

The formation of on-resonant Raman-like pump and probe solitons is examined to extract its characterizing features and define the physical condition suitable for its occurrence.

I. INTRODUCTION

Using self-consistent computational methods developed previously [1], we have investigated the **on-resonance** propagation of two pulses, one strong pump and one initially weak probe in a homogeneously broadened three-level atomic systems[2]. This analysis reveals the essential physical mechanism. We have establish that **solitary wave behavior[3] can occur provided the medium nonlinearity balances the linear diffraction for each pump and probe transition.**

In this study we examined how we can modify the soliton evolution by tailoring the field-matter physical parameters. The goal of this study is to understand how the threshold of the on-resonance coherent probe and pump soliton evolution depends on a number of physical parameters. Among those pertinent characteristics are the population relaxation $\mathbf{T_1}$ and polarization dephasing $\mathbf{T_2}$ times, the initial **populations**, beams diffraction lengths κ, Beer's lengths α, Beer's length Fresnel numbers $\mathbf{F_g}$, the initial **temporal coincidence** of the probe and the pump peaks and the role of the dispersion-induced phase. This sheds light on the inertial properties of the medium.

It should help us secure conditions that insure distortionless co-propagation in an all-optical-communication system.

II. NUMERICAL RESULTS

i. Probe Soliton

By modifying the gain-to-loss ratio, one controls the soliton emergence. A different threshold for the **probe area asymptotic stabilization occurs whenever diffraction is varied.**

To modify the gain we select physical situations in which the medium characteristics are varied. As an example, we calculate the effect of changing the **ratio of relaxation times (T_1 and T_2) to pump duration τ_p** on both on-axis probe area (graph a) and pump beam width (graph b) (Fig. 1). As T_1 decreases the effective Beer's length and the soliton formation is slower. The pump ceases to narrow and even lengthens for short T_1. The on-axis probe area stabilization evolves like an oscillator that is under-, over-, and/or critically damped. It occurs earlier or later. At other instances, the probe saturation peak disappears and its area stabilizes immediately. The coherent stabilization corresponds to self-trapping.

To see the importance of **gain-to-loss** ratios in the probe soliton formation, calculation of gain g dependence and appropriate **Fresnel number** F, have been carried out simultaneously. As the gain increases (while maintaining the diffraction loss as is) the pump beam narrowing decreases and the probe buildup is stronger (Fig. 2). One varies

† Supported by ARO, AFOSR, NSF, and ONR.

the diffraction coupling by operating at another wavelength, λ, or by selecting a different beam width r_p. The dependence on F_g the Beer's length Fresnel number, is shown in Fig. 3. The ratio F_g/F is the total gain $g\ell$. As the Beer's length and the beam input widths increase the pump narrows less and the probe soliton formation evolves more slowly. Graphs a and b respectively display the probe and pump axial areas. When the input pump area increases (Fig. 4) one encounters pump pulse breakup à la Self-Induced-Transparency (SIT). The probe on-axis stabilization is strongly influenced.

For the first time one can control the interplay of pump depletion, diffraction and on-axis optical area stabilization in a long three-level vapor Raman-like amplifier and modify the probe asymptotic stabilization.

The effect of **dispersion** through the probe detuning, reduces the probe gain since $\tilde{\alpha}_b(\Delta\omega_b)<\alpha_b(\Delta\omega_b=0)$. The probe grows at a smaller rate causing the pump to deplete less. The cross-amplification, changes from its maximum on-resonance to a smaller off-resonance.

The probe magnification spectrum is symmetrical in the UPW (or for large F) regime while it is **not** when diffraction (for small F) is accounted for. The off-resonant phase buildup is independent of the detuning sign. It can either interfere constructively with or be cancelled by the diffraction-induced phase. The peak of the frequency response occurs off-line center for narrow beams.

The **initial populations** can accelerate the probe growth. **By establishing an initial population inversion in the probe transition,** the probe begins to grow earlier or later.

By modifying the **initial temporal coincidence** between pump and probe peaks probe growth can be controlled. By advancing the probe relative to the pump, the probe begins to grow earlier. If the probe peak is delayed with respect to the pump, its growth is decelerated.

Changes in pulse and beam shapes, temporal lengths, spatial widths and strengths of the input pump were analyzed.

These calculations demonstrate that the rate of growth of the probe can be controlled by selecting the input field and matter parameters. The resulting on-resonance probe and pump soliton are altered.

ii. **Pump Soliton**

Further propagation which leads to **pump area stabilization** is presented for a longer metallic vapor cell. The effect of pump input **radial shape** (Gaussian vs. super-Gaussian) has been studied (Fig. 5). The larger the gain the sooner is the build up. The stabilization occurs every time. In Fig. 6 variation of pump area and fluence asymptotic stabilization are seen as diffraction loss changes for both UPW and non-planar regimes[5]. This is the on-resonance equivalent of the Raman soliton[6].

III. CONCLUSION

Control of a probe soliton evolution is accepted as a typical process of light control by light. The concept that **through nonlinear feedback, the controller (pump) itself, can stabilize while the probe depletes in its turn is rather striking.** The interchange of role between pump and probe insures that the two beams do **not** stabilize simultaneously over an extended optical thickness. The dependence of the on-resonance pump and probe solitons on the descriptive pulse and medium parameters have been determined.

REFERENCES

1. (a) F.P. Mattar, Appl. Phys. 17, 53 (1978); (b) F.P. Mattar & M.C. Newstein, Comput. Phys. Commun. 20, 139 (1980); 32, 225 (1984).
2. F.P. Mattar, **Physical Mechanisms in Coherent On-Resonance Propagation of Two Beams in a Three-Level System**, submitted to Progress in Quantum Electronics.
3. F.P. Mattar, **Asymptotic Regime in On-Resonance Reshaping in Coherent Probe and Pump Solitons Evolution in a Three-Level System**, submitted to Progress in Quantum Electronics.
4. S.L. McCall & E. Hahn, Phys. Rev. 183, 457 (1969).
5. F.P. Mattar et al., in **Modeling and Simulation XVII**, ed. W.G. Vogt & M.H. Mickle (Instr. Soc. Am., 1987) pp. 1559-1642.
6. R.G. Wenzel et al., J.Stat. Phys. 39, 615 (1985) & ibid 621 (1985)

FIGURE CAPTIONS

Fig. 1 Variation of relaxation times effect T_1/τ_p, T_2/τ_p.

Fig. 2 g gain dependence with everything else constant.

Fig. 3 Variation of the {g/κ} gain-to-loss ratio (a) UPW & (b) with diffraction on the probe on-axis area.

Fig. 4 Influence of the input pump area.

Fig. 5 Radial shape variation on the probe area during the pump soliton regime formation.

Fig. 6 Probe and pump areas while varying F_g (a) UPW and (b) non-planar regimes.

CURVE	LR2CFG	T1INVA	T1INVB	T2INVA	T2INVB	T2INVC
1	1252	.0025	.0025	.00285714	.00285714	.00285714
2	1210	.0125	.0125	.0142857	.0142857	.0142857
3	1249	.0625	.0625	.0714285	.0714285	.0714285
4	1250	.125	.125	.142857	.142857	.142857
5	1251	.25	.25	.285714	.285714	.285714

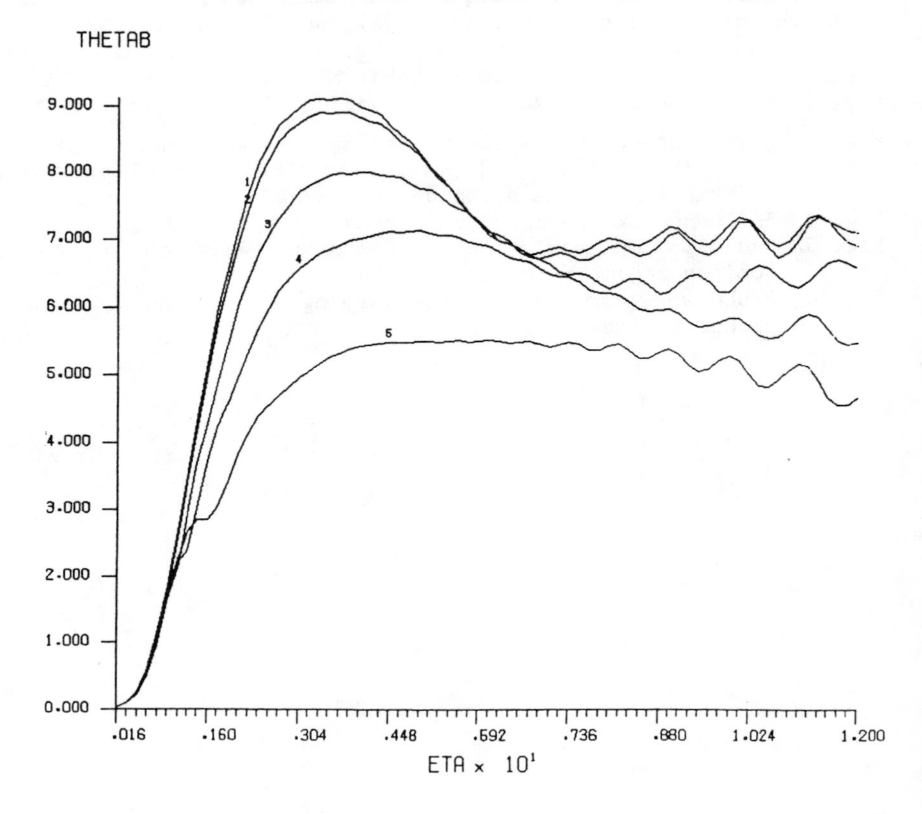

Fig. 1

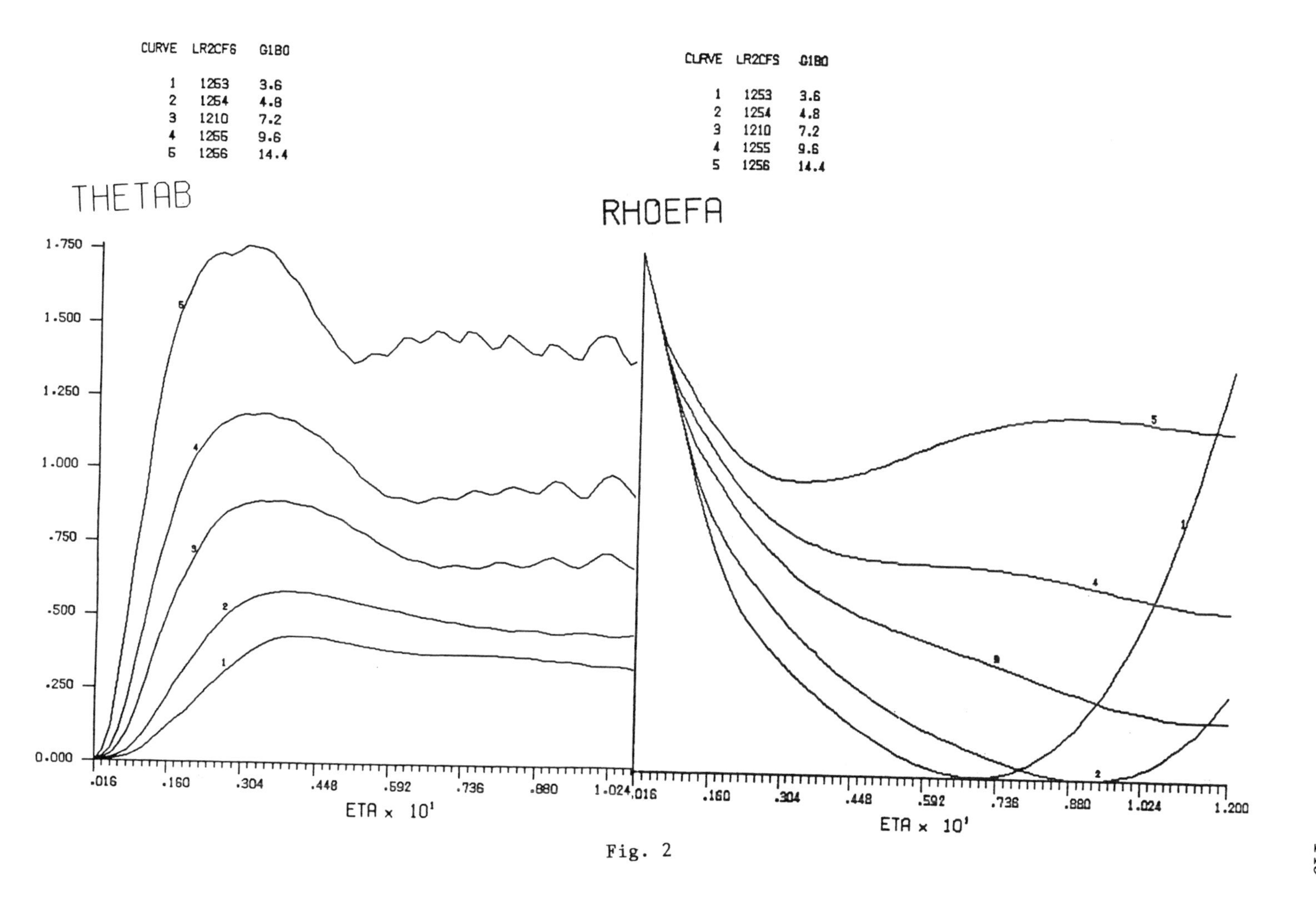
CURVE LR2CFS G1BO
1 1253 3.6
2 1254 4.8
3 1210 7.2
4 1255 9.6
5 1256 14.4
THETAB
1.750
1.500
1.250
1.000
.750
.500
.250
0.000
.016 .160 .304 .448 .592 .736 .880 1.024
ETA × 10¹
CURVE LR2CFS G1BO
1 1253 3.6
2 1254 4.8
3 1210 7.2
4 1255 9.6
5 1256 14.4
RHOEFA
.016 .160 .304 .448 .592 .736 .880 1.024 1.200
ETA × 10¹

Fig. 2

CURVE	LR2CFS	G1A0	G1B0	TBRHOA	TBRHOB	TBRHON
1	1210	.24	7.2	7.0	7.0	7.0
2	1211	.21	6.3	9.0	9.0	9.0
3	1257	.18	5.4	11.0	11.0	11.0

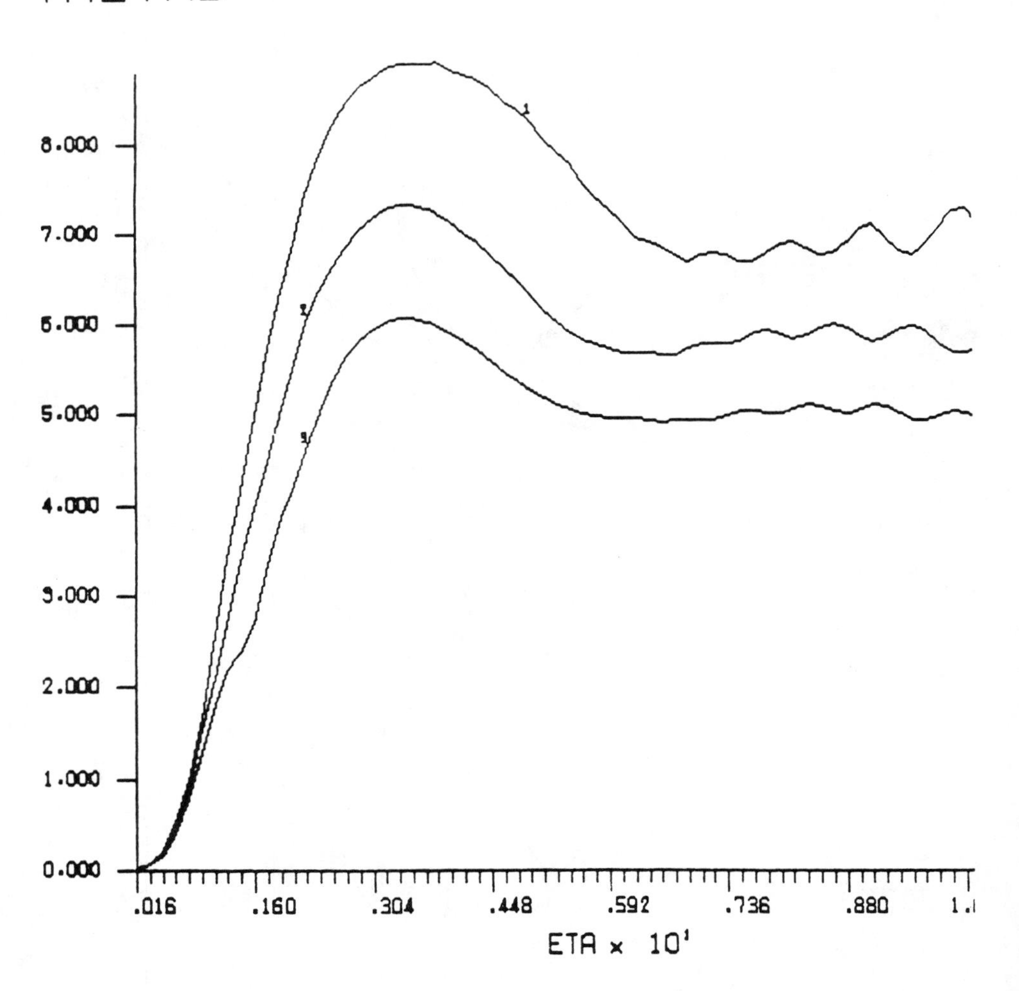

Fig. 3

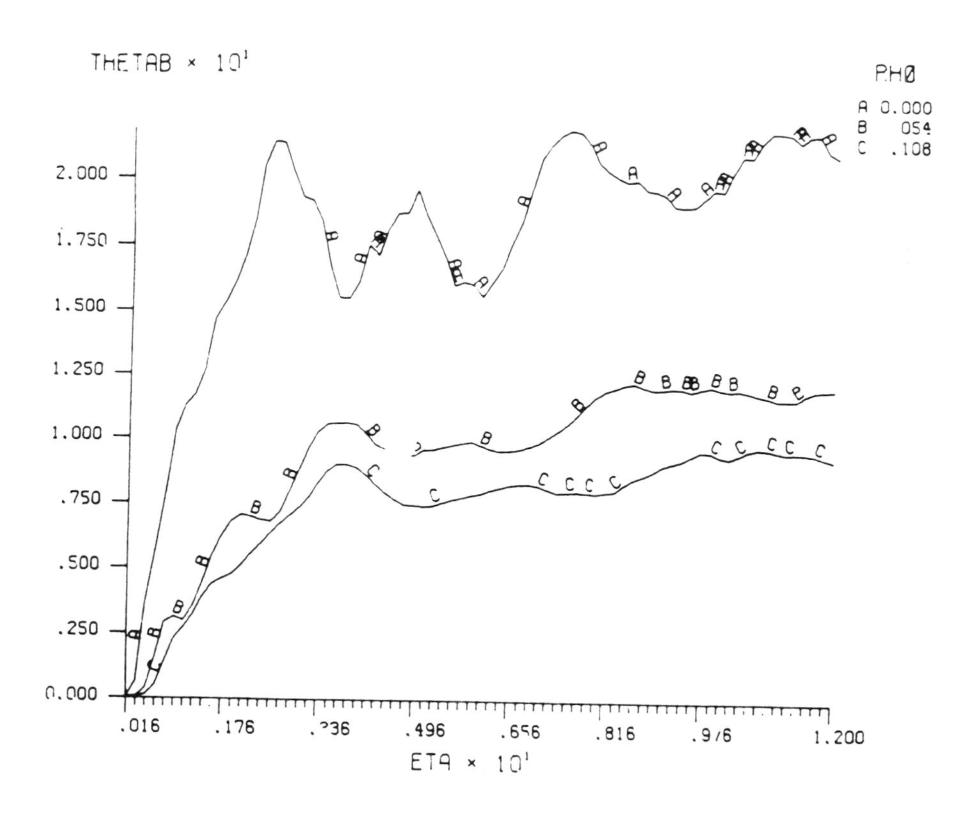

THETAB

SCALE GLØBALE
MIN = .34401910E-50
MAX = 21.843092

AXIS	TØT.PT	SYM	RANGE	NB.SEL	RANGE ØF VALUES
PØW	1	NØ	1	1	0.0
RHØ	32	NØ	(1.7)	3	(0.0.0.10839)
ETA	75	NØ	(1.75)	75	(0.16.12.0)

Fig. 4

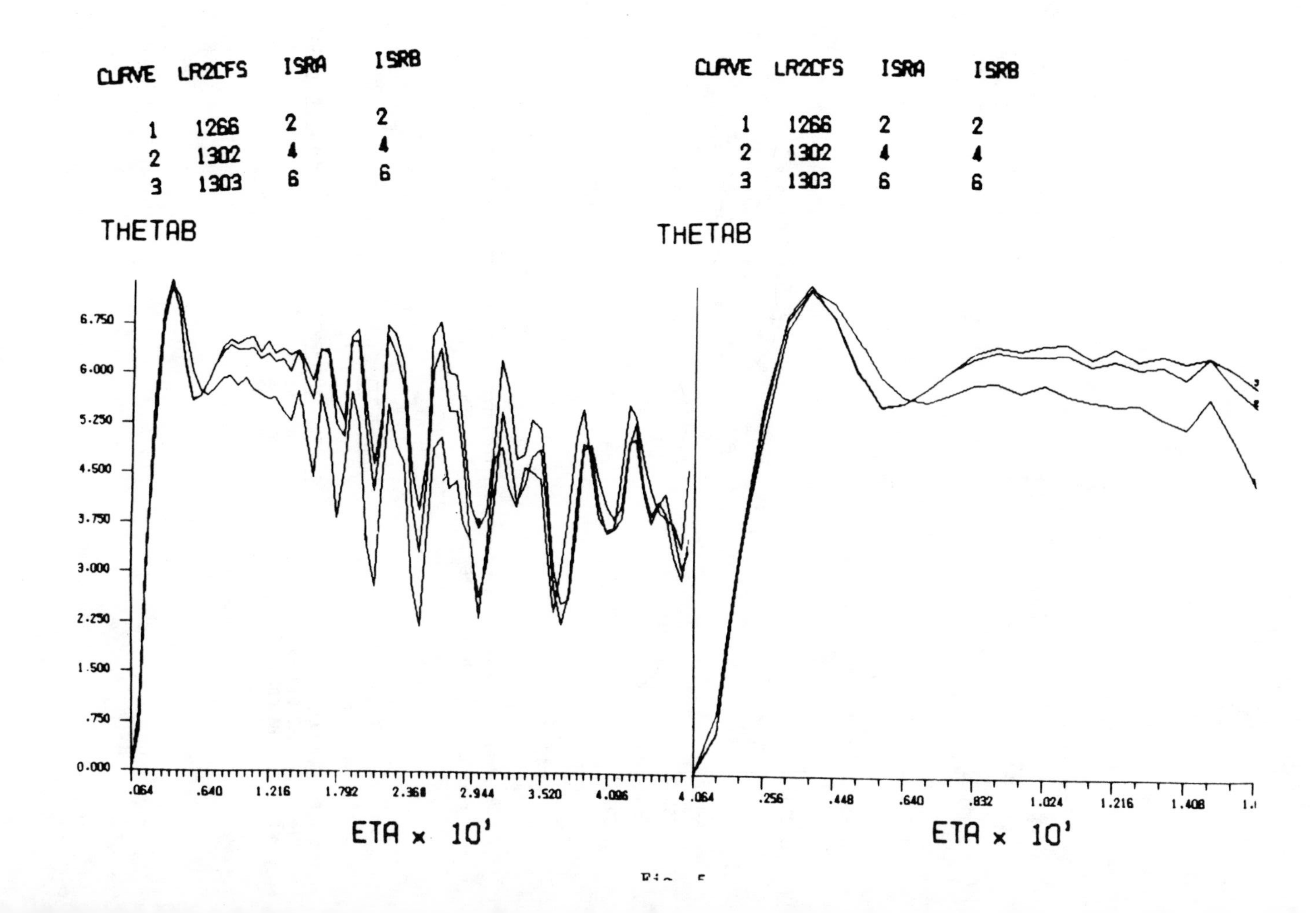
CURVE LR2CFS ISRA ISRB
1 1266 2 2
2 1302 4 4
3 1303 6 6
THETAB
6.750
6.000
5.250
4.500
3.750
3.000
2.250
1.500
.750
0.000
.064
.640
1.216
1.792
2.368
2.944
3.520
4.096
ETA x 10¹
CURVE LR2CFS ISRA ISRB
1 1266 2 2
2 1302 4 4
3 1303 6 6
THETAB
.064
.256
.448
.640
.832
1.024
1.216
1.408
ETA x 10¹

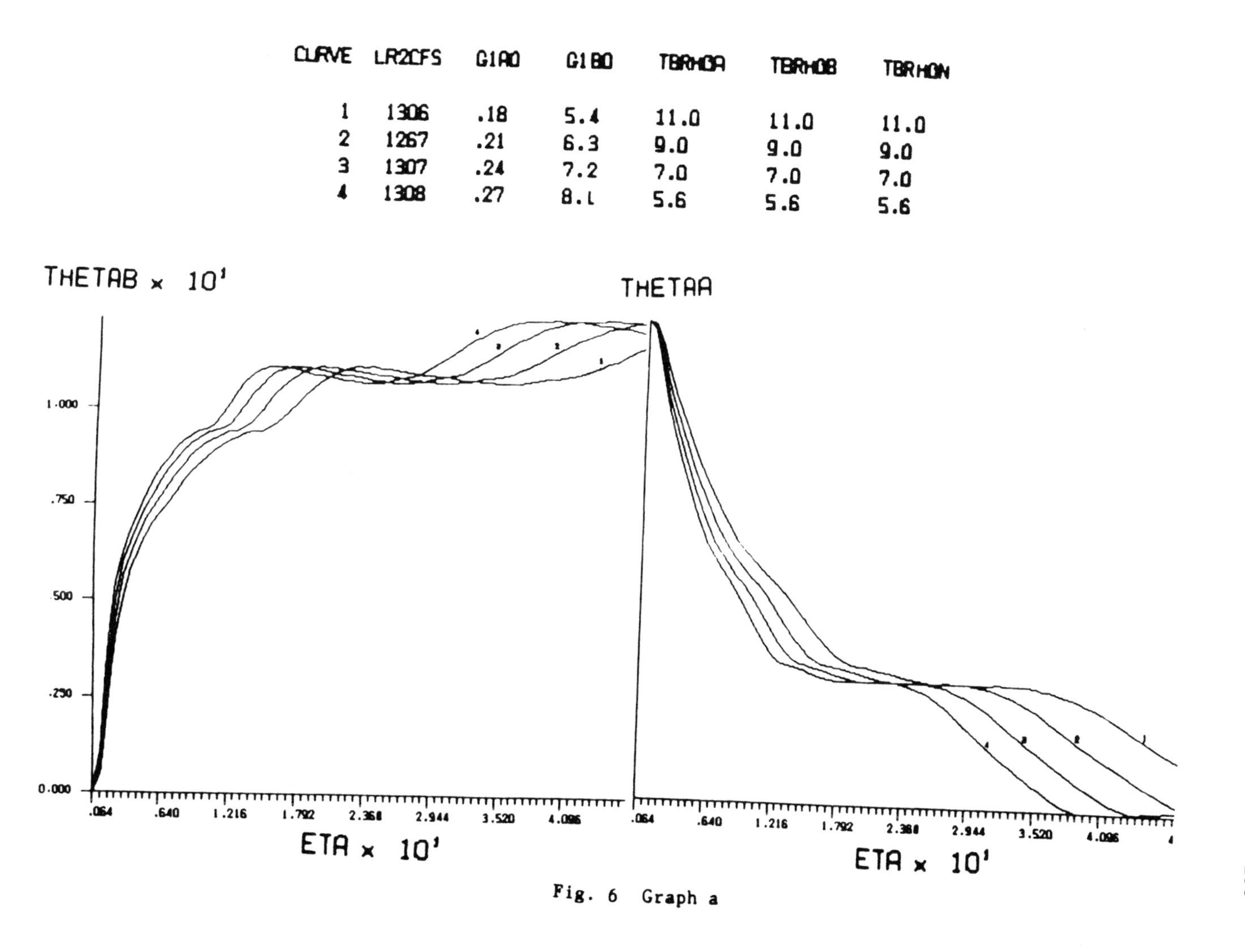
CURVE LR2CFS G1AO G1BO TBRHOA TBRHOB TBRHON
1 1306 .18 5.4 11.0 11.0 11.0
2 1267 .21 6.3 9.0 9.0 9.0
3 1307 .24 7.2 7.0 7.0 7.0
4 1308 .27 8.1 5.6 5.6 5.6
THETAB × 10¹
THETAA
1.000
.750
.500
.250
0.000
.064
.640
1.216
1.792
2.368
2.944
3.520
4.096
ETA × 10¹
ETA × 10¹

Fig. 6 Graph a

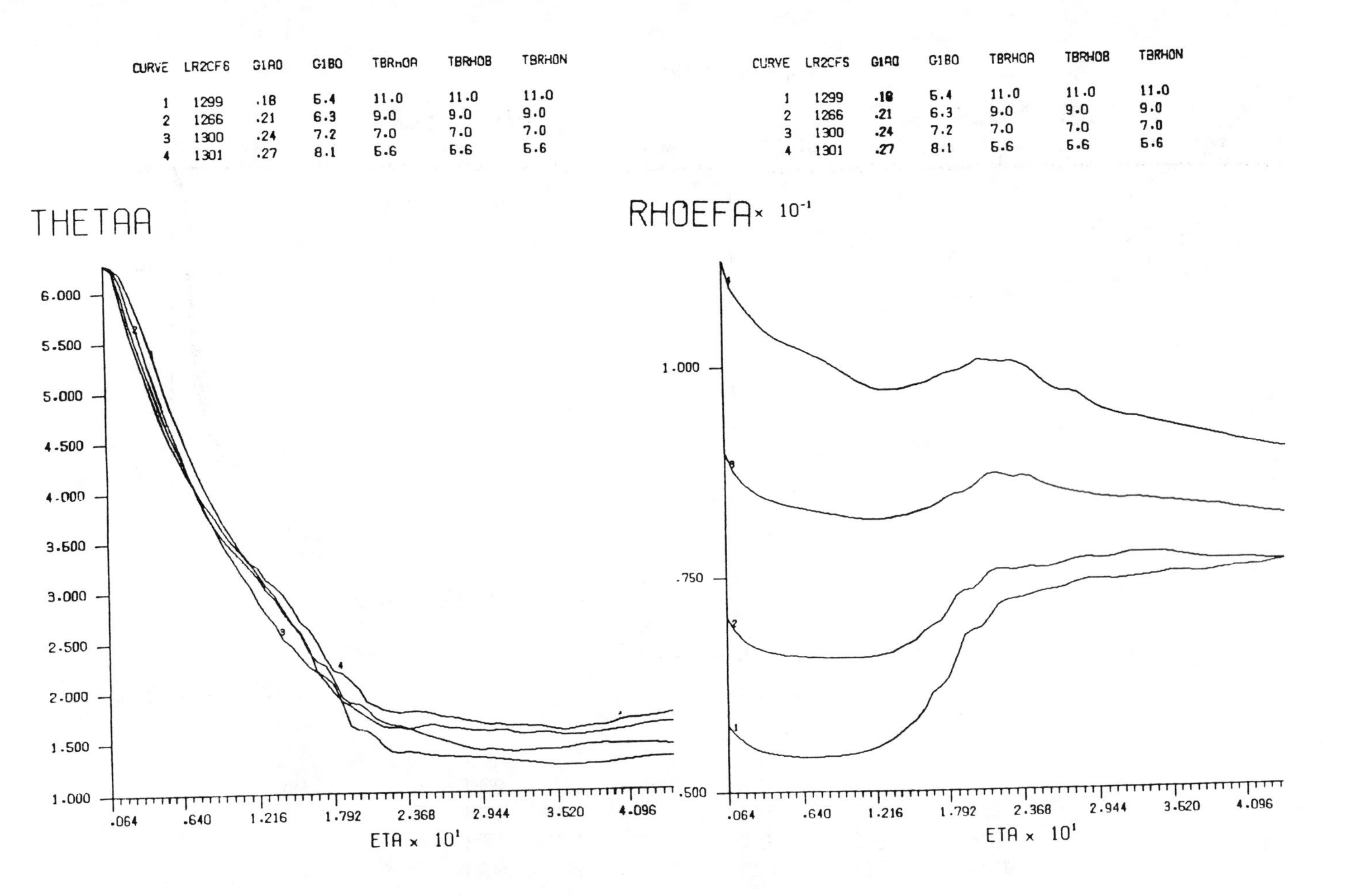
CURVE LR2CF6 G1A0 G1B0 TBRhOA TBRHOB TBRHON
1 1299 .18 6.4 11.0 11.0 11.0
2 1266 .21 6.3 9.0 9.0 9.0
3 1300 .24 7.2 7.0 7.0 7.0
4 1301 .27 8.1 6.6 6.6 6.6
THETAA
6.000
5.500
5.000
4.500
4.000
3.500
3.000
2.500
2.000
1.500
1.000
.064
.640
1.216
1.792
2.368
2.944
3.620
4.096
ETA × 10¹
CURVE LR2CFS G1A0 G1B0 TBRHOA TBRHOB TBRHON
1 1299 .18 6.4 11.0 11.0 11.0
2 1266 .21 6.3 9.0 9.0 9.0
3 1300 .24 7.2 7.0 7.0 7.0
4 1301 .27 8.1 6.6 6.6 6.6
RHOEFA × 10⁻¹
1.000
.750
.500
.064
.640
1.216
1.792
2.368
2.944
3.620
4.096
ETA × 10¹

Fig. 6 Graph b

ATOMIC RESPONSE TO BICHROMATIC CROSSED-BEAM FIELD

Wilhelmus M. Ruyten, Lloyd M. Davis, Christian Parigger, Dennis R. Keefer
University of Tennessee Space Institute
Center for Laser Applications, Tullahoma, TN 37388

ABSTRACT

We treat the excitation–fluorescence response of an atom to a bichromatic, crossed-beam field, both in terms of a time-dependent physical spectrum, and optical Bloch equations. Agreement is found between narrowband illumination for a broad velocity distribution, and broadband illumination in phase sensitive fluorimetry. For strong fields, essential features of the Rabi structure are shown to remain even if the velocity distribution exceeds the homogeneous linewidth. Advantages of a crossed-beam experiment are pointed out.

INTRODUCTION

In a proposed laser fluorescence velocimeter[1], two beams of equal intensity with frequencies ω_L and $\omega_L+\omega_B$, obtained by passing a frequency stabilized dye laser beam through a high efficiency Bragg cell, are crossed at a small angle θ, to produce moving interference fringes. Bichromatic, single-photon[2] excitation of tracer atoms gives rise to fluorescence, which is collected from a sub-fringe sized volume element, and the component at the modulation frequency ω_B is detected through a lock-in amplifier. From the depth-of-modulation M, and/or the phase-shift Φ relative to the exciting field, the atom's velocity component $v_\perp$ along the fringe velocity vector can be derived. Expressions for $M(v_\perp)$ and $\Phi(v_\perp)$ in Ref.1 were obtained solely by considering the atomic *fluorescence* response, as is customary in phase-sensitive fluorimetry with broad band illumination. Here, we consider in detail the roles of atomic linewidth, Doppler shift, and field strength, resulting from the *excitation* response. We present results for the limiting cases of narrow and broad velocity distributions in the beam direction. For the latter, we find complete agreement with the previous work.

WEAK FIELD

For weak fields, the excitation is linearly proportional to the spectral intensity of the exciting light. To take into account the atomic linewidth, we use Eberly *et al.*'s expression for a time-dependent physical spectrum (TDPS)[3], as measured by a Lorentzian absorber with linecenter ω_0 and FWHM 2γ:

$$S(\omega_0,t;\gamma) = 2\gamma \int_{-\infty}^{t} dt' \int_{-\infty}^{t} dt''\, e^{-2\gamma(2t-t'-t'')} e^{-i\omega_0(t'-t'')} \langle v^*(t')v(t'')\rangle. \qquad (1)$$

Upon substituting the bichromatic field $v(t) \propto e^{-i\omega_1 t} + e^{-i\omega_2 t}$, and introducing the scaled average detuning $\bar{\Delta} = (2\omega_0 - \omega_1 - \omega_2)/2\gamma$, the scaled difference frequency $\delta = (\omega_2 - \omega_1)/2\gamma$, and a normalized time $\tau = \gamma t$, the TDPS becomes:

$$S(\bar{\Delta},\delta;\tau) \sim 1 + M\cos(2\delta\tau - \Phi), \qquad (2)$$

where the modulation M and phase shift Φ are each functions of both $\bar{\Delta}$ and δ:

$$M = \left[1 - \left(\frac{2\bar{\Delta}\delta}{1+\bar{\Delta}^2+\delta^2}\right)^2\right]^{\frac{1}{2}}; \qquad \Phi = \tan^{-1}\left(\frac{2\delta}{1+\bar{\Delta}^2-\delta^2}\right). \tag{3}$$

Through the Doppler shift, $\bar{\Delta}$ and δ become functions of the parallel ($v_{\|}$) and perpendicular ($v_{\perp}$) velocity components of the atoms (see Fig.1):

$$\bar{\Delta} = (\omega_L - \omega_0 + \tfrac{1}{2}\omega_B)/\gamma - \cos\tfrac{\theta}{2}\,\bar{k}\,v_{\|}/\gamma; \qquad \delta = \omega_B/2\gamma + \sin\tfrac{\theta}{2}\,\bar{k}\,v_{\perp}/\gamma, \tag{4}$$

where $\bar{k}$ is the mean wavenumber. Thus, both $v_{\perp}$ and $v_{\|}$ can be determined from M and Φ in principle. However, since the angle θ is small (typically 20 mrad), there is a large asymmetry in the respective velocity dependences. Consider for example velocity distributions significantly wider than 2γ, yet narrower than $2\gamma/\theta$. These map onto thin, extremely elongated "lines" in $\bar{\Delta},\delta$-space, as shown in Fig.1. By comparison, a collinear experiment ($\theta = 0$) would map the same distributions onto the same truly one dimensional line. Thus, in a crossed-beam experiment different δ's can be accessed for a fixed ω_B, merely by varying θ. Note that the modulation frequency detected in the laboratory frame is just ω_B, independent of the atom's velocity.

Figure 1: Mapping of velocity space onto atomic rest frame parameters.

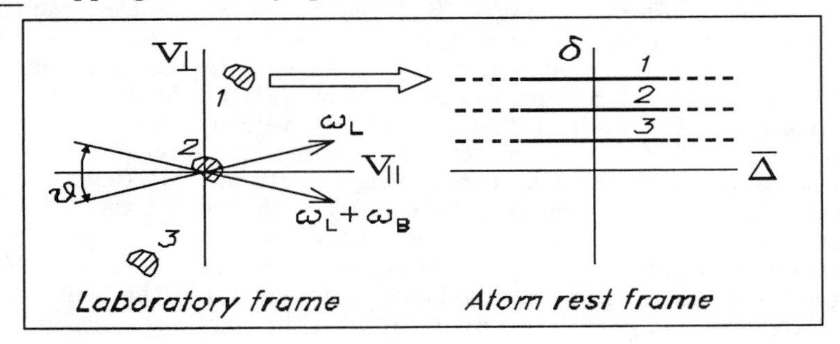

The composite modulation and phase-shift are obtained by integrating the TDPS over the velocity distribution. In the case of Fig.1, the detailed structure of the distribution is irrelevant, and the resulting integration $\int_{-\infty}^{\infty} S(\bar{\Delta},\delta,\tau)d\bar{\Delta}$ may be performed analytically to give (see Fig.2a also):

$$M' = (1+\delta^2)^{-\frac{1}{2}}; \qquad \Phi' = \tan^{-1}(\delta) = \tfrac{1}{2}\Phi(\bar{\Delta}=0). \tag{5}$$

As it turns out, the same results are obtained by neglecting the atomic excitation response, and assuming that M' and Φ' are merely the convolution of the sinusoidally varying excitation field with the exponential fluorescence decay. Thus the results of Ref.1 and the usual assumptions of frequency domain techniques for measuring fluorescence lifetimes are reproduced.[4] It must be emphasized that for a distribution with negligible spread in $v_{\|}$, different results are obtained, as indicated by the dotted line in Fig.2a (obtained by setting $\bar{\Delta} = 0$ in Eq.3).

STRONG FIELD

To treat the strong field case,[4] we write the optical Bloch equations in terms of the inversion R_z and the polarization components R_1, R_2, as:

$$\begin{pmatrix} \dot{R}_z \\ \dot{R}_1 \\ \dot{R}_2 \end{pmatrix} = \begin{pmatrix} -2 & 2\frac{|V|}{\gamma}\cos\delta\tau & 0 \\ -2\frac{|V|}{\gamma}\cos\delta\tau & -1 & \bar{\Delta} \\ 0 & -\bar{\Delta} & -1 \end{pmatrix} \begin{pmatrix} R_z \\ R_1 \\ R_2 \end{pmatrix} + \begin{pmatrix} -2 \\ 0 \\ 0 \end{pmatrix}. \quad (6)$$

The rotating wave approximation has been made, and radiative damping has been assumed for simplicity. Indeed, the weak field results obtained from the TDPS are reproduced. In general, a Floquet expansion leads to a solution in terms of continued fractions.[5] By numerical evaluation, we again obtain the modulation and phase shift (see Fig. 2b). For broad velocity distributions, it is particularly interesting that many features of the Rabi structure at the modulation component remain, including the positions of the resonances. This feature should alleviate the requirement of ultracold atomic beams[6] in fundamental studies of the field–atom interaction.

Figure 2: Depth-of-modulation and phase shift for increasing interaction strengths, in the limits of narrow (···) and broad(–) velocity distributions.

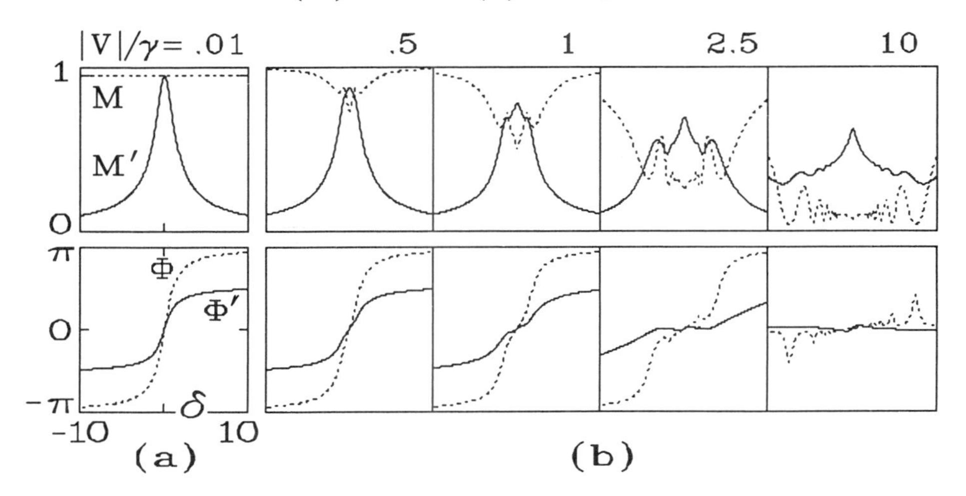

REFERENCES

1. D. Keefer, Appl. Opt. **26**, 91 (1987).
2. L. Davis, Opt. Soc. Am. 1987 annual meeting, paper TUI3.
3. J.H. Eberly, K. Wódkiewicz, J. Opt. Soc. Am. **67**, 1252 (1977).
4. R. Saxena, G.S. Agarwal, J.Phys.B, **12**, 1939 (1977).
5. L.W. Hillman *et al.*, J. Opt. Soc. Am. **B2**, 211 (1985).
6. H. Chakmakjian *et al.*, Opt. Soc. Am. 1987 annual meeting, paper WF8.

OBSERVATION OF ANTI-STOKES-STIMULATED RAMAN SCATTERING AND TWO-PHOTON EMISSION IN LITHIUM AND SODIUM VAPORS

F.Z. Chen and C.Y. Robert Wu
University of Southern California, Los Angeles, CA 90089-1341

T.S. Yih and H.H. Wu
National Central University, Chung-Li, Taiwan 32054, ROC

Y.C. Hsu and K.C. Lin
Institute of Atomic and Molecular Sciences
Academia Sinica, Taipei, Taiwan, ROC

ABSTRACT

We have observed anti-Stokes-stimulated Raman scattering (ASRS) and two-photon emission (TPE) in Li and Na metal vapors. We found that the TPE can be generated from most of the excited even-parity states of both atoms but not ASRS. In the present work, the "pump laser lines" which stimulate the ASRS and TPE are those optically-pumped stimulated emissions (OPSE) generated in the same alkali metal vapor. Since ASRS and TPE processes take place from the same metastable state, strong competitions between both processes are expected. Enhanced outputs due to the presence of a resonant intermediate state in the ASRS and TPE processes have also been observed.

INTRODUCTION

The metastable states of atoms and molecules have been considered as energy storage media,[1] two-photon laser amplifier,[2-4] and frequency up-conversion and down-conversion.[5] If the metastable states are coherently prepared by two-photon resonant excitation, then the overall mechanism falls into the general scheme of four-wave mixing, namely, sum- and difference-frequency generation, third harmonic generation (THG) and parametric oscillations, etc. In the present discussion, the stimulated emissions generated from the interaction of an injected laser beam ω with metastable states (at ω_m above the ground state) are termed[3] anti-Stokes-stimulated Raman scattering (ASRS) ω_A and two-photon emission (TPE) ω_T. The relations can be given by:

$$\omega_A = \omega_m + \omega \tag{1}$$

$$\omega_T = \omega_m - \omega \tag{2}$$

Several cases with simplified energy levels are illustrated in Fig.1. Toschek and his colleagues[4,6] have experimentally studied the two-photon emissions as shown in Figs.1(a) and (d). Carman[3] reported a theoretical study on both ASRS and TPE of case (a) and (d). Narducci et al[2] have discussed the propagation of two-photon amplifier.

In many systems there exist intermediate state(s) which may enhance and thus favor one of the competing processes, e.g., Fig.1(b) and 1(c). We have encountered many such cases in the Li and Na metal vapors. The experimental setup has previously been

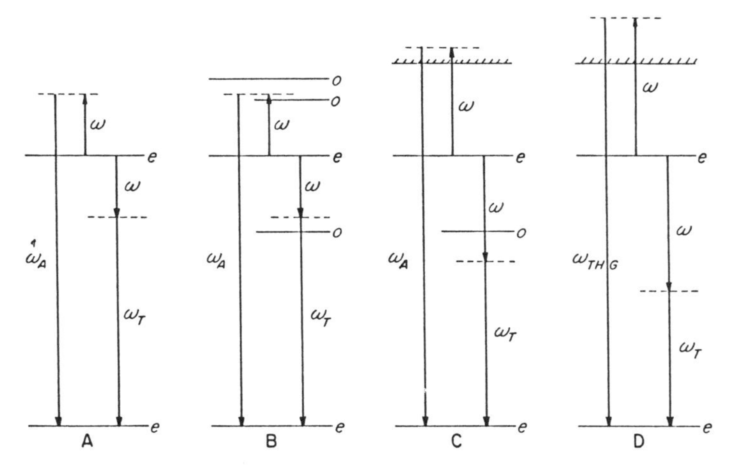

Figure 1. The typical ASRS and TPE schemes. (e and o for even and odd parity, respectively.)

Table 1. Observation of ASRS and TPE in the Li and Na Metal Vapors

Atom	Metastable State	ω	Obs (Å)	Cal (Å)	T°K	Buffer gas/ Pressure(Torr)
Li	4s	ω_{2p-3s}	- 4402	2113 4402	850-1,050	He/4
Na	5s	ω_{4s-4p}	2649 3486	2650 3487	690	He/8
		ω_{5s-4p}	2767 3302	2767 3303	690	He/8
		ω_{3d-4p}	2915 3115	2915 3114	690	He/8
		ω_{3p-4s}	- 4092	2382 4093	450-580	He/8
	4d	ω_{3d-4f}	- 3430	2502 3432	450-580	He/8
		ω_{4s-4p}	- 3330	2558 3331	450-580	He/8
		ω_{4p-4d}	- 3302	2575 3303	450-580	He/8
		ω_{3d-4p}	2805 2988	2805 2989	740-930	Ar/50

described in detail.[7]

RESULTS

Table 1 summarizes the results. The metastable states, the heatpipe temperature, the buffer gas and its pressure are included. The injected wave is the optically pumped stimulated emission (OPSE) generated in the same metal vapor. The observed wavelengths and the expected ones calculated according to Eqs.(1) and (2) are also listed in the Table. The upper value refers to the ASRS while the lower value to the TPE. Many ASRS were searched for but were not detected. The detection limit is mainly due to the noise levels in the present experimental setup. We have noted the following observations:

- The thresholds for producing TPE is lower than those for the corresponding ASRS.
- The output of TPS is stronger than the corresponding ASRS before saturation effects become important.
- In the low vapor density regime, the intensity of TPE grows faster than that of ASRS as a function of metal vapor density.
- The ASRS output exhibits approximately a I^2 dependence whereas the TPE shows a less than ideal power square low. This suggests that the power saturation for the TPE may be considerable lower than that for the ASRS.
- The quantum efficiency for generation of ASRS decreases as ω increases.

CONCLUSION

We have found that TPE always dominates in the cases studied. This is because the TPEs are enhanced by resonant intermediate states which have large transition moment. As a special case, the generation of TPE at 3302Å in Na has a high quantum efficiency[8] of 5×10^{-2}.

ACKNOWLEDGMENTS

This work was partially supported by the USC Faculty Research and Innovation Fund.

REFERENCES

1. e.g., D.J. Ehrlich and R.M. Osgood, Jr., IEEE QE-15, 301 (1979).
2. L.M. Narducci, W.W. Eidson, P. Furcinitti, and D.C. Eteson, Phys. Rev. A 16, 1665 (1977) and references theirin.
3. R.L. Carman, Phys. Rev. A 12, 1048 (1975).
4. B. Nikolaus, D.Z. Zhang and P.E. Joschek, Phys. Rev. Lett. 47, 171 (1981).
5. e.g., L.J. Zych, J. Lukasik, J.F. Young and S.E. Harris, Phys. Rev. Lett. 40, 1493 (1978).
6. K.-J. BOLLER, H.-J. Lau, J. Sparbier and P.E. Tosehek, Bulletin Amer. Phys. Soc. 32, 1625 (1987), paper TuL7.
7. H.H. Wu, T.C. Chu and C.Y.R. Wu, Appl. Phys. B 43, 225 (1987).
8. W. Hartig, Appl. Phys. 15, 427 (1978).

NEAR IR EMISSION IN SODIUM VAPOR UNDER MULTIPLE EXCITATIONS

T.S. Yih, H.H. Wu
National Central University, Chung-Li, Taiwan 32054, R.O.C.

Y.C. Hsu, K.C. Lin
Academia Sinica, Taipei, Taiwan, R.O.C.

INTRODUCTION

Here we use a tunable yellow coherent source to pump sodium vapor and generate an intense near ir radiation at about 840.3nm which might be due to an atomic nonlinear optical wave-mixing process. A similar result[1] in lithium has been reported before by Nikolaus et al.

EXPERIMENTAL SET-UP

The experimental set-up is similar to that reported before[2] except a Lambda Physik EMG202-C excimer laser is used to pump a FL3002E dye laser. A Rhodamine 6G dye solution is used in the dye laser. Sodium is confined in the crossed heatpipe with argon buffer gas pressure up to about 50 Torr. The trasmitted uv and ir emissions are focused by a quartz f=200mm lens into the Acton Research Corp. 0.5m monochromator. A Hamamatsu R666 multialkali cathode PMT detects the dispersed emissions and converts them to electric current collected by Keithley 404A picoammeter. The spectrum then is recorded by a HP7015B x-y recorder.

RESULTS AND DISCUSSION

Fig.1 shows the wavelength dispersed spectrum of this near ir emission around 840nm under various laser wavelength pumping. The peaks move into the shorter wavelength region as the pump laser wavelength gets longer.

Listed in Table I are the near ir emission peak wavelengths, laser wavelengths and the calculated summation of their frequencies. The pump laser wavelength 578.7nm is near the Na 3s-4d two photon absorption resonance. This two photon absorption process is enhanced by the presence of the intermediate state 3p which is $301cm^{-1}$ less than that of the laser photon. The 4d and 4p states could be also populated through Raman process[2,3] state mixing under laser induced plasma field[4] and molecular enhancement[2,5] as well as collision assisted processes.[4] The succesive decay to 3d state is obviously a strong process since the 3d-3p transition appears in high intensity in the near ir spectrum. We assume then the high power laser beam induces the 4p-3d stimulated emission which mixes with the laser photons to induce a two photon emission process from 3d state. Wang et al.[6] recently suggest this is a six-wave mixing process. Our results have no discrepancy with their interpretation.

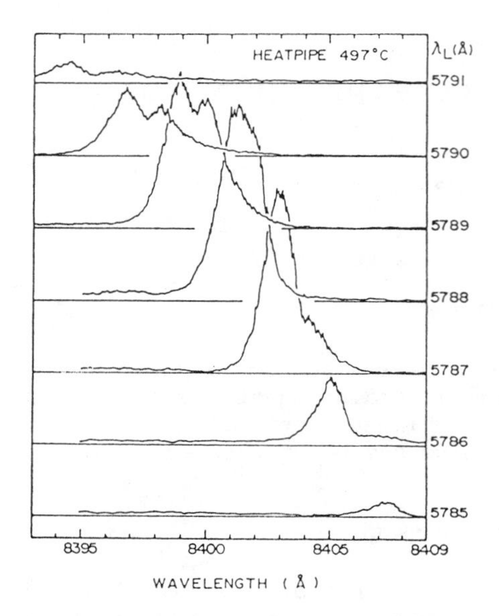

Fig.1 Dispersed spectrum of the near ir emission under several laser wavelength excitations. Sodium density is about 4.6×10^{16} cm^{-3} and the monochromator bandwidth is 0.05nm.

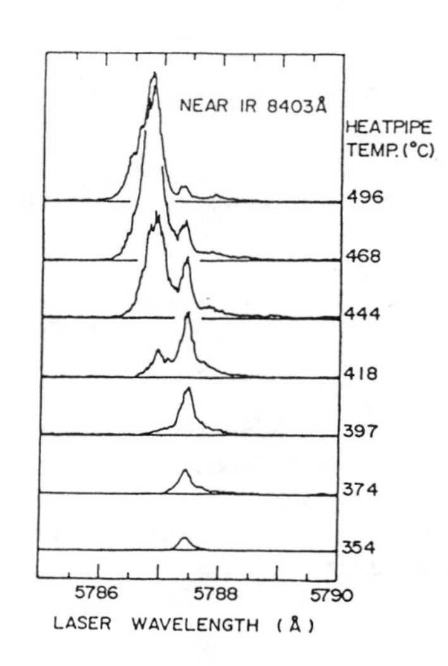

Fig.2 Excitation function of the 840.3nm emission at various heatpipe temperatures. The excitation function is not normalized to the laser output power.

Fig.2 shows the excitation function of this 840.3nm emission at various heatpipe temperatures. For heatpipe temperatures up to 374°C there is only one effective excitation peak around 578.74nm. For temperatures higher than 397°C the peak of the excitation function gets higher and extends to shorter wavelengths. There is an absorption dip at 578.72nm which corresponds to the 3s-4d two photon resonance. When the pump laser is tuned to this wavelength, there are many other nonlinear processes which outweigh the 840.3nm generation mechanism such as the very intense ultraviolet emissions at 330nm[2,7] and 343nm[2] have been observed. The fact that the peaks get wider at higher tempearature might be due to collisions between sodium atoms since its vapor density is higher. Detailed study of the buffer gas pressure dependence is planned.

The power dependence has a slope of about 3 which may suggest the involvement of photons in other competing nonlinear processes. Above the density of $3 \times 10^{16} cm^{-3}$ the absorption of the pump beam is sufficient to cause the ir generation process to be reduced. The density dependence of this ir emission has a slope of about 3 and could be explained as one Na(3s) atom involved in populating the 4d state or 4p state, and the final nonlinear wave-mixing process gives a quadratic dependence on the Na(3s) density.

Table.I. Near ir emission frequency in sodium vapor at density of about 4.6×10^{16} cm^{-3}.

λ_{Laser} (nm)	λ_{IR} (nm)	$\omega_{Laser} + \omega_{IR}$ (cm^{-1})
578.47	840.75	29172.66
578.57	840.51	29173.07
578.68	840.30	29172.76
578.69	840.25	29173.10
578.79	840.15	29171.60
578.89	839.89	29172.23
578.99	839.44	29172.29
579.07	839.44	29173.31

CONCLUSION

We have observed this near ir emission at 840nm which is tunable associated with the changes of pump laser wavelength. It is found that this photon energy is in complement with the laser photon energy to give the sodium 3d-3s energy difference. Intensity of this near ir emission is about an order of magnitude higher than other four wave mixing generated uv emissions such as 280nm and 298.9nm.[8] The high resonance effect occurs at 578.7nm indicating that this nonlinear mixing process is actually enhanced by the increased population of Na(3d) which is indirectly evidenced by the increased population of Na(4p) at this laser wavelength.[7,9]

ACKNOWLEDGEMENT

We thank the referee who has acquainted us with the recently published work by Wang et al.[6] Also we thank National Science Council of R.O.C. for financially support this study under the grant #NSC76-0208-M008-23.

REFERENCES

1. B. Nikolaus, D.Z. Zhang and P.E. Toschek, Phys. Rew. Lett. 47, 171 (1981).
2. C.Y.R. Wu, J.K. Chen, D.L.Judge and C.C. Kim, Proc. of the Lasers'83, ed. by R.C. Powell, 641 (1983).
3. C.Y.R. Wu and J.K. Chen, Proc. of the Lasers'85, 20, (1985).
4. T.B. Lucatorto and T.J. McIlrath, Phys. Rev. Lett. 37, 428 (1976).
5. S.G. Dinev, I.G. Koprinkov and I.L. Stefanov, Appl. Phys. B39, 65 (1986).
6. Z.G. Wang, H. Schmidt, and B. Wellegehausen, Appl. Phys. B44, 41 (1987).
7. W. Hartig, Appl. Phys.15, 427 (1978).
8. T.S. Yih, H.H. Wu, Y.C. Hsu and K.C. Lin, to be submitted.
9. J.R. Taylor, Opt. Comm. 18,504 (1976).

TRANSIENT RESPONSE OF A SATURATED TWO-LEVEL SYSTEM IN ABSORPTION AND STIMULATED EMISSION TO A WEAK PROBING FIELD

C. J. Hsu
University of New Brunswick, Saint John Campus
Saint John, N.B., Canada E2L 4L5

ABSTRACT

The transient response of a resonant saturated system in steady state to a weak probing near resonant field was analyzed by using density metrix formulism. We found that this perturbation alters the saturation system significantly, which exhibits absorption and stimulated emission with decaying and modulation related to the Rabi frequency due to the strong field.

When a saturated system in steady state interacts with a weak probing field, beside the strong saturating resonant field, the interaction is treated as a perturbation to the saturated system. In this paper I shall report a theoretical investigation of the transient effect from a resonantly saturated atom in response to the switch-on of a weak perturbing field. Since the strong resonant field will redress[1] the atom, upon the application of weak field, the transient effect of this origin can be sizeable and the saturated system is not stable.

The atomic system in consideration has two levels designated as $|0>$ (lower) and $|1>$ (upper), and the electric fields are given classically.

$$\vec{E} = \vec{E}_R + \vec{E}_p$$

$$\text{and} \quad \vec{E}_R = [\tfrac{1}{2}\epsilon_R e^{i\omega_R t} + \text{c.c.}]\hat{e}_1 \tag{1}$$

$$\vec{E}_p = [\tfrac{1}{2}\epsilon_p e^{i\omega_p t} + \text{c.c.}]\hat{e}_2$$

$\vec{E}_R$ and $\vec{E}_p$ are the strong resonant field and weak probing field respectively; C.C.'s are complex conjugates, and $\hat{e}_1$ and $\hat{e}_2$ ar unit vectors. The amplitudes are constants. In a dipole approximation, the equation of motion of the density matrix of the atom can be given as follows:

$$(\frac{d}{dt} + \gamma + i\omega_{10})\,\rho_{10} = \frac{i\vec{u}_{10}}{\hbar}\cdot\vec{E}\,(\rho_{00} - \rho_{11}).$$

$$(\frac{d}{dt} + \gamma')\,\rho_{00} = \gamma'\bar{\rho}_{00} + \frac{i\vec{u}_{10}}{\hbar}\cdot\vec{E}\rho_{10} + \text{c.c.} \tag{2}$$

$$(\frac{d}{dt} + \gamma')\,\rho_{11} = \gamma'\bar{\rho}_{11} - \frac{i\vec{u}_{10}}{\hbar}\cdot\vec{E}\rho_{01} + \text{c.c.}$$

Where the notations are conventional[1]. Let us consider E_p giving rise to the perturbation and the first order equation can be obtained. For simplicity, we assume the atoms are stationary and $\gamma=\gamma'$. In a similar procedure[1], in a rotating wave approximation, the equation for the amplitudes of the density matrix elements can be given as:

$$
\begin{aligned}
&\left(\frac{d}{dt} - i\Delta\omega' + \gamma\right) \Delta\rho_{01}^{(1)} - 2i\Omega_0^{*} \rho_{10}^{(1)}(t,\omega) + 2i\Omega_0 \rho_{01}^{(1)}(t, 2\omega_R-\omega) = 2n\Omega_p \Omega_0^{*}/z^{*} \\
&\left(\frac{d}{dt} + i(\Delta\omega' + \Delta\omega) + \gamma\right) \rho_{10}^{(1)}(t,\omega) - i\Omega_0 \Delta\rho_{01}^{(1)} = i\Omega_p n \qquad (3)\\
&\left(\frac{d}{dt} - i(\Delta\omega' - \Delta\omega) + \gamma\right) \rho_{01}^{(1)}(t, 2\omega_R-\omega) + i\Omega_0^{*}\Delta\rho_{01}^{(1)} = 0
\end{aligned}
$$

$$\text{where}\quad \Omega_0 = \frac{u_{10}\epsilon_R}{2\hbar} \qquad \Omega_p = \frac{u_{10}\epsilon_p}{2\hbar}$$

$$\Delta\omega' = \omega_p - \omega_R \qquad \Delta\omega = \omega_R - \omega_{10} \qquad z = -i\Delta\omega + \gamma$$

$$n = \rho_{00}^{(0)} - \rho_{11}^{(0)} = (\bar{\rho}_{00} - \bar{\rho}_{11})/[1 + 4|\Omega_0|^2/(\Delta\omega^2 + \gamma^2)]$$

It is worth noting that, these equations contain not only the perturbation terms due to probing field, but also terms from the interaction with the strong saturating field which is thus redressing the system. We shall show that the probing field may be absorbed or amplified due to the redressing of the strong resonant field. The change of the probing field is assumed to be moderate and we shall adopt independent field approximation (IFA)[3]. We shall see later that this redressing triggered by the probing field alters the saturated system significantly. The term $\rho_{10}^{(1)}(t,\omega)$ gives rise to absorption and stimulated emission through the association with the oscillation of population difference and four-photon mixing. By employing Laplace transform, we have obtained a steady state solution equivalent to Mollow's[3] and a transient solution for $\rho_{10}^{(1)}(t,\omega)$. The imaginery part of this transient is equal to $n\Omega_p \cdot \mathrm{Real}(F_t)$. In the limit of $\Omega_0 \gg \gamma$, the function Real (F_t) is given as:

$$
\begin{aligned}
\mathrm{Real}(F_t) = (-\tfrac{1}{2})\, e^{-\gamma t} \Big[&\frac{\gamma \cos\Delta\omega' t - \Delta\omega' \sin\Delta\omega' t}{\Delta\omega'^2 + \gamma^2} |Z_0| \\
&+ \frac{\gamma \sin(\Delta\omega'+\Omega)t - (\Delta\omega'+\Omega)\cos(\Delta\omega'+\Omega)t}{(\Delta\omega'+\Omega)^2 + \gamma^2} |Z_+| \qquad (4)\\
&+ \frac{\gamma \sin(\Delta\omega'-\Omega)t + (\Delta\omega'-\Omega)\cos(\Delta\omega'-\Omega)t}{(\Delta\omega'-\Omega)^2 + \gamma^2} |Z_-| \Big]
\end{aligned}
$$

where $Z_0 = \frac{4|\Omega_0|^2}{\Omega^2} \frac{\gamma}{z^*}$; $Z\pm = (1 \pm \frac{\Delta\omega}{\Omega}) - \frac{2|\Omega_0|^2}{\Omega^2} \cdot \frac{\mp i\Omega + \gamma}{z^*}$

and $\Omega = \sqrt{4|\Omega_0|^2 + \Delta\omega^2}$.

A gain function[2] can be found for a typical heterodyne experiment, thus we have:

$$g = \frac{4\pi N}{\hbar} \frac{\omega_p}{c} n|\mu_{10}|^2 .l.\text{Real}\ (F_t)$$

where N is the number density of atoms, and l is the interaction length. This function describes the variation of absorption (or stimulated emission) with respect to time, frequency detuning and the Rabi frequency due to the strong saturating field. The following are the important features in this investigation.

(1) There are three characteristic frequencies in this study. Since the probing field may be detuned so that $\Delta\omega' \gg \gamma$, and E_p may be allowed to be increased to a value comparable to E_R without invalidating the perturbation, and these signals can therefore be comparable to ordinary nutation signal.

(2) The line shape of the first component is resembling the Ramsey line shape. This shape has been studied in time delay spectra.[2] However, the other two components are different from a typical Ramsey shape, because of phase leading and phase lagging of approximately $\pi/2$ in the second and the third terms respectively, which gives rise to the dispersive part to be dominated at the beginning so that the detuning near the value $\pm\Omega$ is very sensitive to the switch-on of the probing field at an earlier stage.

Besides the transient effect in absorption and emission, the investigation of the related four-photon mixing, which has been of interest to many researchers, in the transient regime is underway.

REFERENCES

1. C.J. Hsu, Optics Lett. 7, 102 (1982).
2. M. Ducloy, J.R.R. Leite and M. Feld, Phys. Rev. A17, 623 (1978).
3. B.R. Mollow, Phys. Rev. A5, 2217 (1972).

WAVE-MIXING PROCESSES IN SODIUM-POTASSIUM VAPOR

B. K. Clark, M. Masters and J. Huennekens, Lehigh University, Bethlehem PA 18015

ABSTRACT

We report observations of 4-wave and 6-wave mixing processes based upon two-photon pumping of the potassium 4S → 6S and 4S → 4D transitions in a sodium-potassium mixture. Additionally we observe coherent emissions at the potassium 3D → 4P transition frequencies for pump wavelengths between 725 and 760 nm. The excitation spectra for these emissions show several broad peaks, each of roughly 10 nm width. A series of sharp dips is superimposed on the most intense peak due to depletion of the 3D population by laser-induced absorption to Rydberg F levels. Although the 3D → 4P emissions can be observed in both the forward and backward directions, they are ~50 times more intense in the forward direction. We believe the forward emission is primarily due to 6-wave mixing while the backward emission can most likely be attributed to amplified spontaneous emission following two-photon molecular photodissociation. However, at present, the exact mechanisms responsible for these 3D → 4P emissions are not fully understood.

THE EXPERIMENT

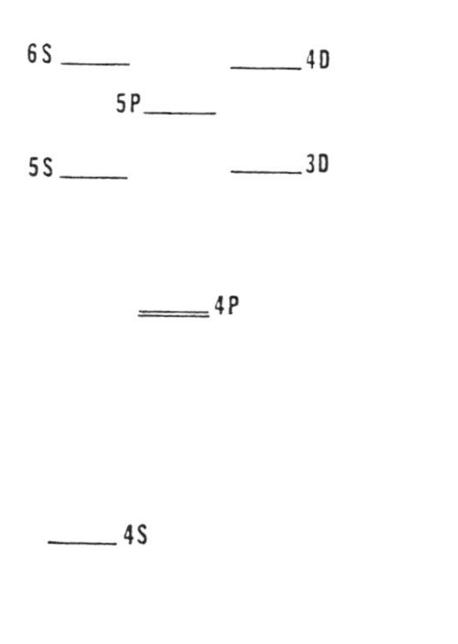

Fig. 1 - Potassium energy levels

The experiment was carried out in a crossed heat-pipe oven containing a 2:1 mixture of potassium and sodium and operated at ~435°C with an argon buffer gas pressure of ~2 Torr. The alkali vapor was excited with a Nd:YAG pumped dye laser operating in the 690 to 760 nm range with LDS 722 dye. The laser typically delivered ~10 mJ in a 10 ns pulse, and the beam was in all cases unfocussed. Forward, backward and side fluorescence was imaged onto a monochromator equipped with either a photomultiplier or an intrinsic germanium detector.

RESULTS AND DISCUSSION

With the dye laser wavelength tuned to the 4S → 6S two-photon transition at 728.4 nm we observed K_2 and NaK molecular fluorescence in the side direction along with various sodium and potassium atomic lines. In the forward direction we observed coherent emission at the potassium 5P → 4S, 3D → 4P and 5S → 4P transition wavelengths (see Fig. 1) and at 1280 and 1370 nm which do not correspond to any atomic transitions. In the backward direction, only the 3D → 4P and 5S → 4P emissions were observed. The 5P → 4S forward emission was also observed when pumping the 4S → 4D two-photon transition. The various coherent emissions can be understood as resulting from the following processes:

4-wave mixing

$$\omega_{5P\to4S} = 2\omega_{laser} - \omega_{6S\to5P(4D\to5P)} \quad , \quad 2\omega_{laser} = \omega_{4S\to6S(4S\to4D)}$$

6-wave mixing

$$\omega_{1280(1370)} = 2\omega_{laser} - (\omega_{6S\to5P} + \omega_{5P\to3D(5P\to5S)} + \omega_{laser})$$

amplified spontaneous emission (ASE)

$$\omega_{3D\to4P(5S\to4P)} = 2\omega_{laser} - (\omega_{6S\to5P} + \omega_{5P\to3D(5P\to5S)} + \omega_{4P\to4S})$$

The ASE emissions are characterized by approximately equal intensities in the forward and backward directions, while the wave-mixing processes result only in forward emission due to phase-matching requirements. These various processes have either been observed previously in potassium vapor,[1-2] or can be understood by analogy to similar observations made in other alkalis.[3-5]

We also observe strong forward and backward emission at 1170 and 1180 nm (potassium 3D → 4P) when pumping the vapor over a broad range between 725 and 760 nm. The excitation spectrum for the 1170 nm line is shown in Fig. 2. The forward to backward ratio is approximately 55 for a pump wavelength of 752 nm and an argon pressure of 1.8 Torr. As can be seen, the excitation spectrum consists of 3 broad peaks: a very strong one centered at 752 nm, a somewhat weaker one at 741 nm, and a very weak one at 730 nm. The most intense peak has a series of sharp dips superimposed.

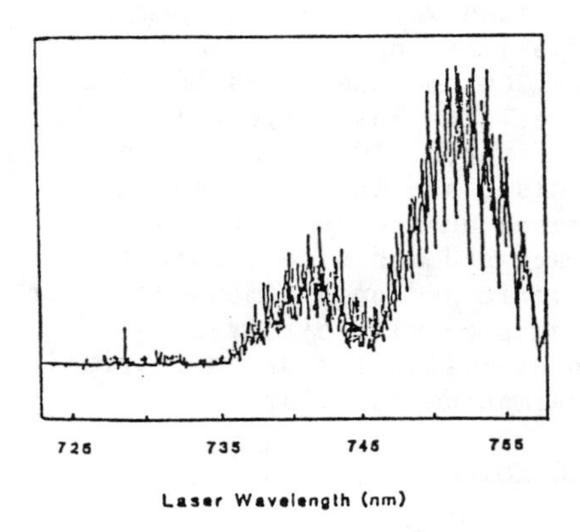

Fig. 2 - 3D → 4P excitation spectrum

Measurements of the 1170 nm forward and backward intensity as a function of argon pressure and laser power were carried out. As the pressure increased from 2 to 25 Torr, the forward signal strength decreased by about a factor of 2 for a pump wavelength of 752 nm. Over the same range the 742 nm peak almost completely disappeared. Similar but not identical observations were made for the backward emission. Laser intensity dependences showed log-log slopes of ~.7 and 1.3 for the backward and forward 1170 nm emission, respectively. These observations support the idea that the forward and backward emission are caused by different mechanisms.

An investigation of the positions of the sharp dips in the 752 nm peak of the excitation spectrum show that they coincide with 3D → nF transitions with n ranging from 21 to 28. Thus at these pump wavelengths the laser depopulates the 3D state by inducing transitions to high-lying Rydberg states, and thereby reduces the 1170 nm emission. Similar excitation spectra for the 1180 nm line, which probe the $3D_{5/2}$ fine-structure level, show dips shifted by the expected 2 cm^{-1}.

It is worth noting that the 3D → nF dips are significantly more pronounced in the backward direction. This reflects the greater sensitivity of the 3D - 4P population inversion required for ASE compared to total 3D population responsible for forward wave mixing. We now believe the 3D → 4P backward emission can be explained as ASE probably following a two-photon molecular photodissociation process. The forward emission is most probably due primarily to 6-wave mixing involving 2 laser photons and one photon at 1170 or 1180 nm. Also involved in the process are 4P → 4S resonance photons (which are highly trapped and therefore unobservable), and two IR photons which lie beyond our detector response.

We must emphasize that these explanations are still somewhat speculative at present, and additional data is needed before final conclusions can be reached.

ACKNOWLEDGEMENTS

We thank R. N. Compton, W. R. Garrett and M. A. Moore for several helpful discussions and suggestions concerning this work. This work was supported by the Army Research Office under grant DAAL03-86-K-0161.

REFERENCES

1. P. Agostini, P. Bensoussan and J. C. Boulassier, Op. Commun. 5 293 (1972).
2. P.-L. Zhang, Y.-C. Wang and A. L. Schawlow, J. Opt. Soc. Am. B 1 9 (1984).
3. W. Hartig, Appl. Phys. 15 427 (1978).
4. Z. G. Wang, H. Schmidt and B. Wellegehausen, Appl. Phys. B 44 41 (1987).
5. S. M. Hamadani, J. A. D. Stockdale, R. N. Compton and M. S. Pindzola, Phys. Rev. A 34 1938 (1986).

Nonlinear Optical Properties of Strained GaAs-$In_xGa_{1-x}As$ Multiple Quantum Well Structures Using a Titanium Sapphire Laser Probe

K. Aron, G. Hansen, R. Stone, R. Lytel and W.W. Anderson

Lockheed Palo Alto Research Laboratory
3251 Hanover Street, Palo Alto, California 94304-1191

ABSTRACT

The nonlinear saturation of the excitonic absorbtion in a GaAs-$In_xGa_{1-x}As$ multiple quantum well sample consisting of 80 wells was investigated in the vicinity of the excitonic resonance at 991nm using a tunable titanium sapphire laser probe. This characterization yields a saturation intensity of less than 50 kW/cm^2 and a maximum off-resonance nonlinear refractive index of 6.42×10^{-7} cm^2/W. This dispersion can lead to practical optically bistable etalons fabricated from these multiple quantum well structures.

1. INTRODUCTION

Saturable absorbers have long been used in intracavity and extracavity applications. They also form the basis for an important class of optical bistable devices employing semiconductor multiple quantum well structures (MQWs) operating at room temperatures [1]. The nonlinear optical properties of MQWs have been the subject of much recent interest [2]. In these systems, the binding energy of the free excitons is increased due to the quantum confinement, allowing a sharp atomic-like transition peak to be observed even at room temperature due to the exciton resonance. These exciton transitions have been found to saturate at low light intensities, so that very efficient nonlinear processes have been observed in the vicinity of these resonances. Because of the resonant enhancement of these nonlinear optical properties, $\chi^{(3)}$ values of 10^{-3} esu, or 10^9 times larger than CS_2, have been reported to date.[3]

The work reported herein represents the first nonlinear absorption and saturation experiments in InGaAs-GaAs MQWs. It has been the aim of this study to characterize the nonlinearity of this system in order to obtain design parameters for optically bistable switches fabricated by surrounding the MQW structure with reflectors, thereby creating a Fabry - Perot etalon. The physics of bistability in such etalons is well known and has been especially studied for GaAs-AlGaAs MQWs.[4] The extremely large nonlinear absorption and dispersion arising from semiconductor MQWs permits micron - sized etalons to be fabricated and switched with relatively low intensities (10 - 100 kW/cm^2) and high speeds (1 psec on, 200 psec off), and even operated cw with multimode diode lasers with 100 mW power.

This paper includes both a description of the Titanium-Sapphire (Ti:SAP) laser-based apparatus used in probing the InGaAs-GaAs MQWs as well as a discussion of the MQWs spectra under low power (linear) and saturated conditions. This is followed by a calculation of the nonlinear absorption coefficient and dispersion of the

MQWs from the observed spectral data. It will be shown in future work how these results dictate the design parameters for the InGaAs-GaAs - based Fabry -Perot switches.

EXPERIMENTAL

MQWs samples of strained layer InGaAs-GaAs were provided by Profs. Harry Wieder and W. Chang of the University of California at San Diego (UCSD). This system was selected for our device work because the GaAs on which the MQWs are grown is transparent at the operating wavelength (near 1 micron) and need not be etched away, as in GaAs-AlGaAs structures. In addition, InGaAs-GaAs is free of the deep level traps associated with GaAs-AlGaAs. These samples were tested as received and were not optimized for optical absorption and saturation measurements.

The samples consisted of eighty 200 A thick wells of GaAs-InGaAs (15% indium concentration), or a total MQW length of 1.6 microns. The MQW was capped at one end with 200 A of n-GaAs, and grown on a 500 micron p-GaAs substrate, with a 1000 A GaAs buffer layer between the MQW and the substrate. After receiving the first such samples, it was discovered that the p-GaAs substrate doping density was unusually high --9 x 10^{18} cm^{-3} -- producing an overall absorption coefficient for the sample of 40 - 60 cm^{-1}, and an overall sample absorption loss of 85 - 95%. This high absorption created experimental problems, as is discussed below, and will have to be eliminated by substitution with a more lightly doped substrate before device work is begun.

The high substrate absorption is apparent in Figure 1, where the total sample transmission is only a few percent even after correcting for Fresnel losses. This figure was obtained using a Nd:YAG-pumped Ti:SAP laser. The Ti:SAP laser was tuned using a four plate birefringent tuning element (linewidth 3 A) and could be tuned from 700 nm past 1 micron, with appropriate mirrors. The Ti:SAP output pulse of 80 - 100 nsec FWHM was "sliced" electro-optically to generate a 30 nsec square pulse. This eliminated the need to deconvolute the input laser intensity for a temporal Gaussian input, providing essentially a cw input over a 30 nsec gate. The 1 mm radius (1/e points) laser beam was incident unfocussed onto the MQWs, and deconvolution of the input intensity due to the spatial Gaussian was carried through using standard techniques for nonlinear observables.[5]

The MQWs exhibit a well resolved, saturable room temperature absorption peak at 991 nm due to creation and subsequent ionization of excitons. This is clearly seen in Figure 1 by comparing the low power scan [a] (61 W/cm^2) to the high power scan [b] (6.1 kW/cm^2). At low powers the sharp excitonic transition is apparent below the band gap energy; as the input laser intensity is increased, the transmission at the resonance frequency also increases as the peak recedes into the background. (It is important to note again that, as discussed above, as much as 95% of the light is lost due to absorption in the p-GaAs substrate. It is for that reason that the *total* sample transmission is in the few percent range given in Figure 1.).

Figure 2 details this effect of input laser intensity on the saturation of the exciton absorption. It was generated by increasing the input laser intensity while maintaining the excitation wavelength at the excitonic resonance at 991 nm. Here laser power is plotted against the *saturable* part of the absorption coefficient, having numerically eliminated the background contribution to the absorption coefficient α due to the heavily doped substrate.[6] The saturable absorption measurements were obtained by injecting the input beam into the MQW side of the sample, measuring the intensity transmitted through the entire sample as a function of the input intensity, and calculating the total absorption through the sample at each input intensity. The total unsaturable background absorption of the sample (MQW plus substrate) was then subtracted to produce the figure. Both the MQW and the substrate contribute to the unsaturable background intensity, but the two contributions could not be independently measured or calculated because no sample of the substrate material *alone* was available. For the device fabrications planned, clearly a more transparent (in this case, lightly doped) substrate will be grown.

It is seen from Figure 2 that the saturable contribution to the absorption coefficient has a marked intensity dependence. α_{sat} is seen to be falling steadily even at the lowest input intensities of a few tens of watts per square centimeter. This data must be modeled in order to extract the nonlinear characteristics.

DISCUSSION

The spectra obtained in this experiment yield sufficient information to carry through preliminary calculations of design parameters for optically bistable devices. Future experiments will incorporate a pump/ probe (two laser) experimental scheme, so that the physical processes underlying the excitonic saturation can be more rigorously understood. In addition, with a more lightly doped (transparent) substrate higher input intensities can be used without the danger of the onset of optical damage which constrained these experiments. In spite of these shortcomings, reasonable designs can be obtained.

The data in Figure 2 can be fit using a saturable absorption of the form $\alpha_{sat} = \alpha_1 [1+I/I_{sat}]^{-1}$, where I is the average intensity through the MQW portion of the sample. When $I=I_{sat}$, the absorption is reduced to half of its value at I=0. The fit yields $\alpha_1=3115 \pm 32\ cm^{-1}$ and $I_{sat} = 12.5 \pm 0.9\ kW/cm^2$. This result implies that the maximum off-resonance nonlinear refractive index is $n_2 = 0.026\ \alpha_1\ \lambda/I_{sat} = 6.42 \times 10^{-7}\ cm^2/W$. The value of n_2(max) is smaller than expected for InGaAs-GaAs, but is still seven orders of magnitude larger than the best nonresonant third order materials.

The value of I_{sat} calculated above is also higher than expected. Previous work[7] in similar semiconductor MQWs systems had found I_{sat} values ranging from 70 W/cm^2 to approximately 2.5 kW/cm^2. An approximate form for the excitonic saturation intensity is[1]:

$$I_{sat} = h\omega / 2\pi\, a_x^2\, \tau\, \alpha\, L_z \qquad (1)$$

where $h\omega$ is the photon energy, a_x is the exciton radius, and τ is the lifetime. To account for the high satuaration intensity observed in this work, the exciton lifetime τ would have to be anomalously short. Since there is no reason a priori to expect this to be the case, a definitive value for I_{sat} must await a sample with better substrate transmission properties.

CONCLUSIONS

Nonlinear optical materials may be fabricated into optically bistable switches by surrounding the material with reflectors to make a Fabry - Perot etalon. In future work we will show how the nonlinear properties described herein can de used to design and fabricate such devices from InGaAs - GaAs MQWs.

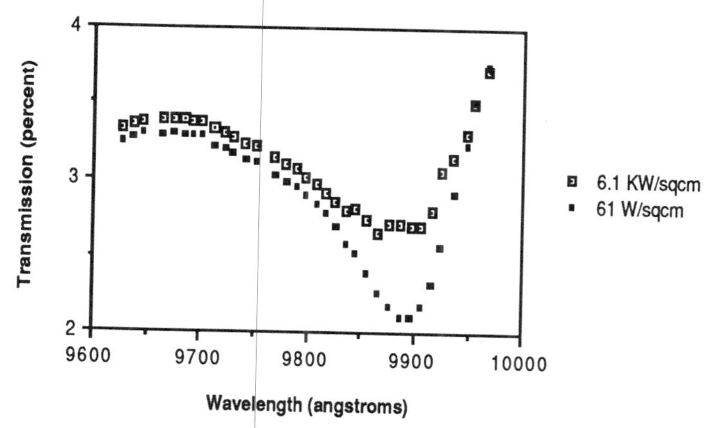

Figure 1. Transmission vs. Wavelength

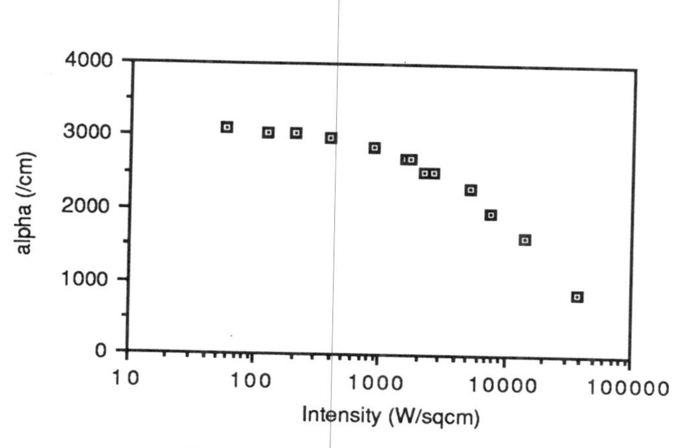

Figure 2. Saturable Absorption vs. Intensity

REFERENCES

1. H.M. Gibbs, Optical Bistability: Controlling Light with Light (Academic, New York, 1985).
2. D.S. Chemla and D.A.B. Miller, J. Opt. Soc. Am. **B 2**, 1155 (1985).
3. D.A.B. Miller, D.S. Chemla, D.J. Eilenberger, P.W. Smith, A.C. Gossard and W.T. Tsang, Appl. Phys. Lett. **41**, 679 (1982).
4. D.S. Chemla, D.A.B. Miller, P.W. Smith, A.C. Gossard and W. Wiegmann, IEEE J. Quant. Elect. **QE-20**, 265 (1984).
5. P. Kolodner, H.S. Kwok, J.G. Black and E. Yablonovitch, Opt. Lett. **4**, 38 (1979).
6. H.C. Casey Jr., D.D. Sell and K.W. Wecht, J. Appl. Phys. **46**, 250 (1975).
7. A.M. Fox, A.C. Maciel, M.G. Shorthose, J.F. Ryan, M.D. Scott, J.I. Davies, and J.R. Rifatt, Appl. Phys. Lett. **51**, 30 (1987) and references therein.

Amplified Spontaneous Emission in Surface-Emitting Semiconductor Lasers

J. Khurgin,
Philips Laboratories,
North American Philips Corporation
Briarcliff Manor, NY, 10510
and
D.A. Davids,
Polytechnic University
333 Jay St. Brooklyn, NY 11201

Abstract

Theories of amplified spontaneous emission in surface-emitting semiconductor lasers are advanced in which spatial and spectral inhomogeneities of gain are taken into account. Experiments to determine the dependence of laser threshold and differential efficiency on the diameter of an excited circular region demonstrate good agreement with the threshold predictions of the linear theory, however, the lack of agreement in differential efficiency suggest that a nonlinear phenomenological theory be considered. Data for such calculations are provided from experiments on saturation of spontaneous emission. The resultant nonlinear theory accounts well for the observed dependence of differential efficiency on the diameter of the excited region.

I Introduction

Attempts have been made by Basov, et. al. [1] and Lavrushin [2] to quantitatively predict the influence of Amplified Spontaneous Emission (ASE) on the threshold of lasing in surface-emitting electron-beam pumped semiconductor lasers. Both studies employed a one-dimensional model with either plane or cylindrical waves.

A two-dimensional approach is reported by Bogdankevitch et. al. [3] in which it is assumed that ASE is formed by zig-zag rays caused by multiple reflections of the spontaneous radiation between resonator mirrors, rather than by lateral waves as in references 1 and 2. This approach yields a lower estimate of the influence of ASE on the threshold of lasing which is closer to experimental observation, but the model is not physically meaningful in that it is not reasonable to neglect lateral waves in transverse directions as they experience the highest gain.

Experiments conducted by Packard et. al. [4] show much weaker dependence of the threshold of lasing on the diameter of the excited region than predicted in reference 1. Also, Packard et. al. [5] report that the lasing threshold had been achieved for diameters of excited region that according to one-dimensional theory of reference 2 make lasing impossible.

A possible explanation of the discrepancies in the articles cited is that besides not using a three-dimensional model, none of them account for the spectral shapes of gain and spontaneous radiation. It is assumed that all the spontaneous radiation emitted is amplified with the same value of gain-per-unit-length, equal to the value at threshold. In fact, the value of gain is maximum at the wavelength of lasing and drops off quite sharply at the shorter wavelengths, where the spontaneous radiation is concentrated. Thus, taking into account the spectral dependencies of gain and spontaneous radiation should provide better agreement with experiment.

II Linear Theory of ASE

Consider the geometry of the laser region shown in Fig.1. It is a cylinder with radius R and height H. The cylinder is pumped axially with a pumping power density $P(r)$. The balance equations for the photon density in the laser mode $I_\lambda(r)$ and electron-hole pair density $N(r)$ above threshold are

$$\frac{dI_\lambda(r)}{dt} = I_\lambda(r)\times[cG(E_\lambda,N(r)) - \frac{1}{\tau_c}] = 0 \tag{1.a}$$

$$\frac{dN(r)}{dt}=P(r)-\frac{N(r)}{\tau}-c\sum_\lambda G(E_\lambda,N(r))I_\lambda(r)-c\int_0^\infty G(E,N(r))I_{ase}(r,E)dE=0 \tag{1.b}$$

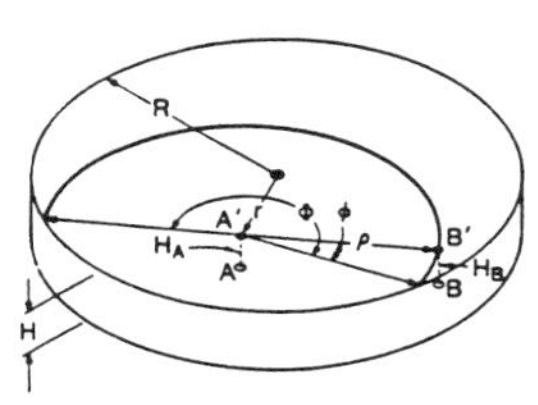

Figure 1 The geometry of the excited region of a surface-emitting laser

where c is the speed of light; $G(E,N(r))$ is a density dependent gain at energy E; E_λ is an energy corresponding to the laser mode with wavelength λ; τ is the spontaneous lifetime of electron-hole pairs; τ_c is the photon lifetime in the laser cavity; and $I_{ase}(r,E)$ is the photon number density of amplified spontaneous emission at energy E. The total output power density of the laser is given by

$$I_{out} = ½cTh\nu \sum_\lambda I_\lambda = \frac{T}{T+L}\, h\nu H\, [P(r) - P_{th}(r)], \tag{2}$$

where $P_{th}(r)$ is the threshold power density, and L internal loss per pass. It follows that

$$P_{th}(r) = P_{th}^0 \times \gamma(r) \tag{3}$$

where

$$P_{th}^0 = \frac{(L+T)}{2H\sigma(E_\lambda)\tau} \tag{4}$$

is the threshold power density for the small active region where the influence of amplified spontaneous emission is negligible and

$$\gamma(r) = 1 + c\tau\int_0^\infty \sigma(E)I_{ase}(r,E)dE \tag{5}$$

is the factor that represents the increase in the threshold power density due to the presense of amplified spontaneous emission.

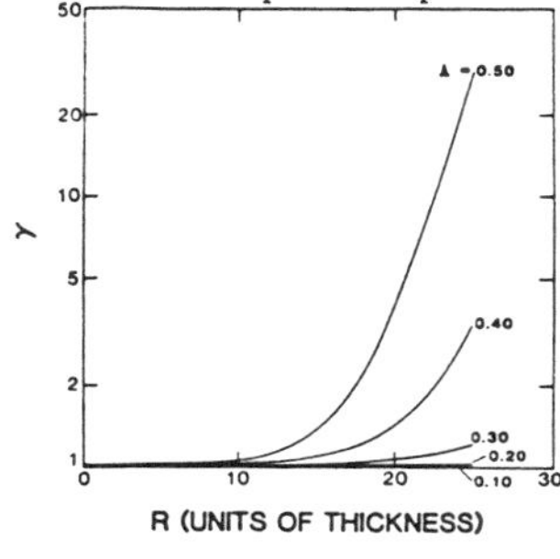

Figure 2 Relative increase in threshold power due to ASE as a function of radius-to-thickness ratio for various values of the longitudinal gain Λ.

The factor $\gamma(r)$ can be separated into spectral and geometrical components given by

$$\gamma(r) = 1 + F_E \times F_{r,R}, \tag{6}$$

where the spectral factor is

$$F_E = \int_{E_g}^{\Delta E_f} \kappa(E)g(E)\frac{e^{\Lambda R g(E)}}{e^{\Lambda R}}dE, \tag{7}$$

the geometrical factor is

$$F_{r,R} = \Lambda \int_0^{R+r} \frac{\Phi(r,\rho,R)}{4\pi}\,\frac{\rho e^{\Lambda\rho}}{\rho^2 + \frac{1}{6}}d\rho, \tag{8}$$

$\kappa(E)$ and $g(E)$ are normalized spectral shapes of

spontaneous emission and gain respectively, and Λ is the single pass loss.

The value of γ for a threshold carrier density N_{th} equal to $5\times10^{17}cm^{-3}$, which corresponds to a current density of $5A/cm^2$ in plotted in Fig.2. If one assumes a 30 - 50% round-trip nonresonant loss L and an output mirror with a transmission coefficient of 0.2, the single pass loss Λ is 0.25 - 0.35. For this transmission coefficient, the excited region radius must be about 20 times as large as its thickness for ASE to cause an increase of threshold of 10%. Because the penetration depth of 35kV electrons is about 3–4μm, no change in threshold due to ASE can be observed for beam diameters of up to 180 μm. For output mirrors with transmission transmission coefficients of 0.5, ($\Lambda = 0.4$–0.5), one should be able to see the increase in threshold once the beam diameter reaches 75 μm.

III Laser Experiment

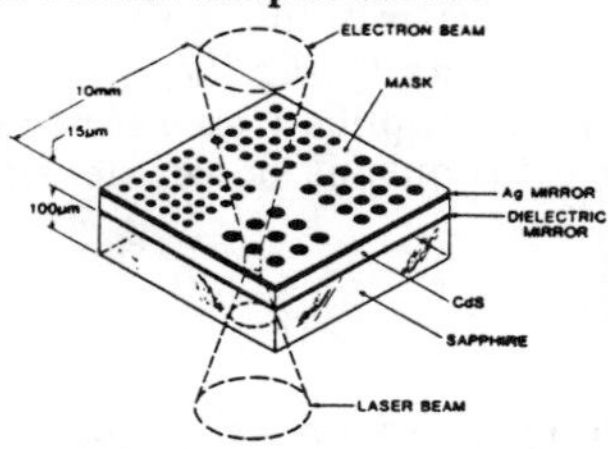

Figure 3 Geometry of surface-emitting laser experiment.

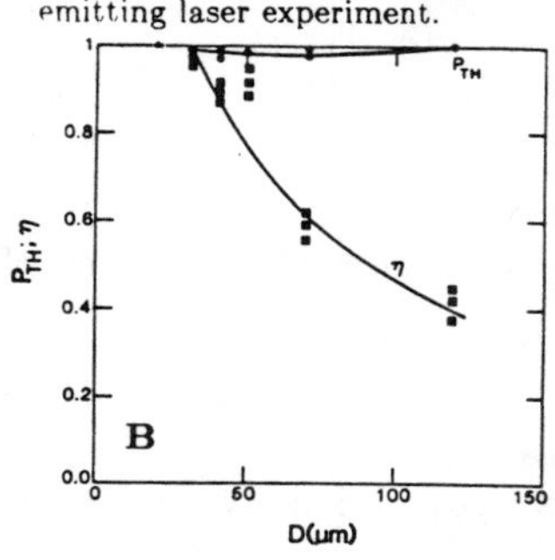

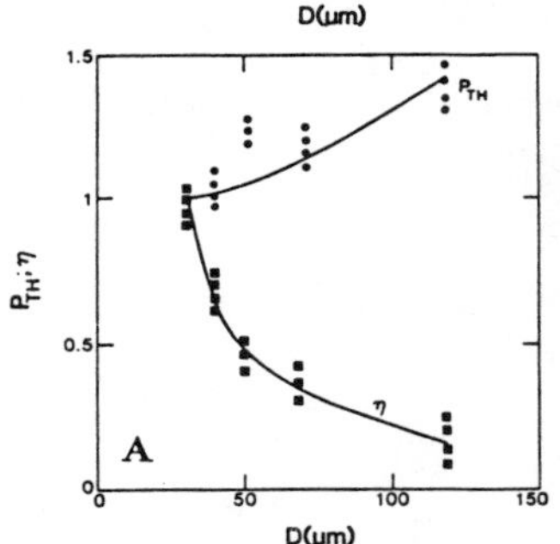

Figure 4 Dependencies of lasing threshold power and differential efficiencies η on the excited region diameter for (A) 0.8 and (B) 0.5% reflective output mirrors.

The laser experiments used several samples obtained by polishing bulk-grown CdS crystals attached to 100 μm thick sapphire substrates (Fig.3). The upper surfaces of the sapphire substrates were coated with dielectric films having reflection coefficient of either 0.5 or 0.8. The upper surface of the CdS was coated with a Ag film to provide a reflection coefficient of 0.95. The samples were cooled to 100K and excited by a 35kV electron beam. The electron beam pulse length was 15 *ns* and the repetition rate was chosen to be the order of 1 *KHz* in order to limit thermal effects. To vary the diameter of the excited region without changing the current density, the beam was defocused to a 150 μm spot and a mask used to limit the diameter of the excited region. The mask was formed from a 15 μm thick molybdenum square with four sets of holes having various diameters. In order to reduce the influence of sample inhomogeneity, four sets of measurements were performed with the mask rotated by 90 degrees for each successive set. The results of the four measurements for each hole diameter were averaged, giving four values of threshold and slope efficiency for the four excited region diameters. Another value was obtained using the defocused beam without the mask.

The results of the experiments are plotted in Figs.4a and 4b for output mirror reflectivities of 80% and 50% respectively. The curves were obtained by the cubic spline interpolation of the data points. As can be seen for the 80% reflectivity mirrors case, no change in threshold occurs for diameters of up to 150 μm, while for the 50% reflectivity case, there is a small difference in threshold caused by ASE. This agrees well with calculations, especially the fact that the threshold dependence is stronger for the higher-loss (larger Λ) laser.

The most interesting aspect of the results is that while threshold does not appear to depend strongly on the excited region diameter in the tested range, the slope efficiency appears to fall-off rapidly with an increase in diameter. The linear theory developed by us above, assumes "clamping" of the quasi-Fermi levels at their threshold values, and thus a constant differential efficiency $T/(T+L)$ can not explain this dependence. This observation prompted an investigation into the the influence of such factors as spatial and spectral inhomogeneity of the gain and losses on the characteristics of lasing.

IV A Nonlinear Theory of ASE

If gain broadening is not homogeneous or there are non-lasing regions in the pumped volume due to spatial variation of losses, the carrier densities and the quasi-Fermi levels are not clamped to their threshold values but continue to grow thereby increasing the spontaneous radiation and hence the ASE. In order to take this effect into account , the carrier density $N(r)$, the spectral factor F_E and the geometrical factor $F_{r,R}$ are made to follow a linear phenomenological relation as for example

$$F_E = F_{E,th}(1 + a_E[P(r) - P_{th,0}(r)]). \tag{9}$$

The output power then becomes

$$I_{out} = \frac{T}{T+L} h\nu H[1 - P'_{th}(r)]\,[P(r) - P_{th,0}(r)], \tag{10}$$

where

$$P'_{th}(r) = \frac{\partial P_{th}(r)}{\partial P(r)} = a_{sp} + \frac{\gamma(r)}{1+\gamma(r)}[a_E + a_R] \tag{11}$$

is the rate of change of threshold with pump power and a's are phenomenological coefficients introduced above. The first component of threshold increase is due to the increase in spontaneous emission and therefore does not depend on the dimensions of the excited region, while the second and third components are due to a variation of the ASE. As can be seen, an increase in carrier density and therefore in spontaneous and amplified spontaneous emission above threshold does not affect the apparent threshold taken from the input-output curves, but it does reduce the differential efficiency by the factor $P'_{th}(r)$.

V Saturation of Spontaneous Emission Measurement

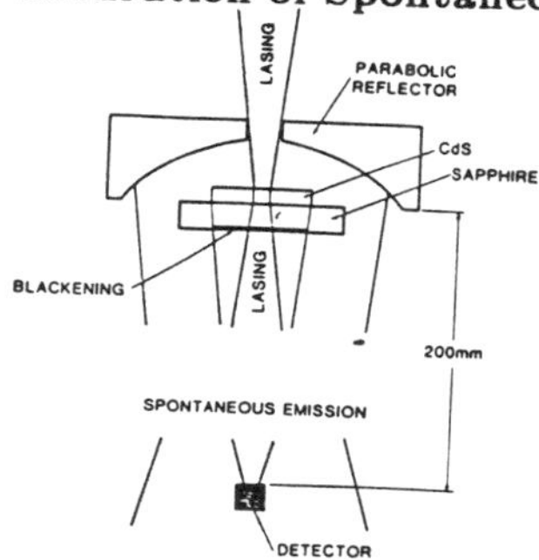

Figure 5 Geometry of the spontaneous emission saturation experiment

To estimate changes in spontaneous and amplified spontaneous emission above threshold, the following experiment was performed. A CdS sample was prepared with semitransparent coatings on either side (Fig.5). It was placed close to the focus of a parabolic reflector having a round opening of 3mm diameter. The opening served the dual purpose of giving the electron beam access to the sample, while at at the same time preventing the laser emission from being collected by the reflector. The laser emission emerging from the lower semi-transparent coating is partially absorbed by black paint to the degree that there remains only enough laser power density for monitoring purpose, but far less than the

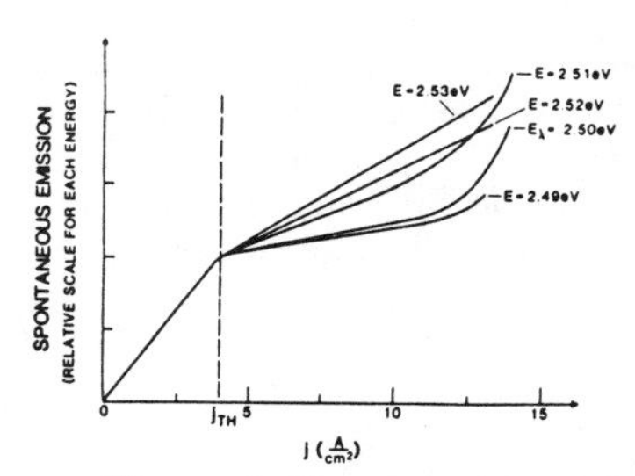

Figure 6 The dependency of spontaneous emission on pumping current density below and above threshold for the energies around lasing energy E_λ.

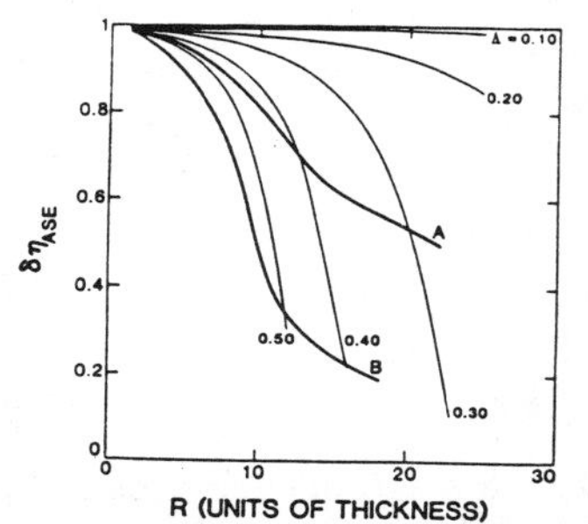

Figure 7 Differential efficiency decrease as a function of excited region radius-to-thickness ratio for (A) 0.8 and (B) 0.5% reflective output mirrors.
Thin lines - calculated for various values of longitudinal gain Λ
Thick lines - experimental

cathodoluminescence focused by the reflector. This permits us to measure the peak power of cathodoluminescence below, at and above lasing threshold by means of a photodiode, and to observe its power spectrum.

The results of the experiment are shown in Fig.6. It can be seen that below lasing threshold, which occurs at .8 A/cm^2, the spontaneous emission grows linearly over the entire range of energies. At threshold, the input-output curves bend for all energies, although the curvature is far greater for energies below the lasing energy E_λ. The fact that for energies above lasing, the spontaneous emission continues to grow means that the quasi-Fermi levels continue to move into their respective bands thereby enhancing the population of the upper states. This effect can be attributed to the fact that the population of the bands is not thermalized and there exists a dynamic equilibrium between intraband relaxation and band-to-band nonradiative recombination. The increase in spontaneous emission above threshold for lasing energy E_λ and energies below, can follow from an assumed spatial inhomogeneity of lasing, which means that for some parts of the excited region the local gain is higher than the threshold value. At pumping densities far above threshold, spectral narrowing is observed, which can be attributed to either ASE or stray laser emission reaching the photodetector. These results agree qualitatively with observations of GaAlAs heterojunction injection lasers by Somers [6] and Hakki and Paoli [7].

Phenomenological constants a_{sp}, a_E and a_R were determined from these results. The overall result in the relative decrease in differential efficiency $\eta_{ASE}(R)$ caused by ASE is plotted in Fig.7, where

$$\eta_{ASE} = \frac{1 - P_{th}'(R)}{1 - P'_{th}(0)}. \tag{12}$$

As one can see, for a 0.2 transmission loss, or equivalently, Λ equal to 0.25 - 0.35, there is a reduction in differential efficiency by a factor of 0.95 at R equal to 10 for a beam diameter of $70\mu m$. For 0.5 transmission loss ($\Lambda = 0.4 - 0.5$) the differential efficiency is reduced by the factor ½ for the same beam diameter. A cubic spline approximation to the experimental points is included in Fig.7 to demonstrate at least qualitative agreement between theory and experiment.

A significant disagreement between this theory and experiment is that according to the theory, the differential efficiency for large radii can become less than zero, which means that for an increase in pumping power the corresponding increase in threshold due to increased ASE is larger than the increase in the pumping power itself, so that the threshold value can never be reached . This should happen only when threshold is already very high, say 10-50 times above threshold for an excited region with vanishingly small radius. In these experiments such high excitation density was never reached. However, rather than observing a continuous decrease in differential efficiency for large radii, an asymptotic approach to zero has been observed. One possible explanation for this discrepancy is that some non-lasing regions act as absorbers, possibly due to crystal defects, so they isolate the lasing regions surrounded by them from the influence of ASE. In general, the question of whether the non-lasing regions have gain above or below threshold is unclear and requires further investigation.

VII Conclusions

A theory concerning amplified spontaneous emission and its influence on the threshold and differential efficiency of a surface-emitting semiconductor laser has been developed. The linear theory takes into account the spectral distribution of both gain and spontaneous emission. Previous work failed to deal with this dependence, and as a result the influence of ASE on laser threshold had been overestimated.

Experiments were conducted to observe the threshold and differential efficiency of an electron-beam-pumped CdS lasers for different diameters of excited region. The results show a weak dependence of threshold on diameter, in good agreement with the linear theory. The strong dependence of differential efficiency on excited region diameter cannot be explained by the linear theory. Moreover, the experiment on saturation of spontaneous emission above lasing threshold has shown that saturation is far from being complete, implying that a linear ASE theory is not valid above threshold.

Based on the results of the spontaneous emission saturation experiments, a phenomenological nonlinear theory of ASE is advanced that predicts a sharp decrease in the differential efficiency due to an increase in ASE above threshold which demonstrates qualitative agreement with the laser experiments.

References

1. N.G.Basov, O.V.Bogdankevich, A.N.Pechenov, A.S.Nasibov, and K.P.Fedoseev, So.Phys.-JETP, **28**, 900 (1969).
2. B.M.Lavrushin, Tr. Fiz. Inst. Akad. Nauk SSSR **59**, 123 (1971).
3 O.V.Bogdankevich, S.A.Darznek, M.M.Zverev, and V.A.Ushakhin Sov. J. Quant. Electron., **5**, 953 (1975).
4 J.E.Packard, W.C.Tait and G.H.Dierssen, "Two Dimensionally Scannable Electron-Beam Pumped Laser ", Appl.Phys.Lett., **19**, 338 (1971).
5 J.R.Packard, W.C.Tait, and D.A.Campbell, "Standing Waves and Single Mode Room Temperature Laser Emission from Electron-Beam-Pumped CdS", IEEE J. Quant. Electron., **QE—5**, 44 (1969).
6 H.S.Sommers Jr, "Experimental Studies of Injection Lasers: Spontaneous Spectrum at Room Temperature", J. Appl. Phys., **43**, 4067, (1972).
7 B.F.Hakki and T.L.Paoli "Gain Spectra in GaAs Double-Heterostructure Injection Lasers", J. Appl. Phys., **46**, 1299, (1974).

1.06μm Nanosecond Laser Amplification via Degenerate Multiwave Mixings in Silicon

I. C. Khoo, R. R. Michael and T. H. Liu
Department of Electrical Engineering
The Pennsylvania State University
University Park, PA 16802

ABSTRACT

Degenerate multiwave mixing and optical phase conjugation in semiconductor silicon are reexamined both theoretically and experimentally. It is found that, in general, the often neglected diffracted beams influence considerably both the forward as well as the backward phase conjugated signal gain.

INTRODUCTION

Recently, with the emergence of highly nonlinear materials, degenerate optical wave mixings have received renewed attention from both fundamental and applied standpoints. Studies in photorefractive crystals, liquid crystals, multiple quantum wells and semi-conductors have shown that efficient wave mixing effects can be realized with low power cw or pulsed lasers. In particular, in a typical configuration as depicted in Figure 1 (ref.1) where a strong pump and a weak probe beam are incident on the nonlinear medium, several orders of diffracted beams are generated. As the crossing angle between the pump and the probe beams are decreased, thereby reducing the phase mismatches amongst the beams, the number of diffracted beams, as well as their relative intensities (compared to the probe beam) also increase.

Since these diffracted beams can be very strongly coupled with the strong pump beam, and scatter light into or out of the probe beam, they obviously play very important roles in the transmitted probe beam intensity as well as the "reflected" phase conjugated signal in an interaction configuration involving a retro-reflected pump beam or another strong beam.
The effect of the diffracted beams are particularly significant in the case involving very weak probe, when the first diffracted order on the pump beam side can be of comparable magnitude to the probe beam. Accordingly, we have developed theories[2,3] for these forward and backward phase conjugation processes. In particular, we have studied the gain, and the phase conjugated probe beam reflection, by taking into account explicitly the roles played by the diffracted beams (in both forward and backward directions). Our theories also include all the other relevant mechanisms such as self-phase modulation effects, pump-probe beam ratio, pump-retroreflected pump beam asymmetry, intensity dependent phase modulations, frequency shifts, phase mismatches, sample thickness, etc. These theories provide quantitatively accurate description of experimental results obtained in liquid crystals and in the

semiconductors silicon.

In our study involving silicon, the laser used is a linearly polarized 20 ns Nd:Yag 1.06μm single longitudinal pulse. The laser is divided into a strong and a weak probe, and they are then overlapped on a silicon wafer (0.4mm thick) at Brewster angle (to reduce reflection losses). In the configuration involving a counterpropagating strong beam , the beam is derived by a 50% beam splitter from the pump beam. The angle of crossing between the pump and the probe are varied.

Detailed experimental results of this study and their comparison with theories are relegated to a longer article elsewhere. In summary, we have observed very large probe beam amplification (up to 1200%) for moderate pump laser pulse energies (≈ a few tens of milli-joules/cm^2). We have also observed the theoretically predicted gain dependence on the pump-probe beam ratio, as well as the dependence of the probe beam phase conjugation reflectivity on the pump-probe beam ratio (c.f., Figures 1a and 1b respectively) which are similar to results obtained in liquid crystals with cw lasers.[4] This dependence of the phase conjugation reflectivity on the pump-probe beam ratio is particularly important in current studies of self-pumped phase conjugation, where the probe beam originates from scattering noise (from the strong pump beam). As a matter of fact, we have shown that by taking into account the diffracted beam, the phase conjugation self-oscillation threshold is significantly lower than the usual threshold value obtained without accounting for the diffracted beam.

REFERENCES

1. I. C. Khoo and T. H. Liu, IEEE J. Quant. Electron. QE23, 171 (1987).

2. T. H. Liu and I. C. Khoo, IEEE J. Quant. Electron. QE23, 2020 (1987).

3. I. C. Khoo and Y. Zhao, IEEE J. Quant. Electron. (submitted Dec. 1987).

4. I. C. Khoo, P. Y. Yan, R. Michael and G. M. Finn, J. Opt. Soc. Am. (Feb. 1988).

FIGURES

Fig. 1a Dependence of probe beam gain on the pump-probe beam ratio. ○ circles.Solid is theoretical fit using ref.1.

Fig. 1b. Dependence of phase conjugated reflection on the pump-probe beam ratio. ● Solid circles.

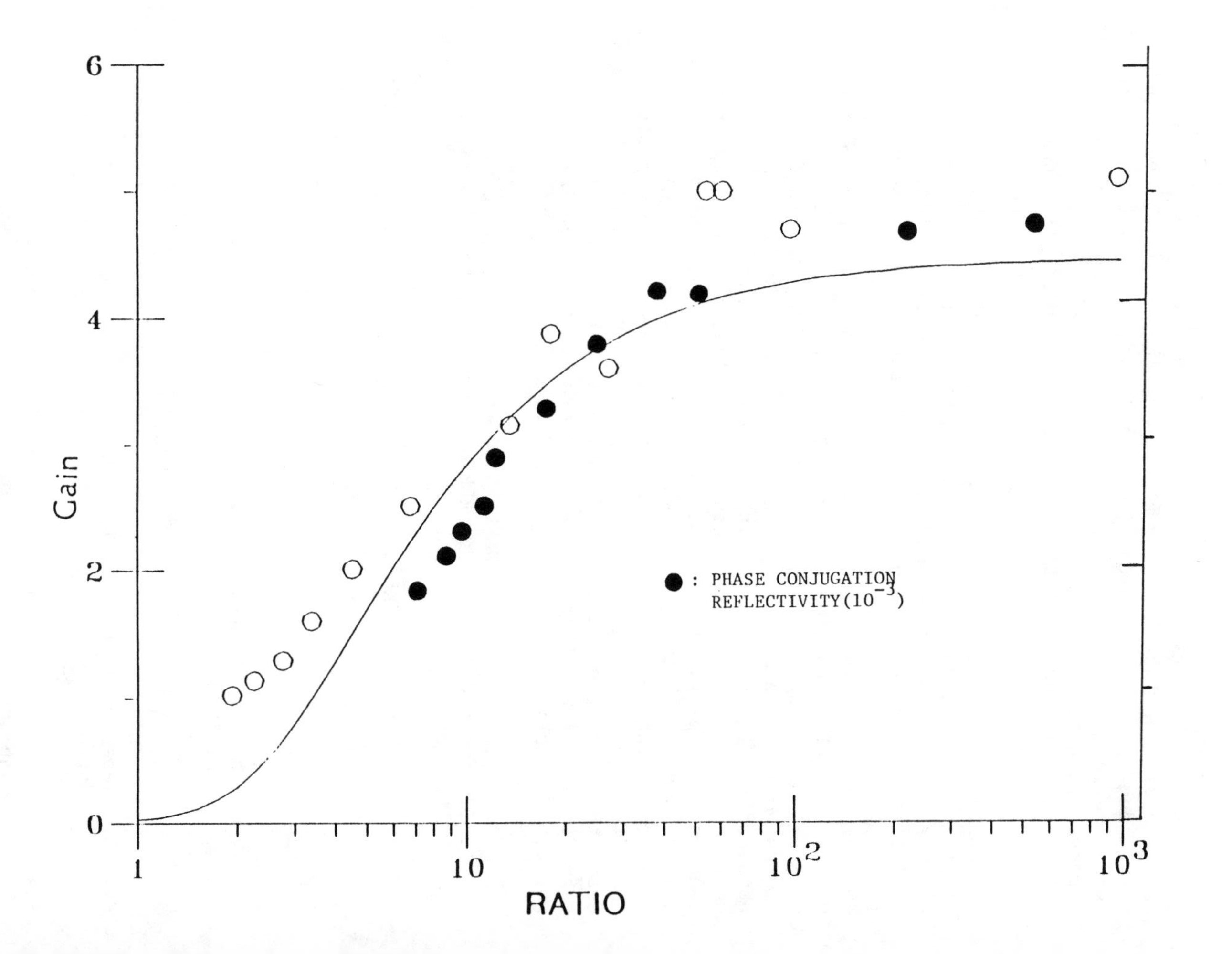
6
4
2
0
Gain
1
10
10^2
10^3
RATIO
● : PHASE CONJUGATION
REFLECTIVITY(10^-3)

REAL TIME PHONON KINETICS IN CALCITE BETWEEN 300 K AND 5 K

R. Dorsinville, P. J. Delfyett, and R. R. Alfano
Institute for Ultrafast Spectroscopy and Lasers,
Department of Electrical Engineering,
The City College of New York , New York, New York 10031

ABSTRACT

The relaxation of the 1086 cm^{-1} A_{1g}-mode of calcite was observed in real time within a temperature range from 300 K to 5 K. A theoretical fit to the experimental data suggests that the temperature dependence of the phonon dephasing time is due to three and four phonon splitting processes.

We have measured phonon lifetimes in calcite as a function of temperature in **real time** using a new method which combines the Raman induced phase conjugation (RIPC) geometry with streak camera techniques.[1] In this method, the generated phase conjugate pulse using RIPC is passed into a streak camera and a video computer system to record its time profile. The rise and decay times of the generated pulse is directly related to the phonon formation and decay times. The decay time of the A_{1g} -mode of calcite was investigated. The free relaxation of this vibration was observed in real time within a temperature range from 300 K to 5 K.

The RIPC experimental set-up was previously described in detail.[2] Two pulses of different frequencies, one derived from a nonlinear supercontinuum at $w-\Omega$ =562 nm (where Ω = 1086 cm^{-1} corresponds to the symmetric molecular vibration to be investigated) and a second harmonic pulse at w = 530 nm from an Nd-YAG laser (25 ps, 0.5 mJ, 0.5 Å bandwidth) interact in a calcite crystal at an angle to create a spatial modulation of the optical polarizability of the medium. A large crossing angle was used to limit the size of the excitation volume. For the same reason the interacting beams were focused into the calcite crystal with 25 cm focal length lenses. As a result the excitation time was determined by the pulse duration of the shortest pulse. In the absence of electronic resonance and in the presence of phonon resonance at Ω the medium optical polarizability modulation is due mainly to a spatially modulated coherent phonon population with a relatively small non resonant electronic contribution. The picosecond pulse continuum is produced by focusing a 2.5 mJ, 25 ps second harmonic pulse into a 5 cm liquid D_2O cell. The frequency bandwidth of the continuum is typically 3000 cm^{-1}. A 10 nm spectral window at the appropriate frequency ($w-\Omega$) is selected from the continuum with narrow band filters. The pulse duration of the continuum pulses after spectral selection is about 3 ps when measured with a 2 picosecond streak camera, with an average energy of 20 nJ/nm.

Since the phonon population at freqency Ω is coherently driven only when both the second harmonic laser pulse (at w) and the continuum pulse (at w-Ω) are present the phonon excitation should last for less than 3 ps. Once the excitation has ended the sample relaxes according to the characteristic dephasing time of the excited phonon. A third weak (.2 mJ) probe pulse of frequency w = 530 nm and 25 ps duration is scattered off the spatial modulation in the phase conjugate direction. The scattered signal frequency is w-Ω and its amplitude and duration depend directly on the coherent phonon population lifetime. The intensity of the phase conjugate signal as a function of time is monitored directly in **real time** using a Hamamatsu model C1587 2 ps resolution streak camera and temporal analyzer.The intensity of the signal reaching the streak camera at time t is given by[1]:

$$I_{pc}(t) = |E_{pc}(t)|^2 \sim$$
$$\sim |E_{pr}(t)\, E_{po} Q_o \exp(-t/T_\phi)|^2 \sim I_{pr}(t) e^{-2t/T_\phi} \quad (1)$$

where E_{pc} is the phase conjugate field, E_{pr} is the probe pulse, T_ϕ is the phonon dephasing time. We have assumed a very short continuum ($E_c(t') \approx \delta(t')$) and a very long pump ($E_p(t') \approx E_{po}$ = const). All three beams coincide at the sample site. For a long probe pulse of any shape the phonon response function is obtained by dividing the phase conjugate signal by the probe intensity. For short decay times this correction does not affect substantially the measured lifetimes and the temporal shape of the phonon response function of the material can be obtained in real time.

Measurements of the temperature dependence of the phonon lifetime between 300K and 5K were carried out using a helium Dewar. Fig. 1 gives the observed relaxation times as a function of temperature.The mesured values of the dephasing time agree with spontaneous Raman linewidth data in the same temperature range.[3] The measured values vary from $T_\phi/2 = 5 \pm 0.5$ ps at room temperature to $T_\phi/2 = 10 \pm 1$ ps at 5 K. The experimental dephasing time T_ϕ can be written:

$$2/T_\phi = 1/\tau_{ph} + 1/T_1 \quad (2)$$

where τ_{ph} is the pure dephasing time due to scattering from impurities, defects, free carriers, etc.. This leads to a **k**-vector change along the phonon branch without significant loss of energy since the dispersion of the internal mode A_{1g} is very small. T_1 is the population decay time due to energy interchanges with other lattice modes through collisions due to anharmonicity of the lattice forces. The temperature dependence of phonon-phonon scattering processes can be calculated and compared with the experimental values to evaluate the relative contribution of both mechanisms (dephasing or energy relaxation) to the phonon decay and identify the relaxation process.

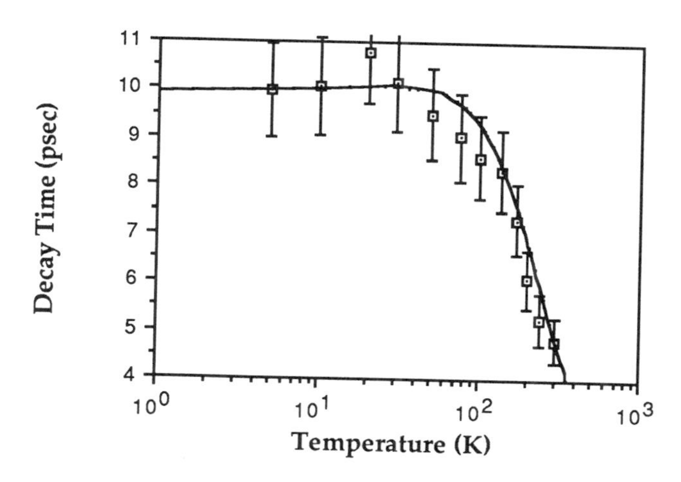

Fig. 1. Experimentally measured decay times ($T_\phi/2$) versus temperature. The solid curve represents the best fit for a combination of three phonon and four phonon splitting processes.

In general, the two main phonon-phonon scattering decay schemes are: 1) splitting of an optical phonon into two or more acoustic phonons, 2) splitting of optical phonon into two or more optical phonons. In calcite the density of states of acoustic phonons is relatively small and decay into optical phonons are more probable. Other schemes involving the production of phonon of higher energy than the A_{1g} mode are possible but are important only at high temperatures.The best fit to our experimental data is obtained with a combination of a three phonon splitting process and a four phonon splitting process. This is shown in fig.1. This seems to indicate that the dephasing is mostly due to phonon splitting processes and both **three** and **four** phonon splitting processes are reponsible for the temperature dependence of the dephasing time of the 1086 cm^{-1} mode in calcite.

This research is supported by ASFOR-84-01H4B.

---

[1] P. J. Delfyett, R. Dorsinville, R. R. Alfano, Optics Letters, 12, 1002 (1987).
[2] R. Dorsinville, P. Delfyett, R. R. Alfano, Applied Optics, 17, 3655 (1987).
[3] K. Park, Phys. Lett., 22, 39 (1966).

NONLINEAR ABSORPTION OF CO_2 LASER RADIATION IN p-TYPE InSb*

R. B. JAMES
Sandia National Laboratories, Livermore, CA 94550

Y. C. CHANG
University of Illinois, Champaign-Urbana, IL 61801

D. L. SMITH
Los Alamos National Laboratory, Los Alamos, NM 87545

ABSTRACT

Calculations of the intervalence-band absorption saturation in p-type InSb by high-intensity CO_2 laser light are reported. The intensity-dependent absorption coefficient is found to be well described by an inhomogeneously broadened two-level model, and values of the saturation intensity are presented.

In many p-type semiconductors, direct intervalence-band transitions are responsible for the absorption of light with a wavelength near 10 μm. For intensities exceeding approximately 1 MW/cm^2, the absorption due to these transitions has been found to decrease with increasing intensity in p-Ge[1-2] and p-GaAs[3]. The saturation property has been exploited for the generation of passively mode-locked CO_2 laser pulses of subnanosecond duration,[4] and to provide interstage isolation of high-power oscillator-amplifier stages of CO_2 laser systems.[2] In this paper we report theoretical results for the intensity dependence of the free-hole distribution in p-type InSb for light with a wavelength of 10.6 μm and apply the results to a calculation of the absorption cross section. The calculations predict that the absorption will saturate due to a state-filling effect, in which the occupation probabilities of resonantly coupled states begin to approach one another at high pump fluences.

The absorption of 10.6-μm radiation in p-InSb is dominated by direct free-carrier transitions between the heavy- and light-hole bands.[5] Since both energy and wave vector are conserved in the optical transitions, only holes in a narrow region of **k**-space can directly participate in the absorption process. At low light intensities the relaxation mechanisms maintain the hole distribution near its equilibrium value. However, for sufficiently large laser intensities, scattering can no longer maintain the equilibrium distribution, and the occupation probability for a state in the resonant region decreases in the heavy-hole band and increases in the light-hole band. Since the absorption is governed by the population difference of these resonant states, this redistribution leads to a reduced absorption coefficient.

The intervalence-band absorption coefficient is given by[6]

$$\alpha_p(\omega,I) = \frac{4\pi^2 e^2}{(K_\infty)^{1/2} m\omega c} \sum_{\mathbf{k}} \left\{ (f_h(I)-f_l(I)) |\eta\cdot P_{hl}|^2 \frac{\hbar\Gamma_{hl}(\mathbf{k})/\pi}{(\hbar\Omega(\mathbf{k})-\hbar\omega)^2+(\hbar\Gamma_{hl}(\mathbf{k}))^2} \right\}, \quad (1)$$

where the subscripts h (l) designate the heavy- (light-) hole band, N_h is the density of holes, K_∞ is the high-frequency dielectric constant, m is the free-electron mass, $\hbar\omega$ is the photon energy, $f_i(\mathbf{k})$ is the intensity-dependent probability that a hole state with wave vector $\mathbf{k}$ is occupied in band i, $\mathbf{P}_{bc}(\mathbf{k})$ is the momentum matrix element between Bloch states in bands b and c, η is the polarization of the light, $\hbar\Omega$ is the energy difference $[E_h(\mathbf{k})-E_l(\mathbf{k})]$ where $E_i(\mathbf{k})$ is the valence subband energy for a state with wave vector $\mathbf{k}$ in band i, and $\Gamma_{hl}(\mathbf{k})$ is the average scattering rate for holes in states $(h,\mathbf{k})$ and $(l,\mathbf{k})$.

The scattering rate of holes occurs on a picosecond time scale, so for laser pulses of nanosecond or longer duration (the typical experimental situation), transient effects are damped out. We calculate the steady-state hole distribution functions by solving equations which allow for intervalence-band optical transitions between the heavy- and light-hole bands and the scattering of holes by phonons, ionized impurities and other free holes.

Phonon scattering was calculated on the basis of the deformable potential model. The valence-band deformation parameters were chosen by fitting the temperature dependence of the calculated hole mobility to the experimental data.[7] Following Ref. (8), we neglect the angular dependence of the phonon matrix elements and take the scattering rate to be the same function of energy for the heavy- and light-hole bands. The fit to the mobility data gives acoustical and optical deformation potential constants of 2.32 and 4.64 eV, respectively.

For acceptor and hole concentrations greater than about 5×10^{13} cm^{-3}, the scattering of holes by ionized impurities and other free holes becomes important. The hole-ionized impurity scattering rate is given in Ref. (9), and the rate of hole-hole scattering as a function of the carrier energy is shown in Ref. (10). In the calculation we consider only uncompensated samples of p-InSb, where the acceptors are all shallow and ionized at a temperature of 77 K (i.e., $N_I=N_h$).

The free-hole energies $E_h(\mathbf{k})$ and $E_l(\mathbf{k})$ and the momentum matrix elements $\mathbf{P}_{hl}(\mathbf{k})$ are determined by degenerate $\mathbf{k}\cdot\mathbf{p}$ perturbation theory.[11] The cyclotron resonance parameters and spin-orbit splitting of Lawaetz[12] are used in the calculation.

Figure 1 shows the computed values for the distribution function (multiplied by the effective density of states) versus k^2 for $\mathbf{k}$ in the [100] direction and a lattice temperature of 77 K. The solid curves illustrate the equilibrium values, and the dashed curves show the modified distribution for states in the resonant region due to excitation by a pulse with an intensity of 30 kW/cm^2 and a wavelength of 10.6 μm. The calculation was performed for a free-hole density of 10^{10} cm^{-3}, in which case the hole-ionized impurity and hole-hole scattering rates are negligible compared to the hole-phonon scattering. The large dip in the occupation probabilities of the heavy-hole band and the large increase in the light-hole band at $k^2=5.36\times10^{-4}$ $Å^{-2}$ are caused by the direct optical transitions.

The intensity-dependent distribution functions are then used in the numerical integration of Eq. (1). The absorption coefficient was calculated for intensities in the range of 5-500 kW/cm^2, and the results are found to be well described by the equation

$$\alpha_p(N_h,\omega,I) = \alpha_o(N_h,\omega)\,[1 + I/I_s(N_h,\omega)]^{-1/2} \quad , \qquad (2)$$

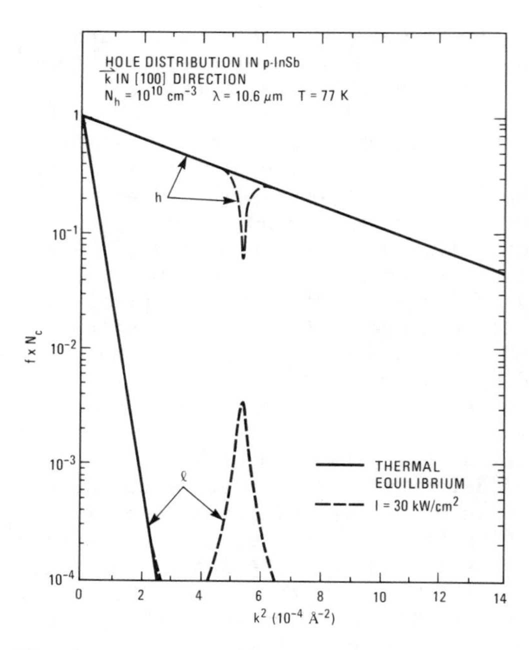

Figure 1. Calculated one-hole distribution functions in p-InSb versus k^2 for **k** in the [100] direction. N_c is the effective density of states.

where α_o is the intervalence-band absorption coefficient at low intensity, and I_s is the saturation intensity.

At higher free-hole concentrations, ionized impurity and free-carrier scattering become more important. The inclusion of these scattering rates increases the total scattering rate and introduces a concentration dependence in the saturation behavior. Figure 2 shows the calculated values for I_s as a function of the hole density for 10.6-μm radiation at a temperature of 77 K. We note that I_s is substantially independent of the hole density for concentrations less than about $5x10^{13}$ cm^{-3}, that is, in the region where the hole-phonon scattering mechanism is dominant. For hole concentrations greater than about $5x10^{13}$ cm^{-3}, the saturation intensity begins to increase monotonically with increasing hole density due to the increased scattering rate of the free holes participating in the optical absorption process.

For both applications and comparison of experiment with theory, it is important that the dynamic range in the saturable absorption be as large as possible. The maximum reduction of the total absorption coefficient is, in general, determined by the value of I/I_s and the threshold for the two-photon absorption (TPA) process. The TPA process increases the free-carrier absorption via the generation of non-equilibrium electron-hole pairs. The two-photon absorption coefficient, K_2, at 77 K is measured to be about 0.2 $cm\text{-}MW^{-1}$ for 10.6-μm radiation[13-14], so for intensities greater than several tens of kW/cm^2, two-photon absorption across the band gap must be considered. Thus, an experimental test of the free-hole absorption saturation should be performed on lightly doped p-type samples (i.e., the samples should have a relatively small value for I_s), so that a significant reduction in the free-hole absorption cross section occurs prior to the onset of two-photon absorption. This condition is satisfied in p-InSb when $[\alpha_p(I=0)-\alpha_p(I)] \gg K_2 I$.

In conclusion, we have presented theoretical results for the intensity dependence of the intervalence-band absorption cross section. For lightly doped p-type InSb at a temperature of 77 K, we find that the absorption begins to saturate at intensities less than 10 kW/cm^2. The values for the saturation intensity, I_s, for p-InSb are found to be between one and two orders of magnitude smaller than

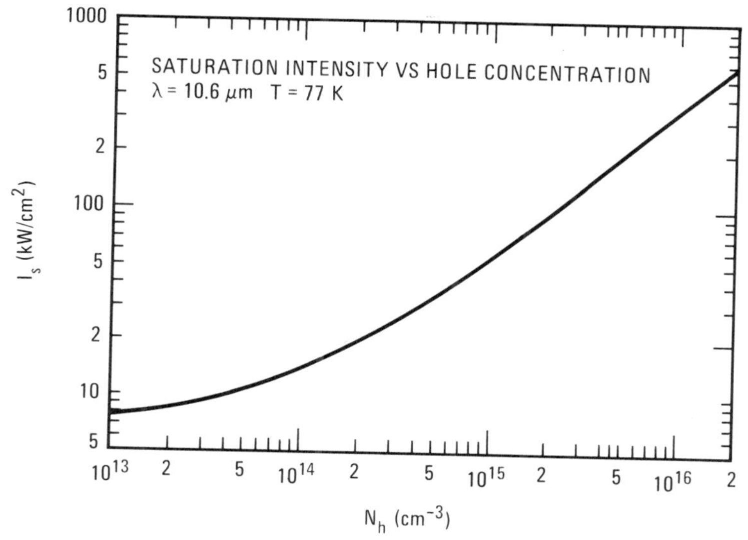

Figure 2. Computed values of the saturation intensity as a function of the hole concentration.

the measured values of I_S for p-type Ge and GaAs. Given the interests in saturable absorbers of 10.6-μm radiation, these relatively small values of I_S for p-InSb may be of significant practical importance.

* Work supported by the U. S. Department of Energy.

REFERENCES

1. F. Keilmann, IEEE J. Quantum Electron. 12, 592 (1976).
2. C. R. Phipps, Jr. and S. J. Thomas, Opt. Lett. 1, 93 (1977).
3. R. B. James, W. H. Christie, R. E. Eby, B. E. Mills and L. S. Darken, Jr., J. Appl. Phys. 59, 1323 (1986).
4. A. J. Alcock and A. C. Walker, Appl. Phys. Lett. 25, 299 (1974).
5. S. W. Kurnick and J. M. Powell, Phys. Rev. 116, 597 (1959).
6. R. B. James and D. L. Smith, Phys. Rev. Lett. 42, 1495 (1979).
7. R. B. James and D. L. Smith, J. Appl. Phys. 51, 2836 (1980).
8. E. Conwell, J. Phys. Chem. Solids 8, 236 (1959).
9. H. Brooks, Advances in Electronics and Electron Physics, Vol. 7, edited by L. Marton (Academic, New York, 1955), p. 85.
10. I. N. Yassievich and I. D. Yaroshetskii, Sov. Phys. Semicond. 9, 565 (1975).
11. E. O. Kane, J. Phys. Chem. Solids 1, 82 (1956).
12. P. Lawaetz, Phys. Rev. B4, 3460 (1971).
13. A. F. Gibson, C. B. Hatch, P. N. D. Maggs, D. R. Tilley and A. C. Walker, J. Phys. C: Solid State Phys. 9, 3259 (1976).
14. The equations used in Ref. (13) to obtain a value for K_2 assume the intervalence-band absorption cross section to be independent of intensity. The inclusion of our results for $\alpha_p(I)$ in the equations of Ref. (13) results in a significantly increased value for K_2.

HIGH-EFFICIENCY FREQUENCY CONVERSION OF IR LASERS WITH $ZnGeP_2$ AND $CdGeAs_2$

Yu.M. Andreev, P.P. Geiko and V.V. Zuev
Institute of Atmospheric Optics SB USSR Academy of Sciences, Tomsk, 634055, U S S R

The problem on developing effective, reliable parametric frequency converters (FC) for overlapping the middle IR by the converted spectrum of the available lasers has not been solved as yet. This is caused by low nonlinear susceptibility of nonlinear crystals widely used and by relatively low optical quality, damage levels and small length of the new promising crystals.

We think that insignificant improvement of technology of growing new nonlinear ternary semiconductor crystals, in particular, $CdGeAs_2$, $ZnGeP_2$ and Tl_3AsSe_3, and optimization of nonthreshold FC of such lasers as CO_2 and CO enables one to solve this problem.

The crystals mentioned above are of interest here due to high birefringence and figure of merit equaling +0.09, +0.04 and -0.18, respectively. The Tl_3AsSe_3 crystals allow SHG and mixing of frequencies of all available middle IR lasers. The same can be made using both $CdGeAs_2$ and $ZnGeP_2$ crystals. The FC spectrum thus obtained can overlap the 2-17 µm region several times. In the simplest case, this can be carried out using one two-frequency laser or two single-frequency CO_2 lasers. The $CdGeAs_2$ and $ZnGeP_2$ crystals have second after Te figure of merit. But the peculiarities of Tl_3AsSe_3 spectral characteristics allow one to use longer crystals thus compensating the lower figure of merit.

The technological advances allowed us to obtain sufficiently large optically qualitative monocrystals bulls of $ZnGeP_2$ (Ø 20-25 mm, length up to 150 mm) and Tl_3AsSe_3, $CdGeAs_2$ monocrystals with the volume up to 2-3cm^3. This makes it possible to conduct experimental studies of one- and two-cascade nonthreshold FC of CO_2, CO and NH_3 laser radiations with these crystals. Types and energy efficiencies of some of the constructed FC, such as SHG, fourth harmonic generators (FHG) and sum frequency generators (SFG) are presented in Table I.

In all the cases, collinear I type three-frequency interactions were investigated. The $ZnGeP_2$ FC have been studied in more detail. Our experiments have not had as its objective the optimization of converters, but the works N 1,5 and 9 in accordance with Table I.

It should be noted that the hybrid CO_2 laser radiation had the energy degree of contrast 1:10. When investigating a two-cascade fourth harmonic generator of the above laser radiation a 3 mm crystal $ZnGeP_2$ was used in the first cascade and a 7 mm one was used in the second cascade. The damage intensity was about 1 GW/cm^2 or slightly higher. The efficiency of the second cascade was limited by plasma formation on the surface of a LiF band filter. The SH wave front control using the photographic method with light sensitization has shown the absence of distortions even at predamage pumping intensities. At SHG of the TEA laser and of the CO and CO_2 Q-switched lasers radiation the crystals of 10 mm length were used. At chemical-dynamic polishing of surfaces the pumping damage intensity was 60 MW/cm^2 for the TEA laser radiation. For the cw CO_2 laser radiation it was about 200 kW/cm^2. Frequency doubling of 4.3 um band of CO_2 laser radiation was made in the antireflection coated up to 87.5% 7 mm crystal. A 9.2mm crystal was used with the SHG of cw CO laser radiation, and a 7 mm crystal was used with the Q-switched CO laser radiation. The damage intensity for cw radiation was slightly higher than that for CO_2 laser radiation. The Q-switched CO laser pulses were of 22-1 us duration, the pulse repetition rate was 10-200 Hz, energy per pulse was 2-0.1mJ. Maximum mean power efficiency of doubling exceeded 3% and was obtained at 84.5 mW pumping power with f = 75 Hz. Maximum value of SH mean power being equal to 4 mW was obtained at 194 mW pumping with f = 89 Hz. The power of SH cw radiation did not exceed 10 mW.

Radiation frequencies of cw lasers were summarized in the 3.1 mm crystal at 0.5° convergence angle. The CO_2 and CO laser radiation powers were 5.7 W and 4.7 W, respectively. The Tl_3AsSe_3 frequency doublers have been tested at present in the lidar. The efficiency of Q-switched CO_2 laser radiation doubling was $2.2 \cdot 10^{-2}$ % and TEA CO_2 laser radiation was 2.3% with a significant surface destruction of the first shot. The crystal length was 4.65 mm, the optical losses were $\alpha = 0.2$ cm^{-1}. Stable efficiency of doubling in the last case was 0.8 %. The FC efficiencies and energy characteristics of radiation generated in most cases were limited by low power of pumping radiation and by the absence of crystal antireflection coating.

It should be noted that with doubling of the 4.3 µm emission of CO_2 laser, whose average power did not exceed 10 mW, and pulse-periodic CO laser emission with average power 200 mW, the converted power was measured in milliwatt.

For SHG with $ZnGeP_2$ of CO_2 laser operating in selective regime the angular phase-matching width exceeded 4°, and for CO lasers it was 2°06'. For SHG of CO laser operating in nonselective regime and for mixing of CO_2 and CO lasers it was 2°50'. The angular width of phase matching for SHG of CO_2 laser radiation in $CdGeAs_2$ and Tl_3AsSe_3 was 3° and 1.8°, respectively. For SHG of the radiation in $ZnGeP_2$ the above spectral width of phase matching was 4.9 cm^{-1} for a 10 mm crystal, and at SHG in Tl_3AsSe_3 it was of the same order of magnitude. The CO laser radiation frequency tuning did not cause the necessity for adjustment of $ZnGeP_2$ monocrystals to phase-matching angle, including the heating up to temperature of 100°C. The large 40-50°C temperature range of phase matching is typical for SHG of CO_2 laser in $ZnGeP_2$. For SHG in Tl_3AsSe_3 monocrystals it was 6°.

Resistance to humidity and thermal impact of $ZnGeP_2$ was determined by periodic placing of crystals into boiling water and liquid nitrogen. After two-three tens of cycles the crystals were not cracked, the characteristics of surfaces were invariable. Thus, high efficiency of conversion and excellent operation parameters of frequency converters with $CdGeAs_2$, $ZnGeP_2$ and Tl_3AsSe_3 enable one to create the coherent sources of 2-17 μm range which can be competitive with another known types of sources.

Table I. Efficiencies of frequency converters

FC type	Crystal	Pumping laser	Laser parameters λ, μm	τ, s	Efficiency (internal), %
SHG	$ZnGeP_2$	Hybrid CO_2	9.28	2.10^{-9}	49(83.5)
	$ZnGeP_2$	Hybrid CO_2 SH	4.64	$1.5.10^{-9}$	14(22)
	$ZnGeP_2$	TEA CO_2	9.2-10.8	2.10^{-7}	9.3
	$ZnGeP_2$	Q-switch.CO	5.3-6.1	4.10^{-5}	3.1(5.6)
	$ZnGeP_2$	Q-switch.CO	4.3	$3.3.10^{-7}$	8.4(10.1)
	$CdGeAs_2$	pulsed NH_3	11.7	$1.5.10^{-7}$	2 (5.2)
	Tl_3AsSe_3	Q-switch. CO_2	9.2-10.8	10^{-3}	2
FHG	$ZnGeP_2$	Hybrid CO_2	9.28	2.10^{-9}	1.4
SFG	$ZnGeP_2$	Q-switch. double-wave CO_2	= 4.3 =10.4	3.10^{-7} 6.10^{-7}	20% of 4.3 μm radiation
	$ZnGeP_2$	CW CO and CO_2	=5.3-6.1 = 10.6	- -	0.6 mW

GAS ANALYSIS USING CO_2 LASER FREQUENCY CONVERTERS

Yu.M.Andreev, P.P.Geiko, V.V.Zuev, O.A.Romanovskii
Institute of Atmospheric Optics, Siberian Branch
USSR Academy of Sciences, Tomsk, 634055, USSR

It is known that the 2-15 μm region of the middle IR holds considerable promise for sensing the atmospheric gas content by the differential absorption method. On the other hand, the spectral region and the method considered are not widely used in practice for the lack of the required radiation sources. To construct such sources one should solve the following problems: overlapping of a wide spectral interval by frequency tunable radiation; formation of a narrow spectral line width with high energy parameters; continuous control for the output-radiation spectral characteristics; and provision for reliability of operation and efficiency of generation.

The goal of our research is to study the possibilities and ways of developing multi-purpose gas-analyzers based on CO_2 and CO low-pressure lasers and a set of frequency converters of their radiations: harmonic, sum-frequency and difference-frequency generators based on the nonlinear $ZnGeP_2$ and Tl_3AsSe_3 crystals developed.

The spectral interval between individual lines of molecular low-pressure lasers is up to 2 cm^{-1}, thus the problem of spectral selection and frequency tuning is simplified. The 2 to 17 μm spectral region can be overlapped by frequency converted spectrum of single CO_2 laser emission with a set of two-cascade frequency converters. Two CO_2 or CO and CO_2 lasers can be used here. Using combinations of different emission line pairs or of these lines and their second harmonics, one can obtain very dense, with up to 10^{-3} cm^{-1} step, overlapping of this spectral region. A narrow 10^{-2} - 10^{-3} cm^{-1} spectral line width and the spectral line center position are specified by physical parameters of an active medium. That is why the line center positions and those of converted frequencies are known with high accuracy. This enables one to overcome the problem on formation and control of spectral characteristics. Such a radiation source, used under tropospheric conditions, extracts the information on the atmospheric parameters not worse than continuously tunable lasers do. High efficiency of frequency converters developed, tenths to tens per cent, depending on the type of the pumping lasers used, allows one to develop the trace meters operating with topographic targets. Due to this fact, the remote measurements with the use of large-scale optics, at least in the range to 5.5 μm, can be carried out. These converters have high reliability of operation.

For the first step, a gas analyzer with a line tunable CO_2 laser was modernized. It made it possible in field conditions to measure, in addition to concentrations of ethylene, water vapor, ammonia, ozone, the CO concentration using $ZnGeP_2$ frequency doublers [1]. The CO concentration sensitivity was 4 ppb at the 2 km path length. Under the laboratory conditions the possibility of measuring N_2O and H_2O vapor concentrations with sensitivity sufficient for measurements in winter was shown [2]. The CO absorption line centered at frequency ν = = 2154.596 cm^{-1} and used for measurements is strongly perturbed by the H_2O vapor absorption line. The CO concentration can be correctly measured in this case either at low atmospheric humidity ($\leqslant 1$ g/m^3), or at multifrequency sounding, where the inverse incorrect problem is solved using the Tikhonov's regularization method. The main disadvantage of the $ZnGeP_2$ frequency doubler was the necessity for crystal heating to temperature above 100°C for frequency doubling with the wavelength 10.3 um.

The gas analyzer was further modernized by replacement of a $ZnGeP_2$-crystal frequency doubler by a Tl_3AsSe_3 doubler operating at room temperature. The CO absorption line, free of the H_2O vapor perturbation effect, with the center at 2086.3266 cm^{-1} was found, Fig.1. It coincided with the second harmonic of the 9P(24) $^{12}C^{16}O_2$ laser line and had the same 29.8 $cm^{-1}atm^{-1}$ cross-section. The measurement results of time variations of CO concentration measured based on the found coincidence are presented in Fig.2. The path length was 250 m. A plane mirror was used as a reflector. The results presented were averaged over 6 min intervals.

When CO concentration was measured with a topographical target (a plywood removed at 25 m distance), the signal-to-noise ratio was not lower than 10, if CO concentration was higher than 0.5 ppm. A Q-switched CO_2 laser at pulse duration 100 μs, repetition rate of about 100 Hz, peak power 150 W and MDS-structure of the oxidized n-InSb were used here.

Then tunable CO_2 and CO lasers, a Tl_3AsSe_3 SHG and a set of different $ZnGeP_2$ frequency converters were added. Practically all atmospheric gases can be measured using this gas analyzer. A detailed analysis of the N_2O absorption spectrum, including cell measurements, revealed the possibility of measuring the N_2O background and higher concentrations with the use of the absorption line at 2186.002 cm^{-1}. The emission at this frequency was obtained by summarizing frequencies of the main-isotope CO_2 laser line 9R(40) and the isotope $^{12}C^{18}O_2$ laser line 9P(18). The absorption cross-section of this line was 9.8 $cm^{-1}atm^{-1}$ that allowed measurements of N_2O background concentrations to be made with the same accuracy as the CO measurements were made. The N_2O concentration measurements with the use of a topographic target were also successful.

The results obtained show that our technique is quite promising.

1. Yu.M.Andreev, V.G.Voevodin, A.I.Gribenyukov et al. Sov.J. Applied Spectroscopy 47, N1, 15-20 (1987).
2. Yu.M.Andreev, V.G.Voevodin, A.I.Gribenyukov et.al. VIIIth All-Union Symposium on Laser and Acoustic Sounding of the Atmosphere. Tomsk, part 1, p.277-280.

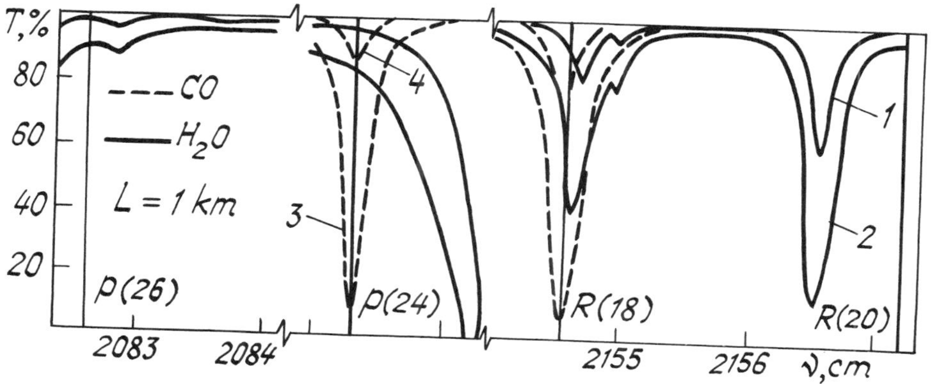

Fig.1. The atmospheric transmission spectrum for the ground 1 km path in the CO_2 laser SH region used for sounding CO 3(0.1 ppm), 4(1 ppm), with water vapor concentrations 14 g/m^3 (1) and 3.5 g/m^3 (2).

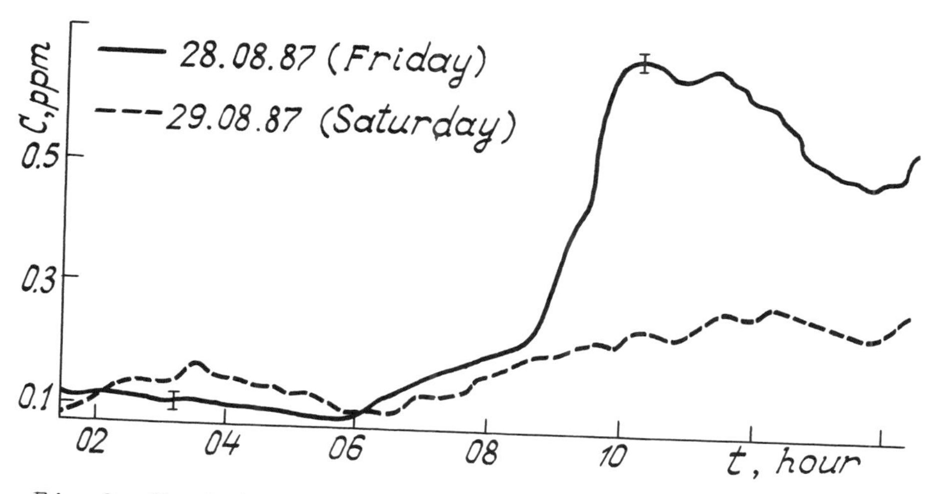

Fig.3. Variation of CO concentration in the suburb of an industrial center.

OBSERVATIONS ON SECOND HARMONIC GENERATION IN MEROCYANINES

R.G.S. Pong,* C.L. Marquardt, and James S. Shirk
Naval Research Laboratory, Washington, DC 20375-5000

ABSTRACT

Merocyanine samples were produced photochemically from six derivatives of spiro[[2H]-1-benzopyran-2,2'-[2H]indole] by precipitation from methylcyclohexane or ethanol solutions. Several of the photolyses were performed in an electric field to evaluate orientational effects. Second harmonic generation efficiencies relative to quartz were measured at 1.06 μm using a powder method. Infrared absorption spectra and spectral dichroism were also investigated. It was found that many merocyanines could be prepared in a noncentrosymmetric form without admixture of residual spiropyran. Application of an electric field during photolysis was found to produce macroscopic alignment of the precipitate but not an alignment of molecular dipoles.

INTRODUCTION

Merocyanines have been investigated extensively as potential nonlinear optical materials because of their high molecular polarizabilities.[1] Production of a bulk merocyanine material for second harmonic generation (SHG) has been achieved in alternating Langmuir-Blodgett films,[2] but the only x-ray diffraction studies reported for pure merocyanine crystals have shown them to be centrosymmetric.[3] Observations of visible dichroism and SHG in merocyanine crystallites produced by photolysis in an applied field of 10^3 V/cm have been reported as evidence of molecular alignment.[4] Present experiments were undertaken to further investigate techniques for producing noncentrosymmetric bulk merocyanines. The results demonstrate that many merocyanines can be prepared in a noncentrosymmetric form by photolysis in methylcyclohexane solution. Application of an electric field during photolysis is shown to produce macroscopic alignment of the precipitate, but not an alignment of molecular dipoles.

EXPERIMENTS

Materials studied were the merocyanine forms of six derivatives of spiro[[2H]-1-benzopyran-2,2'-[2H]indole]:

(MET) 3',3'-dimethyl-1'[2-methacroyl-oxyethyl]-6-nitro(<u>spiro-[[2H]-1-benzopyran-2,2'-[2H]indole]</u>)
(6NT) 6-nitro-1',3',3'-trimethyl-(_)
(BMN) 5-bromo-8-methoxy-6-nitro-1',3',3'-trimethyl-(_)
(CDN) 5'-chloro-6,8-dinitro-1',3',3'-trimethyl-(_)
(BMP) 5-bromo-8-methoxy-6-nitro-1',3',3',5',6'-pentamethyl-(_)
(CNT) 5'-chloro-6-nitro-1',3',3'-trimethyl-(_)

*Sachs/Freeman Associates, Landover,MD 20785

MET is the compound in which alignment by electric field was previously reported.[4] 6NT has a relatively simple molecular structure; it was included in order to facilitate interpretation of the IR spectra.

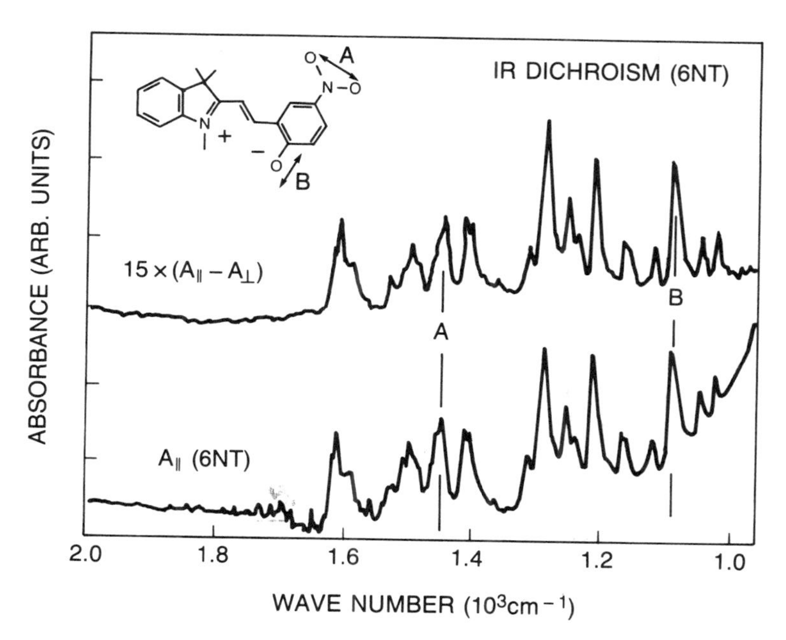

Fig.1 IR absorption spectrum of an aligned sample of 6NT for light polarized parallel to the alignment axis (A//). Spectral dichroism (A// - A⊥) is displayed in 15 X magnification.

Each merocyanine sample was prepared by UV photolysis of the parent spiropyran in methylcyclohexane (or 95% ethanol solution) and subsequent evaporation of the solvent. UV irradiation for photolysis was provided by a UV multiline Ar^+ laser (351-363nm), a Nd:YAG laser (3rd harmonic, 353nm), or a Hg/Xe arc lamp with filters (350-390nm). Samples for IR spectroscopy were deposited on CaF_2 substrates with electric fields ranging from 0-10 kV/cm applied during photolysis. These samples were also evaluated for SHG using a Q-switched Nd:YAG laser (1.06 μm). More quantitative SHG measurements were made on powder samples prepared by photolysis of larger quantities of parent solution without application of an electric field.

RESULTS AND DISCUSSION

Samples produced by photolysis of methylcyclohexane solutions in electric fields $\geq$ 1 kV/cm exhibited properties previously reported.[4] They deposited on the substrate in the form of "strands" aligned along the field direction. These aligned samples had greater optical absorbance for visible light polarized parallel to the alignment direction than for light polarized perpendicular to the alignment direction (A// > A⊥). To determine whether these aligned samples consisted of arrays of oriented molecules, we measured the spectral dichroism(A// - A⊥) in the IR. Results were qualitatively the same for all compounds; they are illustrated in Fig. 1 for 6NT. Bands labelled A and B are identified as the O-N-O asymmetric stretch (1470 cm^{-1}) and the C-O stretch (1100 cm^{-1}) which have their respective transition moments parallel and perpendicular to the molecular dipole. If the molecular dipoles were oriented along the alignment direction, the dichroism(A// - A⊥) would be positive for band A, but negative for band B. The observed dichroism is positive throughout the entire visible and IR spectral regions, and is probably wholly attributable to differential scattering.

Small SHG signals and enhancement of the SHG signals by subsequent application of electric fields were observed in the aligned samples as previously reported.[4] However, experimentally indistinguishable effects were also observed in samples prepared under identical conditions but in zero electric field. These results, together with those of the dichroism experiments, indicate that macroscopic alignment of merocyanines produced by photolysis in an electric field does not result in alignment on the molecular level.

Table I Powder SHG relative to Quartz

Compound	SHG Powder Signal
MET(methylcyclohexane)	2.2×10^{-4}
6NT(ethanol)	2.2×10^{-4}
BMN(methylcyclohexane)	1.2×10^{-2}
CDN(methylcyclohexane)	3.7×10^{-2}
BMP(methylcyclohexane)	5.5×10^{-2}
6NT(methylcyclohexane)	3.0×10^{-1}
CNT(methylcyclohexane)	1.5×10^{0}

Powder measurements of SHG from all six merocyanine compounds are presented in Table I. SHG efficiencies found for MET, and for 6NT precipitated from ethanol solution, are only slightly above the detection limit in our experiment ($10^{-4} \times SHG_{quartz}$). This is consistent with previous evidence that these crystal structures are centrosymmetric;[3] the sizes of the SHG signals are approximately as anticipated for small crystallites with large surface area and internal imperfections. In the remaining samples the SHG signals are 2-4 orders of magnitude larger, far too large to be attributable to surfaces and defects. This result implies that these five merocyanine compounds form noncentrosymmetric structures when precipitated from methylcyclohexane.

REFERENCES

1. D. J. Williams, Angew. Chem. Int. Ed. Engl. 23, 690 (1984)
2. I. R. Girling, Opt. Comm. 55-4, 289 (1985)
3. S. M. Aldoshin, Izv. Akad. Nauk. Ser. Khim. 12, 2720 (1981)
4. G. R. Meredith, J. Phys. Chem. 87, 1697 (1983)

NONLINEAR OPTICAL EFFECTS DUE TO IR INDUCED PHOTOISOMERIZATION AND PHOTOORIENTATION

by

James S. Shirk
Naval Research Laboratories
Washington, D.C. 20375

ABSTRACT

A class of organic molecules will photoisomerize when irradiated into resonances in the OH and CH stretching (2.7-3.4 micron) region. These reactions, which have been studied principally in low temperature solid hosts, are single photon IR reactions. These photoisomerizations can lead to high quantum efficency saturable IR absorbers. The resulting near resonant nonlinear refractive index of an example system is reported.

INTRODUCTION

The number of materials with large optical non-linearities in the 3-5 and 8-14 micron region of the infrared is limited. Many non-linear materials, including the organic non-linear materials absorb in this region. The absorptions are due to vibrational transitions which are generally not useful for resonant non-linearities because it is difficult to saturate vibrational transitions; the v=1 to v=2 and higher transitions are usually almost degenerate with the v=0 to v=1 transition.

Recently it has been discovered that some organic molecules will photoisomerize when irradiated into resonances in the 2.7-3.4 micron region.[1,2] These photoisomerizations are single photon processes. The photoinduced reaction is a rotation about a single bond, i.e. a conformational change. The barrier for these reactions is generally 10 to 40 kJ/mole. If the photoisomerization is carried out with polarized light in solid hosts, an oriented dichroic sample results.[3] Spectroscopy of such photooriented samples can provide considerable information on the dynamics of the solid state isomerizations.[4]

RESULTS AND DISCUSSION

IR photoisomerizations are shown to give rise to infrared photochromism. The photochromism has been used to make materials which are easily saturable IR absorbers at cryogenic temperatures.

As an example of this effect, the optical non-linearities resulting from saturation of the 2.7 micron band of 2-fluoroethanol in solid Ar have been calculated from measured quantum yields and absorption cross sections. The near resonant non-linear refractive index in this case is given as:

$$n_2 = 7 * 10^{-2} \phi T ((\delta+i)/(1+\delta^2)^2) \text{ esu} \quad (1)$$

where ϕ is the quantum yield and T is the response time (in sec) for the non-linearity, which is also the rate of the back isomerization from the photochemically produced conformation, $\delta = (\omega - \omega_0)/\Delta\omega$.

The quantum yield for this photoisomerization was found to depend upon the vibrational mode excited. It varied from 10^{-2} to near 1. For the 2.7 micron band the quantum yield was found to be 0.18. The randomization time has been measured, it is temperature dependent and follows an Arrhenius law. It varies from many hours near 10 K to an estimated 50 nsec at room temperature.

We have demonstrated that in a solid host at low temperatures saturation can be achieved for 2-fluoroethanol with laser fluences of a few hundred mJ/cm^2.

Most of our measurements have been carried out at low temperatures where the optical properties are easy to measure because the recovery time is slow, hence time resolution is not critical. Room temperature studies in liquid hosts are in progress.

REFERENCES

1. A.E. Shirk and J.S. Shirk, Chem. Phys. Lett. 97, 549 (1983)
2. H. Frei and G.C. Pimentel, Ann. Rev. Phys. Chem. 36, 491 (1985)
3. W.F. Hoffman, A. Aspiala, and J.S. Shirk, J. Phys. Chem. 90, 5706 (1986)
4. Z.K. Kafafi, C.L. Marquardt, and J.S. Shirk, Paper # TuK-6 Third Int'l Laser Science Conf., Atlantic City, N.J. (1987)
5. W.F. Hoffman and J.S. Shirk, J. Phys. Chem. 89, 1715 (1985)

OPTICAL GENERATION OF COHERENT SURFACE ACOUSTIC WAVES:* A NEW PROBE OF SURFACE DYNAMICS AND STRUCTURE

J.J. Kasinski, L. Gomez-Jahn and R.J.D. Miller
Dept. of Chemistry, University of Rochester, Rochester, NY 14627

ABSTRACT

The optical generation of coherent surface acoustic waves (SAW) is shown up to frequencies as high as 2 GHz on semiconductor surfaces. The frequency response of the acoustic coherence is directly related to thermal relaxation processes within the surface layer. Propogation of the optically generated SAW is shown to be sensitive to phase restructuring that occurs at solid-liquid interfaces.

INTRODUCTION

Previous studies of optical coupling to a single frequency SAW via photo thermal effects used optical pulses that were too long to excite the SAW coherently.[1] Consequently, the coupling was less efficient than if coherence had been achieved. We have extended these earlier studies using much shorter pulses and have used SAWs to explicitly probe surface processes of fundamental importance to surface chemistry. The particular system under study was n-TiO_2, which can photodissociate water into oxygen and hydrogen with high efficiency.[2] The first reaction step in this very complicated surface reaction sequence is that of interfacial electron transfer. The physics of the electron transfer step have recently been questioned. The highly quantized nature of the space charge region of the semiconductor, which occurs upon the formation of a liquid junction, should dramatically alter carrier thermalization processes from single phonon cascades to much slower multiphonon relaxation processes.[2] The timescale for this is thought to be on the hundred picosecond timescale which would enable hot carrier transfer processes to dominate the interfacial electron transfer process. In addition to this consideration, the static potential of the TiO_2 surface is expected to align water molecules at the atomic surface layer. The electrostatic interaction between the polar water molecules and the surface could change the phase structure of the water layer for several atomic layers and significantly reduce the dielectric relaxation rate of the aqueous phase to a charge separation event. The latter process is an integral component to the overall dynamics of interfacial electron transfer. Optically generated SAWs can address both thermal relaxation processes and phase restructuring at the surface.

RESULTS AND DISCUSSION

The experiment is essentially a transient grating (described previously) where above band excitation pulses are used to write an optical interference pattern on the semiconductor surface and a below band gap pulse is used to probe a surface phase grating. The SAW is excited within the surface layer by thermal relaxation of the photogenerated electron-hole carriers and lattice expansion which exactly mimics the optical interference pattern. This thermal expansion is highly constrained to the surface by the short optical penetration depth of the grating excitation pulses (~1000Å) such that a surface wave rather than a bulk wave is selectively excited. A detailed theoretical analysis of the photothermal coupling, which will be given elsewhere,[3] demonstrates that either the pulse duration or the thermal relaxation processes should be faster than one quarter the acoustic period or else a decrease in the depth of signal modulation would be observed. The excitation pulse durations (100 psec) were less than this in the present experiment. The frequency response of SAW coherence enables a direct measurement of energy relaxation processes within the surface layer. Experimental results are shown in figure 1. Under the high excitation conditions used (10^{20} carriers/cm^3), Auger recombination dominates the non-radiative relaxation processes and increases the overall energy thermalized into the lattice so as to increase the SAW amplitude. The SAW signature is the distinctive oscillations in the diffracted probe signal superimposed on top of a population carrier phase grating. The initial fast decay is due to Auger recombination of the carriers and disappears at lower intensities. From the optical fringe spacing (i.e. acoustic wavelength) and the measured standing wave frequency of figure 1A, the speed of sound is measured to be 4.7×10^5 cm/sec which is within 5% of the known SAW velocity of TiO_2. All other acoustic mode velocities are considerably higher. Figure 1A clearly shows the coherent optical excitation of a single frequence SAW. A striking difference was observed when this same crystal was used to construct a liquid junction. The data shown in Figure 1B, rather than showing a single frequency SAW, shows a beat frequency corresponding to three different wave velocities. The intrinsic TiO_2 SAW component (4.7×10^5 cm/sec) is observed as above, along with an interfacial aqueous acoustic component (1.5×10^5 cm/sec) which is expected in this non-traction free case for SAW propogation. However, there is a third component (4.3×10^5 cm/sec) which corresponds to a solid phase velocity component. Since this component is only observed in the presence of the aqueous phase, it must be related to restructuring of either the TiO_2 or H_2O interface. We attribute this high velocity component to an ice like structure for the tightly bound water layer at the TiO_2 surface. The acoustic mode, which propagates along this layer and modulates the surface deformation, must be excited from thermal energy transfer across the TiO_2/H_2O interface and not by the SAW itself.

Lower intensity studies (3 X 10^{18} carriers/cm^3) have enabled a measurement of thermal relaxation within the highly quantized space charge layer of this n-TiO_2/aqueous junction. AT 2 GHz, there is no difference in the SAW coherence or depth of signal modulation when a junction is present or not. The SAW frequency range will be extended and lower intensities used but this result indicates that thermalization is occurring within the space charge layer on time-scales faster than 300 psec. This result demonstrates that previous measurements of hole minority carrier transfer processes (~400 psec dynamics) were involving thermalized hole vacancies at the atomic surface layer. The SAW frequency range can be extended up to 50 GHz giving a factor of 25 improvement (~10 psec resolution) on characterizing the thermalization processes within a highly quantized surface space charge layer.

*This work was supported in part by the Department of Energy Grant #81-049.

REFERENCES

1. a) R.E. Lee and R.M. White, Appl. Phys. Lett. 12, 12 (1968).
   b) G. Cahier, Appl. Phys. Lett. 17, 419 (1970).
2. D.S. Boudreaux, F. Williams and A.J. Nozik, J. Appl. Phys. 51, 2158 (1980.
3. J.J. Kasinski, L. Gomez-Jahn, S. Gracewski and R.J.D. Miller, to be submitted to Opt. Lett.

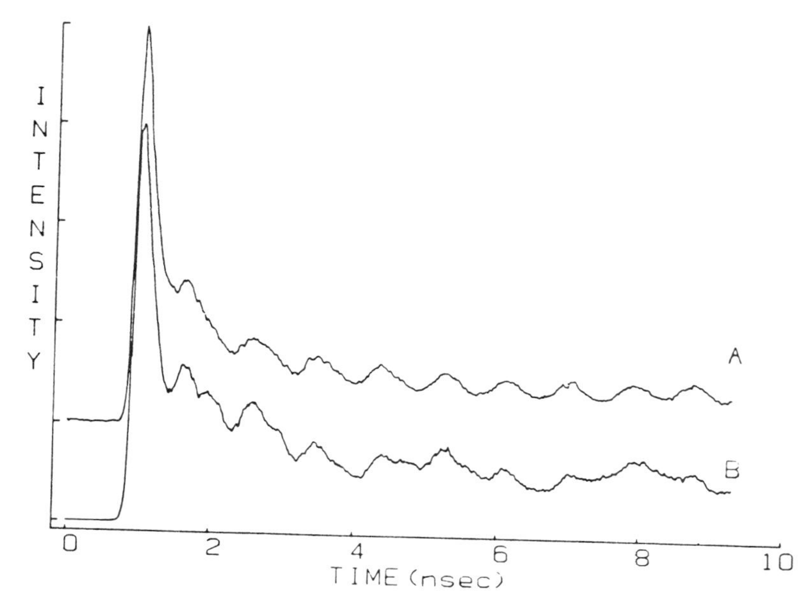

Figure 1. Optical SAW generation; n-TiO_2 in air (A), and in the presence of an aqueous liquid junction (B).

QUANTUM FLUCTUATIONS IN THE STIMULATED RAMAN SCATTERING LINE WIDTH

D.C. MacPherson and J.L. Carlsten
Physics Dept., Montana State University
Bozeman, Montana 59717

ABSTRACT

Results of measurements of the Raman linewidth in an H2 Raman generator at 10 and 32 atmospheres using a Fabry Perot interferometer are presented. The results show that the measured linewidth can be much narrower than the linewidth predicted from gain narrowing. In addition the measured power spectrum exhibits large frequency fluctuations which may be related to observed soliton decay.

INTRODUCTION

The stimulated Raman scattering effect amplifies the statistics of spontaneous emission to a measurable macroscopic field. The conversion of a pump laser pulse to a frequency shifted Stokes pulse in a Raman cell is essentially a single pass laser. However unlike a conventional laser the Raman laser does not have a cavity and therefore the Stokes power spectrum is not influenced by cavity modes.

Numerical models have been quite successful in modeling the growth of a known Stokes seed in a Raman amplifier. However many questions remain as to how to properly model the transition from vacuum fluctuations to a small but real Stokes field[1]. We present preliminary data from an experiment which will be used to measure the statistics of fluctuations in the Stokes power spectrum for comparison with those from computer models.

Fluctuations in the Stokes pulse energy have previously been observed by Raymer et al.[2] and recently frequency fluctuations in a cavityless dye laser have also been observed. These fluctuations appear to be similar to what we have observed but for a broader linewidth system.

EXPERIMENTAL CONDITIONS

For our experiment a frequency doubled, single mode, Nd:YAG laser operating at 532nm was used to pump a multi pass Raman cell. A 1.5m cell was centered between two concave mirrors with a focal length of 50cm. The pump laser beam entered the multi pass cell through a hole in one mirror and did not retrace its path before exiting through a hole in the opposite mirror. The beam had a confocal parameter of 38cm and made 15 passes through the Raman cell. The laser pulse had a nearly gaussian temporal profile with a FWHM of 26ns and approximately 1mJ of energy.

The Stokes pulse grew from spontaneous emission and at the

exit of the cell had fully depleted the central portion of the pump laser pulse. A typical Stokes pulse had a fast rise and fall time and a FWHM of 12ns. The Stokes beam was then expanded and passed through a parallel plate Fabry Perot interferometer with a plate spacing of 15cm and a finesse of approximately 45. The divergence of the Stokes beam was adjusted to obtain two orders of rings. A cross section of the rings was measured using a photodiode array which was interfaced to a computer. The computer scaled the data to produce a power spectrum of the Stokes pulse.

To ensure stability in the frequency of the laser pulse its power spectrum was also monitored using a second Fabry Perot interferometer. Between thermally induced mode jumps, when the Stokes pulse was measured, the laser frequency was found to be quite stable. Thus our experimental setup allows us to make a power spectrum measurement on a single shot.

EXPERIMENTAL RESULTS

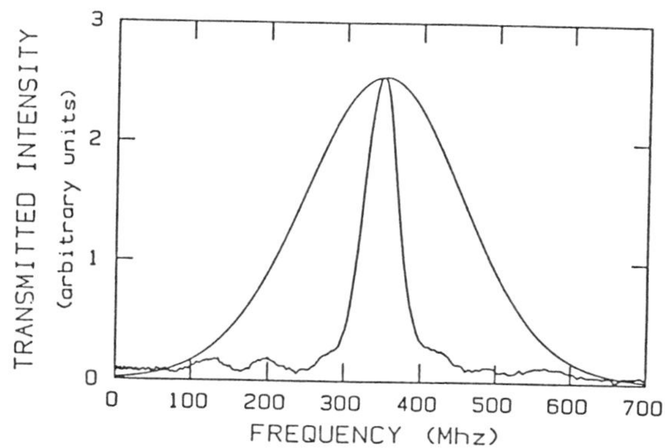

Fig.(1) Power Spectrum

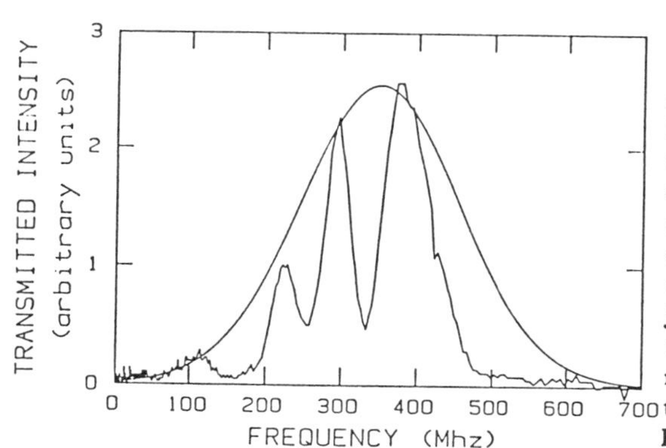

Fig.(2) Power Spectrum

Our first measurements were made with 10 atm. of H2 in the Raman cell. At 10 atm. the collisionally broadened Raman linewidth is 525MHz. Due to gain narrowing we expect this linewidth to be narrowed by approximately a factor of 7 resulting in a 75MHz linewidth FWHM[3]. The measured linewidth was 60MHz for some shots and for roughly 1 in 10 shots the power spectrum consisted of two well resolved spikes. Fluctuations in the location in frequency of the power spectrum could also be seen. The Fourier transform of a typical Stokes pulse had a FWHM of 57MHz. It appears that the pulses are transform limited but the central frequency jumps from shot to shot to build up the gain narrowed linewidth. As a test the pressure in the Raman cell was increased to 32atm. to give a gain narrowed linewidth of 240MHz. The power spectrum from two shots at this pressure are shown in figures 1 and 2, along with the predicted gain narrowed linewidth. The location of the gain narrowed line was visually centered about the measured power spectrum because its actual location has not been determined. On some shots the Stokes pulse is still essentially transform limited as indicated in figure 1. On other shots the power spectrum breaks up into two or more peaks, somewhat resembling cavity modes. We have not yet made an accurate measurement of the average linewidth, however from Fabry Perot scans using a collimated Stokes beam our estimates appear to be quite close.

DISCUSSION

We believe the statistics of these large fluctuations in the Raman power spectrum will provide a good test for the modeling of the growth of spontaneous emission into a macroscopic field. A model being used by Englund and Bowden treats the initiation of the Stokes pulse by injecting a real Stokes seed which has a randomly fluctuating phase and gaussian amplitude statistics with the pump pulse. Through the nonlinear interaction with the medium this random seed develops substantial coherence. Our experiment will test wether this model predicts the correct amount and fluctuations of the coherence.

Finally the frequency jumps in the Stokes line may be responsible for the observed decay of the Raman soliton[4]. These solitons are formed by electrooptically inducing a π phase shift in the Stokes beam between a Raman generator and amplifier[5]. Frequency jumps would cause the Stokes beam to be off resonance in the amplifier which computer models show would lead to soliton decay.

REFERENCES

1. J.C. Englund and C.M. Bowden, Phys. Rev. Lett. 57, 2661 (1986) : C.M. Bowden and J.C. Englund, Optics Comm. Special Issue (to be published)

2. I.A. Walmsley and M.G. Raymer, Phys. Rev. Lett. 50, 962 (1983) : C. Radzewicz, Z.W. Li and M.G. Raymer (to be published)

3. M.G. Raymer, J. Mostowski, and J.L. Carlsten, Phys Rev. A19, 2304 (1979)

4. K.J. Druhl, Proceeding of the Third International Laser Science Conference (1987).

5. D.C. MacPherson, J.L. Carlsten, and K.J. Druhl, J.Opt.Soc.Am.B 4,1853,(1987)

ENHANCED NONLINEAR PROCESSES BY MICROPARTICLE SURFACE MODES IN NONLINEAR MEDIA

T.P. Shen and D.N. Rogovin
Rockwell International Science Center
1049 Camino Dos Rios
Thousand Oaks, CA 91360

INTRODUCTION

It is well known that resonant excitation of microparticle surface modes will give rise to enhanced nonlinear processes of molecules localized in the immediate vicinity of the particle surface. For these processes interest focus's on the behavior of the field at the microparticle surface and little attention is given to its detailed behavior for distances far from the particle surface.

Here we examine the nonlinear electrodynamics of a two-component system consisting of small metallic microparticles embedded in a nonlinear medium. In particular, we examine second harmonic generation (SHG) in a composite medium, which consists of a silver microsphere embedded in a KDP crystal. The silver sphere is irradiated by a laser light of 0.4413 μ wavelength, which is resonant with the Fröhlich mode of the silver sphere. In turn this enhances the scattered field for distances on the order of several times the microparticle dimension from the particle surface and thereby increases the nonlinear polarization generated by the nonlinear medium. As a consequence, the detailed spatial structure of the surface-enhanced electric field plays an important role in the nonlinear electrodynamics of the composite.

THEORY

The surface enhancement of SHG is most readily appreciated by studying a situation in which the nonlinear medium is so configured that harmonic generation can occur only in the presence of the microparticle. For example, a KDP[1] crystal irradiated by a coherent light which is linearly polarized with respect to its optic axis (z-axis) and propagating along the y-axis. The electric field of the incident radiation at ω is given by

$$\vec{E}_0(\vec{r},t) = \hat{z}\, E_0(\vec{r},t) \quad , \tag{1}$$

where $E_0(\vec{r},t) = 1/2\, E_0\, e^{i(Ky-\omega t)} + c.c.$, $K = (\omega/c)\sqrt{\varepsilon_0(\omega)}$, and $\varepsilon_0(\omega)$ is the linear dielectric constant[1] of KDP at ω.

In the Raleigh regime, the total field outside the particle is given by

$$\vec{E}(\vec{r},t) = \vec{E}_0(\vec{r},t) + \delta\vec{E}(\vec{r},t) \quad , \tag{2}$$

where the scattererd field is given by

$$\delta\vec{E}(\vec{r}) = \{3f(\omega)\,\frac{z}{r}\,\hat{r} - f(\omega)\hat{z}\}\,\frac{r_0^3}{r^3}\,E_0(\vec{r},t) \quad . \tag{3}$$

r_0 is the radius of the particle and $f(\omega)$ is the local field factor.[2] The scattered field gives rise to radiation with polarization components orthogonal to the incident wave and will generate a nonlinear polarization within the composite which will emit second harmonic light.

The second harmonic field generated by the microsphere embedded in KDP is given by

$$\vec{E}^{(2)}(\vec{r}) = \frac{e^{ikr}}{r} \vec{f}(\vec{r},k) \quad , \tag{4}$$

where $k = (2\omega/c)\sqrt{\varepsilon_0(2\omega)}$, $\varepsilon_0(2\omega)$ is the linear dielectric constant[1] of KDP at 2ω and the scattered amplitude is given by

$$\vec{f}\,(\vec{r},k) = \left(\frac{2\omega}{c}\right)^2 \int_V d^3r'\, e^{i\eta} \int dt\, e^{-2i\omega t}\, \vec{P}(\vec{r},t) \quad , \tag{5}$$

the integration volume v is the volume of KDP excluding the microsphere, the phase factor $\eta = kr'Tt\cdot r$, and $t = (\Delta y - r)/T$, $T = [\Delta^2 + 1 - 2\Delta \sin\theta \sin\phi]^{1/2}$ and $\Delta = [\varepsilon_0(\omega)/\varepsilon_0(2\omega)]^{1/2}$. Then the differential cross section is given by

$$\frac{d\sigma}{d\Omega} = \frac{288\pi^2}{c}\, g G(kr_0;\, \theta,\phi)\, \sigma_0 I_0 \quad , \tag{6}$$

where $\sigma_0 = \pi r_0^2$, I_0 is the incident laser intensity, g is the appropriate second order optical coefficient of KDP and

$$G(kr_0;\, \theta,\phi) = \frac{|f(\omega)|^2}{|\varepsilon_0(2\omega)|^2}\, \frac{\cos^2\theta}{\sqrt{\varepsilon_0(\omega)}\, T^6} \left(1 - \frac{\cos^2\theta}{T^2}\right) \times \left(\frac{\sin kr_0T}{T} - \cos kr_0T\right)^2 \quad . \tag{7}$$

In Eq. (7), the contributions from the short range components (r^{-6}) in the nonlinear polarization are small compared to the long range (r^{-3}) ones, and have been omitted. The second harmonic radiation pattern is virtually independent of ϕ for small polar angles (θ). However, as one approaches the xy-plane, the radiation pattern becmoes strongly peaked in the forward dielection, namely, along positive y-axis and$(d\sigma/d\Omega) \to 0$ as $\theta \to 90°$.

As an example, we estimate about 100 photons/sec generated at second harmonic frequency for the incident light of wavelength $\lambda = 0.4413\ \mu$, $I_0 = 1\ \mathrm{MW/cm^2}$ and the radius of the silver sphere $r_0 = 220.6$Å. This represents an enhancement of over 100, namely, $|f(\omega)|^2 = 101.2$.

REFERENCES

1. F. Zernike and J.E. Midwinter, Applied Nonlinear Optics (John Wiley and Sons, New York 1973).
2. C.J.F. Bottcher, Theory of Electric Polarization, (Elsevier, Amsterdam, 1973) p. 78.

THEORY OF NONLINEAR OPTICS IN DROPLETS

G. Kurizki, Chemical Physics Department,
Weizmann Institute of Science, Rehovot, Israel 76100

The experiments by Chang et al. on lasing[1] and stimulated Raman scattering (SRS)[2] in spherical droplets with radii of 30-40 μm have revealed certain intriguing features. In particular: 1) The thresholds for lasing and SRS are drastically lowered compared to bulk or even compared to high-quality Fabry-Perot resonators. 2) The peaks of stimulated emission are shifted from the Mie scattering resonances of the inactive droplet medium. The theory[3] outlined below is capable of explaining these features. Conventional laser theory[4], formulated for a Fabry-Perot resonator, describes stimulated emission by means of normalized internal modes in a cavity with a prescribed Q-value. In contrast, the lasing fields inside and outside the sphere need be treated on equal footing, in order to take account properly of Mie scattering resonances in the emission. This is achievable by judicious use of scattering theory, which is not employed in conventional laser theory.

The coupled semiclassical Maxwell-Bloch equations[4] yield in the small-signal lasing regime the following equation for the radial amplitude g(r) of a given transverse electric (TE) field mode, on assuming a time-independent, <u>isotropic</u> [5] (but radially nonuniform) pumping field[6]

$$[d^2/dr^2+(2/r)d/dr - \ell(\ell+1)/r^2+k_\lambda^2-U_0-\Delta U]g_\lambda = 0 \qquad (1)$$

Here the mode and its frequency $k_\lambda c$ are labelled by two indices $\lambda=(\ell,\nu)$, ℓ denoting the polar-angle symmetry of the mode. The dominant "potential" term in this equation is $U_0=k_\lambda^2(1-\varepsilon_\lambda)\theta(a-r)$, where ε_λ is the frequency-dependent complex dielectric index accounting for light refraction and absorption in the medium, and $\theta(a-r)$ is a step function for a sphere of radius a. The other "potential" term therein is $\Delta U(r)=4\pi k_\lambda^2\chi_\lambda^{(1)}(r)$, where $\chi_\lambda^{(1)}(r)$ is the first-order susceptibility proportional to the inverted (pumped) population density $N^{(0)}(r)$.

Consistently with the small signal regime, let us assume that $|\Delta U|<<|U_0|$, i.e., that susceptibility perturbs weakly the dielectric index. The solution of (1) to zeroth order in the susceptibility is obtainable from the Mie theory[6]

$$g^{(0)}_\ell(r<a) = -(i/k_\lambda a M_\ell)j_\ell(\varepsilon_\lambda^{1/2}k_\lambda r);$$
$$g_\ell^{(0)}(r\to\infty)=(e^{i\delta_\ell(0)}/k_\lambda r)\sin(k_\lambda r-\ell\pi/2+\delta_\ell^{(0)}) \qquad (2)$$

Here $M_\ell(\varepsilon_\lambda^{1/2}k_\lambda a)$ is the ℓ'th TE-wave Mie denominator measuring the enhancement of the field inside the sphere relative to the incident wave. The phase shift $\delta_\ell^{(0)}$ determines the scattering amplitude

$\exp(i\delta_\ell^{(0)})\sin\delta_\ell^{(0)}$, i.e. the ratio of the amplitude of the outgoing wave to that of the incident wave. To first order in the susceptibility, Eq. (2) is modified near a Mie resonance frequency (a value of $\rho_\lambda=\varepsilon_\lambda^{1/2}k_\lambda a$ such that M_ℓ is minimized) in that[3] $\delta_\ell^{(0)}$ is replaced by

$$\delta_\ell^{(1)}\simeq\delta_\ell^{(0)}-\alpha_\lambda^{(1)}, \text{ where}$$

$$\alpha_\lambda^{(1)} = 4\pi\varepsilon_\lambda^{-1/2}[\rho_\lambda M_\ell(\rho_\lambda)]^{-2}\chi_\lambda^{(1)}(\langle N_\ell^{(0)}\rangle) \tag{3}$$

$\chi_\lambda^{(1)}(\langle N_\ell^{(0)}\rangle)$ being the bulk susceptibility for the expectation value of inverted population density

$$\langle N_\ell^{(0)}\rangle=\int_0^a d(k_\lambda r')^3 N^{(0)}(r')j_\ell^2(k_\lambda r').$$

The shift of a Mie resonance frequency from the value dictated by U_0 is determined by the condition $\mathrm{Re}\alpha_\lambda^{(1)}\simeq-\cot\delta_\ell^{(0)}$. The lasing (amplification) condition is $\mathrm{Im}\delta_\ell^{(0)}<\mathrm{Im}\alpha_\lambda^{(1)}$ which is tantamount to the requirement that the scattering amplitude $|\exp(i\delta_\ell^{(1)})\sin\delta_\ell^{(1)}|$ exceed 1 near resonance. In what follows the implications of this condition will be considered.

In the case of near-surface population inversion

$$N^{(0)}(r)\simeq\bar{N}\delta(r-a),$$

the amplification condition reads, on using (3)

$$\mathrm{Im}\delta_\ell^{(0)}<4\pi\varepsilon_\lambda^{-1/2}[j_\ell(\rho_\lambda)/M_\ell(\rho_\lambda)]^2\ \mathrm{Im}\chi_\lambda^{(1)}(\bar{N}) \tag{4}$$

where $\mathrm{Im}\delta_\ell^{(0)}$ is of the order of the bulk absorption constant times the radius a. It follows from (4) that the factor $[j_\ell(\rho_\lambda)/M_\ell(\rho_\lambda)]^2$ is the effective Q-value for lasing in this case. For typical experimental conditions $\rho_\lambda\gtrsim300$, $M_\ell(\rho_\lambda)\lesssim10^{-5}$ and $j_\ell(\rho_\lambda\simeq\ell)\gtrsim10^{-2}$, this Q-factor may well exceed 10^6. This explains the finding of Chang et al.[1] that the lasing threshold in spherical droplets is reduced well below that of Fabry-Perot resonators.

The above treatment is easily adaptable to stimulated Raman scattering (SRS) at the Stokes frequency $ck_{\lambda1}\simeq ck_{\lambda0}-\omega_v$, on replacing $\chi_\lambda^{(1)}$ by the susceptibility for third order Stokes polarization $\chi_{\lambda1}^{(3)}$ multiplied by $|E_{\lambda0}|^2_{ref}$. Here $(E_{\lambda0})_{ref}$ is the amplitude of the reflected pump field (the field inside the sphere) which, according to Eq. (2), is related to that of the incident pump field by $|(E_{\lambda0})_{ref}/(E_{\lambda0})_{inc}|=\varepsilon_\lambda^{1/2}j_\ell(\varepsilon_\lambda^{1/2}k_\lambda r)/\rho_\lambda M_\ell(\rho_\lambda)$. For pumping nearly confined to the surface, the threshold pumping intensity $|E_{\lambda0}|^2_{inc}$ is then found by a procedure analogous to that leading to (4), to be lowered by a factor of

$$(k_{\lambda1}a)^{-2}[j_\ell(\rho_{\lambda0})/M_\ell(\rho_{\lambda0})]^2[j_\ell(\rho_{\lambda1})/M_\ell(\rho_{\lambda1})]^2$$

as compared to bulk. Here the Q-factor of (4) is replaced by product of factors pertaining to the pump and Stokes fields. For the parameter values used above this product of factors can exceed 10^7

or 10^8, i.e., threshold lowering in SRS should be even more spectacular than in lasing.

If $(E_{\lambda 1})_{ref}$ undergoes amplification, then it can serve, after accumulating sufficient intensity, as a pump for the second Stokes-order frequency $ck_{\lambda 2} \simeq ck_{\lambda 0} - 2\omega_v$. The close spacing of Mie resonances $k_{\lambda n}$ up to n=14 Stokes orders in the experiment of Ref. 2 corresponds to the availability of small $M_\ell(\rho_{\lambda n})$ in the threshold-reduction factor given above. This explains the observed[2] effectiveness of multiorder SRS processes, in which a $k_{\lambda n}$ mode pumps the consecutive $k_{\lambda n+1}$ mode.

In order to interpret quantitatively the cited experimental results, as well as obtain the angular and spectral features of SRS and four-wave mixing in spherical microparticles, it is necessary to consider couplings among modes of different angular symmetry, caused either by pumping anisotropy or the E_λ^2 (saturation) terms in population-inversion factors. Work on such multimode theory is currently underway.

References

1. H.M. Tzeng, K.F. Wall, M.B. Long and R.K. Chang, Opt.Lett. 9, 499 (1984).
2. S.X. Qian and R.K. Chang, Phys.Rev.Lett. 56, 926 (1986).
3. G. Kurizki and A. Nitzan, Phys.Rev.A (submitted).
4. H. Haken, Laser Theory (Springer, Berlin, 1983).
5. T. Baer, Opt.Lett. 12, 392 (1987).
6. J.A. Stratton, Electromagnetic Theory (McGraw-Hill, New York, 1941).

SURFACE-POLARITON ENHANCED SMITH-PURCELL RADIATION FROM DOPED SEMICONDUCTORS

N.E. Glass
Rockwell International Science Center, 1049 Camino Dos Rios
Thousand Oaks, CA 91360

ABSTRACT

Calculations of the Smith-Purcell diffraction radiation from an electron beam traveling over a grating, on the surface of doped GaAs, show large enhancements in the infrared, due to excitation of surface polaritons. These enhancements will be important for ascertaining the feasibility of a solid state Smith-Purcell laser.

INTRODUCTION

A charged particle traveling at constant velocity over a conducting grating surface will emit Smith-Purcell diffraction radiation. The incoherent radiation from a beam of such particles is weak, but when feedback in a resonator is achieved, a free-electron laser -- sometimes called an Orotron -- can be built.[1] Recent theories, which for the first time have correctly considered the finite conductivity of the underlying medium, have shown that the particle beam can excite surface polaritons (SP), thus leading to large enhancements in the radiation.[2,3] These theories have taken the conductor to be a metal, so that the SP enhancement (due to surface plasmons) was calculated to be in the UV. For solid-state integration, it would be interesting to see if SP enhanced Smith-Purcell radiation can be achieved using semiconductors rather than metals, and to see if the frequency range of the enhanced radiation can be extended (to the IR and even mm wavelengths). Given the inherent weakness of the underlying mechanism, these SP enhancements could be crucial in ascertaining the feasibility of a semiconductor Smith-Purcell laser. Thus, in the present work, the theory of Ref. 3 was used to calculate the incoherent Smith-Purcell radiation from electrons traveling over a classical grating (periodic in one direction) on the surface of doped GaAs.

THEORY

Reference 3 considered particles of fixed velocity $\vec{v}$, with an arbitrary orientation relative to the axes of either a bigrating (periodic in two different directions) or a classical grating surface, separating vacuum from a medium characterized by a complex dielectric function $\varepsilon(\omega)$. The theory begins with the method of Toraldo di Francia[4] and van den Berg,[5] wherein the Fourier transform of the electric field of a moving point charge is found, and then each Fourier component is considered as an incident wave on the (bi)grating. The frequency ω and wavevector (parallel to surface) $\vec{k}$, for these Fourier component incident fields, are related by $\omega = \vec{v} \cdot \vec{k}$. The scattered field for each incident wave is determined with the bigrating scattering theory of Ref. 6. From the total scattered field, the Poynting vector is determined, and then an incoherent superposition gives the total power radiated from the entire beam.

Here, the particles will have velocity $\vec{v} = v\hat{x}$ transverse to the grooves of a classical grating (in the xy surface plane), whose profile is given by $z = h \cos(2\pi x/a)$. For the GaAs medium

$$\varepsilon(\omega) = \varepsilon_\infty + (\varepsilon_0 - \varepsilon_\infty)\omega_T^2/(\omega_T^2 - \omega^2) - \varepsilon_\infty \omega_p^2/\omega(\omega + i\gamma); \quad (1)$$

where $\varepsilon_\infty = 10.9$, $\varepsilon_0 = 12.9$, the transverse phonon frequency is $\hbar\omega_T = 0.0334$ eV, and for a donor electron concentration of 2.7×10^8 cm^{-3} the plasma frequency is $\hbar\omega_p = 0.0682$ eV (550 cm^{-1}), with a collision frequency $\gamma = 10^{13}$ s^{-1}.[7] The zeros of $\varepsilon(\omega)$ are given by $\hbar\omega_{L-} = 0.0324$ and $\hbar\omega_{L+} = 0.07$ eV. The long-dashed lines in Fig. 1 are the dispersion curves for transverse bulk light waves (where $\varepsilon(\omega) > 0$, i.e., $\omega > \omega_{L+}$ and $\omega_{L-} < \omega < \omega_T$). The solid lines in Fig. 1, in the two regions where $\varepsilon(\omega) < 0$, are the dispersion curves for surface polaritons (coupled plasmon - LO-phonon). The dash-dot line is the dispersion curve $\omega = vk$ for the "incident waves," where $v = 0.7\ c$ (c = light speed). At the two points where this line $\omega = vk$ intersects the two branches of the SP curves, the "incident wave" resonantly excites the SPs. These points are at $\omega/2\pi c = 261.2$ and 518.0 cm^{-1}.

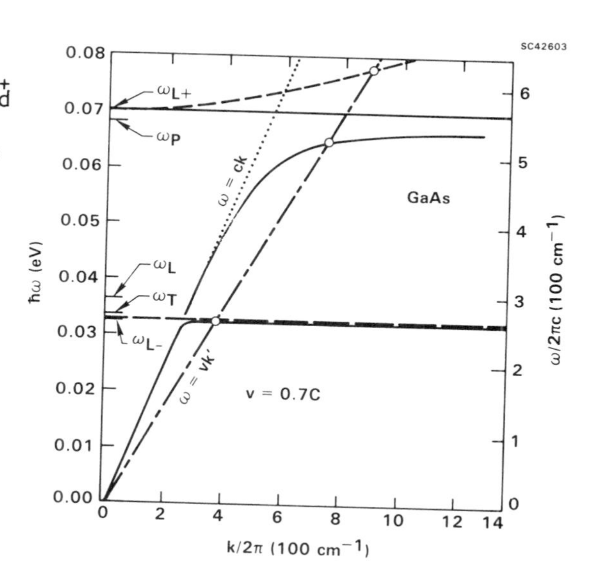

Fig. 1 Dispersion curves for GaAs.

Figure 2 shows R, the power radiated in a given direction (in the xz plane at angle η measured from the z-axis -- the surface-normal), per unit solid angle, per unit beam-area parallel to the surface, per unit current density. Positive η corresponds to + x. Figure 2a, for h = 1.2 μm and a = 20 μm, shows two enhancement peaks, corresponding to the excited SP (the large peak is for the mostly plasmon SP and the small peak is for the phonon SP). The relation between η, plotted on the lower axis, and the emitted frequency, on the upper axis, is $\omega = c(\sin\eta - c/v)^{-1}\,2\pi m/a$, for $m = -1$ (first order coupling). The peaks are not precisely at the frequencies of the intersection points of $\omega = vk$ and the SP dispersion curves in Fig. 1, because the latter are for a flat surface, and the grating shifts the SP frequencies. The structure indicated (by the insert) around the lower peak is due to the grating induced splitting (and gap formation) in the dispersion relation at the first Brillouin zone of the grating.

With the period "a" chosen such that $2\pi/a = K_{SP}$, the wavevector of the excited SP, the enhancement peak must be in the direction $\eta = 0$ (a convenient geometry for achieving feedback). In Fig. 2b, "a" was chosen so as to put the major peak at $\eta = 0$. In that case the lower peak vanishes: for the lower branch SP, $K_{SP} - 2\pi/a$ crosses a light-line and becomes nonradiating. In Fig. 2b the height h was adjusted to keep $h/a \approx 0.06$, as in Fig. 2a. The effect

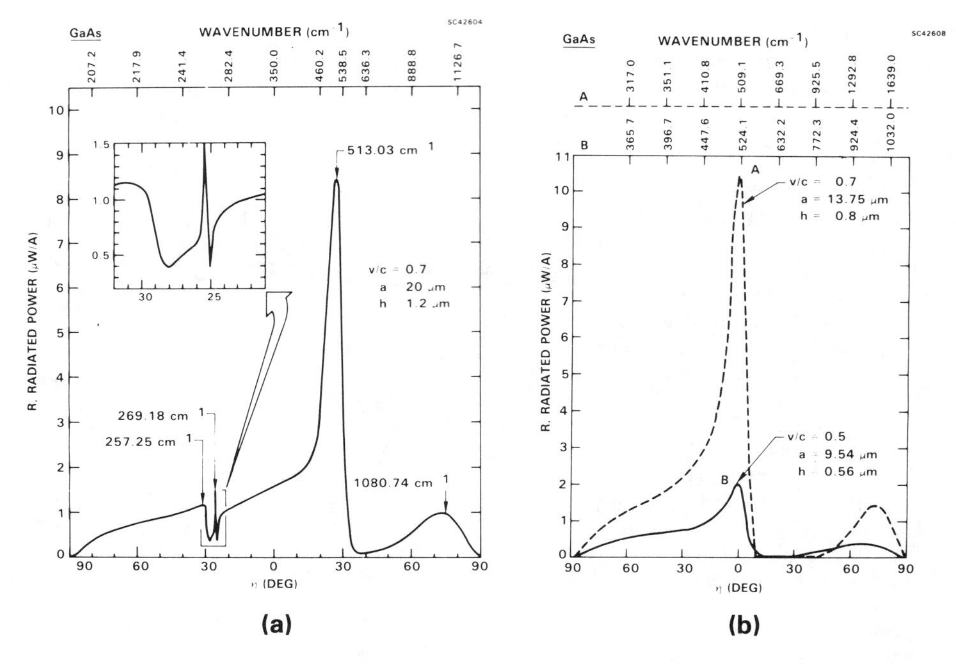

Fig. 2 Power radiated in direction η, per unit solid angle, vs η.

of lowering v to 0.5 c is also shown in Fig. 2b, for the same h/a. The radiated power, R, is on the order of μW/A in all these plots and is thus much lower than the mW/A levels from Ag reported in Ref 3.

Reference 3 showed that there is an optimal value of h/a to maximize the enhancement, viz., h/a ≈ 0.08. It was also shown that a bigrating can simultaneously decouple two SP (at the same ω and η), yielding an added enhancement over the classical grating. One might therefore expect to be able to double the enhancement peaks shown here, by careful optimization. Moreover, one could lower the frequency, of the larger peak, to around 100 cm^{-1}, by lowering the donor concentration.[7]

REFERENCES

1. R.P. Leavitt, D.E. Wortman, and H. Dropkin, IEEE J. Quantum Electron. QE-17, 1333 (1981).
2. S.L. Chung and J.A. Kong, J. Opt. Soc. Am. A1, 672, 1984.
3. N.E. Glass, Phys. Rev. A36 to appear (Dec. 1987).
4. G. Toraldo di Francia, Nuovo Cimento 16, 61 (1960).
5. P.M. van den Berg, J. Opt. Soc. Am,. 63, 1588 (1973).
6. N.E. Glass, A.A. Maradudin, and V. Celli, J. Opt. Soc. Am. 73, 1240 (1983).
7. A. Mooradian and A.L. McWhorter in Light Scattering Spectra of Solids, edited by G.B. Wright (Springer, N.Y., 1969), p. 297.

MULTI-FREQUENCY CONVERSIONS USING ONE BIAXIAL CRYSTAL $KTiOPO_4$

Sun Decai Wu Xing and Yao Jianquan
Dept. of Precision Instrument Engineering, Tianjin University, PRC

ABSTRACT

By calculation of phase match parameters and effective nonlinear coefficients, several different nonlinear optical processes using one biaxial crystal $KTiOPO_4$ are considered. The nonlinear optical processes include 1, SHG (Type II) of 1.064→0.532 μm. 2, SFG of 1.064+0.6→0.383μm. 3, DFG of 1.064—0.6→1.375μm.

SUMMARY

The application of biaxial crystal $KTiOPO_4$ in nonlinear optical processes is very popular recently for its good optical qualities. But usually only one process (such as SHG of 1.064→ 0.532 μm) can be provided. In this paper we discuss the technique of multi-frequency conversions using one crystal material which possesses the low symmetry property of $KTiOPO_4$.

From the equation of refractive indices

$$\frac{k_x^2}{n^{-2}-n_{x,w}^{-2}}+\frac{k_y^2}{n^{-2}-n_{y,w}^{-2}}+\frac{k_z^2}{n^{-2}-n_{z,w}^{-2}}=0$$

for the process of$K_1//K_2//K_3$, we can get following equations[1]:

$$x_1^2+b_1x_1+c_1=0$$

$$x_2^2+b_2x_2+c_2=0$$

$$x_3^2+b_3x_3+c_3=0$$

Solving these equations we get

$$n_{w_q,i}=\sqrt{2}/\sqrt{-B_q\pm\sqrt{B_q^2-4C_q^2}}$$

As i=1 or 2, the symbols under the square root sign take on plus or minus values respectively, where q=1,2,3.

1. For the process of SHG of 1.064→0.532μm:
   Type I PM: $n_{w_{1.2}}=n_{w_{2,1}}$
   Type II phaase macth (PM): $\frac{1}{2}(n_{w_{1,1}}+n_{w_{1,2}})=n_{w_{2,1}}$,

   Through the calculation, we get two curves of phase match angles (Fig.1, 1 and 2).

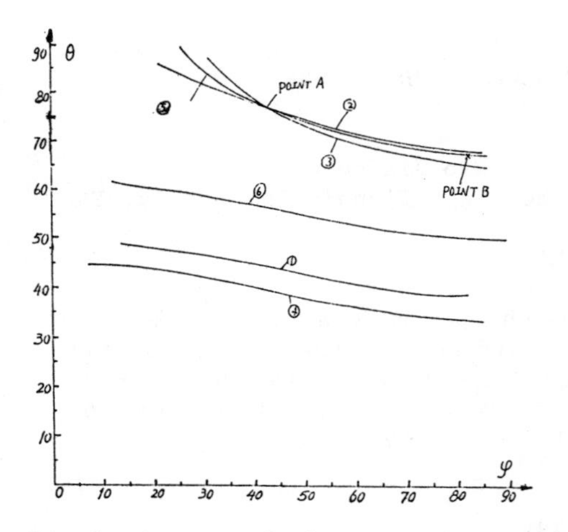

Fig.1. Curves of phase match angles

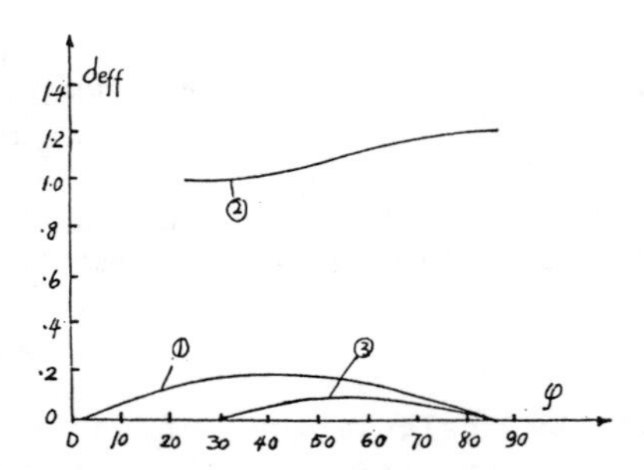

Fig.2. Curves of effective nonlinear coefficient

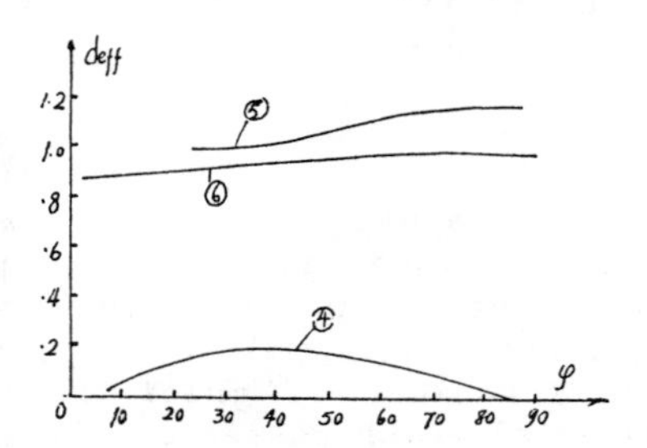

Fig.3. Curves of effective nonlinear coefficient

2. For the process of SFG of $1.064+0.6 \rightarrow 0.383\mu m$:
   Type I PM:

   $n^{e2}\ w_1 + n^{e2}\ w_2 = n^{e1}\ w_3$,

   (w_1, w_2 — e2 light, w_3 — e1 light).

That is $n_{w_1,2} w_1 + n_{w_2,2} w_2 = n_{w_3,1} w_3$,

Type II PM: $n_{w_1,2} w_1 + n_{w_2,1} w_2 = n_{w_3,1} w_3$, ... (1)

and $n_{w_1,1} w_1 + n_{w_2,2} w_2 = n_{w_3,1} w_3$ ... (2)

Equations 1 and 2 have no solutions, we obtained curves of PM angles only for Type I (Fig. 1, 3).

3, For the process of DFG of $1.064 - 0.6 \rightarrow 1.375\mu m$.
It is equivalent to process of SFG of $1.064 + 1.375 \rightarrow 0.6\mu m$. Using the same method, we get three curves (Fig.1,4,5,6).

Then we calculate the effective nonlinear coefficients (ENC) for the six processes and obtain six ENC curves (Fig. 2.1,2,3 and Fig. 3.4,5,6).

From Fig. 1, we find that curves 2 and 3 have a common point A corresponding to large ENC values and a space direction $\vec{A}$: $\theta=76.79°$, $\varphi=43.09°$. Through the symmetry characteristic of the vbiaxial crystal and calculation, we find a point B on the curve of DFG (PM 2A) which has a space direction $\vec{B}$: $\theta=67.15°$, $\varphi=82.59°$, vertical to A.

Depending on the space direction $\vec{A}$ and $\vec{B}$, we can process a piece of crystal KTP with two surfaces vertical to each other and get three wavelengths: 0.532, 0.383 and 1.375 μm respectively, by experiment.

Above technique can be applied to biaxial crystal materials which have a low symmetry property, and can get three or more nonlinear optical processes.

References

[1] Q. Yao et al., Appl. Phys. 55(1), 65, 1984.

COUPLED PAIR OF PARTICLES IN A MEAN FIELD FOR A CUBIC PARTICLE COMPOSITE

T.P. Shen
Rockwell International Science Center
1049 Camino Dos Rios
Thousand Oaks, CA 91360

ABSTRACT

A self-consistent variational calculation including both multipole and multi-site interactions to account for fluctuation effects around an arbitrary particle in a composite is developed for the macroscopic dielectric function of a statistically homogeneous two-component small particle composite. This work systematically improves on simple mean field theories. It is formally an exact expansion of which mean field theory is the zeroth term.

THEORY

The composite sample which is considered consists of two kinds of particles with dielectric functions ε_1 and ε_2. The particles are situated in a simple cubic lattice and random in their dielectric properties. The shape of particles is the Wigner-Seitz cell of simple cubic lattice, viz cube. The quasi-static approximation is used, as long as the size of particles are small compared to the wavelength of incident radiation.

For a cube in a uniform external field $\vec{E}_o$, the induced polarization inside the cube is not uniform. Moreover, electrostatic interactions among the particles give rise to spatially varying reaction fields at a given particle and hence to the variations of the induced polarization inside each cube, too. The induced polarization is expanded in terms of multipoles and, in particular, the dipole and quadrupole terms for the ith particle are given by

$$\vec{P}_i^{(0)}(\vec{r}) = \vec{P}_i = (P_{ix}, P_{iy}, P_{iz}) \quad , \tag{1a}$$

$$\begin{aligned} \vec{P}_i^{(1)}(\vec{r}) &= \frac{2}{\sqrt{3}} Q_{ix}(-2x,y,z) + \frac{2}{\sqrt{3}} Q_{iy}(x,-2y,z) \\ &+ \frac{2}{\sqrt{3}} Q_{iz}(x,y,-2z) + \sqrt{6}\, Q'_{ix}(y,x,o) \\ &+ \sqrt{6}\, Q'_{iy}(o,z,y) + \sqrt{6}\, Q'_{iz}(z,o,x) \quad , \end{aligned} \tag{1b}$$

and subject to the constraint $Q_{ix} + Q_{iy} + Q_{iz} = 0$, where $\vec{P}_i$ and $(\vec{Q}_i, \vec{Q}'_i)$ are the dipole and quadrupole strengths of the ith particle, and $\vec{r}$ is the position vector from the center of the ith particle. The multipole strengths are the variational parameters in the total electrostatic energy of the composite. Using the variational principle, one obtains the following set of linear equations for the multipole strengths:

$$\vec{P}_i = \gamma_i \vec{E}_0 + \gamma_i \sum_{j\neq i}^{N} \{\overleftrightarrow{J}_{ij}\vec{P}_j + \overleftrightarrow{K}_{ij}\vec{Q}_j + \overleftrightarrow{K}'_{ij}\vec{Q}'_j\} \quad , \tag{2a}$$

$$\vec{Q}_i = \bar{\gamma}_i \sum_{j\neq i}^{N} \{\overleftrightarrow{K}^+_{ij} \vec{P}_j + \overleftrightarrow{V}_{ij}\vec{Q}_j + \overleftrightarrow{S}_{ij}\vec{Q}'_j\} \quad , \tag{2b}$$

$$\vec{Q}'_i = \bar{\gamma}'_i \sum_{j\neq i}^{N} \{\overleftrightarrow{K}'^+_{ij} \vec{P}_j + \overleftrightarrow{S}^+_{ij} \vec{Q}_j + \overleftrightarrow{V}_{ij}\vec{Q}'_j\} \quad , \tag{2c}$$

where $i = 1, \ldots, N$, N is the total number of particles in the composites, and each particle is coupled to the rest of particles in the composite. The interaction matrices, $\overleftrightarrow{J}_{ij}$, $\overleftrightarrow{K}_{ij}$, etc., represent the reaction fields from other particles and are purely geometric factors. The matrices are straightforward to work out and will not exhibit here. In Eqs. (2), the coupling constants are the dipole (γ_i) and quadrupole ($\bar{\gamma}_i, \bar{\gamma}'_i$) polarizabilities, which are given by

$$\gamma_i = \frac{3}{4\pi} \frac{\varepsilon_i - 1}{\varepsilon_i + 2} \quad , \tag{3a}$$

$$\bar{\gamma}_i = \frac{\gamma_i}{1 + 6.9252\ \gamma_i} \quad , \tag{3b}$$

$$\bar{\gamma}'_i = \frac{\gamma_i}{1 + 3.3858\ \gamma_i} \quad , \tag{3c}$$

where ε_i is the dielectric function of the ith particle, which can be either ε_1 or ε_2. The quantities in Eqs. (2) and (3) for a statistically homogeneous composite shall be averaged over an ensemble of realizations of the given composite, which is taken to be the volume average.

For any composite, it is impossible to solve Eqs. (2) exactly. Instead, attention is focused on a cluster of (z+1) particles, which consists of an arbitrary central particle and z neighbors, and the interactions for (z+1) particles will be virially expanded into single-site, two-site, ..., terms. The remaining particles outside the cluster are approximated by an effective medium[1] of dielectric function $\bar{\varepsilon}$, which is also the macroscopic dielectric function of the composite and is going to be determined by a self-consistent condition. The condition requires that on the average no additional dipole moment exists for the central particle of the cluster embedded in the effective medium, where the fluctuation effects from the cluster have been incorporated. The self-consistent condition will be solved virially by iteration, where a mean field value is used as an initial guess.

The effective coupling constants for particles embedded in the effective medium are those given in Eqs. (3) with the replacement of ε_i by $\varepsilon_i/\bar{\varepsilon}$, which is the rescaled dielectric function of the ith particle. This scaling law and the self-consistent condition will yield a vanishing effective dipole polarizability and small effective quadrupole polarizabilities. This implies that the dipole-dipole interactions between the particles in the cluster vanish on the average and their multipole interactions are decreasing very rapidly as the separations between the particles increase. Hence it is sufficient to keep the z neighbors to the nearest neighbors and at most the second nearest neighbors for a chosen particle. The virial expansion for a chosen cluster shall converge nicely, since the n-site interaction in this expansion is at least proportional to the n products of the effective coupling constants, which are either zero or small. Finally, we have to check for the convergence of the multipole expansion used in the total electrostatic energy. The convergent results have obtained by dividing each cube into eight sub-cubes and including both dipole and quadrupole terms.

One of the important tests for a theory of small particle composites is the predicted percolation limit for a metal-dielectric composite. It is known that a suspension of fine metallic particles mixed with dielectric particles will become effectively metallic, if the volume fraction, f, of the metallic component is larger than a percolation limit, f*. A simple effective medium theory[2] using a single site approximation predicts f* = 1/3. An example of numerical network simulation[3] with no correlation among the particles gives f* = 1/4. Our calculation predicts f* = 0.307 for a simple cubic lattice of random mixture of metallic and dielectric cubes by including the dipole and quadrupole terms, two- and three-site interactions and up to third nearest neighbors for an arbitrary particle in the composite.

REFERENCES

1. J.S. Smart, Effective Field Theories of Magnetism, (W.B. Saunders Co., Philadelphia and London, 1966).
2. D.A.G. Bruggeman, Ann. Phys. 24, 636 (1935).
3. I. Webman, J. Jortner and M.H. Cohen, Phys. Rev. B15, 5712 (1977).

RESULTS ON SEVERAL METHODS OF FREQUENCY MIXING IN ß-BaB_2O_4 TO ATTAIN THE WAVELENGTH REGION 197 NM TO 189 NM

W. Muckenheim, P. Lokai, B. Burghardt, D. Basting, Lambda Physik GmbH; M. F. Essary, Lambda Physik Inc.. Lambda Physik, D-3400 Goettingen, FRG

ABSTRACT

ß-BaB_2O_4 has been used to obtain wavelengths in the range from 188.9 to 197 nm by Type-I sum frequency generation. The fundamental beams were supplied by a pulsed dye laser tuned between 780 and 950 nm and a second laser operating at 248.5 nm. Four different sources for the 248.5 nm wavelength were used: In all cases output at the ArF excimer wavelength of 193 nm was demonstrated.

INTRODUCTION

The easiest way to generate tunable uv pulses is to use frequency doubling or mixing in nonlinear crystals. In order to attain the lowest wavelengths, a new crystal ß-BaB_2O_4 (BBO) has proved to be most suitable. Frequency doubling yields a lower limit of about 205 nm [1-3], and 197.3 nm has been reached [4, 5] by frequency mixing. Of special interest would be to attain the wavelength 193 nm, i.e. to match the gain regime of the ArF excimer laser.

The first attempt to obtain 193 nm radiation involved cooling the BBO crystal. This was not successful [3 5] and it could be estimated that the minimum wavelength is 194.4 nm even at zero degrees. But, according to the Sellmeier equations given by Chen et al. [6] and also by Kato [1], sum frequency generation (SFG) should be possible to the end of the transparency range below 190 nm by exploiting two very different initial wavelengths. This conjecture has been verified.

EXPERIMENTAL SET-UP AND RESULTS

In each of the following experiments we used one laser to produce the fixed wavelength λ_1 = 248.5 nm. The second wavelength λ_2 was obtained from a pulsed dye laser (FL 3002, Lambda Physik) which could be tuned between 780 and 950 nm using the dyes Rhodamine 800, Styryl 9, HITCI and IR 125. The relevant operating parameters of the dye laser were: pulsewidth 12 ns, beam cross section at the crystal 0.1 cm^2, and bandwidth 0.2 cm^{-1}. The total output energy per pulse is displayed in Fig. 1. The beam was not linearly polarized; the percentage which was suitably polarized for our experiment (type-I phase matching) is also indicated in Fig. 1. From the given values, a power density of 2 (+/-1) MW/cm^2 over the main part of the tuning range can be calculated. In particular at λ_2 = 865 nm we obtain (for HITCI) 2.5 MW/cm^2.

For frequency mixing a BBO crystal of 8 mm length was used, cut under 80° with respect to the optical axis. The resulting beam was separated using a quartz prism.

TWO DYE LASERS

In the first experiment we employed as a source of λ_1 = 248.5 nm, a pulsed dye laser, (FL 3002, Lambda Physik), operated at 497 nm and frequency-doubled in a second BBO crystal. Each frequency-doubled pulse contained 4.5 mJ energy. The pulse duration was 15 ns, the other parameters were comparable to those above, in particular the power density, 3 MW/cm^2.

Both beams were made collinear by means of a quartz plate coated for high reflectance at 248.5 nm and high transmittance in the infrared.

The actual pulse energies in the resulting beam are given in Fig. 1. The width of the SFG pulse was 12 ns. At 193 nm the pulse energy output was 95 μJ, leading to a power density of 0.08 MW/cm^2. Thus, the SFG efficiency in terms of power density $\eta = P(\omega_1 + \omega_2)/\sqrt{P(\omega_1)P(\omega_2)}$ was nearly 3% at 193 nm. (1)

As Fig. 1 shows, even 189 nm can be reached with 8 μJ output. Below 189 nm the energy drops immediately below 1 μJ. This effect is attributed to absorption.

STANDARD EXCIMER LASER AND DYE LASER

In the second experiment the frequency-doubled dye laser was replaced by a standard excimer laser operating with KrF (EMG 103 MSC, Lambda Physik). The power density of the excimer laser was attenuated to about 5 MW/cm^2 with half of this value being suitably polarized for type-I phase matching. The effective interaction area was restricted to the beam cross section of the dye laser, i.e. 0.1 cm^2. At 193 nm we observed an SFG power density of 1 kW/cm^2, corresponding to an SFG efficiency $\eta = 0.04\%$. This low value is due to the broad linewidth of the excimer laser (0.3 nm corresponding to 50 cm^{-1}).

BANDWIDTH-NARROWED EXCIMER LASER AND DYE LASER

In order to enhance the SFG efficiency, the bandwidth of the excimer laser was reduced to 0.4 cm^{-1}. Further parameters of the linearly polarized 248.5 nm pulses were: pulse energy 1.5 mJ, pulse duration 15 ns, beam cross section 0.3 cm^2. The resulting 12 ns long SFG pulses at 193 nm contained an energy of 2 μJ (0.1 cm^2 beam cross section). The SFG efficiency is according to (1) $\eta = 0.1\%$.

Because of the high damage threshold of BBO, it would be possible to increase the power density to 5 MW/cm^2 or more without destroying the crystal. This should result in an SFG efficiency of nearly 0.5%.

PICOSECOND EXCIMER LASER AND DYE LASER

In a fourth experiment, λ_1, was supplied by a linearly polarized picosecond excimer laser (PSL 4000, Lambda Physik) delivering a power density of 20 MW/cm^2. The bandwidth was 16 cm^{-1} with the central wavelength at 248.5 nm. The pulse duration of 30 ps was measured with a streak camera. Due to a lack of appropriate optics the duration of the SFG pulse at 193 nm could not be measured in this way. However,the pulse shapes of the picosecond pulse (ω_1) and the SFG pulse ($\omega_1 + \omega_2$) were measured using a fast photodiode as shown in Fig. 2. Assuming a pulse duration of 30 ps for the SFG pulse, the power density was 0.5 MW/cm^2. Thus, the SFG efficiency amounts to $\eta = 7\%$.

CONCLUSION

It has been demonstrated that SFG in BBO opens the spectral range between 189 and 197 nm. In particular, the important wavelength 193 nm can be generated by several methods. By means of a picosecond excimer laser, picosecond output at 193 nm has been observed with an SFG efficiency of 7%. The presented method should be well-suited to applications in the femtosecond range [7, 8].

[1] K. Kato, IEEE J. Quantum Electron. QE-22, 1013 (1986).
[2] K. Miyazaki, H. Sakai, T. Sato, Opt. Lett. 11, 797 (1986).
[3] P. Lokai, B. Burghardt, D. Basting, W. Muckenheim, Laser and Optoelectronik 19, 296 (1987).
[4] W.L. Glab, J.P. Hessler, IQEC 1987, Baltimore, Postdeadline Paper PD5
[5] P. Lokai, B. Burghardt, W. Muckenheim, Sum frequency generation in a cooled β-BaB_2O_4 crystal, submitted to Appl. Phys. B.
[6] C. Chen, B. Wu, A. Jiang, G. You, Sci. Sinica B 28, 235 (1985).
[7] S. Szatmari, B. Racz, F.P. Schafer, Optics Comm. 62, 271 (1987).
[8] B. Dick, S. Szatmari, B. Racz, F.P. Schafer, Optics Comm. 62, 277 (1987).

Figure Captions

Fig. 1(a): Pulse energies, $E(\omega_2)$, of the IR-dye laser. Only the percentage indicated in parentheses was useful for SFG. (b): SFG, pulse energies $E(\omega_1+\omega_2)$ versus wavelengths λ_2 and λ_1, respectively.

Fig. 2(a): Pulse shape of the IR dye laser. (b): The picosecond excimer laser pulse superimposed upon that of the IR dye laser. (c): Pulse shape of the SFG pulse at 193 nm.

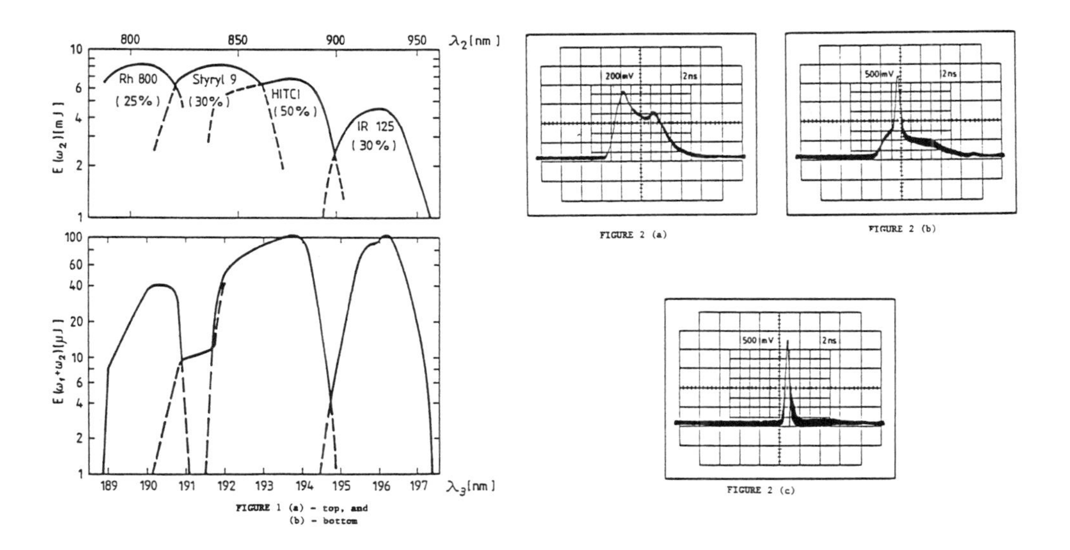

FIGURE 1 (a) - top, and (b) - bottom

FIGURE 2 (a)

FIGURE 2 (b)

FIGURE 2 (c)

COLLISION-INDUCED GRATINGS IN FOUR-WAVE MIXING OSCILLATORS

Gilbert Grynberg
Laboratoire de Spectroscopie Hertzienne
de l'École Normale Supérieure
Université Pierre et Marie Curie - 75232 Paris Cedex 05, France

ABSTRACT

Collision-induced effects for a set of two level atoms can often be explained in term of collision-induced gratings. In particular, we interpret the Bloembergen PIER 4 resonances and the collision-induced gain in two-wave mixing using these gratings. In the case of a ring atomic parametric oscillator, we show that these gratings can lead to a non reciprocal effect between the two directions of propagation. Experimentals results obtained with a cell containing sodium and helium show that a potential application to optical gyros can be imagined.

INTRODUCTION

The effect of dephasing collisions in non linear optics is often far from being intuitive. For example, it is well known that collisions can induce the apparition of new resonances in four-wave mixing. These resonances occur for a Bohr frequency associated with excited states which are not resonantly excited [1]. It has also been demonstrated that collisions can induce an energy transfer between two non resonant light beams [2]. Soon after the first experimental observation of these collision-induced effects in non linear optics, it has been shown that these effects can generally be interpreted in terms of collisionnally-aided excitation of the upper states [3]. This type of picture is particularly simple and useful in the case of a set of two-level atoms. We show in this paper that several preceeding experimental results obtained in the case of nearly degenerate exciting beams can be easily interpreted in terms of collision induced gratings. This picture also suggests new type of experiments. For example, we present here the results obtained on a nearly degenerate sodium vapor parametric oscillator [2] when a buffer gas is introduced in the sodium cell. We show that for a ring oscillator, a non reciprocal behaviour between the two counterpropagating waves can be observed in presence of collisional damping.

COLLISION AIDED EXCITATION AND COLLISION INDUCED GRATINGS

We consider a set of two-level atoms (ground state g, excited state e) interacting with a non resonant light beam of frequency ω_1. In presence of collisional damping the atom can be brought in the excited state e in a process involving a photon absorption and a collision (Fig. 1).

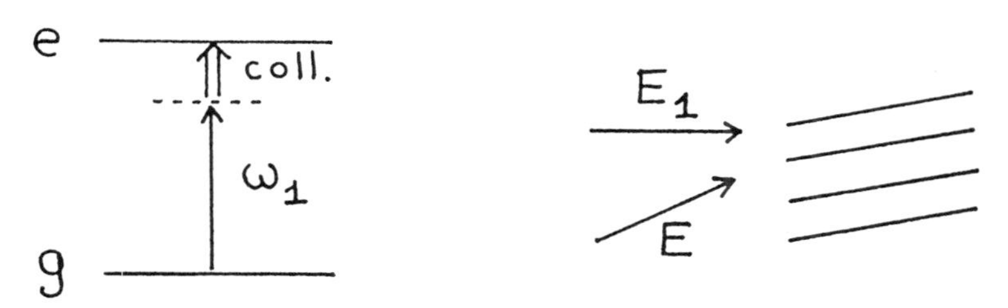

Figure 1 : Collision-aided excitation of the upper state.

Figure 2 : Light grating created by the beams E and E_1.

If the intensity of the incident light is spatially modulated, the probability of collision-aided excitation is also spatially modulated. If the atoms interact with two plane waves E and E_1 propagating in different directions (Fig. 2), the excitation probability will be larger at an antinode than at a node of the standing wave. Because of the collision-aided process of Fig. 1, we create a grating of excited state atoms starting from a light grating. Because of the lifetime of the excited state, the grating of excited state atoms has a lifetime $1/\Gamma_e$ where Γ_e is the natural width of the upper state. This point has an important consequence when the beams E and E_1 have different frequencies ω and ω_1. In this case the light grating move in the cell with a velocity proportional to $(\omega_1 - \omega)$. Because of the finite lifetime of the excited state, the atomic excitation does not follow instantaneously the variations of the light intensity. The grating of excited state atoms follow the light grating with a delay (Fig. 3). Furthermore, when $|\omega_1 - \omega| \gg \Gamma_e$, the light grating moves so rapidly inside the cell that the atoms are almost uniformly excited. In this case, the modulation depth of the grating of excited state atoms is almost 0 (Fig. 3.c).

DIFFRACTION OF THE PROBE BEAM ON THE COLLISION-INDUCED GRATING

In the following, we assume that the beam E_1 (pump beam) is more intense than the beam E (probe beam). We study the diffraction of E_1 on the collision-induced grating created by E and E_1. If we just consider the diffracted beams of lowest order, we find three beams at the exit of the cell (fig. 4). In the following, we concentrate our interest on the beams E' and E" of the frequencies ω and $2\omega_1 - \omega$.

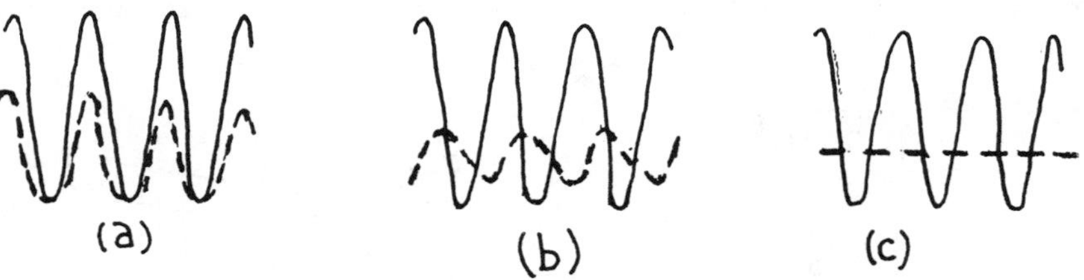

Figure 3 : Light grating (full lines) and atomic grating (dashed lines) in the case $\omega - \omega_1 = 0$ (a), $|\omega - \omega_1| \sim \Gamma_e$ (b), $|\omega - \omega_1| \gg \Gamma_e$ (c).

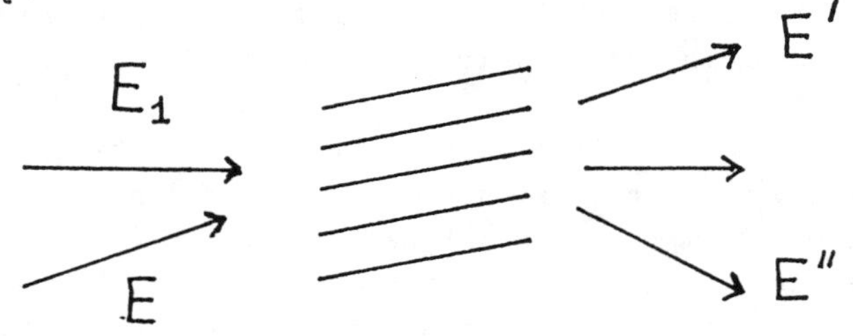

Figure 4 : Diffraction of the pump beam E_1 on the atomic grating created by interaction of the atoms with the beams E_1 and E.

COLLISION-INDUCED RESONANCES IN NEARLY DEGENERATE FOUR-WAVE MIXING

We consider here the intensity variations of E" as a function of $(\omega_1 - \omega)$. This situation is very similar to the one experimentally considered by Bloembergen and his coworker [1]. The intensity of the diffracted beam E" increases with the modulation depth of the collision-induced grating. When $\omega - \omega_1 = 0$, this modulation attains its largest value and the diffracted intensity is maximum. When $|\omega - \omega_1| \gg \Gamma_e$, the modulation is very small and the diffracted intensity reaches its asymptotic value (which is not zero because of the mixing of the wavefunctions induced by the electromagnetic field, this term being collision-free). The four-wave mixing generation at $2\omega_1 - \omega$ thus appears to be maximum when $\omega_1 - \omega = 0$ and the width of the collision-induced enhancement is of the order of Γ_e.

COLISION INDUCED GAIN IN TWO-WAVE MIXING

Let us now consider the diffracted beam E' which has the same frequency than the incident beam E and propagates in the same direction. To find the total field after the cell, we have to add the transmitted beam tE (where t is the transmission coefficient of the medium) and E'. As it has been shown earlier [2], $|tE + E'|$ can be larger than E. The net effect is then a gain for beam E. This gain process does occur neither when $\omega_1 - \omega = 0$ because in this case the diffracted beam E' and tE are in quadrature not when $|\omega_1 - \omega| \gg \Gamma_0$ because the modulation depth of the collision-induced grating is vanishly small in this case. In fact, the variation of the gain with $(\omega_1 - \omega)$ is a dispersive curve and the maximum gain is expected for $|\omega_1 - \omega| \sim \Gamma_e$ (in fact, if we have gain for $\omega_1 - \omega = \Gamma_e$, there is a maximum extra absorption for the opposite value $-\Gamma_e$ of $\omega_1 - \omega$).

COLLISION-INDUCED PHASE SHIFTS IN ATOMIC PARAMETRIC OSCILLATORS

In the preceeding example, we have considered the amplitude of the beam of frequency ω after the cell. However, it should also be noticed that the phase of $tE + E'$ differs from the phase of tE. Another effect of the diffraction of E_1 on the collision-induced grating is thus to induce a phase-shift on the transmitted beam. Such a phase-shift is particularly important to understand the behaviour of the atomic parametric oscillator [2].

We present in Fig. 5 the ring oscillator that we have used. A cell containing sodium and 3 Torrs of helium at temperature T = 155°C is enclosed inside a ring cavity. The sodium atoms interact with two pump beams E_1 and E_2 of same frequency ω. The four-wave mixing interaction leads to an oscillation in this cavity with two counterpropagating waves of frequency ω_+ and ω_-. The gain mechanism being associated with the absorption of two pump photons and the emission of two photons of frequencies ω_+ and ω_-, we have $\omega_+ + \omega_- = 2\omega$. Let us now consider the difference $\omega_+ - \omega_-$. If we call Ω_+ the eigenfrequency of the cavity which is closest to ω_+ for a beam rotating in the anticlockwise direction (a similar definition is used for Ω_- in the clockwise direction) we find using the boundary conditions on the mirrors of the cavity that $\omega_+ - \omega_- = \Omega_+ - \Omega_-$.

If the intensities I_1 and I_2 of the two pump beams are different, the efficiencies of the diffraction of E_1 on the grating created by E_+ and E_1 and of E_2 on the grating created by E_- and E_2 are different. The consequence is that the phase-shift experienced by E_+ and E_- are different and thus Ω_+ and Ω_- are also different. The difference $\Omega_+ - \Omega_-$ can also be calculated using the non linear susceptibilities [4] and we find

$$\Omega_+ - \Omega_- = \chi_0 \frac{\beta p}{\Gamma_a} \frac{\Omega_2^2 - \Omega_1^2}{\delta^2} \omega \frac{l}{L} \tag{1}$$

where χ_0 is the linear susceptibility, β is the pressure broadening coefficient, p is the pressure of buffer gas, $\Omega_i = dE_i/\hbar$ is the Rabi frequency associated to wave E_i, $\delta = \omega_0 - \omega$ is the frequency detuning from resonance, l is the length of the atomic cell and L is the length of the cavity.

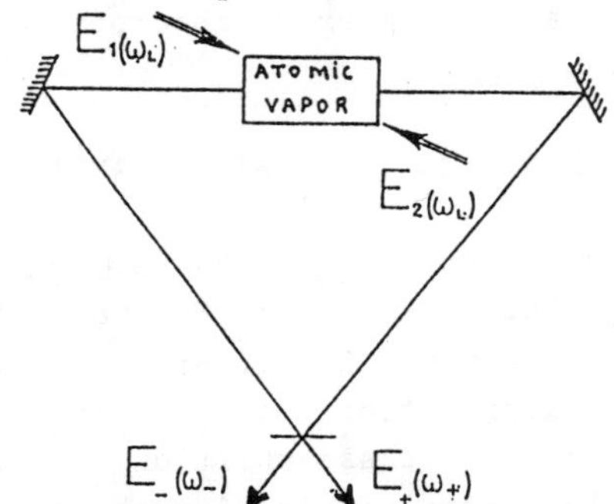

Figure 5 : Scheme of the ring atomic parametric oscillator.

We deduce from (1) that the oscillation should be non degenerate when $I_1 \neq I_2$. In fact, the preliminary experimental results (Fig. 6) show that this is not true if I_1 and I_2 are very close. More precisely, we have measured the beat frequency between E_+ and E_- as a function I_2, I_1 remaining constant. We see that for $0.8 < I_2/I_1 < 1$, ω_+ remains exactly equal to ω_-. When I_2/I_1 becomes smaller than a value of the order of 0.8, we find a law which agrees fairly well with the theoretical formula (1).

The curve of figure 6 is very similar to the one obtained in a laser gyro. When Ω_+ and Ω_- are very closed, the retrodiffusion of light on the optical elements inside the cavity couple the two directions of propagation and a frequency locking occurs.

Laser gyros are based on the Sagnac effect and they are used to measure the angular velocity Ω of the frame in which the laser is fixed. The difference between Ω_+ and Ω_- being proportional to Ω, there can be some difficulty to measure Ω when $\Omega_+ - \Omega_-$ is smaller than the dimension of the dead zone. To overcome this problem, one usually uses a mechanical activation. We see that in the case of the four-wave mixing oscillator, it is possible to replace the mechanical activation by an optical activation using pump beams I_1 and I_2 of different intensities. The results reported in figure 6 are of course very preliminary and it is not possible to deduce the limiting possibilities of this system from this curve.

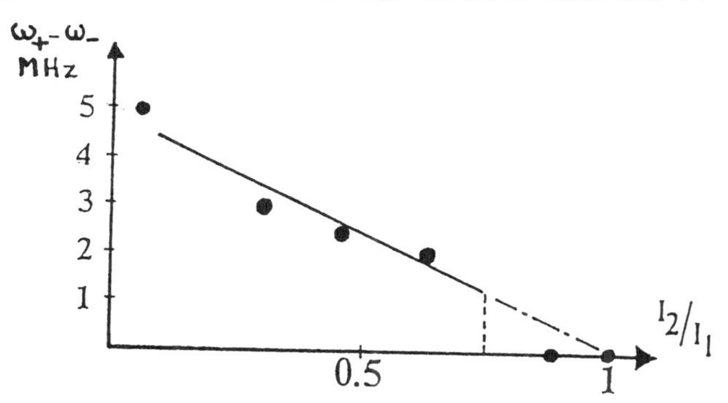

Figure 6 : Experimental variation of $\omega_+ - \omega_-$ as a function of I_2/I_1 (I_1 remaining fixed).

CONCLUSION

We have shown that collision-induced gratings can be useful to understand several non linear optical phenomena for a set of two-level atoms. One of the main interest of this interpretation is that it provides a close link with the effects obtained in solid state physics and particularly in the case of photorefractive materials where many effects can be understood using photocarriers gratings.

REFERENCES

[1] N. Bloembergen in Laser Spectroscopy IV (Springer, Berlin 1979) p. 340, in Laser Spectroscopy V (Springer, 1981)p. 157.

[2] D. Grandclément, G. Grynberg and M. Pinard, Phys. Rev. Lett. 59, 40 (1987) ; Phys. Rev. Lett. 59, 44 (1987).

[3] G. Grynberg, J. Phys. B14, 2089 (1981).

[4] G. Grynberg and M. Pinard, Phys. Rev. A32, 3772(1985).

FOUR-WAVE MIXING SPECTROSCOPY OF STATE SELECTIVE COLLISIONS IN GASES AND SOLIDS*

Juan F. Lam
Hughes Research Laboratories
Malibu, California 90265

ABSTRACT

The quantum evolution from a closed to an open system, induced by collisional processes, is presented in the framework of nearly degenerate four-wave mixing (NDFWM) spectroscopy. We show that an open system manifests itself by the appearance of a subnatural linewidth in the spectrum of the phase conjugate signal. Examples are described for sodium vapor in the presence of buffer gases and in Nd^{+3} doped β''-Na-Alumina for high concentration of Nd^{+3} ions.

INTRODUCTION

The discovery of real time phase conjugate optics by Stepanov et al[1] and Woerdman[2] has provided the foundation for the use of four-wave mixing processes as novel spectroscopic tools. The inherent advantages of backward degenerate four-wave mixing over saturated absorption techniques are the existence of Doppler-free spectrum combined with their nearly background-free signals, and the simultaneous measurement of the longitudinal and transverse relaxation times in a single spectral scan. The Doppler free feature was first pointed out by Liao and Bloom[3] in their investigation of resonantly enhanced phase conjugate mirrors. While the simultaneous presence of the longitudinal and tranverse relaxation times in the spectrum were described by Lam et al[4].

Recent studies of collisional processes using four-wave mixing techniques have shed new insights on how collisional processes affect the spectral lineshape[5]. A point in question was the measurement of ground state behavior in spite that the laser was resonant to an optical transition. This fact illustrates the complexity involved in our understanding of collision-induced lineshapes. The objective of this review is to provide an up-to-date account of how the technique of nearly degenerate four-wave mixing probes the evolution of a quantum system in the presence of perturbers[6].

THE TECHNIQUE OF NDFWM SPECTROSCOPY

The technique of NDFWM spectroscopy[4] involves the generation of a travelling wave excitation in the medium (with an atomic resonance ω_o) by the interference of two nearly co-propagating radiation fields oscillating at frequency ω and $\omega+\delta$; respectively. The nearly phase conjugate field is produced by the scattering of a counter-propagating read-out beam off the travelling wave excitation. The nearly conjugate field oscillates at frequency $\omega-\delta$ and travels in opposite direction to the input beam oscillating at $\omega+\delta$.

The small signal spectrum of the phase conjugate fields takes on distinct behavior depending on whether the resonant medium is homogenously or inhomogenously broadened. For the case of a two-level homogenously broadened system, the theoretical spectral lineshape[7] for $\omega=\omega_o$ exhibits a resonance at $\delta=0$ with linewidth given by $.41\gamma$, where γ is the spontaneous decay rate, in the absence of a buffer gas. In the presence of a buffer gas, the NDFWM lineshape shows an effective narrowing in the linewidth. This phenomenon arises from the formation of a long-lived ground-state population excitation formed by the interference of two input radiation fields. The limiting factor is ultimately determined by the transit time of an atom traveling across a laser beam.

The spectral lineshape takes on a similar behavior for the case of an inhomogenously broadened system when $\omega=\omega_o$. It describes the evolution of the spectral lineshape as buffer gas is added to the quantum system. For $\omega=\omega_o$, the lineshape contains two resonances located at $\delta=-2\Delta$ and $\delta=0$. However the linewidths have different behavior in the presence of buffer gases[4]. The resonance line located at $\delta=-2\Delta$ experiences phase interrupting collision, which tends to broaden the linewidth. The resonance line at $\delta=0$ experiences a decoupling of the ground state from the excited state and the width of the line is determined by the lifetime of the ground state (in this case it is the transit time). Δ is the detuning of the pump from resonance.

The spectral behavior of the nearly phase conjugate field illustrates two important points. First, the spectrum contains simultaneous information on the energy and dipole relaxation times, when $\omega=\omega_o$. And second, the spectrum provides a direct measurement of the ground state lifetime in the presence of a buffer gas.

COLLISION-INDUCED RESONANCES IN SODIUM VAPOR

Sodium vapor provides a testing ground for the concept outlined above. We consider the excitation of the D_2 line of sodium using two correlated cw lasers having a bandwidth of approximately 1 Mhz. Since the spectral lineshape depends on the difference of frequency between the two correlated lasers, phase fluctuations are automatically eliminated and the fundamental limitation to the sensitivity of our measurement technique is the time of flight of the atom across the optical beam or the laser linewidth, whichever is larger. Figure 1a shows the spectral behavior of the nearly phase conjugate field for equally polarized input fields and for ω tuned to the $3\ S_{1/2}$ (F=2) to $3\ P_{3/2}$ (F=3) transition[4]. It depicts a

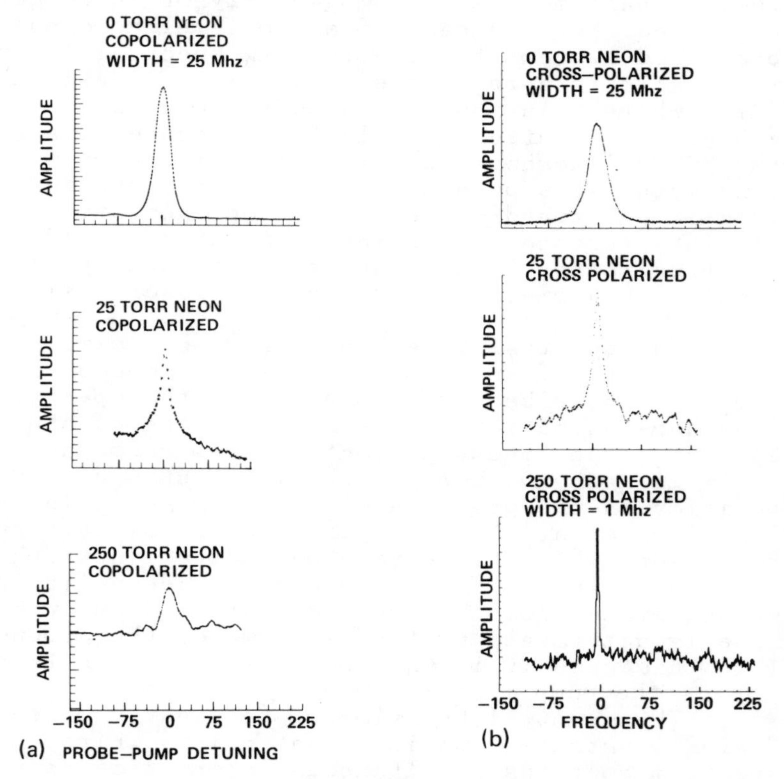

Figure 1. Spectral lineshape of the NDFWM of sodium for (a) co-polarized radiation fields, (b) cross-polarized radiation fields.

narrowing of the linewidth, from 20 Mhz to 1 Mhz, as argon buffer gas is added to the sodium cell. It can also be interpreted as the appearance of a new resonance with a much narrower linewidth. Hence it provides a direct measurement of the effective lifetime of the 3 $S_{1/2}$ state. As the buffer gas pressure increases to 250 torr, the linewidth broadens up due to the excited state.

Figure 1b shows the spectral lineshape for the case where the counterpropagating fields, oscillating at ω, is orthogonally polarized with respect to the field oscillating at $\omega + \delta$. Again as the buffer gas pressure increases, the linewidth decreases from 20 Mhz to 1 Mhz, even beyond the 250 torr regime. The use of cross polarized radiation fields induces a Zeeman coherence grating rather than a population grating[8]. The collision cross section for Zeeman coherences is relatively small, which accounts for the 1 Mhz linewidth even at a pressure of 250 torr. These data provides a glimpse of the simplicity of collisional processes on the NDFWM spectrum. However, Khitrova and Berman[9] has predicted that additional ultranarrow features are also present even in the absence of buffer gases. These features are the result of optical pumping processes that exist when the atomic multipoles are produced by the interference of two orthogonally polarized light beam. A detail account of the theory and experiment related to these novel features are presented by Steel et al.[10].

OPTICAL PAIR INTERACTION IN Nd^{+3}-β''-Na-Alumina

The study of cooperative processes in solids has been a subject of great interest since they determine the practical limitations of solid state optical devices. The technique of NDFWM spectroscopy has provided an initial step toward understanding the role of ion-ion interaction in the spectral lineshape[11]. An interesting candidate material is Nd^{+3} doped β-Na-Alumina[12,13]. For this material, the four-wave mixing process takes place at 575 nm between the $^4I_{9/2}$ and $^4G_{7/2}$ states. However due to the fast nonradiative relaxation rate of the excited state $^4G_{7/2}$, the NDFWM spectrum shows a ultranarrow resonance whose width is determined by radiative decay rate at 1.06 μm (which is of the order of few hundreds μsec). The transition at 1.06 μm is between the $^4F_{3/2}$ and $^4I_{11/2}$ states of Nd^{+3}.

Figure 2a depicts the spectral lineshape of the nearly phase conjugate field for three different pump intensities. The curve labeled by 25 mW provides a direct measurement of the spontaneous decay rate at 1.06 μm even though the laser used in the experiments has a wavelength of 575 nm. At higher value of the laser intensity, the linewidth broadens up due to pump-induced saturation. Rand et al[11] showed that the intensity behavior of the NDFWM linewidth is consistent with the system being inhomogenously broadened. Figure 2b describes the dependence of the saturation intensity and the linewidth as a function of Nd^{+3} concentration. For densities larger than .5 x 10^{21} cm^{-3}, both physical parameters experience a decrease. Lam and Rand[14] has shown that this behavior arises from the increase in the pair interaction which creates an additional channel for the electron located in $^4F_{3/2}$ to escape, leading to such a behavior.

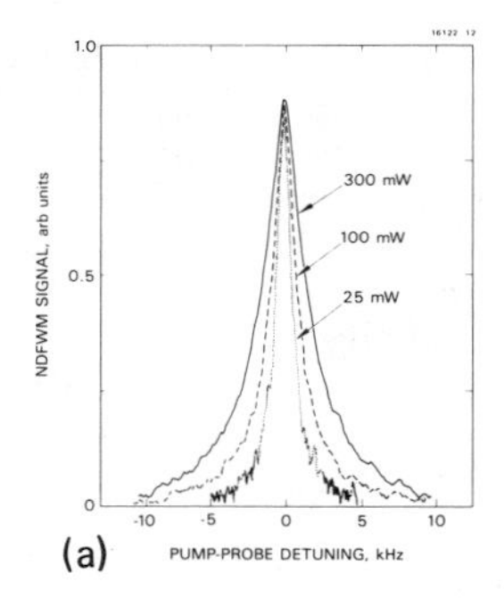

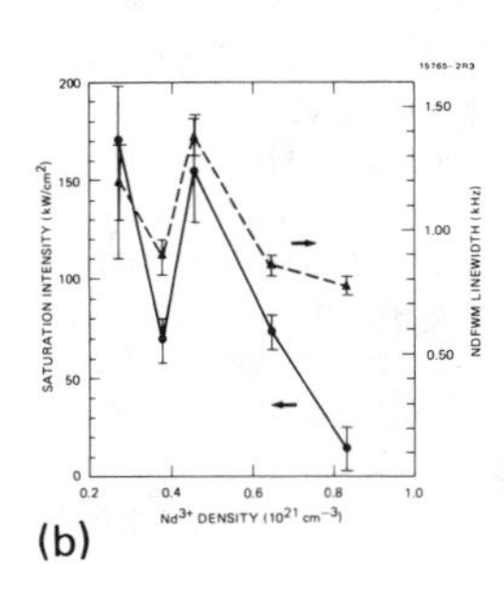

Figure 2. (a)Spectral lineshape for different intensities. (b) Behavior of the saturation intensity and linewidth as a function of dopant density.

SUMMARY

We have described some of the subtle issues associated with spectral lineshape of NDFWM in the presence of collisional processes. The results have demonstrated that collisions do indeed provide a means of separating the ground state from the excited state. The ultranarrow linewidth of the NDFWM spectrum is a manifestation of the ground state effective lifetime.

* Work supported by the United States Army Research Office under contract #DAAL03-87-C-0001 and by the Air Force Office of Scientific Research under contract #F49620-85-C-0058

ACKNOWLEDGEMENT

The author gratefully acknowledge the important contributions made by Drs. Duncan Steel, Steve Rand, Ross McFarlane, Steve Turley and Oscar Stafsudd.

REFERENCES

1. Stepanov, B.I., E.V. Ivakin and A.S. Rubanov,, Sov. Phys. Dokl. Tech. Phys. 15, 46 (1971)
2. Woerdman, J.P., Opt. Commun. 2, 212 (1971)
3. Liao, P.F., and D.M. Bloom, Opt. Lett. 3, 4 (1978)
4. Lam, J.F., D.G. Steel and R.A. McFarlane, Phys. Rev. Lett. 49, 1628 (1982)
5. Rothberg, L., in PROGRESS IN OPTICS, ed. by E. Wolf, North Holland, Amsterdam (1987)
6. Berman, P.R., in NEW TRENDS IN ATOMIC PHYSICS, Les Houches, ed. by G. Grynberg and R. Stora, North Holland, Amsterdam (1984)
7. Lam, J.F., D.G. Steel and R.A. McFarlane, in LASER SPECTROSCOPY VII, ed. by T.W. Hansch and Y.R. Shen, Springer Verlag, Berlin (1985)
8. Lam, J.F., D.G. Steel and R.A. McFarlane, Phys. Rev. Lett. 56, 1679 (1986)
9. Khitrova, G. and P.R. Berman, to be published (1988)
10. Steel, D.G., J. Liu, G. Khitrova and P.R. Berman, unpublished (1988).
11. Steel, D.G. and S.C. Rand, Phys. Rev. Lett. 55, 2285 (1985)
12. Boyd, R.W., M.T. Gruneisen, P. Narum, D.J. Simkin, B. Dunn and D.L. Yang, Opt. Lett. 11, 162 (1986)
13. Rand, S.C., J.F. Lam, R.S. Turley, R.A. McFarlane and O.M. Stafsudd, Phys. Rev. Lett. 59, 597 (1987)
14. Lam, J.F. and S.C. Rand, (unpublished 1988)

PHASE CONJUGATION CALCULATIONS AND EXPERIMENTS

K. Leung, R. Holmes, and A. Flusberg
Avco Research Laboratory, Inc.
Everett, Massachusetts 02149

INTRODUCTION

SBS phase conjugation has been a useful means for improving laser beam-quality in many applications.[1] Further, many authors have discussed the conditions necessary for obtaining high-fidelity phase conjugation.[2] Experimental results show that stimulated Brillouin scattering can phase conjugate a laser effectively under transient as well as steady-state phonon-response conditions.[3] A broadband laser may also be conjugated.[4] Waveguide- and focussed- geometries both may be useful for laser beam-quality control. However, some questions nevertheless remain. In this paper we attempt to answer several questions related to the fidelity. First, is there a loss of phase conjugation fidelity when the laser is both broadband and the phonon response is transient? Second, is there a difference in fidelity with SBS in a liquid compared to a gas? Third, is there any advantage to the use of a waveguide in improving the conjugation fideltiy? The first two of these questions are answered experimentally, the third question is answered theoretically. The numerical portion of the theoretical work is compared to the experimental results.

EXPERIMENT

The experimental setup is shown in Figure 1. The experiment used a Nd:YLF laser equipped with glass amplifiers. The laser beam passed through a $\lambda/4$ plate and a polarizer for isolation of the laser from the backscattering. The laser beam then passed through several pickoff windows, and then through an aberrator if desired. The beam was then focussed into the cell containing the Brillouin media. The SBS return was observed in reflection from the pickoff windows. The temporal profile of both the laser beam and the SBS return was monitored. The far-field and near-field profiles , and near-field shearing interferograms were recorded. The laser pulse length was measured at 40 $\pm$ 5 nsec, and the phonon decay time was 45 and 20 nsec, respectively, for the liquid CS_2 and the gaseous SF_6. the SF_6 gas was kept at 17 atm. Hence, the SBS in both the SF_6 and CS_2 was definitely transient, although not with the same degree of transiency. Another key parameter is the laser bandwidth, which was measured at .5 cm^{-1}. The inverse of this bandwidth is to be compared to the length of the scattering region. For our experiment, the scattering volume is equal to the focal region, so for our F/50 focussing system the length of the scattering volume is roughly 1.1 cm for an unaberrated beam. Thus the length of the scattering volume was somewhat less than the coherence length. The threshold energy for the observation of SBS was approximately 5 and 30 ($\pm$ 1)mJ respectively for CS_2 and SF_6. The corresponding laser intensity at the cell center was about 5 and 30 GW/cm^2 respectively.

The observations were conducted at threshold, so the transient gain G was roughly equal for all observations and approximately equal to 30.

Low frequency aberrations were introduced by a phase-plate as well as by tilting the focussing lens. These aberrations were mostly of low spatial frequency. The quality of conjugation for these aberrations were compared with the quality of conjugation when no aberrations were introduced. The results are described in the following section.

EXPERIMENTAL RESULTS

The experiment had diagnostics for both the temporal and spatial profiles. Both temporal and spatial profiles showed a tendency for the SBS to correlate with the pump, as is predicted by the theories.[2] However, there are important qualifications to this remark.

The temporal profile of the SBS and the pump was observed on an oscilloscope. The profiles were similar within the resolution of the oscilloscope, which was operating at 20 nsec per division. The pump fluctuations on the order of 1 to 2 nsec were reproduced on the SBS return. However, the SBS return pulse width was somewhat (20 percent) shorter than the pump pulse, with a rise time much faster than the pump beam. Also, the SBS decay was somewhat faster than the pump beam.

The spatial profiles of the SBS return pulses usually indicated that a reasonable quality of spatial reproduction of the pump was obtained. The far-field width of the SBS return signal was within 50 percent of that for the unaberrated pump. However, the width was consistently observed to be less than the pump when no aberrations were introduced. This result may arise from the so-called gain-narrowing effect, in which the central portion of the pump exceeds threshold at focus in the Brillouin cell but the surrounding regions do not. The far-field profile is the image of the focal spot in the SRS cell when no aberrator is present. Hence, the far-field should be narrower when the spatial gain-narrowing is present.

On the other hand, gain-narrowing may also be used to explain our results when aberrators were introduced in the beam path. When aberrators were used the SBS far-field profile was usually affected somewhat. In this SBS-threshold experiment, the higher spatial frequencies away from the focus in the SBS cell did not have sufficient gain to exceed threshold. Hence, we saw a notable degradation of phase conjugation fidelity for aberrated beams.

Other observations which are relevant are the near-field profiles. These profiles indicate that the SBS process need not introduce significant intensity modulation in the near-field. This result contrasts somewhat with some previous numerical calculations in which depletion is included.[5]

Finally, it is interesting to observe that the CS_2 and the SF_6 return profiles exhibited the same qualitative behavior as discussed above. Hence, at our fluence and power levels, the liquid did not introduce additional interfering nonlinear effects.

COMPARISON BETWEEN EXPERIMENT AND THEORY

Numerical calculations were performed for a geometry similar to the experiment. The results are not directly comparable to experiment since the calculations pertain to the steady-state phonon regime. However, the similarity of results warrant comparison. As shown in Figure 2, results are qualitatively comparable. The SBS intensity and phase profiles are the bottom left and bottom right plots, respectively. The pump intensity and phase plots are the upper left and upper right profiles, respectively. The plots are the profiles of the azimuthally-symmetric quantities which are used and calculated in our code. Hence, two-dimensional effects are included in the code. These azimuthally-symmetric results show that the SBS return does have ≃20 percent intensity modulation, and additional defocus is present on the backscattered light. This is observed experimentally when aberrations are introduced on the laser beam. Both the code and experiment show the effect of spatial-narrowing of the gain region by SBS at threshold. This effect is seen experimentally in the far-field profiles. Note that the phase profile of the pump is not well reproduced on the Brillouin return, but some similarities are seen despite the difference in scale of the ordinates for the plots of pump and SBS phase profiles.

Those results lead us to conclude that there is qualitative agreement between the code and the experiment.

COMPARISON OF ANALYTICAL THEORY FOR WAVEGUIDES WITH EXPERIMENTAL AND NUMERICAL RESULTS

The theory of fidelity of phase conjugation for waveguides is well known, but unresolved issues still remain in the literature. A controversy regarding the applicability of waveguide results to the focussed geometry has persisted.[6] These waveguide results indicate that the fidelity of phase conjugation at threshold should improve as the angular divergence of the laser is increased at the waveguide entrance. We have found experimentally and numerically that the conjugation fidelity at threshold in focussed geometries may actually decrease as the angular divergence at or near focus is altered by aberrations in the vicinity of the focussing lens. The trend of decreasing fidelity with increasing aberration at threshold intensity in the focussed geometry suggests that the waveguide theory is not applicable to the focussed geometry. A physical explanation is as follows: in the focussed geometry, sufficient medium polarization for observable backscatter is induced only at the central peak of the focal spot. On the other hand, in the waveguide, gain gratings are created as light is reflected or refracted from the waveguide walls. These gain gratings allow parametric amplification of the spatial sidelobes in the waveguide geometry. Hence, the spatial sidelobes are more completely reproduced in the waveguide geometry relative to the focussed geometry. Hence, conjugation fidelity is expected to improve with the use of a waveguide.

SUMMARY AND CONCLUSIONS

Experiments confirm that phase conjugation may work despite the simultaneous presence of two potentially disruptive conditions. The first condition is that the laser pulsewidth is less than the phonon buildup time and the phonon lifetime. The second condition is that the laser bandwidth exceeds both the Brillouin linewidth and the Brillouin shift. Despite the presence of these two conditions, the SBS signal is present and conjugates the pump.

Experiment also shows that stimulated Brillouin scattering in both a gas and a liquid may yield reasonable conjugation fidelity.

Finally, the experiment indicates that a highly aberrated beam may be difficult to conjugate accurately in a focussed geometry at threshold, and theory indicates that it may be easier to conjugate the same beam more precisely in a waveguide geometry.

REFERNCES

1. R.A. Fisher, "Applications of Phase Conjugation," in "Optical Phase Conjugation," ed. by R.A. Fisher, (Academic Press, New York, 1983).
2. B. Ya. Zel'dovich, N.F. Pilipetsky, and V.V. Shkunov, "Principles of Phase Conjugation," (Springer-Verlag, New York, 1985).
3. V.F. Efimkov, et.al., "Inertia of Stimulated Mandel'shtam-Brillouin Scattering and Nonthreshold Reflection Short Pulses with Reversal of the Wavefront", Sov. Phys. JETP 50, 267 (1979).
4. G.C. Valley, "A Review of Stimulated Brillouin Scattering Excited with a Broad-Band Pump Laser", IEEE J. Quant. Electron. QE-22, 704(1986).
5. R.H. Lehmberg, "Numerical Study of Phase Conjugation in Stimulated Backscatter with Pump Depletion," Opt. Commun. 43, 369 (1982).
6. P. Suni and J. Falk, "Theory of phase conjugation by stimulated Brillouin scattering," J. Opt. Soc. Am. B 3, 1681 (1986).

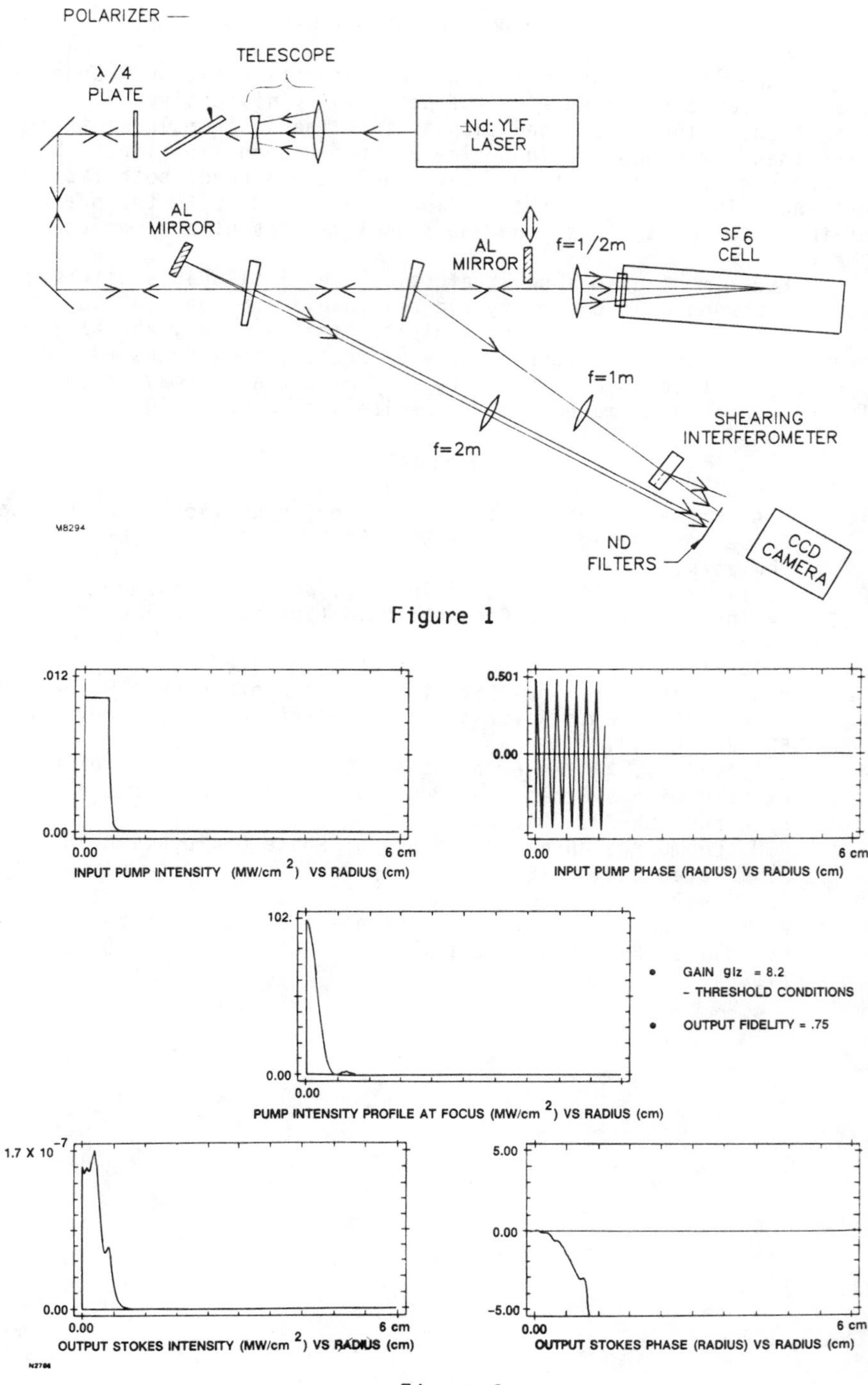

Figure 1

Figure 2

SELF-SEEDED BRILLOUIN PULSE COMPRESSION

GABRIEL G. LOMBARDI, H. KOMINE, and S. J. BROSNAN
Northrop Research and Technology Center
One Research Park, Palos Verdes Peninsula, CA 90274

ABSTRACT

A one joule, 60 ns, XeCl laser pulse was compressed to 6 ns (FWHM) by a phase-conjugated, stimulated-Brillouin-scattering compressor in SF_6 with an efficiency of 50%. The Stokes seed beam was generated by focussing the pump beam into a Brillouin cell.

INTRODUCTION

Self-seeded pulse compression was demonstrated using a phase-conjugated, XeCl laser with 60 ns long (FWHM) optical pulse. Figure 1 is a schematic diagram of the apparatus. The master oscillator (MO) consisted of a XeCl unstable resonator, with an output energy of 10 mJ, which was injection-seeded by a frequency-doubled dye laser to narrow the emission linewidth to 1-2 GHz. The linearly polarized MO beam was amplified by two passes through a power amplifier. The amplifier output beam was separated from the MO beam by a polarization-sensitive beamsplitter (PS). A quarter-wave plate (QW) rotated the polarization of the beam by 90° so it would be reflected, rather than transmitted, by PS. The amplifier output was phase conjugated by the seed generator after passing through the compressor cell. This backward-travelling wave served as the Stokes seed for the SBS compressor, which contained 13 amagat of SF_6. As the Stokes wave passed through the pump wave in the compressor cell, energy was transferred from the pump to the Stokes wave. The pump and Stokes waves were separated with a PS as in the power amplifier.

The length of the folded optical path through the compressor was 800 cm, approximately corresponding to half the length of the optical pulse. This is the minimum length required in order for the pump pulse to be depleted over its entire duration. With 13 amagat of SF_6, the cell had a transmittance loss of approximately 25%. At least a part of this loss (9%) is attributable to Rayleigh scattering by SF_6. Because of the uncertainty in the measurement, it is possible that the dominant loss is attributable to scattering.

The use of phase conjugation of the pump beam to generate the compressor seed has two important advantages over other pulse-compression schemes. First, the pump and seed beams are automatically aligned with respect to one another. Second, the effects of optical aberrations caused by the compressor and other elements in the optical train are cancelled.

Two effects lead to the temporal narrowing of the Stokes wave in a self-seeded compressor. First, only the leading edge of the Stokes wave is amplified by undepleted pump; this results in enhanced amplification of the leading edge and consequent narrowing of the Stokes wave. Second, the Stokes seed is itself shortened because it is produced by the depleted pump wave, the trailing edge of which has a smaller amplitude than the leading edge.

RESULTS

Pulse shapes were recorded with vacuum photodiodes connected to transient digitizers having a time resolution of 5 ns. The pulse shape of the amplified Stokes wave was also recorded on a storage oscilloscope with a 400 MHz bandwidth. The energy of the compressed pulse was measured with a calorimeter. In addition, the pump input and depleted pump photodiodes were calibrated, with a calorimeter, to provide absolute energies for each pulse. Some results obtained with this apparatus are given in Fig. 2. The pump input and compressed Stokes output waveforms are shown in Figs. 2 (a) and (b), respectively. For this shot, the conversion efficiency, deduced from the fractional pump depletion, was 40% at a calculated compressor gain of between 3 and 5. In general, the efficiency ranged from 32 to 53%. Variations in the conversion efficiency are attributed to shot-to-shot variations in the bandwidth and spectral distribution of the laser output, with a consequent change of the gain of the compressor. These data have been compared to the results of calculations made assuming a plane-wave model for the compressor. The results of efficiency and pulse shape simulations agree closely with the data.

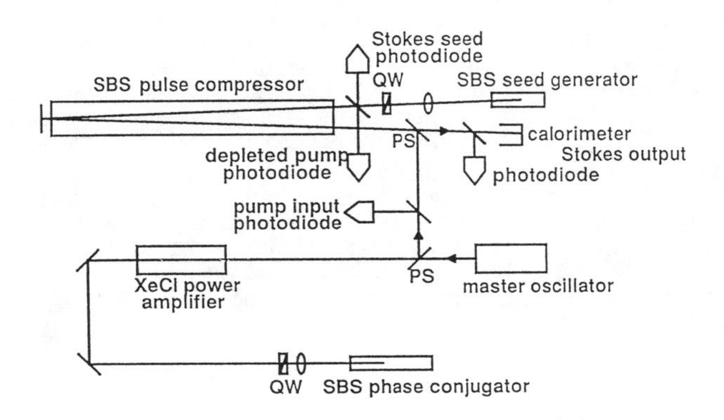

Fig. 1. Schematic diagram of the apparatus

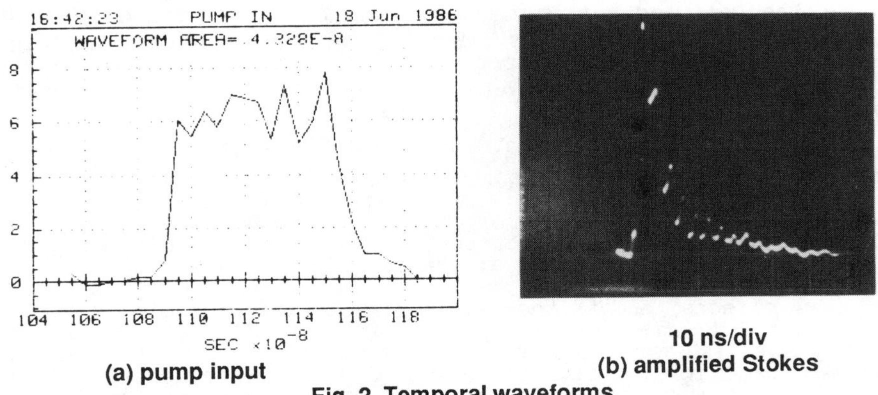

10 ns/div

(a) pump input

(b) amplified Stokes

Fig. 2. Temporal waveforms

STATISTICAL LIMITS ON BEAM COMBINATION BY STIMULATED BRILLOUIN SCATTERING

Joel Falk, Morton Kanefsky, Paul Suni*
University of Pittsburgh, Pittsburgh, PA 15261

ABSTRACT

We investigated the phase difference ϕ between the SBS outputs produced by two beams pumping slightly different portions of the same active medium. SBS is narrow-band amplified noise and thus has a statistical nature. The parameter ϕ must be characterized by a probability distribution which can be calculated. A time average of $\cos\phi$ can be evaluated theoretically and measured by interfering the two SBS outputs and using integrating detectors. For imperfect overlap of the pumping beams there is a significant probability that $\cos\phi<1$..

INTRODUCTION

In recent years there has been a growing interest[1-4] in finding ways to combine the output beams from multiple lasers into a single high brightness beam. This interest stems from a need to overcome limitations on the power which can be extracted from a single gain medium. For beam combination to be most useful the beams which are combined should be coherent and in phase with each other in both space and time.

In the work reported here two high power laser pump beams are focused into the same region of a stimulated Brillouin scattering (SBS) medium. We have shown that, since the Stokes beams produced are derived from noise, the phase difference between generated beams must be described by a probability distribution. This distribution shows a significant probability of a nonzero phase even when pumping beams are fairly well overlapped. We have developed an experimental method to measure the distribution of the time average of $\cos\phi$. The distribution functions recorded for experimental runs of 10^3-10^4 pulses, agree well with those predicted.

THEORETICAL RESULTS

If we focus two beams into two separate spatial regions of the SBS medium then the two generated Stokes beams will have a completely random phase difference when they emerge. This occurs because they are generated by two uncorrelated random noise sources and this has been experimentally verified by Basov et al.[5]. The only way to force the phase difference of the Stokes beams to be exactly zero is to focus the input beams in such a way that they

*Present address: Department of Physics, University of Southampton, SO9 5NH, United Kingdom.

are driven by exactly the same thermal noise field. This means in practice that the beams must be exactly overlapped near their focal regions. If they are not exactly overlapped then the degree of phase locking is found to depend on the ratio of the Stokes pulse width T to the phonon lifetime τ_p.

Two SBS beams are generated by amplifying two noise sources which, because of the partial overlap of the SBS pump beams, are correlated with correlation coefficient ρ. The noise sources are described in terms of correlated, narrow-band, Gaussian processes each symmetric about the Stokes radian frequency.

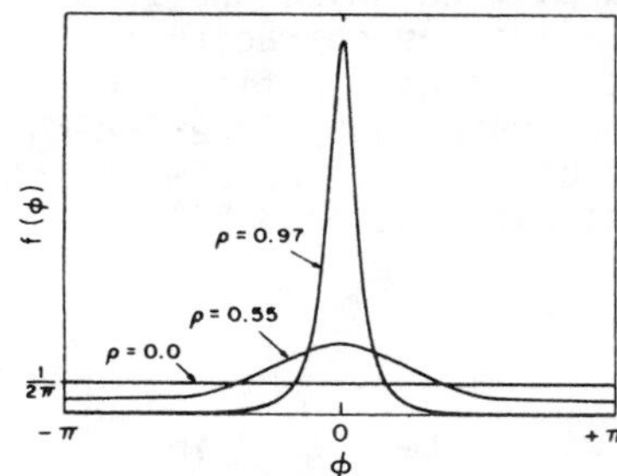

Fig. 1. The distribution of phase.

If the Stokes pulse width T is much shorter than the phonon lifetime τ_p, then ϕ and consequently $\cos\phi$ remains constant throughout the pulse. For this case we have derived closed form solutions for the probability distribution of ϕ. These distributions are shown in Figure 1 for ρ=0, 0.5 and 0.97.

Figure 1 shows that even for highly correlated noise sources (ρ = 0.97) there is a sizeable probability that $\phi \neq 0$ and $\cos\phi \neq 1$.

If T/τ_p is not <<1, which is the case for the experiments we have performed, then the temporal fluctuations of ϕ must be taken into account. These fluctuations come about because of the finite Brillouin bandwidth. In this case a computer simulation can be used to find the probability distribution of $S \equiv (1/T)\int \cos\phi(t)dt$. Computer simulations and a comparison with laboratory experiments are described in the following section and are illustrated[4] in Figures 2.

EXPERIMENTAL RESULTS

In our measurements of S we used a single-longitudinal mode frequency doubled Nd:YAG laser operating at 10 pps with a pulse width of approximately 7 ns. We used a phase grating to split the incident 5 nsec 532 nm laser beam into many orders. The ± 1 grating orders were focused into ethanol. The two SBS return signals were recombined in the grating. The output 0, +1 and -1 order energies from the grating were measured on a large number (10^4) of pulses. The zeroth grating order recorded a coherent addition of the two SBS field amplitudes, i.e. $\int_0^T |E_x + E_y e^{j\phi}|^2 dt$, where the integration time is over the Stokes pulse duration. The detectors which recorded the +1 and -1 orders produced signals proportional

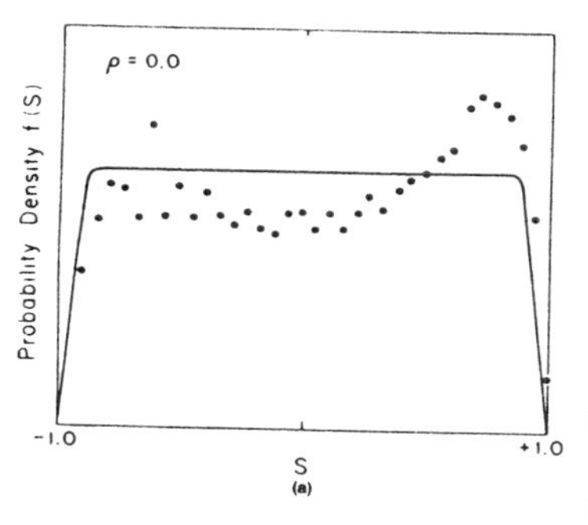

to $\int|E_x|^2$ dt and $\int|E_y|^2$ dt. From these measurements the value of $\int\cos\phi$ dt can be calculated for each laser pulse.

Figures 2 shows distributions (ρ = 0, 0.55, 0.97) of S found from the computer simulations as well as experimentally measured distributions for three separate crossings of the two pump beams (9 mm and 3 mm from the beams' focal plane, and at the beam's focal planes). Figures 2 show that the distribution of S is well described by the correlation coefficient between the two noise sources. We note from Figure 2 that even with the pump beams crossing at their focal point (ρ = 0.97) there is a significant probability of finding the phase difference not near zero degrees. Further details of our experimental work are given in reference 4.

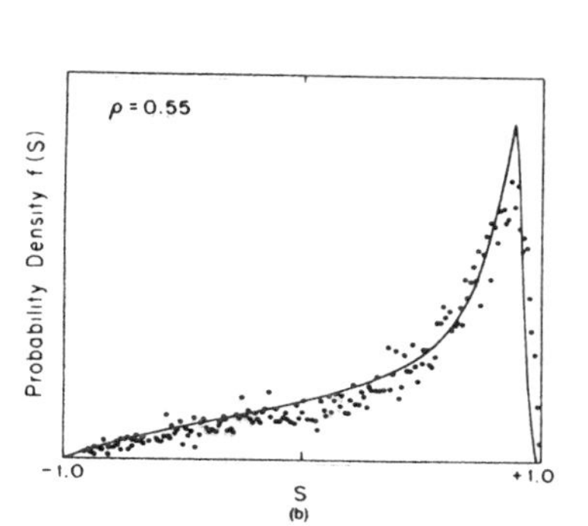

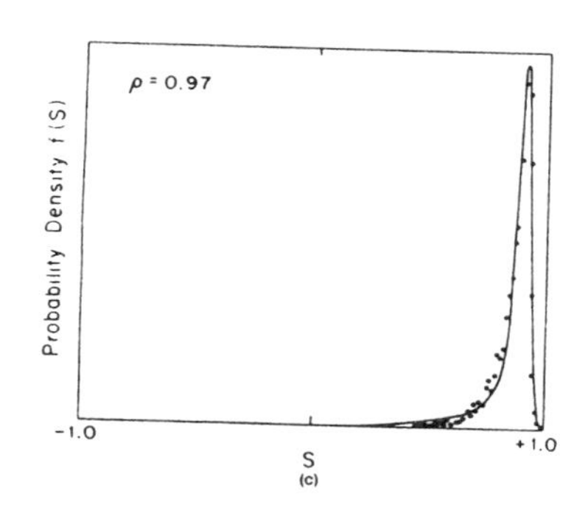

Figure 2. The probability distribution f(s).

REFERENCES

1. M. Valley, G. Lombardi and R. Aprahamian, J. Opt. Soc. Am. B 3 1492, 1986.
2. T.R. Loree, D.E. Watkins, T.M. Johnson, N.A. Kurnit and R.A. Fisher, Optics Letters, 12, 178, 1987.
3. D.A. Rockwell and C.R. Giuliano, Optics Letters, 11, 147 (1986).
4. J. Falk, M. Kanefsky and P. Suni, to be published, Optics Letters, January 1988.
5. N.G. Basov, I.G. Zubarev, A.B. Mironov, S.I. Mikhailov and A. Yu. Okolov, JETP Letters 31, 645, 1980.
6. J. Falk, M. Kanefsky and P. Suni, submitted to JOSA B, 1987.

Beam Coupling and Self-Pulsations in Self-Pumped $BaTiO_3$

Putcha Venkateswarlu, H. Jagannath, M. C. George and A. Miahnahri
Department of Physics, Alabama A&M University
Huntsville, AL 35762

ABSTRACT

Barium Titanate crystal is self-pumped at 5145, 4880, 4765 and 4580Å by two incoherent beams A1 and A2 with variable power ratios. In the first set of experiments, the beams cross in the crystal while in the second, they cross before reaching the crystal surface. The phase conjugate beams travelling in the reverse directions of A1 and A2 can be expressed in general by $[\overleftarrow{A1}(t) + \hat{A}2(t) - \Delta 1(t)]$ and $[\overleftarrow{A2}(t) + \hat{A}1(t) - \Delta 2(t)]$ where $\overleftarrow{A1}(t)$ and $\overleftarrow{A2}(t)$ are self-pumped phase conjugates of A1 and A2, $\hat{A}2(t)$ and $\hat{A}1(t)$ are Bragg diffracted components of A2 and A1 in the reverse directions of A1 and A2 respectively. The presence of $\hat{A}1(t)$ and $\hat{A}2(t)$ is more obvious in the second set of experiments, while $\Delta 1(t)$ and $\Delta 2(t)$ are more obvious in the first set.

EXPERIMENT AND RESULTS

Self-pumped phase conjugation and beam couplings have been discussed by different workers.[1-5] Eason and Smout reported recently their work on incoherent beam coupling[6] while we have reported our preliminary work on coherent beam coupling in Barium Titanate[7]. We report here our results on beam coupling in self-pumped electrically poled Barium Titanate crystal using two mutually incoherent beams A1 and A2 from a multimode 5w Ar ion laser with a path difference of 150cm. A1 and A2 make angles of about 17° and 15° with the normal to the c axis of the crystal (Fig. 1a).

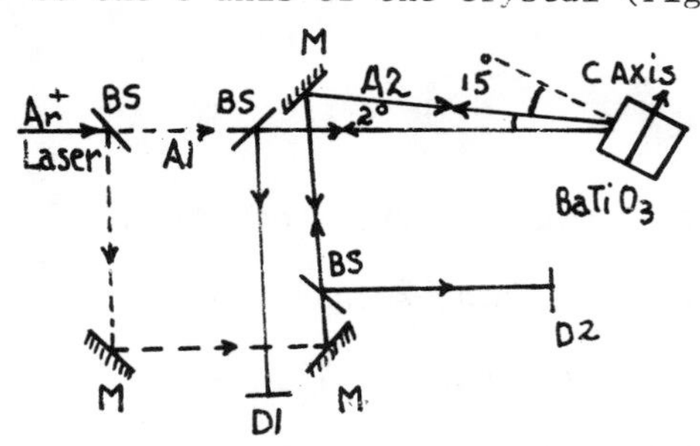

Fig. 1a. Experimental arrangement. A1, A2 are incoherent. Path difference 150cm. Multimode Ar ion laser

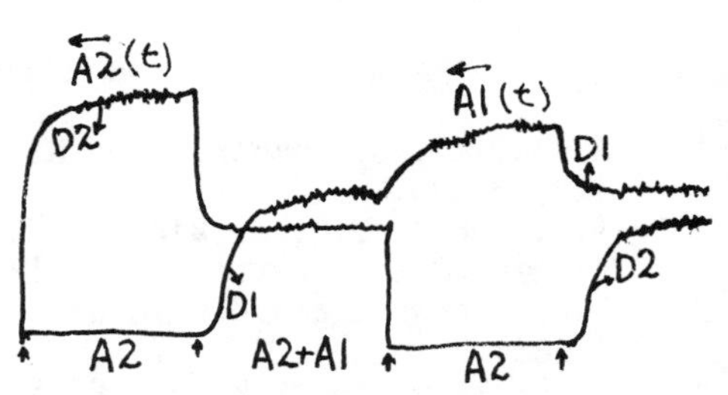

Fig. 1b. Signals at D1, D2 versus time. A1=10mw, A2=4mw. Beams cross in crystal in Figs. 1b, 2 and 3. Total time 7 min.

In the first set of experiments, the beams meet in the center of the crystal. Keeping the power in A1 two to three times more than that in A2 (4mw), their respective self-pumped phase conjugate beams $\overleftarrow{A1}$ (t) and $\overleftarrow{A2}$ (t) could be detected at D1 and D2 (Fig. 1a) when individually pumped. When pumped simultaneously by both the beams, the strengths of the signals at D1 and D2 may be represented by $\overleftarrow{A1}$ (t) + $\hat{A}2$ (t) - $\Delta 1$ (t) and $\overleftarrow{A2}$ (t) + $\hat{A}1$(t) - $\Delta 2$(t) respectively. $\hat{A}2$ (t) and $\hat{A}1$ (t) represent the parts of the beams A2 and A1 getting Bragg diffracted (due to fanning) into the reverse direction of A1 and A2 respectively.[6,7] $\Delta 1$ (t) and $\Delta 2$ (t) are the erasure effects on $\overleftarrow{A1}$(t) and $\overleftarrow{A2}$(t) by the beams A2 and A1 respectively. The signal at D1 during simultaneous pumping is greater or less than that under individual pumping depending upon whether $\hat{A}2$ (t) is greater or less than $\Delta 1$ (t), and similarly with the signal at D2. (See Fig. 1b).

With a ratio of about 4:1 in the powers of A1 (16.5mw) and A2 (4mw), $\overleftarrow{A1}$ (t) at D1 shows self-oscillations while $\overleftarrow{A2}$ (t) at D2 remains steady when individually pumped (Fig. 2). The frequency of oscillations increases with the wavelength of the laser and its power. With simultaneous pumping, the signal at D1 becomes stable (or less oscillatory) while the one at D2 is zero showing that in this case $\overleftarrow{A2}(t) = \Delta 2(t) - \hat{A}1(t)$ and $\overleftarrow{A1}(t) > [\Delta 1 - \hat{A}2(t)]$. With the beam powers nearly equal (6-17mw), $\overleftarrow{A1}$(t) and $\overleftarrow{A2}$(t) do not coexist and each beam effectively erases the other beam's grating (Fig.3.). Here $\overleftarrow{A2}$ (t) = $\Delta 2$ (t) - $\hat{A}1$ (t) and $\overleftarrow{A1}$ (t) = $\Delta 1$ (t) - $\hat{A}2$ (t).

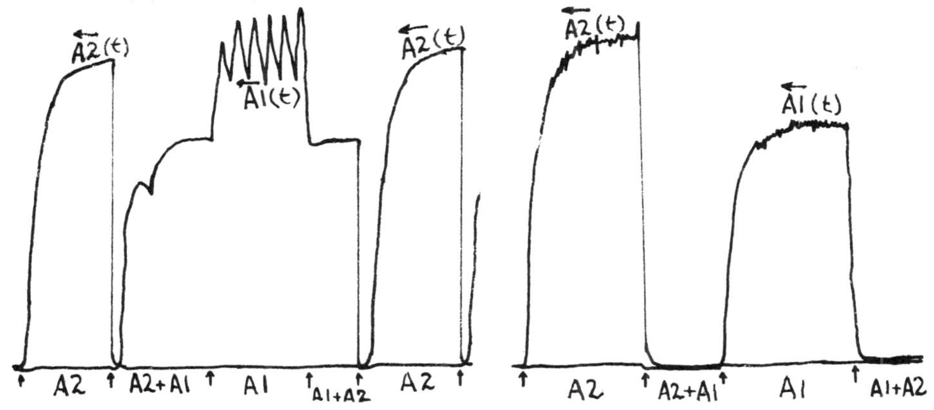

Fig. 2. Signals D1, D2 vs. time. A1 = 16.5 mw, A2 = 4 mw. 11 min.

Fig. 3. Signals D1, D2 vs. time. A1 = 9.5 mw, A2 = 9 mw. 12 min.

In the second set of experiments the beams A1 and A2 with equal powers (6-17mw) cross before reaching the crystal surface. $\overleftarrow{A1}$ (t) and $\overleftarrow{A2}$ (t) show oscillations under individual pumping while they are nearly stable with minor coherent pulsations under simultaneous pumping (Fig. 4). If after simultaneous pumping A1 is shut off, the signal at D2 shows a sharp drop to zero while that at D1 shows an exponential decay. Similarly when A2 is shut off, D1 shows a sharp drop to zero while D2 shows an exponential decay (Figs. 4a,b). These results clearly indicate the existence of the cross coupled beams

$\hat{A}1$ (t) and $\hat{A}2$ (t) in the directions of $\overleftarrow{A2}$ (t) and $\overleftarrow{A1}$ (t) respectively[5,6]. As indicated earlier the signals at D1 and D2 under simultaneous pumping are given by $\overleftarrow{A1}$ (t) + $\hat{A}2$ (t) - Δ1 (t) and $\overleftarrow{A2}$ (t) + $\hat{A}1$ (t) - Δ2 (t) respectively. Thus when A2 is shut off, $\overleftarrow{A2}$ (t) and Δ2 (t) go to zero and $\hat{A}1$ (t) decays exponentially (Fig. 4b) depending on the life of the transient grating from which A1 gets Bragg diffracted to get into the opposite direction of A2. One can see that the signal at D2 shoots up slightly before it exponentially decays (Fig. 4b). This increase comes up probably because the grating erasure effect on $\overleftarrow{A1}$ (t) by A2 goes to zero when A2 is shut off. The signal at D1 goes to zero abruptly when A2 is shut off probably because $\hat{A}2$ (t) goes to zero and so also does $\overleftarrow{A1}$ (t) - Δ1 (t). After the signal at D1 sharply goes to zero, it rises again to the saturated value of $\overleftarrow{A1}$ (t) as the crystal is under the pumping by A1 only.

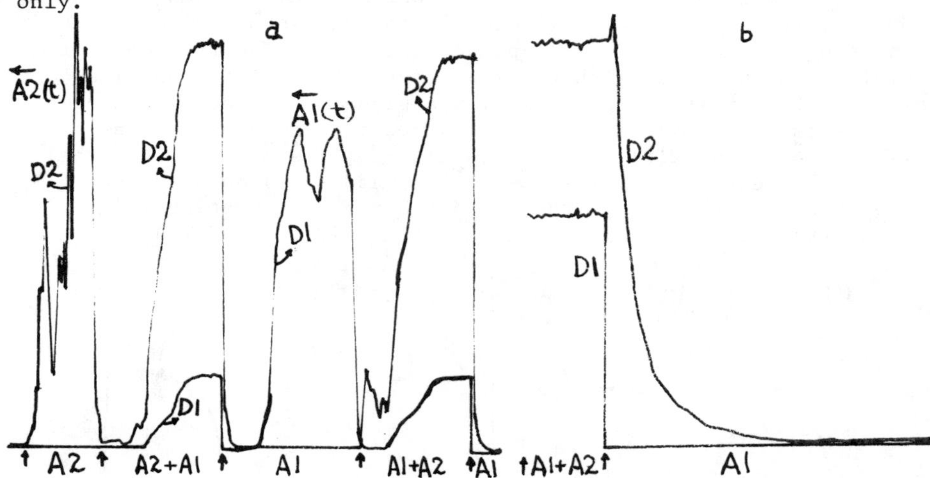

Fig. 4. Signals at D_1, D_2 vs time. $A_1 = A_2$ = 11 mw. Beams cross before reaching the crystal surface. Total time (a) 15 min, (b) 3 sec.

This work is supported by ARO Grant #DAAL03-G-0078

REFERENCES

1. J. Feinberg, Opt. Lett. 7, 486 (1982); 8, 480 (1983).
2. P. Gunter, E. Voit, M.Z. Zha and J. Albers, Optics. Comm. 55, 210 (1985).
3. A. M.C. Smout, R. W. Eason and M. C. Gower, Optics, Comm. 59, 77 (1986).
4. S. Sternklar, S. Weiss, M. Segev and B. Fischer, Opt. Lett. 11, 528 (1986).
5. S. Sternklar and B. Fischer, Opt. lett. 12, 711 (1987).
6. R. W. Eason and A. M. C. Smout, Opt. Lett. 12, 51 and 498 (1987).
7. P. Venkateswarlu, H. Jagannath, M. C. George and A. Miahnahri, J. Opt. Soc. Am. A4, (13), 24 (1987).

FIELD AND PRESSURE INDUCED FOUR-WAVE MIXING LINE SHAPES

W. M. Schreiber, N. Chencinski, A. M. Levine, and A. N. Weiszmann
The College of Staten Island of The City University of New York,
Staten Island, New York 10301

Yehiam Prior
Weizmann Institute of Science, 76100 Rehovot, Israel

ABSTRACT

A closed density matrix formalism for four-wave mixing applicable to strong fields is discussed. The resulting line shapes are analyzed for weak, intermediate, and strong intensities.

INTRODUCTION

Recently,[1,2] we have presented a strong field theory for four-wave mixing. A wave function technique was used, and dissipation was not included. The procedure involved identifying an effective Hamiltonian for each of the respective diagrams and solving it exactly to all orders. This was accomplished by transforming the problem to an equivalent time independent one, and then using eigenvalue techniques to obtain the solution. Equivalently, Laplace transform methods can be employed.

Labeling the four atomic levels in the order of increasing energy, the nonzero dipole moments are p_{13}, p_{14}, p_{23}, and p_{24}. These are the selection rules associated with the 3S-3P transitions of sodium. There are 24 effective Hamiltonians[2] to be associated with the 48 diagrams corresponding to the lowest order ($\chi^{(3)}$) results for four-wave mixing.[3-5]

Referring to the classification delineated in Prior,[5] 12 Hamiltonians are associated with the respective parametric diagrams 1-12, and 12 Hamiltonians are associated with the nonparametric diagrams 13-48. The nonparametric diagrams are ordered in groups of three, and a single effective Hamiltonian is assigned to each group.

FORMALISM

We discuss a density matrix treatment of the corresponding model. Both longitudinal and transverse decay are included in the density matrix equation. For each of the 24 effective Hamiltonians, the corresponding equation has a time independent supermatrix.[6] This is accomplished with the same respective diagonal unitary transformation matrix T(t) employed in the wave function treatment.[2]

In terms of the transformed density matrix

$$\rho' = T\rho T^{+}, \qquad (1)$$

the steady state expectation value of the polarization is

$$\langle P \rangle = \sum_{i,j=1}^{4} p_{ij} \rho'^{s}_{ji} T_{ii}(t) T^{*}_{jj}(t). \qquad (2)$$

In Eq. (2), ρ'^{s}_{ji} is the time independent part of ρ'_{ji}. In particular, Eq. (2) yields the polarization component $\mathcal{P}_{\omega_p}$ at the frequency

$$\omega_p = \omega_a + \omega_b - \omega_c, \qquad (3)$$

where ω_a, ω_b, and ω_c represent the respective frequencies of the three incident fields. Specifically,[2]

$$T_{ii}(t) T^{*}_{jj}(t) = e^{-i\omega_p t} \qquad (4)$$

according to the following prescription: i=1,j=4 for the first set of parametric diagrams (1-6); i=3,j=1 for the second set of parametric diagrams (7-12); i=4,j=2 for the first set of nonparametric diagrams (13-30); i=2,j=3 for the second set of nonparametric diagrams (31-48). From Eq. (2), the respective polarization components at ω_p are

$$\mathcal{P}_{\omega_p} = \begin{cases} p_{14}\rho'^{s}_{41} & (1\text{-}6). \\ p^{*}_{13}\rho'^{s}_{13} & (7\text{-}12). \\ p^{*}_{24}\rho'^{s}_{24} & (13\text{-}30). \\ p_{23}\rho'^{s}_{32} & (31\text{-}48). \end{cases} \qquad (5)$$

RESULTS

In many four-wave mixing experiments, two of the frequencies (ω_a,ω_b) are equal and the third frequency (ω_c) is swept through a resonance peak. The method described in this work can be used to predict the real and imaginary parts of the polarization at arbitrary field strengths and input frequencies for any resonance.

For input field strengths much smaller than the appropriate dissipation rates, the exact calculations agree completely with traditional perturbation theory. By including proper dephasing, the usual PIER4[7] resonance line shape can be predicted. A typical line shape is shown in Fig. 1, where the real and imaginary parts of the polarization are plotted as a function of ω_c. The peak to peak amplitudes of the real and imaginary parts are equal, and the resonance occurs at the usual frequency $\omega_c = \omega_{jk} + \omega_a$, where ω_{jk} is the transition frequency between the unpopulated upper two levels. The linewidth is determined by the dissipation rates.

When the fields are large compared to the dissipation rates, the results of the density matrix calculation agree with the strong field theory previously published.[1,2] The amplitudes and the widths are insensitive to the dissipation rates. This is true both for the ordinary resonances and the extra resonances. In Fig. 2, a typical line shape for a field induced extra resonance[2,8] is shown. The imaginary part is very small, while the real part exhibits a dispersive shape of opposity polarity to the PIER4 shown in Fig. 1. The resonance occurs at the frequency $\omega_c = \omega_{jk} + \omega_a + \Delta\omega_a/2 - [(\Delta\omega_a/2)^2 + V_a^2]^{\frac{1}{2}}$, where V_a is the field stength and $\Delta\omega_a$ is the detuning of the input at ω_a. The linewidth is determined by the field strength.

At intermediate field strengths,both pressure and field induced effects are present. A typical line shape is shown in Fig. 3, where the asymmetric shape is indicative of the competition between field induced and pressure induced phenomena.

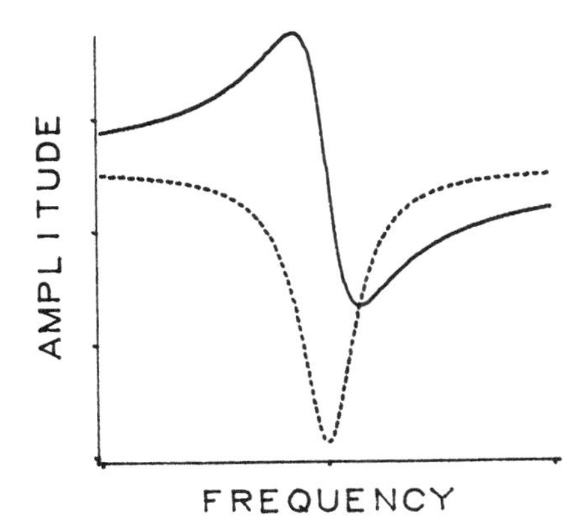

Figure 1. A pressure induced extra resonance line shape. The real (——) and imaginary (---) parts of the polarization are shown as a function of the frequency ω_c.

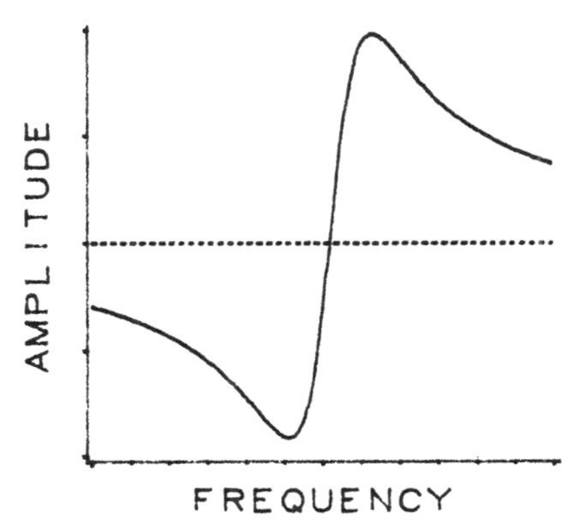

Figure 2. A field induced extra resonance line shape.

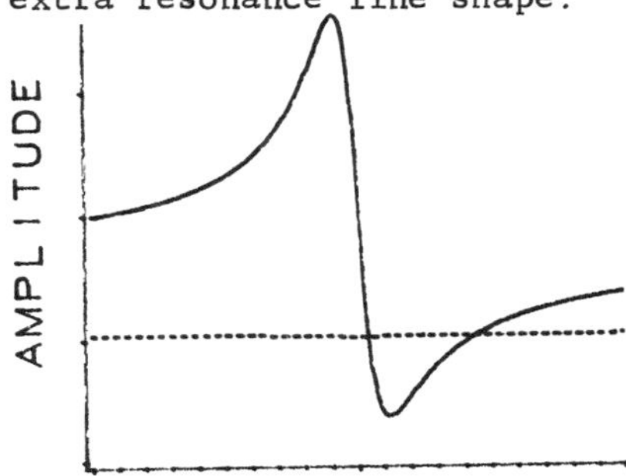

Figure 3. An intermediate field strength showing the competition between pressure and field induced effects.

REFERENCES

1. W. M. Schreiber, N. Chencinski, A. M. Levine, and A. N. Weiszmann in Methods of Laser Spectroscopy, edited by Y. Prior, M. Rosenbluh, and A. Ben-Reuven (Plenum, New York, 1986).
2. A. M. Levine, N. Chencinski, W. M. Schreiber, A. N. Weiszmann, and Y. Prior, Phys. Rev. A 35, 2550 (1987).
3. N. Bloembergen, H. Lotem, and R. T. Lynch, Jr., Ind. J. Pure Appl. Phys. 16, 151 (1978).
4. S. Y. Yee and T. K. Gustafsen, Phys. Rev. A 18, 1597 (1978).
5. Y. Prior, IEEE J. Quantum Electron. QE-20, 37 (1984).
6. B. Dick and R. M. Hochstrasser, Chem. Phys. 75, 133 (1983).
7. Y. Prior, A. Bogdan, M. Dagenais, and N. Bloembergen, Phys. Rev. Lett. 46, 111 (1981).
8. H. Friedmann and A. D. Wilson-Gordon, Phys. Rev. A 26, 2768 (1982); ibid. A 28, 302 (1983).

ACKNOWLEDGMENTS

We acknowledge the support of the CUNY/UCC and the academic computer center of The College of Staten Island. This research was supported by Grant No. 666470 from the PSC-CUNY Research Award Program of The City University of New York.

FOUR-WAVE MIXING AND THERMAL NOISE

R. McGraw and D.N. Rogovin
Rockwell International Science Center
1049 Camino Dos Rios
Thousand Oaks, CA 91360

ABSTRACT

In this paper, we present a statistical thermodynamic analysis of thermal noise in four-wave mixing. For degenerate four-wave mixing, we obtain a static susceptibility relation between the Kerr coefficient ε_2 (related to the signal strength) and light scattering thermal fluctuations (related to the noise). The signal-to-noise power ratio (SNR) is given in the weak coupling limit.

INTRODUCTION

In a recent paper,[1] we established a fundamental connection between light scattering fluctuations in an artificial Kerr medium (a Brownian suspension of microspheres) and thermal noise. An expression for the root-mean-square (RMS) fluctuations in conjugate wave intensity was obtain which appears to be the first to incorporate light scattering noise into four-wave mixing. Its predictions are in remarkably good agreement with the measurements of Smith et al[2] and account quantitatively for a previous discrepancy between theory and experiment.

In the present paper, we extend the theory of thermal noise fluctuations in nonlinear optics materials initiated in Ref. 1, to general Kerr media. The analysis is based on a novel application of the fluctuation-dissipation theorem to laser-induced gratings and yields a relation between the Kerr coefficient, ε_2, and the light scattering fluctuations, $\delta\varepsilon$. A new and more general expression for the SNR in four-wave mixing is obtained upon making the connection between these light scattering fluctuations and thermal noise.

Thermal Noise Fluctuations in Kerr Media

Figure 1 depicts a four-wave mixing configuration through which a field-induced grating (solid curve) is written in a Kerr medium by the left pump and probe beams. The open arrows represent grating motion for $\Omega \neq 0$, in which case power is dissipated in the medium.[3] Here, we limit discussion to degenerate four-wave mixing ($\Omega = 0$), so that only stationary gratings are formed. The material grating response consists of a spatial modulation of the dielectric constant. For incident plane waves $\Delta\varepsilon = \varepsilon_2 \mathscr{E}_1 \bullet \mathscr{E}_p \cos(\vec{Q} \bullet r) = a_1^{eq} \cos(\vec{Q} \bullet r)$, where ε_2 is the Kerr efficient, $\mathscr{E}_j$ is the amplitude of wave j, and $\vec{Q}$ is the wavevector of the grating. Given the nonlinear polarization $\vec{P}_{NL} = \Delta\varepsilon\vec{E}/4\pi$, where $\vec{E}$ is the total field, the thermodynamic analysis of Landau and Lifshitz[4] can be adapted to obtain the free energy of grating formation in a dielectric medium. Based on a detailed analysis given elsewhere,[5] we find:

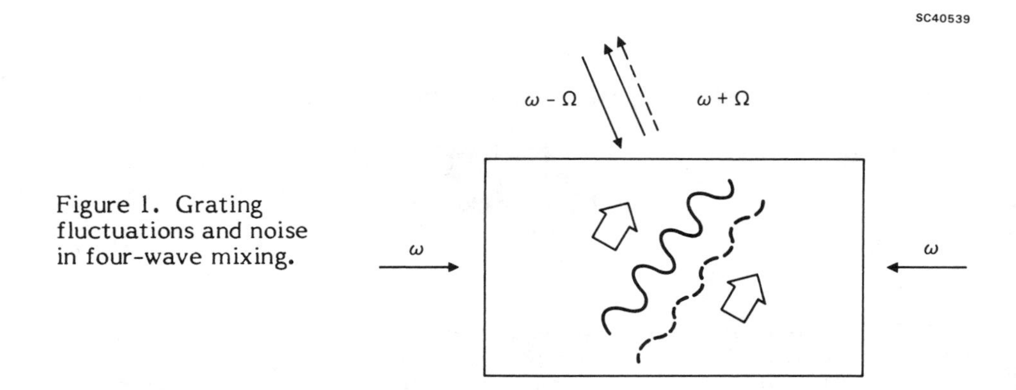

Figure 1. Grating fluctuations and noise in four-wave mixing.

$$d\langle F\rangle = -\langle S\rangle\, dT - \frac{\varepsilon_2}{32\pi}\, d(\mathscr{E}_1 \bullet \mathscr{E}_p)^2 \tag{1}$$

where F is the Helmholtz free energy density, S is the entropy density, and the angular brackets denote averaging over a unit volume which is large compared to a grating spacing.

Equation (1) yields conjugate variables in terms of which the fluctuation-dissipation theorem can be applied. For $x = \mathscr{E}_1 \bullet \mathscr{E}_p$, we find the conjugate variable y with time average $\overline{y} = V_S \varepsilon_2(\mathscr{E}_1 \bullet \mathscr{E}_p)/16\pi = V_S\, a_1^{eq}/16\pi$, where V_S is the beam interaction volume. Following standard procedure, one can immediately write a static susceptibility relation for the fluctuations in y which, in turn, can be used to obtain the fluctuations of $a_1 = a_1^{eq} + \delta a_1$ about its time average value. Referring to Fig. 1, a_1^{eq} is the amplitude of the field-induced grating, $\Delta\varepsilon$, and $\delta a_1(t)$ is the amplitude of a spontaneous fluctuation grating (dashed curve) of the same wavevector, i.e., $\delta\varepsilon = \delta a_1 \cos(\vec{Q} \bullet r)$. The static susceptibility relation gives[4]

$$8\pi\, kT\, \varepsilon_2 = V_S \langle|\delta\varepsilon|^2\rangle = V_S \langle |\delta a_1|^2/2 \tag{2}$$

which is the result we have been seeking. The first equality is a relation between the Kerr coefficient and the light scattering fluctuations $\delta\varepsilon$. An equivalent relation was obtained previously by Hellwarth,[6] but the derivation outlined here based on the fluctuation-dissipation theorem appears to be new.

A heuristic argument, motivated by Fig. 1, suggests that the SNR is given by a ratio of the field-induced (signal) to the spontaneous (noise) gratings:[1]

$$\frac{S}{N} = \frac{\langle|a_1^{eq}|^2\rangle}{\langle|\delta a_1|^2\rangle} = \frac{V_S\, \varepsilon_2\, (\mathscr{E}_1 \bullet \mathscr{E}_p)^2}{16\pi\, kT} \tag{3}$$

where in the second quality, Eq. (2) has been used. For the special case of a microparticle suspension medium with $\delta\varepsilon = 4\pi\alpha\delta n$ and $\varepsilon_2 = 2\pi\alpha^2 \langle n\rangle/kT$, Eq. (3) reduces to our previous result which was found to be in excellent agreement with experiment.[1,2] Here α is the particle polarizability and n is the particle number density. A more complete analysis than can be given here[7] shows that Eq. (3) is exact in the weak coupling regime.

REFERENCES

1. R. McGraw and D. Rogovin, SPIE Vol. 739, Phase Conjugation and Beam Combining and Diagnostics (1987), pp. 100-104.
2. P.W. Smith, P.J. Maloney and A. Askin, Opt. Lett. 7, 347 (1982).
3. R. McGraw and D. Rogovin, Phys. Rev. A35, 1181 (1987).
4. L.D. Landau and E.M. Lifshitz, Electrodynamics of Continuous Media, Pergamon, New York (1960).
5. R. McGraw, Phys Rev. A, submitted for publication.
6. R.W. Hellwarth, J. Chem. Phys. 52, 2128 (1970).
7. R. McGraw and D. Rogovin, in preparation.

HIGHER-ORDER SQUEEZING OF QUANTUM ELECTROMAGNETIC FIELD IN NON-DEGENERATE PARAMETRIC DOWN CONVERSION

Xizeng Li , Ying Shan
Department of Physics, Tianjin University, Tianjin, P.R.China

ABSTRACT

It is found that tne process of Non-degenerate Parametric Down Conversion exhibits higher-order squeezing. The degree of squeezing increases with the order N, and the higher-order squeeze parameter q_N may approach -1.

With the development of techniques for making higher-order correlation measurements in quantum optics, Hong and Mandel introduced a new concept of higher-order squeezing in 1985, which is the natural generalization of the usual second-order squeezing.[1,2]

In the usual approach to squeezing in the context of quantum optics, the real field $\hat{E}$ is decomposed into two quadrature components $\hat{E}_1$ and $\hat{E}_2$ which are canonical conjugates. Then the state is squeezed to the Nth order in $\hat{E}_1$ (N=1,2,3,......)if there exists a phase angle ϕ such that $\langle(\Delta\hat{E}_1)^N\rangle$ is smaller than its value in a completely coherent state of the field.

By using the Campbell-Backer-Hausdorff identity Hong and Mandel readily obtain the relation

$$\langle(\Delta\hat{E}_1)^N\rangle = \langle:(\Delta\hat{E}_1)^N:\rangle + \frac{N^{(2)}}{1!}\left(\frac{1}{2}C\right)\langle:(\Delta\hat{E}_1)^{N-2}:\rangle + \frac{N^{(4)}}{2!}\left(\frac{1}{2}C\right)^2\langle:(\Delta\hat{E}_1)^4:\rangle + \cdots + (N-1)!!\,C^{N/2}, \quad \text{if N is even.}$$

Here $N^{(r)}$ stands for N(N-1)$\cdots$(N-r+1), the commutator $C=\frac{1}{2i}[\hat{E}_1, \hat{E}_2]$. Now the normally ordered moments $\langle:(\Delta\hat{E}_1)^N:\rangle$ all vanish for a coherent state. It follows that the state is squeezed to any even order N if

$$\langle(\Delta\hat{E}_1)^N\rangle < (N-1)!!\,C^{N/2} .$$

Hong and Mandel have applied this concept of higher-order squeezing to several physical situations, such as Degenerate Parametric Down Conversion, Second Harmonic Generation , etc. In this paper, we consider the process of Non-degenerate Parametric Down Conversion, in which a strong, classical field of complex amplitude ($v=|v|e^{i\theta}$) and frequency ω_3 is incident on a nonlinear crystal. The frequency is ω_1 for signal mode 1 and ω_2 for idler mode 2, where $\omega_1+\omega_2=\omega_3$. The simplest Hamiltonian for this problem has the form

$$\hat{H} = \hbar\omega_1\hat{a}_1^+\hat{a}_1 + \hbar\omega_2\hat{a}_2^+\hat{a}_2 + \hbar g(ve^{-i\omega_3 t}\hat{a}_1^+\hat{a}_2^+ + \text{H.C.}) .$$

Here g is the mode coupling constant, $\hat{a}_j$ and $\hat{a}_j^+$(j=1,2) are annihilation and creation operators respectively of a photon in the jth mode.

It is convenient to use the more slowly varying variables

$$\hat{A}_j = \hat{a}_j e^{i\omega_j t} \qquad (j=1,2) ,$$

then we get the Heisenberg equations of motion for $\hat{A}_1$ and $\hat{A}_2$

$$\dot{\hat{A}}_1 = -igv\hat{A}_2^+ ,$$
$$\dot{\hat{A}}_2 = -igv\hat{A}_1^+ .$$

Their solutions are

$$\hat{A}_1(t) = \mu\hat{a}_1(0) - \nu\hat{a}_2^+(0) ,$$
$$\hat{A}_2(t) = \mu\hat{a}_2(0) - \nu\hat{a}_1^+(0) ,$$

where
$$\mu = \cosh(g|v|t) ,$$
$$\nu = e^{i\theta}\sinh(g|v|t) .$$

It can be verified that the fields of $\hat{A}_1$ mode and $\hat{A}_2$ mode do not exhibit higher-order squeezing, so we consider the combined mode

$$\hat{B}(t) = \frac{\hat{A}_1(t) + \hat{A}_2(t)}{\sqrt{2}} .$$

We then define the quadrature component $\hat{E}_1$ by

$$\hat{E}_1 = \hat{B}e^{-i\phi} + \hat{B}^+e^{i\phi} ,$$

where ϕ is some phase angle that may be chosen at will.

Let $\hat{B}_{in} = \dfrac{\hat{a}_1(0) + \hat{a}_2(0)}{\sqrt{2}}$, $\hat{B}_{in}^+ = \dfrac{\hat{a}_1^+(0) + \hat{a}_2^+(0)}{\sqrt{2}}$.

Then
$$\hat{E}_1 = h\hat{B}_{in} + h^*\hat{B}_{in}^+ ,$$
where
$$h = \mu e^{-i\phi} - \nu^* e^{i\phi} .$$

We now apply C-B-H identity in the form

$$e^{(\Delta\hat{E}_1)x} = \langle :: e^{(\Delta\hat{E}_1)x} :: \rangle e^{\frac{1}{2}x^2C} ,$$

but with the normal ordering applicable to the $\hat{B}_{in}$, $\hat{B}_{in}^+$ operator, and with the commutator C therefore identified with $|h^2|$, we then obtain, after equating coefficients of $x^N/N!$,

$$\langle(\Delta\hat{E}_1)^N\rangle = \langle::(\Delta\hat{E}_1)^N::\rangle + \frac{N^{(2)}}{1!}\left(\tfrac{1}{2}|h|^2\right)\langle::(\Delta\hat{E}_1)^{N-2}::\rangle + \frac{N^{(4)}}{2!}\left(\tfrac{1}{2}|h|^2\right)^2\langle::(\Delta\hat{E}_1)^{N-4}::\rangle + \cdots + (N-1)!!\,|h|^N \quad (N \text{ even}).$$

Where :: :: denotes normal ordering with respect to $\hat{B}_{in}$, $\hat{B}_{in}^+$.

Now we take the initial quantum state to be $|\xi\rangle_1|0\rangle_2$ which is a product of the coherent state $|\xi\rangle_1$ for $\hat{a}_1(0)$ mode and the vacuum state for $\hat{a}_2(0)$ mode, then

$${}_2\langle 0|_1\langle\xi|::(\Delta\hat{E}_1)^N::|\xi\rangle_1|0\rangle_2 = {}_2\langle 0|_1\langle\xi|::(h\Delta\hat{B}_{in} + h^*\Delta\hat{B}_{in}^+)^N::|\xi\rangle_1|0\rangle_2$$
$$= \sum_{r=0}^{N}\binom{N}{r}{}_2\langle 0|_1\langle\xi|(\Delta\hat{B}_{in}^+)^r(\Delta\hat{B}_{in})^{N-r}|\xi\rangle_1|0\rangle_2\, h^{*r}h^{N-r}$$
$$= \frac{1}{(\sqrt{2})^N}\sum_{r=0}^{N}\binom{N}{r}{}_2\langle 0|_1\langle\xi|[\Delta\hat{a}_1^+(0) + \Delta\hat{a}_2^+(0)]^r[\Delta\hat{a}_1(0) + \Delta\hat{a}_2(0)]^{N-r}|\xi\rangle_1|0\rangle_2$$

$$\cdot h^{*r} h^{N-r} = 0$$

Hence the higher-order moments

$$\langle (\Delta \hat{E}_1)^N \rangle = (N-1)!! \, |h|^N$$
$$= (N-1)!! \, [\cos(2g|v|t) - \sin(2g|v|t)\sin(2\phi-\theta)]^{N/2} .$$

If ϕ is chosen so that

$$2\phi - \theta = \frac{\pi}{2} , \qquad \sin(2\phi-\theta) = 1 ,$$

then the above equation leads to the result

$$\langle (\Delta \hat{E}_1)^N \rangle = (N-1)!! \, e^{-Ng|v|t} .$$

As the right-hand side is less than (N-1)!! , which is the correspon-ding Nth order dispersion for a coherent state. We see that the field of the combined mode of the two down converted light beams in NPDC exhibits higher-order squeezing to all even orders.

Moreover, in order to determine whether the squeezing is intrinsically of higher-order in the sense defined by Hong and Mandel[2], we obtain the normally ordered moments, for any even N,

$$\langle :(\Delta \hat{E}_1)^N: \rangle = (N-1)!! \, (-1)^{N/2} \, [1 - e^{-2g|v|t}]^{N/2} .$$

It follows that $\langle :(\Delta \hat{E}_1)^N: \rangle < 0$ when N/2 is odd, therefore there is intrinsic Nth order squeezing for all values of N for which N/2 is odd, viz. for N=2,6,10,14,... .

Finally, we calculate the squeeze parameter q_N for measuring the degree of Nth order squeezing

$$q_N = \frac{(N-1)!! \, e^{-Ng|v|t} - (N-1)!!}{(N-1)!!} = e^{-Ng|v|t} - 1$$

We find that $|q_N|$ increases with N, and the higher-order squeeze parameter may approach -1. So the higher-order squeezing of the quantum electromagnetic field has potential applications in optical communication, interferometry, spectroscopy and gravitation-wave detection.

This research was supported by National Science Foundation of China.

REFERENCES

1. C.K.Hong, L.Mandel, phys. Rev. Lett. 54 , 323(1985).
2. C.K.Hong, L.Mandel, phys.Rev.A 32, 974(1985).

COLLISION, FLUCTUATION, AND FIELD-INDUCED EXTRA RESONANCES

Yehiam Prior

Department of Chemical Physics
Weizmann Institute of Science
Rehovot, Israel 76100

ABSTRACT

Four wave mixing experiments are normally performed near atomic (molecular) resonances, when some of the input frequencies or their combinations are at resonance with the atomic transitions. If dephasing processes are present, several new resonances appear, that may be traced directly to the influence of the dephasing process. The origin of the "dephasing" may vary from collisions and other material relaxation processes, to field fluctuations, to higher order interactions in the wave mixing process. The source of these extra resonances is discussed, and specific results are shown for the various cases.

INTRODUCTION - PRESSURE INDUCED RESONANCES

Many experiments in Four Wave Mixing (FWM) have dealt with the question of interferences between two or more contributions to the induced polarizability. In particular, Bloembergen Lotem and Lynch[1] concluded that most of these observations can be explained within the third order susceptibility theory of nonlinear wave mixing if the phenomenological relaxation rates Γ_{ij} for the various transitions are handled in a consistent way. Thus, one may derive a 48 term complete expression for the third order susceptibility $\chi^{(3)}$, which includes all contributions and all interferences to third order. This expression may also be derived from time ordered, two sided Feynman diagrams, and such diagrams were introduced[2,3]. When the perturbation theory treatment of FWM is done properly, in addition to the normal resonances, several "extra resonances" are observed, namely, resonances which occur when none of the input frequencies or any of their combinations matches a material frequency.

The first experimental demonstration of "extra resonances" was done in Na vapor[4]. In these experiments, resonances were observed between the excited 3P states, when the difference of two input frequencies matched the 17 cm^{-1}. Additional resonances were observed between the ground hyperfine transitions, and between Zeeman magnetic sub levels, with or without magnetic fields[5]. In all these cases, the "standard" theory of the nonlinear third order susceptibility $\chi^{(3)}$ was sufficient to explain the observed results. An intuitive explanation is possible, which will be applied again in the case of other dephasing processes. If more than one intermediate level participates in a resonant excitation, or even more generally, if several terms appear in the expression for the nonlinear susceptibility, they interfere coherently. If, however, some dephasing process is present, the two terms are no longer added coherently, and the interference between them may be removed. If the interference had been destructive, a new resonance will now appear, manifesting itself in the observed spectrum of the FWM.

In particular, for a three level system consisting of a **ground state** a and two excited states b,c, three relaxation rates exist $\Gamma_{ba}, \Gamma_{ca}, \Gamma_{bc}$, which, in the absence of proper dephasing, obey the relationship:

$$\Gamma_{bc} = \Gamma_{ba} + \Gamma_{ca} \tag{1}$$

This relation is unique, and may break down for many different reasons. Equ.(1) is responsible for the destructive interference in the expression for FWM, and if for any reason it does not hold - one may expect the appearance of extra resonances. In particular, this equality breaks down if proper dephasing is present. Two mechanisms were shown to cause equ.(1) to break down: collisions in the Na vapor case, or phonon interactions in molecular crystals where the temperature determined the dephasing rate[6].

The model suggested by the third order perturbation theory was verified in several respects, and its predictions and observations may be summarized as follows:

a. Resonances between excited states, which are not populated, and where the observed signal is not due to population building up in the excited state.
b. Resonances between equally populated ground states (states of fine or hyperfine structure).
c. Hanle resonances between degenerate ground states.

In all these cases, it was recognised quite early that other theoretical approaches are fruitful, and the dressed atom approach was shown to provide an intuitive explanation to the otherwise "mysterious" extra resonances[7].

FLUCTUATIONS

Pressure Induced resonances are created by the dephasing of the pseudopolarization vector due to collisions with other atoms. By considering the Bloch picture it becomes obvious that dephasing may occur either when the polarization vector is displaced, or when the rotating frame itself is displaced, as in a phase change of the laser. Agarwal, in several early papers[8], has considered the effect of laser phase fluctuations on the extra resonances. The model treated by Agarwal, was the Phase

Diffusion Model (PDM), mostly for ease of calculations, and they predicted quantitatively the combined effect of pressure and phase fluctuations (within the PDM). That work, while very detailed, did not include possible correlation between the input fields. We have considered the PDM[9], and included correlations between the input fields. A FLS (Few Level System) was treated for correlated phase diffusing fields. One can write an equation similar to equation (1), and show that even in the absence of collisions, when the lasers are correlated a Stochastic Fluctuation Extra Resonance (SFIER4) exists. The results were corroborated by computer simulations[10], and possible preliminary experimental observation of this effect was reported[11].

More recently[12], a unified approach was presented for the treatment of stochastic fluctuations in four wave mixing processes. The theory allows for phase as well as amplitude fluctuations and covers Markovian and non Markovian (non Lorentzian) input lineshapes. Uncorrelated input fields give rise to resonances similar to those described previously by Agarwal, while correlated fields cause the extra resonances. Partial crosscorrelations were also treated showing oscillations in the intensity of the generated FWM signal as a function of the degree of crosscorrelation. In this work the induced polarization is written in terms of the three input fields, and the six point correlation function of the input fields is treated. For uncorrelated fields, the 6-point correlation function separates into three functions for each of the fields, with the resulting resonances predicted by Agarwal. These resonances are depicted in fig. 1, where all the possibilities of fluctuation induced resonances for uncorrelated fields are shown.

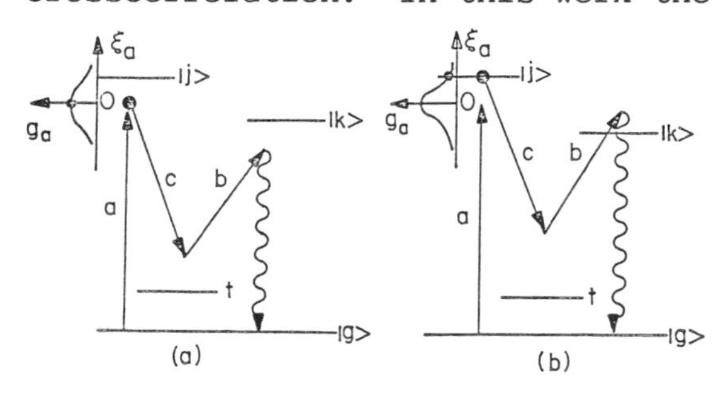

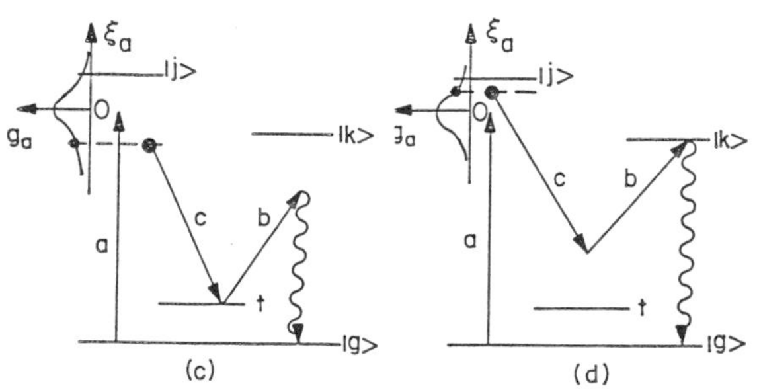

Fig.1
Ordinary (a) and fluctuation induced resonances (b,c,d)

These resonances can be easily understood: for monochromatic inputs, only a delta function output is predicted and observed, whereas for input fields of finite spectral width, the bandwidth of an input field (or the combination of such bandwidths) may overlap one of the atomic transition frequencies - resulting in a resonance. These resonances occur at the atomic frequency, they are regular resonances, and their intensity is proportional to the spectral power at the frequency component responsible for the overlap with the atomic line.

When the input fields are correlated, the 6-point correlation function cannot be broken to independent functions. In this case, there are two contributions to the observed FWM signal, one from the uncorrelated part, which is identical to the one discussed above, and one from the correlated part which may be written as

$$I_c = C \left[1 + \frac{a'}{\Delta^2 + \Gamma_{jk}^2} \right] \qquad (2)$$

where $a'=(2\nu_b+\Gamma_{kg}+\Gamma_{jg})^2-\Gamma_{jk}^2$, and C is a constant. Note that in PIER4 the pressure determined the width of the extra resonance (through Γ_{jk}), while in SFIER4 ν_b does not appear in the denominator, giving rise to a strong, sharp line.

By far, the novel and interesting case is of partial correlation. Partial correlation may be achieved by deriving the field ε_c from ε_b from the same laser, with appropriate delay: $\varepsilon_c(t) = R\ \varepsilon_b(t-T)$. The two fields are fully correlated for T-->0, and are uncorrelated for T-->∞. Analytic results were obtained for partially correlated fields, both for T<0 and T>0. The observed intensity as a function of the degree of cross correlation is depicted in fig. 2.

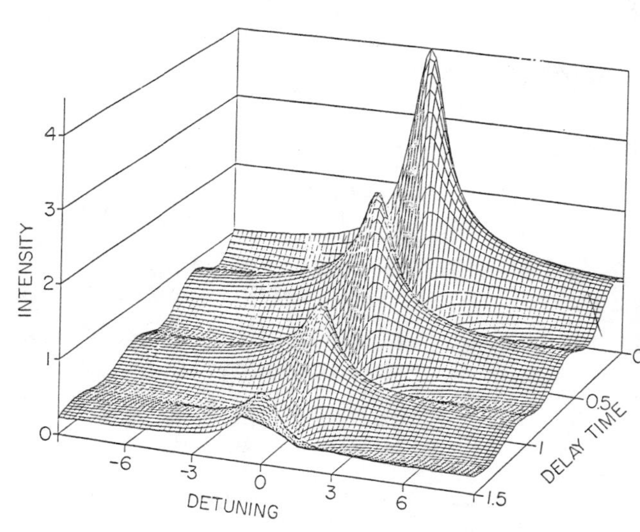

Fig.2
The observed intensity as a function of the delay between the input pulses.

This result is related to the white light coherent measurements of Yajima and of Hartmann[13]. The first laser sets up the polarization vector which precesses and returns to its original position at frequency $\Delta\omega_1$ where 1 stands for the first laser. The second laser pumps the system coherently, with a definite phase relationship to the first one, and if the delay between them corresponds to the precession frequency, the two lasers will add in phase, hence, oscillations at $\Delta\omega_1$.

FIELD-INDUCED RESONANCES

It can be shown that strong fields will cause the existence of extra resonances. Several theoretical treatments had been presented for handling strong fields[14], but none has been adequate for treating all input fields as strong. In this context, a strong field is a field that transfers significant population from the ground to the excited state.

Recently we have presented[15] a theory suitable for the case where all fields are strong. The theory had been developed in a wave function formalism, where material relaxations were not included, but had been generalized since to include relaxations, both longitudinal and transverse[16]. By choosing the right transformation to a generalized rotating frame, one may remove all fast time dependencies (in a manner analogous to what is done for the two level system), and by numerically diagonalizing the resulting time independent matrix, an expression is obtained for the induced polarization. The solutions obtained by this method are not perturbative, and the fields may be arbitrarily strong. In the case of one strong field, each resonant line is split into two components in a way very similar to the AC Stark effect (or Rabi splitting) with the appropriate changes due to the FWM. Under these conditions, one may identify the extra resonance peak resulting from the strong field, and trace its origin. If more than one field is strong, a new type of behavior appears, namely "stirring" of the components of the induced doublet. The first strong field induces a splitting, and if the Rabi frequency of the second field is comparable to this splitting, the two components are mixed (or exchange narrowed), resulting in one spectral feature. This is a clear signature of the strong fields, and should be identifiable in experiments.

In figure 3 the calculated induced third order susceptibility $\chi^{(3)}$ is plotted against the power of one of the lasers (V_a) in a FWM experiment, where all lasers are nearly resonant with their corresponding transitions. In this calculation, the input power is varied over a very large dynamic range, in order to demonstrate the three separate regions: a) Low power region ($-8<\log V_a<-4$), where perturbation theory holds, a PIER4 effect is predicted and the polarization is linearly proportional to the power V_a. The PIER4 signal reaches saturation at the expected level- as determined by the pressure and other relaxation parameters; b) High power region ($\log V_a>-2$), where the Field Induced Resonance (FIRE) is observed, the third order susceptibility $\chi^{(3)}$ is proportional to the square of V_a, and reaches saturation at a higher level; c) Intermediate power level ($-4<\log V_a<-2$), where both pressure and field effects are present, and the observed lineshapes will manifest the competition between these two effects.

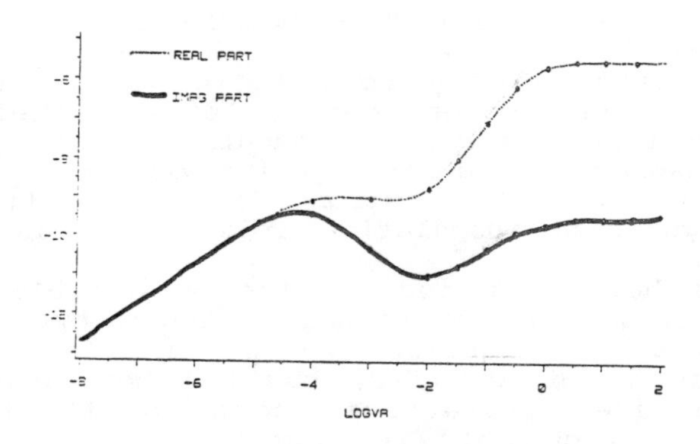

Fig.3
The real and imaginary parts of the calculated third order susceptibility $\chi^{(3)}$ as a function of the power of input field a. Three regions are apparent.

In another paper in this volume[17] we present detailed lineshapes for the different cases.

CONCLUSIONS

The phenomena of extra resonances in Four Wave Mixing has been briefly overviewed in this paper. It is shown that several different physical effects may give rise to the observed "extra resonances", and in fact these resonances are not as rare as originally thought. In the process of handling these new resonances, a full non perturbative theory has been developed for the treatment of strong fields in FWM. Stochastic fluctuations may also cause apparent dephasing in the coherent process, and should be considered as a viable source for extra resonances. A unified approach has been presented for the treatment of correlated, partially correlated and uncorrelated field fluctuations in coherent nonlinear wave mixing, and extra resonances as well as new oscillations were discussed.

ACKNOWLEDGEMENTS

I wish to thank people who collaborated with me on various parts of this work. These are A. M. Levine, and W.M. Schreiber, A.N. Weiszmann, and N. Chencinski from Staten Island, and A. G. Kofman of the WIS. This work was partially supported by grants from the Volkswagenwerk Stiftung, and the US-Israel binational Science Foundation.

REFERENCES

1. Bloembergen N. H. Lotem and R.T. Lynch, Jr., Ind.J.Pure and Appl. Phys. **16**, 151 (1978).
2. Yee S.Y., T.K. Gustafson, S.A.J. Druet and J.-P. Taran, Opt. Comm. **23**, 1 (1977). S.A.J. Druet, B. Attal, T.K. Gustafson and J.-P.E. Taran, Phys. Rev. **A18**, 1529 (1978); J. Borde and Ch. Borde, J. Mol. Spec.**78**, 353 (1978); A. Yariv, IEEE J. Quant. Elect. **QE-13**, 943 (1984).
3. Y. Prior, IEEE J-QE, **QE-20**, 37 (1984).
4. Prior Y., A. Bogdan, M. Dagenais, N. Bloembergen, Phys. Rev. Lett. **46**, 111 (1981).
5. Rothberg L. Progress in Optics, To Be Published.
6. J.R. Andrews and R.M. Hochstrasser, Chem. Phys. Lett. **83**, 427 (1981); J.R. Andrews, R.M. Hochstrasser and H.P. Tommsdorf, Chem. Phys. **62**, 87 (1981).
7. G. Grynberg, J. Phys. B **14**, 2089 (1981).
8. Agarwal G.S. and J. Cooper, Phys.Rev.A **26**, 2761 (1982); Agarwal G.S. and C.V. Kunasz, Phys.Rev.A **27**, 996 (1983); Agrawal G.P., Phys Rev.A, **28**, 2286 (1983).
9. Prior Y., I. Schek and J. Jortner, Phys.Rev.A 31,3775 (1985).
10. Y. Prior and P.S. Stern, in **Methods of Laser Spectroscopy**, eds. Y. Prior, A. Ben Reuven and M. Rosenbluh (Plenum Press, NY 1986), p. 286.
11. Y.H. Zou and N. Bloembergen, Phys. Rev. A. **34**, 2968 (1986)
12. A.G. Kofman, A. M. Levine and Y. Prior, Phys. Rev. A (February 1988, In Press).
13. N. Morita and T. Yajima, Phys. Rev. A. 30, 2525 (1984); R. Beach, D. DeBeer and S.R. Hartmann, Phys. Rev. A **32**, 3467 (1985) and references therein.
14. Harter D.J., and R.W. Boyd,Phys. Rev. A **29**, 739 (1984) and references therein; Oliveria F.A.M., Cid B. de Araujo, and J.R. Rios Leite, Phys. Rev. A **25**, 2430 (1982); Wilson-Gordon A.D., R. Klimovsky-Barid, and H. Friedmann, Phys. Rev. A **25**, 1580 (1982); Agarwal G.S. and N. Nayak, J. Opt. Soc. Am. B, **1**, 164 (1984). Dick B. and R.M. Hochstrasser, Chem. Physics **75**,133 (1983).
15. A.M. Levine, N. Chencinski, W.M. Schreiber, A.N. Weiszmann and Y. Prior, Phys. Rev A **35**, 2550 (1987);
16. A.M. Levine, N. Chencinski, W.M. Schreiber, A.N. Weiszmann and Y. Prior, Phys. Rev A (to be published).
17. W.M. Schreiber, N. Chencinski, A.M. Levine and Yehiam Prior, Field and Pressure Induced Four Wave Mixing Line Shapes, this volume.

RESONANCES IN NONLINEAR MIXING DUE TO DEPHASING INDUCED BY PUMP-PROBE FLUCTUATIONS

G.S. Agarwal
University of Hyderabad, Hyderabad - 500 134, India

ABSTRACT

I show that the temporal fluctuations of the pump-probe are an important source of dephasing and these can lead to additional resonances in four wave mixing signals. I demonstrate the existence of fluctuation-induced Hanle resonance both in forward and phase conjugate geometries. I show the existence of resonances for two different models of the pump and probe fluctuations thereby establishing the general nature of the fluctuation-induced resonances.

It is now well known that the collisional dephasing leads to additional resonances (PIER) in a variety of nonlinear mixing experiments[1]. These additional resonances have also been predicted and observed both in fluorescence and nonlinear absorption measurements[2]. Even the effects of cross relaxation[2] on PIER studied. Some examples of these resonances are:

(a) Two Level System: Consider the four-wave-mixing signal at the frequency $2\omega_1 - \omega_2$ and in the direction $2\vec{k}_1 - \vec{k}_2$ produced by pump and probe fields with frequencies and wave vectors ω_1, $\vec{k}_1$ and $\omega_2, \vec{k}_2$ respectively. The signal will exhibit a resonance at $\omega_1 = \omega_2$ if the longitudinal and transverse relaxation constants T_1, T_2 are such that $T_1 \neq T_2/2$.

(b) Three Level System: Consider a transition involving two excited states $|1\rangle$ and $|2\rangle$ connected to a state $|3\rangle$ by dipole transition and not connected otherwise. The four-wave mixing signal can exhibit extra resonance, say at $\omega_1 - \omega_2 = \omega_{12}$ (energy separation between the two excited states) provided that the relaxation Γ_{ij} of the

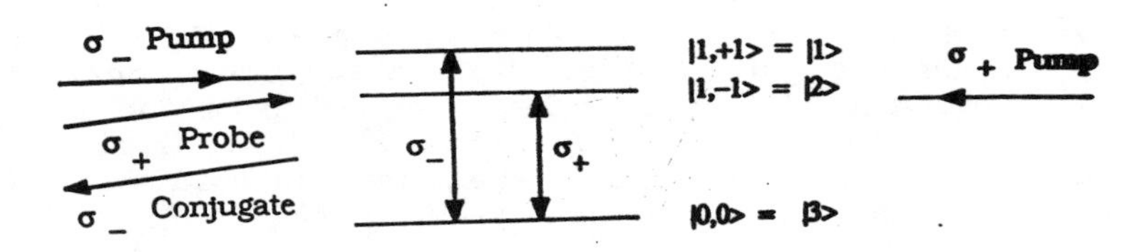

Fig.1 Schematic diagram of the four wave mixing.

coherences is such that $\Gamma_{12} \neq \Gamma_{13} + \Gamma_{23}$. For suitable polarizations of the pump and probe, Grynberg [3] predicted the existence of pressure-induced Hanle resonances at $\omega_{12} = 0$ for the case when the two excited states correspond to zeeman levels (cf Fig.1) and ω_{12} is scanned by changing the magnetic field.

A question that has received considerable attention is - Is collisional dephasing the only one leading to additional resonances; i.e. are there some other mechanisms that can produce such resonances even in a system without collisions. Two independent mechanisms have been recently pursued. These concern the existence of additional resonances due to (a) dephasing induced by the fluctuations[4-6] of the pump and probe (b) saturation effects[7]. In this lecture I concentrate on the effects of dephasing arising from the temporal fluctuations of the source. For a system with radiative relaxation and no collisional broadening, such resonances may be termed as fluctuation-induced extra resonaces, FIER. We consider two different models of the fluctuations, viz.(i) chaotic[8] pump and probe, (ii) the fields with phase fluctuations given by the phase-diffusion model. The work for the latter model has been done in collaboration with C.V. Kunasz and J. Cooper of the University of Colorado, Boulder.

We start with a very simple problem that demonstrates how the fluctuations of the pump can create new resonances. Consider the resonance fluorescence produced by a two-level optical transition driven by a weak coherent field of frequency, ω_ℓ . The spectrum of resonance fluorescence is $\alpha\,\delta\,(\omega - \omega_\ell)$ for a monochromatic field. However, for a source with width, γ_c, the spectrum $S(\omega)$ for large detuning, Δ, has the form

$$S(\omega) \sim \frac{\varepsilon^2 \quad \gamma_c}{\Delta^2[(\omega - \omega_\ell)^2 + \gamma_c^2]} + \frac{\varepsilon^2 \quad \gamma_c}{\Delta^2[(\omega - \omega_0)^2 + \gamma^2]} \,. \qquad (1)$$

Thus the fluorescence peak at the natural frequency ω_0 is induced by the fluctuations. We also point out that the spectrum differs from (1) if the dephasing due to collisions is there - The spectrum continues to have a coherent component $\delta\,(\omega - \omega_\ell)$ even if collisional width $\gamma_p \neq 0$ and a fluorescence component with a width $(\gamma + \gamma_p)$ showing that the details of the spectrum depend on the model of dephasing. Fluctuation-induced resonances can also occur in modulated fluorescnece[9]. For example the fluorescence produced by $j = 1$ to $j = 0$ transition showed a fluctuation induced resonance at modulation frequency = zeeman splitting. We next turn our attention to the question of the existence of fluctuation-induced resonances in nonlinearly generated coherent radiation. We consider two different models of fluctuations of the pump and probe so that the existence of FIER's is established under rather general conditions.

I. CHAOTIC MODELS[8]

Consider a system interacting with a pump field $\varepsilon_\ell(t)e^{-i\omega_\ell t}$ + c.c. and a probe field $\varepsilon_s(t)e^{-i\omega_s t}$ + c.c. According to third order perturbation theory, the nonlinear polarization responsible for four wave mixing can be written as

$$P(\omega) = 3\int d(\omega)\ \chi^{(3)}(\omega_1 + \omega_\ell, \omega_2 + \omega_\ell, -\omega_3 - \omega_s)\varepsilon_\ell(\omega_1)$$
$$\varepsilon_\ell(\omega_2)\varepsilon_s^*(\omega_3)\ \delta(\omega_1 + \omega_2 - \omega_3 + 2\omega_\ell - \omega_s - \omega)\ . \tag{2}$$

If ε's are fluctuating, then $P(\omega)$ becomes a random process. The FWM will be determined by the spectrum of the fluctuations of P. This can be calculated in terms of the sixth-order correlation function of the applied fields. Such a correlation function can be calculated for chaotic fields in terms of the spectral shapes

$$<\varepsilon_i(\omega)\ \varepsilon_i^*(\omega')> = I_i(\omega)\ \delta(\omega - \omega') \tag{3}$$

and the cross correlation $C(\omega)$ between the pump and probe

$$<\varepsilon_1(\omega)\ \varepsilon_s^*(\omega')> = C(\omega)\ \delta(\omega - \omega')\quad . \tag{4}$$

Our calculations show that the spectrum $S(\omega)$ of the FWM signal is

$$S(\omega) = 9.2\int d(\omega) I_\ell(\omega_1 - \omega_\ell)\ I_\ell(\omega_2 - \omega_\ell)\ I_s(\omega_3 - \omega_s)$$
$$\delta(\omega - \omega_1 - \omega_2 + \omega_3)|\ \chi^{(3)}(\omega_1, \omega_2, -\omega_3)|^2$$

$$+36\ I_\ell[\omega - (2\omega_\ell - \omega_s)]|\int d\omega_2 C(\omega_2)\ \chi^{(3)}(\omega + \omega_s - \omega_\ell, \omega_2 + \omega_\ell, -\omega_2 - \omega_s)|^2 . \tag{5}$$

The last term in (5), which we denote by S_c, arises from the cross correlation between pump and probe. Agarwal and Kunasz[4] examined the consequences of the first term in (5) and showed that the fluctuations lead to additional resonances in four-wave-mixing experiments. However, such FIER were shown to occur at an overall frequency which was different from $2\omega_\ell - \omega_s$. In the present work we show how the cross correlation can lead to FIER at $2\omega_\ell - \omega_s$.

For simplicity consider the case of fluctuation-induced Hanle resonance in $j = 0$ to $j = 1$ transition. The degenerate ($\omega_\ell = \omega_s$) FWM signal is scanned as a function of the magnetic field. The term S_c now simplifies to

$$S_c \alpha \left| \int C(\omega_2 - \omega_\ell)\, \chi^{(3)}(\omega_\ell, \omega_2, -\omega_2) d\omega_2 \right|^2 . \qquad (6)$$

The terms like

$$(\Gamma_1 + i\Delta_1)^{-1}(\Gamma_2 - i\Delta_2)^{-1}(i\delta + \Gamma_1 + i\Delta_1)^{-1} + 1 \leftrightarrow 2$$

in the Raman susceptibility $\chi^{(3)}(\omega_1, \omega_2, -\omega_2)$ lead to

$$S_c \alpha \frac{I^3}{\Delta^6} \left| 1 + \frac{2\gamma_c(\Gamma_1 + \Gamma_2)}{\omega_{12}^2 + (\Gamma_1 + \Gamma_2)^2} \right|^2 + \text{background terms} , \qquad (7)$$

if we assume a Lorentzian form for the cross correlation $C(\omega)$. Note that for the Hanle experiment the pump and probe are fully correlated since both are derived from the same laser. From (7) we see the existence of the fluctuation-induced Hanle resonance in four wave mixing. We have also proved two other important results[8] the simpler form of which we quote without proof:

(a) Consider degenerate FWM with pump and probe derived from the same source. Next consider the spectrum of the signal produced in an homodyne experiment. Such a spectrum turns out proportional to S_c. Thus S_c can be studied in an homodyne experiment.

(b) The cross correlation also determines the outcome of a FWM experiment in which pump and probe are delayed with respect to each other. In the degenerate FWM the delay dependence of the signal is determined by S_c i.e. by

$$I_\ell(\omega - \omega_\ell) \left| \int d\omega_2 C(\omega_2 - \omega_1)\, \chi^{(3)}(\omega_\ell, \omega_2, -\omega_2) e^{-i(\omega_2 - \omega_\ell)\tau} \right|^2 .$$

This can be used to determine the dephasing times if the pump has a very short correlation time.

II. PHASE DIFFUSION MODELS

We next consider the case when the input fields have stabilized amplitudes but with phases $\varphi_i(t)$ such that $\varphi_i^s(t)$ are Gaussian, markovian and delta correlated: $\langle \varphi_i(t)\varphi_j(t')\rangle = 2\gamma_{ij}\delta(t-t')$. For such models exact results[4,5] can be obtained i.e. even the saturation effects can be included. The fluctuations of the fields lead to a correlation between different atoms. Hence one has to solve for the ensemble average of single atom and two atom density matrices. Moreover since there is considerable redistribution of the emitted radiation, we have to separate out the contribution responsible for FWM. This can be done from an analysis of the spectrum of the emitted radiation. The details of the procedure can be found Refs.4,5 which discuss the existence of FIER in a number of FWM situations. Here we briefly consider two cases of Hanle resonances.

(A) Forward FWM - Consider the degenerate FWM in the forward goeometry in a three level system j = 0 to j = 1 transition. The pump is assumed to be linearly polarized and the probe is σ_+ polarized. The FWM is dominated by σ_- polarized generated radiation We assume radiative relaxation. In fig.2. We show the existence of fluctuaion-induced Hanle resonances in the forward geometry. In contrast, the pressure-induced Hanle resonance in the correspon-ding geometry is difficult to see as the collisional dephasing also adds to the background.

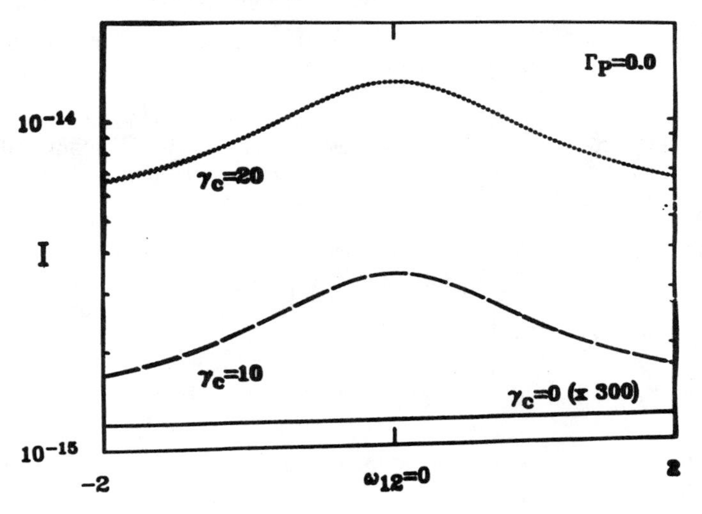

Fig.2: Fluctuation induced Hanle resonance in forward FWM.

(B) Phase Conjugation Geometry - (Fig.1)

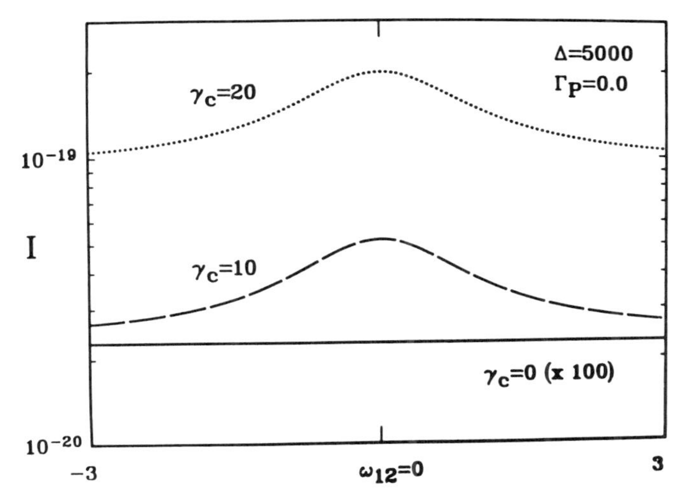

Fig.3: Fluctuation induced Hanle resonance in phase conjugation.

We show in fig.3 the fluctuation - induced Hanle resonance in phase conjugation geometry. We assume that the pump - atom detuning is much bigger than Doppler broadening which in turn is much bigger than natural width. Thus, in conclusion, we have shown that the correlated fluctuations of the pump and probe leads to new resonances in a variety of nonlinear phenomena, just like the phenomena associated with other sources of dephasing. The detailed character of the new resonances, however, depends on the particular dephasing mechanism. For the observation of FIER it is perhaps best to use sources where fluctuations can be controlled.

The author is grateful to the National Science Foundation USA for the award of a travel grant and to the Department of Science and Technology, Government of India, for partially supporting this work.

REFERENCES

1. Y.H. Zou and N.Bloembergen, Phys. Rev. A 34, 2968 (1986) and refs. therein.
2. G.S. Agarwal, Opt. Commun. 57, 129 (1986); M.S. Kumar and G.S. Agarwal, Phys. Rev. A 35, 4200 (1987);W. Lange, Opt. Commun. 59, 243 (1986) and D. Grandolement, G. Grynberg and M. Pinard, Phys. Rev. Lett. 59, 40 (1987).
3. G. Grynberg, Opt. Commun. 38, 439 (1981).
4. G.S. Agarwal and C.V. Kunasz, Phys. Rev.A 27, 996 (1983).
5. G.S. Agarwal, C.V. Kunasz and J. Cooper, Phys. Rev.(a) A 36, 143 (1987); (b) in press; (c) in press.
6. Y. Prior, I. Schek and J. Jortner, Phys. Rev. A 31, 3775 (1985); Y. Prior, A.M. Levin and A. Weiszmann, to be published.
7. G.S. Agarwal and N. Nayak, J. Opt. Soc. Am B 1, 164 (1984); Phys. Rev. A 33, 391 (1986);H. Friedmann, A.D. Wilson-Gordon, Phys. Rev.A 36, 1333 (1987).
8. G.S. Agarwal, Phys. Rev. A in press.
9. R. Saxena and G.S. Agarwal, Phys. Rev. A 25, 2123 (1982).

DEPHASING-INDUCED COHERENCES: HIGHER-ORDER SELECTION RULES AND EXPERIMENTS IN A MAGNETIC FIELD

Larry A. Rahn and Rick Trebino
Combustion Research Facility
Sandia National Laboratories
Livermore, California 94550

ABSTRACT

We observe dephasing-induced third- and higher-order wave-mixing spectra of Zeeman and hyperfine coherences in a sodium-seeded flame in a magnetic field. As the perturbation parameter approaches unity, we compare our experimental spectra with theoretical higher-order selection rules.

Dephasing-induced phenomena are a novel class of effects that are ordinarily forbidden because the two or more relevant quantum-mechanical amplitudes add coherently and cancel out.[1] When dephasing is present, however, these amplitudes no longer cancel, and previously unseen effects can be observed. A number of novel dephasing-induced phenomena have been seen.[2-5] Of particular interest, both for fundamental reasons and for diagnostic purposes, are (third-order) dephasing-induced resonances between equally populated levels of a ground electronic state,[6,7] specifically, Zeeman and hyperfine coherences. Working with a sodium-seeded flame, we have observed these latter effects[8] and, at higher intensities, subharmonics of them. The subharmonics resulted from higher-order dephasing-induced processes (see Fig. 1) from susceptibility orders as high as $\chi^{(13)}$. In this work, we report additional experimental studies in the presence of a magnetic field to determine the applicable selection rules. Because selection rules appear to distinguish true perturbative nonlinear effects from other high-intensity effects,[9] these studies allow us to verify the perturbative nature of these resonances and to determine the deviations from perturbation theory.

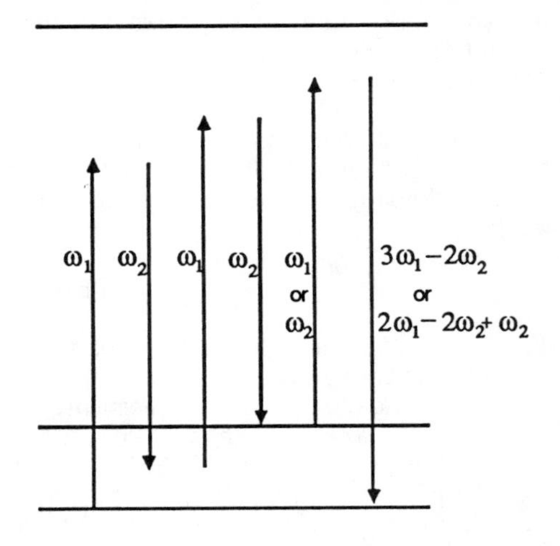

Fig. 1. Energy-level diagram for 6WM showing a four-photon $[2(\omega_1-\omega_2)]$ subharmonic resonance. The $3\omega_1-2\omega_2$ process requires a 6WM geometry, while the $2\omega_1-2\omega_2+\omega_2$ process is phase-matched in $2\omega_1-\omega_2$ 4WM geometries. Adding ω_1-ω_1 or ω_2-ω_2 successively can generate arbitrarily high-order processes and additional subharmonics.

In our experiments, two pulse-amplified, single-mode, cw dye lasers provide the ~590-nm

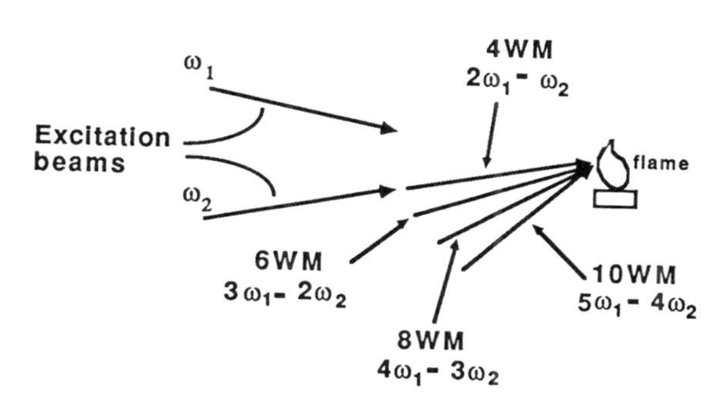

Fig. 2. Planar wave-mixing beam geometries. Higher-order beam geometries exclude lower-order processes by phase-matching. In our studies, the excitation beam at ω_1 is z-polarized, while that at ω_2 is y-polarized.

light necessary to nearly resonantly excite the D_1 line of sodium (detuning $\approx 2\ cm^{-1}$). We split one of these beams into two, labeled by $(\omega_1, \mathbf{k_1})$ and $(\omega_1, \mathbf{k_1}')$, respectively. The other beam, labeled by $(\omega_2, \mathbf{k_2})$, is frequency-tunable. We use a non-planar four-wave-mixing (4WM) geometry and a planar 6WM geometry (see Fig. 2), having k-vector conservation equation $\mathbf{k_{sig}} = 2(\mathbf{k_1} - \mathbf{k_2}) + \mathbf{k_1}'$. The vertically directed (z), 320 Gauss magnetic field is transverse to the beam propagation direction (x). Accordingly, we refer to the polarization arrangements as "zyz" or "zyy," according to whether the probe beam (with k-vector $\mathbf{k_1}'$) is polarized in the z- or y-direction. All spectra are necessarily dephasing induced because initial population densities of all ground electronic sublevels are equal.

Using the zyz 4WM geometry, we obtain the dephasing-induced 4WM spectrum shown in Fig. 3. It shows unresolved multiplets of Zeeman resonances (at $\omega_1 - \omega_2 \approx \pm 0.01\ cm^{-1}$) and hyperfine resonances [at $\pm(\omega_1 - \omega_2)$ ranging from 0.04 cm^{-1} to 0.09 cm^{-1}]. This spectrum agrees with that of Rothberg and Bloembergen.[7] Figure 4 shows the dephasing-induced spectrum using the zyy 6WM geometry and a somewhat higher intensity. This spectrum is plotted on an expanded scale so that subharmonics appearing at one-half the frequencies of the 4WM resonances can be compared with the 4WM resonances.

To understand higher-order spectra, we have derived higher-order selection rules. (Two-photon selection rules for Zeeman and hyperfine coherences in 4WM in sodium were given by Rothberg and Bloembergen.[9]) In N-wave mixing, an M-photon resonance must simultaneously obey both M-photon and (N-M)-photon selection rules. (In $\chi^{(3)}$, this is not obvious because M=N-M=2.) In addition, there are many more photon permutations to consider, allowing a process that is forbidden in one permutation to be allowed in another. We find the following selection rules for four-photon resonances in the ground state of sodium: $\Delta F = 0, \pm 1$ and $\Delta m = 0, \pm 2$ for two pairs of orthogonally polarized beams; $\Delta F = \Delta m = 0$ for two pairs of parallel-polarized beams, and $\Delta F = 0, \pm 1$ and $\Delta m = \pm 1$ for one of each pair. Using these rules, we find that zyy 6WM allows subharmonics of all of the zyz 4WM resonances. On the other hand, zyz 6WM forbids them. These selection rules appear to be confirmed by the spectra shown in Fig. 4. The hyperfine subharmonics at ~0.045 cm^{-1} and ~0.037 cm^{-1} and Zeeman subharmonics at ~.005 cm^{-1} in this zyy 6WM spectrum appear at half the corresponding frequencies in the 4WM spectrum shown in Fig. 3. The weaker pair of hyperfine subharmonics are not apparent in Fig. 4 and may be too weak to see. Alternatively, higher-order resonances may be interfering with them.

We have also measured zyz 6WM spectra (not shown) which are completely different in appearance.

We have found, however, that the 8WM processes that can occur in the zyz 6WM geometry are similar to those measured in a zyz 8WM geometry. These observations are also consistent with the selection rules discussed above.

In conclusion, we find that selection rules based on perturbation theory accurately predict which lines appear in higher-order collision-enhanced spectra. The applicability of perturbation theory is clearly limited, however. It is evident, for example, that, at the values of the perturbation parameter required to observe these effects (i.e., near unity), saturation effects will be important in determining line shapes and line widths.

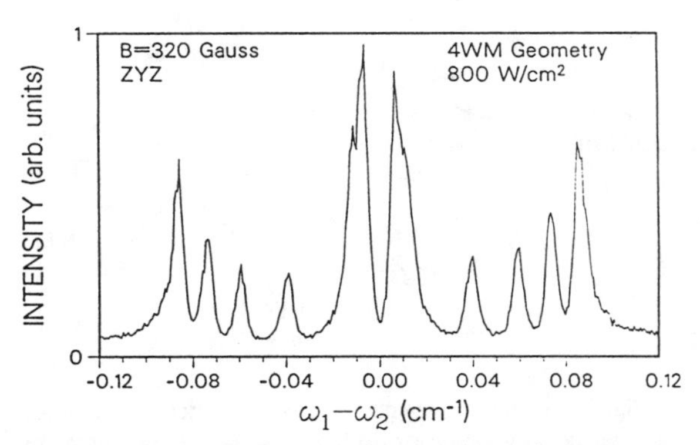

Fig. 3. Dephasing-induced spectrum of Zeeman and hyperfine coherences in sodium using a 4WM geometry at low intensity. Input polarizations are zyz. All resonances result from four-wave mixing. The quadruplets of lines at $\omega_1-\omega_2\approx\pm0.04$, ±0.06, ±0.075, and ±0.086 cm^{-1} are from hyperfine coherences, while the unresolved triplets of lines at $\omega_1-\omega_2\approx\pm0.01$ cm^{-1} are from Zeeman resonances.

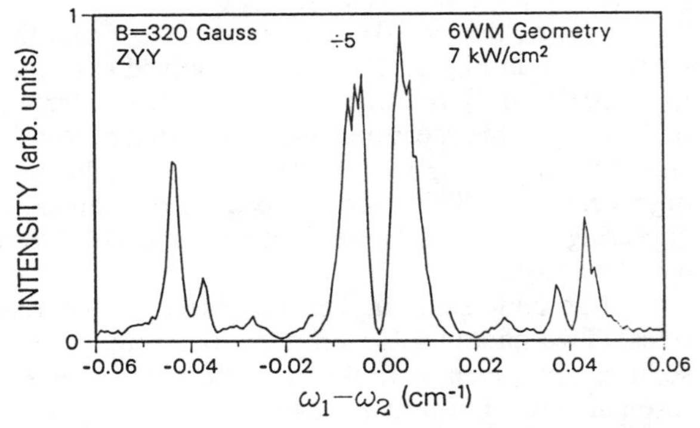

Fig. 4. Dephasing-induced spectrum of Zeeman and hyperfine coherences in sodium using a 6WM geometry. Input polarizations are zyy. All resonances result from six- or higher-order wave mixing. The frequency scale is expanded by a factor of two to illustrate the similarities with the 4WM spectrum above.

Acknowledgements

This work was supported by the U.S. Department of Energy, Office of Basic Energy Sciences, Chemical Sciences Division.

References

1. N. Bloembergen, H. Lotem, and R.T. Lynch, Ind. J. Pure and Appl. Phys. **16**, 151 (1978).
2. Y. Prior, A.R. Bogdan, M. Dagenais, and N. Bloembergen, Phys. Rev. Lett. **46**, 11 (1981).
3. A.R. Bogdan, Y. Prior, and N. Bloembergen, Opt. Lett. **6**, 82 (1981).
4. D. Grandclément, G. Grynberg, and M. Pinard, Phys. Rev. Lett. **59**, 40 (1987).
5. G. Grynberg and P. Verkerk, Opt. Commun. **61**, 296 (1987).
6. A.R. Bogdan, M.W. Downer, and N. Bloembergen, Opt. Lett. **6**, 348 (1981).
7. L.J. Rothberg and N. Bloembergen, Phys. Rev. A **30**, 820 (1984).
8. R. Trebino and L.A. Rahn, Opt. Lett. **12**, 912 (1987).
9. G.S. Agarwal, "Fractional Raman Effect in Nonlinear Mixing," submitted to Opt. Lett.

A NEW NONLINEAR-OPTICAL PERTURBATION EXPANSION AND DIAGRAMMATIC APPROACH

Rick Trebino
Combustion Research Facility
Sandia National Laboratories
Livermore, California 94550

ABSTRACT

A new nonlinear-optical perturbation expansion and diagrammatic approach are derived yielding expressions for higher-order dephasing-induced phenomena.

Dephasing-induced nonlinear-optical resonances have zero strength in the absence of dephasing because the relevant quantum-mechanical amplitudes add coherently and cancel out.[1] In the presence of dephasing, however, these amplitudes do not cancel out, yielding resonances whose strength increases with pressure. An example of a dephasing-induced effect is a resonance between initially unpopulated states. Algebraic manipulation of the third-order terms that contain these unpopulated-state resonances reveals their strengths to be proportional to trinomials of dephasing rates:[1,2]

$$\Gamma_{ijk} \equiv \Gamma_{ij}+\Gamma_{jk}-\Gamma_{ik} \tag{1}$$

Typically, this quantity vanishes when pure dephasing can be neglected, i.e., when the pressure is zero. Recently, researchers[3] have extended the range of dephasing-induced phenomena to higher order by observing subharmonics of hyperfine

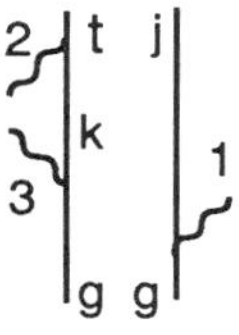

$$\frac{(-1)\,\mu^{(3)}_{gk}\mu^{(2)}_{kt}\mu^{(0)}_{tj}\mu^{(1)}_{jg}\rho^{(0)}_{gg}}{(\omega_{gj}-\omega_1)(\omega_{kj}-[\omega_1-\omega_3])(\omega_{tj}-\omega_0)}$$

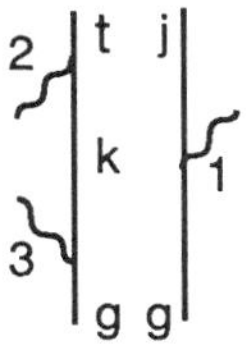

$$\frac{(-1)\,\mu^{(3)}_{gk}\mu^{(2)}_{kt}\mu^{(0)}_{tj}\mu^{(1)}_{jg}\rho^{(0)}_{gg}}{(\omega_{kg}+\omega_3)(\omega_{kj}-[\omega_1-\omega_3])(\omega_{tj}-\omega_0)}$$

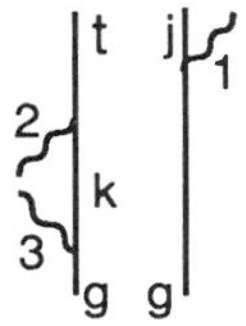

$$\frac{(-1)\,\mu^{(3)}_{gk}\mu^{(2)}_{kt}\mu^{(0)}_{tj}\mu^{(1)}_{jg}\rho^{(0)}_{gg}}{(\omega_{kg}+\omega_3)(\omega_{tg}-[\omega_2-\omega_3])(\omega_{tj}-\omega_0)}$$

Fig. 1.
Three double-sided diagrams and the terms obtained from them for the $\chi^{(3)}$ process, $\omega_0=\omega_1+\omega_2-\omega_3$. The states are labeled g, k, t, and j; $\mu^{(\gamma)}_{\alpha\beta}$ is the dipole matrix element between states α and β for the polarization of the beam at ω_γ; and $\rho^{(0)}_{gg}$ is the population density in the state g. Detailed descriptions of the construction and interpretation of double-sided diagrams are given in references 4 and 5.

resonances, which appear to be due to dephasing-induced nonlinearities as high-order as $\chi^{(13)}$. It would thus be useful to derive expressions for dephasing-induced phenomena for all orders of the perturbative nonlinear susceptibility.

Expressions for each order of the susceptibility can be obtained using double-sided diagrams[4,5] (see Fig. 1), but such expressions require significant algebraic manipulation before yielding trinomial-dephasing terms. Consequently, the required manipulation has only been performed for second and third orders.[1] We have performed this manipulation in all orders, yielding, as a result, a new perturbation expansion. In this expansion, the nonlinear-optical susceptibility is written as the sum of: (1) "principal terms," obtainable from single-sided diagrams,[5,6] (see Fig. 2) and (2) "correction terms," all of which are proportional to trinomials of dephasing rates, Γ_{ijk}. This new expansion is equivalent to the cur-

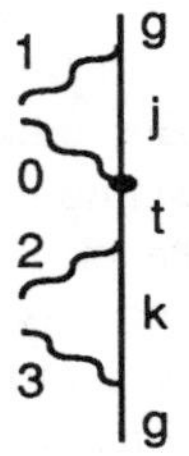

$$\frac{(-1)\,\mu^{(3)}_{gk}\mu^{(2)}_{kt}\mu^{(0)}_{tj}\mu^{(1)}_{jg}\,\rho^{(0)}_{gg}}{(\omega_{kg}+\omega_3)(\omega_{tg}-[\omega_2-\omega_3])(\omega_{gj}-\omega_1)}$$

Fig. 2.
A single-sided diagram for the $\chi^{(3)}$ process, $\omega_0=\omega_1+\omega_2-\omega_3$. A detailed description of the construction and interpretation of single-sided diagrams is given in reference 5.

rently used expansion, valid in the impact approximation, assuming isolated lines, monochromatic radiation, steady-state operation, and no initial system coherence. It reveals dephasing-induced phenomena with amplitudes proportional to trinomials of dephasing rates, Γ_{ijk}, in all orders and has the advantage that it explicitly displays the terms proportional to Γ_{ijk} in all orders.

To write down and work with the latter terms, we have defined a new diagrammatic approach (see Appendix). This approach uses single-sided diagrams to yield the principal terms and a new type of double-sided diagram, which we call a "trinomial-dephasing diagram" (see Fig. 3) to yield the correction terms. The latter type of diagram yields dephasing-induced effects proportional to a trinomial of dephasing rates directly and immediately *in any order,* without the need for algebraic manipulation.

One result[7] that follows easily using these diagrams is that any term proportional to Γ_{ijk} contains a resonance between the states i and k and is proportional to $\rho^{(0)}_{jj}$. Thus, while a trinomial-dephasing effect involves dephasing rates between all pairs of a set of three states, at least one of them must be populated initially. Other results that follow easily are:[7] in all orders, resonances between initially unpopulated states only occur when induced by dephasing, and for two-photon resonances between initially unpopulated states, the states i and k in the factor, Γ_{ijk}, must represent the two unpopulated states. Finally, as the order increases, the fraction of the terms that are trinomial-dephasing terms approaches 100%. Thus, the computational simplicity resulting when these terms can be neglected can be significant.

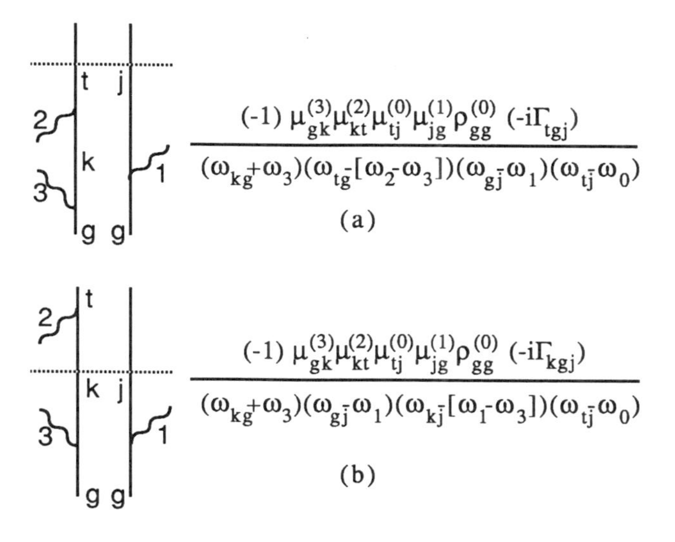

Fig. 3.
(a) A trinomial-dephasing diagram and its corresponding term for the four-wave-mixing process, $\omega_0 = \omega_1+\omega_2-\omega_3$. Observe the factor, $-i\Gamma_{tgj}$, in the numerator, indicating the dependence of this term on dephasing.

(b) The additional trinomial-dephasing diagram (obtained by sliding an interaction vertex from below to above the dashed horizontal line in a) and its corresponding term for the same four-wave-mixing process. Figs. 1, 2, and 3 illustrate the main result of this work: the sum of the terms in Figs. 2 and 3 is the same as the sum of the terms in Fig. 1.

APPENDIX: TRINOMIAL-DEPHASING DIAGRAMS

Trinomial-dephasing diagrams (see Fig. 3) look like double-sided diagrams, except for a horizontal line, which separates trinomial-dephasing diagrams into two regions. At and above the line, the rules for interpreting double-sided diagrams apply. Below the line (at least one interaction must be below the line on each side), however, rules for interpreting single-sided diagrams apply. Thus, relative heights of interactions on opposite sides below the line are not important. Another distinguishing feature of the interpretation of trinomial-dephasing diagrams is that a factor of $-i\Gamma_{ijk}$, where j is the state at the base of the diagram and i and k are the states at the line, multiplies each term corresponding to a trinomial-dephasing diagram. Reference 7 gives detailed instructions for writing down terms in the Nth-order susceptibility corresponding to trinomial-dephasing diagrams.

ACKNOWLEDGEMENTS

I would like to thank Larry Rahn for his insight and many suggestions. This work was supported by the U.S. Department of Energy, Office of Basic Energy Sciences, Chemical Sciences Division.

REFERENCES

1. N. Bloembergen, H. Lotem, and R.T. Lynch, Ind. J. Pure and Appl. Phys. **16**, 151 (1978).
2. G. Grynberg, J. Phys. B: Mol. Phys. **14**, 2089 (1981).
3. R. Trebino and L.A. Rahn, Opt. Lett. **12**, 912 (1987).
4. T.K. Yee and T.K. Gustafson, Phys. Rev. A **18**, 1597 (1978).
5. Y. Prior, IEEE J. Quant. Electron. **QE-20**, 37 (1984).
6. G.L. Eesley, J. Quant. Spectrosc. Radiat. Transfer **22**, 507 (1979).
7. R. Trebino, submitted to Phys. Rev. A.

III. ATOMIC, MOLECULAR, AND IONIC SPECTROSCOPY

UNDERSTANDING THE DYNAMIC BEHAVIOR OF MOLECULAR VIBRATIONAL STATES

T.G. Kreutz and G.W. Flynn
Department of Chemistry and Columbia Radiation Laboratory,
Columbia University, New York, NY 10027

ABSTRACT

Experimental studies of dynamic molecular processes are described which use a powerful new infrared diode laser probe technique. The combination of the very high spectral resolution of the diode laser (0.0003 cm^{-1}) and a temporal resolution of 100 ns allows detailed measurements to be made of state specific energy deposition in small polyatomic molecules (such as CO_2) which are produced by processes such as collisional energy transfer, photofragmentation, or chemical reactions. Such in depth probing of the vibrational, rotational, and translational excitation in product molecules provides a detailed picture of the underlying dynamical events.

INTRODUCTION

State-to-state dynamics has traditionally provided one of the most detailed, experimentally accessible descriptions of fundamental molecular interactions such as collisional energy transfer and chemical reactions.[1] Practically, however, state-to-state experimental data remains extremely difficult to obtain, usually requiring molecular beams, lasers, and substantial fortitude! As a result, many elegant schemes have been devised to acquire dynamical information which is somewhat less detailed, but nonetheless provides insight into collision/reaction dynamics. The most prominent of these techniques involves the measurement of final quantum states of molecules which have undergone gas phase dynamical events, such as collisions, reaction, or photodissociation, with minimal initial state selection.[2] Laser induced fluorescence (LIF) has been one of the most useful detection methods in these experiments, finding wide application to the measurement of state distributions for diatomic molecules, whose excited state spectroscopy is often well known.[3] In this paper, a complimentary technique, infrared diode laser absorption spectroscopy, which can be used to conveniently measure vibrational, rotational, and translational energy disposal in somewhat larger molecules, is described. This method is applicable to small polyatomics and other IR active species, such as molecules with complicated, unassigned or predissociative electronic spectra, which are less amenable to LIF detection.[4] The diode laser probe technique has been successfully employed in state specific measurements in a variety of dynamically interesting processes which are described below.

THE DIODE LASER PROBE TECHNIQUE

The power of the diode laser absorption spectroscopy arises from the high spectral resolution (~0.0003 cm^{-1}) and continuous tunability of the cw diode laser radiation, which allows transitions of the type:

$$CO_2(m,n^l,p;J) \; + \; h\nu(\lambda{\sim}4.3\ \mu m) \; \rightarrow \; CO_2(m,n^l,p+1;J\pm1)$$

to be probed. In the equation above, m, n, and p refer to the CO_2 quantum numbers for the symmetric stretching (ν_1=1388 cm^{-1}), bending (ν_2=667 cm^{-1}), and asymmetric stretching (ν_3=2349 cm^{-1}) vibrational modes. The IR absorption lines corresponding to these transitions are well separated due to the anharmonicity of the molecule, and in addition, they all share the strong oscillator strength of the ν_3 fundamental. As a result, the population in virtually any ro-vibrational state of the CO_2 molecule can be measured by simply monitoring the appropriate transition. The diode laser (Laser Analytics) is a low power (~1 mW), spectrally bright source of cw infrared radiation which is tunable over a range of ~50 cm^{-1}, centered anywhere in the infrared spectral region from 3-30 µm. The temporal resolution of a time domain absorption probe experiment is governed by the bandwidth of the IR detector, and ranges from 1 to 100 ns.

Time domain diode laser absorption spectroscopy has a number of significant advantages over traditional IR chemiluminescence detection. Paramount, of course, is the 100,000 fold increase in spectral resolution which allows ro-vibrational state probes of small polyatomic molecules. In addition, by focusing the coherent diode laser radiation onto a small IR detector, an order of magnitude can be gained in detection speed without loss of sensitivity. Finally, states whose emission is difficult or impossible to detect, such as the CO_2 bend (01^10) or symmetric stretch (10^00), can easily be probed with diode laser absorption spectroscopy. In short, this technique represents such a significant advance over previous infrared probing methods, that virtually all previous IR fluorecense experiments are worth repeating with a diode laser! Fine kinetic details of dynamically interesting processes such as bimolecular reactions, collisional energy transfer, and photodissociation can be easily studied with ro-vibrational state specificity. In addition, the high resolution of the diode laser permits sub-Doppler probing of the absorption lineshapes, which in turn provides information about the translational degrees of freedom, as well as correlations between the velocities, angular momenta, and electric dipoles of the interacting species.[5,6]

The time domain absorption spectroscopy technique used in our laboratory employs a double resonance configuration, in which a cw diode laser is used to monitor time dependent changes in the ro-vibrational state populations of some species of interest (such as CO_2) following a fast perturbation within the gas mixture by a pulsed excitation dye or excimer laser. In the typical experimental apparatus shown in Figure 1, the pulsed laser beam and the diode laser radiation are propagated co-linearly through a long sample cell containing a flowing gas mixture. Upon exiting the cell, the diode radiation passes through a monochromator to discriminate against competing spatial and longitudinal laser modes, and is detected with a cooled IR detector. Time resolved changes in the transmitted intensity of the diode light are acquired with a signal averager.

In order to monitor the population in a single CO_2 ro-vibrational level, the diode frequency is fixed to the peak of the appropriate absorption line throughout the duration of the experiment. This is accomplished by splitting off a portion of the beam (~8%) before the sample cell and directing it through a reference cell into a second monochromator/IR detector. The reference cell absorption signal is sent to a lock-in detector whose (derivative) output is fed back into the diode laser current controller for frequency stabilization. In order to probe transitions which have minimal or zero population at room temperature, a CO_2 discharge reference cell, which consists of a high voltage DC discharge applied to a low pressure mixture of CO_2, N_2, and He, is employed. The discharge generates a steady state, high temperature distribution of vibrationally excited CO_2 molecules, providing frequency references for thousands of high lying rovibrational lines which are normally inaccessible at room temperature. Typical reference cell absorption spectra are shown in Figure 2; note the dramatic increase of available absorption lines when the discharge is turned on.

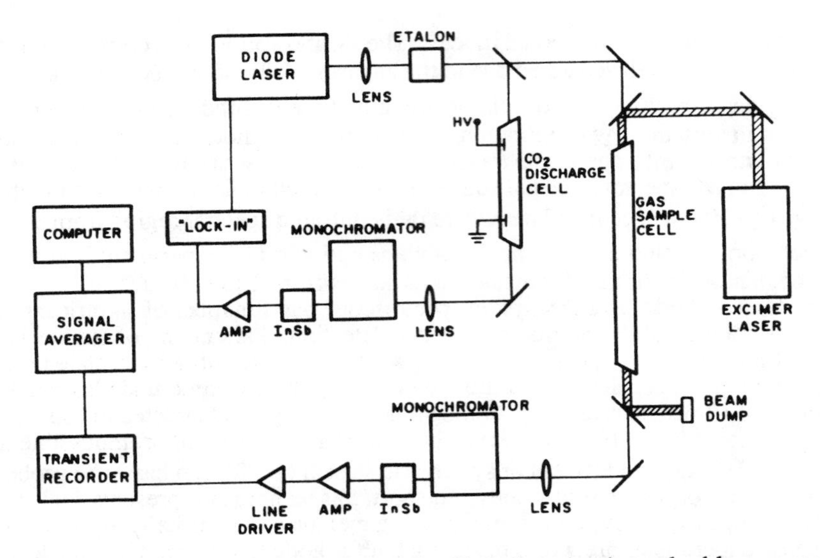

Figure 1. A schematic diagram of the diode laser/excimer double resonance apparatus used for time domain diode laser absorption spectroscopy.

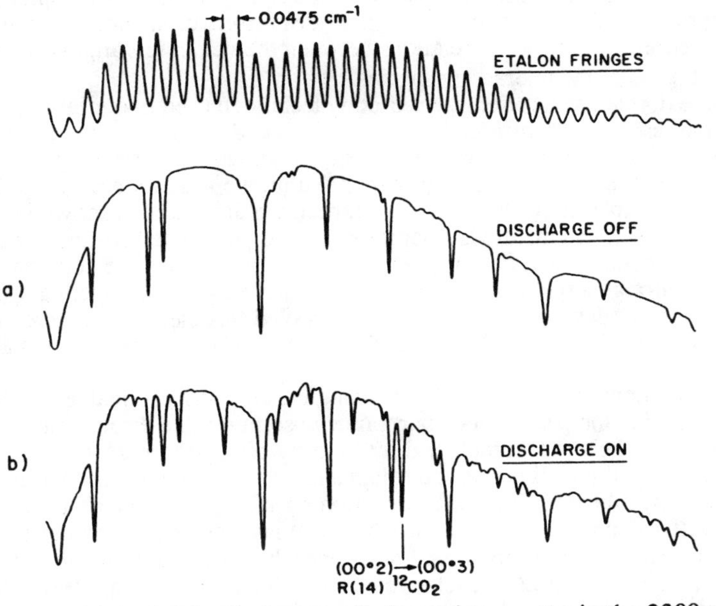

Figure 2. Typical CO_2 discharge cell absorption spectra in the 2300 cm^{-1} region with the discharge a) turned off, and b) ignited (25 mA). Note that the intensity of the $(00^02)\rightarrow(00^03)$ R(14) line at 2310.035 cm^{-1} is increased by nine orders of magnitude when the discharge is turned on.

COLLISIONAL ENERGY TRANSFER BETWEEN POLYATOMIC MOLECULES AND HOT ATOMS

The collision and reaction dynamics between hydrogen atoms and other species has been of substantial experimental and theoretical interest for many years.[7] A series of studies of the H*/CO_2 system have been performed using a diode laser to measure the the probability of ro-vibrational excitation of the CO_2 molecule.[8-16] These experiments are initiated by the 193 nm excimer laser photolysis of a flowing CO_2/H_2S gas mixture:

$$H_2S + h\nu(193\ nm) \rightarrow HS + H^* \qquad \text{(Hot Atom Production)}$$

which produces hydrogen atoms having approximately 2.5 eV of translational energy, corresponding to a temperature of nearly 24,000 K. The hot atoms collide with CO_2 and excite it to a variety of ro-vibrational levels

$$H^* + CO_2(00^00;J) \rightarrow H + CO_2(mn^lp;J') \qquad \text{(Collisional Excitation)}$$

whose populations are then probed with a diode laser. These experiments are performed at a pressure which is sufficiently low that the hydrogen atoms are lost rapidly by diffusion from the probe beam. As a result, only the first collision with a hot hydrogen atom is important in the vibrational scattering process.[10] The rotationally resolved excitation of CO_2 has been measured in a number of vibrational levels as described below. The relative efficiency of vibrational excitation is found by integrating the rotationally resolved results.

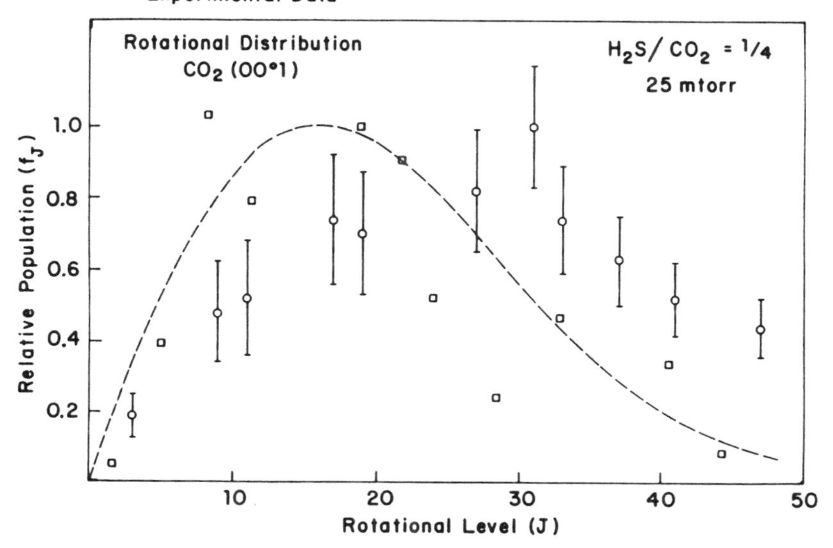

Figure 3. Relative rotational excitation of CO_2 (00^01) from collisions with hot hydrogen atoms.

The rotational distribution for the 00^01 state, shown in Figure 3, is peaked near J=31, approximately 17 units of angular momentum higher than the peak of the initial ground state distribution (300 K Boltzmann).[9,10] There is more than enough collision energy to produce much higher rotational levels in 00^01, but the available angular momentum appears to be limited by the by the small impact parameter required in order to excite the ν_3 vibration (2349 cm^{-1})[10]. Support for this interpretation comes from a study of the relative probability of exciting CO_2 (00^01) by collisions with hot deuterium versus hot hydrogen atoms, in which D* is found to be more efficient than H* in exciting the (00^01) level as J increases.[11]

Rotationally resolved measurements of the excitation of the first excited bending state (01^10) exhibit a broad peak near J=42, a finding which is again consistent with a 'constrained angular momentum' model.[12] The degree of rotational excitation in this case is significantly higher than that found in the 00^01 level as a result of its smaller vibrational energy (667 cm^{-1}). In addition, the cross section for 01^10 vibrational excitation is expected to be substantially larger than that for 00^01, not only because of the energetic effects, but also because CO_2 is longer than it is wide!

A particularly interesting feature of the 01^10 rotational distribution is that the nascent population in the odd J levels is markedly larger than that in the even J states.[13] This effect is clearly related to the symmetry of the the initial and final scattering states; note that 00^00 has only even J states, while 01^10 has both even and odd rotational levels. The intensity alternation strongly suggests that ro-vibrational scattering from the ground state is more important than pure rotational scattering from within the 01^10 manifold (which has 8% of the CO_2 population at 300 K). In the 01^11 bend-stretch combination level, the observed alternation is reversed (i.e. odd J < even J), as would be predicted by the change in rotational state symmetry for this level compared to that for 01^10. This effect has been predicted theoretically by Alexander and Clary for low energy He/CO_2 scattering,[14] but has never been observed experimentally.

Preliminary data has been obtained for the excitation of the first excited symmetric stretch level 10^00, which is in Fermi resonance with the 02^00 state.[15] The rotational distribution and total cross-section for excitation into this state are similar to that observed for the 00^01 level. Pure rotational scattering into high rotational levels (J=54 to J=74) of the ground state has also been observed, and the probability of excitation found to decrease monotonically with increasing J.[16] As in the excitation of 00^01, D* is more efficient than H* at exciting high J states.

In general, the rotational distributions observed in these experiments can be qualitatively understood using a simple ellipse model.[17] Moreover, an analysis based upon the infinite order sudden (IOS) approximation, which allows the extraction of state-to-state rates from this data, appears to work quite successfully. The IOS approximation is well justified at the high collision energies and low rotational constant of CO_2 characteristic of these experiments. The most ambitious theoretical study of the H/CO_2 system has been by Schatz and co-workers[18] who used a quasi-classical trajectory study with a full potential surface to calculate distributions of final states. Comparisons of our data with these calculations have revealed problems with their potential surface and have lead to improvements in the surface. Further theoretical analysis of our data should reveal the extent to which our detailed measurements shed light upon the potential surface governing this important process.

PYRUVIC ACID PHOTOFRAGMENTATION DYNAMICS

Time domain diode laser absorption spectroscopy is particularly suited to measurements of final product state distributions of polyatomic fragments produced in molecular photodissociation. In a series of studies, the extent of vibrational, rotational, and translational excitation in the CO_2 photofragment from the 193 nm photolysis of pyruvic acid has been measured.[19-21] In these experiments, pulsed excimer laser radiation at 193 nm is collinearly propagated with a cw diode laser beam through a 2 m sample cell containing flowing pyruvic acid vapor at ~25 mtorr. Pyruvic acid is photolysed by the excimer pulse and produces excited CO_2 fragments:

$$CH_3COCOOH + h\nu(193\ \text{nm}) \rightarrow CH_3CHO^* + CO_2\,(nm^lp;J)$$

whose relative rotational and vibrational populations are probed via time domain diode laser absorption spectroscopy.

Measurements of the relative vibrational excitation among the CO_2 photofragments were made in 10:1 mixtures of argon:pyruvic acid.[20] The argon promotes rotational equilibration within the response time of the detector, but has a negligible effect on CO_2 vibrational relaxation. The nascent vibrational populations, shown in Fig. 4, indicate that the vast majority of CO_2 fragments (~97%) are found with no vibrational excitation, while the remaining 3% are vibrationally hot. These results suggest a two channel photodissociation mechanism, with a major channel (~97%) which produces vibrationally cold photofragments on an electronically excited surface, and a minor channel (~3%) which yields vibrationally excited fragments following internal conversion of the excited pyruvic acid to the electronic ground state surface. Since this mechanism predicts that the acetaldehyde carries away the vast majority of the 6 eV photolysis energy, evidence of further fragmentation has been sought. Indeed, diode laser probing has revealed the evolution of both CH_4 and CO upon the 193 nm irradiation of pyruvic acid.

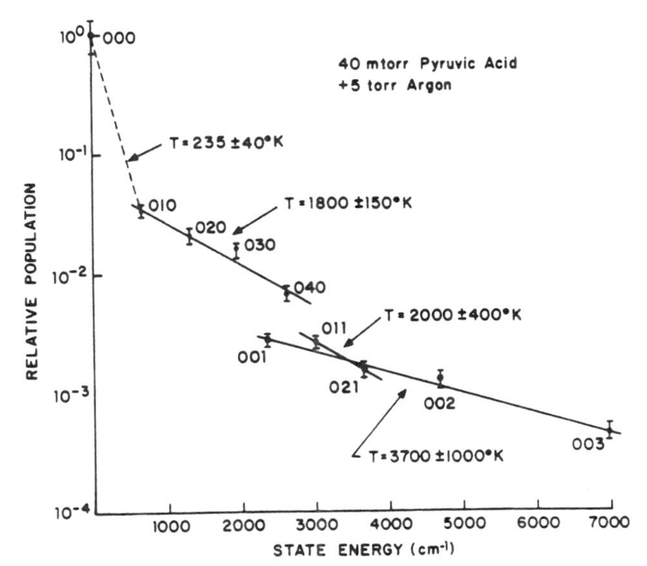

Figure 4. Nascent distribution of vibrational energy in the CO_2 photofragments generated from the 193 nm photolysis of pyruvic acid.

Measurements of rotational and translational excitation in the nascent CO_2 molecules have also been performed following the 193 photolysis of pure pyruvic acid at low pressures.[21] A high degree of rotational excitation was found in the 02^20 level, giving a nascent rotational distribution which can be roughly fit to a 1300 K Boltzmann distribution. In contrast, little excess translational energy (as measured by examining the nascent CO_2 lineshapes) was found in CO_2 fragments for either the 01^10 or 02^20 vibrational levels. These results might possibly be explained by the presence of an internal hydrogen bond in gas phase pyruvic acid.[22] As the C-C bond breaks, the residual hydrogen bond imparts substantial rotational motion to the separating photofragments. A high efficiency for such a process might reconcile the substantial rotational excitation observed in the CO_2 fragment with its more moderate translational excitation. These results appear to be consistent with a photodissociation mechanism which involves rather low velocities along the 'reaction' coordinate, as suggested by the observed lack of vibrational excitation. The deposition of rotational excitation within the 00^00 and 01^10 levels is being investigated to see if differences in the two proposed dissociation channels will be reflected in the rotational distributions of these states. These studies will help to further determine the photofragmentation mechanism and perhaps provide clues to the nature of the potential surfaces which govern the photodissociation process.

CATALYTIC OXIDATION OF CARBON MONOXIDE ON A PLATINUM SURFACE

While time domain studies make up the bulk of our experimental efforts, the diode laser is also extremely useful in its more traditional role as a source for high resolution IR spectroscopy. A diode laser spectrometer has been used to measure vibrational excitation in *steady state* CO_2 produced from the catalytic oxidation of carbon monoxide on a platinum surface:[23]

$$CO + 1/2\, O_2 \rightarrow CO_2(mn^lp;J).$$

Aside from its enormous practical importance, this reaction is of particular interest because the CO_2 product does not fully accommodate with the surface before desorbing, as is thought to be typical of chemical reactions on catalytic surfaces.[24-26] Instead, CO_2 molecules leave the surface with translational, vibrational, and rotational temperatures far in excess of the surface temperature, and thus may be probed for mode specific energy disposal in the internal degrees of freedom. Such data provides a wealth of information about the reaction dynamics and the activated complex on the surface. These experiments are being carried out in collaboration with Steve Bernasek and Larry Brown of Princeton University who have observed high levels of vibrational excitation in the CO_2 product by measuring IR chemiluminescence from the asymmetric stretching vibrations.[25]

The experimental apparatus consists of a low pressure flow tube in which a mixture of CO, O_2, and argon flows through a platinum mesh which is electrically heated to 800K. CO is oxidized on the surface to form ro-vibrationally excited CO_2 which is probed just downstream of the mesh by diode laser radiation. Relative line intensities are used to determine the ro-vibrational state populations. Under the present experimental conditions, the CO_2 suffers enough collisions before it is probed to complete translational, rotational and *intra*mode vibrational relaxation. As a result, only separate vibrational mode temperatures can be determined for each of the three CO_2 vibrational modes. Preliminary results show that the high frequency asymmetric stretch mode is substantially hotter than the low frequency bend in essential agreement with previous molecular beam/FTIR and IR chemiluminescence studies.[26] Since the previous experiments all probed IR emission, the diode laser experiments appear to be the first in which the low energy bends have been studied directly.

These results may be indicative of a reaction transition state in which the incipient CO_2 molecule has highly asymmetric bond lengths (leading to excitation in the ν_3 mode) but is only slightly bent. Alternatively, they may simply reflect a greater degree of surface accommodation by the relatively low frequency bending mode as compared with the asymmetric stretch. This point will be further studied by measuring the relative efficiency of vibrational excitation in these two modes during collisions of gas phase CO_2 with a heated platinum mesh. It should be noted that while the measured mode temperatures are quite disparate, the vibrational *energy* per mode is approximately equal for the bends and the stretches. Further experiments will measure the vibrational mode temperatures as a function of surface temperature and reactant coverage. In addition, angular resolved diode laser measurements of this reaction on a single crystal platinum surface are planned for the new UHV molecular beam apparatus at Princeton University. Such detailed probing of the state specific deposition of reaction exothermicity in the CO_2 product is expected to lead to a substantially enhanced understanding of the transition state and the nature of the reaction coordinate(s) for this important catalytic reaction.

BIMOLECULAR CHEMICAL REACTION

Time domain diode laser spectroscopy provides an excellent method for the study of state specific energy disposal in bimolecular chemical reactions which can be initiated with an optical pulse. One particularly interesting bimolecular process is the chemical vs physical quenching of $O(^1D)$ atoms by CO_2 (ref. 27)

$$^{16}O_3 + h\nu(248\ \mathrm{nm}) \rightarrow {}^{16}O_2 + {}^{16}O(^1D) \qquad \text{(Initiation)}$$

$$^{16}O(^1D) + C^{18}O_2 \rightarrow {}^{18}O(^3P) + C^{16}O^{18}O(mn^lp;J) \qquad \text{(Reaction)}$$

$$\rightarrow {}^{16}O(^3P) + C^{18}O_2(mn^lp;J). \qquad \text{(Quenching)}$$

The electronic quenching of $O(^1D)$ is known to proceed rapidly (at a rate approximately equal to the gas kinetic collision rate[28]), but the relative efficiency of chemical versus physical quenching has never been determined.[27] There is, in addition, an intermediate $(CO_3)^*$ species which has been isolated in a matrix, whose structure has been the subject of a number of theoretical and experimental investigations.[29] Since the two isotopic forms of CO_2 allow a separation between the chemical and physical quenching channels, measurements of state specific energy deposition in each species should lend insight into the nature of this intermediate species. Since the isotope shifts are quite large and the spectroscopy of these species well known, such measurements are straightforward with a diode laser probe.

In contrast to the expectation of strong, prompt vibrational excitation of the CO_2 molecules, our preliminary results indicate that *none* of the relatively low lying CO_2 rovibrational states (< 5000 cm^{-1}) which have been studied so far show direct excitation. Instead, substantial excitation has been observed in the 00^01 level at a rate (~20 gas kinetic collisions) which is substantially slower than the ν_3 intramode vibration-to-vibration (V-V) equilibration rate. This result may indicate direct excitation into high lying ro-vibrational levels (particularly ν_3) followed by V-V equilibration among the CO_2

molecules. However, our analysis of such a 'harvesting' process suggests an enormous deficit between the electronic energy which is fed into the CO_2 molecule (~2 eV per $O(^1D)$) and that which shows up as vibrational excitation. Such a finding might be explained by the intervention of some intermediate, metastable electronic state of CO_2, possibly a triplet, since the rapidity of this 'spin forbidden' quenching process has long been a puzzle.[30] Experiments are presently under way to probe very high lying levels of CO_2 such as 00^05, 00^06, and 00^07 in order to search for states which are directly populated by the quenching of $O(^1D)$. These experiments should help to elucidate the key dynamical features of this atmospherically important process.

CONCLUSIONS

The powerful technique of diode laser absorption spectroscopy has been used to study state specific energy disposal in small molecules involved in a variety of dynamically interesting processes such as bimolecular chemical reactions, collisional energy transfer, photodissociation, electronic-to vibration (E-V) energy transfer, and chemical reactions on catalytic surfaces. Detailed measurements of vibrational, rotational, and translational excitation in CO_2 molecules formed by these processes have provided a great deal of insight into the mechanisms and underlying dynamics which govern their behavior.

ACKNOWLEDGEMENTS

We are indebted to our colleagues at Columbia University, Brookhaven National Laboratories, and Princeton University, who have participated in the many experimental studies described here, for their inspiration and insight. This work was supported by the Department of Energy. Equipment and administrative support was provided by the National Science Foundation and the Joint Services Electronics Program of the Depatment of Defense (U.S. Army, U. S. Navy, and U.S. Air Force)

REFERENCES

1. See for example, State-to-State Dynamics, edited by P.R. Brooks and E.F. Hayes (Americal Chemical Society, Washington, 1977).
2. For example, B.E. Holmes and D.W. Setser, in Physical Chemistry of Fast Reactions, Vol. 2, edited by I.W.M. Smith (Plenum, New York, 1980).
3. M.A.A. Clyne and I.S. McDermid, in Dynamics of the Excited State, edited by K.P. Lawley (Wiley, New York, 1982); Adv. Chem. Phys. 50, 1 (1982).
4. R.N. Zare, Faraday Discuss. Chem. Soc., no. 67, 7 (1979).
5. R.N. Zare, Mol. Photochem. 41, 1 (1972).
6. P.L. Houston, J. Phys. Chem. 91, 5388 (1987).
7. G.W. Flynn and R.E. Weston, Jr., Ann. Review of Physical Chemistry, 37, 551 (1986).
8. J. O. Chu, C. F. Wood, G. W. Flynn, and R. E. Weston, Jr., J. Chem. Phys. 80, 1703 (1984); 81, 5533 (1984); J.A. O'Neill, J.Y. Cai, G.W. Flynn, and R.E. Weston, Jr., J. Chem. Phys. 84, 50 (1986);
9. J.A. O'Neill, C.X. Wang, J.Y. Cai, G.W. Flynn, and R.E. Weston, Jr., J. Chem. Phys. 85, 4195 (1986);
10. J.A. O'Neill, J. Y. Cai, C. X. Wang, G.W. Flynn, and R.E. Weston, Jr., J.

Chem. Phys., 88 0000 (1988).

11. S.A. Hewitt, J.F. Hershberger, G.W. Flynn, and R.E. Weston, Jr., J. Chem. Phys., 87, 1894 (1987).
12. .A. Hewitt, J. Chou, J.F. Hershberger, G.W. Flynn, and R.E. Weston, Jr., "Rotational Profiles in the Excitation of the First Bending State of CO_2 by Hot Deuterium Atoms," manuscript in preparation.
13. J.F. Hershberger, S.A. Hewitt, G.W. Flynn, and R.E. Weston, Jr., "Observation of Symmetry Effects in CO_2/Hot Deuterium Atom Scattering," manuscript in preparation.
14. D.C. Clary, J. Chem. Phys. 78, 4915 (1983); M. H. Alexander and D. C. Clary, Chem. Phys. Lett., 98, 319 (1983).
15. J. Hershberger, J. Chou, A. Hewitt, G. Flynn, and R.E. Weston, Jr., "Rotational State Distribution of the Bend/Stretch Fermi Mixed Level 10^00 of CO_2 Produced by Collisions with Hot Hydrogen Atoms," manuscript in preparation.
16. J. Hershberger, A. Hewitt, G. Flynn, and R.E. Weston, Jr., "Pure Rotational Scattering in Hot Atom CO_2 Collisions," manuscript in preparation.
17. S. Bosanac and U. Buck, Chem. Phys. Lett. 81, 315 (1981); S. Bosanac, Phys. Rev. A22, 2617 (1980).
18. G.C. Schatz, M.S. Fitzcharles and L.B. Harding, Faraday Discussions, to be published.
19. C.F. Wood, J.A. O'Neill, and G.W. Flynn, Chem. Phys. Lett., 109, 317 (1984).
20. J.A. O'Neill, T.G. Kreutz, and G.W. Flynn J. Chem. Phys., 87, 4598 (1987).
21. T. Kreutz and G. Flynn, "Rotationally Resolved Product State Distribution of CO_2 Produced by the Excimer Laser Photodissociation of Pyruvic Acid," manuscript in preparation.
22. W.J. Ray, J.E. Katon, and D.B. Phillips, J. Mol. Struct. 74, 75 (1981).
23. T.G. Kreutz, J.A. O'Neill, G.W. Flynn, L.S. Brown, and S.L. Bernasek, "Tunable Diode Laser Probe of Vibrationally Excited CO_2 Formed by Catalytic Oxidation on a Platinum Surface", manuscript in preparation.
24. T. Engel and G. Ertl, Advan. Catal. 28, 1 (1979);C.A. Becker, J.P. Cowin, L. Wharton, and D.A. Auerbach, J. Chem. Phys. 67, 3994 (1977); R.L. Palmer and J.N. Smith Jr., J. Chem. Phys. 60, 1453 (1974).
25. S.L. Bernasek and S.R. Leone, Chem. Phys. Lett. 84, 401 (1981); L.S. Brown and S.L. Bernasek, J. Chem. Phys. 82, 2110 (1985).
26. D.A. Mantell, S.B. Ryali, B.L. Halpern, G.L. Haller, and J.B. Fenn, Chem. Phys. Lett.81, 185 (1981); D.A. Mantell, S.B. Ryali, and G.L. Haller, Chem. Phys. Lett. 102, 37 (1983).
27. H. Yamazaki and R.J. Cvetanovic, J. Chem. Phys. 40, 582 (1964); D.L. Baulch and W.H. Breckenridge, Trans. Faraday Soc. 62, 2768 (1966).
28. G.E. Streit, C.J. Howard, A.L. Schmeltekoph, J.A. Davidson, and H.I. Schiff, J. Chem. Phys. 64, 57 (1976), and references therein.
29. D. Katakis and H. Taube, J. Chem. Phys. 36, 416 (1962); E. Weissberger, W.H. Breckenridge, and H. Taube, J. Chem. Phys. 47, 1764 (1967); M. Arvis, J. Chim. Phys. 66, 517 (1969); W.B. DeMore and C. Dede, J. Phys. Chem. 74, 2621 (1970); N.G. Moll, D.R. Clutter, and W.E. Thompson, J. Chem. Phys. 45, 4469 (1966); M. Cornille and J. Horsley, Chem. Phys. Lett. 6, 373 (1970).
30. J.C. Tully, J. Chem. Phys. 62, 1893 (1975).

CONTINUOUS STOPPING AND TRAPPING OF NEUTRAL ATOMS*

G.P. Lafyatis#, V.S. Baganato##, A.G. Martin, K. Helmerson, and D.E. Pritchard
Massachusetts Institute of Technology, Cambridge, MA 02139

ABSTRACT

We are now able to hold billions of laser cooled sodium atoms for minutes in a magnetic neutral trap. Physics experiments that might be done in this type of trap are outlined. The trap itself is described and results which characterize the loading and trapping of the atoms are presented. Finally, we report very recent work in which R.F. transitions between magnetic hyperfine levels have been driven.

I shall discuss some of the results we have obtained in our work at M. I. T. to slow, trap, and cool neutral sodium atoms. In this work we use magnetic fields to trap atoms by interacting with the magnetic dipole moment of the unpaired spin in a ground state sodium atom. Atoms in the M_F=+2 level in the ground state Zeeman manifold are trapped around a minimum in the magnetic field. Laser light is used to slow and cool atoms so that they will be confined in our relatively weak trap.

Before I present specific results I would like to make several introductory comments to put into perspective our work. First, the most significant conclusion of this talk is that the trapping field is at the point were we can begin to look beyond "trapology" -- the study of traps for their intrinsic interest --and start using traps as powerful tools for novel physical studies. This past year has seen orders of magnitude improvement in numbers of trapped particles and confinement times of neutral traps. We can now (almost routinely) trap a fair number of atoms -- literally billions and billions. We are able to confine these atoms for minutes and cool them to millikelvin temperatures.

This is not to say trap development is finished: many interesting experiments require orders of magnitude more atoms. Presently, trapping times seem to be limited only by the background pressure of the apparatus and, in principle, there appears to be no fundamental reason we cannot hold atoms indefinitely; however this remains to be seen. A major direction of our current research is to cool atoms to temperatures

*Supported by the Office of Naval Research

#Permanent address: Department of Physics, Ohio State University, Columbus, Ohio 43210.

##Permanent address: IFQSC-USP, São Paulo, Brazil.

corresponding the the recoil "kick" an atom receives when it scatters a single photon; for sodium this is a couple of microkelvins.

General fields in which neutral atom traps may be particularly useful include the study of very low energy atomic collisions, observations of quantum collective effects, and precision atomic spectroscopy. As will be described in other papers in this session, interesting new physics is predicted and has begun to be observed in collisions among atoms which have been cooled to millikelvin temperatures. Traps may prove useful for studying collective behavior among dense, ultracold samples of atoms. The most frequently discussed effect is Bose condensation. However, we are still over ten orders of magnitude in atom density or, alternatively, about seven orders of magnitude in sample temperature from seeing a Bose condensate of sodium atoms. Neutral atom traps share virtues of ion traps for precision spectroscopy experiments: long interaction times may be used to interrogate spectroscopic features (in contrast, for example, to the "transit time" limitations in a Cs beam) and second order Doppler shifts may be virtually eliminated by laser cooling the atomic sample. The major advantage of atom trap experiments over corresponding ion trap experiments is that, because neutrals interact very weakly compared to ions, a neutral experiment may use many more atoms and hence should have better signal to noise. A disadvantage of the neutral trap is that the trapping field itself shifts the frequencies of the atomic resonances.

Fig. 1 shows, schematically, the apparatus used for this work. I shall describe it briefly, a more detailed description is found in Ref. 1. A thermal (550 C) beam of sodium atoms is brought to a stop by using the "Zeeman tuned slowing" technique developed by Phillips and Metcalf[2]: atoms in a beam are slowed as they absorb photons from a counterpropogating laser beam, which is tuned to the 589 nm sodium D_2 line. A specially tapered magnetic field keeps the atoms in resonance with the laser as they slow by compensating the changing Doppler shift of the resonant frequency with an equal and opposite Zeeman shift. In our experiment two stages of "Zeeman tuned slowing" are used. The first uses the "slowing laser" and "slower magnet" to decelerate the atomic beam to an intermediate velocity -- between 100 m/s and 400 m/s. The second stage uses the

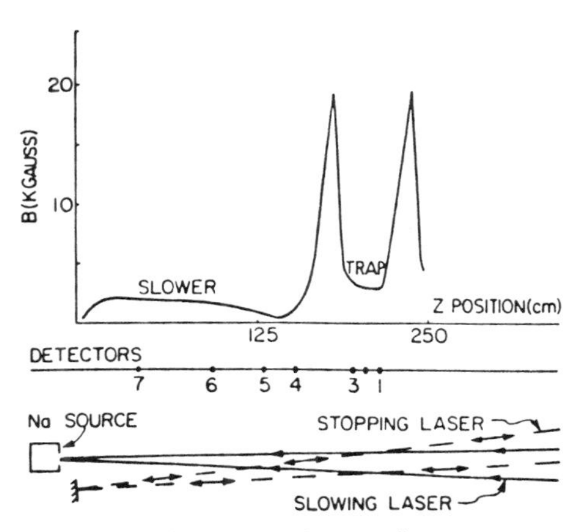

FIG. 1 The Experimental Apparatus

"stopping laser" to bring atoms to a stop within the trap itself. The second laser is retroreflected. This is to keep atoms from being pushed by the laser back out of the trap after they have been stopped. Retroreflecting the beam creates, in addition, a one dimensional "optical molasses" in which it was expected that the atoms would be laser cooled to around one millikelvin.

The magnetic field minimum, which constitutes the trap, is created by solenoid magnets which produce the axial field shown in Fig. 1 and an octopole magnet whose pole pieces are parallel to the axis of the apparatus and extend (roughly) between the two axial field peaks. The octopole magnet confines atoms radially. All fields are created by superconducting magnets which are operated in a presistent mode. Typical operating currents in the magnets produce a 100 mK deep trap. The vacuum wall of the trap was actually the inside surface of the LHe dewar that contained the magnets. This cryogenic environment provided an excellent vacuum for the experiment, $P<10^{-11}$ torr.

As they are slowed, stopped, and trapped, the atoms are diagnosed by observing their laser excited fluorescence in the photodiodes at the positions numbered in Fig. 1. The results I report here are, for the most part, based on measurements made with detector 1, which is located at the minimum of the trap.

The two stage Zeeman tuned slowing scheme enabled us to continuously stop atoms and load them into our trap. This is a significant improvement over the only other reported magnetic neutral trap which was loaded with pulses of slowed atoms[3]. Continuous loading allowed us to fill our trap with over five orders of magnitude more atoms than was previously reported.

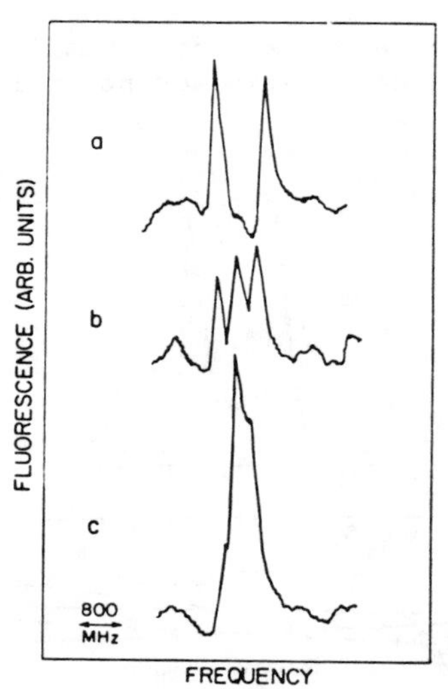

FIG.2 Continuous stopping measurements

Initially, we investigated continuous stopping. By varying the frequency of the slowing laser, we can vary the velocity of the atoms exiting the first stage of slowing. It is important to adjust this velocity for optimal trap loading: if the atoms are moving too fast when they enter the trap region, the second stage of slowing cannot stop them. Fig. 2 demonstrates this. The three curves in Fig. 2 represent measurements for three different frequencies of the slowing laser. The data were obtained by scanning the frequency the stopping laser and observing the fluorescence in detector 1. In Fig. 2(a), the initial slowing stage produces 280 m/s atoms. These are moving too fast to be stopped and, consequently, two peaks in the fluorescence signal are seen: the first occurs when the direct beam from the stopping laser is resonant with the 280 m/s atoms (the atoms are moving toward the direct beam and see it Doppler shifted to the blue) and the second when the retroreflected beam is resonant. In Fig. 2(b) 200 m/s atoms leave the slower. These are slow enough to be stopped. The outer two peaks in Fig. 2(b) correspond to unslowed atoms as in Fig. 2(a). The third, central peak occurs when the stopping laser is tuned correctly for the second stage of Zeeman tuned slowing to bring the atoms to a halt at the bottom of the trap. This peak represents the continuous stopping of atoms from an atomic beam. In Fig. 2(c) the peaks overlap as the slowing laser is tuned to provide still slower atoms (130 m/s).

A convincing experiment to demonstrate that one is indeed confining atoms is to load the trap, block the lasers and the atomic beam, wait a while, and then turn a laser back on and see if any atoms are left. In Fig. 3 we performed this experiment for different periods of darkness. The peaks show the fluorescence from the atoms remaining in the trap. The loss of atoms from the trap is fairly well described by an exponential decay with a time constant on the order of two minutes. This is consistent with a loss mechanism in which the trapped atoms are scattered out of the trap when they collide with background gas molecules in the trap's vacuum.

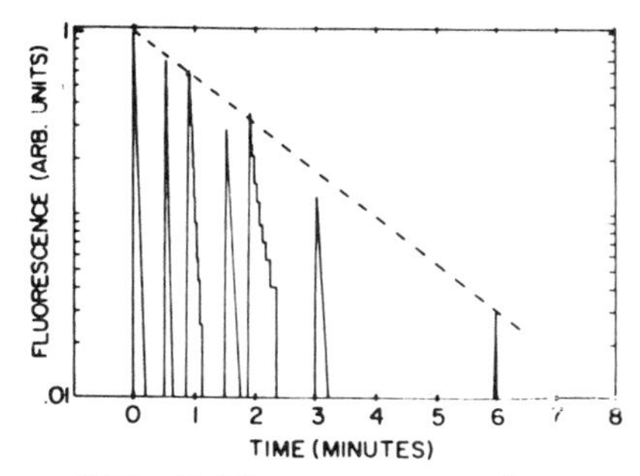

FIG. 3 Fluorescence of atoms remaining in trap. The trap was loaded at t=0.

Very recently we have been able to study trapped atoms by measuring their absorption of a probe laser beam. Absorption is a useful technique for studying the atoms because extremely weak probe beams may be used and hence problems in which the probe beam changes the sample that is being studied are reduced. In addition, absorption results are more reliable than fluorescence ones in determining the number of atoms in the trap. These measurements are made using a greatly attenuated probe beam from the stopping laser and measuring absorption in the retroreflected beam. We observe peak absorptions (down and back through the trap) of sixty per cent. This confirms that several billion atoms are being trapped.

Trapping times of minutes are sufficient to begin experimenting with the trapped atoms. Initially, we undertook a double resonance (R.F.-optical) experiment to investigate magnetic dipole transitions between different Zeeman levels in the hyperfine manifold of the ground state. Such transitions have been proposed to cool atomic samples below the Doppler cooling limit[4]. The loading process produces a trapped atom population that is largely in the M_F=+2 state. This is the most energetic state in the manifold and we decided to first try to drive transitions to the next state down, an M_F=1, (M_J=1/2) state.

Transitions are detected by observing optical excitations (we have used both fluorescence and absorption) from the various levels. We put an antenna in the trap and supplied enough R.F. signal power to yield a Rabi frequency between 100 and 1000 hz (on resonance). The transition frequency is a function of magnetic field, however, for the field at the bottom of our trap the nominal resonance frequency is about 330 MHz. The actual experiments consisted of applying R.F. for several seconds and then measuring the relative populations of the two levels with a probe laser. Fig. 4 plots the ratios of the two level populations for a series of runs in which the R.F. frequency was varied. The sharp low frequency edge occurs at the resonant frequency for atoms at the bottom of the trap. The signal at the higher frequencies results from the more energetic atoms in the trap which sample larger magnetic fields. These data may be used, then, to diagnose the trapped atoms. Indeed, a very preliminary analysis suggests that the temperature of the atomic sample is about 25 mK -- significantly larger than the Doppler limit, 240 microkelvin, that we expected to approach with the "optical molasses" at the bottom of the trap.

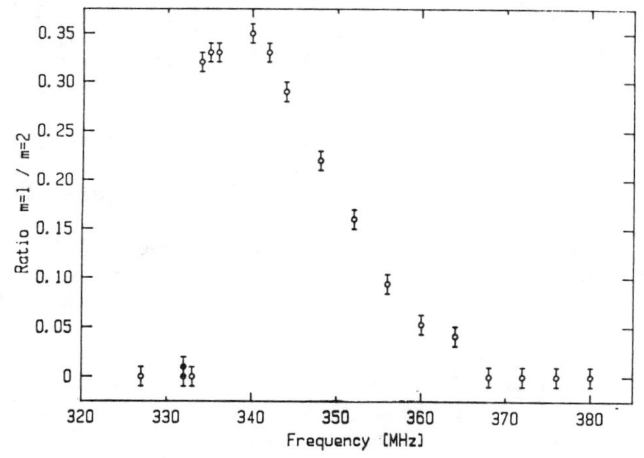

FIG. 4 Ratio of populations of $M_F=1$ to $M_F=2$ as a function of R.F. drive frequency

REFERENCES

[1]V.S. Bagnato, G.P. Lafyatis, A.G. Martin, E.L. Raab, R.N. Ahmad-Bitar, and D.E. Pritchard, Phys. Rev. Lett. 58, 2194 (1987).

[2]W.D. Phillips and H.J. Metcalf, Phys. Rev. Lett. 48, 596 (1982).

[3]A.L. Migdall, J.V. Prodan, W.D. Phillips, T.H. Bergeman, and H.J. Metcalf, Phys. Rev. Lett. 54, 2596 (1985).

[4]D.E. Pritchard, Phys. Rev. Lett. 51, 1336 (1983).

OBSERVATION OF ASSOCIATIVE IONIZATION OF ULTRA-COLD LASER-TRAPPED SODIUM ATOMS

P.L. Gould, P.D. Lett, W.D. Phillips, and P.S. Julienne
National Bureau of Standards, Gaithersburg, MD 20899

H.R. Thorsheim and J. Weiner
Chemistry Department, University of Maryland, College Park, MD 20742

ABSTRACT

We have observed associative ionization of laser cooled and trapped sodium atoms. The measured rate coefficient for the process at a temperature of (0. 75 ±0. 25) mK is $(1.1^{+1.3}_{-0.5})\times10^{-11}$ cm^3/sec, which implies a cross section of 8.6×10^{-14} cm^2. This is three orders of magnitude larger than the cross section measured in previous experiments at higher temperatures.

The possibility of studying collisions between ultra-cold neutral particles has interested researchers since the development of laser cooling and trapping techniques. This low energy regime is novel for several reasons: (1) The collision dynamics can be dominated by long range, weakly attractive potentials that are unimportant at higher energies. (2) The de Broglie wavelength of the particles may be long compared to the range of the interaction potential making the concept of a classical trajectory invalid. (3) Significantly fewer partial waves contribute to the reaction than at room temperature. We report the first identification and measurement of a specific collisional process in such a unique sample.

We study associative ionization (AI) of sodium:

$$Na^* + Na^* \rightarrow Na_2^+ + e^- \quad (1)$$

where $Na^* = Na(3P_{3/2})$. This reaction has been investigated in both vapor cells[1] and beams[2]. The lowest collision energy studied previously was 20 K. We extend the knowledge of this reaction to the ultracold mK region.

The experiment uses sodium atoms that are laser cooled and confined in a new hybrid laser trap[3]. This trap uses both the spontaneous radiation pressure force and the dipole force for confinement. Two counterpropagating, circularly polarized (σ+), Gaussian laser beams are brought to separate foci (separated by approximately one confocal parameter) such that each beam is diverging at the other's focus. With the laser tuned below resonance, radial confinement is provided by the dipole force. Axial confinement about the midpoint of the two foci results from the spatial variation in the radiation pressure of the beams. The trap has a time averaged radial well depth of ~10 mK and an axial depth (to each focus) which is on the order of 1K. Dipole heating[4] is avoided by alternating the trap beams in time. Each trap beam is on for 3 μs. The trapping cycles are also alternated with Doppler cooling periods[5] to damp the atomic motion. The cooling is provided by "optical molasses" [6,7,8] which surrounds the trap. Slow atoms emanating continuously from a laser-cooled atomic beam are first captured and confined by the molasses and then load the trap.

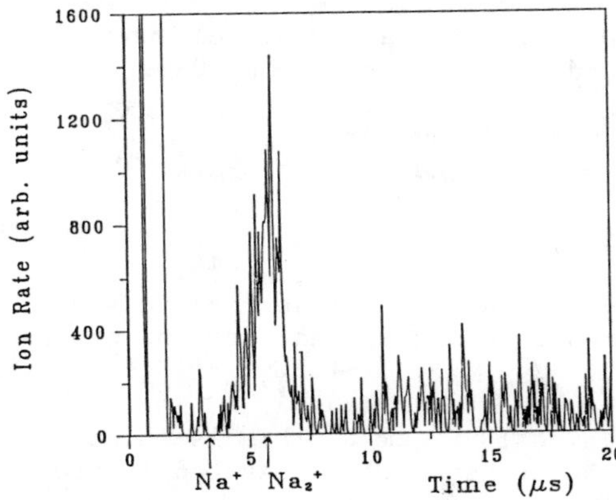

Fig. 1 Time-of-flight mass spectrum of ions produced in trap. Arrows indicate the calculated arrival times for Na^+ and Na_2^+.

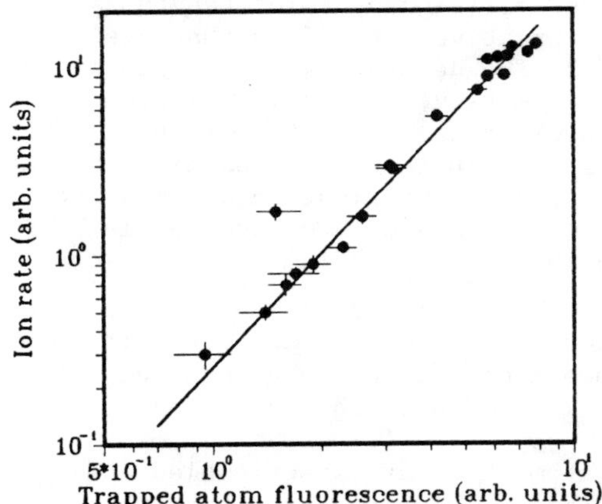

Fig. 2 Log-log plot of ion signal vs. trapped atom fluorescence. The solid line has a slope of 2.0.

Fluorescence from the excited atoms in the trap is collected and imaged using a calibrated CCD video camera. The FWHM of the trap is ~40 μm in the radial direction and ~850 μm in the axial direction. We derive the three dimensional excited state density profile from the two dimensional video image of the excited atoms. Ions are detected with a focused mesh electron multiplier.

In order to verify that the ions are produced in AI collisions, we performed two tests. First a time of flight mass spectrum (see Fig. 1) is taken by applying a pulsed electric field to extract the ions which accumulate over many trapping/cooling cycles. The predominant ion formed is Na_2^+, with the signal to noise ratio allowing us to to determine that the Na^+ fraction is <20%.

Secondly, the dependence of the ion rate on the excited state density is measured (see Fig. 2). The density of the trapped atoms is changed by varying the flux of the slow atoms, thereby changing the loading of the trap without modifying the trap parameters. On a log-log plot, a binary collision would yield a straight line with a slope of two. The data support the hypothesis that the ions are formed by a binary process, and not a single body process, which would have a slope of one.

The absolute rate coefficient for associative ionization is calculated from absolute measurements of the ion production rate and the number and distribution of excited atoms. We typically measure an ion rate of 1.1×10^3 ions/sec from 2×10^4 excited atoms in an effective volume of 4.0×10^{-6} cm^3, implying an effective trap density of 5.0×10^{-9} cm^{-3}. This yields a rate coefficient of $(1.1 {}^{+1.3}_{-0.5})$ $\times10^{-11}$ cm^{3}sec^{-1}. At a temperature of (0.75 ±0.25)mK, this corresponds to a cross section of $(8.6 {}^{+10.0}_{-3.8}$ x 10^{-14} cm^2. The uncertainty in this value is mainly due to the determination of the spatial distribution of trapped atoms. The temperature is determined from the distribution of atoms in the calculated trap potential.

Although the cross section we obtain is three orders of magnitude larger than that previously observed at higher temperatures, it is not unreasonable. The maximum cross section, assuming unit probability of inelastic scattering leading to AI would be $\sigma_{max}=\pi\lambda\!\!\!^{-2}(\ell_{max}+1)^2$ where $2\pi\lambda\!\!\!^{-}$ is the de Broglie wavelength and ℓ_{max} is the maximum partial wave that can contribute to the cross section. If only s-waves participate in the reaction, the maximum cross section is ~$7\text{x}10^{-13}$cm^2. Thus the large measured value does not violate unitarity, even for only s-wave scattering.

In conclusion, we have observed associative ionization in a cold, dense sample of sodium atoms. The cross section we measure is much larger than what has been previously observed at higher temperatures. The large cross section implies that this and other collisional processes[9] may dominate the lifetime of optical traps (as opposed to losses due to collisions with background gas) as higher densities are achieved. This experiment represents the first use of laser-cooled neutral atoms to study atomic physics.

This work was supported in part by the Office of Naval Research and the National Science Foundation.

REFERENCES

1. J. Huennekens and A. Gallagher, Phys. Rev. A 28, 1276 (1983); V.S. Kushawaha and J.J. Leventhal, Phys. Rev. A 22, 2468 (1980); A. Klyucharev, V. Sepman, and V. Vuinovich, Opt. Spectrosc. 42, 336 (1977).
2. R. Bonnano, J. Boulmer, and J. Weiner, Comments At. Mol. Phys. 16, 109 (1985); Phys. Rev. A 28, 604 (1983); M.-X. Wang, J. Keller, J. Boulmer, and J. Weiner, Phys. Rev. A 35, 934 (1987)
3. A. Ashkin, Phys. Rev. Lett. 40, 729 (1978).
4. J.P. Gordon and A. Ashkin, Phys. Rev. A 21, 1606 (1980).
5. J. Dalibard, S. Reynaud, and C. Cohen-Tannoudji, Opt. Commun. 47, 395 (1983).
6. P.L. Gould, P.D. Lett, and W.D. Phillips, in Laser Spectroscopy VIII, W. Persson and S. Svanberg, editors (Springer-Verlag, Berlin, 1987), p.64.
7. W.D. Phillips, J.V. Prodan, and H.J. Metcalf, J. Opt. Soc. Am. B 2, 1751 (1985).
8. S. Chu, L. Hollberg, J.E. Bjorkholm, A. Cable, and A. Ashkin, Phys. Rev. Lett. 55, 49 (1985).
9. P.S. Julienne, S.H. Pan, H.R. Thorsheim, and J. Weiner, in these proceedings.

QUANTUM OPTICS EXPERIMENTS WITH A SINGLE ION

R. G. Hulet*, J. C. Bergquist, W. M. Itano, J. J. Bollinger,
C. H. Manney and D. J. Wineland
National Bureau of Standards, Boulder, Colorado 80303

ABSTRACT

Several experiments performed at NBS using a single (or few) trapped atomic ions are described. We recently observed the "quantum jumps" in the fluorescence from a single ion of Mg^+ due to spontaneous Raman transitions between the energy levels of the ion. In this system, a *single* laser is used to induce transitions between Zeeman sub-levels of the ion. As with other realizations of quantum jumps, the dynamics of the population evolution is governed by an effective two-level rate equation. A novel coherence phenomenon between excited state sub-levels makes this system ideally suited for quantitatively testing the quantum jump theory. Our data are consistent with theory to within the measurement precision of 2%. Images of two to four ions which have been laser-cooled until they "crystallize" are displayed and several other recent single ion experiments are mentioned.

INTRODUCTION

The fluorescence emitted by one or a few atoms or atomic ions can exhibit phenomena which are unobservable in the fluorescence emitted by a collection of many atoms. A classic example is that of photon anti-bunching,[1] where the correlation in the time interval between detected photons reveals the photons to be "anti-bunched" in time. This purely quantum effect was first observed in the fluorescence from a weak atomic beam[2] and later, from a single ion confined in an ion trap.[3] A more recent demonstration of the quantum nature of the fluorescent radiation emitted by only a few emitters is the observation of the "quantum jumps" of an electron between its atomic energy levels. In 1985, three groups reported the observation of this phenomenon in a single confined ion[4–6] and a fourth group inferred quantum jumps from the time correlation of fluorescence from a weak atomic beam.[7] In the National Bureau of Standards experiment, anti-bunching was demonstrated without directly detecting, and therefore, destroying, the anti-bunched photons.[5]

These are examples of quantum optics experiments which have been performed using a single trapped ion. Ion trap and laser cooling technology enables the experimenter to "look" at a single ion for continuous periods of time which can extend to many minutes in duration. In this paper, we will discuss several new single/few ion experiments performed at N.B.S. Primarily, we will describe a very recent experiment which has quantitatively verified the theory of

*Present address: Rice University, Physics Department, Houston, Texas 77251.

quantum jumps to high precision.[8] Also, a brief description of the observation of "Coulomb clusters" is presented.[9]

QUANTUM JUMPS

As an introduction to "quantum jumps" we describe the "prototype" quantum jump experiment, shown in Fig. 1. It consists of three atomic energy levels in the "V" configuration. The ground state (0) is strongly coupled to an excited state (1) which has a spontaneous decay rate A_1, and is weakly coupled to another excited state (2) which decays at a rate A_2 ($A_1 >> A_2$). Two radiation sources separately drive the $0 \leftrightarrow 1$ and $0 \leftrightarrow 2$ transitions while the fluorescence from level 1 is monitored. On a time scale long compared to A_1^{-1}, but less than A_2^{-1}, the fluorescence displays periods of constant intensity interrupted by periods of zero intensity while the atomic electron is "shelved" in level 2. The on-off switching of the fluorescence indicates a discontinuous jump, or quantum jump, between level 2 and the strongly fluorescing system consisting of level 0 and level 1. This prototype experiment is based on Dehmelt's "electron shelving" scheme for the detection of a transition between the weakly coupled levels,[10] and was first analyzed by Cook and Kimble.[11] There have been several subsequent theoretical analyses.[12–21]

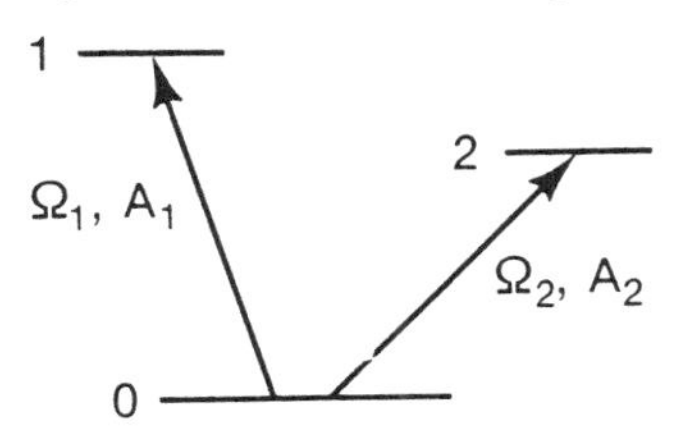

Figure 1. V configuration of energy levels used in "prototype" quantum jump experiment. Level 2 is metastable, so that $A_1 >> A_2$.

The population dynamics of the three levels, and therefore, the statistics of the emitted photons, can be reduced to those of an effective two-level system for times which are long compared to A_1^{-1}.[18] Let R_+ denote the rate of excitation out of the strongly fluorescing level 0-level 1 system and into level 2, while R_- is the rate back out of level 2. The rate equation analysis yields the mean fluorescence on-time $<T_{on}> = R_+^{-1}$ and off-time $<T_{off}> = R_-^{-1}$.

PRECISE TEST OF QUANTUM JUMP THEORY

We have recently observed quantum jumps in an atomic system which is quite different than those in which quantum jumps had been previously observed: a *single* laser is used to induce spontaneous Raman transitions into and out of a ground state shelving level.[8] As we shall show, this new system has particular advantages for precisely testing the statistical predictions of the quantum jump theory. The process is illustrated in Fig. 2. The single radiation source is tuned between a pair of Zeeman levels of an atom or ion with a $^2S_{1/2}$ ground state and an excited $^2P_{3/2}$ state. The six Zeeman sub-levels in Fig. 2 are labelled from 1 to 6 for convenience. The frequency ω of the radiation is tuned near the level 1 $\leftrightarrow$ level 3 transition frequency ω_o. The atom cycles nearly continuously between these levels since the dipole selection rules allow spontaneous decay only to the original ground level. A steady stream of fluorescence photons, which are readily detected, is emitted by the atom during this period. However, if the radiation is linearly polarized perpendicular to the direction of the magnetic field, the 1 ↔ 5 transition is also allowed, although it is far from resonance. This transition is indicated by the dashed arrow in Fig. 2. A spontaneous decay from level 5 can then leave the atom in the "other" ground level (i.e. level 2). This spontaneous Raman transition into level 2 takes the atom out of the 1 ↔ 3 cycling loop, causing the emitted fluorescence to suddenly stop. The off-resonant 2 → 4 → 1 spontaneous Raman transition (not shown in Fig. 2) will return the atom to the cycling loop where it will resume scattering. Consequently, the detected fluorescence will alternate between periods of "on" and "off".

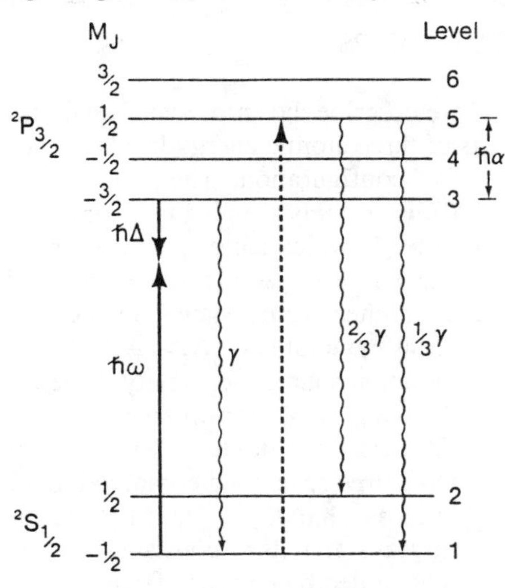

Figure 2. The energy level structure of the $^2S_{1/2}$ and the $^2P_{3/2}$ states of an atom in a magnetic field. Off-resonant excitations from level 2 and spontaneous decay from levels 4 and 6 are not shown.

We have used a single $^{24}Mg^+$ ion confined by the static magnetic and electric fields of a Penning trap to demonstrate this process.[8] The ring electrode of the trap was split into two halves to provide a large collection solid angle (> 1% 4π sr) for collecting the fluorescent photons.[22,8] The magnetic field in the trap was 1.39 T, which gives $\alpha \approx 1200\gamma$, where $\hbar\alpha/2$ is the energy separation between adjacent excited state sublevels and $\gamma = (2\pi)$ 43 MHz is the excited state spontaneous decay rate (see Fig. 2). The $3S_{1/2}$ to $3P_{3/2}$ transition wavelength is 280 nm for which we generated up to 200 μW by frequency doubling the output of a dye laser. The laser beam was focused near the center of the trap with a beam waist $w_o \approx 45$ μm. The ions were laser cooled[23] to a fraction of 1 K when the detuning $\Delta<0$ ($\Delta = \omega-\omega_o$), and since the beam was directed at an angle of 74° relative to the magnetic field axis all of the ion's degrees of motion were cooled directly. Under these conditions, the Doppler broadening of the transition frequency was much less than γ and the ion was confined to dimensions much less than w_o. The fluorescent photons were collected in a direction perpendicular to both the magnetic field and laser beam directions. Because of the large collection solid angle, we were able to detect 2×10^5 photons/s. from a single ion.

Data exhibiting quantum jumps are displayed in the inset of Fig. 3. The horizontal axis is divided into time intervals of 0.5 ms duration and the vertical axis is a measure of the number of photons counted per time interval. The

CONCLUSIONS

Ion trap techniques have yielded new investigations of the quantum nature of the radiation emitted by a single atomic ion. These include the observation of quantum jumps and, more recently, a test of quantum jump theory which exploits a coherence phenomenon to quantitatively verify the theory to a precision of 2%. The statistics of quantum jumps have also been used to precisely measure radiative decay rates of metastable states of an ion.[26] Finally, it was reported that a "gas" of several atomic ions has been observed to crystallize when laser cooled to below a critical temperature.

The work was supported by the U.S. Air Force Office of Scientific Research and the U.S. Office of Naval Research. R.G.H. thanks the National Research Council for support.

REFERENCES

1. c.f., J. D. Cresser, J. Häger, G. Leuchs, M. Rateike and H. Walther in Dissipative Systems in Quantum Optics, Vol. 27 of Topics in Current Physics, edited by R. Bonifacio (Springer-Verlag, Berlin, 1982) and references therein.
2. H. J. Kimble, M. Dagenais and L. Mandel, Phys. Rev. Lett. 39, 691 (1977).
3. F. Diedrich and H. Walther, Phys. Rev. Lett. 58, 203 (1987).
4. W. Nagourney, J. Sandberg and H. Dehmelt, Phys. Rev. Lett. 56, 2797 (1986).
5. J. C. Bergquist, R. G. Hulet, W. M. Itano and D. J. Wineland, Phys. Rev. Lett. 57, 1699 (1986).
6. Th. Sauter, W. Neuhauser, R. Blatt and P. E. Toschek, Phys. Rev. Lett. 57, 1696 (1986).
7. M. A. Finn, G. W. Greenlees and D. A. Lewis, Opt. Commun. 60, 149 (1986).
8. R. G. Hulet, D. J. Wineland, J. C. Bergquist and W. M. Itano, submitted for publication.
9. D. J. Wineland, J. C. Bergquist, W. M. Itano, J. J. Bollinger and C. H. Manney, Phys. Rev. Lett., to be published.
10. H. G. Dehmelt, J. Phys. (Paris) 42, C8-299 (1981).
11. R. J. Cook and H. J. Kimble, Phys. Rev. Lett. 54, 1023 (1985).
12. T. Erber and S. Putterman, Nature (London) 318, 41 (1985).
13. J. Javanainen, Phys. Rev. A 33, 2121 (1986).
14. A. Schenzle, R. G. DeVoe and R. G. Brewer, Phys. Rev. A 33, 2127 (1986).
15. C. Cohen-Tannoudji and J. Dalibard, Europhys. Lett. 1, 441 (1986).
16. D. T. Pegg, R. Loudon and P. L. Knight, Phys. Rev. A 33, 4085 (1986).
17. A. Schenzle and R. G. Brewer, Phys. Rev. A 34, 3127 (1986).
18. H. J. Kimble, R. J. Cook and A. L. Wells, Phys. Rev. A 34, 3190 (1986).
19. P. Zoller, M. Marte and D. F. Walls, Phys. Rev. A 35, 198 (1987).
20. G. Nienhuis, Phys. Rev. A 35, 4639 (1987).
21. M. Porrati and S. Putterman, Phys. Rev. A 36, 929 (1987).
22. M. H. Prior and H. A. Shugart, Phys. Rev. Lett. 27, 902 (1971).
23. W. M. Itano and D. J. Wineland, Phys. Rev A 25, 35 (1982).
24. R. G. Hulet and D. J. Wineland, Phys. Rev. A 36, 2758 (1987).
25. F. Diedrich and H. Walther, Phys. Rev. Lett., to be published.
26. W. M. Itano, J. C. Bergquist, R. G. Hulet and D. J. Wineland, Phys. Rev. Lett. 59, 2732 (1987).

PROPOSAL FOR MAGNETIC TRAPPING OF NEUTRAL 2^3S HELIUM

Harold Metcalf
Physics Dept., S.U.N.Y. Stony Brook, N.Y. 11790

Optical cooling and magnetic trapping of metastable 2^3S helium (He*) is very attractive for many reasons. First, the narrowness of the cooling transition ($2^3S \rightarrow 2^3P$ at λ_o=1.083 μm) results in a VERY low Doppler cooling limit $T_D \equiv \hbar\gamma/2k_B \sim 40\mu K$ where k_B=Boltzmann constant and $\gamma/2\pi$=natural width=1.6 MHz. Second, its magnetic moment μ is two Bohr magnetons giving a trapping coefficient μ/k_B=1.2 Kelvin/Tesla, twice the value for alkali atoms. Third, it would enable fundamental studies in atomic spectroscopy such as a far better, direct measurement of the metastable lifetime. Fourth, collisions of very cold He* atoms may yield new information about their potentials, especially at long distances and with isotopic mixtures. Fifth, because of its small mass, the quantized energy levels of trapped He* are more widely separated than those of alkalis, enabling studies of trap spectra and Majorana transitions out of the lowest quantum states, thus providing access to some of the fundamental problems of magnetic trapping of neutral atoms. Sixth, formation of aggregations could lead to the study of dimers and clusters.

A three stage method for producing ultra-cold He* atoms is proposed. First, He* is made in the cold gas above a very small amount of rapidly pumped liquid helium by an rf discharge that is quenched when most of the He gas is pumped away. Typical discharge conditions of $10^{14}/cm^3$ atoms (mass M) produce about $10^9/cm^3$ He*. The inelastic electron collisions ($\Delta E \sim 20$ eV) that produce the excitations also transfer kinetic energy $(m_e/M)\Delta E$, about 30 K, that constitutes heating. The 30 K He* atoms are cooled to the 1.5 K cryostat temperature by collisions with the high density, cold ground state atoms (aided by metastability exchange collisions).

The second stage begins when the discharge stops. Ground state atoms continue to be pumped away while He* atoms are held (and optically cooled) by a low-gradient magneto-optical radiation force trap[1] of depth $T_o \sim$few K. For pumping speed S on N ground state atoms in volume V we find $N(t)=N(0)\exp(-t/\tau)$ where $\tau \equiv V/S$. We define $\tau^* \equiv V/v\sigma^*$ where σ^* is the cross section for collisions that eject He* from the trap and depends on T_o, and v is the atomic speed. The number of trapped metastables N* obeys $dN^*/dt=-N^*N(t)/\tau^*$ so $N^*(t)= N^*(0)\exp[N(0)Q(t)\tau/\tau^*]$ where $Q(t)\equiv[\exp(-t/\tau)-1]$ approaches -1 for $t \gg \tau$. If $x \equiv N(0)\tau/\tau^* \sim 1$, most He* atoms remain in the trap and are not ejected by collisions during pumpout of the ground state atoms. Therefore we seek conditions that satisfy $x \sim 1$.

We can estimate x for both molecular and viscous flow domains during the pumpout process. Using standard formulae for pumping speed, viscosity, and mean free path λ, we find that a pumpout tube of length L and diameter D gives $x=(64VL/\pi D^4)(\sigma^*/\sigma)$ for viscous flow $(D>\lambda)$ and a factor $\sim D/5\lambda$ smaller for molecular flow ($D<\lambda$ so this is always <1). The ratio of σ^* to the kinetic cross section σ is estimated to be $\exp(-T_o/T)$ where T is the temperature of the background gas atoms that can eject trapped metastables.

In order to find T_0 we next discuss magneto-optical trapping (MOT) which is used to contain the He* during pumpout. MOT depends on both inhomogeneous magnetic fields and radiative cooling to exploit both optical pumping and the radiative force for the purposes of cooling and confinement. The idea was suggested by J. Dalibard and demonstrated by Raab et al.[1] We consider atoms in a linearly varying magnetic field $B=B(z)\equiv Az$ and neglect the Doppler shift for the moment. A simple atomic transition (J=0 to J=1) has three Zeeman components, excited by each of three polarizations, whose frequencies tune with field (and therefore with position) as shown below. We employ two oppositely directed laser beams of opposite circular polarization, each detuned below the zero field atomic resonance by δ as shown. Atomic resonance can then only occur at the two points $z=\pm z'$ where the Zeeman tuning of each transition corresponds to the laser frequency, and because of the polarizations, it can only occur with the σ^+ beam at $z=-z'$ and with the σ^- beam at $z=+z'$. Consequently, an atom at $+z'$ will be driven toward $-z'$ and conversely, so that atoms bounce back and forth between these turning points, which can be moved either by tuning the laser or by changing A. This scheme is readily extended to three dimensions using two opposed magnet coils[2].

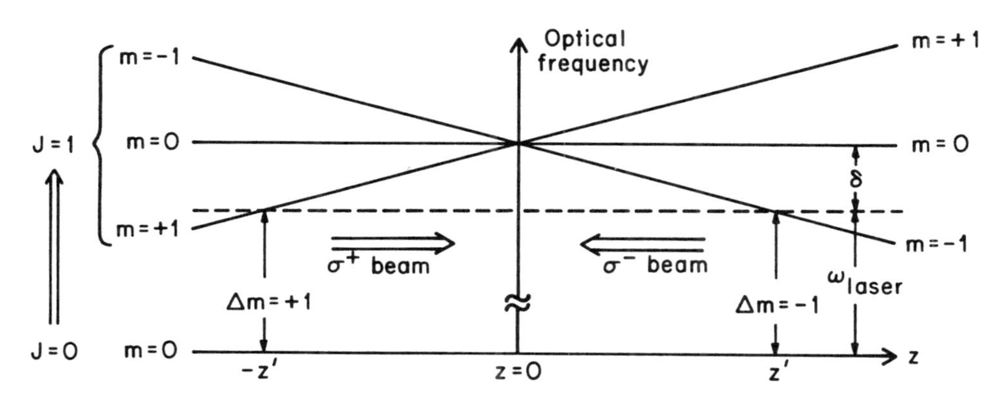

The Doppler effect of a moving atom changes the figure by shifting the optical frequencies in the atomic rest frame. The Zeeman and Doppler shifts then combine to produce resonance with the $\sigma^{\pm}$ beams when $\delta=\pm(|k|v+\mu' Az/\hbar)$, where k is the optical wave vector and μ' is the Zeeman tuning coefficient. This enables resonance over an extended distance because the changing Doppler shift of the decelerating atom can be compensated by the changing Zeeman shift[3]. Furthermore, the red tuning ($\delta<0$) and the choice of polarization combine to produce a radiative force that opposes the motion of an atom for a wide range of atomic velocities.

The maximum value of the radiative force is $\hbar k\gamma/2$, and it can stop an atom of maximum speed $v=v_0$ when resonance with such a fast atom at one end of the trap is preserved until it is stopped at the other end. The resonance conditions for $v=v_0$ and $v=0$ at opposite ends of the trap combine to give $\delta=-kv_0/2$ and $A=\hbar\delta/\mu' z'$. Fast atoms are therefore decelerated along their entire trip across the trap, and slower atoms have gentle (inelastic) collisions with extended magneto-optical walls instead of hard bounces at $\pm z'$. The maximum height of the soft walls is approximately $\hbar k\gamma z'=3k_BT_o/2$.

With this relation between T_o and trap size we can now calculate the conditions for $x\sim 1$ that confines most of the He* during pumpout. We choose the reasonable pumpout tube parameters of length L=25 cm and diameter D=2.5 cm, and then we find z'=3.2 cm which gives $V=250\ cm^3$, $T_o\sim 10.2$ K, $v_o\sim 250$ m/s, and $\sigma^*/\sigma\sim 4\times 10^{-4}$ for 1.3 K gas. We also find $\delta/2\pi\sim 115$ MHz and A~0.25 T/m.

Because of the red tuning, the decrease in the speed of an atom moving toward its turning point is larger than the increase moving away from it, back toward z=0, so that MOT also produces efficient cooling. The rebound speed of an atom from the inelastic collision with the soft magneto-optical wall is described by a non-linear differential equation, but it can be estimated as follows. For a moving atom, the position of resonance is displaced from $\pm z'$ by $-\hbar|k|v/\mu'A$, and for an accelerating atom it therefore moves at speed $-\hbar|k|a/\mu'A$. Approximating the radiative force on an atom by $\hbar k\gamma/2$ within a distance $\hbar\gamma/\mu'A$ of the resonance position, and zero outside, we readily find that the receding wall accelerates atoms away from their stopping point to a speed of about $2\gamma/k\sim 3.6$ m/s, corresponding to temperature $T_\gamma\equiv 4M\gamma^2/3k_Bk^2\sim 2$ mK. At this temperature, a purely magnetic trap[2,4] less than 20 gauss deep can confine He*.

The third stage consists of two parts. First, the sample is isothermally compressed by first sweeping the laser frequency close to $\delta=-\gamma$ and then raising the field gradient to A'~5, the maximum allowed by experimental constraints of power, cooling, etc. The MOT holds the sample at T_γ during this compression to a size of $\pm\hbar\gamma/\mu'A'\sim 22\ \mu m$. (The maximum allowable gradient for MOT occurs for $\mu'A_m\lambda_o=2\hbar\gamma$, which gives $A_m\sim 230$ T/m, not attainable by any practical magnet.) Then the laser beams are turned off and those atoms in the $M_J=+1$ state (1/3 of them) are held in the dark by purely magnetic trapping. In the absence of the strong radiative force of MOT, the sample expands to approximately $\pm 3k_BT_\gamma/2\mu A'\sim 450\ \mu m$. Second, with the lasers off, the atomic sample is adiabatically expanded by lowering the field gradient. We have shown that truly adiabatic expansion in this linear potential preserves the quantity $TV^{2/3}$ just as for an ideal gas. If the sample is expanded to its original size of 3.2 cm, the final temperature is lowered to $\sim T_\gamma/5000\sim 0.4\mu K$.

Because the $^3S_1$ to $^3P_{0,1,2}$ transition at 1.08 μm in He* is not J=0 to J=1, the simple MOT scheme described above is not rigorously correct. However, we have done a numerical simulation that includes optical pumping among the multiple magnetic sublevels, Doppler shifts, power broadening, etc., and it has shown that He* atoms still can be contained by MOT, and be rapidly cooled to T_γ. Furthermore, the process produces copious fluorescence that can be used as a non-destructive probe of the trapped He*.

The computer model has also shown that once the trap is loaded and the atoms cooled to T_γ (end of stage two), further cooling occurs when the laser frequency is swept toward the blue ($\delta\rightarrow-\gamma$). The minimum temperature that can be obtained in this optical molasses before the beginning of the third stage is $T_D\sim 40\ \mu K$, where the sample would be confined to the same 22 μm size as in MOT. Then the expansion phase of the subsequent third stage would then produce a temperature decrease by a factor of 3×10^6, resulting in an immeasurably small final temperature.

It is planned to produce the λ_o=1.083 μm laser light for MOT using solid state, single mode lasers (LNA crystals), optically pumped by efficient diode lasers, and to conduct the light into the cryostat by optical fibers. This results in a highly reliable, very efficient, all solid state system for this important helium resonance frequency. There are now published descriptions of at least two lasers[5] that would make MOT in He* feasible. Because the saturation intensity of the He* transition is about 1 mW/cm^2, the power required for the 6.5 cm size MOT is about 40 mW, easily achievable with these lasers.

He* atoms carry about 20 eV of internal energy and are therefore easily detected. Position sensitive multi-channel plate detectors could be used in conjunction with various timings to make ballistic measurements that would enable determination of the atomic motion. Nevertheless, there are at least two limitations to these speculative experiments at temperatures well below T_D. First, the large cross section (~10^{-14} cm^2) for Penning ionization collisions between He*'s suggests that they would not survive for very long if crowded into a small volume. (However, this might be a way to produce a single trapped atom.) Second, He* atoms distributed over a region of space with vertical extent are subject to a gravitational energy spread of mg/k_B, about 5 nk/μm, thereby producing a temperature-size limitation. Finally, it is useful to remark that the deBroglie wavelength of an atom with kinetic energy corresponding to T_D is 130nm, still very much smaller than the 20 μm size orbits of magnetically trapped atoms, so quantization would have a small effect on atomic motion. The adiabatic expansion will neither change the effect of quantization on atomic motion nor will it improve upon the conditions appropriate to observe Bose condensation.

The author wishes to acknowledge many helpful conversations with several temporary colleagues at Ecole Normale Superieure in Paris. This work was supported by NSF and ONR.

REFERENCES

1. E. Raab et al., Phys. Rev. Lett. 59, 2631 (1987).
2. A. Aspect, private communication.
3. W. Phillips and H. Metcalf, Phys. Rev. Lett. 48, 596 (1982).
4. T. Bergeman et al., Phys. Rev. A35, 1535 (1987).
5. L. Schearer et al., IEEE J. Quant. Elect. QE22, 713 (1986);
   L. Schearer et al., IEEE J. Quant. Elect. QE22, 756 (1986).

THEORY OF COLLISION-INDUCED OPTICAL TRAP LOSS

P. S. Julienne and S. H. Pan
National Bureau of Standards, Gaithersburg, MD 20899

H. R. Thorsheim and J. Weiner
Chemistry Department, University of Maryland, College Park, MD 20742

ABSTRACT

Quantal and semiclassical calculations have been carried out for the rate coefficient of the collision-induced emission process in which a ground and excited state sodium atom collide, emit a red-shifted photon, and separate with enough kinetic energy to escape the trap. The large rate coefficient implies that this process will limit the densities which can be achieved in optical traps.

THEORY

Both excited and ground state atoms will be present in optical atom traps[1] or optical "molasses".[2] When a ground and excited atom collide, a photon can be emitted in the far wings of the pressure broadened line,

$$Na(^2P_{3/2}) + Na + \varepsilon_i \rightarrow Na + Na + \varepsilon_f + h\upsilon \ . \qquad (1)$$

The photon is frequency shifted by $\Delta = \upsilon_0 - \upsilon$ from the atomic transition υ_0. This is a free-free molecular transition of the Na_2 molecule in which the kinetic energy ε_f of the separating atoms is related to the incident kinetic energy ε_i by energy conservation, $\varepsilon_f = \varepsilon_i + h\Delta$. Here red shifts are positive. If ε_f is large enough that $\varepsilon_f > 2\kappa T_D$, where T_D is the trap depth, the atoms (which share center of mass kinetic energy equally) can escape the trap if the mean free path is larger than the trap dimension.

We have estimated the rate coefficient for (1) using known Na_2 potentials and the following assumptions:
(1) Spontaneous emission can be treated as a perturbation on the collision.
(2) Effects of the trapping laser radiation field on the collision are ignored.
(3) The adiabatic molecular Ω potentials[3] found by diagonalizing the electronic + spin-orbit Hamiltonian give excited state potentials, which vary at long range as R^{-3}, for calculating the free-free transition matrix elements $<\varepsilon_i J|\mu_{if}(R)|\varepsilon_f J>$, where J is the rotational quantum number and μ_{if} is the transition dipole.
(4) Only the attractive 0_u^+, 0_g^-, 1_u, and 1_g states contribute.

The most severe assumptions are (1) and (2). Including the neglected effects will modify but not substantially change our conclusions.

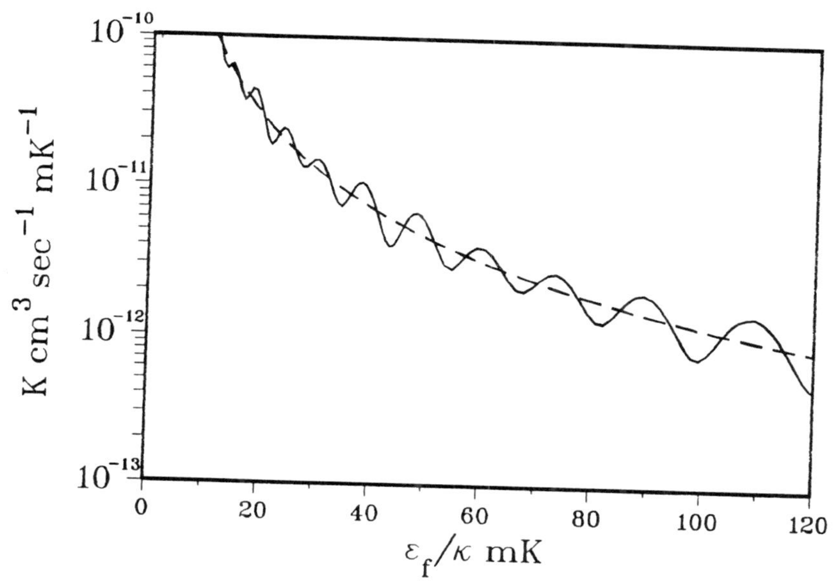

Figure 1. Calculated quantal(solid) and semiclassical(dashed) $K(\varepsilon_i, \varepsilon_f)$ versus ε_f/κ for $\varepsilon_i/\kappa = 1.4$ mK

Fig. 1 compares the calculated quantal and semiclassical spectral rate coefficient for (1) (i.e., the rate coefficient per unit interval of ε_f/κ) for the $0_u^+ \rightarrow 0_g^+$ transition in Na_2. The semiclassical spectrum is found by applying the stationary phase approximation to the free-free matrix element and the random phase approximation to the sum over contributing J values.[4] Here it is necessary to cut off the sum over J due to long range centrifugal barriers with maxima $\geq \varepsilon_i$ at distances well outside the point of stationary phase. This truncation causes the semiclassical rate to be much less than the usual semiclassical expression.

If T is the temperature of the trapped particles and κT_D is the trap depth, the total rate coefficient ($cm^3 sec^{-1}$) for a ground-excited state collision which results in trap escape is

$$K(T,T_D) = \int_{2\kappa T_D}^{\infty} < K(\varepsilon_i, \varepsilon_f) > d\varepsilon_f/\kappa \quad . \tag{2}$$

The brackets imply a thermal average of ε_i over T. The semiclassical expression for (2) summed over contributing transitions can be obtained analytically for $T_D \gg T$ in terms of the formula,

$$K(T,T_D) = 1.3\text{x}10^{-11}/\ (T^{1/6}T_D^{5/6})\ \ cm^3 sec^{-1}. \tag{3}$$

If we assume T = 1mK and T_D = 20mK, then we estimate K to be about 10^{-9} $cm^3 sec^{-1}$. The coefficient is large because the very long range of the excited $1/R^3$ potential allows about 25 J values to contribute even for such a low collision energy. We anticipate that effects due to assumptions (1) and (2) will reduce the rate coefficient. The amount of reduction will depend on the actual trap parameters. The greatest reduction will occur for conditions of small detuning and Rabi frequency, comparable to the natural linewidth. This is because the colliding atoms decouple from the radiation field at very long range due to the detuning from the laser frequency caused by the long range $1/R^3$ molecular interaction energy. The rate coefficient must be multiplied by a survival factor which accounts for the probability of spontaneous emission decay of the excited atom as the two atoms approach each other between the decoupling distance and the point of stationary phase. We estimate a reduction in the rate coefficient of between one and two orders of magnitude due to this effect. A much less severe reduction is expected for traps using strong laser fields where the Rabi frequency is larger than detuning, and both are much larger than $\kappa T/h$.

Even after a reduction in magnitude from Eq. (3), the predicted rate coefficient remains large and will lead to upper limits on the density of atoms possible in optical traps under conditions in which enough excited state atoms are present. If the fraction of excited atoms is small compared to the ground state density, the lifetime of the trap relative to this process is $1/(KN_e)$, where N_e is excited state density. If we assume $K = 10^{-10}$ $cm^3 sec^{-1}$ and $N_e = 10^{10}$ cm^{-3}, the trap would decay in 1 sec. Although the optical trap reported by Chu, et al.,[1] has high total density, $> 10^{11} cm^{-3}$, the excited state fraction was very small because detuning was very large relative to the Rabi frequency. The excited state fraction was high during the molasses cycle of this trap, but the smaller rate coefficient and the additional cooling for molasses conditions inhibit the trap loss process during the molasses cycle. Although the collision-induced loss process (3) was not dominant for Chu's trap, it will be an important loss process for optical traps with sufficiently high excited state density. The process is a generic one for any pair of like atoms and is expected to have a large rate coefficient, typically in the range 10^{-12} to 10^{-10} $cm^3 sec^{-1}$.

REFERENCES

1. S. Chu, J. E. Bjorkholm, A. Ashkin, and A. Cable, Phys. Rev. Lett. 57, 314(1986).
2. S. Chu, L. Holberg, J. E. Bjorkholm, A. Cable, and A. Ashkin, Phys. Rev. Lett. 55, 48(1985).
3. M. Movre and G. Pichler, J. Phys. B 10, 1(1977).
4. P. S. Herman and K. M. Sando, J. Chem. Phys. 68, 1153(1978).

WAVELENGTH AND ISOTOPE SHIFTS IN THE Mg I RESONANCE LINE

N. Beverini, E.Maccioni, D.Pereira*, F.Strumia, G.Vissani,Wang Yu-Zhi*[1]
Dipartimento di Fisica dell' Università - piazza Torricelli , 2 - Pisa Italy
and INFN Sezione di Pisa

ABSTRACT

The isotopic shifts and the wavelength of the Magnesium resonance line $^1S_0 - {}^1P_1$ at 285.2 nm has been measured in an atomic beam.

INTRODUCTION

The spectroscopic properties of the Mg atom are of interest for astrophysics, atomic frequency standards and laser cooling[1]. Accurate measurements of isotopic shifts, Landè factors, and fine structure transition frequency are available for the triplet metastable states. On the contrary a few measurements are available for the singlet resonance transition $^1S_0 - {}^1P_1$ at 285.2 nm. The isotopic shift was mesured by comparison of the line centers from isotopically enriched hollow cathode lamps, cooled at the liquid nitrogen temperature[2]. The hyperfine structure of the ^{25}Mg was measured by a level crossing experiment [3] and found to be smaller than the natural line width (79 MHz). In the present experiment sub-Doppler resolution has been obtained by irradiating orthogonally a Mg atomic beam with a frequency doubled CW tunable dye laser.

EXPERIMENTAL APPARATUS AND RESULTS

The experimental apparatus is shown in fig.1. A thermal beam of natural Mg is produced from a thermostatized oven and collimated to about 3×10^{-3} in order to obtain a residual Doppler width of 26 MHz , smaller than the natural linewidth The UV radiation is generated by a Coherent 699/21 ring dye laser with an intracavity frequency doubling crystal. The CW output power is of the order of a few mW, and the frequency is actively stabilized with a residual jitter lower than 2 MHz at UV frequency. The absolute wavelength is calibrated by observing the absorption from a Iodine cell of the fundamental laser radiation at 570.4 nm. The frequency scale around the Mg absorption line is calibrated by means of a confocal Fabri-Perot with 592(1) MHz of free spectral range.

---

* Fellows of the International. Center for Theoretical Phys. - Trieste, Italy

The laser radiation excite the atoms orthogonally to the beam direction and the fluorescence light is observed with a photomultiplier as a function of the

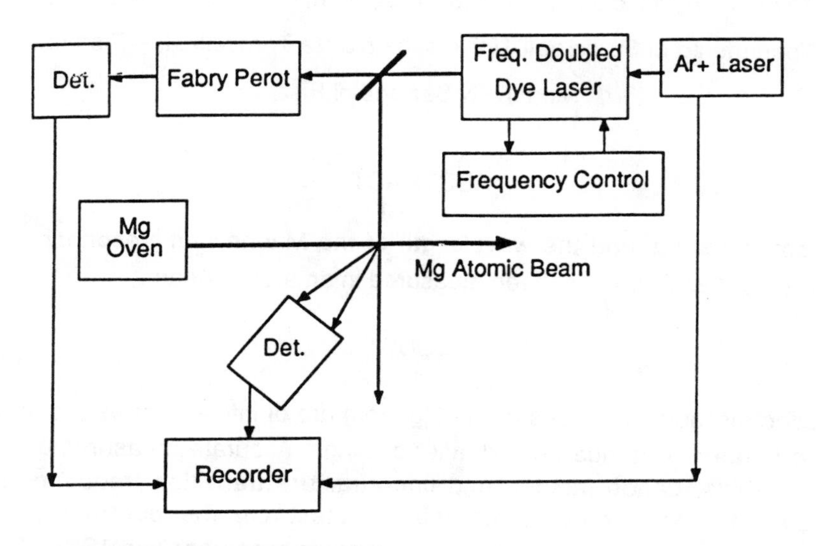

Fig. 1 - Experimental apparatus

laser frequency tuning as shown in fig. 2. The Mg line was found to be close to the absorption of a strong I_2 line in the fundamental at 570 nm. This reference line is reported two time in the I_2 atlas of ref 4 (line Nº 3140 in part II, and line Nº 115 in part III, with an average wavenumber of 17525.6522(20) cm^{-1}). Natural Mg consist of a mixture of three stable isotopes : ^{24}Mg (78.70%), ^{25}Mg (10.13 %), and ^{26}Mg (11.17%). The even isotopes have no nuclear spin and no hyperfine structure. With reference to the I_2 line the wavenumber of the ^{24}Mg was found to be

$$^{24}\mathrm{Mg}\ (^1S_0 - {}^1P_1) = 35051.273 \pm 0.004 \pm 0.005\ \mathrm{cm}^{-1}$$

This value is smaller than that reported in the NBS Tables (35051.36 cm^{-1}). The isotope shifts were measured with reference to the peak of the fluorescence signal and the laser frequency was calibrated against the reference interferometer. We discovered that the laser tuning was not linear. A fit with a small quadratic correction was necessary to obtain agreement with the interferometer transmission peaks The frequency scale was corrected in accordance and the best values for the isotopic shifts are given in the table. The agreement with the values of ref.2 is within the experimental errors. Future improvements are expected.

Table I - Isotopic shift (in MHz) of the Mg resonance line at 285

Isotopes	25 – 24	26 – 24
This work	743.8 ± 6	1415.3 ± 9
Hallstadius	728.5 ±12	1412.0 ± 21

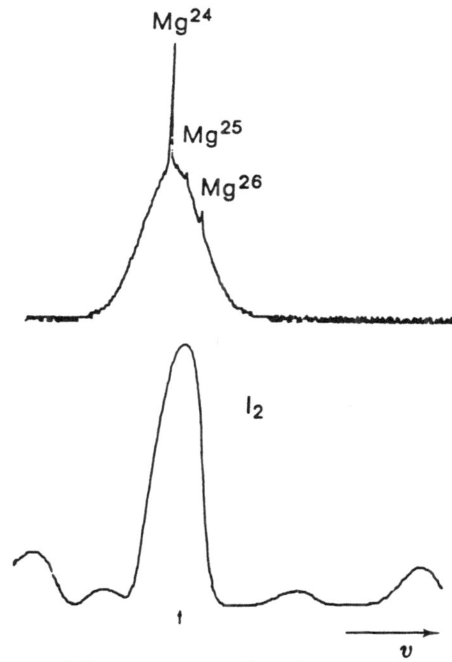

Fig. 2 - Observed fluorescence signal

REFERENCES

1. F.Strumia : " Application of laser cooling to the atomic freque standards", in Laser science and technology (A.N.Chester, S.Martellucci eds. , Plenum Press 1988)
2. L.Hallstadius : Z.Physik, A291, 203, (1979)
3. H.Jurgen Kluge, H.Sauter : Z.Physik, 270, 295, (1979)
4. S.Gerstenkorn, P.Luc : Atlas du spectre d'absorption de la mole d'iode, Editions du CNRS, Paris 1978 ; Rev. Phys. Applique, 14. (1979)

A NEW TECHNIQUE TO STUDY RYDBERG STATES BY MULTIPHOTON IONIZATION SPECTROSCOPY

R.D. Verma and Alak Chanda
Department of Physics, U.N.B., Fredericton, N.B., Canada

ABSTRACT

A new technique to study the Rydberg states of the Ba atom has been developed. In this technique a Multiphoton Ionization signal is detected by selective excitation of the ground state ion (6s) to an excited state (6p), which results in a collimated Amplified Spontaneous Emission (ASE) signal at the 6p→5d transition of Ba^+. Discrete Rydberg states, 6snℓ (ℓ=0,2), as well as autoionizing Rydberg states, 5dnℓ (ℓ=0,2) and 6pnℓ (ℓ=0,2) are observed by this novel but very simple method.

INTRODUCTION

Traditionally, single photon spectroscopy[1] in the vacuum region has been used extensively to study the discrete as well as the autoionizing Rydberg states. These studies were, of course, limited in their scope to the extent that from a given ground state only certain angular momentum and parity states could be excited. In recent years, laser multiphoton spectroscopy has overcome these difficulties. Currently, there are the following important modes of Multiphoton laser spectroscopy in use to study Rydberg states: (i) Multi-step photo-ionization spectroscopy[2], (ii) Multiphoton-ionization mass-spectroscopy[3], (iii) Multiphoton-ionization photo-electron spectroscopy[4], and (iv) Optogalvanic spectroscopy[5].

Recently, Bokor et al.[6] showed that when two-step photo-excitation was applied to excite the Ba atom from its ground state 6s to $6p_{3/2}12p$ autoionizing states, an Amplified Spontaneous Emission (ASE) signals were observed at 493 nm and 650 nm. These are the transitions from the $6p_{1/2}$ state to the 6s and $5d_{3/2}$ states, respectively, of Ba^+. They result because the autoionizing state $6p_{3/2}n\ell$ decays ($t=10^{-12}$s) selectively to $6p_{1/2} + e^-$. The energy difference is taken up in the kinetic energy of the ejected electron, and a population inversion is created between $6p_{1/2}$ and the low lying $5d_{3/2}$ and 6s states of the ion.

The above work of Bokor et al. provided an incentive to develop a new technique to detect the ionizing signal by selectively exciting the ion state and thereby producing a collimated ASE beam along the direction of the exciting photon.

In this paper, we describe this novel but simple and very sensitive method to study discrete as well as autoionizing Rydberg states. Basically, the method differs from the other methods mentioned earlier only in the detection of the ionization signal. Here, multiphoton ionization is observed through the ASE signal produced by selective excitation of the ground state ion via optical pumping. The technique is illustrated in detail using the Ba atom as a test case and the present results are compared with known results for Ba.

EXPERIMENTAL PROCEDURE

The experimental set up is shown in Fig. 1. Two dye lasers were pumped by an excimer laser (Lumonics). Both dye lasers delivered energy of up to 3-6 mJ per pulse with line width 0.003-0.004 nm and pulse duration of ≃ 10 ns. Two dye laser pulses were brought from opposite direction into an oven containing Ba metal in Ar (p=15 Torr) buffer and were loosely focussed by 1m and 2m focal length lenses. Pump, probe and exciting lasers, as they are called in Fig. 2. The probe laser was delayed by a few nanoseconds with respect to the pump laser. Because the probe and exciting photons belonged to the same single laser pulse, they had the same frequency. This way only two lasers were required. The probe laser intensity was cut down considerably by neutral density filters, and in later experiments this laser was used without focussing. It was important that the beams overlapped at the focal spot of the pump laser.

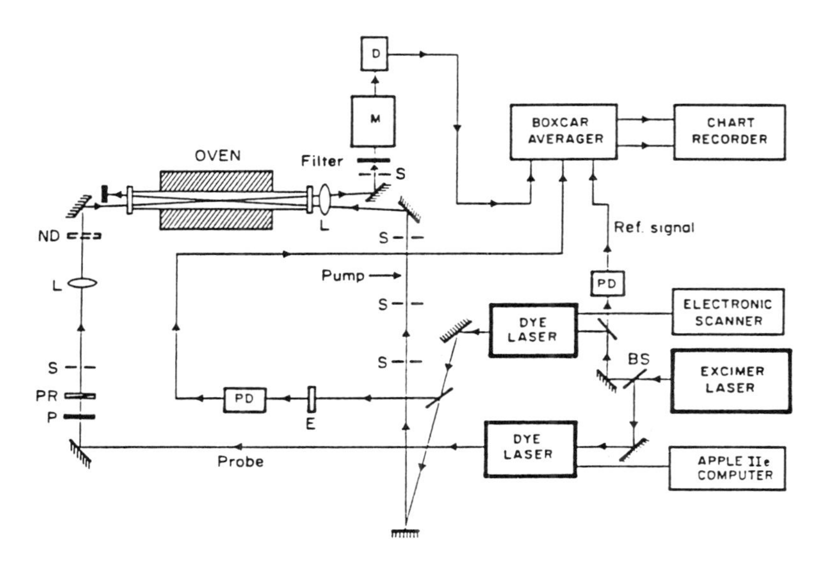

Fig. 1. The experimental set up.

Barium vapour was produced in a steel tube, which was connected to a glass tube at both ends via a water cooled jacket. The barium was heated by a commercial furnace (Lindberg) to a temperature of 850°C - 900°C. The heated zone was 120 cm long and 50 mm in diameter.

The ASE signal at 614 nm or 650 nm was isolated from the exciting photon by a filter and detected by a monochromator-photomultiplier system. The signal was processed by a boxcar averager at the pulse rate of 10 Hz. An optogalvanic spectrum from a commercial uranium H.C. lamp provided the calibration standards.

RESULTS AND DISCUSSION

The underlying principle of this technique is best explained by referring to Fig. 2. A two photon resonance of the pump laser

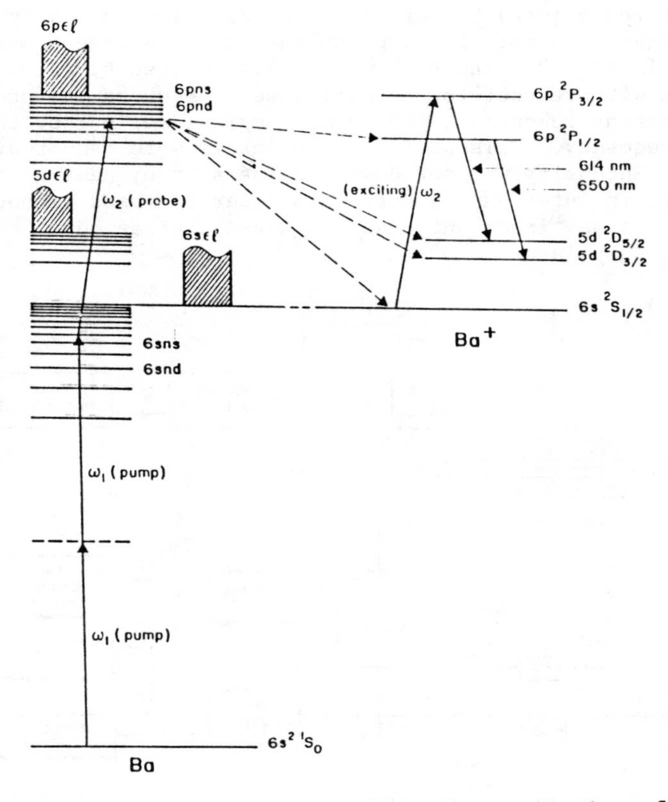

Fig. 2. The process involved in the excitation of the spectrum of the discrete Rydberg states.

populates the discrete Rydberg states, $6sn\ell$ (ℓ=0,2). A few nanoseconds later, the probe laser set at 455 nm corresponding to the transition $6s \rightarrow 6p_{3/2}$ of Ba^+ excites the Ba atom from $6sn\ell$ to $6p_{3/2}n'\ell$ autoionizing states. As discussed earlier by Tran et al.[2] and Mullins et al.[7] in this transition the Rydberg electron $n\ell$ simply behaves like a spectator, so that $n \simeq n'$ and there is only a slight adjustment in the quantum defect. Thus the $6sn\ell \rightarrow 6p_{3/2}n'\ell$ transition has essentially the same frequency and strength as the $6s \rightarrow 6p_{3/2}$ transition of Ba^+. The autoionizing effect is likely to cause the following decay processes: $6p_{3/2}n'\ell \rightarrow 6p_{1/2}, 5d_{3/2}, {}_{5/2}, 6s + e^-(\Delta E)$. The kinetic energy, ΔE, of the electron corresponds to the energy difference between the Ba atom in the $6p_{3/2}n'\ell$ state and the final ionic state.

The fact that the $6sn\ell \rightarrow 6pn\ell$ transition frequency of the Ba atom is nearly the same as that of the $6s \rightarrow 6p$ transition Ba^+ allowed us to use the 455 nm probe for excitation of the ion as well. Since the ionization process is very fast ($t \simeq 10^{-12}$ s) and the exciting laser pulse length is long ($t \simeq 10^{-8}$ s), the same pulse which acts as a probe for the atom can pump the ground state ion, 6s, up to an empty $6p_{3/2}$ state. This results in a population inversion between the $6p_{3/2}$ and $5d_{5/2}$ states of Ba^+, causing an intense collimated ASE radiation at 614 nm along the path of the exciting pulse. The ASE signal was isolated and processed as described in the preceding Section.

Probe and pump intensity were reduced until we obtained a Gaussian line shape for our signal. Also, it was checked that the signal corresponded to both the pump and the probe excitation together and not to a multiple excitation of one laser. Both lasers were linearly polarized.

When the pump laser was then scanned with the probe laser kept fixed at 455 nm, two-photon resonances at various $6sn\ell$ (ℓ=0,2) states yielded enhancements in the ASE signal providing the spectrum of the discrete Rydberg states, 6sns and 6snd, as shown in Fig. 3. The present spectrum compares very well with the ionization-current spectrum[7]. However, the signal detection in the present case is very simple and cheaper.

When the two photon frequency of the pump laser exceeded the first ionization limit, the spectrum of the $5dn\ell$ (ℓ=0,2,4) autoionizing states was obtained. Here two photon excitation drives the ground state atom to the autoionizing levels $5dn\ell$, which decay to the 6s ground state of the Ba ion. The exciting laser fixed at 455 nm then pumps the 6s ion to the $6p_{3/2}$ state causing an ASE signal at 614 nm.

Identical spectra were obtained when the above experiments were repeated but with the probe and exciting laser now set at the 493 nm line corresponding to the $6s \rightarrow 6p_{1/2}$ transition of Ba^+, and the signal being detected at 650 nm, the $6p_{1/2} \rightarrow 5d_{3/2}$ transition line.

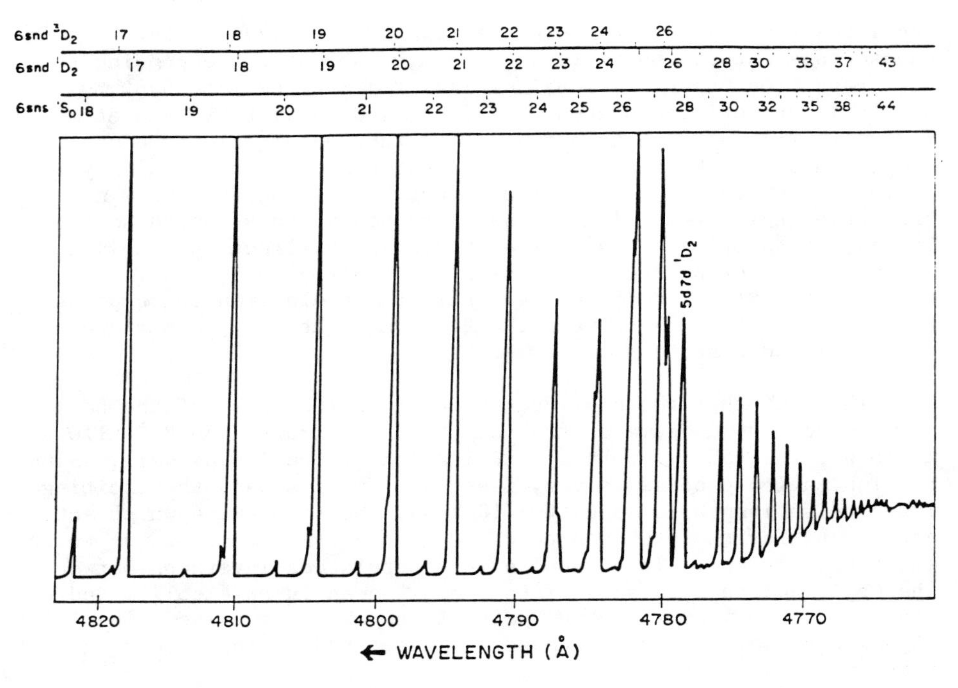

Fig. 3. The spectrum of 6snℓ series of Ba recorded by the new technique.

REFERENCES

1. See, for example, W.R.S. Garton and K. Codling, Proc. Phys. Soc. London 75, 87, 1960; or for recent work on Ba, C.M. Brown and M.L. Ginter, J. Opt. Soc. Am. 68, 817, 1978.
2. Tran, N.H., Pillet, P. Kachru, R. and Gallagher, T.F., Phys. Rev.A, 29, 2640, 1984.
3. Compton, R.N. and Miller, John C. AIP Conference Proc. No. 90. P.319, 1982.
4. Gallagher, T.F., AIP Conference Proc. No. 90, P.358, 1982 and also see Ref. 3.
5. Camus, P., Dieulin, M. and El Himdy, A., Phys. Rev.A, 26, 379, 1982.
6. Bokor, J., Freeman, R.R. and Cooke, W.E., Phys. Rev.A, 26, 1242, 1982.
7. Mullins, O.C., Yifu Zhu and Gallagher, T.F., Phys. Rev.A, 32, 243, 1985.

HIGH PRECISION CW LASER MEASUREMENT OF THE 1S-2S INTERVAL IN ATOMIC HYDROGEN AND DEUTERIUM

E.A. Hinds
Physics Department, Yale University, New Haven, Ct. 06520, U.S.A.

M.G. Boshier, P.E.G. Baird, C.J. Foot, M.D. Plimmer, D.N. Stacey, D.A. Tate and G.K. Woodgate
The Clarendon Laboratory, University of Oxford, Parks Road, Oxford OX1 3PU, United Kingdom

J.B. Swan
Department of Physics, University of Western Australia, Nedlands, Western Australia 6009, Australia

D.M. Warrington
Physics Department, University of Otago, P.O. Box 56, Dunedin, New Zealand

ABSTRACT

We have excited the 1S-2S transition in atomic hydrogen and deuterium by two-photon absorption of cw 243 nm light. The transition frequency has been measured by comparison with calibrated lines in the spectrum of the $^{130}Te_2$ molecule, providing new precise values for the 1S Lamb shifts or the Rydberg constant.

INTRODUCTION

The 1S-2S transition of atomic hydrogen has attracted considerable attention because of its extremely narrow natural width (1 Hz). High resolution Doppler-free spectra of this transition, obtainable by two-photon absorption of 243 nm light, offer the prospect of precise tests of bound-state quantum electrodynamics (QED) and more accurate values for fundamental constants [1]. Early studies of this transition took advantage of the high peak powers available from pulsed laser sources [2,3], but the resolution of these measurements was limited by instrumental effects associated with their use. A large increase in precision has now been obtained following the development of continuous-wave (cw) 243 nm sources [4,5]. We report here recent results from our cw experiment [5].

DESCRIPTION OF THE EXPERIMENT

The major difficulty in exciting the 1S-2S transition by two-photon absorption is the generation of sufficient cw 243 nm radiation (~1mW) in a nonlinear crystal. Our experiment uses frequency-doubling to generate the 243 nm light because it provides a direct link to a visible wavelength (486 nm), permitting the use of accurate heterodyne calibration techniques and also facilitating a future direct comparison with the hydrogen Balmer-β transition [2]. However, until very recently, frequency-doubling at 486 nm has been difficult because of the lack of a suitable nonlinear crystal. We have developed an intra-cavity frequency-doubling system using lithium formate monohydrate (LFM) [6], but this was not adequate for the hydrogen experiment because of damage to the crystal rapidly induced by the UV. Better results were obtained with urea, especially after it became possible to polish the crystal surfaces well enough to dispense with an index-matching fluid. Urea also suffers from UV-induced damage, but we were nevertheless able to make the initial observation and preliminary measurements of the 1S-2S transition using it. We have now obtained a crystal of β-barium borate (BBO) which is superior to urea in almost all respects: BBO is hard, has good optical quality and can be polished well. Further, it provides good conversion efficiency and, most important, it does not suffer from any UV-induced damage.

The primary quantities measured in our experiment are the frequencies of the 1S-2S transitions in hydrogen and deuterium relative to nearby accurately-calibrated lines in the $^{130}Te_2$ spectrum [7]. A simplified schematic diagram of the apparatus used for these measurements is shown in Fig. 1.

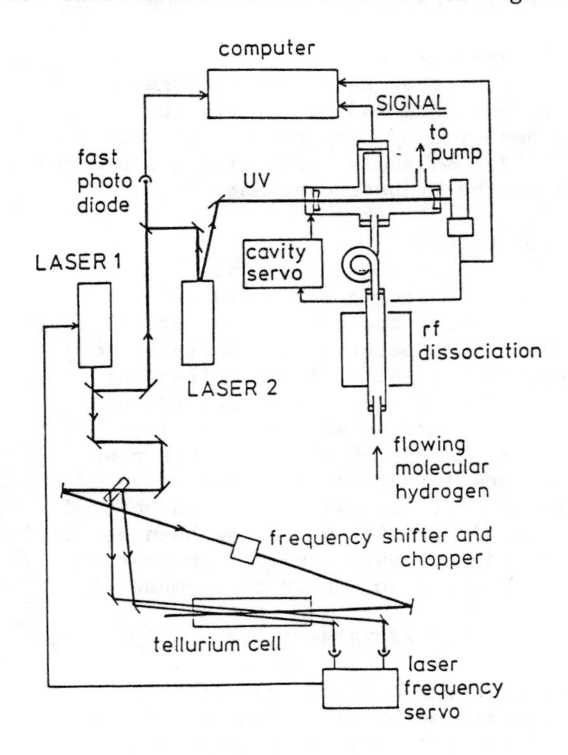

Fig. 1: Schematic diagram of the apparatus. Both lasers are krypton-ion pumped C102 dye lasers.

The 243 nm light from the frequency-doubled ring dye laser (typically 2 mW) is enhanced by about a factor of 12 in a standing-wave enhancement cavity, which is windowless to avoid loss and scattering. The "walk-off" distortion of the uv beam profile can be fairly well approximated by an elliptical gaussian beam, and hence compensated for with an appropriate cylindrical lens. In this way more than 65% of the total uv power can be coupled into the fundamental TEM_{00} mode of the cavity. The two-photon excitation is detected by monitoring collisionally-induced Lyman-α fluorescence. A second dye laser is locked to the appropriate tellurium transition, detected by saturated absorption. Radiation from each dye laser at 486 nm is mixed on a fast photodiode and the resulting beat (~1400 MHz for hydrogen, ~4240 MHz for deuterium) is measured to provide frequency calibration. The measurements reported here were made with respect to the original tellurium cell calibrated by the NPL [7]. We used this cell after discovering a ~1 MHz discrepancy between it and our own (nominally identical) cell. The reason for the difference has not yet been found.

RESULTS AND DISCUSSION

A typical two-photon signal is shown in Fig. 2. It was recorded with an intra-cavity 243 nm power of ~20 mW, focussed to a waist size of 100 μm in the interaction region. The cell contained 230 mTorr of a $5\%H_2$ / $5\%D_2$ / 90%He mixture. The linewidth at this pressure was 2.5 MHz (FWHM at 486 nm), and at lower pressures this decreased to 1.6 MHz. The dominant source of broadening is the finite transit time of the atoms through the laser beam.

In order to measure the pressure shift in the hydrogen cell, we have made several measurements and extrapolated to zero pressure. Measurements have been made with pure hydrogen, pure deuterium and the helium mixture mentioned above which has been found to give a smaller pressure shift [4].

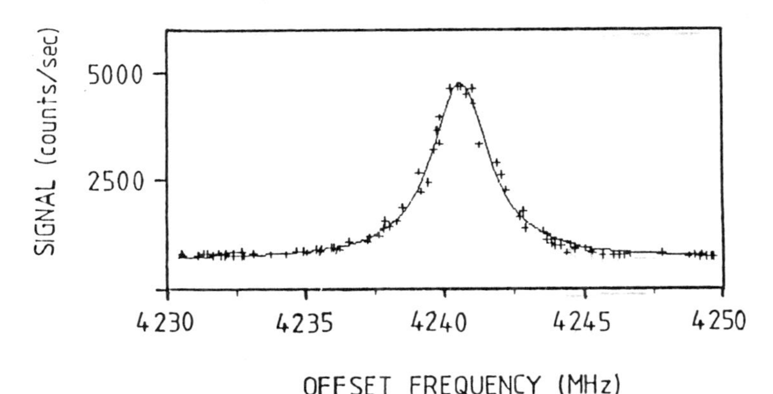

Fig. 2: F=3/2 to F=3/2 component of the 1S-2S transition in deuterium. The offset frequency is measured relative to the b_1 line of tellurium. The solid line is a least squares fit of a Lorentzian profile to the data (crosses).

The final values for the frequency of the 1S-2S transition centroid are 2466061414.08(75) MHz for H and 2466732408.45(69) MHz for D. We then use the recent value of the Rydberg constant [8] and the known 2S-2P Lamb shift to extract a value for the 1S Lamb shift. We find the values 8172.98(85) MHz for hydrogen and 8184.11(80) MHz for deuterium, in excellent agreement with theory [9]. Our definition of the Lamb shift is the same as that of JOHNSON and SOFF [9], i.e. the sum of all QED contributions plus the finite nuclear size correction. Our results can also be used to provide a value for the Rydberg constant by using the theoretical value of the 1S Lamb shift. In this way we obtain R_∞ = 109737.315733(34) cm^{-1} from hydrogen and 109737.315724(31) cm^{-1} from deuterium. The mean value of R_∞ = 109737.31573(3) cm^{-1} compares well with other recent determinations [8 and references therein] and is as precise as the best. Finally, we obtain an H-D 1S-2S isotope shift of 670 994.4(9) MHz, also agreeing well with the theoretical value.

The precision of our determination of the 1S Lamb shift is limited by the accuracies of the tellurium standard and the Rydberg constant. These sources of uncertainty will be removed in the next stage of this work, which will involve the direct comparison of the 1S-2S and 2S-4S transitions, excited in atomic beams. The natural width of the 2S-4S transition is less than 1 MHz and we expect to make the comparison to about 25 kHz (at 486 nm). This will provide a value of the 1S Lamb shfit with a precision than approaches that of rf 2S Lamb shift measurements.

ACKNOWLEDGEMENTS

It is a pleasure to acknowledge the contribution to this work of Mr. G. Read, Mr. C.W. Goodwin and Dr. S.H. Smith. We are grateful to Dr. A.I. Ferguson for lending us the calibrated tellurium cell. One of us (M.G.B.) acknowledges the support of the Royal Commission for the Exhibition of 1851 and Christ Church, Oxford. This work is supported by the Science and Engineering Research Council which also awarded a Visiting Fellowship to E.A.H., a Senior Fellowship to D.N.S. and studentships to M.D.P. and D.A.T. C.J.F. holds a Royal Society Research Fellowship.

REFERENCES

1. E.V. Baklanov and V.P. Chebotaev, Opt. Spectrosc. 38, 215 (1975).
2. C. Wieman and T.W. Hänsch, Phys. Rev. A22, 192 (1980).
3. J.R.M. Barr, et al., Phys. Rev. Lett. 56, 580 (1986).
4. R.G. Beausoleil, et al., Phys. Rev. A35, 4878 (1987).
5. M.G. Boshier, et al., Nature 330, 463 (1987).
6. C.J. Foot, et al., Opt. Comm. 50, 199 (1984).
7. J.R.M. Barr, et al., Opt. Comm. 54, 217 (1985).
8. P. Zhao, et al., Phys. Rev. Lett. 58, 1293 (1987).
9. W.R. Johnson and G. Soff, At. Data Nucl.Data Tables 33, 405 (1985).

THE AUTLER-TOWNES EFFECT IN HYDROGEN AT LOW PRESSURE

Albert M. F. Lau
Combustion Research Facility, Sandia National Laboratories, Livermore, CA 94550

ABSTRACT

The Autler-Townes effect (also known as ac, optical or dynamical Stark splitting) on the Balmer-α transition of hydrogen in 0.5 - 5 torr of helium is investigated. In the double optical resonance scheme, H(3p) is excited by absorption of two 243-nm photons and a 656-nm photon via the 2s resonant intermediate level. Calculated results show that in a Doppler-free or a "Doppler-reduced" beam configuration, the required intensity of the 656 nm laser (0.2 cm^{-1} bandwidth) for observing the Autler-Townes doublet has to be greater than 2×10^4 W/cm^2 and 2×10^5 W/cm^2 for a 243-nm laser bandwidth of 0.1 cm^{-1} and 0.5 cm^{-1}, respectively.

Recently the Autler-Townes effect has been observed[1] in the Balmer-α line of hydrogen produced in an air/H_2 diffusion flame at atmospheric pressure and at $\approx$ 2500K with a laser intensity of 4×10^6 W/cm^2. Calculations[2] showed that the spectra are mainly broadened by collisions and the Doppler effect due to the high pressure and high temperature. Owing to the lack of measured or calculated collisional quenching and dephasing cross sections of H excited states by N_2, O_2 and H_2O present in the flame, the theoretical fit was done using estimated values of these quantities.

It appears that the Autler-Townes effect, whether as doublets in absorption/excitation spectra or as triplets in fluorescence specta,[3] has not been observed in hydrogen at low pressure or in a beam. This paper reports our calculations[4] of the effect in the Balmer-α transition of hydrogen at low helium pressure. For this system, most of the needed collisional cross sections have been measured,[5,6] so that the input parameters in the calculations are expected to be quite reliable.

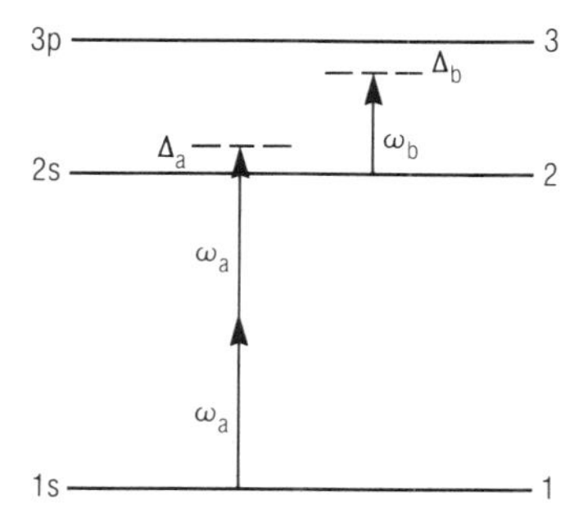

Fig. 1 Schematic energy diagram of hydrogen showing the three-photon excitation of the 3p level by UV ($\lambda_a = 243$ nm) and visible ($\lambda_b = 656$ nm) laser radiation.

The multiphoton excitation scheme of H in our calculation is shown in Fig.1, where $\hbar\omega_i$ (i = 1, 2 and 3) are the energies of the three principal levels 1s , 2s, and 3p respectively; and ω_a and ω_b are the frequencies of the UV (λ_a = 243 nm) and the visible (λ_b = 656 nm) lasers respectively. The UV laser detuning Δ_a is defined as $2\omega_a - (\omega_2 - \omega_1)$, and the visible laser detuning Δ_b is defined as $\omega_b - (\omega_3 - \omega_2)$. The parameters of the lasers (such as frequencies, intensities and bandwidths) used in the calculations are achievable with the lasers in earlier experimments.[1] In the model for hydrogen, we neglect the fine structure in the 3p level since the laser (FWHM) bandwidths γ_a(0.1 - 0.5 cm^{-1}) and γ_b (0.2 cm^{-1}) are broader than the fine-structure separation. The hydrogen is subject to 0.5 - 5 torr of He at 310 K, like the system described in Weber.[5] At 310 K, the Doppler width in the 1s - 2s transition is 1 cm^{-1} and that in the 2s - 3p transition is 0.2 cm^{-1}. The spectral linewidths are predominantly due to the laser bandwidths and Doppler widths.

One way of detecting the Autler-Townes effect is to monitor the Balmer-α or the Lyman-β (102 nm) fluorescence as a function of the UV frequency detuning at fixed visible laser intensity and detuning. Therefore we calculate the population density in the 3p level. Consider the situation of two counter-propagating UV beams with the visible beam propagating parallel to them. This eliminates the Doppler broadening ($\approx$ 1 cm^{-1}) in the absorption of the two UV photons while the Doppler broadening ($\approx$ 0.2 cm^{-1}) in the visible transition remains. This "Doppler-reduced" beam geometry greatly improves the resolution of the Autler-Townes doublet.[4] To calculate the actual excitation spectrum (represented by σ_{33}) as a function of Δ_a, we use Eqs.(2) - (7) in Ref. 2 in which the laser bandwidths are included phenomenologically (rather than using the phase diffusion model, as stated erroneously in Ref. 2). From these relations, we obtain analytic expressions for calculating the threshold value I_b' of the visible laser intensity I_b required for resolving the doublet.[4,7] The results are (a) $I_b' > 2 \times 10^4$ W/cm^2 for γ_a = 0.1 cm^{-1}; and (b) $I_b' > 2 \times 10^5$ W/cm^2 for γ_a = 0.5 cm^{-1}. Figures 2 and 3 show the Doppler-averaged excitation spectra at various I_b values for γ_a = 0.1 cm^{-1} and 0.5 cm^{-1} respectively, confirming the above threshold intensities. It may also be pointed out that these "Doppler-reduced" spectra are indistinguishable from the Doppler-free spectra, indicating that the residual Doppler contribution of the 2s - 3p transition is insignificant.

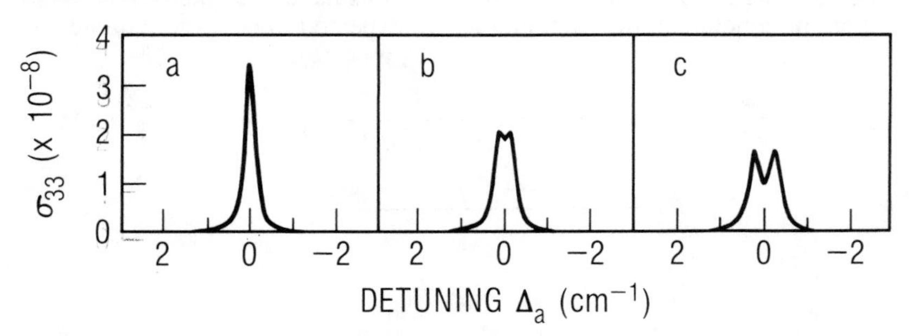

Fig. 2 The population density in the 3p level vs the UV laser detuning Δ_a for the visible laser intensity I_b equal to (a) 10^4 W/cm^2; (b) 4 x 10^4 W/cm^2 and (c) 10^5 W/cm^2. The visible laser detuning Δ_b is fixed at 0. The UV laser bandwidth is 0.1 cm^{-1}. The two UV laser beams are assumed counter-propagating, with the visible beam propagating parallel to them. Helium pressure is 1 torr.

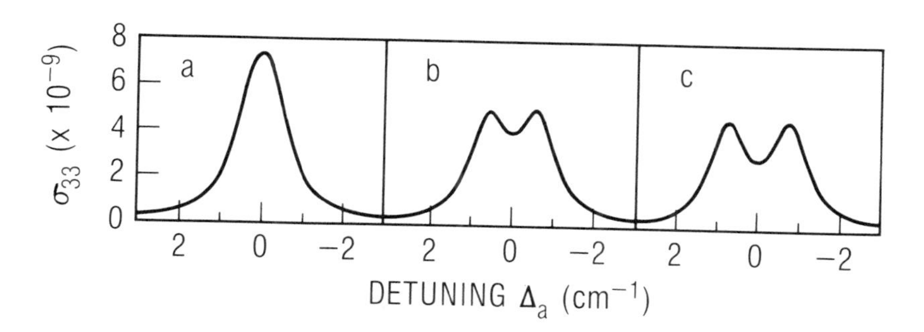

Fig. 3 Same as Fig.2 except that the UV laser bandwidth is 0.5 cm^{-1} and the visible laser intensity is (a) 10^5 W/cm^2 ; (b) 5 x 10^5 W/cm^2 ; and (c) 10^6 W/cm^2.

Similar calculations show that for helium pressure in the range 0.5 - 5 torr, the numerical results do not differ significantly from the above results for 1 torr.

In conclusion, our calculations show that with the Doppler-free and the above-described "Doppler-reduced" beam configurations, the Autler-Townes splitting can be observed in the 3p-2s transition of hydrogen in 0.5 - 5 torr of helium at 656-nm laser (bandwidth = 0.2 cm^{-1}) intensities greater than (a) 2 x 10^4 W/cm^2, and (b) 2 x 10^5 W/cm^2 for the 243-nm laser bandwidth of 0.1 cm^{-1} and 0.5 cm^{-1} respectively. These threshold values are one to two orders of magnitude lower than that (4 MW/cm^2) required in the high-temperature and high-pressure environment of flames.

This work was supported by the U.S. Department of Energy, Office of Basic Energy Sciences, Division of Chemical Sciences.

REFERENCES

1. J. E. M. Goldsmith, Opt. Lett. 10, 116 (1985).
2. A. M. F. Lau, Phys. Rev. A 33, 3602 (1986).
3. P. L. Knight and P. W. Milonni, Phys. Rept. 66 21 (1980) for a review of related processes and references.
4. A. M. F. Lau, J. Phys. B 20, L369 (1987).
5. E. W. Weber, Phys. Rev. A 20, 2278 (1979).
6. D. Bloch, R. K. Raj and M. Ducloy, Opt. Commun. 37, 183 (1981).
7. A. M. F. Lau, Opt. Commun. 64,144 (1987).

ENHANCEMENT OF HIGH–ORDER MANY–PHOTON ABSORPTION PROCESSES BY BROADBAND NOISE

J. E. Bayfield and D. W. Sokol
University of Pittsburgh, Pittsburgh, Pennsylvania 15260

ABSTRACT

Highly excited hydrogen atoms have been exposed to combined strong sinewave and noise microwave electric fields. When the rms noise field strength exceeds 30% of the sinewave strength, the addition of the noise can increase state-changing and ionization probabilities by amounts larger than that for a same increment in the sinewave field.

INTRODUCTION

At high field strengths, a pulse of single-frequency microwaves can induce high-order photon absorption in highly excited hydrogen atoms. The absorption is characterized by fairly smooth final bound-state quantum number distributions and nonzero many-photon ionization probabilities (1). Such photon absorption can occur when the microwave coupling of adjacent atom energy levels is comparable to their spacing, for many levels. The time evolution of the system then can involve a strong coupling of many free-atom states. An experimental signature of this regime is observable adjacent-level n-changing probabilities at sinewave frequencies away from two-state multiphoton resonances (2). This permits a transient semiclassical quantum diffusion of microwave energy into the atom (3). As the diffusion occurs during the strong microwave pulse, it is not inhibited by the energy-conservation requirements of individual two-state quantum multiphoton transitions. In the diffusion regime, the physics of the time evolution of the sinewave-driven bound electron is a quantized deterministic nonlinear particle dynamics problem (1,3).

A similar global state-changing regime would be possible for simultaneous application of sinewave microwaves at all the adjacent-level resonance frequencies. Another way to achieve this would be to apply microwave noise of a sufficiently large bandwidth. This would add randomness to the system via the incoherence of the noise. In macroscopic nonlinear systems, external noise can induce shifts in deterministic critical points, and can induce the appearance of new critical points otherwise not present because of deterministic constraints due to symmetry (4). The theory of randomizing effects in quantized nonlinear systems is under development (5).

NOISE EXPERIMENTS

The experiments were carried out with electrically polarized n=63 hydrogen atoms in sinewave microwaves within the range 12-18 GHz. The atoms saw a microwave electric field pulse about 100 periods long and with a sinefunction pulse envelope. The ratio of sinewave microwave frequency to initial electron orbit frequency ranged between 0.45 and 0.68, within the region where initial two-state adjacent-level n-changing energetically requires the absorption of two or three photons. The additive microwave noise was generated by a traveling wave tube amplifier with its input terminated, and arose from shot noise in tube electron gun emission with added randomness due to thermal electron velocity distribution. The tube noise bandwidth was clipped by a combination of waveguide cutoff at the low frequency end and a low pass filter at the high frequency end. The resulting noise power spectrum was observed to be in the range 9.5-12.5 GHz, with a 1% tail extending to 18 GHz.

Figure 1 shows measured down n-changing probabilities as a function of sinewave frequency, with and without large amounts of incident waveguide noise power. With no noise, 2 mW of sinewaves produced only resonant 2-photon n-changing down to n=62; no semiclassical n-changing was observed. With 8 mW of noise only, semiclassical n-changing was observed. For both fields together, 2-photon transitions involving one photon from each field occurred for sinewave frequencies capable of producing resonance with some component of the noise spectrum.

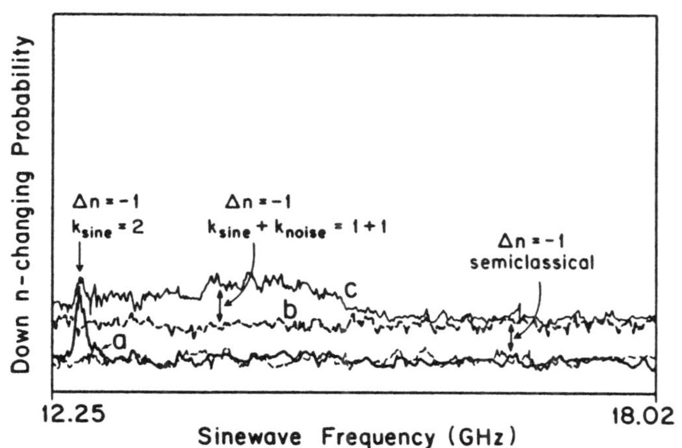

Figure 1. Microwave down n-changing probabilities in (n,n_1,m) = (63,0,0) hydrogen atoms, as a function of sinewave microwave frequency, for sinewave, noise powers of (a) 2,0; (b) 0,8 and (c) 2,8 mW.

Figure 2 shows n-changing final state distributions for fields in the semiclassical regime. 20 mW of sinewaves alone produced simultaneous n-changing only to states n=62 and 64. Replacing 2 mW of the sinewave power with noise rms power produced no observable change. However, replacing most of the sinewave power with noise power resulted in an extended n-changing distribution characteristic of the diffusion regime. This observation might be due to a noise-induced shift in the threshold for diffusion arising from underlying classical chaos. An increased effectiveness of noise over sinewaves might be expected because the noise amplitude spectrum has large components that play a disproportionate role in nonlinear processes.

Microwave ionization was observed to behave similarly to the semiclassical n-changing, see Figure 3. As ionization is believed to be the final outcome of the diffusion of microwave energy into the atom, this is to be expected. Again low amounts of noise power produced no changes, while replacement of a sizeable fraction of sinewaves with noise enhanced the ionization probability, quite strongly near and below the sinewave power ionization threshold.

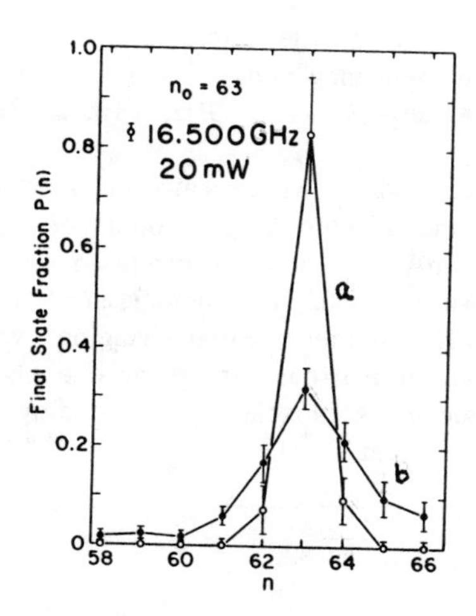

Figure 2. Medium-power n-changing final state distributions (a) for 20 mW of sinewave power and negligible noise power, and (b) 4 mW of sinewave power and 16 mW of noise power, adding to a total of 20 mW.

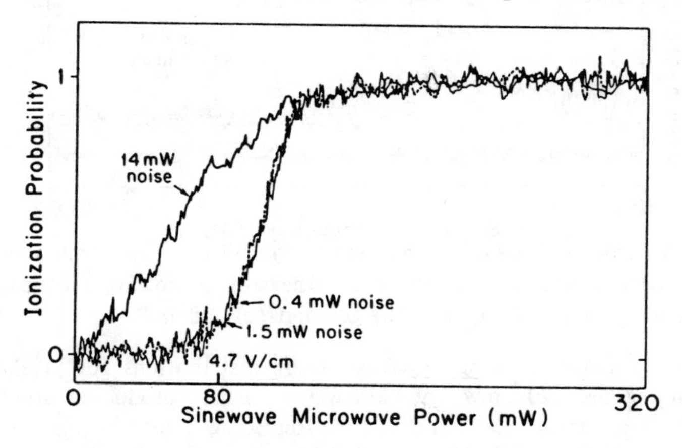

Figure 3. Microwave ionization probability as a function of sinewave microwave power at a microwave frequency of 17.501 GHz, for three powers.

1. J. E. Bayfield, in Quantum Measurement and Chaos, E. R. Pike and S. Sarkar, editors, Plenum Press, 1987, pages 1-33.
2. J. N. Bardsley, B. Sundaram, L. A. Pinnaduwage and J. E. Bayfield, Phys. Rev. Lett. 56, 1007 (1986).
3. G. Casati, I. Guarneri and D. L. Shepelyansky, Phys. Rev. A 36, (5),1987.
4. W. Horsthemke and R. Lefever, Noise-Induced Transitions, Springer-Verlag, Berlin, 1984, page 164.
5. T. Dittrich and R. Graham, Europhys. Lett. 4, 263 (1987).

TWO-PHOTON (VUV & VISIBLE) RESONANT IONIZATION SPECTROSCOPY OF ATOMS AND MOLECULES

M. P. McCann, C. H. Chen, and M. G. Payne
Chemical Physics Section, Oak Ridge National Laboratory
Oak Ridge, Tennessee 37831-6378

Two-photon absorption spectroscopy is the simplest nonlinear optical method. Two-photon spectroscopy offers several advantages over conventional one-photon absorption. In atoms and molecules with a center of symmetry, two-photon transitions can access states which are not allowed by one-photon transitions.[1] The two-photon transition energies and the two-photon rate constants can be used to determine rotational and vibrational constants along with potential energy surfaces. Two-photon spectroscopy can be used to examine selected regions of a bulk sample. Thus, surface properties or bulk properties can be studied exclusive of each other depending on where the two laser beams intersect. If the two laser beams are linearly polarized, then the symmetry of the excited state can be determined in a randomly oriented sample such as a gas or a liquid. Two counter-propagating laser beams can be used to produce spectra from a high pressure gas cell that is free of Doppler broadening.

In order to do two-photon spectroscopy of argon, krypton, and molecular hydrogen, it was decided to use one high energy photon at a fixed wavelength and a tunable photon of lower energy. The first photon was produced by third harmonic generation using xenon as the nonlinear medium.[2] Coherent VUV light was produced at 118 nm from the third harmonic of a Nd:YAG laser (355 nm). A tunable dye laser was used to produce the second photon.

The two photons were absorbed in a high pressure ($\leq$ 200 torr) cell which was filled with the sample gas (Ar, Kr, H_2). The excited atom or molecule then absorbed a UV photon (355 nm) and was subsequently ionized. The photoelectrons were detected by parallel plates. By tuning the dye laser, spectra were produced. With knowledge of the VUV and visible photon flux, the sample pressure and the signal size, the two-photon rate constant could be calculated.[3] The results are shown in Table I. It is interesting to note the the two-photon transition to the $v'=7$ in the E, F potential is just below the hump that separates the two wells in the E, F potential. The Franck-Condon factor for this transition should be very unfavorable as none of the lower vibrational levels in the outer well were observed. This points to a tunneling effect. The two-photon rate constant compares well with a previous measurement of the ground state to $v'=0$ of the E, F state at different wavelengths.[4]

Table I

Experimental Transition Energies and Two-Photon Rate Constants for Ar and Kr and H_2

Transition	Energy (cm^{-1})	K_2 (cm^4 s)
Argon		
$3p^5 4p\,[5/2]_{J=2} \leftarrow 3p^6$	105,611	1.6×10^{-46}
$3p^5 4p\,[3/2]_{J=1} \leftarrow 3p^6$	106,083	2.5×10^{-47}
$3p^5 4p\,[3/2]_{J=2} \leftarrow 3p^6$	106,235	5.8×10^{-47}
$3p^5 4p\,[1/2]_{J=0} \leftarrow 3p^6$	107,051	2.3×10^{-47}
$3p^5 4p'\,[3/2]_{J=1} \leftarrow 3p^6$	107,128	6.2×10^{-48}
$3p^5 4p'\,[3/2]_{J=2} \leftarrow 3p^6$	107,287	1.3×10^{-47}
$3p^5 4p'\,[1/2]_{J=1} \leftarrow 3p^6$	107,495	6.6×10^{-48}
Krypton		
$4p^5 4f\,[3/2]_{J=2} \leftarrow 4p^6$	105,961	3.7×10^{-47}
$4p^5 4f\,[5/2]_{J=2} \leftarrow 4p^6$	106,017	6.2×10^{-48}
$4p^5 4f\,[1/2]_{J=2} \leftarrow 4p^6$	107,410	9.5×10^{-48}
Hydrogen		
$E,F^1\Sigma_g^+, v' = 3 \leftarrow X^1\Sigma_g^+, v'' = 0$	101,486	2.7×10^{-49}
$E,F^1\Sigma_g^+, v' = 6 \leftarrow X^1\Sigma_g^+, v'' = 0$	103,554	3.1×10^{-49}
$E,F^1\Sigma_g^+, v' = 7 \leftarrow X^1\Sigma_g^+, v'' = 0$	103,833	1.6×10^{-49}
$E,F^1\Sigma_g^+, v' = 9 \leftarrow X^1\Sigma_g^+, v'' = 0$	105,377	4.2×10^{-49}
$E,F^1\Sigma_g^+, v' = 10 \leftarrow X^1\Sigma_g^+, v'' = 0$	105,960	4.1×10^{-49}
$E,F^1\Sigma_g^+, v' = 11 \leftarrow X^1\Sigma_g^+, v'' = 0$	106,709	1.0×10^{-48}
$E,F^1\Sigma_g^+, v' = 12 \leftarrow X^1\Sigma_g^+, v'' = 0$	107,421	6.1×10^{-49}
$E,F^1\Sigma_g^+, v' = 13 \leftarrow X^1\Sigma_g^+, v'' = 0$	108,096	1.3×10^{-48}
$E,F^1\Sigma_g^+, v' = 14 \leftarrow X^1\Sigma_g^+, v'' = 0$	108,790	5.1×10^{-49}

REFERENCES

1. R. G. Bray and R. M. Hochstrasser, Mol. Phys. 31, 1199-1211 (1976).
2. A. H. Kung, J. F. Young, and S. E. Harris, Appl. Phys. Lett. 22, 301-302 (1973).
3. M. P. McCann, C. H. Chen, and M. G. Payne, Chem. Phys. Lett. 138, 250-256 (1987).
4. E. E. Marinero, C. T. Rettner, and R. N. Zare, Phys. Rev. Lett. 48, 1323-1326 (1982).

PRECISE MULTIPHOTON SPECTROSCOPY OF H_2

E.E. Eyler, J.M. Gilligan and E. McCormack
Department of Physics, Yale University
217 Prospect Street, New Haven, Connecticut 06511

ABSTRACT

This paper summarizes our recent progress in the determination of absolute energy levels in the H_2 molecule using two- and three-photon excitation. Several vibrational bands of the E,F←X transition have been measured using two-photon excitation near 220 nm, to an accuracy of 0.01 cm^{-1}, an improvement by nearly an order of magnitude. Three-photon transitions to the B(2pσ) and C(2pπ) states have been measured to 0.06 cm^{-1}. The three-photon measurements are somewhat less precise because they are affected by shifts arising from the AC Stark effect and from interference between three-photon excitation and third harmonic generation. Finally, transitions have been measured from the E,F state to the dissociation limit yielding H(1s) + H(2s or 2p), providing a preliminary new value of 36118.1 ± 0.2 cm^{-1} for the dissociation energy of H_2.

TWO-PHOTON SPECTROSCOPY OF E,F←X TRANSITIONS

The hydrogen molecule is the only neutral molecule for which theoretical calculations of the dissociation energy and ionization potential have been performed with an accuracy better than 0.1 cm^{-1}. To test these calculations, absolute wavelengths must be measured for transitions from the ground state to the electronically excited states. The lowest excited state accessible using two photons is the E,F $^1\Sigma_g^+$ state, with a total energy of about 99,000 cm^{-1}. This state, which has a double-minimum potential curve, is of particular interest because it can be used as a convenient intermediate level for subsequent excitation to higher states.

We have observed transitions to the v=0 and v=1 levels of the inner E state well using two-photon resonant, three-photon ionization. To accomplish this, a laser system operating near 210 nm was used to excite a beam of vibrationally excited ground state H_2. The vibrational excitation was created by a discharge between electrodes placed just outside the exit of a pulsed supersonic nozzle. The laser radiation was obtained by frequency doubling the output of a pulse-amplified cw ring dye laser in a crystal of β-BaB_2O_4. Pulses about 10 nsec in duration with energies of 200-500 μJ were obtained, with a bandwidth of 0.005 cm^{-1}. Collimation of the molecular beam reduced the residual Doppler width to about 0.01 cm^{-1}

The wavelengths of 22 transitions in the (0,1), (0,2) and (1,2) bands were measured to an absolute accurcy of 0.01 cm^{-1} by comparison with the absorption spectrum of a tellurium cell. A complete table of the results appears in Ref. 1. The accuracy was limited principally by nonlinearity in the laser scans. Minor improvements in the apparatus should result in an improvement to about 0.001-0.002 cm^{-1}.

These new measurements of the E,F state are in good agreement with most previous determinations. They disagree slightly with a recent measurement made using a Raman shifted dye laser, but subsequent work indicates that the Raman shifting process introduces small shifts that probably account for the discrepancy.[2]

By combining these results with other measurements, a value of 124417.4 ± 0.15 cm^{-1} is obtained for the ionization potential of H_2.[1] This value is in good agreement both with theory and with other recent experiments, and of about the same accuracy. A value accurate to 0.01 cm^{-1} should be attainable in the near future.

THREE-PHOTON SPECTROSCOPY OF B←X AND C←X TRANSITIONS

An alternative arrangement for determining vacuum ultraviolet intervals in H_2 is the measurement of three-photon transitions. Preliminary measurements of transitions to the B and C states were described at this conference a year ago.[3] It was clear from this initial work that measurements to an accuracy of 0.05 cm^{-1} or better are feasible, but that at this level the measurements can be affected by line broadening and shifts that depend both on the laser power and the number density in the molecular beam apparatus. During the past year we have confirmed that most of the frequency shift arises from interference between third harmonic generation and three photon excitation. This process has been observed previously, mainly in atomic systems, and a detailed theoretical treatment has recently appeared.[4]

To confirm this explanation we observed the three-photon excitation with a pair of counterpropagating laser beams. Some of the transition amplitude must then arise from the absorption of two photons from one laser beam and of one from the other, a process that cannot be cancelled by third harmonic generation because momentum conservation is impossible. Thus a large enhancement in the signal size is expected. In the experiment, one of the C←X transitions was studied under conditions such that when only the main beam was present, the line was shifted about 0.2 cm^{-1} to the blue. When the counterpropagating beam was introduced, we observed an enhancement by a factor of 20 of the signal at the unshifted resonance position. We have also observed the dependence of the line shift on number density by varying the voltage that opens the pulsed nozzle.

The obtainable accuracy should be much better for measurements to a level from which single-photon decay is impossible, so that third harmonic generation cannot occur. To investigate this possibility, we are currently undertaking measurements of three photon intervals from the ground state of H_2 to the 4f and 5f Rydberg states.

PHOTODISSOCIATION OF THE E,F STATE

The region near the second dissociation limit of H_2, to H(1s) + H(2s or 2p), can be studied with very high resolution by exciting the E,F state and using an additional laser to scan the region near the dissociation threshold. Hydrogen atoms in the 2p state can be detected by their Lyman α fluorescence, and 2s state atoms can be quenched through the 2p state by applying a small electric field. Figure 1 shows preliminary experimental data for this transition, obtained by exciting the v=2 level of the E state using an ArF excimer laser at 193 nm, then observing photodissociation produced by a pulse-amplified cw laser operating near 680 nm. Although this data suffers from a very poor signal to noise ratio, the top trace, showing the signal produced by H(2s) atoms, clearly shows the onset of the threshold at a total energy of 118377.2 cm^{-1}. The shape of the threshold is easily resolved, but is partially obscured by noise. The bottom trace was taken with much higher laser powers, near 10^6 W/cm^2, and a different detector arrangement. The signals in this case arise from the absorption of two photons from the E state. The obvious peaks arise from resonances at the one photon level with the highest bound vibrational levels of the B and B' states.

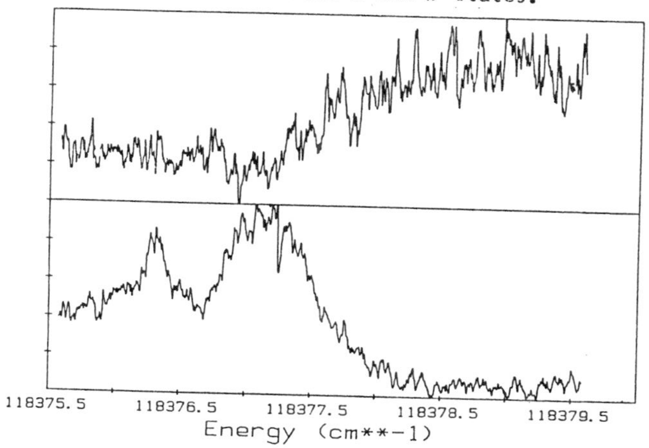

Fig. 1 Top trace: preliminary data showing the H(1s) + H(2s) threshold, observed by Lyman α fluorescence. Bottom trace: Scan at higher laser power showing two-photon transitions with resonances through the highest bound levels of the B and B' states. The peak near 118376.3 cm^{-1} is the R(1) branch of the (39,2) band of B←E. The sharp negative signal near the threshold is an experimental artifact.

The threshold can be determined directly from the top trace, or indirectly by extrapolating the highest B state vibrational levels using the procedure of Stwalley.[5] Finally, it can be determined from the decrease in the two-photon signal above the one-photon threshold. All three methods agree, and yield a preliminary new value of 36118.1 ± .2 cm^{-1} for the ground state binding energy. This value is a factor of 2.5 more precise than previous work, and agrees well both with earlier experiments and theory.

REFERENCES

[1]E.E. Eyler, J.M. Gilligan, E. McCormack, A. Nussensweig and E. Pollack, Phys. Rev. A 36, 3486 (1987).
[2]W.L. Glab and J.P. Hessler, Phys. Rev. A 35, 2102 (1987).
[3]E.E. Eyler, J.M. Gilligan and E. McCormack, in Advances in Laser Science-II, 388 (1987).
[4]M.G. Payne, W.R. Garrett, and W.R. Ferrell, Phys. Rev. A 34, 1143 (1986).
[5]W.C. Stwalley, Chem. Phys. Lett. 6, 241 (1970).

This work was supported by the NSF, grant number PHY-8403324.

HIGH-ORDER MULTIPHOTON IONIZATION PHOTOELECTRON SPECTROSCOPY OF NO

H. S. Carman, Jr., and R. N. Compton
Chemical Physics Section, Oak Ridge National Laboratory*
P.O. Box X, Oak Ridge, Tennessee 37831-6125
and
Department of Chemistry, The University of Tennessee
Knoxville, Tennessee 37996

ABSTRACT

Photoelectron energy and angular distributions of NO following three different high-order multiphoton ionization (MPI) schemes have been measured. The 3+3 resonantly enhanced multiphoton ionization (REMPI) via the $A^2\Sigma^+$ (v=0) level yielded a distribution of electron energies corresponding to all accessible vibrational levels (v^+=0–6) of the nascent ion. Angular distributions of electrons corresponding to v^+=0 and v^+=3 were significantly different. The 3+2 REMPI via the $A^2\Sigma^+$ (v=1) level produced only one low-energy electron peak (v^+=1). Nonresonant MPI at 532 nm yielded a distribution of electron energies corresponding to both four- and five-photon ionization. Prominent peaks in the five-photon photoelectron spectrum (PES) suggest contributions from near-resonant states at the three-photon level.

INTRODUCTION

REMPI-PES has become an established tool for probing the photoionization dynamics of atoms and molecules.[1] In particular, measurements of photoelectron angular distributions (PEAD) in a REMPI process can, in principle, provide information about both the dynamics of the bound-continuum transition and the alignment of an optically-prepared intermediate state.[2a,b,c] Of the molecular systems studied thus far, NO has received the most attention due to its low ionization potential and well-characterized spectroscopy. For details and references to previous REMPI-PES studies of NO and other systems, the reader is referred to a recent review by Compton and Miller.[1] To date, the $A^2\Sigma^+$ state has been probed by either 2+2 or 1+1 REMPI processes, with few PEAD measurements. In this paper we describe PES and PEAD measurements for 3+3 and 3+2 REMPI via the $A^2\Sigma^+$ (v=0) and (v=1) levels, respectively. In addition, we have measured the PES for nonresonant MPI of NO at 532 nm. The apparatus has been described in detail elsewhere[3] and will not be described here.

RESULTS AND DISCUSSION

The spectra of the 3+3 $A^2\Sigma^+ \leftarrow X^2\Pi$ (0,0) and 3+2 $A^2\Sigma^+ \leftarrow X^2\Pi$ (0,1) bands were measured by monitoring the mass-selected NO^+ ion intensity as the laser wavelength was scanned. Both spectra lacked resolved rotational structure and appeared as broad, shifted resonances, indicative of a.c. Stark effects due to the high laser power necessary to affect six- and five-photon ionization. As a result, no rotational state selection was possible and PES and PEAD were measured at arbitrary wavelengths within the resonances where optimum signals were obtained.

Figure 1 shows a representative PES obtained for the 3+3 process at 676.2 nm. Other wavelengths within the resonance produced similar spectra, although peak height ratios varied (especially v^+=2,4). Due to the Rydberg character of the $A^2\Sigma^+$ state, the Franck-Condon princple predicts that direct ionization of the v=0 level would produce only the v^+=0 level of the ion. The appearance of higher v^+ levels thus suggests additional ionization mechanisms, as discussed previously[1] for 2+2 and 1+1 processes. In the present case, possible mixing of states at the fourth- and fifth-photon levels might complicate the dynamics further.

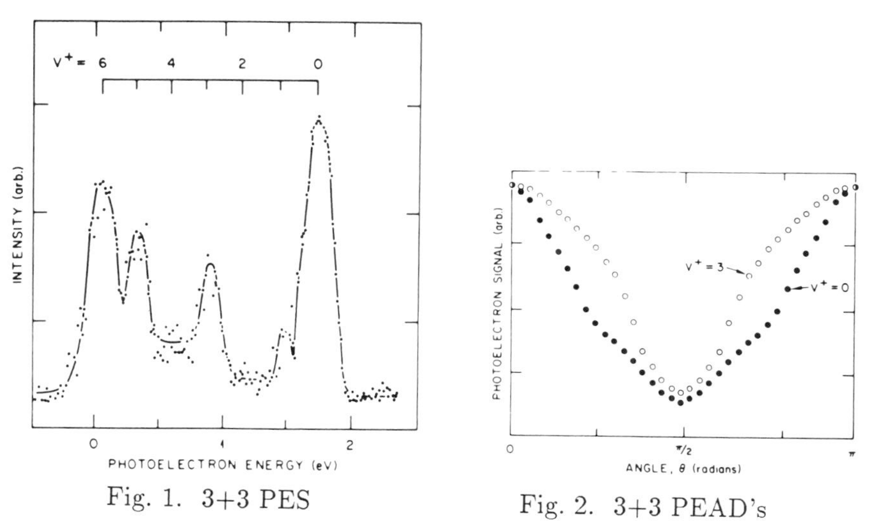

Fig. 1. 3+3 PES

Fig. 2. 3+3 PEAD's

PEAD's measured for electrons corresponding to v^+=0 and 3 are shown in Fig. 2. Both distributions were fit, using a least-squares procedure, to a Legendre polynomial expansion[2a]

$$I(\theta) = \sum_{k=0}^{N} a_{2k} P_{2k}(\cos\theta),$$

where N is the order of the ionization process. It was found that terms up to $P_6(\cos\theta)$ were necessary to fit the data within experimental error:

$$v^+ = 0 : \; I(\theta) = 1 + 1.23(1)P_2 + 0.16(2)P_4 + 0.22(2)P_6$$

$$v^+ = 3 : \; I(\theta) = 1 + 1.07(1)P_2 - 0.34(1)P_4 + 0.13(1)P_6,$$

where uncertainties of the fit are shown in parentheses. This contrasts with recent 1+1 PEAD measurements of the $A^2\Sigma^+$ (v=0) level[4] for which the data were fit by including only the P_2 term. The most dramatic difference between the distributions above is in the sign of the P_4 coefficients. Further theoretical work is necessary, however, to determine if this difference might lead to better understanding of the ionization mechanisms involved.

The 3+2 REMPI of the $A^2\Sigma^+$ (v=1) level at 644.1 nm produced only one electron peak, corresponding to $v^+=1$, indicating direct ionization of this level. In this case, however, the fifth photon energy is only slightly above the $v^+=1$ threshold and channels leading to higher vibrational levels of the ion are not accessible. In order to fit the PEAD, it was necessary to include terms up to P_4:

$$I(\theta) = 1 + 1.54(2)P_2 + 0.32(2)P_4.$$

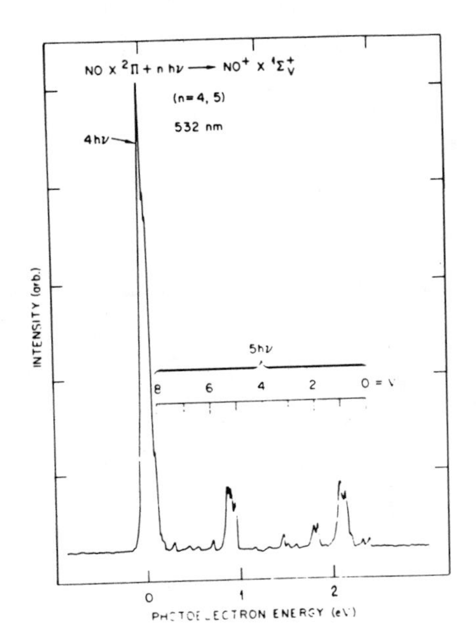

Fig. 3. 532 nm PES

The PES resulting from nonresonant MPI at 532 nm is shown in Fig. 3. Four photons are sufficient to ionize NO at this wavelength (0.06 eV above the $v^+=0$ threshold), and a very intense low energy electron peak is seen in the PES.

In addition, however, several weaker peaks are observed at energies consistent with absorption of five photons, leaving the ion predominantly in the $v^+=1$, 2, and 5 vibrational levels. It is interesting to note that the third photon energy is within ~800 cm^{-1} of the $A^2\Sigma^+$ (v=5), $C^2\Pi$ (v=2), and $D^2\Sigma^+$ (v=1) levels of NO. The five-photon PES might be reflecting the contributions of these "near-resonant" intermediate states. Experiments are currently in progress to measure the PEAD's of these electrons. More extensive measurements of PEAD's for various n+1 REMPI processes are also planned for the future.

REFERENCES

* Operated by Martin Marietta Energy Systems, Inc., under contract DE-AC05-84OR21400 with the U.S. Department of Energy.

1. R. N. Compton and J. C. Miller, in Laser Applications in Physical Chemistry, D. K. Evans, Ed. (Marcel Dekker, Inc., New York) in press.
2. (a) S. N. Dixit and P. Lambropoulos, Phys. Rev. A 27, 861 (1983); (b) S. N. Dixit, D. L. Lynch, and V. McKoy in Multiphoton Processes, P. Lambropoulos and S. J. Smith, Eds. (Springer-Verlag, New York, 1984); and (c) S. N. Dixit and V. McKoy, J. Chem. Phys. 82, 3546 (1985).
3. P. R. Blazewicz, X. Tang, R. N. Compton, J.A.D. Stockdale, J. Opt. Soc. Am. 4, 770 (1987).
4. J. R. Appling, M. G. White, W. J. Kessler, R. Fernandex, and E. D. Poliakoff, J. Chem. Phys., in press.

RESONANCE-ENHANCED MULTIPHOTON IONIZATION SPECTROSCOPY OF $CHCl_2$ AND $CDCl_2$

Jeffrey W. Hudgens and George R. Long
Chemical Kinetics Division, Center for Chemical Physics,
National Bureau of Standards, Gaithersburg, MD 20899

ABSTRACT

Resonance enhanced multiphoton ionization spectra of $CHCl_2$ and $CDCl_2$ were observed between 355-375 nm via a 2 + 1 excitation process. Electronic origins tentatively assigned to 3d Rydberg states were observed at 370.1 nm (ν_{00} = 54,024 cm^{-1}) for $CHCl_2$ and at 370.4 nm (ν_{00} = 53,980 cm^{-1}) for $CDCl_2$. The C-Cl symmetric stretch frequencies are 845 cm^{-1} in $CHCl_2$ and 814 cm^{-1} in $CDCl_2$.

INTRODUCTION

Resonance enhanced multiphoton ionization (REMPI) spectroscopy has become an effective tool for discovering electronic states of free radicals.[1] We report observation and analysis of the only known gas phase spectra of dichloromethyl radicals, $CHCl_2$ and $CDCl_2$.

EXPERIMENTAL

The apparatus[2] used in this study consisted of a flow reactor which produced the free radical species, an excimer pumped dye laser which ionized the radicals, a time of flight mass spectrometer, and a computer/data acquisition system. Free radicals produced in the flow reactor effused into the ion source of the mass spectrometer where they were ionized by a focussed laser beam (Energy=8-15 mJ/pulse; Bandwidth=0.2 cm^{-1} FWHM; fl=250 mm). Pressure within the ion source was about $5x10^{-5}$ torr. The laser generated ions were mass selected and detected by the mass spectrometer. The ion signal was recorded as a function of wavelength to produce the REMPI spectra.

RESULTS AND ANALYSIS

In the flow reactor the hydrogen abstraction reaction of atomic fluorine with dichloromethane produced the dichloromethyl radical. The laser generated ion signals observed from the flow reactor effluent appeared at m/z=83, 85, and 87 with intensity ratios that confirmed that the spectra carrier possessed two chlorine atoms. When fluorine was reacted with CD_2Cl_2, the ion signal appeared at m/z=84, 86, and 88 which showed that the spectral carrier contains one hydrogen. Thus, the spectra arise from $CHCl_2$ and $CDCl_2$ radicals.

Figure 1 shows the REMPI spectra. Each spectrum displays five distinct bands and the pattern of these bands originates from two active radical vibrations. To derive the frequencies of the

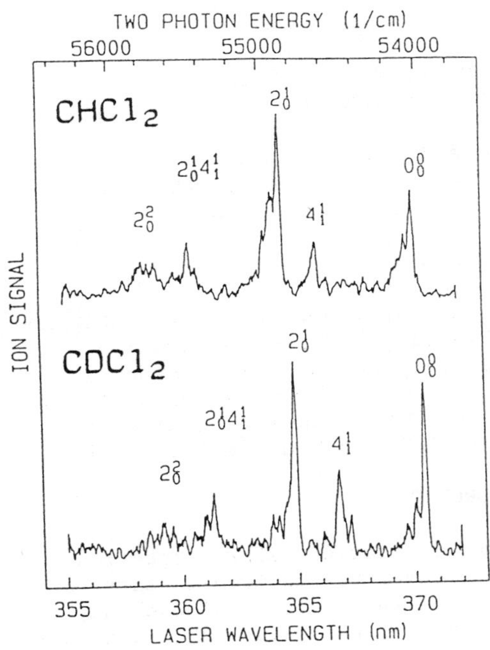

Figure 1. The REMPI spectrum of the $CHCl_2$ radical (upper trace) and the $CDCl_2$ radical (lower trace).

molecular vibrations, these one-photon intervals must be multiplied by the number of photons simultaneously absorbed to produce the REMPI spectra.

The red-most member of each REMPI progression is assigned to the electronic origin. Since the first adiabatic ionization potential of $CHCl_2$ is 8.32 eV,[3] a minimum of three photons are required to ionize $CHCl_2$ between laser wavelengths of 355-375 nm. When we assume that the spectral structure arises from simultaneous two photon absorption which prepares a radical Rydberg state, we can account for the REMPI spectra. In this model the ion signal is generated when the excited Rydberg radical absorbs one more laser photon, i.e. the REMPI signal arises from a 2+1 ionization mechanism.

The type of Rydberg orbital being observed (s,p or d) is determined by solving the Rydberg equation. For two photon absorption, two reasonable solutions are obtained corresponding to a 3d Rydberg state (δ=0.1) and the 4s $^2A_1$ Rydberg state (δ=1.1). If these bands originate from 4s $^2A_1$ Rydberg states, then the Rydberg equation predicts that the 3s $^2A_1$ Rydberg state origin should appear at 545 nm through a 2+2 ionization mechanism. No such bands were found.

The vibrational analysis supports the Rydberg state assignment. Because the cation core of Rydberg states of radicals have the same electron configurations as the cation, the vibrational spacings observed in Rydberg states and cations are usually similar. In $CHCl_2$ the frequency interval of the three membered vibrational progression, assuming a two photon transition, is 845(10) cm^{-1}. This interval is identical to the frequency of the $\upsilon_2''(a_1)$=860(30) cm^{-1} C-Cl stretch of $CHCl_2^+$.[3] In the REMPI spectrum of $CDCl_2$ the two photon vibrational progression is 814(10) cm^{-1}, which is also nearly identical to the $\upsilon_2''(a_1)$=790(30) cm^{-1} C-Cl symmetric stretch observed in $CDCl_2^+$.[3] Thus, we assign the observed transition to a two photon resonant REMPI process, probably through a 3d Rydberg state. The three membered vibrational progression observed in each REMPI spectrum is assigned to transitions from the vibrationless ground state to the $\upsilon_2'(a_1)$=1,2 levels in the Rydberg state (Figure 1).

REMPI spectra of $CHCl_2$ are expected to show bands associated with the $\upsilon_4(b_1)$ out-of-plane bending mode because of changes in the shape of the out-of-plane bending potential between the nearly planar ground state geometry into the planar Rydberg state geometry. Since the $\upsilon_4(b_1)$ out-of-plane bending mode (in C_{2v} symmetry and in the Born-Oppenheimer approximation) is governed by the selection rule, $\Delta v_4=0, \pm 2, \pm 4...$, the REMPI bands in $CHCl_2$ at 366.8 nm and in $CDCl_2$ at 366.00 nm are either 4_1^1 hot bands or 4_0^2 overtone bands. Evidence favors the 4_1^1 hot band assignment. Upon deuteration the ~366 nm bands exhibit a 10% increase of intensity relative to the origin. A relative intensity increase is expected of hot bands because the thermal $v_4''=1$ population in $CDCl_2$ is larger than the thermal $v_4''=1$ population in $CHCl_2$. The observation of 4_1^1 hot bands in the present, ambient temperature experiments implies that the υ_4'' (b_1) modes lie at very low frequencies.

The assignment of the 4_1^1 transition shows that the out-of-plane bending frequency in $CHCl_2$ is 605 cm^{-1} greater in the Rydberg state than in the ground state. This frequency increase also conforms to expectation. Previous studies have established that the out-of-plane bending modes of other methyl radical systems are very sensitive to bonding changes associated with the loss of the radical electron. The out-of-plane bending frequency increases by factors of 2 to 5 between the ground state and Rydberg state radicals,[1] e.g. CH_2F (260 cm^{-1} to 1259 cm^{-1})[4] and CH_3 (606.5 cm^{-1} to 1334 cm^{-1}).[5,6] Thus, our assignment of the ~366 nm bands to the 4_1^1 hot bands appears reasonable. Adopting this band assignment, the bands at 360.40 nm in $CHCl_2$ and at 359.20 nm in $CDCl_2$ are assigned as the $2_0^1 4_1^1$ transitions (Figure 1). *Ab initio* calculations[7] also support the present interpretation of these spectra.

REFERENCES

1) J. W. Hudgens, Advances in Multi-photon Processes and Spectroscopy, Vol. 4, S. H. Lin, ed. (World Scientific Publishing Co., Singapore, in press).
2) G. R. Long and J. W. Hudgens, J. Phys. Chem. 91, 5870 (1987).
3) L. Andrews, J. M. Dyke, N. Jonathan, N. Keddar, and A. Morris, J. Chem. Phys. 79, 4650 (1983).
4) J. W. Hudgens, C. S. Dulcey, G. R. Long, and D. J. Bogan, J. Chem. Phys. 81, 4546 (1987).
5) J. W. Hudgens, T. DiGiuseppe, and M. C. Lin, J. Chem. Phys. 79, 571 (1983).
6) M. Jacox, Chem. Phys. 59, 213 (1981).
7) S. A. Kafafi and J. W. Hudgens, in press.

Resonance-Enhanced Multiphoton Ionization Spectra of SiCl between 430 - 520 nm

Russell D. Johnson III, and Jeffrey W. Hudgens
Chemical Kinetics Division, National Bureau of Standards
Gaithersburg, Maryland 20899

ABSTRACT

SiCl radical is observed from 430 to 520 nm by resonance enhanced multiphoton ionization. The spectra arise from two photon transitions to the C $^2\Pi_r$, D $^2\Sigma^+$, and E states. Absorption of a third laser photon formed the cation. Both higher vibrations of the C $^2\Pi_r$ state (v'=3-7) and the $Si^{37}Cl$ spectrum of the C $^2\Pi_r$ state are observed. The $Si^{35}Cl$ C $^2\Pi_r$ state vibrations can be fit with constants of ω_e=682.7 ± 3.8 cm^{-1} and $\omega_e x_e$=3.8 ± 0.5 cm^{-1}.

INTRODUCTION

This paper demonstrates the first detection of SiCl radical by resonance enhanced multiphoton ionization (REMPI) spectroscopy. Throughout the spectral region of this study very strong $SiCl^+$ ion signals were observed which suggests that REMPI spectroscopy can be used to sensitively detect relatively small concentrations of SiCl radicals. SiCl appears as one of the intermediates in the plasma reduction of $SiCl_4$ to produce semiconductor Si crystals[1]. A sensitive method for detecting SiCl radicals, which REMPI provides, may further the understanding of such Si growth mechanisms.

Resonance enhancement originates from two-photon preparation of the C $^2\Pi_r$, D $^2\Sigma^+$, and E-states. Part of the present work confirms bands previously observed with the traditional emission and absorption spectroscopies, but the REMPI spectrum revealed many more, previously unreported vibrational bands. The sensitivity and mass selectivity of REMPI also enabled the first measurement of the band positions of the $Si^{37}Cl$ isotopic radical. This isotopic information ascertained the vibrational numbering assignments of the REMPI bands. Based on this larger assigned data set, we report spectroscopic constants, ω_e and $\omega_e x_e$, for the C $^2\Pi_r$ state.

EXPERIMENTAL METHODS

The apparatus is similar to one previously described[2]. Briefly, it consists of a flow reactor in which Cl atoms reacted with silane (SiH_4) to produce SiCl radicals. A portion of the flow reactor effluent leaked through a skimmer and into the ion optics of a time-of-flight mass spectrometer. The SiCl radicals were ionized by the focused light from an excimer laser pumped tunable dye laser. The ions were extracted into the time-of-flight mass spectrometer and the mass resolved m/z 63 or 65 (corresponding to $Si^{35}Cl$ or $Si^{37}Cl$) ion currents were recorded. The vacuum chamber which enclosed the mass spectrometer was operated at a pressure of ~10^{-5} torr. Based upon our calculations, which assume 100% conversion of SiH_4 into SiCl, we can detect concentrations as low as 10^9 radicals/cm^3 in the ionization region of the mass spectrometer with a single laser shot.

RESULTS AND DISCUSSION

REMPI spectra of SiCl carried by m/z 63 and m/z 65 were observed over the laser wavelength range of 430-520 nm. Congruous

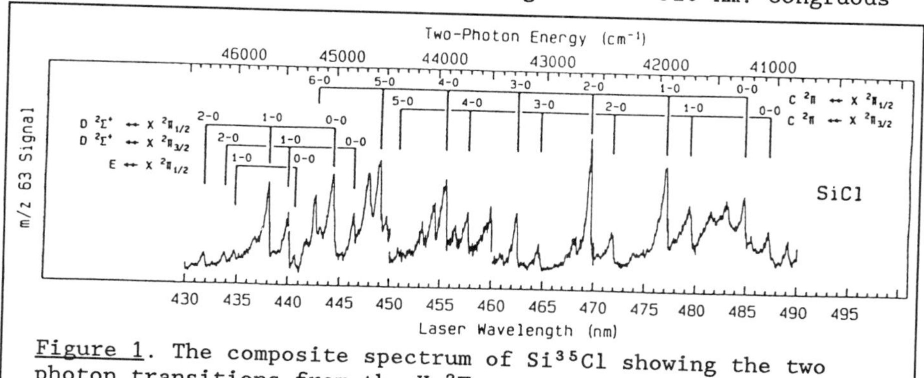

Figure 1. The composite spectrum of $Si^{35}Cl$ showing the two photon transitions from the X $^2\Pi_r$ ground state to the C $^2\Pi_r$, D $^2\Sigma^+$, and E states. The many evident hot bands are not indexed.

with the relative $^{35}Cl/^{37}Cl$ isotopic abundance, the intensity of the m/z 63 signal ($Si^{35}Cl$) was three times greater than the m/z 65 signal ($Si^{37}Cl$). In addition to the ion signals that originated from SiCl radicals, m/z 28 ion signals corresponding to transitions to known states of atomic silicon were also observed.

Long progressions are observed in the region of 420 to 520 nm (Fig. 1). Between 445-490 nm one progression of five doublets shows peak intensity ratios of ~2:1 which arise from the difference in populations between the X $^2\Pi_{3/2}$ and X $^2\Pi_{1/2}$ spin-orbit states.

Between 40,000-43,000 cm^{-1} Jevons[3] has reported the UV emission spectrum of the C $^2\Pi_r$ (v'=0,1,2) → X $^2\Pi_r$ bands. In a two-photon spectrum these bands will lie between 465-490 nm. In the REMPI spectrum the average two-photon laser frequency separation between each pair of doublets is ~200 cm^{-1}. This separation is the same as is expected of a two-photon transition from the ground-state (A"=206.6 cm^{-1}) to the C $^2\Pi_r$ state (A'=10.9 cm^{-1})[4], i.e. A"-A'=195.7 cm^{-1}.

Jevons[3] observed v'=0,1,2 → v"=0 vibrational bands which exhibited an interval of ~665 cm^{-1}. Our present two-photon resonant REMPI spectrum shows these same bands as well as members up to v'=6 from v"=0. These higher vibrational bands overlap with bands from the D $^2\Sigma^+$ ←← $^2\Pi_r$ system. The fact that these new bands belong to the C-state was determined by the vibrational spacing and the isotope shifts. This paper is the first to report the C $^2\Pi_r$ (v'=3-6) vibrational bands.

A least-squares fit of thirty-three upper-state vibrational spacing differences (using v'=0 through v'=7) yielded C $^2\Pi_r$ state spectroscopic constants of ω_e=682.7 ± 3.8 cm^{-1} and $\omega_e x_e$=3.8 ± 0.5 cm^{-1}. These can be compared with the values reported by Jevons[3] of ω_e=674.2 cm^{-1} and $\omega_e x_e$=2.2 cm^{-1}, who used v'=0 through v'=2.

The REMPI spectrum of the $Si^{37}Cl$ radical was measured by recording the m/z 65 signal. For transitions involving v'=0-4 very precise isotopic shifts were measured from positions of the sharp O-branches. Spectral congestion thwarted O-branch measurements of the REMPI bands involving transitions to v'>4 or v">0. For these bands the isotopic shifts were measured from band maxima. The measured shifts confirm the vibrational numbering of the bands and also confirm the specific C- and D-state band assignments for the spectral region in which the two states overlap.

All evidence (vibrational intervals, electronic origin, and doublet splitting) supports the assignment of the majority of the REMPI spectrum to two-photon C $^2\Pi_r$ ↔ X $^2\Pi_r$ transitions.

The C $^2\Pi_r$ state possesses an outer shell electron configuration[4] of the type : $...(z\sigma)^2(y\sigma)^2(w\pi)^4(x\sigma)^2...(4p\pi)$. From this configuration the lowest energy ionization process will form the ground-state ion. For an SiCl ionization potential of 7.2 eV, laser ionization is possible from the two-photon prepared C $^2\Pi_r$ state by absorption of one photon more energetic than 517 nm, i.e. a 2+1 REMPI mechanism.

Between 431-447 nm the REMPI spectrum (Fig. 1) shows a progression of two-photon D $^2\Sigma^+$ (v'=0,1,2) ↔ X $^2\Pi_r$ (v"=0) bands which agree with those previously reported.[3,5] The positions of these lines are listed in Table 1. The vibrational interval in the D $^2\Sigma^+$ state is ~650 cm^{-1}. Hot bands originating from D $^2\Sigma^+$ (v'=0,1,2) ↔ X $^2\Pi_r$ (v"=1) transitions are also seen. Bands originating from the D $^2\Sigma^+$ ↔ X $^2\Pi_r$ system that lie to the red of the origin at 447 nm (i.e. v'=0 ↔ v">0) are obscured by the strong C $^2\Pi_r$ ↔ $^2\Pi_r$ signals. The D-state electron configuration is[6]: $...(z\sigma)^2(y\sigma)^2(w\pi)^4(x\sigma)^2...(4p\sigma)$. From this configuration the lowest energy ionization process will form the ground-state ion. Laser ionization can proceed through a 2+1 REMPI mechanism.

REFERENCES

1) G. Bruno, P. Capezzuto, G. Cicala, F. Cramarossa, Plasma Chem. Plasma Process. 6, 109 (1986).
2) C. S. Dulcey and J. W. Hudgens, J. Phys. Chem. 87, 2296 (1983).
3) W. Jevons, Proc. Phys. Soc. London 48, 563 (1936).
4) F. Melen, Y. Houbrechts, I. Dubois, B. L. Huyen, and H. Bredohl, J. Phys. B. 14, 3637 (1981).
5) G. A. Oldershaw and K. Robinson, J. Mol. Spec. 38, 306 (1971).
6) H. Bredohl, R. Cornet, I. Dubois, and F. Melen, J. Phys. B. 15, 727 (1982).

THREE-PHOTON IONIZATION IN LASER-EXCITED SODIUM BEAM

A.Vellante, L.J.Qin,* F.Giammanco, E.Arimondo
Department of Physics, University of Pisa, Pisa 56100,Italy

ABSTRACT

The ionization spectrum of a sodium beam has been investigated in the 5400Å-6100Å region at different laser intensity and sodium density conditions, in the aim of elucidating the various processes taking place in the weak plasma created by the ionization.

INTRODUCTION

During the last few years many papers reported the formation of Na_2^+ molecular ions and Na^+ atomic ions when a intense laser irradiates sodium vapor in a cell or in an atomic beam [1-5]. A possible process to produce Na_2^+ ions is multiphoton ionization(MPI) of sodium molecules[3,4]. In this case one would expect the following relations:

$$Na_2 + 3h\nu \longrightarrow Na_2^+ + e$$
$$i(Na_2^+) \propto [Na_2] \cdot I^3 = n \cdot I^3 \qquad (1)$$

where $i(Na_2^+)$ represents the detected molecular ion yield,n the molecular density and I the laser intensity. The complex molecular MPI spectrum will be characterized by the bound molecular constants of both ground and excited states of the neutral molecule.

For what concerns the atomic photo-ionization, strong ions signal should be found whenever the energy of one or two absorbed photon matches the energy separation from ground state to high-lying sodium levels. From the high-lying sodium levels, the ionization may be produced by absorption of another laser photon

* Fellow of the International Center for Theoretical Physics, Trieste, Italy, Permanent Address: Department of Physics, East China Normal University, Shanghai, China.

or blackbody absorption depending on the level position and the experimental conditions.[5] In the present experiment we have observed atomic ion signals when the atomic beam was irradiated by intense laser radiation at the 5497Å, 5787Å corresponding to two photon resonant transitions from 3S ground state to 6S and 4D levels respectively, and at 5890Å, 5896Å and 5688 Å wavelengths corresponding to one photon transitions from 3S to $3P_{1/2}$, $3P_{3/2}$ and from $3P_{3/2}$ to 4D respectively[2]. If the ionization from the excited upper levels takes place through laser photon absorption,the atomic ion signal depends on the laser intensity I through the following relations:

$$Na + 3h\nu \longrightarrow Na^+ + e$$
$$i(Na^+) \propto n \cdot I^3 \qquad \text{-------- (2)}$$

We report here the ionization spectrum of a sodium beam in the 5400Å-5915Å region and analyze the different mechanisms leading to atomic or molecular ion formation.

EXPERIMENTAL SET-UP

We used an excimer pumped dye laser , working with coumarin 153 (5220Å-6100Å),to irradiate a sodium beam. The atomic beam was created by an oven placed inside the vacuum chamber. The density of atomic beam was $\sim 10^{11}$ at/cm^3 at 200^{0}C oven temperature.

The ions produced synchronously with the laser pulse were accepted by a microchannel amplifier at the end of a 30 cm time-of-flight drift tube. Signal output from microchannel amplifier was amplified and fed to a gated boxcar integrator.

RESULTS AND DISCUSSION

A typical measured spectrum of three photon ionization irradiated by a dye laser in the 5400Å-5915Å region is shown in Fig.1. This Na^+ ionization spectrum was recorded with a laser intensity of 25 MW cm^{-2}(at maximum dye efficiency) and for atomic sodium density 10^{11} at/cm^3 The peaks labelled 1,2,4 in Fig.1

correspond to one-photon absorption resonant with the transitions $(3S-3P_{1/2})$, $(3S-3P_{3/2})$, $(3P_{3/2}-4D)$ respectively, and the 3,5 peaks correspond to two-photon absorptions resonant with the transitions (3S-4D), (3S-6S) respectively, followed by one photon ionization.

We measured the yield of ions as a function of atomic density n, which exhibits a linear dependence, following the relation (2).

The measured laser power dependences are in agreement with the the predicted behaviors. The 5890Å resonance exhibits a slope 2, because the first step (3S-3P) in the ionization process is saturated at pump power density 2.5 MW/cm^2. The same behavior was observed for the 5896Å line. The peak of the "band" near the 3S-4D resonance at 5777Å exhibits a slope 3 . The line at 5787 Å (3S-4D) exhibits at low power a slope 3, and at the saturation of the two photon transition a slope 1. Finally, after the saturation, a decreasing of the ion signal was observed.

The wide peaks 6,7,8 and the background of the peaks 1,2,4 could be explained as molecular ionization followed by a molecular dissociation.

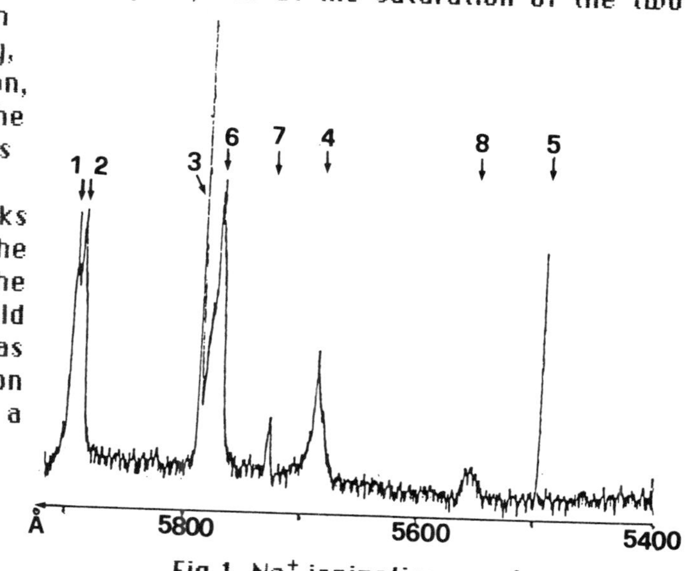

Fig.1, Na^+ ionization spectrum

REFERENCES

1, M.Allegrini, W.P.Garver, V.S.Kushawaha and J.J.Leventhal, Phys.Rev. A28 199 (1983)

2, I.M.Betrov, N.V.Fateev, Opt.Spec.(USSR) 54(6), (1983)

3, J.Keller and J.Weiner, Phys.Rev. A30, 213(1984)

4, F.Roussel, P.Breger and G.Spiess, Mol.Phys. 18, 3769 (1985)

5, C.E.Burkhardt, R.L.Corey, W.P.Garver, J.J.Leventhal, M.Allegrini and L.Moi, Phys.Rev. A34, 80(1986)

STATE MIXING IN COLLISIONS INVOLVING HIGHLY EXCITED BARIUM ATOMS

M. Allegrini†, E. Arimondo§, E. Menchi§, C. E. Burkhardt, M. Ciocca, W. P. Garver, S. Gozzini† and J. J. Leventhal, University of Missouri-St. Louis, St. Louis, Missouri 63121

J. D. Kelley
McDonnell Douglas Research Labs, St. Louis, Missouri 63166

ABSTRACT

Using two-step laser excitation it is shown that the state distributions of highly excited barium atoms are substantially altered by inelastic collisions with barium atoms. For "pure" Rydberg states the initially produced ns or nd eigenstates are rapidly mixed with nearly degenerate n'ℓ' states. For initially produced states that may be viewed as admixtures of Rydberg and doubly excited independent electron eigenstates, collisions convert the distribution to one with nearly pure, albeit state-mixed, Rydberg character. Broadening of the absorption profiles of the highly excited states is also observed.

INTRODUCTION

Over the past decade or more, many properties of Rydberg atoms have been studied[1] both theoretically and experimentally. While much of the experimental work has been performed using "one-electron" alkali-metal atoms, the two-electron alkaline-earth atoms[2] have also received considerable attention. Most previous studies have been directed toward elucidation of the intrinsic properties of these atoms. In contrast, the work reported here was initiated to study collisional effects on the Rydberg states of barium, a two-electron atom with high-lying energy levels readily accessible with available laser wavelengths.

It has been shown that the various states that result from a nominal 5d7d electronic configuration of atomic barium serve as perturbing states that alter the regularity of the nd Rydberg series.[2] For principal quantum numbers n above 20, the only significant perturber[3] is the one designated 5d7d $^1D_2$ which perturbs the J=2 Rydberg levels in the vicinity of n=26. This is illustrated in the partial term diagram shown in Figure 1. the fact that the eigenstates have varying relative Rydberg-valence state compositions in the vicinity of the perturber has important consequences with respect to the effects of collisions on these states. In this paper we report and discuss data that illustrate some of these effects.

Just as for the alkalis we expect collisional state changing,

†Permanent address: IFAM, Pisa, Italy.
§Permanent address: University of Pisa, Italy.

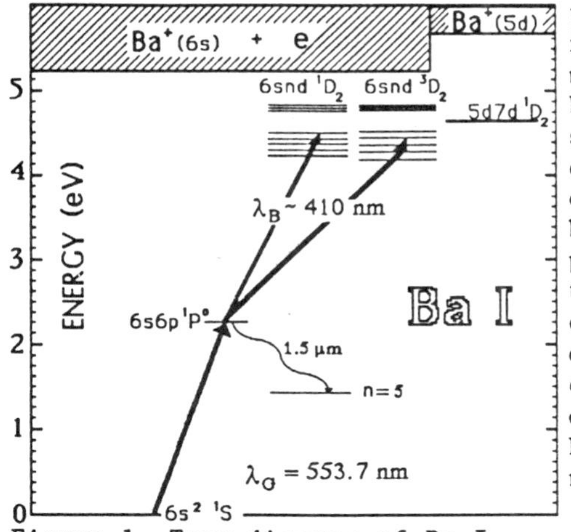

Figure 1. Term diagram of Ba I illustrating the effects of the 5d7d state on the regularity of the Rydberg series.

primarily $n\ell \rightarrow n'\ell'$. Indeed, we find this to be the case, but, not unexpectedly, only for the Rydberg component of the excited state wavefunction. As a consequence, we find that a collection of barium atoms which has been laser excited to a perturbed level with a substantial 5d7d fraction is rapidly converted, by collisions with other barium atoms, to a collection with a distribution over many nearly degenerate Rydberg states and essentially no 5d7d character.

In addition to state changing we also report our observation that the excitation profiles of the highly excited states broaden dramatically with increasing excited atom density, an effect that we deduce to be the result of quasistatic interaction between highly excited atoms during the laser excitation.[4]

The experiments were performed using a new apparatus employing a well-collimated beam of barium atoms and two grazing incidence dye lasers pumped by a single Nd:YAG laser. The collinear linearly polarized laser beams, one green and the other blue, intersected the atomic beam producing highly excited barium atoms, Ba^{**}, in two steps as illustrated in Figure 1. The atom density N in the beam was varied over the approximate range 10^8-10^{13} as estimated from the oven temperature and vapor pressure curves.

Two methods for detecting highly excited atoms were employed. At relatively low N, ~10^8-10^{10} cm^{-3}, field ionization with a preset delay time, 200 ns to 2μs, produced Ba^+ which were then detected with a CuBe particle multiplier. At the higher densities field ionization could not be used because of electrical breakdown of the vapor. The Rydberg states of interest in these experiments are however photoionized by room temperature blackbody radiation. The resulting Ba^+ were detectable allowing this method to be used over the entire range of atom densities. Only the BBPI results will be discussed here.

The laser beam intensities were 5 kW/cm^2 and 2 kW/cm^2 (maximum) respectively for the green and blue lasers. At these levels the green resonance transition was saturated, but the blue transition to the highly excited states was not. The intensity of the blue laser beam was reduced when necessary using neutral density filters. The bandwidth of the green laser beam was about

0.8 cm^{-1}. For most of the experiments the bandwidth of the blue laser was also 0.8 cm^{-1}. For experiments requiring careful comparison of absorption profiles, a reflection grating was used to decrease the bandwidth to 0.2 cm^{-1}.

RESULTS AND DISCUSSION

We can examine the effects of heavy body collisions on the excited state populations by comparing BBPI spectra acquired at different values on N. Fig. 2 shows three such spectra. Comparison of the spectrum in Fig 2(a) with a calculated spectrum

for the collision-free case shows that ionization at this lowest N-value is due primarily to BBPI with no collisional effects.

As N is increased [Fig. 2(a) - Fig. 2(c)], the d-state signal increases with respect to the s-state signal. this result is attributed to collisional state mixing and d-state "lifetime lengthening". That is, increasing N causes the initially prepared d (and s) states to be collisionally distributed over the complete ℓ-state manifold. Increasing the ℓ-value increases the radiative lifetime, allowing a larger fraction of initially prepared d-states to be observed. The effect enhances the d-to-s-state signal ratio because the d-state lifetimes are shorter than the s-state lifetimes. For the J=2 states around n=26 with substantial perturber (5d7d) fraction, this collisional state mixing converts such states to a distribution with essentially no perturber contricution.

Although these data suggest that the state-changing collisions that cause lifetime lengthening are between the highly excited barium atoms Ba^{**} and ground state, $6p^1P$ or $5d^1D$ barium atoms (the last three of which we designate as Ba) additional experiments were undertaken to investigate the possible effects of Ba^{**}-Ba^{**} collisions. First, we fixed the atom density N at a value comparable with that in Fig. 2b and varied the blue laser power density from maximum to 0.1 maximum and finally to 0.02 maximum, acquiring BBPI spectra at each laser power density. The three spectra were identical.

Since the <u>excited</u> atom density was varied by a factor of 50 in these constant N experiments, we infer that Ba^{**}-Ba^{**} collisions do not produce lifetime lengthening. We next reduced N by a factor of 50 while keeping the laser power density of the blue beam at maximum. This effectively reduces both N^{**} and N by a factor of 50. The resulting BBPI spectrum, exhibited increased s to d ratios, indicative of a tendency toward collision-free conditions. This clearly establishes Ba as the major state-changing (lifetime-lengthening) collision partner.

Examination of Fig. 2 shows that as N is increased the absorption profiles broaden. In order to assign the source of this

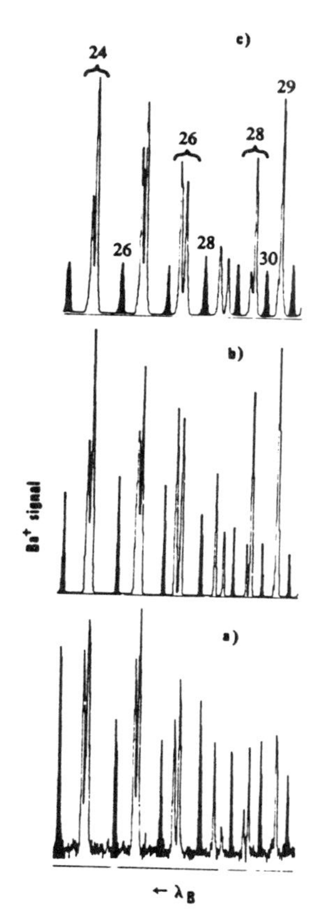

Figure 2. BBPI spectra taken at different values of N; 10^{11} cm^{-3}, 10^{12} cm^{-3} and 10^{13} cm^{-3}, for (a - c). The peaks for s-state excitations are shaded.

broadening we performed additional experiments using a narrow band (0.2 cm^{-1}) blue laser. States in the vicinity of n=40 were prepared with $N^{**} \sim 10^{11}$ cm^{-3} and $N \sim 10^{12}$ cm^{-3}. The linewidths were measured, and the blue laser power was decreased progressively to 0.1, the original level. N was kept constant. As the laser power was decreased, the linewidths decreased from about 0.6 cm^{-1} to about 0.2 cm^{-1}.

In a second set of experiments, N^{**} was kept constant, as measured by the BBPI signal, by increasing the blue power while decreasing N. The linewidths remained constant. Thus, the broadening is independent of N, but does depend on laser power for fixed N. These observations clearly show that the observed broadening mechanism is probably not collisional, but rather a quasistatic Rydberg-Rydberg interaction via the transition-dipole coupling matrix elements.[4]

ACKNOWLEDGEMENTS

This work supported by NSF grants PHY-8418075 and INT-8318024, DOE grant DE-FG02-84ER1327, the McDonnell Douglas Independent Research and Development Program, and the University of Missouri Weldon Spring Fund and US-NSF/Italy-CNR Bilateral Research Programs n. 19931 (M.A. and S.G.) and n.88.01280.02 (E.A. and E.M.) One of us (S.G.) wishes to acknowledge a Fulbright Fellowship. The authors wish to thank Professor T. F. Gallagher for useful discussions.

REFERENCES

1. For a review see Rydberg States of Atoms and Molecules, edited by R. F. Stebbings and F. B. Dunning (Cambridge Univ. Press, Cambridge, 1983).
2. See for example, M. Aymar, Phy. Rep. 110, 163 (1984).
3. T. F. Gallagher, W. Sandner and K. A. Safinya, Phys. Rev. A 23, 2969 (1981).
4. J. M. Raimond, G. Vitrant and S. Haroche, J. Phys. B 14, L655 (1981).

POLARIZATION AND DOPPLER MEASUREMENTS ON CN RADICALS FORMED IN THE 210 NM PHOTOLYSIS OF BrCN

A. Paul, I. McLaren, W.H. Fink and W.M. Jackson
Department of Chemistry, University of California, Davis, California 95616

ABSTRACT

$CN(X^2\Sigma^+)$ radicals have been produced from the photolysis of BrCN using a polarized 210 nm laser. The polarization of the individual rotational lines was determined and a slightly positive value is observed. The Doppler width of the individual rotational lines was also measured. Both of these observations are used to discuss the dynamics of the overall photodissociation process at this wavelength.

INTRODUCTION

Polarization and Doppler measurements of ground state fragments produced in the photolysis of simple molecules are useful for determining how the fragments are oriented during the dissociation process.[1,2] To date most of the work that has been reported on triatomic molecules has been with H_2O,[3] and ICN.[4,5,6] In the present paper the results obtained from polarization and Doppler measurement of the CN radical produced in the 210 nm photolysis of BrCN will be reported.

EXPERIMENTAL

The apparatus used for these experiments is schematically shown in Fig. 1.

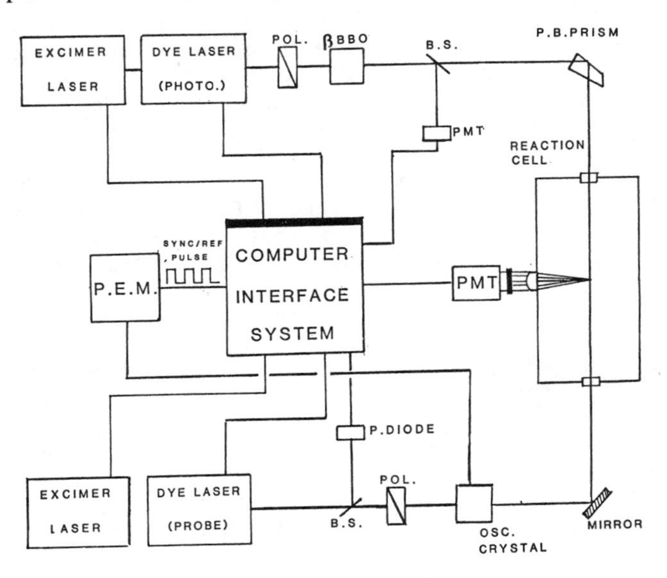

Figure 1. Schematic Diagram of the Experimental Apparatus

The output from a Lambda Physik excimer pumped dye laser is passed through a polarizer and then doubled in a beta barium borate crystal. Part of the output is sampled with a beam splitter and is used to normalize the signal on a shot to shot basis in a Laser Interface computer control system. Most of the output from the doubling crystal is analyzed by a Pellin Broca prism and passed through the observation cell. The analysis light source is generated with a Lambda Physik EMG 101 MSC excimer laser pumping a Lumonics Hyperdye 300 dye laser. Some of its output is also separated with a beam splitter and is used to normalize the signal on a shot to shot basis. The rest of the light goes through a polarizer and then into the oscillator crystal of a Hinds photoeleastic modulator. This crystal oscillator modulates the light at its natural frequency which is about 50 kHz. Since the lasers cannot be pulsed at this frequency we sense the sync signal from the photoelastic modulator and the computer interface system fires the lasers so that we can determine the LIF signal for a given number of shots in the parallel direction and then for the same number of shots in the perpendicular direction. The accuracy of the computer timing is such that we are able to reproduce the zero phase displacement within 1/3 of a degree and the lambda/2 displacement within 1/100 of a degree. The parallel and perpendicular intensities for each rotational line are obtained by integrating the area under the line for the parallel and perpendicular settings respectfully.

The Doppler line shapes are determined by scanning slowly over each line for a given polarization. Some of the Doppler line shape measurements were determined at delays long enough so that the radical had undergone a few collisions. The laser line width was then determined by assuming that the observed line shape was due to the triple convolution of two Gaussians. One of the Gaussians is due to the Maxwellian velocity of the CN radical and the other is due to the line shape of the laser. Once the experimental laser line shape was determined it was used in modeling all of the other lines.

RESULTS AND DISCUSSION

The polarization results are summarized in Fig. 2. From this Figure we see that

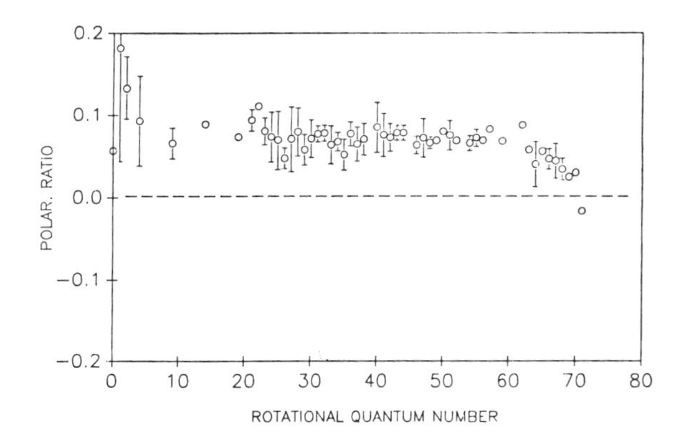

Figure 2. Polarization ratio of CN radicals produced in the 210 nm photolysis of BrCN.

we have only a slightly positive polarization ratio. A slightly positive polarization ratio indicates a bias toward the parallel intensity for our geometrical configuration. The polarization that we observe is similar in sign and magnitude to the polarization observed earlier in ICN.[5,6] The positive polarization response for BrCN extends to larger rotational quantum numbers than it does for ICN and we observe a sharp decrease in the polarization at N equal to 65. We also observe a sharp cutoff in the rotational populations of the v"=0 and v"=1 levels at about the same N" value indicating that there is an angular momentum constraint rather than an energy conservation constraint in these rotational populations. The nature of these constraints will be discussed in a later paper.

The Doppler profile of a translationally cooled line is shown in Fig. 3a. The fit to the line is a convolution of two Gaussian Distributions. One corresponding to a Gaussian line shape for the laser, and the other to the Maxwellian velocity distribution of the CN radical. The shapes of the cooled lines have been used to determine that the FWHM of the laser line is 0.15 cm^{-1}.

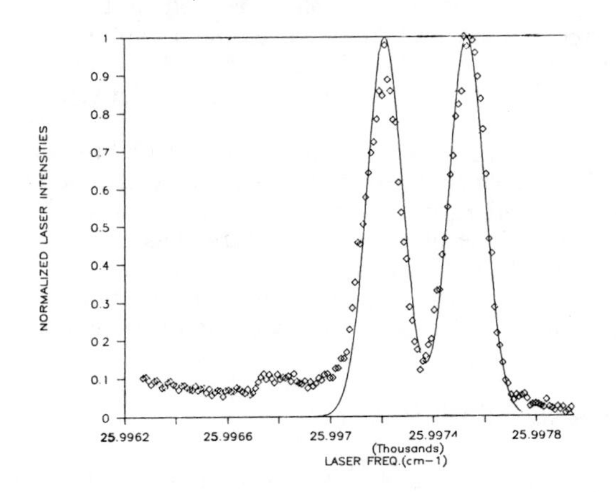

Figure 3a.

Doppler profile of the R-32 line of CN produced in the 210 nm photolysis of BrCN after thermolization via collision with argon. The solid line is the fit to the experiment described in the text. The error bars on the data are about the size of the diamonds.

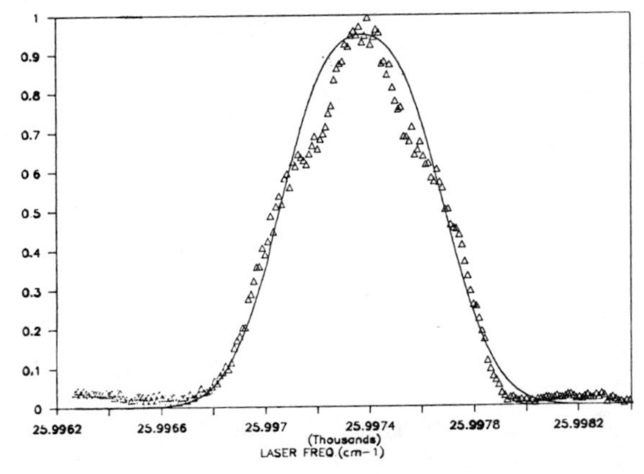

Figure 3b.

A nascent Doppler profile of the R-32 line. The solid line is a convolution of three Gaussians. The error on the data is about the same as in Fig. 3a.

Gaussian distributions may be used to fit the cooled lines but, as Fig. 3b shows, they cannot be used to fit the nascent line shape. This is reasonable because we expect that the nascent line will have fixed velocities corresponding to dissociation processes which form bromine atoms in the $^2P_{1/2}$ or $^2P_{3/2}$ states.

Qualitatively we have observed that these shapes require some mixture of a perpendicular transition in order to obtain the observed line shapes. Because there is an anisotropy in the photodissociation process, a transformation has to be made from the body centered coordinates to the laboratory coordinates.[2,7,8] Our analysis is further complicated by the fact that the F_1 and F_2 components are not equal and completely resolved. We initially tried to account for this by translationally cooling the CN radical. We have found that this is not valid at 210 nm. It may be that the recoil velocity is high enough so that collisions start to equilibrate these components. As a result of this we are in the process of modeling the nascent lineshapes where we vary the ratio of the F_1 and F_2 components. We will determine the ß from the anisotropy of the line and only use the recoil velocity and the F_2/F_1 ratio as adjustable parameters.

SUMMARY

The polarization and Doppler profiles of individual rotational lines produced in the photolysis of CN radicals in the 210 nm photolysis of BrCN have been measured. Both the lineshapes and the polarization ratios are consistent with an anisotropy in the photodissociation process. Modeling studies are currently underway to quantitatively determine the parameters required to fit these lineshapes.

Acknowledgments: William M. Jackson and Albert Paul gratefully acknowledge the support of NSF under grant number CHE-85-20029.

REFERENCES

1. J. P. Simons, J. Phys. Chem. 91, 5378 (1987).
2. P. L. Houston, J. Phys. Chem. 91, 5388 (1987).
3. H. J. Krautwald, L. Schneider, K. H. Welge, M. N. R. Ashfold, Faraday Discuss. Chem. Soc. 81, in press.
4. I. Nadler, D. Mahgerefteh, H. Reisler and C. J. Wittig, J. Chem. Phys. 82, 3885 (1985).
5. G. E. Hall, N. Sivakumar, P. L. Houston, J. Chem. Phys. 84, 2120 (1986).
6. M. A. O'Halloran, H. Joswig and R. N. Zare, J. Chem. Phys. 87, 303 (1987).
7. C. H. Green and R. N. Zare, J. Chem. Phys. 78, 6741 (1983).
8. R. N. Dixon, J. Chem. Phys. 85, 1866 (1986).

STATISTICAL ANALYSIS OF THE MICROWAVE-OPTICAL DOUBLE RESONANCE SPECTRA OF NO_2: ERGODICITY WITHOUT LEVEL REPULSION?

Kevin K. Lehmann
Department of Chemistry, Princeton University, Princeton, NJ 08544

Stephen L. Coy
Department of Chemistry, Harvard University, Cambridge, MA 02138

ABSTRACT

Current theory has predicted that in the semiclassical limit, spectra of molecules should have statistical properties for both the transition energies and eigenvalues which depend only upon the nature of classical dynamics for the same Hamiltonian. It is predicted that both spectral ergodicity (complete breakdown of approximate selection rules) and eigenvalue repulsion should be signatures of classical chaos. In order to observe these statistical properties, spectra must not contain an inhomogeneous distribution of initial levels. By using the technique of microwave detected, microwave-optical double resonance we have obtained spectra of NO_2 starting in known lower states, and thus have been able to determine both the statistical properties of both the transition intensities and for the eigenvalue positions. We find, that in conflict with the predictions of currently available theory, the spectrum shows spectral ergodicity without significant eigenvalue repulsion.

INTRODUCTION

In recent years, there has been much interest in what, if any, are the quantum mechanical analogs of chaos in classical mechanics. While the field initially generated much controversy, it has become almost universally accepted that in the semiclassical limit, the statistical properties of fluctuations in both eigenvalue position and transition intensity depend only upon the nature of the classical dynamics at that energy. Numerical experiments at limited levels of excitation have tended to support these theoretical predicitons but such studies can never establish the universal character of the predictions.

While most of the tests of these new ideas have come from computer calculations, it is possible to use molecular spectroscopy as a way of obtaining highly accurate eigenvalues and transition strengths at much higher excitation levels than is currently possible with available computer methods. The difficulty is that molecular spectra have great complexity do to the population of a large number of different levels at room temperature. This complexity hides statistical properties that one would like to look for in the spectrum. The most powerful method of removing this inhomogeneous effect, is to use double resonance spectroscopy, where one can obtain a spectrum of a single quantum eigenstate at a time, and thus determine the essential or homogeneous complexity in the spectrum.

We choose to use the technique of microwave detected, microwave optical double resonance to study the NO_2 optical spectrum[1]. The detailed experimental setup has been published, so we will only describe the essential features of the technique. NO_2 molecules are exposed to strong, cw microwave radiation which is resonant with a transition between two well defined rotational levels in the ground vibronic state. The sample is then pumped with the output of a pulsed dye laser operating at ~590 nm. If this laser pumps either of the two levels connected by the microwave radiation, then following the laser pulse will be an optical nutation signal which under our experimental conditions lasts about 1 μsec. This nutation will produce microwave absorption or emission depending upon whether the upper or lower microwave level is pumped. The nutation signal is detected with a gated intergrator and then read by a computer which scans the dye laser. In this way, the difference spectra of the two sampled levels is obtained and because the spectra are well resolved, few lines are lost because of blending in the spectra. We can detect NO_2 optical transitions as weak as 10^{-7}/cm/torr.

Figure 1 shows the stick spectra of the observed transitions from four lower levels. Also compared with the stick spectra are the vibronic band origions observed by Smalley *et al* [2] in the jet spectrum of NO_2. If normal rotational selection rules held, we would expect to see only two lines/band from the $8_{0,8}$ and $10_{0,10}$ states. As is easily seen, the number of observed lines is considerably more. Careful consideration of the number of observed lines, combined with double resonance measurements that have been published, establishes that we observe essential all rovibronic levels of the correct overall symmetry to have allowed transitions from the ground state[3,4].

This result establishes that the spectrum shows the qualitative feature of 'spectral ergodicity' but it is important to try to be more quantitative about this. Eric Heller[6] has proposed a statistic, F, which is computed from the number and distribution of the intensity of lines in the spectrum. This number should be near zero for a 'regular' spectrum (one with approximate selection rules) and 1/3 for the idea 'chaotic' spectrum, where the transition amplitudes have a Gaussian distribution. We calculate values between .25-.30 for our observed spectra when we make corrections for only week mixing of spin rotation levels of NO_2. Thus by the statistical properties of the observed intensity of transitions, we can assert that NO_2 is close to the chaotic ideal for spectral ergodicity. As a word of caution, some of the lines in the optical spectrum are strongly saturated, but we have established that the qualitative shape of the intensity distribution does not change when corrections are made for saturation.

We will now look at the statistics of the energy of the eigenstates. If we have a 'regular' quantum spectrum, it is predicted that the local spacing of eigenvalues should be Poisson, i.e., the same as random numbers. It is believed that for a chaotic classical Hamiltonian, the eigenvalues should have the same statistics as those of an ensemble of matrices, whose matrix elements are Gaussian random numbers, known as the GOE The statistical properties of this ensemble have been worked out in great detail. It is known that this GOE spectrum displays two features, level repulsion and spectral rigidity. Level repulsion refers to the fact that the distribution of energy spacings shows few near degeneracies, the levels 'push' apart. Spectal rigidity is more subtle, and refers to the fact that the fluctuations in eigenvalue density grow much more slowly with energy width than for a random spectrum. The most rigid spectrum would be a picket fence of levels where the fluctations in number of levels is at most one level. In contrast a regular spectrum, which has Poisson statistics shows fluctuations of the square root of the averge number of levels. The GOE, spectrum shows fluctuations that go as the square root of the log of the average number of levels and is thus much more like a rigid fence than a random spectrum.

We have studied our NO_2 spectrum and looked for both level repulsion and spectral rigidity. Much to our surprise, neither was observed, the spectra displayed Poisson statistics, implying a regular quantum spectrum! The effect of missing levels tends to turn a GOE spectrum into a Poisson spectrum, but we have made a careful analysis of the change in the above statistical properties as the number of observed levels is changed, and based on the number of levels we observe, we can rule out GOE statistics for the spectum before partial experimental observation.

Our results are thus inconsistent with the current theory of 'quantum chaos'. Recent work, done by Perch and Demtroder[5] has establshed that the statistics of the vibronic band origins, as opposed to the rovibronic levels probed here, show both level repulsion and spectral rigidity consistent with chaotic motion. The present results appear to imply that the molecule has some type of regular rotational motion despite the lack of any selection rules on the internal rotational quantum number K. This work establishes that the types of 'chaos' that molecules can display are much richer than our current theories allow and that these theories need to be developed further before the statistical behavior of NO_2 can be understood.

References

1. K.K. Lehmann and S.L. Coy, J. Chem. Phys. **81**, 3744 (1984).
2 R.E. Smalley, L. Wharton, and D.H.Levy, J. Chem. Phys. **63**, 4977 (1975).
3 K.K. Lehmann and S.L. Coy, J. Chem. Phys. **83**, 3290 (1985).
4. S.L. Coy, K.K. Lehmann, and F.C. Delucia, J. Chem. Phys. **85**, 4297 (1986).
5. G. Persch and W. Demtroder, to be published in Ber. Der Bunsen-Gesellschaft Fur Physikalische Chemie.
6. E.J. Heller, Faraday Discuss. Chem. Soc., **75**, 141 (1983).

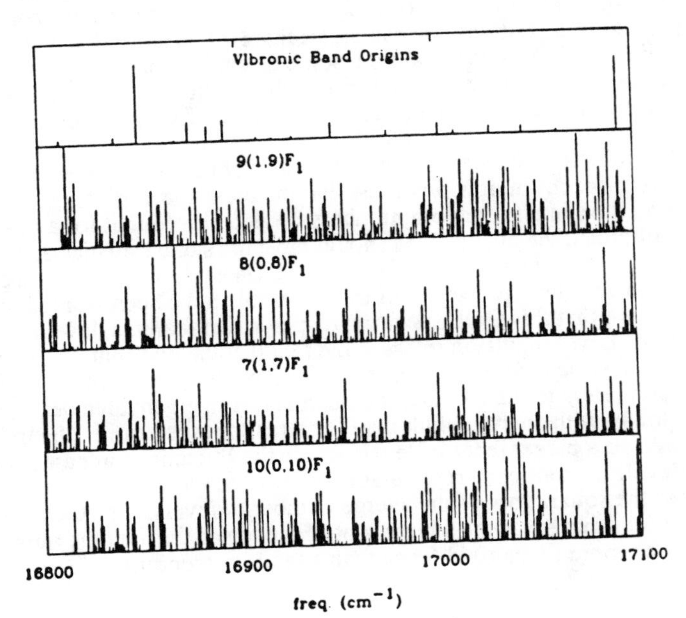

Figure 1. Stick spectrum of observed MODR transitions.

OBSERVATION OF INFRARED ROTATION-VIBRATION TRANSITIONS OF HNO^-

Harold C. Miller
Univ. of Oregon, Eugene, Oregon 97403

John W. Farley
Univ. of Nevada, Las Vegas, 89154

ABSTRACT

We have studied the nitroxyl anion, HNO^-, a bent triatomic molecule, by autodetachment spectroscopy using a color-center laser operating near 3000 cm^{-1}. Over 175 transitions in the 1-0 band of the N-H stretch in the ground $^2A''$ electronic state have been observed with an accuracy of 0.01 cm^{-1}. Our results represent a 10^4-fold improvement in resolution over the best previous results[1], obtained by photoelectron spectrometry. Spin-rotation splittings and asymmetry doubling have been resolved.

INTRODUCTION

Negative molecular ion are important in terrestrial and laboratory plasmas. The electron correlation effects are more prominent in anions than cations or neutrals. The autodetachment of molecular anions is interesting because it results from terms of the Hamiltonian which violate the Born-Oppenheimer approximation. However, there is relatively little detailed information about anion structure, compared to cations or neutral molecules, and there is very little state-resolved information about autodetachment rates.

EXPERIMENT

This experiment employs the same apparatus previously used[2] to make high-resolution measurements of the structure of $^{14}NH^-$ and $^{15}NH^-$, including the first measurement of hyperfine structure in a molecular anion. HNO^- was produced in a hot-cathode discharge in ethyl nitrite (CH_3CH_2ONO) and formed into a 3-kV beam. After mass-selection, the ion beam coaxially overlaps a single-mode infrared laser beam, and the ions are Doppler-tuned into resonance. The laser-excited ionic states autodetach, producing fast neutral particles, which are detected by collisional ejection of secondary electrons. The coaxial beams technique has sub-Doppler resolution (5 MHz), because of a well-known velocity-bunching effect.

We have obtained the first measurement of the N-H

stretch vibrational frequency, and obtained rotational resolution for the first time, making possible accurate measurement of the molecular geometry. Over 175 transitions have been observed in the spectrum in the range 2930-3150 cm^{-1}, with quantum numbers K_a=5←4 through 10←9. We have observed 38 transitions of a ^rQ branch in the 7←6 subband near 2994 cm^{-1}. The spectrum displays the subtle effects of asymmetry doubling and spin-rotation doubling, which allow an independent way to assign the quantum numbers. The new data is a 10^4-fold improvement in resolution compared with the best previous work by Ellis and Ellison[1], using photoelectron spectrometry.

Least-squares fits have been performed to the spectrum using Hamiltonians of varying degrees of sophistication, ranging from a rigid rotor and a symmetric rotor (the molecule is a nearly symmètric prolate top) to an asymmetric rotor. The quantum numbers have been assigned.

Unlike the case of NH^-, the autodetachment widths are all about 10 MHz, corresponding to an autodetachment lifetime of 16 nsec, showing no variation with quantum number. Based on the propensity rules of Simons[3], the autodetachment rate should be greater if the vibration involves an N-O stretch or the bend. Further work along these lines is underway in our lab.

ACKNOWLEDGMENTS

Financial support by NSF grants PHY85-02489 and PHY87-19190 and by AFRPL contract F04611-87-K0021 is gratefully acknowledged.

REFERENCES

[1]H. Benton Ellis and G. Barney Ellison, J. Chem. Phys. 78, 1983.
[2]M. Al-Za'al, H. C. Miller, and J. W. Farley, Chem. Phys. Lett. 131, 56 (1986); H. C. Miller, M. Al-Za'al, and J. W. Farley, Phys. Rev. Lett. 58. 2031 (1987); M. Al-Za'al, H. C. Miller, and J. W. Farley, Phys. Rev. A. 35, 1099 (1987); H. C. Miller and J. W. Farley, J. Chem. Phys. 86, 1167 (1987).
[3]J. Simons, J. Am. Chem. Soc. 103, 3971 (1981).

VIBRATIONAL AUTOIONIZATION AND FANO-ANTIRESONANCE IN RYDBERG SERIES OF NAPHTHALENE

JACK A. SYAGE AND JOHN E. WESSEL

The Aerospace Corporation
P.O. Box 92957
Los Angeles, CA 90009

ABSTRACT

Asymmetric Fano-line profiles have been observed by autoionization for specific vibrational excitation of Rydberg series in jet-cooled naphthalene. The line shapes were analyzed in terms of degenerate resonance interference between optically active discrete and continua states. Observed linewidths were as narrow as 6 cm^{-1} for naphthalene and 3 cm^{-1} for naphthalene-d_8.

INTRODUCTION

In recent years, multiphoton excitation studies in molecular beams have revealed the presence of well-resolved molecular autoionizing Rydberg states in large molecules[1-3]. In each case, the autoionization spectra consisted of relatively sharp and symmetric transitions superimposed on an otherwise continuous background. In the present work, we report clear evidence of asymmetric Fano line profiles observed in autoionizing Rydberg series of jet-cooled naphthalene[4]. These unusual lineshapes, referred to as antiresonances, occur by degenerate resonance interference of a discrete state with a continuum when both carry oscillator strength from a common initial or intermediate state[5].

RESULTS AND ANALYSIS

Our pulsed supersonic molecular beam apparatus and time-of-flight (TOF) mass spectrometer have been described in detail elsewhere[6]. Briefly, the output of two independently tunable dye lasers are frequency doubled to generate ω_1 and ω_2. Naphthalene was introduced into a differentially pumped molecular beam apparatus via a pulsed supersonic nozzle heated to 70°C and under 40 psi He. Two-photon ionization signals arising from one-color excitation by the laser scanning ω_2 were suppressed by operating a shutter that blocked the ω_1 pulse on alternate triggers. The ω_2 only signal was then subtracted from the $\omega_1+\omega_2$ signal, providing the signal-to-noise ratio required to resolve the weak Rydberg series superimposed on the continuua.

A vibrationally selected autoionizing excitation spectrum is presented in Fig. 1 for a region above the adiabatic ionization level, IP_o. Excitation by ω_1 to the $\bar{8}^1 7^1$ nontotally symmetric combination vibration of S_1 followed by ω_2 excitation reveals a strong Rydberg series, R_n, converging to the presumed $^+\bar{8}_1 7_1$ ionization threshold at 1537 cm^{-1} (versus 1422 cm^{-1} in S_1). Additional structure is evident in Fig. 1, which appear to be part of other Rydberg series. The prominence of the $R_n(\bar{8}^1 7^1)$ series by ω_2 excitation is indicative of a propensity for $\Delta v = 0$ transitions. Excitation of the totally symmetric 7^1 vibration of S_1 (at 988 cm^{-1}) by ω_1 followed by ω_2 excitation

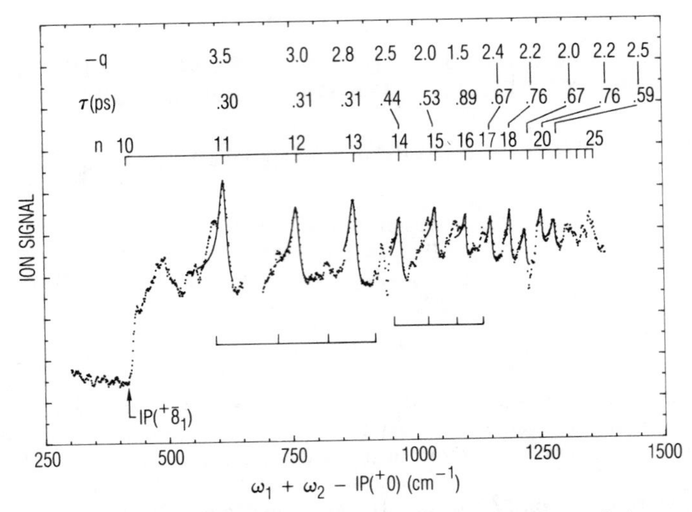

Figure 1. Double resonance autoionization spectrum of napthalene recorded via the intermediate state $S_1(\bar{8}^1 7^1)$. The energy scale is relative to the vertical ionization potential at 65665 cm^{-1}. A major Rydberg series is identified and Fano lineshape analysis represented by the best-fit values for q and τ(ps). Parts of two other Rydberg series are identified but not assigned.

gave a similar Rydberg series for 7^1. Other transitions were studied that did not include ν_7 in S_1, however no Rydberg autoionization was observed. The autoionizing features in Fig. 1 were fitted to the n=11 to 21 members of a Rydberg series that converges to IP_o + 1537 cm^{-1} and has a quantum defect of $\delta = .10$. The low value of δ is consistent with a d series, which in turn is consistent with the large transition moment from the $S_1(\pi, \pi*)$ intermediate state if we consider a qualitative $\Delta l = 1$ selection rule.

The involvement of a smooth ionization continuum in the present case allows us to use a simple formulation[5] of Fano's original theory for the antiresonance lineshape $L(\epsilon)$;

$$L(\epsilon) = \frac{(q+\epsilon)^2}{(1+\epsilon^2)}$$

$$\epsilon = 2\hbar(\omega - \omega_o)/\Gamma \quad ; \quad q = \frac{A\mu_{SR}}{|A|\mu_{Sj}V_{Rj}}$$

where ω and ω_o are the excitation and line center frequency, respectively; Γ is the spectral linewidth. The line profile index q consists of the transition dipole moments from the intermediate S_1 level to the Rydberg R_n and continua j states, μ_{SR} and μ_{Sj}, respectively; the cosine of the angle between the transition dipole vectors, A; and the discrete continua interaction responsible for the vibrational autoionization, V_{Rj}. Calculated Fano-line profiles are superimposed onto the observed data in Fig. 1 and the best fit values for q and the lifetime $\tau = (2\pi\Gamma)^{-1}$ are reported for each member of the Rydberg series R_n.

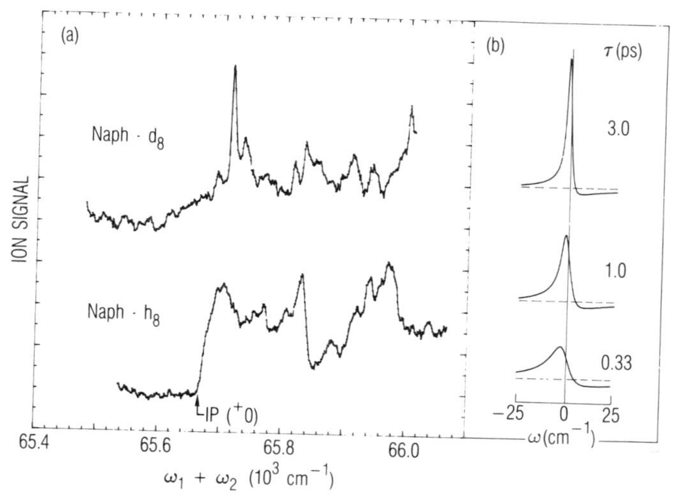

Figure 2. (a) Double resonance autoionization spectra recorded via the intermediate state $S_1(7^1)$ for napthalene-d_8 (ω_1 = 32957 cm^{-1}) and naphthalene-h_8 (ω_1 = 33014 cm^{-1}). (b) Calculated lineshapes are illustrated as a function of the relative change in the coupling term V_{Rj}. Variations in V_{Rj} are represented by the parameter τ(ps) ($\propto 1/V_{Rj}^2$).

The effect of varying the coupling term V_{Rj} was observed by recording autoionizing excitation spectra for naphthalene-d_8. A portion of the spectrum for specific excitation of the $R_n(7_1)$ series is presented in Fig. 2. The most evident features are lineshapes that are narrower and more symmetric than those observed for naphthalene-h_8. These effects are due to weaker vibrational-electronic coupling and are in accord with the predictions of the Fano-model described above. The relative change in the antiresonance lineshape is illustrated in Fig. 2 for variations in the value of V_{Rj}. The observed linewidth of 4 cm^{-1} for the major feature in Fig. 2 corresponds to an autoionization time of greater than 1.3 ps. (Linewidths as narrow as 3 cm^{-1} have been measured for other lines.) This time is undoubtedly larger if we take into account the rotational broadening (~ 2 cm^{-1}) inherent in our expansion conditions.

REFERENCES

1. J. Hager, M. A. Smith, and S. C. Wallace, *J. Chem. Phys.* **84,** 6771 (1986).
2. A. Goto, M. Fujii, and M. Ito, *J. Phys. Chem.* **91,** 2268 (1987).
3. M. Fujii, T. Kakinuma, N. Mikami, and M. Ito *Chem. Phys. Lett.* **127,** 297 (1986); T. Suzuki, N. Mikami, and M. Ito, *Chem. Phys. Lett.* **120,** 333 (1985); M. Fujii, N. Mikami, and M. Ito, *Chem. Phys.* **99,** 193 (1985).
4. J. A. Syage and J. E. Wessel, *J. Chem. Phys.,* **87,** 6207 (1987).
5. U. Fano, *Phys. Rev.* **124,** 1866 (1961); J. Berkowitz, *Photoabsorption, Photoionization, and Photoelectron Spectroscopy,* (Academic Press, New York, 1979).
6. (a) J. A. Syage and J. E. Wessel, *J. Chem. Phys.* **87,** 3313 (1987); (b) J. A. Syage and J. E. Wessel, *Appl. Opt.* **26,** 3573 (1987).

DIODE LASER KINETIC SPECTROSCOPY

D. R. Lander, C. B. Dane, R. F. Curl, G. P. Glass, F. K. Tittel
Rice University, Houston, Texas 77251

Diode laser kinetic spectroscopy has been used to obtain the high resolution spectrum of the CH stretching fundamental of the formyl radical (HCO). The analysis of this band was performed combining the diode data with the difference frequency laser spectroscopy data of C. B. Moore's group at Berkeley on this band.

The diode laser system and its method of operation for normal absorption spectroscopy has been described previously[1]. The diode laser kinetic spectroscopy experimental arrangement is shown in Fig. 1. HCO was produced in the multipass absorption White cell by XeCl (308nm) flash photolysis of acetaldehyde. The excimer laser was introduced just below the "D" mirrors of the White cell and intercepted by a beam block above the notched mirror. Thus the infrared interference from the excimer was small. Within the confines of this geometry the infrared diode probe beam and the excimer photolysis beam were as collinear as possible. The transient absorption signal is detected by acquiring the output of the signal detector immediately before and after the excimer laser flash with a transient digitizer (Biomation 805).

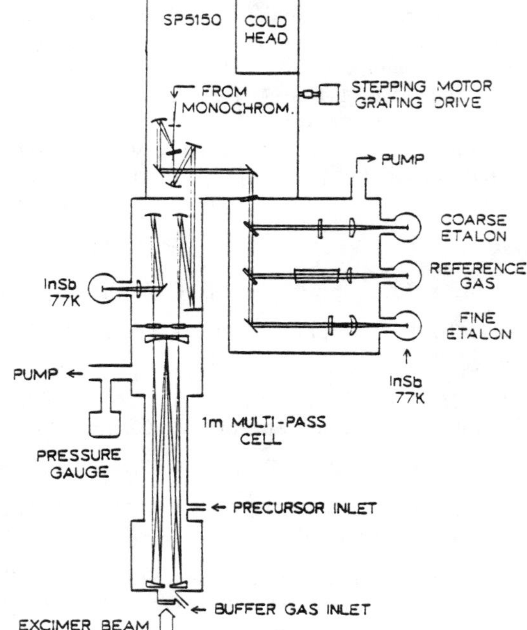

Fig. 1 Experimental set-up. The optical arrangement allows data to be collected from the the experiment as well as the three diagnostic channels in a single scan.

Typical operating conditions were total pressure=7.5Torr, acetaldehyde pressure=2.5Torr, acetaldehyde flow rate=60scc/min, excimer repetition rate=20Hz, and excimer pulse energy=75mJ. The HCO frequencies were calibrated against N_2O in the reference cell. Generally the accuracy of the resulting line positions was limited by the accuracy of the calibration lines which is typically ±0.001 cm^{-1}.

The rotational assignment was made by comparing the intensity of the second to first member of the two obvious a-type Q-branches observed thereby determining their K quantum numbers. The rotational, spin-rotational, and centrifugal distortion constants of the two vibrational states involved in the CH stretch were determined[2] by least squares fitting of ~750 diode laser and difference frequency absorptions. The resulting ν_1 band origin was 2434.47790(24) cm^{-1}. This agrees satisfactorily with the matrix isolation CH stretching frequency[3] of 2483cm^{-1} and the stretching frequency estimated from laser fluorescence[4] of 2432±20cm^{-1}.

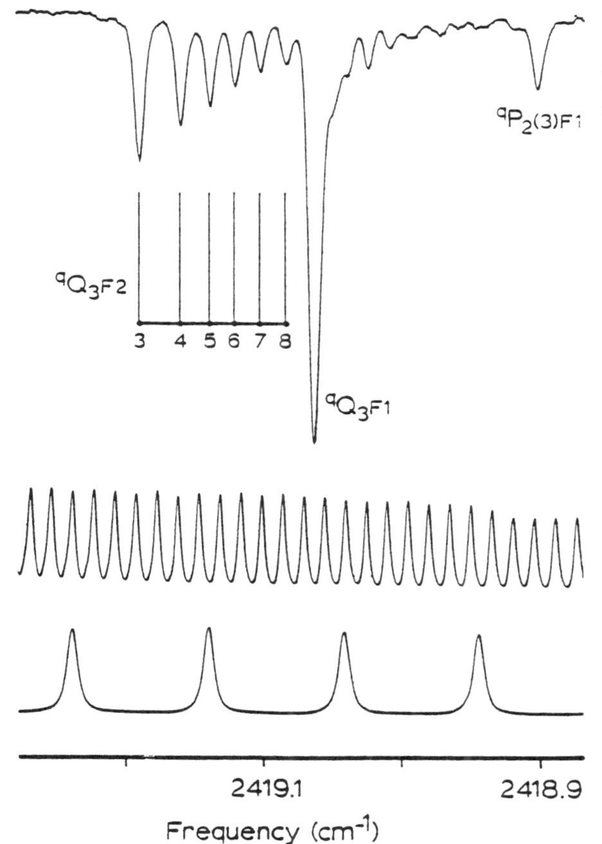

Fig. 2. K=3 parallel Q-branches of the CH stretching fundamental of HCO. 50 transient digitizer points each of 500nsec duration are averaged before and after each excimer shot and subtracted. The laser is stepped 20MHz in scanning with 30 shots averaged per step. Also shown are the 500MHz and 3GHz etalon fringes used in calibration.

The kinetic spectroscopy technique also enables time resolved study of a transient species. The time dependence is used to measure rate constants, and the signal amplitudes are used to measure product yields. Methods have been developed for frequency locking the diode laser to a PZT tunable cavity , which can then be tuned to the peak of an absorption line and the time dependence studied. HCO is relatively long lived as can be seen by the transient digitizer trace of its decay shown in Fig. 3. In our system, the principal decay mechanism is the recombination of two HCO molecules as can be seen by the roughly second order decay curve. The system was tested by measuring HCO decay rates in the presence of varying pressures of O_2. The resulting rate constant for the reaction $HCO+O_2 \rightarrow CO+H_2O$ agrees with the published measurement[5] within experimental uncertainty.

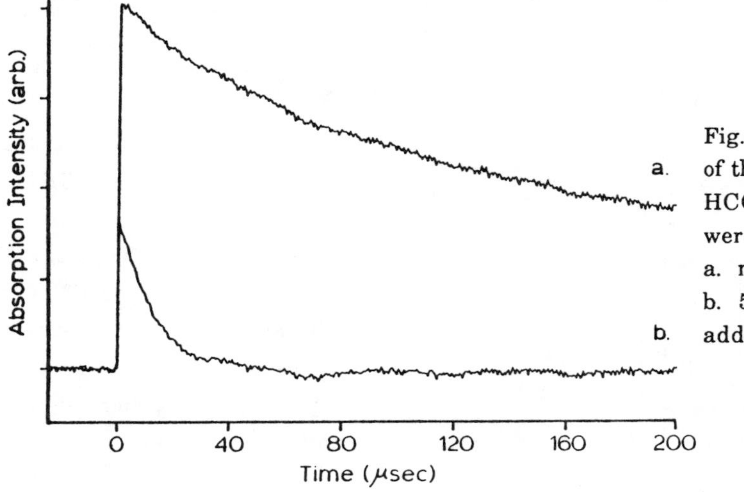

Fig. 3. Time decay of the signal from the HCO. 3000 traces were averaged.
a. no oxygen added.
b. 500mTorr oxygen added.

This work was supported by National Science Foundation Grant CHE-8504171 and by the Robert A. Welch Foundation under Grant C-586.

1. C. B. Dane, R. Bruggemann, R. F. Curl, J. V. V. Kasper, and F. K. Tittel, *Appl. Opt.* **26**, 95 (1987).
2. C. B. Dane, D. R. Lander, R. F. Curl, F. K. Tittel, Y. Guo, M. I. F. Ochsner, and C. B. Moore, accepted for publication in *J. Chem. Phys.*.
3. D. E. Milligan and M. E. Jacox, *J. Chem. Phys.* **51**, 277 (1969).
4. B. M. Stone, M. Noble, and E. K. C. Lee, *Chem. Phys. Lett.* **118**, 83 (1895).
5. R. S. Timonen, E. Ratajczak, and D. Gutman, *J. Phys. Chem.* (in press).

MODE SPECIFIC VIBRATIONAL RELAXATION IN WEAKLY BOUND BINARY COMPLEXES

R. E. Miller
Department of Chemistry
University of North Carolina
Chapel Hill, N.C. 27514

ABSTRACT

Sub-Doppler resolution infrared laser spectroscopy has been used to study a variety of binary and tertiary complexes formed in a free jet expansion, yielding detailed structural and dynamical information. The vibrational predissociation of these complexes is found to be strongly dependent on the nature of the intramolecular vibration initially excited indicating that the process is highly non-statistical. A correlation is observed between the vibrational predissociation lifetime and the vibrational frequency shift associated with complex formation which emphasizes the importance of the stretching dependence of the intermolecular potential in both.

1. INTRODUCTION

The opto-thermal detection method has recently been developed in our laboratory to the point where it can now be used to obtain near infrared spectra, under super-cooled and sub-Doppler resolution conditions, for a wide range of van der Waals and hydrogen bonded complexes [1-4]. Spectra of this type provide not only accurate molecular constants, from which molecular structures can be determined, but also detailed information concerning the dynamics which follows vibrational excitation of the complex. As we will see in later sections of this report, these two aspects are completely complimentary.

The experimental apparatus used in the present study [1] consists of a liquid helium cooled bolometer (minimum detectable power of 10^{-14} W/$\sqrt{Hz}$) which is positioned in order to monitor the total energy of a highly collimated molecular beam formed from a free jet expansion. An F-center laser is crossed with the molecular beam and tuned into resonance with a ro-vibrational transition of the molecules. The result is either an increase in energy delivered to the bolometer when the excited state of the molecule is long lived, or a decrease due to the dissociation of a weakly bound complex in a time which is short with respect to the flight time from the laser crossing region to the bolometer. In the latter case the translational energy of the resulting fragments is removed from the beam. Due to the highly collimated nature of both the laser and molecular beams, as well as the narrow linewidth of the laser, spectra recorded in this way have sub-Doppler resolution.

2. SPECTROSCOPY

In view of its importance in a wide range of fields, including quantum [5] and atmospheric [6] chemistry, the water dimer has been the subject of a number of gas phase microwave [7,8] and infrared [9,10] spectroscopic studies. In all of the latter studies, however, rotational fine structure has not been resolved so that unique species identification and detailed molecular constants could not be obtained. As a result, we have recently undertaken a study of this system using the apparatus discussed above. Several vibrational bands associated with the water dimer have been observed near the "free" O-H stretching region which show well resolved rotational fine structure. Figure 1 shows the spectrum associated with one of these bands which, although not yet fully assigned due to the complications arising from tunneling motions [11], is unambiguously attributable to the water dimer. Results of this type clearly hold considerable promise in obtaining detailed information on the potential surfaces, structure and dynamics of the water dimer, as well as for larger water complexes. In view of the complexities associated with the various bulk phase properties of water, this microscopic approach to the problem should be of considerable help in understanding this uniquely important system.

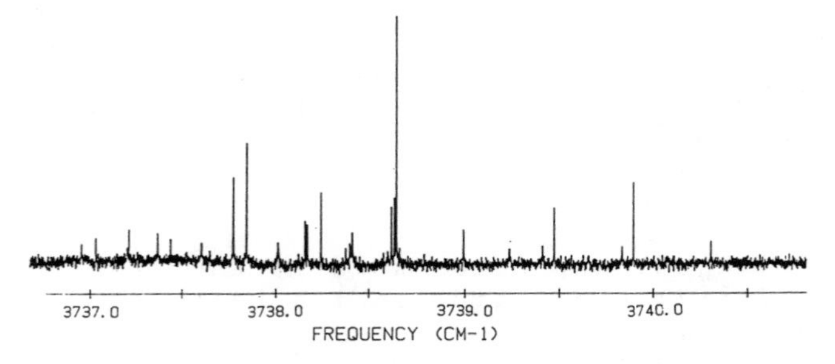

Figure 1. Infrared spectrum of $(H_2O)_2$

The C_2H_2-HF spectrum shown in Figure 2 serves as an excellent illustration of many of the experimental details discussed above. This spectrum corresponds to a perpendicular band of a near prolate symmetric top resulting from excitation of the asymmetric C-H stretch of this T-shaped complex [2]. The calculated spectrum was obtained by fitting the observed transitions to a rigid asymmetric rotor Hamiltonian yielding accurate molecular constants for both the ground and excited vibrational states. It is worth noting here that the three to one nuclear spin statistics resulting from the C_2 axis of this molecule have been included in the spectral simulation. Since the infrared spectra give reliable relative intensities, the effects of nuclear spin degeneracy can often be seen, which aids considerably in the determination of structures for these species. Indeed, only a single isotopic form was needed in order to determine the structure of carbon dioxide dimer [4].

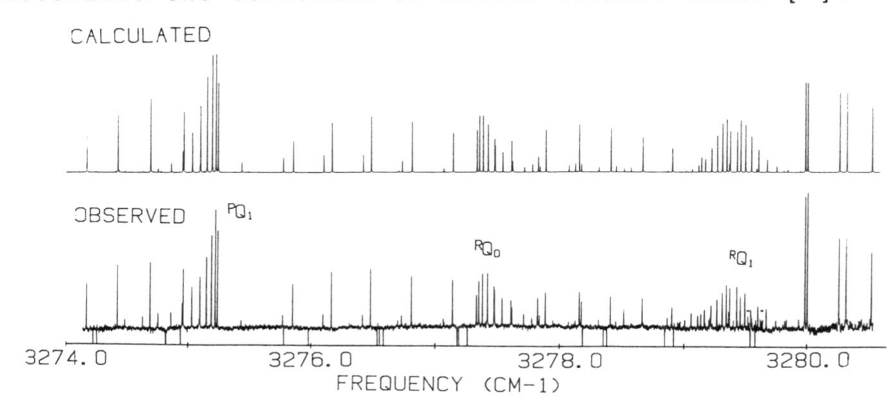

Figure 2. Experimental and calculated spectra of C_2H_2-HF

3. PREDISSOCIATION DYNAMICS

In addition to the C_2H_2-HF transitions seen in Figure 2, acetylene monomer transitions, associated with the ν_3 fundamental and associated hot bands, are also observed, only with opposite sign. From this it is clear that the complex dissociates before reaching the bolometer. Indeed, a careful comparison between the lineshapes for the monomer and complex reveals that the latter are homogeneously broadened due to the finite lifetime of the excited state. The Lorentzian component of the observed lineshape corresponds to a lifetime of 3.6 ns. As discussed elsewhere [12], there is strong evidence to support the association of these linewidths with the predissociation lifetime of the complex.

In an earlier study of the C_2H_2-HF complex [2] we obtained similar results for excitation of the H-F stretch. In this case, the lifetime obtained from the homogeneous linewidth was 0.8 ns. This shorter lifetime can be easily understood if one considers that the HF bond is directly coupled to the weak intermolecular bond while the C-H stretch is essentially decoupled from the intermolecular motion. Even more dramatic mode dependence is observed for the case of the HCN-HF complex [13]. Figure 3 shows two spectra associated with the C-H and H-F stretches of this linear complex. Once again, the HF proton is directly involved in the hydrogen bond, leading to a short vibrational predissociation lifetime upon excitation of the H-F stretch, while for the "free" C-H stretch the lifetime is long. These results demonstrate not only the non-statistical behavior characteristic of these "small" complexes but also the importance of the coupling between the intramolecular and intermolecular motions in vibrational predissociation dynamics. It is obvious from these results that, before quantitative calculations on the vibrational predissociation dynamics can be carried out for these systems we must first determine a potential surface which includes the intramolecular stretching dependence. This fact has been previously emphasized by LeRoy and co-workers [14] in their studies of the hydrogen-rare gas systems.

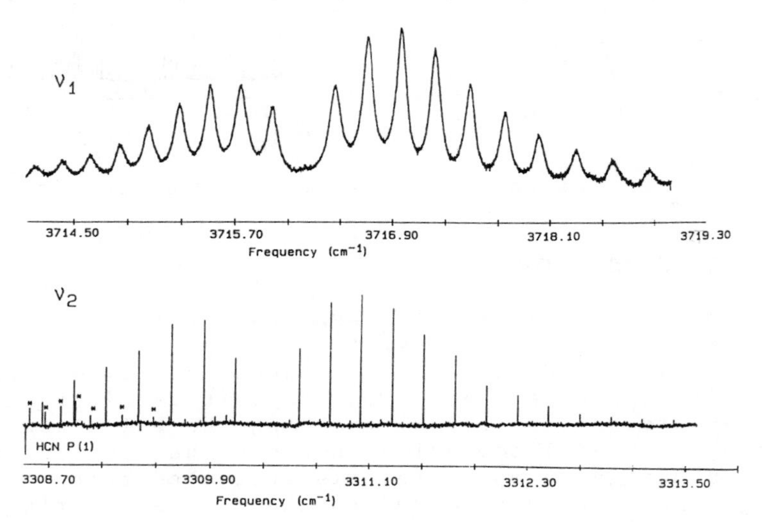

Figure 3. Spectra associated with the H-F (ν_1) and C-H (ν_2) stretch fundamentals of HCN-HF

Although the stretching dependence of intermolecular potentials is difficult to obtain experimentally, the infrared spectra discussed here do provide some information which can be used to at least constrain such a potential. Indeed, (1) the vibrational dependence of the rotational constants is related to the associated change in the position of the potential minimum [15], (2) ground and excited state dipole moments contain information on the vibrational dependence of the anisotropy of the potential [1], (3) fundamental to intermolecular "hot" band frequency shifts are directly related to the vibrational dependence of the potential [16] and (4) the monomer to complex frequency shift depends on the change in the well depth of the intermolecular potential due to vibrational excitation. The future will most certainly see the use of this type of spectroscopic data to constrain the stretching dependence of intermolecular potentials for subsequent use in dynamical calculations.

Molecule	Shift (cm^{-1})	Lifetime (ns)
N_2-HCN	9.5	80.0
OC-HCN	24.0	2.6
C_2H_2-HF	167.1	0.8
C_2H_2-HF	18.5	3.6
C_2H_4-HF	179.7	0.33
OC-HF	117.4	0.9
N_2-HF	43.2	22.0
N_2O-HF	83.2	0.22
ONN-HF	61.4	1.5
$(HCN)_2$	3.1	>120.0
$(HCN)_2$	69.8	6.12
$(CO_2)_2$	0.8	53.0
$(N_2O)_2$	2.4	>80.0
CO_2-HF	55.1	1.1
HCN-HF	1.1	14.0
HCN-HF	245.2	0.058
$(HF)_2$	32.0	12.0
$(HF)_2$	93.0	0.5
NH_3-HF	793.0	0.081
H_2-HF	11.3	27.0
Allene-HF	192.1	0.11
C_3H_6-HF	169.0	0.103

Table 1.

For all of the systems we have studied to date a strong correlation can be seen between the monomer to dimer frequency shift and the lifetime of the complex. More specifically, the vibrational predissociation lifetime of the complex is found to decrease with increasing frequency shift. Although this is qualitatively expected in light of the above discussion,

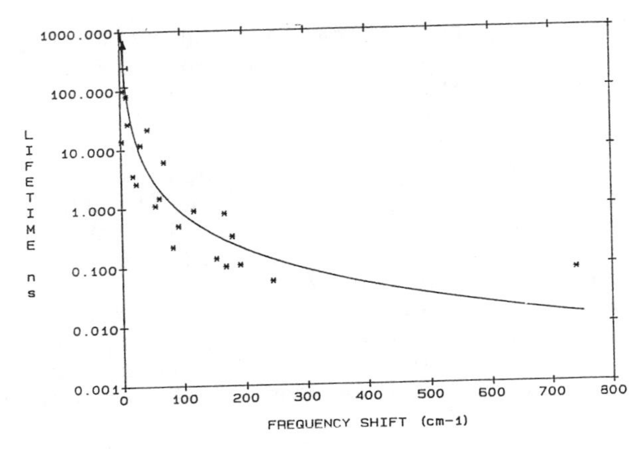

Figure 4.

since both depend of the strength of the coupling between the intramolecular and intermolecular motions, it is perhaps surprising that the correlation is as good as that shown in Figure 4. Each data point in this plot represents a lifetime and frequency shift for a given vibrational mode of a particular complex. As indicated in Table 1, the data covers a wide range of binary complexes. The solid line drawn through the data has the functional form;

$$\tau\ (s) = 8*10^{-6}/(\text{shift}\ (cm^{-1}))^2.$$

Although there is some scatter about this line it clearly gives quite a reasonable description of the data. As a result, a justification of this functional form seems in order. In fact, we have shown elsewhere [17] that this form can be obtained from a Fermi's Golden rule treatment of vibrational predissociation if one assumes that the translational energy of the fragments is small for all of these systems so that the associated DeBroglie wavelengths are long. The implication is, therefore, that low translational energy dissociation channels are readily available in all of the systems shown in Table 1 and that it is not the energy or momentum gaps [18,19] which control the rate of dissociation but rather the strength of the coupling between the intramolecular and intermolecular motions. Although this will certainly not be the case for simpler atom-diatom systems [14,15], where the density of low translation energy exit channels can be quite low (making the momentum gaps large), it most likely does apply to the vast majority of molecular complexes.

In conclusion, infrared spectroscopy of molecular beams has become a powerful new tool for the study of weakly bound complexes. The information obtained from these spectra can be used to determine intermolecular potential surfaces (which in favorable cases include the intramolecular stretching dependence), accurate molecular structures and detailed dynamical information.

4. ACKNOWLEDGMENTS

This research was supported by a grant from the National Science Foundation (CHE-86-03604).

5. REFERENCES

[1] Z.S. Huang, K.W. Jucks and R.E. Miller, J. Chem. Phys. 85 (1986) 3338; 85 (1986) 6905; K.W. Jucks, Z.S. Huang and R.E. Miller, J. Chem. Phys. 86 (1987) 1098.

[2] Z.S. Huang and R.E. Miller, J. Chem. Phys. 86 (1987) 6059.

[3] K.W. Jucks and R.E. Miller, J. Chem. Phys. 86 (1987) 6637.

[4] K.W. Jucks, Z.S. Huang, D. Dayton, R.E. Miller and W.J. Lafferty, J. Chem. Phys. 86 (1987) 4341.

[5] D.J. Swanton, G.B. Bacskay and N.S. Hush, Chem. Phys. 82 (1983) 303 and ref. contained therein.

[6] H.A. Gebbie, W.J. Burrough, J. Chamberlain, J.E. Harris and R.G. Jones, Nature 221 (1969) 143.

[7] T.R. Dyke, K.M. Mack and J.S. Muenter, J. Chem. Phys. 66 (1977) 498.

[8] J.A. Odutola and T.R. Dyke, J. Chem. Phys. 72 (1980) 5062.

[9] R.H. Page, J.G. Frey, Y.R Shen and Y.T. Lee, Chem. Phys. Lett. 106 (1984)373.

[10] D.F. Coker, R.E. Miller and R.O. Watts, J. Chem. Phys. 82 (1985)3554.

[11] T.R. Dyke, "Structure and Dynamics of Weakly Bound Molecular Complexes", ed. A. Weber, NATO ASI Series C, vol 212 (1986)43.

[12] R.E. Miller, Science, to be published

[13] D.C. Dayton and R.E. Miller, Chem. Phys. Lett., in press.

[14] J.M. Hutson, C.J. Ashton and R.J. LeRoy, J. Phys. Chem. 87 (1983)2713.

[15] C.M. Lovejoy, M.D. Schuder and D.J. Nesbitt, J. Chem. Phys. 85 (1986) 4890.

[16] K.W. Jucks and R.E. Miller, J. Chem. Phys., in press.

[17] D.C. Dayton, Z.S. Haung, K.W. Jucks and R.E. Miller, to be published.

[18] G.E. Ewing, J. Chem. Phys. 71 (1979) 3143.

[19] G.E. Ewing, J. Phys. Chem. 91 (1987) 4662.

OPTOGALVANIC SIGNAL IN THE B-X SYSTEM OF HgBr: EFFECT OF DISCHARGE VOLTAGE AND LASER POWER

V. Kumar, A.K. Rai and D.K. Rai
Physics Department, Banaras Hindu University, Varanasi-221005, India

ABSTRACT

Laser Optogalvanic (LOG) signals have been detected from a D.C. discharge in flowing $HgBr_2$ vapour. Different laser lines from a 4 watt Ar^+ laser have been used for excitation. The signals are identified with transitions in the B-X system of this molecule and variation in the signal strength with discharge voltage and input laser power has been investigated. The LOG signal is seen to vary with laser power and the variation is wavelength dependent. An explanation for this dependence is offered.

INTRODUCTION

The spectroscopic study of diatomic mercury bromide has gained in importance due to the observation of laser emission in the visible B-X system. But till now the actual process whereby HgBr is formed in the upper $B^2\Sigma^+$ state from the $HgBr_2$ vapour is not very well understood. The principal excitation mechanism is believed to involve the following steps [1]

$$HgBr_2(^1\Sigma^+) + e \rightarrow HgBr_2(^{1,3}\Sigma_u) + e$$

$$HgBr_2(^{1,3}\Sigma_u) \rightarrow HgBr(B^2\Sigma^+) + Br(^2P)$$

$$HgBr(B^2\Sigma^+) \rightarrow HgBr(X^2\Sigma^+) + h\nu$$

The present work is also supportive of the above mechanism.

Laser optogalvanic (LOG) techniques can accurately monitor the population of atomic/molecular states in a discharge column if the laser frequency can be made to coincide with a pair of states of the atomic/molecular species present in the discharge. It was therefore felt useful to investigate the LOG effect in a D.C. discharge through HgBr vapour.

EXPERIMENTAL

The experimental set up is very similar to the one used in our earlier work [Rai et al [2]]. Pure mercuric bromide (Merch 99.9% pure) in powder form is placed in a small bulk like cavity provided in the discharge tube itself. A reasonably quiet discharge with the characteristic blue colour is achieved at a discharge voltage of around 500 V. Different lines from a 4 watt Ar^+ laser are used for excitation.

RESULTS

The LOG signal (in units of mV) obtained when the discharge through $HgBr_2$ is illuminated by the chopped-radiation from the different lines of the Ar^+ laser have been measured. The change in the LOG signal as the discharge voltage is varied from 540 volts to 690 volts in steps of 10 volts is also measured for each line separately. In addition the effect of varying the laser output power (for each line) on the signals have also been studied. In order to monitor the decay of the B state the simple emission spectrum of the discharge has also been recorded at some typical values of the discharge voltage.

Most of the prominent bands of the B-X system of HgBr coincide with wavelength 4880 A° and 5017 A° of the Ar^+ laser and we find that the LOG signal when 4880 A° and 5017 A° lines are used as exciting radiation first increases with the increase in the discharge voltage and then decrease quite rapidly on further increasing the discharge voltage as shown in fig.1. We find that the LOG signals in both the cases increases with increasing laser power but tends to saturate near the maximum of the output power level that is accessible to us as shown in fig.2.

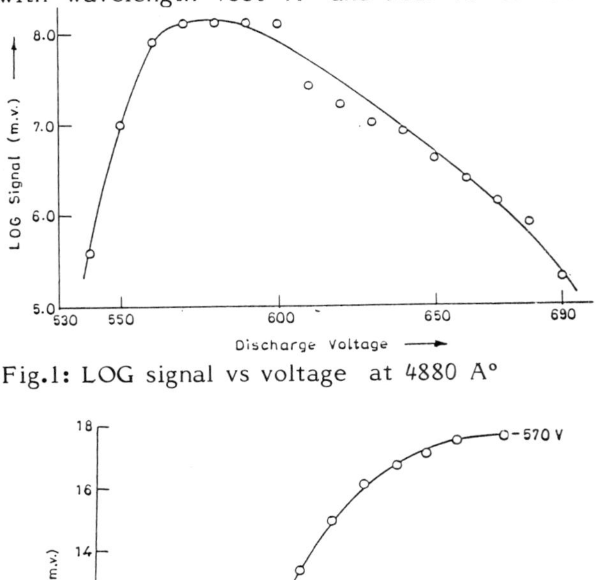

Fig.1: LOG signal vs voltage at 4880 A°

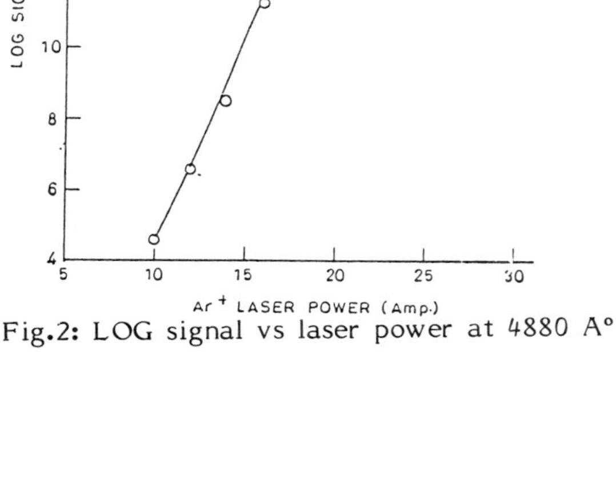

Fig.2: LOG signal vs laser power at 4880 A°

None of other lines of the Ar^+ laser (with wavelength of 4965 4765, 4650 and 4580 A°) coincides with any prominant band head in the B-X system. We found that the change of the discharge voltage does not cause any signicant change in the LOG signal obtained with these lines. The increase in laser power does cause an increase in the LOG signal followed by a tendency to saturate, but the actual signal as well as the rate of increase is relatively small.

DISCUSSION

To explain the observed variation of the LOG signal with applied voltage we make the following proposal :

1. As the applied voltage is increased the mean energy of the electrons in the discharge would increase and this increase may effect the rate of production of HgBr from $HgBr_2$. The HgBr is assumed to be in the B(v = 0) level (very rapid vibrational relaxation of the B-state is assumed).

2. Increasing the discharge voltage has the effect of increasing the intensity of the B — A emission [Kumar et al[3]]. Therefore the effective population in the B state is lowered due to this transition so that the laser radiation will lead to reduced stimulated emission and this will decrease the LOG signal.

We have compare the relative efficiencies of the 4880 A° and the 5017 A° lines of the Ar^+ laser in producing LOG signal in HgBr. We find that the latter line is approximately 2.5 times as efficient as the former in producing an LOG signal. This is partly attributed to the fact that the (0-22) band which coincides with the 5017 A line has a much large visually estimated intensity as compared to the (0-18) band coincident with the 4880 A line.

CONCLUSION

Our study provides indirect additional support for the mechanism of formation of HgBr $(B^2\Sigma^+)$ molecules by the dissociation of $HgBr_2$. The utility of the LOG technique to monitor the population of the $B(^2\Sigma)$ level is demonstrated. The relatively greater efficiency of the 5017 A line as compared to the more intense 4880 A line in producing a LOG signal may provide an explanation for the fact that lasing is observed near 502 nm but not near 4880 A°.

REFERENCES

1. W.L. Nighan and R.T. Brown, J. Appl. Phys. 53, 7201 (1982).
2. A.K. Rai, S.B. Rai, S.N. Thakur and D.K. Rai, Chem. Phys. Lett. 138, 215 (1987).
3. V. Kumar, A.K. Rai, S.N. Thakur and D.K. Rai, Chem. Phys. Lett. (in press).

COLLISIONALLY INDUCED ABSORPTION IN THE CALCIUM - RARE GAS COMPLEX NEAR 4575 Å

J. Coutts, S. K. Peck and J. Cooper
Joint Institute for Laboratory Astrophysics
University of Colorado and National Bureau of Standards
Boulder, CO 80309-0440, USA

Transitions of higher multipole moment than electric dipole are termed "forbidden transitions."[1] It is possible, however, to induce substantial oscillator strength into a forbidden transition when the radiator undergoes collisions with perturbers. These collisionally induced transitions have been qualitatively discussed for many years[2] and have attracted attention from the laser design community, with regard to proposals for high power laser amplifiers.[3,4] Several theoretical treatments of collisionally induced processes have proven convenient in interpreting observed effects, and indeed the decomposition of the net effects into the various physical mechanisms responsible remains a topic of current interest.[5]

As part of our experimental program to investigate the effect of collisions on forbidden lines,[6] we have probed the collision complex formed between calcium atoms and rare gas atoms with pulsed dye laser light of wavelength close to the 4575 Å $^1S_0 \rightarrow {}^1D_2$ electric quadrupole transition in Ca I. In this paper we concentrate on the collisionally induced absorption spectrum of the Ca/Xe complex; the effects due to other rare gas partners will be briefly addressed.

From the simple theory of, e.g., Gallagher and Holstein,[7] we may estimate the signal levels for a vapor cell based experiment. The collisionally induced absorption coefficient per unit bandwidth can be estimated as

$$k_\omega d\omega \sim [\mathrm{Ca}][\mathrm{Xe}] \times 10^{-39} \mathrm{cm}^{-1}, \qquad (1)$$

and for ND:Yag dye laser technology (roughly 50 μJ of laser light), a calcium density of some $10^{14}\,\mathrm{cm}^{-3}$ and a xenon density of some $10^{19}\,\mathrm{cm}^{-3}$, we expect to create some 10^6 Ca $^1D_2$ states. Unfortunately, these states decay, in free space, via intersystem transitions with typical natural lifetimes of the order of milliseconds, resulting in an unacceptably low count rate. We have therefore designed a laser probe technique, shown in Fig. 1. One pulsed dye laser excites the Ca/Xe collision complex via the collisionally induced process, and then a second pulsed dye laser lifts the free atom Ca $^1D_2$ states to a high lying $^1P_1$ level. A substantial fraction of the $^1P_1$ states cascade down to the $^1P_1$ resonance level, and resonance radiation is observed. The advantages of this scheme are (a) that the frequency of the resonance radiation is far from any laser frequency permitting discrimination against Rayleigh scattering and (b) the Ca $^1D_2$ states give rise to resonance radiation evolving over a timescale of the order of 100 ns. This count rate is easily measurable. The disadvantage of this scheme is that any collisional loss mechanism extant for the Ca $^1D_2$ states reduces net signal. For the Ca/Xe collision partners, the collisional energy

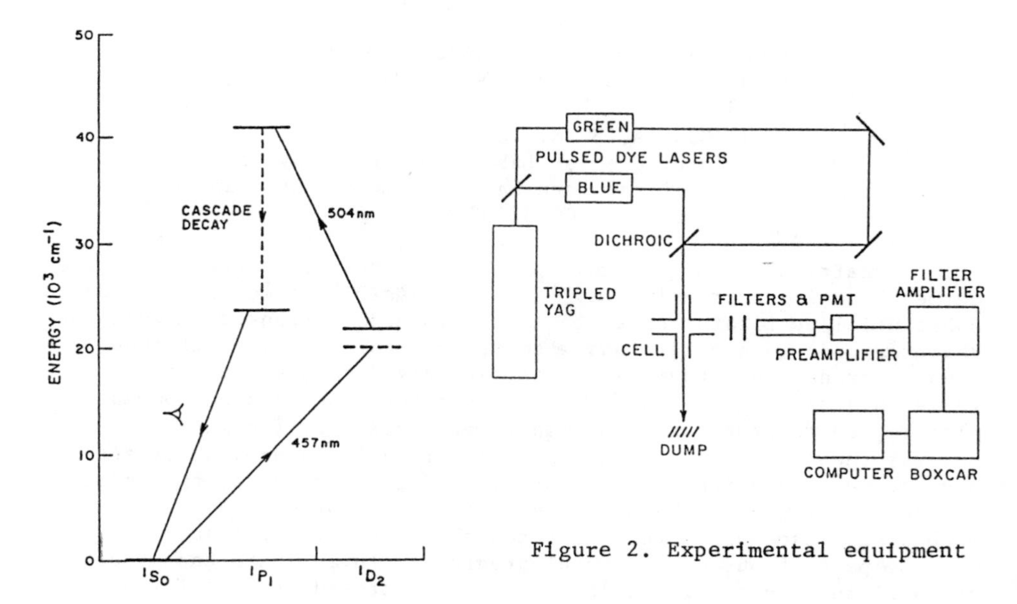

Figure 1. Probe technique

Figure 2. Experimental equipment

transfer cross section for the process $^1D_2 \rightarrow {}^3D_J$ is of the order of 1 Å^2. At buffer gas pressures of the order of 1 atmosphere the $^1D_2 \rightarrow {}^3D_J$ process is the dominant loss mechanism for Ca $^1D_2$ states. (Indeed, the appearance of triplet states can be monitored.) However the lifetime for $^1D_2 \rightarrow {}^3D_J$ transfer becomes comparable to 100 ns at buffer gas pressures of 50 Torr, in the case of xenon.

Reducing the buffer gas pressure to switch off the collisional loss implies a reduction in total signal, of at least one order of magnitude. However, we were still able to measure signals to roughly 5% accuracy with our experimental apparatus, shown in Fig. 2. Briefly, the two colors of dye laser pulse were combined on a dichroic mirror and sent through a well-designed calcium vapor cell, which was run with 50 Torr of buffer gas, at cell/cold finger temperatures of 650°C/625°C respectively. Resonance fluorescence was measured in sidelight via a 420 ± 5 nm interference filter and a photomultiplier (EMI type 9829). Roughly 10 photoelectrons per laser pulse were observed. Preamplified signals were passed to a "timing filter" amplifier (EGG Ortec type 474) and then to a boxcar integrator (SRS type 250). An IBM compatible microcomputer (Wisetek XT) running SRS 265 software logged the data.

By varying the 457 nm blue laser frequency, lineshape data for the collisionally induced process could be detected. We used 401 data point sets, and each point in Fig. 3 is the result of the subtraction of two sets, one with both lasers operating, and one with just the blue laser operating. In this way collisional redistribution into the Ca $^1P_1$ state could be discriminated against. The data shown in Fig. 3 took some 2 hours to acquire and may be subject to a 10% systematic drift, nonlinear in frequency.

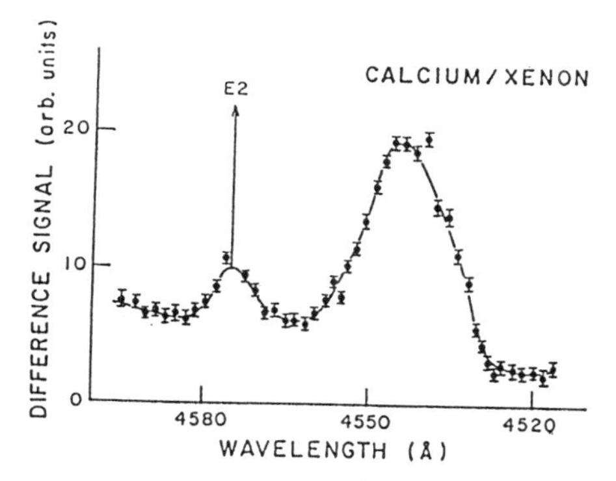

Figure 3. Calcium xenon collisionally induced signal

There are two distinct features in this spectrum. There is a large blue and weak red component extending to detunings of some kT/ħ associated with transient molecular effects. There is also a component that decreases in importance with increasing detuning around the Ca $^1D_2 \leftarrow {}^1S_0$ 4575 Å E2 line. This feature may be explained in terms of a perturbative expansion in terms of adiabatic, time dependent wavefunctions connecting asymptotically to atomic orbitals.[9] Finding such regions has prompted the most comprehensive theoretical investigations into collisionally induced phenomena. This region depends on the rare gas polarizability for its strength, and experiments run with other rare gas partners show only the molecular components surviving.

In conclusion, we have examined the collisionally induced processes associated with calcium rare gas collisions. We can indicate regions in the Ca/Xe spectrum that are the result of molecular transitions as well as regions describable within the framework of line-broadening type calculations. We are now proceeding to investigate the effects of inter-collisional coherences on the collisionally induced oscillator strength.

This work was supported by grant AFOSR 84-0027 to the University of Colorado, and J. Coutts was supported in part by a Lindemann Fellowship for 1986-87, extended for 1987-88, as administered by the English Speaking Union of the Commonwealth.

References

1. R. Garstang, in Atomic and Molecular Processes, edited by D. R. Bates (Academic Press, New York, 1962), p. 1.
2. M. Lapp, Phys. Lett. 23, 523, (1966).
3. J. R. Murray and C. K. Rhodes, J. Appl. Phys. 47, 5041 (1976).
4. P. S. Julienne, M. Krauss and W. Stevens, Chem. Phys. Lett. 38, 374 (1976).
5. G. Alber and J. Cooper, Phys. Rev. A 33, 3084 (1986).
6. J. Coutts, S. K. Peck, R. Stoner and J. Cooper, J. Appl. Phys. in press.
7. A. Gallagher and T. Holstein, Phys. Rev. A 16, 2413 (1977).
8. J. J. Wright and L. C. Balling, J. Chem. Phys. 73, 4, 1617 (1980).
9. K. Ueda and K. Fukuda, J. Phys. Chem. 86, 678 (1982).

FAR WING LASER ABSORPTION AS A PROBE OF REACTIVE COLLISION DYNAMICS

P. D. Kleiber*, A. M. Lyyra†, K. M. Sando†,
A. K. Fletcher*, and W. C. Stwalley*†
Center for Laser Science and Engineering,
University of Iowa
Iowa City, IA 52242-1294.

In a recent paper, we have described our experimental and theoretical studies of the far wing absorption profiles of the MgH_2 collision system. The process may be written symbolically as

$$Mg(3s\ ^1S_0) + H_2 \rightarrow (MgH_2)$$

or

$$(MgH_2) + \hbar\omega_L \rightarrow (MgH_2)^*$$

followed by

$$Mg(3s\ ^1S_0) + H_2 \rightarrow Mg^*(3p\ ^1P_1^0) + H_2$$

or

$$Mg(3s\ ^1S_0) + H_2 \rightarrow MgH(v",J") + H.$$

Here we report studies of the analogous MgD_2 collision complex, probing for isotope effects. The reactive absorption profiles in this case (for both low "J" and high "J" rotational product states) show a pronounced structureless red wing absorption and an unusually strong blue wing absorption, and are essentially identical to our published MgH_2 results[1]. Thus we have observed no essential differences between the MgH_2 and MgD_2 reaction dynamics. These experiments are primarily sensitive to entrance channel effects. The observations are, therefore, consistent with the model for the reaction as being "entrance channel controlled" for both isotopic species.

In this paper we also report studies of the Mg-rare gas 3s $^1\Sigma^+$ - 4s $^1\Sigma^+$ molecular bands. These results will be compared with semiquantitative theoretical predictions based on the model potential curves of Malvern[2] including an assumed transition dipole-moment function. Reasonably good agreement is obtained, supporting our basic interpretation for these collision-induced absorption bands. Finally, we describe studies of the analogous 3s $^1\Sigma^+$ - 4s $^1\Sigma^+$ transition in the MgH_2 reactive collision system. In this case two exit channels are possible: the nonreactive formation of Mg^*(4s $^1S_0$): and the reactive formation of MgH. We have observed evidence

* Also Department of Physics and Astronomy
† Also Department of Chemistry

of the direct reactive formation of MgH as the laser is tuned into the blue wing of the collision-induced molecular absorption band. We believe this to be the first such observation of reactive collision-induced dipole absorption.

The basic experimental arrangement has been described in detail in Refs. 1 and 3. A pulsed frequency doubled dye laser was tuned to the spectral region near the Mg(3s $^1S_0$ - 4s $^1S_0$) strictly forbidden atomic line at 230 nm. The laser was focused into an oven containing Mg metal vapor (~10^{14} - 10^{15}/cm^3) and buffer gas at ~700 Torr. The nonreactive exit channel leading to the formation of Mg* (4s $^1S_0$) was monitored by observing the spectrally and temporally integrated cascade fluorescence on the Mg(3p $^1P_1^0$ - 3s $^1S_0$) resonance transition at 285 nm. The reactive formation of MgH

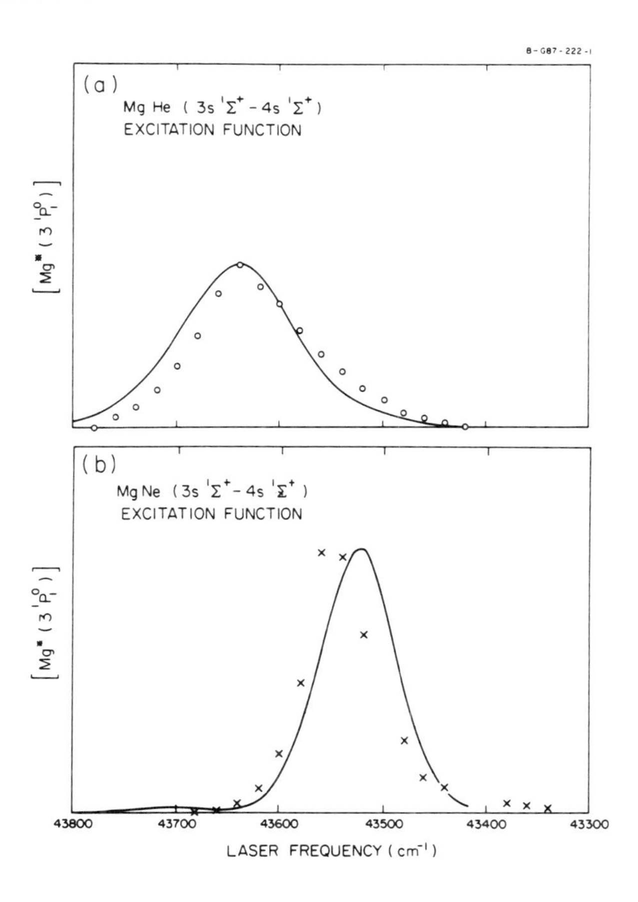

Fig. 1. Comparison of the experimental and theoretical (solid lines) absorption profiles for (a) MgHe and (b) MgNe.

product was detected in a standard pump-probe arrangement by probe laser induced fluorescence.

The absorption profiles for the MgHe and MgNe (3s $^1\Sigma^+$- 4s $^1\Sigma^+$) transitions are shown in Fig. 1. The location of the forbidden atomic transition (3s $^1S_0 \rightarrow$ 4s $^1S_0$) at an energy of 43,503 cm^{-1} is shown with an arrow. In Fig. 1 we also present a quantitative comparison of the observed MgHe and MgNe profiles with theoretical predictions. These predictions (solid curves) are based on the model potential curves of Malvern[2], using a single perturber quantum mechanical calculation for the line profile and an assumed transition dipole-moment function. It is important to note that the theoretical curves are not very sensitive to the form of the assumed dipole-moment function as long as it falls off rapidly as $R \rightarrow \infty$. The position of the absorption profile maximum is thus determined predominantly by the difference potential extremum.

Fig. 2 shows the laser absorption profiles for the Mg + H_2 system. By analogy with our previous work in the Mg-rare gas case, we recognize these molecular absorption bands as arising from a collision-induced dipole process involving collisional mixing with

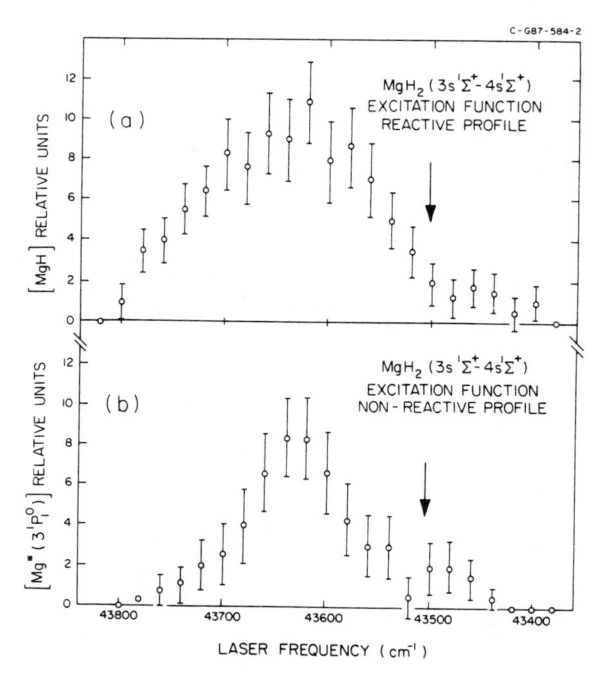

Fig. 2. The collision-induced dipole absorption profiles for MgH_2(3s $^1\Sigma^+$ - 4s $^1\Sigma^+$) molecular bands. (a) The reactive profile and (b) the nonreactive profile.

the lower-lying $MgH_2(3p\ ^1\Sigma^+)$ repulsive state of opposite parity. The essential difference here is the opening of an additional exit channel, namely the reactive formation of MgH. Therefore, Fig. 2 presents the laser absorption profiles leading either to the direct reactive formation of MgH (Fig. 2a), or to the nonreactive formation of $Mg^*(4s\ ^1S_0)$(Fig. 2b).

Under the high-pressure conditions of this experiment, other indirect multistep collisional pathways to form MgH may also be important. It is clear, however, in comparing Figs. 2a and 2b that there are significant differences and, therefore, the reactive profile cannot be entirely dominated by competing multistep processes. In particular, the reactive profile of Fig. 2b shows a distinctly enhanced blue wing relative to the nonreactive profile of Fig. 2b. We believe this blue wing portion of the molecular absorption band is due to the direct reaction channel. These results may be interpreted in terms of nonreactive vs. reactive channel competition in the presence of a reaction barrier; as the laser is tuned farther to the blue ($\geq$ 200 cm^{-1}) of the asymptotic $Mg(3s\ ^1S_0$ - $4s\ ^1S_0)$ resonance, the chemically reactive channel may become energetically allowed. More detailed theoretical calculations based on model potential energy surfaces will be required before a definitive interpretation of these observations can be given.

This work was supported by the National Science Foundation through Grant No. CHE-86-15118.

1 P. D. Kleiber, A. M. Lyyra, K. M. Sando, V. Zafiropulos, and W. C. Stwalley, J. Chem. Phys. 85, 5493 (1986).
2 A. R. Malvern, J. Phys. B. 11, 831 (1978).
3 P. D. Kleiber and K. M. Sando, Phys. Rev. A 35, 3715 (1987).

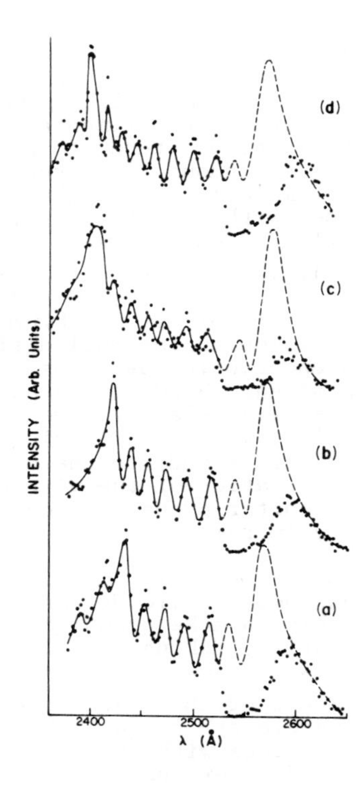

Fig. 1 LIF bands emitted from selectively populated v' levels of the $^1\Pi(\mathrm{Hg}\,^1P + \mathrm{Zn}\,^1S)$ state. (a) v' = 6; (b) v' = 7; (c) v' =8; (d) v' = 9. The wavelengths of the probe radiation were: (a) 5660 Å; (b) 5523 Å; (c) 5586 Å; (d) 5550 Å. The traces suggest the spectral profiles, the dashed lines indicate the probable positions of the components absorbed by Hg.

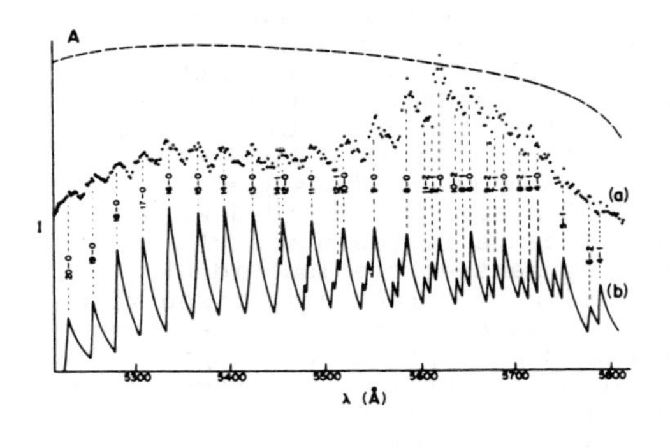

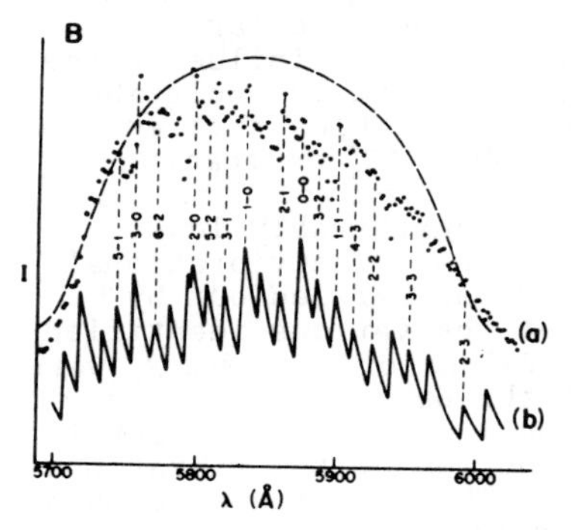

Fig. 2 Traces of the fluorescence-monitored $^1\Pi(\mathrm{Hg}\,^1P + \mathrm{Zn}\,^1S) \leftarrow {}^1\Pi(\mathrm{Hg}\,^1S + \mathrm{Zn}\,^1P)$ excitation band system, showing v' ← v" assignments. A, 5200-5750 Å region; B, 5700-6000 Å region. In each part, the experimental trace is labelled (a) and the computer-modelled spectrum (b); the dashed curves indicate relative dye-laser output power.

TABLE 1 Molecular Constants from the Analysis of the $^1\Pi \leftarrow {}^1\Pi$ Band System.

HgZn State	$\omega_e(cm^{-1})$	$\omega_e x_e(cm^{-1})$
$^1\Pi(Hg\,^1S + Zn\,^1P)$	194 ± 4	0.7 ± 0.2
$^1\Pi(Hg\,^1P + Zn\,^1S)$	117 ± 4	0.5 ± 0.2

REFERENCES

1. J. Supronowicz, E. Hegazi, J.B. Atkinson and L. Krause, Proc. III ILS Conference, p.
2. J. Supronowicz, E. Hegazi, G. Chambaud, J.B. Atkinson, W.E. Baylis, and L. Krause, Phys. Rev. A (to be published).
3. R.J. Niefer, J. Supronowicz, J.B. Atkinson, and L. Krause, Phys. Rev. A. 34, 1137 (1986).
4. J. Supronowicz, R.J. Niefer, J.B. Atkinson, and L. Krause, J. Phys. B 19, L717 (1986); J. Phys. B 19, 1153 (1986).
5. G. Herzberg, Spectra of Diatomic Molecules (van Nostrand, New York, 1950).

LASER-INDUCED FLUORESCENCE OF THE HgZn EXCIMER: THE 460 nm BAND*.

J. SUPRONOWICZ, E. HEGAZI, J.B. ATKINSON
and
L. KRAUSE

Department of Physics
University of Windsor
Windsor, Ontario, Canada N9B 3P4

ABSTRACT

The 460 nm HgZn fluorescence band was excited by pumping a Hg-Zn vapor mixture in a quartz cell with 307.59 nm laser pulses which populated the 4^3P_1 Zn atomic state. The excited HgZn molecules were formed by collisions of the 4^3P_1 Zn atoms with ground-state Hg atoms. The resulting HgZn fluorescent band centred near 460 nm had a half-width of about 80 nm and contained an additional component near 500 nm.

INTRODUCTION

The HgZn excimer is formed by collisions of laser-excited Zn 4^3P_1 with ground-state Hg 6^1S atoms, resulting in the population of the molecular $^3\Sigma$ and $^3\Pi$ ($Hg^1S + Zn^3P$) states and, also, of the $^1\Pi$ ($Hg^1S + Zn^1P$) state which is formed by a curve-crossing mechanism, as may be seen in Fig. 1. The PE curves shown in Fig. 1 are preliminary[1] and the inclusion of spin-orbit coupling effects

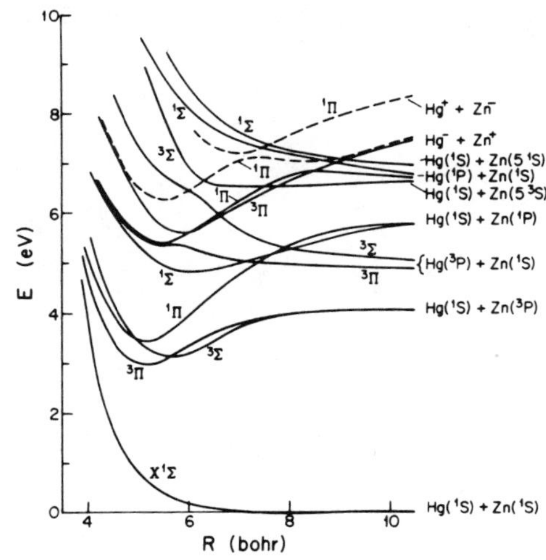

Fig. 1 Partial PE diagram for HgZn [according to Hund's case (a) coupling].

* Supported by the Natural Sciences and Eng. Res. Council of Canada

results in the formation of $\Omega = 0$ and $\Omega = 1$ states in common by the $^3\Sigma$ and $^3\Pi$ states. We believe that the decay of the $\Omega = 1$ ($^3\Sigma$, $^3\Pi$) state to the dissociative $X^1\Sigma$ ground state, gives rise to the broad unresolved 460 nm fluorescence band which has been observed previously.[2]

EXPERIMENTAL

The arrangement of the apparatus has been described previously.[3] The Hg-Zn vapor mixture contained in a quartz fluorescence cell was irradiated with laser pulses and the resulting fluorescence spectrum was observed at right angles to the direction of excitation, analyzed with a monochromator and a transient digitizer and summed in a computer.

The suprasil quartz cell was 7 cm long and 1.8 cm in diameter, and had a 13 cm-long side-arm. It was fitted with two end-windows, perpendicular to the cell axis, and was mounted in a two-compartment oven in which the temperatures of the cell and the side-arm were monitored and controlled separately. The cell containing Zn granules (99.99% purity) was baked under vacuum ($<10^{-5}$ torr) for several days before introducing the Hg sample. The mole fraction of Hg in the Hg-Zn mixture exceeded 0.95 but the quantity of Hg was arranged so that at temperatures above 660 K all Hg was in the vapor phase at a density of $1.9 \times 10^{19}/cm^3$. During the experiments the body of the cell was maintained at 840 K and the side-arm at about 780 K.

The laser beam was produced by pumping an in-house built two-stage dye laser, operated with Rh 640 in methanol, with the second harmonic (532 nm) of a Q-switched Nd:YAG laser. The dye laser output was frequency-doubled with a KDP crystal, to produce 307.59 nm radiation for the excitation of Zn atoms to the 4^3P_1 state. The fluorescence was detected and resolved with a Jobin-Yvon monochromator fitted with an EMI 9816 QB photomultiplier whose output was conveyed to a transient digitizer controlled by a microcomputer.

The profile of the blue fluorescence band, excited with the pump laser, was recorded by scanning the monochromator in the range 380-540 nm and using a "boxcar" time-averaging method. Ten fluorescence pulses following ten pump laser pulses were digitized, time-integrated and accumulated in the computer memory channel corresponding to the wavelength setting of the monochromator. The computer then advanced the monochromator wavelength by 1 nm and the procedure was repeated throughout the wavelength range. The time-resolved measurements were recorded by setting the monochromator at 460 nm and sampling the intensity of the fluorescence signal, following the pump laser pulse, with the transient digitizer which produced a time-spectrum consisting of 1024 channels, each spanning 2 ns.

RESULTS AND DISCUSSION

Figure 2 shows the spectral profile of the 460 nm fluorescence band, which contains an additional and previously unobserved component near 500 nm. We believe that the main band arises from the $^3\Sigma$, $^3\Pi \rightarrow X^1\Sigma$ decay and the 500 nm. component is due to the $^1\Pi(Hg^1S + Zn^1P) \rightarrow X^1\Sigma$ decay.

The time evolution of the band is shown in Fig. 3 which also shows the decay of the $4^3P_1 \rightarrow 4^1S_0$ Zn atomic fluorescence. The band fluorescence decayed exponentially with a decay time of about

6 μs. The rise time of the fluorescence was found to depend on the side-arm temperature (and the HgZn density) and was found at various temperatures to be of the same order as the decay time of the atomic fluorescence.

REFERENCES

1. J. Supronowicz, E. Hegazi, G. Chambaud, J.B. Atkinson, W.E. Baylis and L. Krause, Phys. Rev. A. (Jan. 1, 1988).
2. J.G. Eden, Optics Commun. 25, 201 (1987).
3. R.J. Niefer, J. Supronowicz, J.B. Atkinson and L. Krause, Phys. Rev. A 34, 1137 (1986).

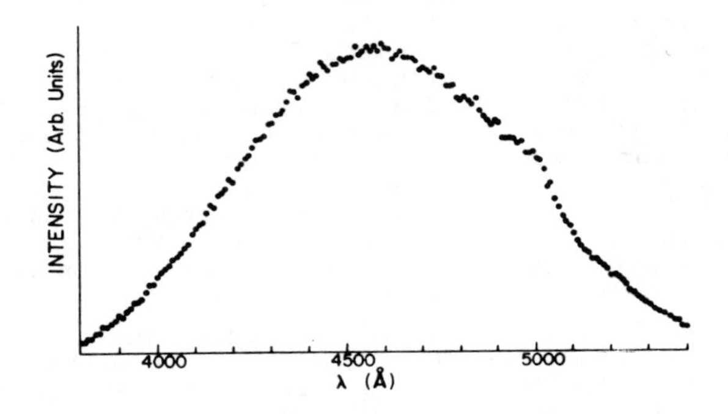

Fig. 2$_o$ HgZn fluorescence band excited with 3075.9A laser radiation.

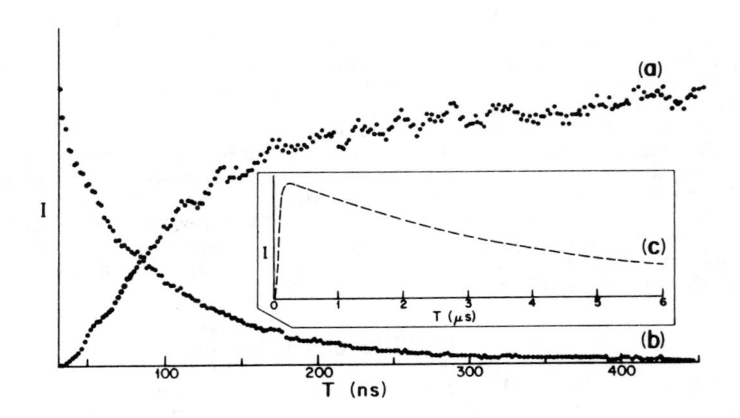

Fig. 3 Time-evolution studies.
(a) Growth and decay of the 460 nm band
(b) Decay of the atomic Zn fluorescence
(c) Extended-time evolution of the 460 nm band.

OBSERVATION OF EMISSION FROM AUTOIONIZING STATES OF Li_2

T.C. Chu and C.Y. Robert Wu
University of Southern California, Los Angeles, CA 90089-1341

ABSTRACT

We report the observation of emission in the 2000 - 2200Å region from lithium molecules. The emission clearly originates from neutral (autoionizing) states above the ionization potential of Li_2. The dispersed spectrum, excitation function and power dependence of the UV emission are presented. The possible mechanisms for producing the observed emission are discussed.

INTRODUCTION

There are neutral states above the lowest ionization potential of molecules which are termed superexcited states[1] or quasi-continuum states. They can be either Rydberg series converging to excited ion states or two-electron (or multiple-electron) excited states. The superexcited states lie in the ionization continuum and therefore will normally decay to the continuum with a characteristic time as short as 10^{-16} sec. The decay process is called autoionization (or pre-ionization). The superexcited states can also dissociate (or predissociate) into fragments with a characteristic time of the order of 10^{-13} sec or longer. Furthermore, the superexcited state can radiatively decay through fluorescence which is typically 10^{-9} sec for electric-dipole allowed transitions. It is thus clear that radiative decays are usually not an efficient process to compete with autoionization and dissociation processes. Only in rare occasions, e.g., due to forbidden coupling between the superexcited state and the ionization continua and dissociation channels, would radiative decays of superexcited states become important processes. Emission from autoionizing levels of H_2[2,3] is the only previous known case. We believe what we have observed in Li_2 represents the second such example. (It is to be noted that in atomic systems emissions from autoionizing states are commonly observed in beam-foil experiments[4] and in the excitation of inner-shell electrons by soft X-ray photons.[5])

EXPERIMENTAL RESULTS

The emission bands were observed under the following experimental conditions: (a) a heatpipe temperature of ~1000K (the Li and Li_2 vapor density is 7.1×10^{15} and $1.9\ \times10^{14}\ cm^{-3}$, respectively) with a helium buffer gas pressure of ~4Torr, (b) the dye laser wavelength tuned near the Li atomic 2s-4s two-photon resonance transition, (c) the dye laser output energy of ~2mJ/pulse, and (d) a solar blind PMT (EMR 541F with spectral response in the 1050 - 3500Å region). The radiation was observed in the forward direction along the laser beam through a quartz prism. The radiation appears to be highly collimated. A 0.3m spectrometer was used to disperse the radiation. The detailed description of the experimental setup can be found in one of our previous publications.[6,7]

Figure 1 shows a typical emission spectrum of Li_2 taken at a laser wavelength of 5711.2Å. The spectrum shows a characteristic blue-shaded feature. Figure 2 displays the excitation function which shows that the emission can only be produced through a narrow laser wavelength range (2.6Å!) in the vicinity of the Li atomic 2s-4s two-photon resonance at 5710.8Å. Using a laser wavelength at 5710.4Å and a laser energy in the range of 0.2 to 2 mJ/pulse, we found the radiation intensity roughly exhibits an I^4 dependence at two different heatpipe temperatures (993 and 1050 K). Further, fixing the laser energy at 1 mJ/pulse, we observed a $n^{1.7}$ dependence for the $Li(Li_2)$ vapor density varying from 1.9×10^{14} (2.0×10^{12}) to 1.6×10^{16} (5.5×10^{14}) cm^{-3}. Unfortunately, both sets of data could not provide conclusive evidence for suggestions of the excitation mechanisms since the ideal behaviors of laser power and vapor density dependences will not hold if (a) a real intermediate state is involved in the multiphoton excitation processes[8] or (b) the stimulated emission or collisional processes are occurring.

DISCUSSION

The possible mechanisms for the excitation of autoionizing states can be the direct three-photon or four-photon excitation of Li_2, collisional association and direct collisional transfer of energy from Li(4s) to Li_2 followed by one photon excitation. These mechanisms can be given in the following:

$$Li_2(X^1\Sigma_g^+,\ \text{low } v'') + 3h\nu \rightarrow Li_2^{**}(^1\Lambda_u) \qquad (1)$$

$$Li_2^{**}(^1\Lambda_u) \rightarrow Li_2(X^1\Sigma_g^+,\ 8 \leq v'' \leq 28) + h\nu' \qquad (1a)$$

$$Li_2(X^1\Sigma_g^+) + 4h\nu \rightarrow Li_2^{**}(^3\Lambda_g) \qquad (2)$$

$$Li_2^{**}(^3\Lambda_g) \rightarrow Li_2(a^3\Sigma_u^+) + h\nu' \qquad (2a)$$

$$Li(2s) + 2h\nu \rightarrow Li(4s) \qquad (3a)$$

$$Li(4s) + Li(2s) + M \rightarrow Li_2^{*}(^1\Lambda_g) + M$$

$$Li_2^{*}(^1\Lambda_g) + h\nu \rightarrow Li_2^{**}(^1\Lambda_u) \qquad (3)$$

$$Li(4s) + Li_2(X^1\Sigma_g^+) \rightarrow Li + Li_2^{*}(^1\Lambda_g)$$

$$Li_2^{*}(^1\Lambda_g) + h\nu \rightarrow Li_2^{**}(^1\Lambda_u) \qquad (4)$$

For a three-photon excitation process the energy of three-photon amounts to $52{,}540cm^{-1}$ which is about $11{,}000cm^{-1}$ and

700cm^{-1} above the IP of Li_2 and the $Li+Li^+$ dissociation limit, respectively. To satisfy the parity selection rule and energy conservation, the observed emission can be assigned to the transition of bound $^1\Lambda_u(\Lambda=\Sigma,\Pi)$ → bound $X^1\Sigma_g^+$ ($v''\cong8$ to 28). The radiation to vibrational levels $v''\leq8$ of $X^1\Sigma_g^+$ will most likely be re-absorbed by the Li_2 molecules as the observed emission shows a high degree of coherence.

The total photon energy of a four-photon process is 28,560cm^{-1} and 18,200cm^{-1} above the IP of Li_2 and the dissociation limit of the lowest ion state, respectively. The observed emission can be assigned to bound ($^3\Lambda_g$, $\Lambda=\Sigma,\Pi$) → free ($a^3\Sigma_u^+$) transitions. The population of $^3\Lambda_g$ is possible because strong spin-orbit coupling between the $A^1\Sigma_u^+$ and $b^3\Pi_u$ states is well known.[9] After absorption of the first visible photon to the $A^1\Sigma_u^+$ state, subsequent multiphoton absorptions from the $b^3\Pi_u$ state to highly excited triplet states are thus possible. The characteristic blue-shaded feature resembles many bound-free transitions observed in alkali-dimers[9-14] making this assignment an attractive one. The observed I^4 dependence appears to support this assignment. However, a serious drawback in this assignment is the large energy, i.e., 18,200 and 3,300cm^{-1}, above the respective $Li(2s)+Li^+$ and $Li(2p)+Li^+$ dissociation limit making an overall four-photon excitation process unlikely.

The collisional processes, i.e., Eqs.(3) and (4), suggest that the excited gerade state(s) $Li_2^*(^1\Lambda_g)$ can be populated through collisional association and/or collisional energy transfer from excited Li(4s) atoms. The $Li_2^*(^1\Lambda_g)$ could subsequently absorb one additional laser photon to the autoionizing state(s). These collisional processes may play a dominate role since the production of the autoionizing emission is effective within a 2.6Å range of the Li 2s-4s two-photon transition. Further, the result of $n^{1.7}$ dependence seems consistent with this picture. Studies of pressure effects with different buffer gases may provide new information to support or repudiate the importance of this process.

The number of photons required in the collisional processes is identical to the three-photon process, but the $Li_2^*(^1\Lambda_g)$ state(s) involved in both methods of excitation may or may not be the same. A VUV single-photon (at 1903Å) excitation might produce a similar result as that by the direct three-photon excitation. This experiment may thus help us in identifying which is the most important mechanism: three-photon, four-photon or collisional processes?

The potential energy curves of the two-electron excited states are highly desired and would be very helpful in the determination of the transitions involved and thus would be very useful in improving our understanding of molecular structures and dynamics above the ionization potential of Li_2.

ACKNOWLEDGMENTS

This work was partially supported by the USC Faculty Research and Innovation fund.

REFERENCES

1. R.L. Platzman, J. Chem. Phys. 38, 2775 (1963).
2. W. Sröka, Phys. Lett. 28A, 784 (1969).
3. P. Borrell, P.M. Guyon and M. Glass-Maujean, J. Chem. Phys. 66, 818 (1977).
4. S. Bashkin, in Progress in Optics XII, ed. E. Wolf (North-Holland, 1974), p.287.
5. e.g., R.G. Caro, P.J.K. Wisoff, G.Y. Yin, D.J. Walker, M.H. Sher, C.J.P. Barty, J.F. Young, and S.E. Harris, in Short Wavelength Coherent Radiation, eds. D.t. Attwood and J. Bokor (American Institute of Physics #147, New York, 1986), p.145.
6. H.H. Wu, T.C. Chu and C.Y.R. Wu, Appl. Phys. B 26, 225 (1987).
7. C.Y.R. Wu, J.K. Chen, D.L. Judge and C.C. Kim, Optics Comm. 48, 28 (1983).
8. C.Y.R. Wu, F. Roussel, B. Carré, P. Breger and G. Spiess, J. Phys. B 18, 239 (1985).
9. X. Xie and R.W. Field, J. Mol. Spectrosc. 117, 228 (1986) and references therein.
10. D.D. Konowalow and P.S. Julienne, J. Chem. Phys. 72, 5815 (1980).
11. G. Pichler, J.T. Bahns, K.M. Sando, W.C. Stwalley, D.D. Konowalow, L. Li, R.W. Field and W. Müller, Chem. Phys. Lett. 129, 425 (1986).
12. J. Huennekens, S. Schaefer, M. Ligare and W. Happer, J. Chem. Phys. 80, 4794 (1984).
13. D.D. Konowalow, S. Milosevic and G. Pichler, J. Mol. Spectrosc. 110, 256 (1985).
14. J.T. Bahns, W.C. Stwalley and G. Pichler (to be published).

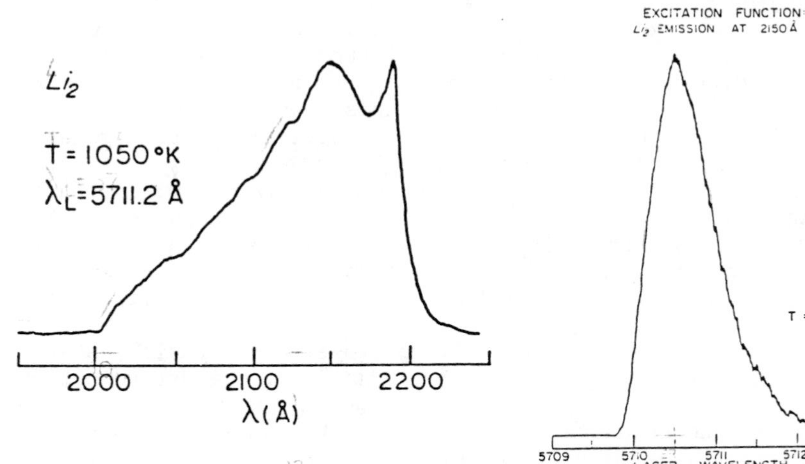

Figure 1. A typical emission spectrum of Li_2 in the 2000 - 2200A region.

Figure 2. The excitation function of the Li_2 emission centered at 2150A with a spectrometer bandwidth of 13A.

A SPECTROSCOPIC STUDY OF THE $1^1\Pi_g$ STATE OF $^7Li_2$

R. A. Bernheim, L. P. Gold, D. A. Miller, P. D. Tripodi
Penn State University, University Park, PA 16802

Previous studies of the excited states of Li_2 in the visible and ultraviolet regions[1] have recently been extended into the infrared. We report here the observation of the $1^1\Pi_g$ state of the lithium dimer by optical-optical double resonance. The associated molecular constants are reported. Together with the recent FT IR work of B. Barakat et al.[2], this study completes the observation of all singlet states correlating with the Li(2s) + Li(2p) dissociation limit.

As the second simplest stable homonuclear diatomic molecule, Li_2 represents a useful system for which the predictions of *ab initio* calculations may be compared. Two series of calculations for the ground and excited states of molecular lithium have recently been reported[3,4] which yield molecular constants and potential curves to within nearly spectroscopic precision. Previous studies in this laboratory have resulted in experimental information for a number of higher excited states including members of three Rydberg series. In the present work we examine the $1^1\Pi_g$ state which belongs to the lowest manifold of excited states. The experimental constants and potential curves may be used as a guide for further refinement of *ab initio* techniques.

The $1^1\Pi_g$ state is observed using optical-optical double resonance excitation of the metal vapor in a heat pipe oven using pulsed lasers. Levels of the $A^1\Sigma_u^+$ state are pumped with one dye laser operating in the red, and subsequent excitation to the $1^1\Pi_g$ state is induced by a probe laser operating in the near infrared. The tunable near infrared radiation is produced by exciting the 2nd Stokes stimulated Raman scattering in a cell filled with H_2 gas. Collisional relaxation transfers some of the excited state population to the adjacent $B^1\Pi_u$ state which has the proper symmetry for direct radiation to the ground state. Transitions may then be detected by monitoring blue-green, B-X, fluorescence as the Raman shifted light is tuned through the range 1.1um -1.3um. Two Nd:YAG pumped pulsed dye lasers provide the sources of excitation. The resulting spectra exhibit the dramatic simplicity that is often obtained in double resonance experiments. A total of 300 lines originating from v = 0 in the A state were easily assigned. Observations cover the first 32 vibrational levels of the $1^1\Pi_g$ state.

A least squares fit of a Dunham expansion to the spectral data requires 17 terms to reproduce the observed line positions. The RKR potential generated from this data corresponds to 85% of the estimated well depth and reveals the $1^1\Pi_g$ state to possess a broad shallow potential with a bond energy of 1426 cm^{-1}, or 17.06 kJ/mol. The differences between the *ab initio* and experimental potentials are shown in Fig. 1.

Table 1

Dunham coefficients for the first 32 vibrational levels of the $1^1\Pi_g$ state of $^7Li_2$. The underlined digits are significant to within one standard deviation.

i	Y_{i0}	Y_{i1}	Y_{i2}
0	21998.251	0.29188837	-1.1901E-5
1	93.35389720	-9.484894E-3	3.74088E-7
2	-1.87400472	2.13376E-4	-2.90145E-8
3	3.257062E-2	-1.8868E-5	5.50366E-10
4	-2.249260E-3	6.60511E-7	
5	7.0570E-5	-7.12661E-7	
6	-7.06769E-7		

Lambda doubling constant, q = 8.5896E-5

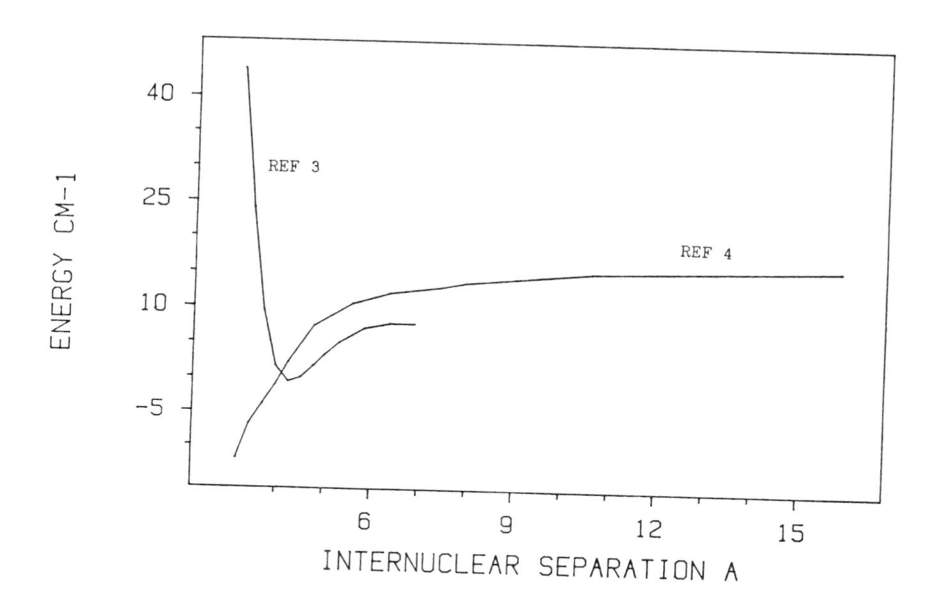

Figure 1. Experimental minus theoretical potential curves.

REFERENCES

1. R. A. Bernheim, L. P. Gold, and C. A. Tomczyk, in Comparison of Ab Initio Quantum Chemistry with Experiment for Small Molecules, Rodney J. Bartlett, ed. (D. Reidel Publ., Boston, 1985) and references therein.

2. B. Barakat et al., J. Mol. Spec. 116, 271-285 (1986).

3. D. D. Konowalow, and J. L. Fish, Chem. Phys. 84, 463-475 (1984).

4. S. Meyer, W. Muller, and W. Meyer, Chem. Phys. 92, 263-285 (1985).

PHOTOELECTRON SPECTRUM OF GAS PHASE Cu_2

A.D. Sappey, J. Harrington, and J.C. Weisshaar
Department of Chemistry, University of Wisconsin–Madison
Madison, Wisconsin 53706

ABSTRACT

We have used resonant two-photon ionization and time-of-flight electron analysis to obtain photoelectron spectra of cold gas phase Cu_2 created by laser vaporization in a pulsed supersonic expansion. The spectra reveal a dense manifold of vibronic bands from 0-1 eV above threshold whose nature is uncertain and a pair of well-resolved $^2\Pi$ excited state spin-orbit components.

INTRODUCTION

Laser vaporization sources[1] of gas phase transition metal clusters permit rapid progress in the chemistry of both neutral (M_n) and cationic (M_n^+) species with $n \leq 30$. Yet our understanding of the electronic and vibrational structure of such species remains primitive, in spite of recent progress in optical, ESR, and photoelectron spectroscopies. A long term goal must be the correlation of size-specific metal structure with reactivity. Cations are well-suited to this goal since they permit mass selection of reactants. However, no spectroscopic data exist even for diatomic transition metal cations.

In this work, we use photoelectron spectroscopy to provide new information about the low-lying states of one such ion, Cu_2^+. Viewed simply, Cu_2 and Cu_2^+ are transition metal analogues of H_2 and H_2^+. Cu_2^+ provides a relatively simple system for calibration of all-electron and core-pseudopotential theoretical techniques as applied to open-shell metal species. The $^1\Sigma_g^+$ ground state of Cu_2 is fairly well described[2] by the single 4sσ-bonded electron $(3d\sigma_g)^2(3d\pi_u)^4(3d\delta_u)^4(3d\pi_g)^4(3d\sigma_g)^4(3d\sigma_u)^2(4s\sigma_g)^2$. Removal of one s- or d-electron gives rise to seven Cu_2^+ molecular terms, all likely chemically bound: the "s-hole" $^2\Sigma_g^+$ term (one-half 4sσ-bond, H_2^+ ground state analogue) and six "d-hole" terms $^2\Sigma^+_{g,u}$; $^2\Pi_{g,u}$; and $^2\Delta_{g,u}$. The d-hole terms correlate to excited state $Cu^+ + Cu$ asymptotes; formally, these have a full 4sσ-bond and a half-bond or a half anti-bond of either σ, π, or δ symmetry. Single configuration self-consistent field (SCF) calculations[3] found the s-hole $^2\Sigma_g^+$ term to be the ground state, with various localized ($C_{\infty v}$ electronic symmetry) d-hole terms at an estimated 2.0 eV excitation energy.

EXPERIMENT AND RESULTS

The experiment uses one-color, resonant two-photon ionization (R2PI) to selectively ionize Cu_2 from a skimmed supersonic beam of cold Cu_n species created by laser vaporization and expanded in He.[4] Time-of-flight analysis yields the photoelectron spectrum (TOF-PES).[4] A frequency doubled, pulsed dye laser (260-267 nm, 5 ns FWHM) ionizes Cu_2 via various v'-0 vibrational bands of the "System V" resonant state,[5] accessing Cu_2^+ states to 1.4 eV.

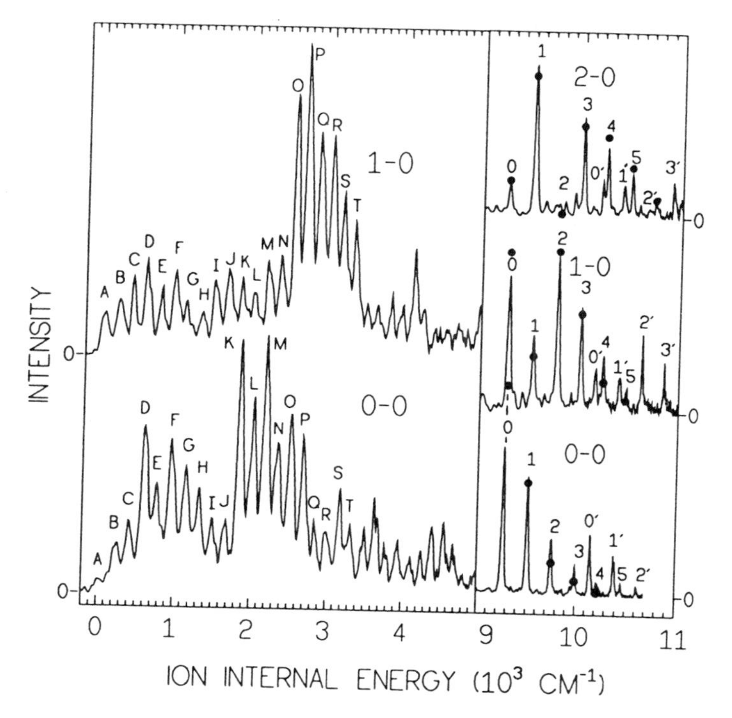

Figure 1. Cu_2 photoelectron spectra obtained by two photon ionization via Smalley's System V, v'-0 resonant bands as indicated. Additional bands occur from 5000-9000 cm^{-1}. The dots are Franck-Condon factors from a Morse potential model.

Figure 1 shows segments of the PESs resulting from 0-0, 1-0, and 2-0 R2PI. The threshold of 7.899 ± 0.005 eV above the Cu_2 ground state agrees with the ionization energy inferred from R2PI band intensities.[5] At higher ion energy, we resolve two excited state vibrational progressions with origins at 1.143 and 1.256 eV; both are well described by $\omega_e^+ = 252 \pm 17$ cm^{-1} and $\omega_e x_e^+ = 1 \pm 3$ cm^{-1}. The frequency is slightly smaller than that of the single-bonded Cu_2 ground state ($\omega_e'' = 266$ cm^{-1}). Ion ← neutral resonant state Franck-Condon factors (FCFs) model all three intensity envelopes reasonably well (dots in Fig. 1). The 908 ± 29 cm^{-1} separation is a sensible spin-orbit splitting for a 3d-hole $^2\Pi$ state.

In the range 0-1 eV above threshold, we observe a dense set of at least 50 moderately well-resolved vibronic states. Within experimental uncertainity, the first 26 band positions can be fit as a <u>single vibrational progression</u> having $\omega_e^+ = 187 \pm 8$ cm^{-1} and $\omega_e x_e^+ = 0.7 \pm 0.2$ cm^{-1}. This frequency is smaller than that of either the 1.2 eV Cu_2^+ excited states or the Cu_2 ground state; it closely matches the Cu_2 $\tilde{A}$ state (192 cm^{-1}). All of the 0-1 eV bands might belong to the anticipated s-hole, half 4sσ-bonded $^2\Sigma_g^+$ state, which would then be the Cu_2^+ ground state.

However, the multi-peaked intensity distribution of the 0-1 eV bands is not easily understood. Taking the 0-0 PES as an example, FCFs between an unperturbed v′ = 0, ω_e' = 288 cm^{-1} resonant state and a single manifold of ion states v^+ with ω_e^+ = 190 cm^{-1} predict a nearly Gaussian-shaped intensity envelope whose energy breadth depends on Δr_e, the ion ← neutral shift in equilibrium bond length. As observed by several groups and recently discussed by Chupka,[6] non-Franck-Condon intensity envelopes are possible in R2PI, although Cu_2^+ would present an extreme example. A strong perturbation could lead to mixed "v′ = 0" resonant state vibrational character. Excitation of long-lived autoionizing states, either diatom bound or repulsive, could permit substantial nuclear motion prior to electron ejection. The System V, 2-0 band is indeed perturbed and split, but the 1-0 and 0-0 bands appear clean.[5] In addition, the success of FCFs in describing the intensity pattern of the 1.14 eV progression for the 0-0, 1-0, and 2-0 PESs argues against dramatic effects from perturbations or autoionization, since each excitation creates an entire PES.

A different possibility is the presence of multiple low-lying d-hole Cu_2^+ electronic states at 0-1 eV, leading to a complicated superposition of Franck-Condon envelopes. The intriguing, reproducible intensity alternation of some bands in the 0-0 PES (note peaks c-d-e-f-g and k-l-m-n-o-p) might then be due to interleaving ~380 cm^{-1} progressions. If such a multi-state explanation proves true, then the d-d bonding in Cu_2^+ is surprisingly important. The fit of the band positions to a single anharmonic progression must then be fortuitous, perhaps assisted by the modest PES resolution. Two-color experiments will provide further information.

We acknowledge the Donors of the Petroleum Research Fund and the National Science Foundation (CHE-8703076) for support of this research.

REFERENCE

1. M.E. Geusic, M.D. Morse, S.C. O'Brien, and R.E. Smalley, Rev. Sci. Instr. 56, 2123 (1985).
2. C.W. Bauschlicher, Jr., S.P. Walch, and P.E.M Siegbahn, J. Chem. Phys. 76, 6015 (1982).
3. E. Miyoshi, H. Tatewaki, and T. Nakamura, J. Chem. Phys. 78, 815 (1983).
4. L. Sanders, A.D. Sappey, and J.C. Weisshaar, J.Chem. Phys. 85, 6952 (1986).
5. D.E. Powers, S.G. Hansen, M.E. Geusic, D.L. Michalopoulos, and R.E. Smalley, J. Chem. Phys. 78, 2866 (1983).
6. W.A. Chupka, J. Chem. Phys. 87, 1488 (1987).

PHOTOELECTRON SPECTRUM OF GAS PHASE TiO

A.D. Sappey, J. Harrington, G. Eiden, and J.C. Weisshaar
Department of Chemistry, University of Wisconsin–Madison
Madison, Wisconsin 53706

ABSTRACT

We have used resonance-enhanced multiphoton ionization (REMPI) of TiO cooled in a supersonic expansion in conjunction with time-of-flight photoelectron spectroscopy (TOF-PES) to refine spectroscopic constants for the $\tilde{X}\ ^2\Delta$ and the $\tilde{A}\ ^2\Sigma^+$ electronic states of TiO^+.

INTRODUCTION

The survey nature of photoelectron spectroscopy should permit general insight into the role of d-electrons in the chemical bonding of simple ligated transition metal species in the gas phase. Dyke and co-workers[1] have used high temperature furnace sources to obtain vacuum ultraviolet photoelectron spectra of gas phase metal oxides including NbO, TaO, BaO, SrO, VO, and TiO. Here we present a different technique, REMPI-PES on TiO cooled in a supersonic expansion, which may eventually yield spectra of a variety of partially ligated transition metal fragments.

EXPERIMENT SECTION

In the source chamber[2] (Fig. 1), a pulsed excimer laser (308 nm, 15 mJ/pulse, 10 pulse/s) focused on a rotating Ti rod creates a packet of high density gas phase Ti species in the throat of a pulsed nozzle expansion (2.2 atm Ar plus 70 torr O_2). The gas phase reaction $Ti + O_2 \rightarrow TiO + O$ creates the desired species. Subsequent collisions during the expansion cool the TiO rotationally to ~50K; the vibrational and electronic degrees of freedom are substantially hotter.

The interaction chamber includes both a time-of-flight mass spectrometer (TOF-MS) and a time-of-flight photoelectron spectrometer (TOF-PES).[2] A pulsed, tunable Nd:YAG pumped dye laser (5 nsec FWHM, 2-5 mJ/pulse, 25 cm focal length lens) intersects the skimmed beam in either the TOF-MS or the TOF-PES. With the laser on resonance, a typical TOF-MS consists of at least 95% TiO^+ with the remainder Ti^+. Gated integration of the TiO^+ TOF-MS signal vs. laser wavelength yields an REMPI spectrum of TiO. From 425-540 nm, many observed REMPI bands can be assigned as nominal one photon resonant, three photon ionizations (one-color, 1 + 2 REMPI). We observe the v'-0 bands of the $\tilde{C}(^3\Delta) \leftarrow \tilde{X}(^3\Delta)$ transition with good intensity for $v' = 0$-5; sequence bands with $v'' = 1$-4 occur as well. The intensities of the bands do not follow $\tilde{C} \leftarrow \tilde{X}$ Franck-Condon factors, suggesting that some of the bands borrow substantial oscillator strength at the two or three photon level. The 3-0 and 4-0 $\tilde{B}(^3\Pi) \leftarrow$ X bands are much weaker.

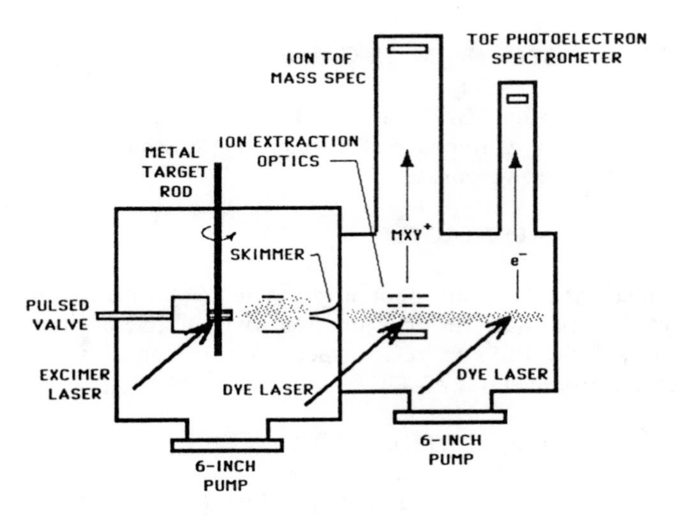

Figure 1. Pulsed molecular beam apparatus.

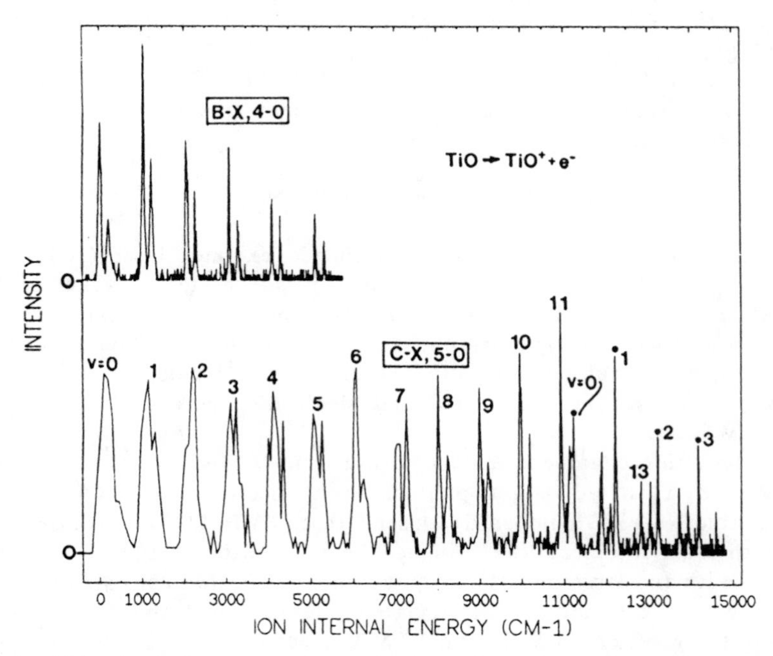

Figure 2. TiO photoelectron spectra obtained from two REMPI bands as shown. The TiO^+ bands indicated are v = 0-13 of the $\tilde{X}^2\Delta$ state and v = 0-3 of the $\tilde{A}^2\Sigma^+$ state (dots).

RESULTS AND DISCUSSION

We study different portions of the TiO^+ spectrum using different resonant bands (different total energy), since the intrinsic energy resolution of the TOF-PES technique varies as $\Delta E \sim E^{3/2}$, where E is electron kinetic energy. Figure 2 shows TOF-PESs obtained via the $\tilde{C} \leftarrow \tilde{X}$ (5-0) and the $\tilde{B} \leftarrow \tilde{X}$ (4-0) resonances. Quantitative results presented here are weighted averages over a number of spectra. At low ion internal energy we observe a progression of spin-orbit pairs of bands. The electronic origin lies 6.818 ± 0.010 eV above the TiO ($\tilde{X}^3\Delta_1$) ground state; we recommend this value as the adiabatic ionization energy of TiO. The spin-orbit splitting is 213 ± 12 cm^{-1}; the first six vibrational states are well fit by the parameters ω_e = 1034 ± 13 cm^{-1} and $\omega_e x_e$ = 0.7 ± 2.0 cm^{-1}. Dyke and co-workers[1b] previously observed this state at a vertical ionization energy of 6.82 ± 0.02 eV and assigned it as the $TiO^+(\tilde{X}^2\Delta)$ ground state based on ab initio calculations. No vibrational structure was resolved. Agreement of the experimental splitting with estimates from atomic Ti(3d) spin-orbit parameters[3] confirms the $^2\Delta$ assignment.

At higher energy, as accessed by the 5-0 and 6-1 bands, we observe a second electronic state whose adiabatic ionization energy is 8.208 ± 0.018 eV, again in close agreement with the vertical energy of 8.20 ± 0.01 eV obtained previously.[1b] Four members of a vibrational progression have average spacing 1032 ± 30 cm^{-1}, in some disagreement with the previous value [1b] of 860 ± 60 cm^{-1}. The absence of spin-orbit partners at our resolution confirms the assignment as the $\tilde{A}^2\Sigma^+$ excited state of TiO^+.

The TiO($^3\Delta$) ground state is well described[4] by the single electron configuration ... $8\sigma^2 3\pi^4 9\sigma 1\delta$, where 8σ and 3π are nearly localized O(2p) orbitals, 9σ is approximately Ti(4s), and 1δ is Ti(3d). The neutral is highly polar, roughly Ti^+O^-. The similarity of the $TiO^+(^2\Delta)$, $TiO^+(^2\Sigma^+)$ and TiO($^3\Delta$) vibrational frequencies (1034, 1030, and 1009 cm^{-1}) suggests similar chemical bonding in all three states. This is consistent with the view that the 9σ and 1δ orbitals in TiO are essentially non-bonding. The charge distribution of the ion is then approximately $Ti^{2+}O^-$, consistent with population analysis of SCF wavefunctions.[1b]

We acknowledge the Donors of the Petroleum Research Fund and the National Science Foundation (CHE-8703076) for support of this research.

REFERENCES

1. (a) J.M. Dyke, A.M. Ellis, M. Feher, A. Morris, A.J. Paul, and J.C.H. Stevens, J. Chem. Soc., Faraday Trans. 2, 83, 1555 (1987) and references therein. (b) J.M. Dyke, B.W.J. Gravenor, G.D. Josland, R.A. Lewis, and A. Morris, Mol. Phys. 53, 465 (1984).
2. L. Sanders, A.D. Sappey, and J.C. Weisshaar, J. Chem. Phys. 85, 6952 (1986).
3. See H. Lefebvre-Brion and R.W. Field, Perturbations in the Spectra of Diatomic Molecules, Sec. 4.3 (Academic, Orlando, Florida, 1986).
4. J.M. Sennesal and J. Schamps, Chem. Phys. 114, 37 (1987).

USING DIODE LASERS FOR ATMOSPHERIC COMPOSITION CONTROL

V.I. Astakhov, V.V. Galaktionov, A.A. Karpukhin,
A.Yu. Tishchenko, V.U. Khattatov
GOSCOMGIDROMET, Central Aerological
Observatory, Dolgoprudny, Moscow Region

The appearance of tunable diode lasers which are now the most effective instruments of IR-spectroscopy stimulates the development of a new class of gas-analyzers with high sensitivity ensured by a significant increase in spectra informativity. Reported are diode laser applications for measurements of carbon monoxide and CFC-12 in the atmosphere.

CARBON MONOXIDE GAS-ANALYZER

To study CO background content it is necessary to have gas-analyzing instrumentation sensitive to slight variations of this minor species in the surface layer and ensuring the required spatial averaging of measurement data. These requirements have been met by a CO long-path gas-analyzer which uses a pulsed diode frequency-tunable semiconductor PbSSe laser operated at 80÷90 K temperature in a 4.7 μm range for sounding the atmosphere's surface layer. Information about CO amount over a sounding path in the atmosphere is obtained by means of a stroboscopic analysis of tunable sounding radiation pulses within the contour of a separate absorption line. The gas-analyzer block-diagram and operation principle are shown in Fig.1. A CO gas-analyzer metrological study has shown that this instrument is insensitive to the presence of water vapor and carbon dioxide, CO_2, over the path, while the maximum sensitivity to CO is 0.25 ppm over a 1-m optical path. The inaccuracy of field CO concentration measurements over paths more than 100 m long does not exceed ±5 ppb. The gas-analyzer measurement data can be recorded both digitally and graphically on different averaging scales.

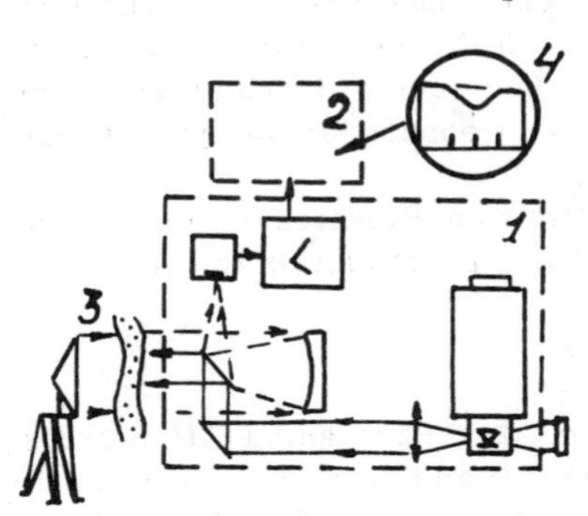

Fig.1. Block-diagram of CO gas-analyzer.
1. Optico-mechanical block. 2. Electronics. 3. Reflector. 4. CO measurement principle.

Its low power consumption of 20 W, liquid nitrogen consumption of 2 liters per hour, and low weight of 30 kg make the gas-analyzer fit for field long-term self-contained measurements.

The gas-analyzer was used to study the surface layer natural CO variability in a number of preserves in the European part of the USSR (Fig.2).

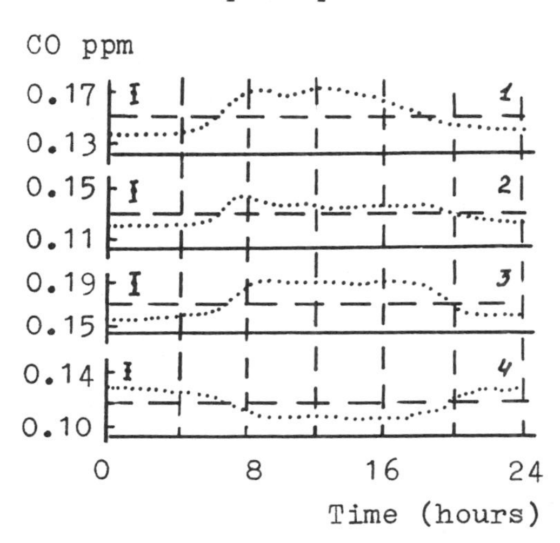

Fig.2. Diurnal CO concentration variabilities in the atmosphere, 1982-1986.
1. 54°54'N, 37°48'E.
2. 55°N, 28°E.
3. 45°N, 34°E.
4. 43°42'N, 44°E.

An intradiurnal cyclic CO concentration variability has been revealed in the surface layer, with a marked maximum in the morning and minimum at night. The 1986 measurements conducted in the Kara-Dag and Ai-Petry Preserves at the Crimean sea coast have revealed the difference between CO concentrations over the land and the sea. The CO concentrations over the land were on the average 1.5 times as high as those over the sea. The diurnal variation amplitude made ±40 ppb and, evidently, was partly of man-made origin.

CFC GAS-ANALYZER

The necessity to control CFCs, whose amount in the troposphere normally does not exceed 1 ppb, is due to their active role as catalysts of the photochemical depletion of stratospheric ozone. Particularly dangerous are the man-made chlorofluorocarbons $CFCl_3$ and CF_2Cl_2.

A tunable diode PbSnSe laser cooled down to 10÷80 K serves in the gas-analyzer as a source of radiation. The gas-analyzer block-diagram is given in Fig.3. The laser operation temperature is maintained with a 10^{-3} K accuracy. The laser generation wavelength is tunable within 10-12 μm range. The gas-analyzer has two identical optical channels, an analytical and a reference ones.

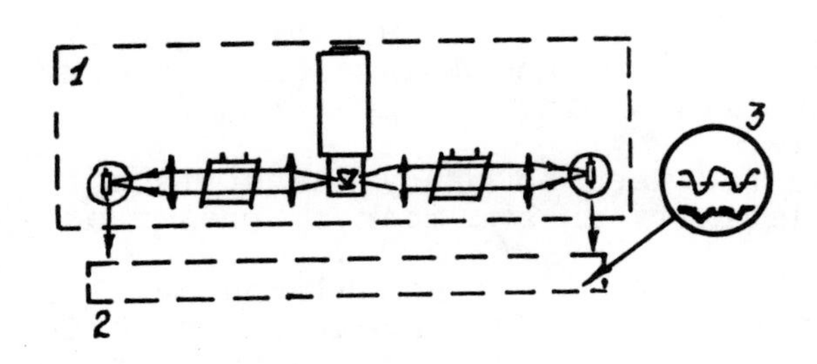

Fig.3. Block-diagram of CFC gas-analyzer.
1. Optico-mechanical block.
2. Electronics. 3. Analogue signal recording.

CFC content in air samples is measured in cells at pressure values as low as 10 Torr. Air is sampled in air-tight cells of stainless steel with a ~10^3-fold cryogenic enrichment of samples. For higher detection sensitivity the gas-analyzer is equipped with a multipath cell with 50-m effective length.

This gas-analyzer was used to measure CF_2Cl_2 amount in the atmosphere. 0.3-3.0 ppb amounts of CF_2Cl_2 were measured in air samples. These preliminary results show that diode laser spectroscopy methods can be applied for the development of the effective means of atmospheric composition control.

In conclusion, the authors express their gratitude to the scientific workers of the Laboratory of Semiconductor Physics of the USSR Academy of Sciences I.I. Zasavitsky,and A.P. Shotov for the provision of the diode lasers.

REFERENCE

1. V.I. Astakhov, V.V. Galaktionov, A.A. Karpukhin, V.U. Khattatov. Modern advances on the application of diode lasers for atmospheric pollution control. In: Instruments and observing methods. Report No. 22, WMO, 1985, pp. 43-56.

IV. CONDENSED MATTER AND SURFACE SPECTROSCOPY

LASER-INDUCED GRATING SPECTROSCOPY OF IONS IN SOLIDS

Richard C. Powell
Oklahoma State University, Stillwater, OK 74078

ABSTRACT

Laser-induced grating spectroscopy is a powerful technique for studying a wide variety of physical properties of solids. We review here the application of this technique in characterizing the spectral dynamics of dopant ions in solid state laser crystals.

INTRODUCTION

When two coherent laser beams with Gaussian profiles intersect at a crossing angle θ, they form an interference pattern in the form of a sine wave. If this occurs inside a solid, the interaction of the electromagnetic fields of the laser beams with matter produces a change in the complex index of refraction of the material. The modulated refractive index will have the same sine wave shape as the light interference pattern and can be modeled as a refractive index grating. A third laser beam incident on this region of the material will be partially diffracted by the laser-induced grating (LIG). This experimental arrangement has been used to investigate many different types of physical processes in a wide variety of materials.[1,2]

Different theoretical formalisms have been developed for treating the effects of interacting laser beams in solids. One approach is the laser-induced grating formalism which is based on the intensity dependent change in the complex index of refraction. This has been especially useful in applications associated with holographic information storage and data processing using photorefractive materials. A second type of formalism is four-wave mixing (FWM) based on the nonlinear polarization and third order susceptibility of the material. This approach has been especially useful in applications associated with nonlinear optical switching and phase conjugation. The LIG and FWM formalisms are semiclassical treatments which are essentially equivalent but different approaches to understanding the same physical phenomena. Full quantum mechanical formalisms have not yet been especially useful in interpreting experimental data. Both the LIG and FWM formalisms are useful in extracting different types of information from experimental results.

There are numerous types of configurations which have been utilized for LIG experiments. Each of these involve two laser beams which produce the modulation of the complex index of refraction (referred to as "pump" or "write" beams) and a beam which interrogates the induced modulation (called a "probe" or "read" beam). The beam resulting from the interaction of the write and read beams is the "signal" beam. These beams can be pulsed or continuous with a variety of different geometric configurations. They can be degenerate in wavelength or all be different. At very high powers the read beam can actually be part of one of the write beams thus producing "self-diffraction". Detection schemes such as hetrodyning and polarization rotation can be used to enhance signal-to-noise ratios. The optimum combination of these experimental details depends on the specific physical properties of interest.

Table I summarizes the variety of experiments that can be carried out with LIG techniques. All of the various measurement techniques and variable parameters are useful in completely characterizing the physical process of interest. Since more than one process is generally occuring simultaneously, it is important to obtain sufficient experimental information to determine the contributions from the different processes.

Table I. Summary of LIG experiments.

EXPERIMENTAL TECHNIQUES	VARIABLE PARAMETERS	PHYSICAL PROCESSES
Scattering Efficiency	Power	Thermal
Scattering Patterns	Temperature	Nonlinear Optical
Signal Dynamics	Crossing Angle	Free Carrier
Beam Mixing:	Wavelength	Exciton
Energy Transfer	Orientation	Trapped Charge Density
Self-Diffraction	Dopant Level	Structural
		Excited State Population

For applications involving characterizing the spectral dynamics of dopant ions in solids, excited state population gratings are of primary interest. The laser write beams are tuned to resonance with an absorption transition of the ions in the host material thus creating a sine wave spatial distribution of the concentration of ions in the metastable state. The difference in the absorption or dispersion properties of the ions when they are in the excited state state versus the ground give rise to the LIG. The grating can decay for two reasons: (1) the relaxation of the excited ions back to the ground state through both radiative and radiationless processes; and (2) the reduction in the modulation depth of the grating due to the migration of the excitation energy from ions in the peak regions of the grating to ions in the valley regions. LIG spectroscopy provides a method for obtaining information about radiationless relaxation, excited state absorption, and energy migration processes. Studies of energy migration (or exciton dynamics) are especially important since LIG spectroscopy is the only technique currently available to characterize long-range energy transport without spectral diffusion.

TWO-LEVEL SYSTEM ANALYSIS

The theoretical model used for analyzing LIG data obtained on ions in solids is based on the interaction of the laser beams with an atomic system having two electronic levels.[1,3-5] For a simple sine wave grating, this model predicts that the scattering efficiency at the Bragg angle will be given by

$$\eta=\exp(-2\alpha L)[\sin^2(d\pi\Delta n/2\lambda)+\sinh^2(d\Delta\alpha/4)] \quad (1)$$

where α and L are the absorption coefficient and sample thickness and d is the grating thickness. For normal conditions the functions in Eq. (1) can be expanded so the LIG scattering efficiency is proportional to the sum of the squares of the products of ($d\Delta n$) and ($d\Delta\alpha$).

For excited state population gratings, the modulation depth of the grating is proportional to the product of the difference in the complex refractive index for the ions in the ground and excited states and the population density of the excited states in the peak region of the grating.[6] Using a simple rate equation analysis for the pumping dynamics of a two-level system, the equilibrium population of the excited state is[7]

$$N_{2p}= NI\sigma_1/\{h\nu/\tau+\sigma_1 I\} \quad (2)$$

where N is the total concentration of ions, σ_1 is the ground state absorption cross section, τ is the excited state lifetime, and ν and I are the frequency and intensity of the laser. For a system of isolated ions the excited state population decays exponentially with the fluorescence decay rate τ^{-1}. Thus,

$$\eta\approx\{-(\pi d/2\lambda)^2[\sigma_1 NI]/[n(h\nu/\tau+\sigma_1 I)]$$
$$\times\Sigma_i f_i[1-\omega/\omega_i]/[4(\omega_i-\omega)^2+\gamma_i{}^2]$$
$$+(d/4)^2[(\sigma_2-\sigma_1)NI\sigma_1]/[h\nu/\tau+\sigma_1 I]\}\exp(-Kt). \quad (3)$$

Here σ_2 is the excited state absorption cross section, f_i is the oscillator strength of the transition centered at frequency ω_i with a width γ_i, and the LIG signal decay rate K is twice the fluorescence decay rate.

In the general two-level system model developed in refs. 3 and 4, the LIG scattering efficiency is determined by two complex coupling constants D_1 and D_2 which mix the four laser beams in the material. D_1 is directly proportional to the complex refractive index of the material and D_2 is proportional to the laser-induced change in this parameter. In this model, the grating modulation depths are derived to be

$$\Delta n=(\alpha c/\omega)D_2{}^r/D_1{}^i \quad (4)$$

$$\Delta\alpha=-2\alpha D_2{}^i/D_1{}^i \quad (5)$$

and the dephasing time of the atomic system is

$$T_2=(2\omega/c)(\Delta n/\Delta\alpha)(\omega-\omega_{21})^{-1}. \quad (6)$$

The latter quantity can be divided into pure dephasing processes associated with phonon scattering and described by a rate parameter γ, and contributions due population decay processes described by a rate T_1,

$$T_2{}^{-1}=(2T_1)^{-1}+\gamma. \quad (7)$$

This theory based on four interacting laser beams in a two-level atomic system is useful for analyzing LIG experimental results to obtain information on the spectral dynamics of ions in solids.

EXPERIMENTAL RESULTS

We are currently using LIG spectroscopy techniques to characterize the spectral dynamics of ions with a $3d^3$ electron configuration in solids.[6,8-10] One purpose of this study is to determine the validity of using the two-level system model described in the proceeding section to analyze data obtained on a real system. Figure 1 compares the schematic picture used for the two-level system model with the configuration coordinate model for the most important levels of $3d^3$ ions in an octahedral crystal field.

Equation (3) predicts that the LIG scattering efficiency should vary quadratically with the power of the laser write beams and saturate at high powers. This type of dependence has been observed when beam depletion is negligible but a smaller than quadratic dependence on power is seen when beam depletion is important.[6,8-10]

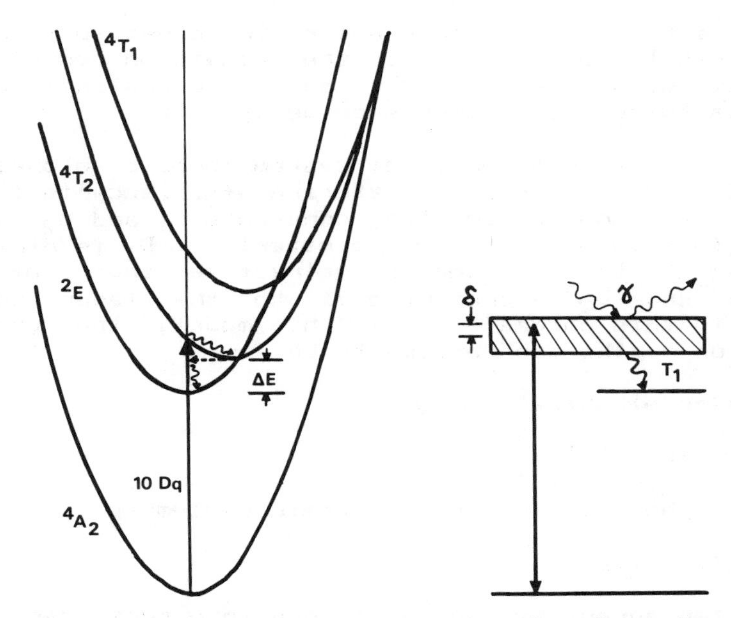

Figure 1. Comparison of the two-level system model with a configuration coordinate diagram.

According to Eq. (3), the temperature dependence of the LIG scattering efficiency should be contained in τ. This has been observed experimentally for laser intensities below saturation.[8]

The prediction of an exponential LIG signal decay with a decay rate equal to twice the fluorescence decay rate has been observed experimentally for situations where no ion-ion interaction is present.[6,8-10] When long range spatial energy migration is present, the signal decay increases as the grating spacing decreases. For incoherent hopping motion the decay dynamics are still exponential but for long mean free path motion nonexponential LIG signals are observed.[6,9-11] A variety of different exciton migration characteristics were observed in different Cr^{3+} doped laser crystals. Ruby showed no long range transfer. Alexandrite and GGG showed a strong energy migration at low temperatures which decreased as temperature was raised.[6,10] Emerald and GSGG exhibited an energy migration which increased as temperature was raised.[10,11] These differences are now being analyzed theoretically in terms of different electron-phonon interactions in the different hosts.

It has been shown that a numerical procedure for fitting the data obtained on the variation of the LIG scattering efficiency with the crossing angle of the write beams can be used to obtain the values of the real and imaginary parts of the coupling parameters.[6] Once these are known, the absorption and dispersion contributions to the grating modulation depths can be determined from Eqs. (4) and (5). For the set of Cr^{3+} doped laser crystals mentioned above, the major contribution to the LIG is the dispersion term although the absorption term is not negligible. These values can then be used in Eq. (3) to obtain the excited state absorption coefficient. The results of doing this for ruby, alexandrite, emerald, and GSGG agree with those reported in the literature obtained directly from excited state absorption measurements.

The values of Δn and $\Delta \alpha$ can also be used in Eq. (6) to determine the dephasing time. For the Cr^{+} doped laser crystals studied here, the write beams are in resonance with the $^4A_2$-$^4T_2$ transition shown in Fig. 1. Under these conditions, the dephasing of the atomic system is dominated by the radiationless relaxation to the bottom of the ^2E potential well. The results obtained on this set of crystals show a pronounced dependence of T_2 on ΔE, the energy splitting between the $^4T_2$ and ^2E levels. These results imply that the relaxation occurs directly in the ^2E potential well instead of first relaxing to the bottom of the $^4T_2$ potential well as is generally assumed.

The results described above show that the simple two-level model can be useful in approximating the multi-level potential wells of ions in solids. In using this model to interpret data obtained through LIG spectroscopy, care must be taken to produce a simple sine wave grating, have negligible beam depletion, and work well below the saturation level. Under these conditions, this type of experimental and theoretical procedure can be extremely useful in obtaining information on exciton dynamics, excited state absorption, and radiationless relaxation processes in solid state laser materials.

This research was supported by the U.S. Army Research Office.

REFERENCES

1. H.G. Eichler, P. Gunter, and D.W. Pohl, "Laser-Induced Dynamic Gratings" (Springer-Verlag, Berlin, 1986).
2. "Optical Phase Conjugation", ed. R.A. Fisher (Academic Press, New York, 1983).
3. A. Yariv and D.M. Pepper, Opt. Lett 1, 16 (1977).
4. R.L. Abrams and R.C. Lind, Opt. Lett. 2, 94 (1987); 3, 205 (1987).

5. H. Kogelnik, Bell Sys. Tech. J. 48, 2909 (1969).
6. A.M. Ghazzawi, J.K. Tyminski, R.C. Powell, and J.C. Walling, Phys. Rev. B 30, 7182 (1984); A. Suchocki, G.D. Gilliland, and R.C. Powell, Phys. Rev. B 35, 5830 (1987).
7. K.O. Hill, Appl. Opt. 10, 1695 (1971).
8. J.M. Allen, A. Suchocki, R.C. Powell, and G. Loiacono, Phys. Rev. B 36, 6729 (1987).
9. G.J. Quarles, A. Suchocki, R.C. Powell, and S. Lai, Phys. Rev. B., to be published.
10. A. Suchocki and R.C. Powell, Chem. Phys., to be published.
11. V.M. Kenkre and D. Schmid, Phys. Rev. B 31, 2430 (1985).

IMPLICATIONS FOR THE $SrF_2:Sm^{2+}$ LASER

$SrF_2:Sm^{2+}$ is known to lase on the $^5D_0 \rightarrow {}^7F_1$ transition at 14,350 cm^{-1}.[13] The absolute efficiency of this laser is determined in part, by the relative values of the ESA (σ_{ESA}) and emission (σ_{em}) cross-sections, since the net effective gain is given by:

$$\sigma_{eff} = \sigma_{em} - \sigma_{ESA} \qquad (6)$$

The ESA cross-section at the laser wavelength can be estimated to be 7×10^{-20} cm^2, while the emission cross-section can be calculated to be about $(8\pm4) \times 10^{-20}$ cm^2.[9,14,15] Although the value of both σ_{ESA} and σ_{em} are clearly estimates, the comparable magnitudes of these cross-sections indicate that ESA is likely to represent a detriment to the effective gain for the $SrF_2:Sm^{2+}$ laser.

CONCLUSION

In conclusion, the first direct experimental measurement of the 4f-5d exchange interaction energy for a rare-earth impurity ion has been made. It was found that Yanase's theory apparently embodies the crucial physical mechanisms, despite the approximations utilized. Lastly, the ESA present in $SrF_2:Sm^{2+}$ was determined to be comparable to the emission cross-section, making efficient operation of this laser somewhat unlikely.

ACKNOWLEDGMENTS

This research was performed under the auspices of the Division of Materials Sciences of the Office of Basic Energy Sciences, U.S. Department of Energy, and the Lawrence Livermore National Laboratory under contract no. W-7405-ENG-48. We are indebted to Gary Wilke, who performed many of the measurements reported in this paper and to Gary Ullery, for maintaining the dye laser in the course of this work.

REFERENCES

1. E. Loh, Phys. Rev. 175, 533 (1968).
2. M.J. Freiser, S. Methfessel, and F. Hultzberg, J. Appl. Phys. 39, 900 (1968).
3. L.L. Chase, Phys. Rev. B 2, 2308 (1970).
4. D.S. McClure and Z. Kiss, J. Chem. Phys. 39, 3251 (1963).
5. S.A. Payne, L.L. Chase, and G.D. Wilke, to be published in Phys. Rev. B 37, (1988).
6. A. Yanase, J. Phys. Soc. Japan 42, 1680 (1977).
7. S.A. Payne, L.L. Chase, W.F. Krupke, and L.A. Boatner, submitted to J. Chem. Phys.
8. P.P. Feofilov and A.A. Kaplyanskii, Opt. and Spectrosc. 12, 272 (1962).

9. D.L. Wood and W. Kaiser, Phys. Rev. 126, 2079 (1962).
10. J.D. Axe and P.P. Sorokin, Phys. Rev. 130, 945 (1963).
11. Wm.R. Callahan, J. Opt. Soc. Am. 53, 695 (1963); S. Johansson and U. Litzen, Phys. Scr. 8, 43 (1973).
12. J. W. Huang and H.W. Moos, J. Chem. Phys. 49, 2431 (1968).
13. P.P. Sorokin, M.J. Stevenson, J.R. Lankarad, and G.D. Petit, Phys. Rev. 127, 503 (1962).
14. V.A. Arkhangelskaya, M.N. Kiselyeva, and V.M. Schraiber, Opt. and Spectrosc. 23, 275 (1967).
15. B. Birang, A.S.M. Mahbub'ul Alam, and B. DiBartolo, J. Chem. Phys. 50, 2750 (1969).

STATISTICAL FINE STRUCTURE IN INHOMOGENEOUSLY BROADENED ABSORPTION LINES IN SOLIDS

W. E. Moerner and T. P. Carter
IBM Research Division, Almaden Research Center, San Jose, CA 95120

ABSTRACT

By using laser FM spectroscopy, we have observed statistical fine structure (SFS) in the inhomogeneously broadened zero-phonon $S_1 \leftarrow S_0$ (0-0) absorption of pentacene molecules in crystals of p-terphenyl at liquid helium temperatures. SFS results from variations in the spectral density of absorbers with optical wavelength due to statistics, illustrating that the inhomogeneous line profile is not a simple, smooth Gaussian function as has been previously assumed but rather contains significant fine structure. Theoretical analysis of the SFS spectra using autocorrelation techniques can be used to estimate the underlying value of the homogeneous linewidth. This new observation not only provides a novel way to study the statistics of inhomogeneous broadening, but it also allows estimation of the homogeneous width without requiring spectral hole-burning or coherent transient techniques.

INTRODUCTION

Inhomogeneous broadening is not only a universal feature of high-resolution laser spectroscopy of defects in solids[1], but it also appears in a fundamental way in other spectroscopies of impurity centers such as nmr, esr, and Mössbauer absorption. For zero-phonon transitions in crystals, the inhomogeneous profile (for a particular orientation or site) is composed of many narrow homogeneous absorption lines with a distribution of center frequencies caused by dislocations, point defects, or random internal electric fields and field gradients. Inhomogeneous broadening also occurs in amorphous hosts, where the center frequency distribution is caused by the large multiplicity of local environments.

Inhomogeneously broadened absorption lines are usually treated as smooth, Gaussian profiles. In recent work we have demonstrated the surprising fact that significant fine structure is a fundamental property of such lines[2]. This structure is static and repeatable for a given probe volume; however, the structure changes completely for different probe volumes. Because we see no correlation between the frequency-dependent structure for different spatial positions, we feel that the source for SFS is statistical variations in the absorber spectral density with optical wavelength. For this reason, we call the effect "statistical fine structure" (SFS). We demonstrate that SFS can be detected in a *high* concentration sample by using a zero-background technique, laser frequency-modulation (FM) spectroscopy[3]. Information about the statistics of the centers and about the underlying homogeneous linewidth γ can be derived from the SFS spectra, without requiring spectral hole-burning [4] or coherent transient techniques[5].

EXPERIMENTAL

Measurements of SFS were performed on the inhomogeneously broadened optical absorption of pentacene molecules in p-terphenyl crystals at liquid helium temperatures. Samples were prepared from mixtures of sublimed pentacene and zone-refined p-terphenyl which were grown into single crystals using Bridgman techniques and cleaved into samples 200-300 μm in thickness. Concentrations ranged between 1×10^{-5} and 2×10^{-7} mole/mole, yielding low temperature optical densities at the peak of the O_1 site absorption[6] between 0.01 and 0.15.

SCALING OF STATISTICAL FINE STRUCTURE

A simulation of expected lineshapes for varying numbers of centers helps in understanding the influence of concentration and sample configuration on the size of the SFS signals. We consider a fixed frequency interval $\Delta\nu$ within the inhomogeneous line that satisfies $\Gamma \gg \Delta\nu \gg \gamma$, where γ is the homogeneous linewidth (full width at half-maximum absorption, or FWHM) and Γ is the FWHM of the inhomogeneous line. We define the spectral density of absorbers in the probed volume $g(\nu)$ by requiring that $g(\nu)d\nu$ be the number of absorbers with center frequencies in $d\nu$ at ν. Then the number of centers per homogeneous linewidth is $N_H \equiv \int g(\nu)d\nu$ where the integral is performed over the spectral range γ, and the variations in $g(\nu)$ and thus in N_H with frequency form the underlying source of SFS. Let ΔN_H , $\overline{N}_H$ and $\Delta\alpha$, $\overline{\alpha}$ signify the rms amplitude and mean value of N_H and α, respectively, over $\Delta\nu$.

We make the crucial assumption that the probability of a given center acquiring a particular center frequency in the range $\Delta\nu$ is constant. This is equivalent to assuming no "microsites" or special frequencies that are more probable than others within the interval. Figure 1 illustrates the variations in absorption coefficient that can then occur due to statistics alone.

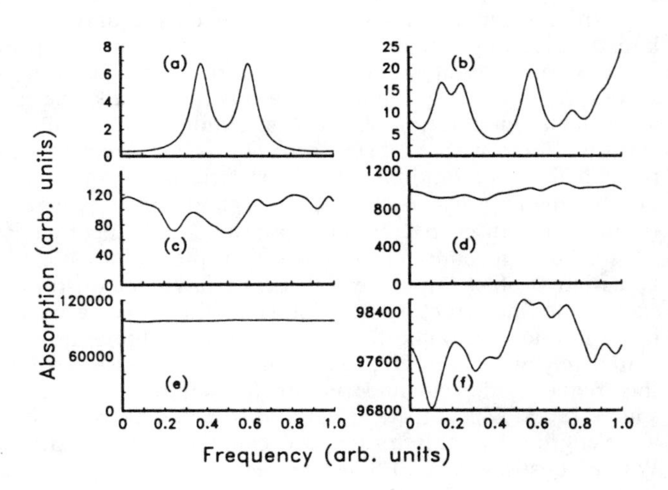

Figure 1. Simulated absorption spectra with different values of $\overline{N}_H$. Traces (a) through (e) correspond to N_H values of 0.2, 1, 10, 100 and 10,000, respectively. Trace (f) is an expanded trace of the same data as in (e).

To show how this figure was generated, let $L_\gamma(\nu) = (\gamma/2\pi)/[\nu^2 + (\gamma/2)^2]$ signify the assumed Lorentzian absorption of FWHM γ for each center, then the full lineshape is

$$\alpha(\nu) = s\int_{-\infty}^{+\infty} g(x)L_\gamma(\nu - x)dx \equiv s\, g * L_\gamma(\nu), \qquad (1)$$

where s is the integrated absorption strength per center and the asterisk signifies convolution. Figure 1(a) shows a possible lineshape if only two centers are in

$\Delta\nu$. Traces 1(b), (c), (d), and (e) show simulated lineshapes for $\overline{N}_H = 1$, 10, 100, and 10^4. Clearly, $\overline{\alpha}$ is growing linearly with $\overline{N}_H$, while the relative fluctuations in absorption $\Delta\alpha/\overline{\alpha}$ are decreasing as $(\overline{N}_H)^{-1/2}$. Therefore, small $\overline{N}_H$ samples ($\overline{N}_H < 10$, for example) would be expected to be optimal for the observation of SFS. However, detecting such a small number of centers in the presence of considerable background from the host matrix is quite difficult[7].

Zero background techniques like FM spectroscopy provide a way around this problem, because the variable measured is $\Delta\alpha$ itself, which is growing as $(\overline{N}_H)^{1/2}$. Trace 1(f) demonstrates that SFS is still present even when $\overline{N}_H$ is large.

FM SPECTROSCOPY DETECTION

Standard FM techniques[3] employing an AD*P electro-optic phase modulator and a Si avalanche photodiode were used to phase-sensitively detect the SFS signal. Complete details of the apparatus will be presented elsewhere[8]. The crucial feature of the FM technique is that the detected signal varying as $\cos(2\pi\nu_m t)$, $F_1(\nu)$, is proportional to

$$F_1(\nu) \sim -MP_0\, e^{-\overline{\alpha}\ell}\, [\alpha(\nu+\nu_m) - \alpha(\nu-\nu_m)]\ell, \tag{2}$$

where ν is the laser frequency, ν_m is the rf modulation frequency, P_0 is the laser power on the sample, M is the modulation index, ℓ is the sample length, and $\overline{\alpha}$ is the background value of α. Thus the FM signal measures the difference in $\alpha\ell$ at the two sideband frequencies. (The component of the photocurrent varying as $\sin(2\pi\nu_m t)$, called $F_2(\nu)$, measures dispersion[8].) $F_1(\nu)$ has two well-defined limits depending upon the ratio of ν_m to the linewidth of the spectral features[3]. When $\nu_m \gg \gamma$, $F_1(\nu)$ consists of two replicas of the Lorentzian line, one positive and one negative, separated by $2\nu_m$. In this regime $F_1(\nu)$ is maximal and independent of ν_m, and it is this regime that we use most often to detect SFS. Using the scaling results of the last section, the FM spectroscopy signal varies as $(\Delta\alpha)\ell = \sigma(\overline{N}_H)^{1/2}/A = \sigma(\rho\ell/A)^{1/2}$, where σ is the peak absorption cross section, A is the beam area, and the volume density per homogeneous linewidth is $\rho = \overline{N}_H/A\ell$. Therefore, $F_1(\nu)$ increases if the concentration of absorbers or the sample length increases, and increases for smaller laser spots. Further, centers with higher cross section lead to larger FM signals.

RESULTS AND DISCUSSION

Figure 2 shows FM spectra of SFS under varying conditions for pentacene in a single crystal of p-terphenyl. These spectra were acquired by repetitively scanning a R6G single-frequency dye laser (2.8 MHz linewidth) over the desired frequency range and averaging 64 scans. Unless stated otherwise, the laser was focused to a 20 μm diameter spot, the laser power was 3 μW, and the sample was immersed in superfluid helium at 1.4 K. The conditions for the various traces are summarized in the caption. Note in particular the following facts: the SFS spectrum for a single spot is highly reproducible (traces (c) and (d)), different spots on the sample show radically different SFS (trace (e)), larger beam areas show smaller SFS (trace (g)), persistent spectral holes can be burned (an unexpected result, trace (h)), and SFS disappears with increasing temperature due to the increase of γ above the fixed value of $\nu_m = 58.1$ MHz (traces (i), (j), and (k)). For the O_1 site of pentacene in p-terphenyl at 1.4 K, γ has been reported to be 7.8 ± 0.6 MHz[9].

Figure 2. FM spectra (F_1) for a single crystal of pentacene in p-terphenyl. (a) No light on the detector. (b) 3 μW on the detector at a wavelength not in resonance with the O_1 site absorption. (c),(d) Spectra at 1.4 K near the peak of the O_1 absorption at 592.3 nm with a focused spot, where $\overline{N}_H \cong 5 \times 10^5$. (e) A new spot on the sample, same spectral range as (c). (f) Laser center frequency offset by 50 MHz from that for (e). (g) Large laser spot (0.75 mm diameter). (h) Persistent hole burned in the spectral range of trace (g) using 11 mW for 30 s (power broadened hole). (i) 1.4 K, focused spot. (j) 5.6 K, same location. (k) 7 K. The vertical scale is exact for (c) and (d); all the other traces have the same scale but are offset vertically for clarity. One volt corresponds to a change in $\alpha\ell$ of 1.1×10^{-3}. The detection bandwidth was 0.1 Hz to 300 Hz and $\nu_m = 58.1$ MHz with M = 0.16. The frequency scale was calibrated by optically observing the rf sideband spacing.

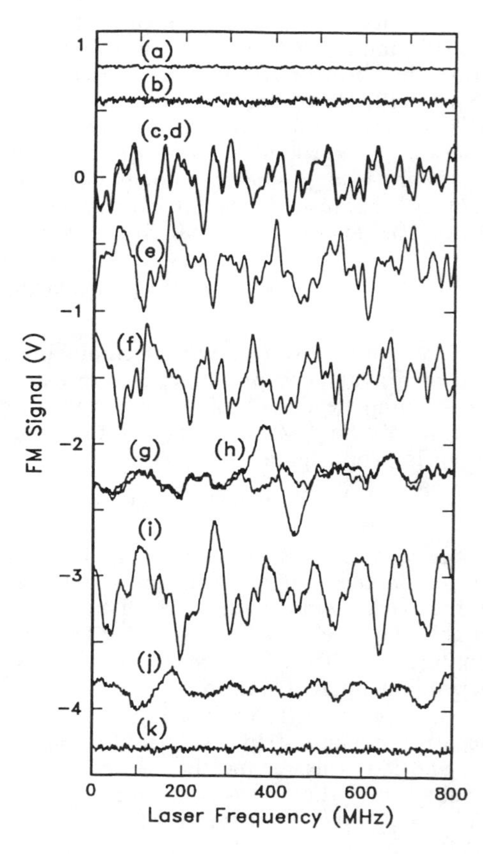

Results similar to those in Fig. 2 were obtained for the O_2 site where $\gamma = 7.3 \pm 0.5$ MHz, but the O_3 and O_4 sites [8] show smaller and broader SFS because γ is larger for O_3 and O_4. In addition, as the laser frequency is moved away from the center of the inhomogeneous line, the amplitude of the SFS continuously decreases, as expected.

SFS provides a new window on inhomogeneously broadened lines with intrinsic detail and complexity. For example, recording of SFS spectra over a large fraction of the inhomogeneous profile may provide new information about the distribution of optical absorption energies available to the impurity centers. Furthermore, the rms amplitude of the SFS spectra should grow as $(\overline{N}_H)^{1/2}$, and a measured dependence of $(\overline{N}_H)^{0.54\pm0.05}$ has been observed experimentally (see Ref. 2). To check for the possibility of "microsites", the SFS spectra can be acquired as a function of position in the sample to see if the spectral structure at one position correlates with that at another position. Figure 3 shows a 3-dimensional plot over a 200 MHz range in frequency and a 200 μm range in laser spot position. No evidence for microsites or departures from the statistical source for the SFS have been observed in this system.

Estimates of γ can be extracted from the SFS spectra by exploiting the properties of autocorrelation functions. We consider the expected autocorrelation of the F_1 spectra, $< F_1 \star F_1 > (\nu)$, in the limit $\nu_m \gg \gamma$. The result[8] is

$$< F_1 \star F_1 > (\nu) \propto -L_{2\gamma}(\nu + 2\nu_m) + 2L_{2\gamma}(\nu) - L_{2\gamma}(\nu - 2\nu_m). \quad (3)$$

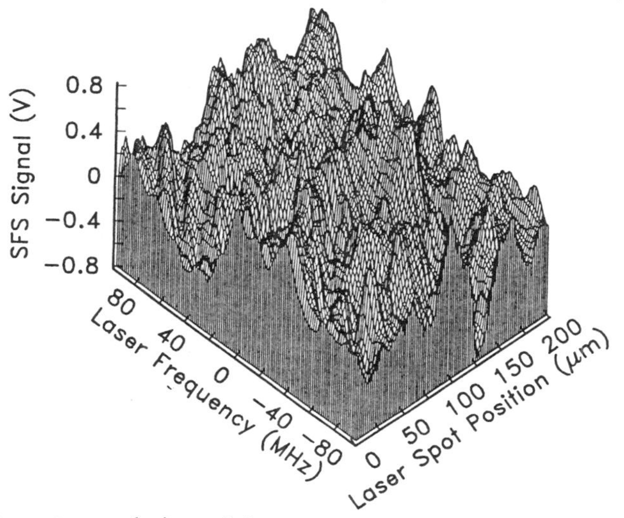

Figure 3. SFS structure versus laser spot position. A sequence of 100 spectra were obtained, moving the 20μm laser spot by 2μm after each spectrum, and the results plotted to show how the SFS structure changes as the laser spot is moved.

The expectation value of the autocorrelation of the FM signal has a FWHM equal to twice that for the underlying homogeneous absorption lines. Thus an estimate for γ can be derived from the FWHM or second derivative of $< F_1 \star F_1 > (\nu)$ at the origin:

$$\gamma \simeq \left[\frac{-2 < F_1 \star F_1 > (0)}{< F_1 \star F_1 > ''(0)} \right]^{1/2}, \quad (4)$$

where the double prime signifies second derivative. (Eqn. (4) is also true for $< F_2 \star F_2 >$, see Ref. 8).

To show how the expectation of the autocorrelation of the SFS signals approaches Eqn. (3), Figure 4 shows examples of measured single autocorrelation functions as well as the average of 10 autocorrelations for both F_1 and F_2.

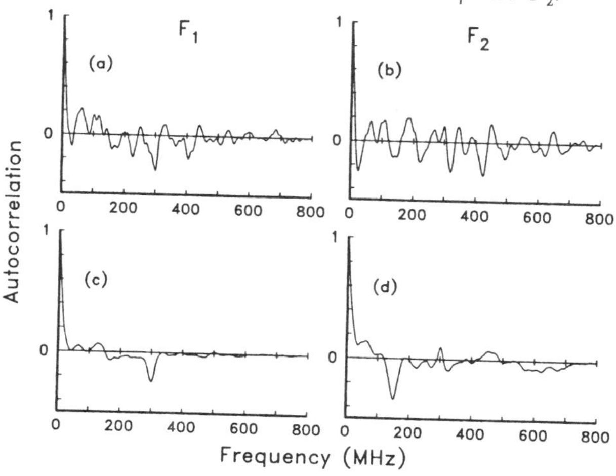

Figure 4. Autocorrelations of SFS spectra in F_1 and F_2 for the O_1 site with $\nu_m = 150$ MHz. Traces (a) and (b) show typical normalized autocorrelations of single SFS spectra. Traces (c) and (d) show the appearance of the peaks described by Eqn. (3) and by a similar equation[8] for F_2 obtained by averaging only ten such spectra.

Long (4.8 GHz) laser scans are best for estimation of homogeneous widths because errors in the autocorrelation due to the limited spectral range of the data are minimized. For a single crystal sample with concentration 3×10^{-7} moles/mole, we find using Eqn. (4) that $\gamma = 7.9 \pm 0.8$ MHz for the O_1 site at 1.4 K by analyzing the autocorrelations of six SFS spectra over a 4.8 GHz spectral range using $\nu_m = 150$ MHz. This value is consistent with the previously reported value [9] of 7.8 ± 0.6 MHz obtained using coherent transient techniques.

CONCLUSIONS

The central result of this work is that for an inhomogeneously broadened line, the spectral density of absorbers $g(\nu)$ is not a smooth Gaussian profile; rather, $g(\nu)$ contains significant statistical fine structure, even for high concentration samples. SFS should be a general feature of all inhomogeneously broadened lines in solids, and even in gases if measurements are performed on a time scale shorter than the collision time. We expect that SFS should be observable for color-centers, ions, and molecular vibrational modes in crystals as well as in amorphous media. We note that SFS imposes a fundamental limit on the detectability of shallow spectral features in inhomogeneous lines.

We acknowledge stimulating discussions with G. C. Bjorklund and N. Pippenger, and the technical assistance of M. Manavi. This work was supported in part by the U.S. Office of Naval Research.

REFERENCES

1. See *Laser Spectroscopy of Solids*, W. M. Yen and P. M. Selzer, eds., Springer Topics in Applied Physics, Vol. 49 (Springer, Berlin, 1981).
2. W. E. Moerner and T. P. Carter, Phys. Rev. Lett. 59 , 2705 (1987); W. E. Moerner and T. P. Carter, Bull. Am. Phys. Soc. 32, 1630 (1987).
3. G. C. Bjorklund, Opt. Lett. 5, 15 (1980); G. C. Bjorklund, M. D. Levenson, W. Lenth, and C. Ortiz, Appl. Phys. B 32, 145 (1983).
4. See *Persistent Spectral Hole-Burning: Science and Applications*, Springer Topics in Current Physics Vol. 44, W. E. Moerner, ed. (Springer, Berlin, Heidelberg, 1988) and references therein.
5. R. L. Shoemaker, in *Laser and Coherence Spectroscopy*, J. I. Steinfeld, ed. (Plenum, New York, 1978), p. 197.
6. R. W. Olson and M. D. Fayer, J. Phys. Chem. 84, 2001 (1980).
7. For a description of background problems for liquid hosts, see D. C. Nguyen, R. A. Keller, and M. Trkula, J. Opt. Soc. Am. B 4, 138 (1987).
8. T. P. Carter, M. Manavi, and W. E. Moerner, submitted to J. Chem. Phys.
9. F. G. Patterson, H. W. H. Lee, W. L. Wilson, and M. D. Fayer, Chem. Phys. 84, 51 (1984).

HIGH-RESOLUTION SPECTROSCOPY OF Cr^{3+} IONS IN SOLIDS

B. Henderson, K.P. O'Donnell and M. Yamaga[†]
Department of Physics and Applied Physics,
University of Strathclyde, Glasgow G4 0NG, Scotland, U.K.

B. Cockayne and M.J.P. Payne
Royal Signals and Radar Establishment, Malvern, England

ABSTRACT

This paper reports measurements of laser-excited fluorescence by Cr^{3+} ions in garnets ($A_3B_5O_{12}$, $A_3B_2C_3O_{12}$) and glasses. The temperature dependence of the photoluminescence spectrum is discussed in terms of the energy splitting, ΔE, between the ^2E and $^4T_2$ excited states. At low temperature ($T \sim 4K$) the ^2E_g - $^4A_2$ emission in garnets shows resolved R-lines and sharply peaked vibronic sidebands. The inhomogeneous broadening of the R-lines in both garnets and glasses is due to structural disorder, which has been probed by fluorescence line narrowing techniques.

INTRODUCTION

The fluorescence spectrum of Cr^{3+} ions in inorganic solids is strongly dependent upon crystal field strength (Dq/B).[1] In the range $2.2 < Dq/B < 2.4$ emission from both ^2E_g and $^4T_2$ levels may be observed with intensities depending upon the sample temperature.[2,3] Since the $^4T_{2g} \rightarrow {}^4A_g$ is parity forbidden whereas the $^2E_g \rightarrow {}^4A_{2g}$ is both spin and parity forbidden there is always some temperature at which the former transition dominates the fluorescence spectrum. This paper reports luminescence spectra of Cr^{3+} ions in garnets and in glasses, where Dq/B varies between _ca_ 1.9 and 2.5. In the garnets structural disorder exists due to ions normally occupying octahedral sites being randomly substituted by ions which normally reside in dodecahedral or tetrahedral sites. The disorder in glasses is associated with the many possible sites occupied by spectroscopic ions which have slightly different crystal field strengths.[4,5] In both material types the R-lines of Cr^{3+} ions are inhomogeneously broadened by the local disorder. We discuss how the R-line transitions may be used to probe this disorder using high resolution spectroscopic techniques.[6]

EXPERIMENTAL METHODS

Studies have been made of Cr^{3+} ions in single crystals of $Gd_3Sc_2Ga_3O_{12}$ (GSGG), $Gd_3Sc_2Al_3O_{12}$ (GSAG), thin film $Y_3Ga_5O_{12}$ and in silicate glasses. Fluorescence measurements were made at 4.2K using an Oxford Instruments MD4 dewar, or in the range 8 - 300K using a cryorefrigerator. The fluorescence was excited using an Ar^+ ion laser and detected through a 1m grating monochromator using phase sensitive

† Permanent address: Department of Information and Computer Sciences, Toyohashi University, Japan.

electronics. High resolution fluorescence measurements were made using an actively stabilized 380D single mode ring laser with linewidths of order 1MHz.

RESULTS AND DISCUSSION

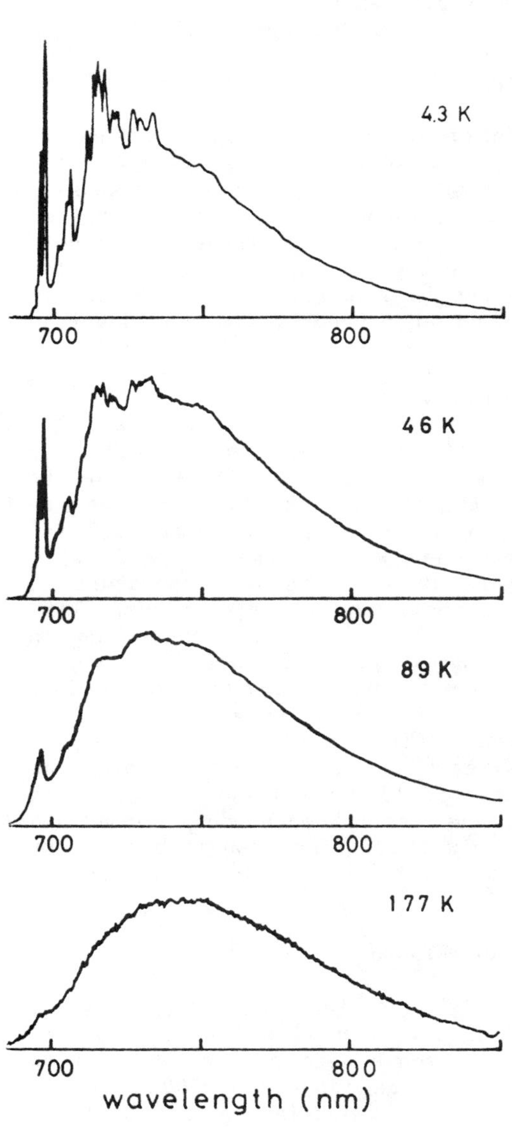

Figure 1 R-line and $^4T_2 \rightarrow {^4A_2}$ emission in Cr^{3+}:GSGG at different temperatures

Fig. 1 presents the luminescence spectra of Cr^{3+} doped GSGG measured at 4.3, 45.7, 88.5 and 176.5K; the prominent features are the broad $^4T_2 \rightarrow {^4A_2}$ emission which is dominant above 150K but which at low temperatures competes for intensity with the $^2E \rightarrow {^4A_2}$ R-lines and their vibronic sidebands. The dominance of the broad band emission at high temperature is due both to increased occupancy of the $^4T_2$ level relative to the lower lying ^2E level and the larger oscillator strength of the $^4T_2 \rightarrow {^4A_2}$ transition. Initially the total emitted intensity arises due to this latter process; however above 170K the total fluorescence output decreases due to competition from multi-phonon non-radiative processes. Similar behaviour is exhibited by both GSAG and YGG except that at low temperature lower emission intensities are observed in the broad band because the splittings, $\Delta E = E(T_{2g}) - E(E_{2g})$ are much larger than in GSGG. Indeed in YGG, where $\Delta E \approx 1100\ cm^{-1}$, the emission below 10K is almost entirely due to the R-line and its vibronic sideband.

In garnet crystals the site occupied by the Cr^{3+} ion is not perfectly octahedral, and the R-line transition is split into two lines, R_1 and R_2 (at shorter wavelength). Generally, at low temperature the R_2 line is not observed. However, as T is increased the

R_2-line intensity increases at the expense of the R_1-line intensity. For GSGG we also detect the appearance of the zero-phonon lines associated with the $^4T_2 \rightarrow {}^4A_2$ emission in the temperature range 15 - 35K. The measured splittings are $\Delta E(R_2-R_1) = 36cm^{-1}$ and $\Delta E(^4T_2-R_1) = 90cm^{-1}$. Above 35K the $^4T_2 - {}^4A_2$ zero-phonon emission lines broaden and are not discernible above ca 45K. However, the most interesting aspect of the R-lines is that both the R_2 and R_1 transitions in GSGG are resolved doublets (Fig. 2), each component of which has a strongly asymmetric lineshape. This behaviour is due to inhomogeneous broadening of the R-line due to disorder associated with the smaller Ga^{3+} ions occupying a fraction of octahedral sites normally reserved for Sc^{3+} ions. By considering the probability that one, two, three etc of the nearest neighbour Sc^{3+} sites are occupied by Ga^{3+} as a function of x in the composition formula $Gd_3Sc_{2-x}Ga_{3+x}O_{12}$ it is possible to show that the sample used for the spectrum shown in Fig. 2 has the composition $Gd_3Sc_{1.9}Ga_{3.1}O_{12}$. This estimate is based purely on the relative intensites of the R_1 and R_1' lines; however, the detailed shape reflects the disorder not solely in the nearest neighbour cation shell but also among the more remote neighbour cations. That a splitting is observed for the R_1 line in GSGG is due to the sensitivity of the ^2E level to changes in Dq. The slope of the $E(E_g)$ versus Dq curve is slightly greater close to the $^2E_g/^4T_{2g}$ cross over point (typical of GSGG) than it is at higher Dq values (typical of GSAG and YGG). In consequence, compositional disorder in GSAG is seen as an unresolved splitting of the R-lines, and in YGG as a strongly assymetric lineshape without splittings.

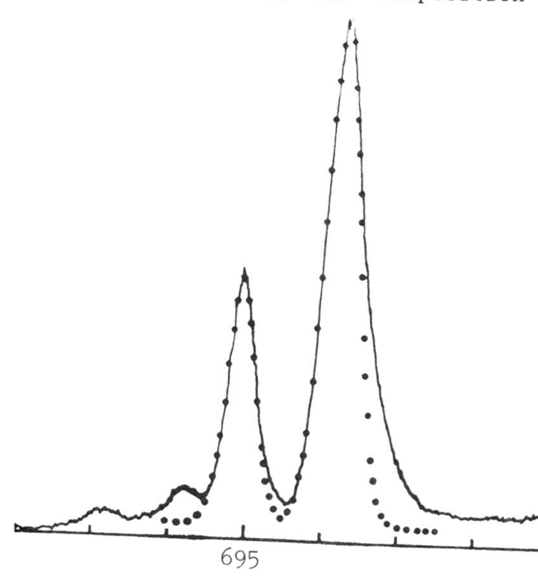

Figure 2 A symmetric shape of the R-lines in Cr^{3+}:GSGG at 30K

The R-line sideband shows poorly resolved structure, associated (mainly) with odd parity phonon modes of T_{1u} and T_{2u} symmetry. Careful comparison of the sideband and the zero-phonon line positions enables one to assign peaks to one, two and (sometimes) 3-phonon processes for the R_1, R_2 and $^4T_2 \rightarrow {}^4A_2$ transitions in Cr^{3+}:GSGG where the odd-parity phonon mode has energy $E/hc \simeq 170cm^{-1}$. Similar measurements yield a mode energy of ca $190cm^{-1}$. However, as we discuss below disorder also broadens vibronic peaks, tending to smear out any structure in the vibronic sideband of the R-line.

High resolution spectroscopic techniques such as optical hole

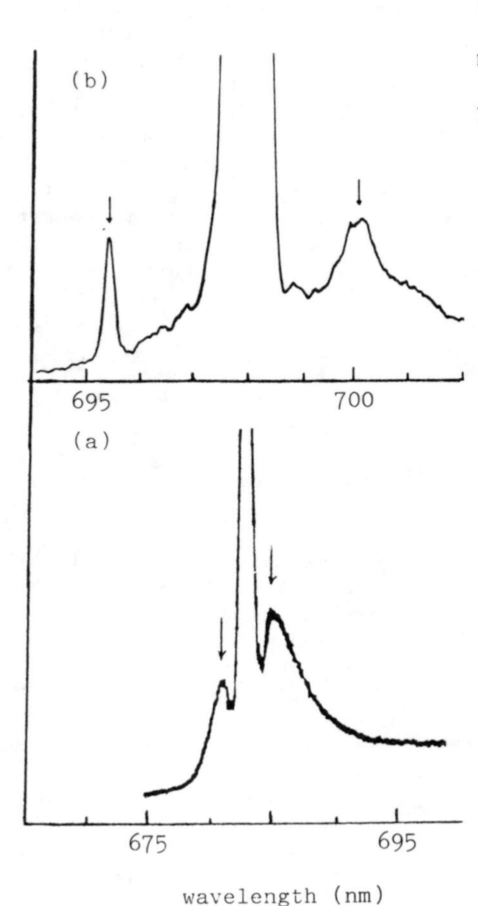

Figure 3 FLN in the R-line of Cr^{3+}-doped (a) glass at 77K and (b) GSGG at 30K

burning (OHB) and fluorescence line narrowing (FLN) are ideal methods with which to study the inhomogeneous broadening produced by structural disorder. The FLN spectra shown in Fig. 3 are for a Cr^{3+}-doped silicate glass (SLb7) and Cr^{3+}:doped GSGG both measured at 77K by resonant excitation in the inhomogeneously broadened R-lines. The line narrowing achieved even at 77K is very marked, the FLN spectra having FWHM of less than $0.1cm^{-1}$ compared with the normal R-line widths of $15cm^{-1}$ for GSGG and $170cm^{-1}$ in silicate glasses. The structure observed at longer and shorter wavelengths relative to the central FLN is due to the R_1 and R_2 components at sites where there are departures from strict octahedral symmetry. As is expected the line at shorter wavelengths (R_2) is not observed at low temperatures (< 60K). When the fluorescence emitted at longer wavelengths under excitation conditions required for FLN in the zero-phonon line is recorded much weaker vibronic sidebands are recorded, especially in the case of silicate glasses. However, in Cr^{3+}:GSGG there is a very dramatic sharpening of the peaked structure in the vibronic sideband (Fig. 4), which allows a very detailed assignment of the various phonon modes to the R_1, R_2 and $^4T_2 \rightarrow {}^4A_2$ transitions. The shifts of these vibronic peaks when varying the excitation wavelength within the R-line profile are initially linear with excitation wavelength. However, when exciting in the long wavelength tail of the R-line the shifts become non-linear presumably reflecting the effects of disorder in more remote shells of octahedrally-disposed cation neighbours of the Cr^{3+} probe ion.

The temperature dependence of the homogeneous width of the FLN spectrum has also been measured to further elucidate the nature of the electron-phonon coupling. In these measurements it is clear that the

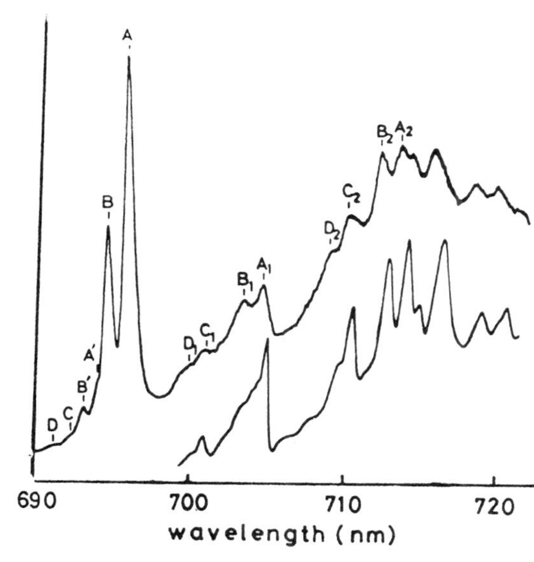

Figure 4 R-line sideband and non-resonant FLN at 30K in Cr^{3+}:GSGG

vibronic behaviour of individual Cr^{3+} ions in crystalline GSGG is different from that of the Cr^{3+} doped SLb7 glass. In glass SLb7 the results are very similar to those reported by Bergin et al.[7] for Cr^{3+} doped ED2 glass. It is found that the temperature dependent broadening of the zero-phonon line is anomalously large in glass at low temperature in comparison with crystal. This is due to the existence of a high density of low frequency "tunnelling modes" in glasses which do not occur in crystals.[7] Experiments are currently underway to probe the vibronic properties of the glass netowrk in detail. The extreme narrowness of the FLN allows direct resolution of the Zeeman effect in modest magnetic fields (0.1 - 0.2T), thus permitting the variation of spin Hamiltonian parameters in the various crystal field sites to be measured. The decay rate of the circularly polarized Zeeman components under pulsed excitation is then just the sum of the radiative (τ_R) and spin-lattice relaxation (T_1) rates. Preliminary measurements of the temperature dependence of T_1 at low temperatures indicate that below 50K a one-phonon direct process dominates the R-line relaxation. Such measurements are being extended to higher temperatures and to a variety of silicate and phosphate glasses.

ACKNOWLEDGEMENTS

The authors are indebted to the Ministry of Defence (U.K.) and Science and Engineering Research Council for support of the research programme of which this report forms a part. One of us (M.Y.) is grateful to the SERC for the award of a Senior Visiting Fellowship during 1986-87.

REFERENCES

1. See e.g. B. Henderson and G.F. Imbusch in Optical Spectroscopy of Inorganic Crystals, Chapter 9, Oxford University Press, (1988), in press.
2. J.C. Walling, O.G. Peterson, H.P. Jensen, R.C. Morris and E.W. O'Dell, I.E.E.E. J QE-16, 1302, (1980).
3. See e.g. G. Huber and K. Petermann (p.11), U. Durr, U. Brauch, W. Knierim and C. Schiller (p.20) in Tunable Solid State Lasers, (Ed.

P. Hammerling, A.B. Bugdor and A. Pinto, Springer-Verlag, Berlin and Heidelberg, 1984).
4. M. Weber in Laser Spectroscopy of Solids I, (Ed. P.M. Selzer and W. Yen, Springer-Verlag, Berlin and Heidelberg, 198).
5. L. Andrews et al. (1981).
6. W. Yen in
7. F.J. Bergin, J. Donergan, T.J. Glynn and G.F. Imbusch, J. Lumin, (1985).

OPTICAL TRANSITIONS IN Cr^{3+}-DOPED $LaMgAl_{11}O_{19}$, A NEW POTENTIAL VIBRONIC LASER MATERIAL

F.M. MICHEL-CALENDINI
Physico-Chimie des Materiaux Luminescents
Université Lyon I - Unité associée au CNRS 442
69622 Villeubanne Cedex - France

ABSTRACT

Electronic structures of Cr^{3+} doped lanthanum hexa-aluminate are determined through a self-consistent molecular orbital method. Cr^{3+} multisites, such as 2a,4f and 4f-4f, are investigated. Energies of optical transitions are calculated and provide a good understanding of the optical spectra. Term energy diagrams, obtained for crystal hosts with various CrO distances, are classified according to the crystal field strength.

INTRODUCTION

The present paper deals with the theoretical properties of some Cr^{3+} doped oxydes. Among them, lanthanum hexa-aluminate $LaMgAl_{11}O_{19}$ (denoted LMA hereafter) has been extensively studied by fluorescence, ESA and EPR techniques [1-2-3]. Such a study appears worthwhile, since in spite of a lot of experimental data relevant to laser materials, few results concerning the electronic structures of the activator ions are found in the litterature[4].

We use a molecular orbital model to get the electronic structure of Cr^{3+} in various crystal hosts through the self- consistent local spin density method, alternative to the $X\alpha$ one. Besides LMA, we investigate ruby, lithium niobate, perovskite oxydes and MgO. Due to the good localisation of Cr 3d levels in these rather wide band gap oxydes, these calculations appear all the more reliable to understand the mechanisms involved into emission and absorption and theirs relations with tunable laser applications. This method has been already applied with some success to investigate the properties of 3d ions embedded in perovskite oxydes (see reference 5 for instance) and part of the results concerning LMA:Cr have been reported recently [6].

THEORETICAL FRAMEWORK AND EIGENVALUE DIAGRAMS

The magnetoplumbite like structure of $LaMgAl_{11}O_{19}$ is well known and multisubstitutional sites are predicted for Cr^{3+} in this material[1-2]. Among

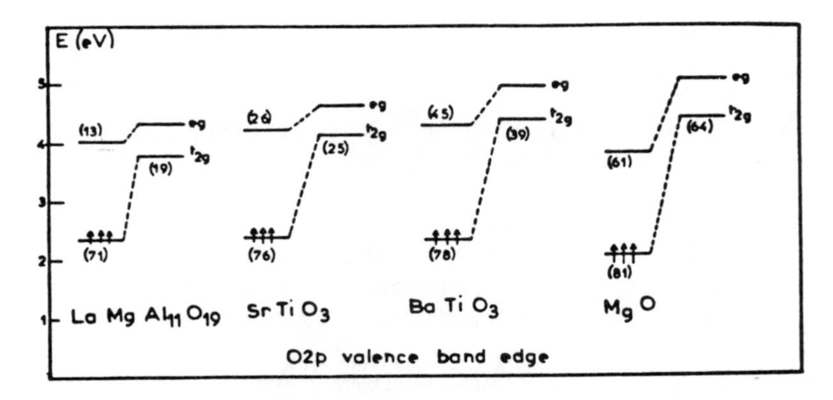

Fig. 1 . 3d eigenvalues of Cr^{3+} in cubic oxydes of increasing unit cell sizes

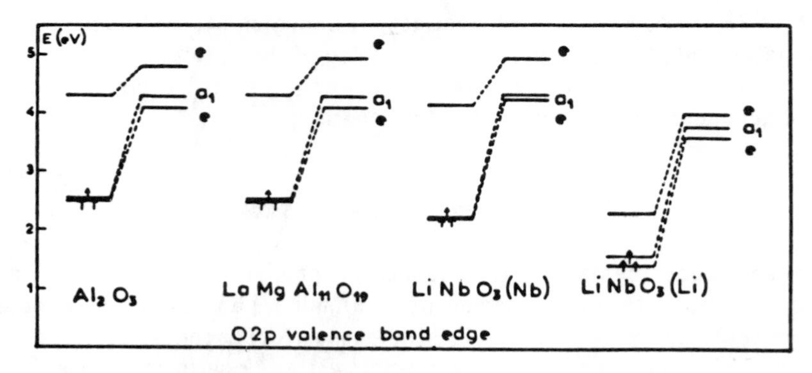

Fig. 2 . 3d eigenvalues of Cr^{3+} at C_{3v} sites in various oxydes

them , we study the perfectly cubic 2a and trigonally distorded 4f ones as well as the Cr–Cr pair assumed to be localized at two adjacent 4f sites . CrO_6^{8-} clusters of O_h and C_{3v} symetries depict the two former centers while the pair is represented by a $Cr_2O_9^{12-}$ cluster of D_{3h} symetry. Similarly , O_h and C_{3v} clusters describe Cr^{3+} in cubic oxydes and $LiNbO_3$ or Al_2O_3 respectively. Two substitutional sites , Li and Nb , are assumed in lithium niobate[7]. Atomic positions associated to the undoped materials are taken from crystallographic data to built up the considered clusters.

Computational details are provided in references 4-6-7. The only difference between earlier SCF Xα calculations and the present ones lies in the use of a local spin density approximation [8] which is more suitable to compute the exchange potentials in large spin polarized systems as Cr^{3+}.

Cr 3d levels associated to the $^4A_2$ ground term are shown in figures 1 and 2 for cubic and trigonal centers. These levels are refered to the highest O2p valence band level, denoted as the valence band edge VB. In figure 1, the CrO distances, d, of the considered crystal hosts are ranging between 1.88 to 2.1 Å, so that the evolutions of the 3d electronic level diagrams, $d\varepsilon$ (t_{2g}) and $d\gamma$ (e_g), may represent also the ones associated to many Cr doped laser materials like YAG or garnets, in a first approximation. It is seen that the $d\varepsilon\uparrow \rightarrow d\varepsilon\downarrow$ splitting increases with d while the $d\varepsilon\uparrow\rightarrow d\gamma\uparrow$ one remains stationary. As it will be seen further, these behaviors are at the origin of the decrease of the crystal field strength Dq/B observed in materials with increasing unit cell sizes. Figure 2 depicts the splittings of t_{2g} levels into e and a_1 ones due to the O_h to C_{3v} symetry lowering. The comparison between the 2a and 4f centers of LMA shows these splittings are less than 0.2 eV as far as the two inequivalent CrO distances remain nearly similar. The situation changes dramatically when these distances go far away as when Cr substitutes the Li site in $LiNbO_3$. The optical properties of this material are then totally different from a site to the other and the comparison with available experimental data allow to consider the Nb site as the more probable[8].

OPTICAL TRANSITIONS

Term energy diagrams have been obtained from transition state computations of the energies of crystal field transition energies associated to the t_2^3 and t_2^2e configurations of d^3. This procedure take into accout the relaxation effects between the initial and final states [5]. Considerations on the determinantal wave functions associated to these configurations give the following relations [9]:

$$2B + 3C = E(t_{2g}\uparrow \rightarrow t_{2g}\downarrow)$$

$$10\,Dq = E(t_{2g}\uparrow \rightarrow e_g\uparrow)$$

between the crystal field parameter Dq, the Racach B, C ones and the transition energies E of cubic levels. These relations are also valid approximatively for small trigonal and tetragonal distortions of the cubic symetry. We assume B/C = 4.5 to get indepandently the B and C values and introduce then the calculated parameters into electrostatic matrices to obtain the complete term energy diagrams which are to be compared to experimental transitions.

Figure 3 shows the theoretical and experimental energies of the first excited states of d^3 in terms of reduced units E/B and Dq/B and allows to localize more specifically the studied compounds into the low, intermediate and high crystal field regions. Experimental parameters are the ones found in the litterature for garnets and YAG [10–12]. For ruby and LMA:Cr, we take B and C from the $^2E \rightarrow {}^4A_2$ emission line assuming B/C = 4.5 and Dq from the maximum of the $^4A_2 \rightarrow {}^4T_2$ absorption. A detailled discussion on the experimental fits for these parameters will be given in a forthcoming paper[9].

As it was expected, ruby belongs to the strong field region but cubic oxydes with d up to 1.95 Å are rather localized in the intermediate field region while larger distances lead to the low field one.

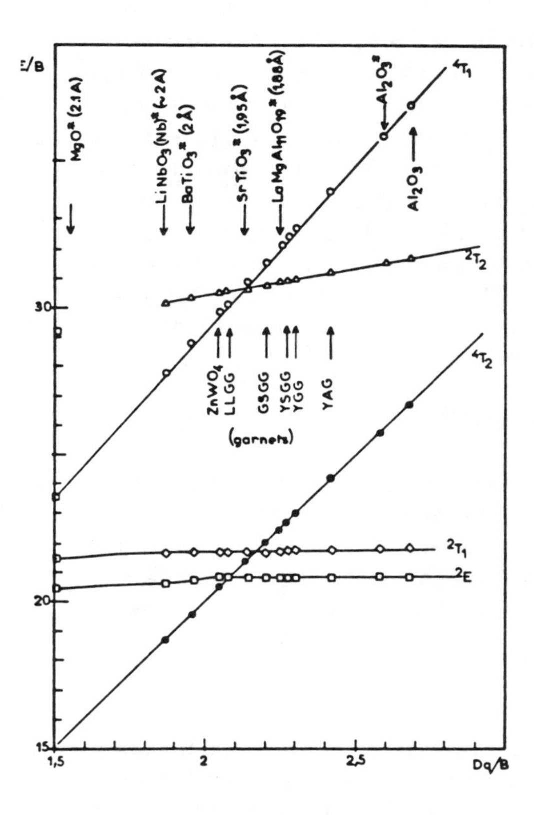

Fig. 3. Theoretical (*) and experimental term diagrams of Cr^{3+} according to the crystal field strength

The calculated B are slightly overestimated relatively to the experimental ones and this discrepancy increases with d while theoretical Dq values follow the experimental trends. Then, the theoretical values of d defining the different crystal field regions must be increased by about 0.5 Å, bringing the hosts with d ≅ 2Å in the zone where ^2E and $^4T_2$ have the same energy. Then, Cr^{3+} substituted at the Nb site in $LiNbO_3$ or at the Ti one in $BaTiO_3$ is expected to give an $^2E \rightarrow {}^4A_2$ emission line smearing in the $^4T_2 \rightarrow {}^4A_2$ band due to the Stokes shift while the absorption line $^4A_2 \rightarrow {}^2E$ lies in the bottom of the phonon enlarged $^4A_2 \rightarrow {}^4T_2$ band ; this may explain why the R line has never been observed for $BaTiO_3$ and in absorption only for $LiNbO_3$.

MULTISITES AND OPTICAL SPECTRA OF $LaMgAl_{11}O_{19}$:Cr

The experimental emission and absorption spectra have been reported elsewhere[1]. The main conclusions of this paper are as follow:

-fluorescence lines at 695 and 688.6 nm are ascribed to $^2E \rightarrow {}^4A_2$ transitions from 2a and 4f Cr sites respectively.

-emissions at 10nm and in the long wavelength range are expected to arize from Cr–Cr pairs and $^4T_2 \rightarrow {}^4A_2$ transitions.

- the absorption spectrum exibit two main structures peaking at 565 and 418 nm, the former one being related to the $^4A_2 \rightarrow {}^4T_2$ absorption.

This latter spectrum is reproduced on the right hand side of figure 4 and compared to the energy diagrams obtained theoretically (a) and experimentally(b) according to the previous section. Table 1 gathers all the informations relevant to the absorption data. Due to the small changes between the 3d eigenvalues associated to the 2a and 4f sites, peaks due to the term splittings induced by the symetry lowering are not differenciated in the broad absorption band. But the decomposition of $^4T_2$ into $^4A_1$ and ^4E states leads to an apparent weakening of the "averaged" Dq for the 4f center. The calculated wavelength associated to $^2E \rightarrow {}^4A_2$ increases from 2a to 4f sites , according to the experimental trends. The theoretical Dq values are between the two experimental estimations obtained , either from the band maximum (1770 cm–1) or from the bottom of the band where the zero-phonon unresolved line is expected to occur (1540–1600 cm^{-1}). The overestimation of about 40 cm^{-1} of calculated B induces a general 50 nm downwards shift of all the calculated wavelengths relatively to the experimental ones , but a general agreement is observed between the two types of data (figure 5 a and b or table 1).

Coming now to the possible positions of the $^4T_2 \rightarrow {}^4A_2$ emission band due to cubic or nearly cubic centers, it is necessary to estimate the Stokes shift between emission and absorption. Experimental results in gadolinium gallium garnets [10] give 120 nm for this shift. If we assume a similar value for LMA:-Cr, the $^4T_2 \rightarrow {}^4A_2$ band would peak around 675 nm . But the only structures observed besides the well defined $^2E \rightarrow {}^4A_2$ peaks occur beyond 710 nm , impling a Stokes shift at least equal to 150 nm in LMA :Cr .

Table 1. Crystal field parameters and wavelengths associated to LMA:Cr for cubic and nearly cubic centers (Dq/B=4.5)

	$B(cm^{-1})$		$Dq(cm^{-1})$	
	2a	4f	2a	4f
a)	734	711	1654	1649
b)	695	688	1770*	

	Absorption wavelengths (nm)				
	^2E	$^2T_1$	$^4T_2$	$^2T_2$	$^4T_1$
a)	655 (2a) 675 (4f)	628-648	604-606	439-452	423-426
b)	689 (2a) 695	675	565* (625-649)**	485	418

* From the absorption maximum ;** from the estimated zero-phonon line
a) calculated ; b) experimental ; limits are obtained with 2a and 4f calculated parameters

Let us now consider the possible effects of the presence of Cr-Cr pair on the emission spectrum. The eigenvalue diagrams associated to Al-Cr and Cr-Cr defects are pictured in figure 5. In fact , the Al-Cr defect is neither else that a 4f Cr center associated to an adjacent Al^{3+} ion , and the electronic levels of Cr ions must be similar to the ones associated to the 4f center in the small cluster representation (figure 2). In fact , the main difference between the two cluster representations of the 4f center lies in the position of the VB edge relatively to the 3d levels which is strongly shifted upwards in the extended $AlCrO_9{}^{12-}$ and $Cr_2O_9{}^{12-}$ clusters. This is a spurious effect on the potentials associated to the middle oxygen plane as observed on extended clusters of the same types [8]. As the relative positions and splittings of $d\varepsilon$ and $d\gamma$ levels remain unchanged between the small and extended cluster representations of 4f center, it remains meaningful to compare the Cr 3d levels in figures 5a and 5b to see the effects of a pair on the 3d optical transitions. First, we observe strong splittings (0.4 eV) of the t_{2g} and e_g levels associated to the two Cr ions. The lowest $d\varepsilon\uparrow \rightarrow d\varepsilon\downarrow$ transition is 0.1 eV below the associated cubic one , that corresponds to an emission peak around 690-700 nm in the emission spectrum which may account for the observed structure at 710 nm. Nevertheless, $d\varepsilon\uparrow \rightarrow d\varepsilon\downarrow$ transitions extend on 0.5 eV, with a transition

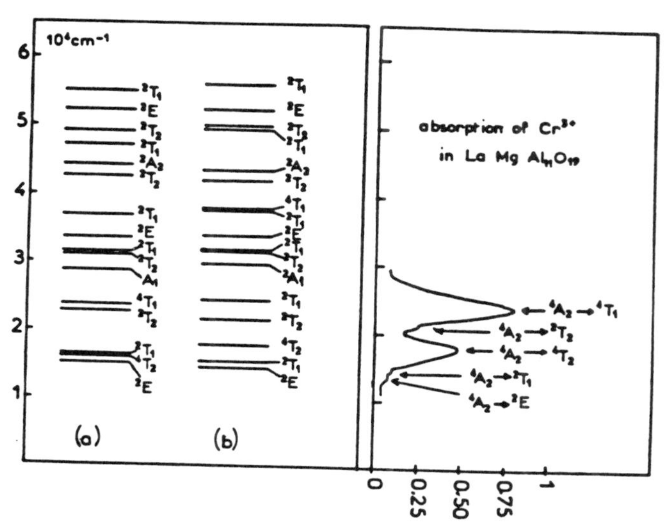

Fig. 4. Theoretical (a) and experimental (b) energy diagrams compared to the absorption spectrum of LMA.

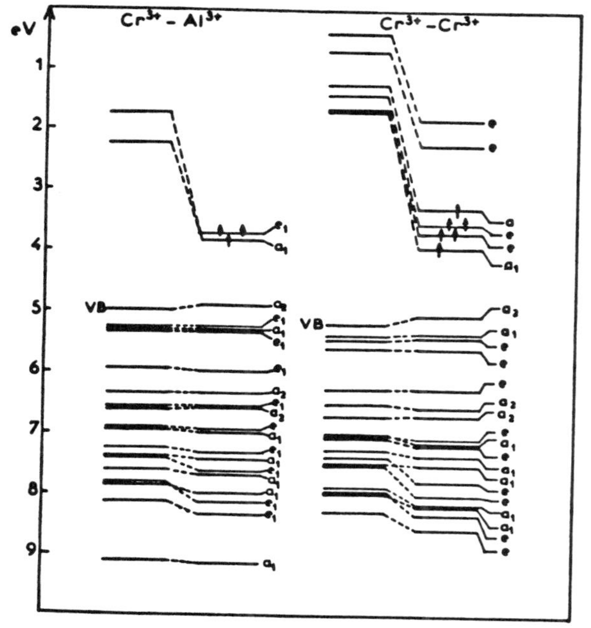

Fig. 5. Eigenvalue diagrams of O2p and Cr3d levels in Cr–Al and Cr–Cr centers at two 4f sites in LMA.

probability probably weak for the higher energies. On the other hand, the lowest $^4A_2 \rightarrow {}^4T_2$ transition is estimated at 700 nm; this implies an emission band from $^4T_2$ associated to the pair localized after 800 nm when a Stokes shift of the same magnitude order as the one of the cubic center is assumed.

The interpretation of the emission spectrum may be sumarized as follow: structures up to 695 nm are due to $^2E \rightarrow {}^4A_2$ emission of cubic and nearly cubic centers. The ^2E emission due to the pair and the $^4T_2$ one of the cubic sites occurs between 710–750 nm while the structures beyond 800 nm are relevant to the $^4T_2 \rightarrow {}^4A_2$ emission of pair centers.

CONCLUSION

Molecular orbital calculations are able to describe complex optical spectra of laser activator ions with different substitutional sites like in LMA. Term energy diagrams show the straight relation between the CrO distance and the crystal field strength Dq/B and define the conditions to obtain tunable properties. Moreover, these diagrams may be efficient tools to understand excited absorption state measurements .

REFERENCES

1. B. Viana, A.M. Lejus, B. Vivien , V. Poncon and G. Boulon, J. Sol. State Chem. 71, 77 (1987)
2. G. Boulon , C. Garapon and A. Monteil, Advances in Laser PhysicsII, 104, Ed Marshall, Stwalley and Kenney-Wallace, N.Y. (1987)
3. C. Pedrini and R. Moncorge, TSSL 1987 , to appear in IEEE/QE (1988)
4. S.J. Till, International Conference Proceedings CLEO (1986)
5. P. Moretti and F.M. Michel-Calendini, Phys. Rev. B34, 8538 (1986)
6. F.M. Michel-Calendini, K. Bellafrough , V. Poncon and G. Boulon, M2P Conference Lyon (1987)
7. S.H. Vosko, L. Wilk and M. Nusair, Can. J. Phys. 58, L1200 (1980)
8. F.M. Michel-Calendini and P. Moretti, LADIC V , Madrid (1986)
9. K. Bellafrough, C. Linares and F.M. Michel-Calendini , to be published
10. A. Monteil, C. Garapon and G. Boulon , J. Luminescence , to appear (1988)
11. W. A. Wall, J.T. Karpick and B. di Bartholo, J. Phys C 4, 3258 (1971)
12. B. Struve and G. Huber, Appl. Phys. B36, 195 (1985)

FLUORESCENCE DYNAMICS IN SOME SOLID-STATE LASER MATERIALS EMITTING IN THE INFRARED REGION : Ho^{3+} DOPED $LiYF_4$ SINGLE CRYSTALS AND FLUORIDE GLASSES*

J. Rubin, A. Brenier, R. Moncorgé, C. Pédrini, B. Moine, G. Boulon
University of LYON I, France

J.L. Adam, J. Lucas
University of RENNES I, France

J.Y. Henry
Centre d'Etudes Nucléaires de Grenoble, France

ABSTRACT

It is reported on the fluorescence mechanisms of Er^{3+} and Ho^{3+} impurity ions in $LiYF_4$ single crystals and in fluoride glasses so-called BATY and BIZYT with the following compositions :

BATY : 20 BaF_2-28.75 AlF_3-22.5 ThF_4- (28.75-x-y) YF_3-xHoF_3-yErF_3
BIZYT : 30 BaF_2-30 InF_3-20 ZnF_2-10 ThF_4-(10-x-y) YF_3-xHoF_3-yErF_3

RESULTS AND DISCUSSION

The fluorescence dynamics in Er^{3+} singly doped $LiYF_4$[1,2] and glasses[3] is governed by very efficient up-conversion and cross-relaxation transfers,the most important of which are indicated in the right part of figure 1. For example, self-quenching mechanisms of the green fluorescence originating from the $^4S_{3/2}$ level in glasses are shown to be very efficient for Er^{3+} concentrations more than few percents[3].

These processes were taken into account in the analysis of the $Er^{3+} \rightarrow Ho^{3+}$ energy transfer mechanisms. The fluorescence dynamics is well described in terms of rate equations[3,4]. It is shown that the transfers strongly depend on the laser excitations in the excited states of Er^{3+} ions and on the Ho^{3+} concentrations. Quantum efficiencies of energy transfers between the lowest excited states of Er^{3+} and Ho^{3+} are calculated and found to be much higher in the crystals than in glasses (table I). The main excitation channels which populate the $^5I_7$ lowest excited state of Ho^{3+} are represented in the left part of figure 1. It should be noted that a fourth weaker deexcitation channel called IR3 describing the transfer $^4I_{9/2} \rightarrow {}^5I_5$ was pointed out in some cases in glasses[3]. The relative contribution

*This work was supported by the DRET grants # 84/071 and # 85/217.

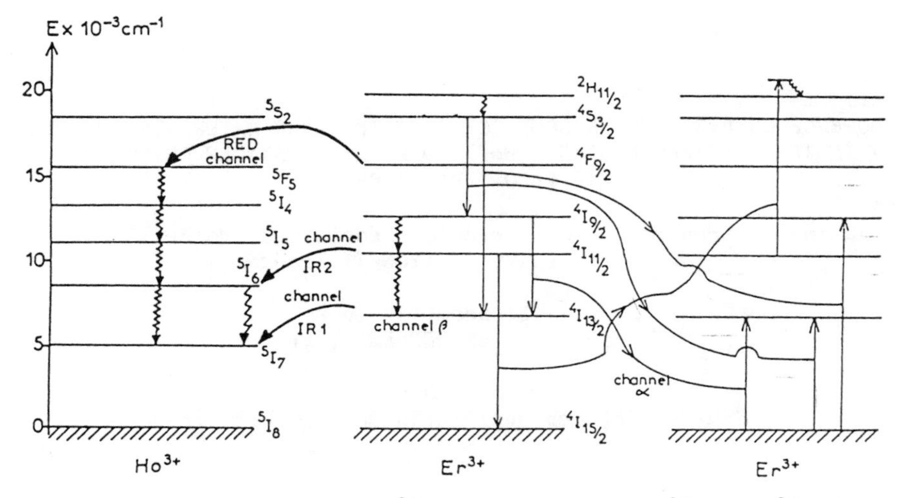

Figure 1. Main channels of Ho^{3+} excitation via $Er^{3+} \rightarrow Ho^{3+}$ energy transfer ans self-quenching mechanisms within Er^{3+} ions.

Table I. Quantum efficiencies of energy transfers.

(i) T_1 : $^4I_{13/2}$, $^5I_8$ ---> $^4I_{15/2}$, $^5I_7$

(ii) T_2 : $^4I_{11/2}$, $^5I_8$ ---> $^4I_{15/2}$, $^5I_6$ (excitation in $^4I_{9/2}$ of Er^{3+})

(iii) $^*T_2$: $^4I_{11/2}$, $^5I_8$ ---> $^4I_{15/2}$, $^5I_6$ (excitation in $^4I_{11/2}$ of Er^{3+})

Samples	Er^{3+}	Ho^{3+}	η_{T_1}	η_{T_2}	$\eta^*_{T_2}$
$YLiF_4$	50	0.5	75	50	24
		2	93	73	38
		5	96	83	50
BIZYT	9	1	23	19	
	5	5	76	46	
	1	9		24	
BATY	27.75	1	55	31	
	10	1	53	53	

Table II. Relative contribution of each channel of excitation to the $^5I_7$ level of Ho^{3+} as a function of the Er^{3+} excitation level in $LiYF_4$: 50 % Er^{3+} : x % Ho^{3+}.

Exc	Ho^{3+} concentration 0.5 %	2 %	5 %
$^4I_{13/2}$	IR_1 100 %	IR_1 100 %	IR_1 100 %
$^4I_{11/2}$	IR_1 + IR_2 57% 43%	IR_1 + IR_2 79% 21%	IR_1 + IR_2 94% 6%
$^4I_{9/2}$	IR_1 + IR_2 92% 8%	IR_1 + IR_2 82% 18%	IR_1 + IR_2 66% 34%
$^4F_{9/2}$	IR_1 + RED ? ?	IR_1 + RED 88% 12%	IR_1 + RED 26% 74%
$^2H_{11/2}$	IR_1 100 %	IR_1 + IR_2 72% 28%	IR_1 + IR_2 88% 12%

of each channel of excitation to the $^5I_7$ level of Ho^{3+} in $LiYF_4$: Er^{3+} : Ho^{3+} are given in table II as a function of the Er^{3+} excitation level. The RED channel is efficient only in the case of excitation of the $^4F_{9/2}$ level of Er^{3+}. In particular, excitation of $^4S_{3/2}$ level does not active this channel owing to the strong above-mentionned cross-relaxation processes. On the other hand, the IR1 and IR2 channels play an important role in the feeding of the $^5I_7$ initial state of the infra-red laser transition $^5I_7 \rightarrow {}^5I_8$ near 2 µm. These results show that the IR1 channel strongly dominates in most cases, and this is a good point since it gives the shortest risetime on $^5I_7$.

REFERENCES

1. J. Rubin, A. Brenier, R. Moncorgé and C. Pédrini J. Lumin 36, 39 (1986)
2. A. Brenier, J. Rubin, R. Moncorgé and C. Pédrini J. Less Common Metals 126, 203 (1986)
3. B. Moine, A. Brenier and C. Pédrini to be published
4. J. Rubin, A. Brenier, R. Moncorgé and C. Pédrini J. Phys. (Paris) 48, 1761 (1987)

RAMAN SCATTERING FROM COUPLED LO PHONON-PLASMON MODES IN p-GaAs

K. Wan, J.F. Young, R.L.S. Devine, W.T. Moore
National Research Council, Ottawa, Canada K1A 0R6

A.J. SpringThorpe and P. Mandeville
Bell-Northern Research Ltd., Ottawa, Canada K1Y 4H7

ABSTRACT

We used non-resonant, allowed Raman scattering to observe the coupled LO phonon-plasmon modes in MBE-grown p-GaAs samples heavily doped with Be. With increasing hole concentration, the single coupled-mode spectrum moves first fom the unscreened LO phonon energy to higher energy, then shifts continuously back towards, and finally asymtotes at, the TO phonon energy. The Raman lineshapes at all densities are consistent with calculations when inter-valence band, intra-light hole and intra-heavy hole band transitions are included in the system's dielectric function.

INTRODUCTION

In polar crystals, longitudinal optical (LO) phonons and plasmons can couple with each other through their associated macroscopic electric fields. The nature of the coupling in n-type, polar crystals is well understood [1]. At low carrier concentration , the high (low) energy coupled-mode, ω^+ (ω^-), has the unscreened LO phonon (plasmon) energy, and with increasing carrier concentration, it monotonically increases and eventually asymtotes at the plasmon (TO phonon) energy. The coupled-mode energies are never between the TO and unscreened LO phonon energies, at any carrier density.

We have studied the Raman spectra of coupled-modes from Be-doped MBE-grown p-GaAs. The concentration dependence of the single coupled-mode is dramatically different from that usually observed in n-type semiconductors (SC); namely, for concentrations between ~ 4 and ~ 10 x 10^{18} cm^{-3}, its peak energy is between the TO and unscreened LO phonon energies. In this paper we demonstrate (i) the differences between the coupled-mode spectra in p-GaAs and those in n-type SC, and (ii) the importance of inter-valence band and intra-light hole band transitions in determining the p-type coupled-mode behavior.

EXPERIMENT

All the Raman spectra shown in this paper were obtained using a standard backscattering geometry from the (100) surface of Be-doped MBE-grown p-GaAs samples at room temperature. The incident and scattered beam polarizations were in the (010) and (001) directions respectively for allowed LO phonon scattering.

RESULTS AND DISCUSSIONS

Fig. 1 shows the Raman spectra of the single coupled-mode observed in our p-GaAs samples (dots), along with the hole concentrations (PH) as determined by Hall measurements. With increasing hole concentration, the asymmetric, sharp peak in 1a develops a distinct high energy shoulder (in 1b) which shifts to the low energy side in 1c. With further increase in hole concentration, a broad structure develops away from the sharp peak, at an energy between the TO phonon and unscreened LO phonon energies (in 1d). Finally, the broad structure evolves to a strong peak at about the TO phonon energy (in 1e). The sharp peak at 291 cm^{-1} never shifts. These spectra are interpreted as follows. The sharp peak at 291 cm^{-1} is attributed to scattering from the unscreened LO phonons in the depletion layer near the sample surface. The shoulders in 1a, 1b and 1c, the broad peak in 1d and the strong peak at about 269 cm^{-1} in 1e, are due to the coupled-modes. That is, with increasing hole concentration, the coupled-mode initially moves to higher energies, then continuously to lower energies, passes the unscreened LO phonon energy and finally asymtotes at the TO phonon energy.

These spectra were fit by a modified version of the calculations by Hon and Faust [2], and Klein et al. [3], with the electron susceptibility in the system's dielectric function replaced by the susceptibilities corresponding to the intra-light hole and heavy hole band transitions, and the inter-light/heavy hole band transitions. A phenomenological damping constant, which corresponds to the damping of the hole plasmons, is appropriately implemented in the intra-band susceptibilities, but is not included in the inter-band susceptibility. Details of this calculation will be published elsewhere.

The dashed lines shown in Fig. 1 represent fits obtained with a fixed phenomenological damping constant of 125 cm^{-1}. The hole concentrations estimated from these fits (PR) are also shown. The major discrepancies between the calculated and the experimental spectra are (i) the unscreened LO phonon peak which is not included in the calculation; and (ii) the strength of the high energy tail, which might be improved by using a better approximation for the inter-band susceptibility. In Fig. 2 we plot the measured peak energy of the coupled-mode vs hole concentration (squares), along with three calculated curves: (i) calculated with all three susceptibilities (solid); (ii) with the inter-valence band term neglected (dash-dot), and (iii) with the intra-light hole band term neglected (dashed). These calculations clearly show that both the inter-valence band transitions and the light holes, which represent only 8% of the total hole concentration, must be included for a correct calculation of the coupled-mode lineshape.

By slightly decreasing the damping constant, a curve similar to the dash-dot curve in Fig. 2 was obtained. This is understandable, since the inter-valence band transitions provide an intrinsic damping mechanism for the hole system. With the damping decreased to about 7 cm^{-1}, the calculation yields the familiar two-mode behavior found in n-type SC. This suggests that the damping of the hole system determines whether the coupled-modes exhibit two-mode or single-mode characteristics.

CONCLUSION

We have presented Raman spectra from the coupled LO phonon-plasmon modes in Be-doped MBE-grown p-GaAs. The single coupled-mode behavior is completely different from that of the ω^+ and ω^- modes in n-type SC. We found that such single-mode behavior is due to the large intrinsic and extrinsic damping mechanisms of the hole system, where the (intrinsic) extrinsic damping corresponds to (inter-valence band transitions) hole-phonon and hole-impurity scattering processes.

REFERENCES

1. M.V. Klein in Light Scattering in Solids I, ed. M. Cardona (Springer, New York, 1983), p. 147.
2. D.T. Hon and W.L. Faust, Appl. Phys. 1, 241 (1973).
3. M. Klein, B.N. Ganguly and P.J. Colwell, Phys. Rev. B6., 2380 (1972).

Figure 1. Raman spectra of coupled LO phonon-plasmon modes in p-GaAs. PH (PR) corresponds to the free hole concentration ($x10^{18}$ cm^{-3}) from Hall (Raman) measurements.

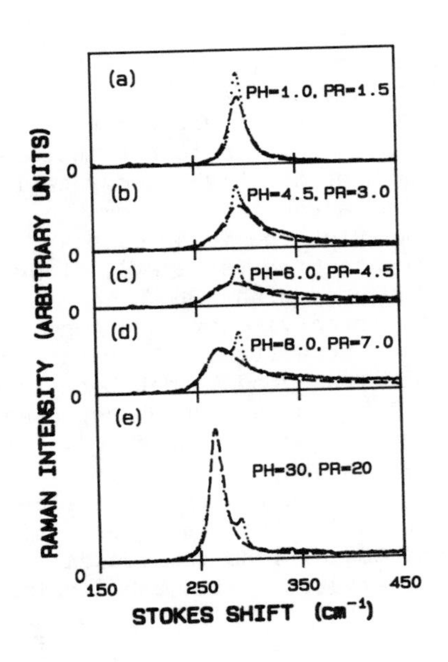

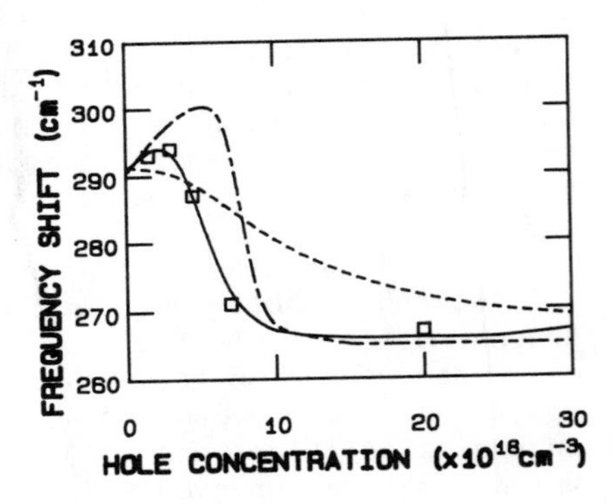

Figure 2. Concentration dependence of the coupled-mode peak energy from (i) experimental data (squares), calculations including (ii) all the intra- and inter-valence band transitions (solid), (iii) only intra-band processes (dash-dot), and (iv) only intra-heavy hole and inter-band processes (dashed).

responsible for the line at $\omega < \omega_D$, which would usually be a Lorentzian. Since there are no phonons with a frequency larger than ω_D, however, photon absorptions with $\omega > \omega_D$ do not occur. Consequently, the line vanishes for $\omega > \omega_D$. Diagrams (b) and (c) explain the line around $\omega \sim \omega_o + \omega_D$, and it follows immediately that the line must disappear for $|\omega-\omega_o| > \omega_D$.

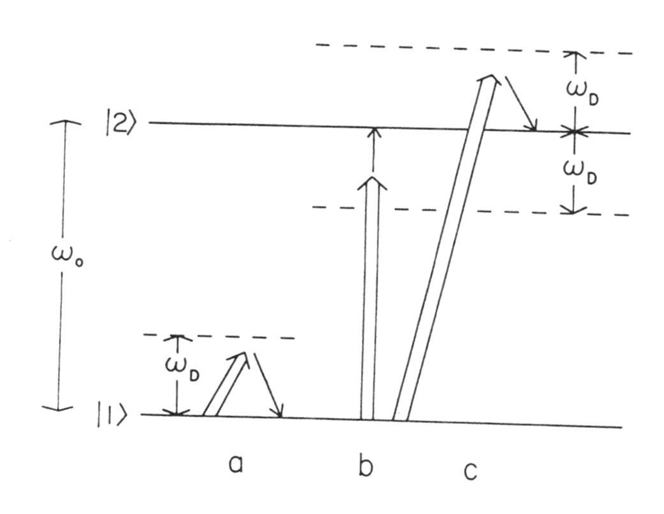

Figure 2. Transition diagrams which are responsible for the profile of Fig. 1. Double arrows indicate photons and single arrows are phonons.

ACKNOWLEDGMENTS

This research was supported in part by the Office of Naval Research and the Air Force Office of Scientific Research (AFSC), United States Air Force, under Contract F49620-86-C-0009.

REFERENCES

1. H. F. Arnoldus and T. F. George, J. Math. Phys., in press.
2. H. F. Arnoldus and T. F. George, Phys. Rev. B, submitted.

A NEW TECHNIQUE FOR THE SPECTROSCOPIC ANALYSIS OF INSULATING CRYSTAL FIBERS

C. E. Byvik and A. M. Buoncristiani †
NASA Langley Research Center, Hampton, VA 23665

ABSTRACT

A simple procedure for extracting the optical absorption and emission spectra of insulating crystal fibers is described. Experimental results are presented for a crystal fiber of Titanium doped Sapphire grown by the laser heated pedestal growth technique. Results of this technique compare well with those obtained from Czochralski grown samples.

INTRODUCTION

The last decade has seen a renewed interest in the development of new materials for solid state lasers[1]. Research efforts are continuing to identify materials which extend the range of solid state laser wavelengths and it is expected that many new materials with useful emission wavelengths remain to be discovered. Optical spectroscopy plays an important role in this search as it is used both to characterize new laser materials and to improve their performance by optimizing crystal properties[2]. Growth of new laser materials by Czochralski, flame fusion, and heat exchange methods is expensive and time consuming. A laser heated pedestal technique for the growth of single crystal fibers with diameters ranging from tens of micrometers to millimeters has been developed at Stanford University[3]. The fiber growth process provides a means to more rapidly and inexpensively survey new laser materials than conventional growth methods. However, their size and optical quality pose a new class of experimental problems.

EXPERIMENTAL

Two measurements are made on the fibers. In one, the laser induced fluorescent radiation is analyzed to give both the absorption and fluorescence spectra (corrected for self absorption) at emission wavelengths. Light from an Argon-ion laser (488 nm) transported by an optical fiber provides a convenient way to deliver excitation radiation which can be accurately positioned along the fiber sample, see Figure 1. The fluorescent emission from the Titanium ions is entrained within the fiber and exits from the fiber ends. The intensity of this radiation, $I(\alpha,\lambda)$, depends upon the distance traveled in the fiber. The sample fiber is side-excited at several points along its length and the spectral intensity of the emerging radiation is recorded as a function of excitation position. This data is used to determine the absorption and emission spectra in the emission region with a model described below.

The second measurement determines the absorption spectra away from the emission region by a transmission measurement. Light from a broadband source is introduced

---

† National Research Council Resident Research Associate

into one end of the fiber and transmitted radiation is analyzed to give an absorption spectrum in the spectral region where there is no fluorescence. The absorption and emission spectra from the two measurements are combined to give the spectral properties of the sample over the entire spectral region.

The Sapphire fiber used in this experiment was grown with its c-axis oriented 60° with respect to the fiber axis. Two parallel faces were polished parallel to the fiber axis of the 1 mm diameter and 5 cm long fiber. Typical output at emission wavelengths is shown in Figure 2.

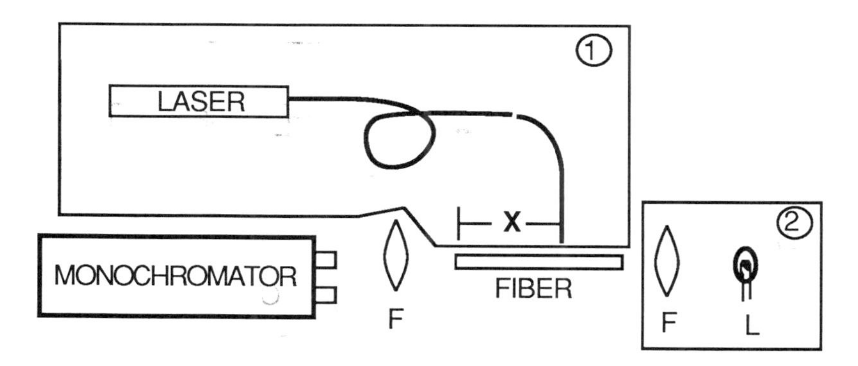

Figure 1. Experimental arrangement. Configuration 1 is used to determine the absorption in the fiber in the emission region. Configuration 2 consists of a broadband source, L, and a lens, F, to focus the source into the sample fiber.

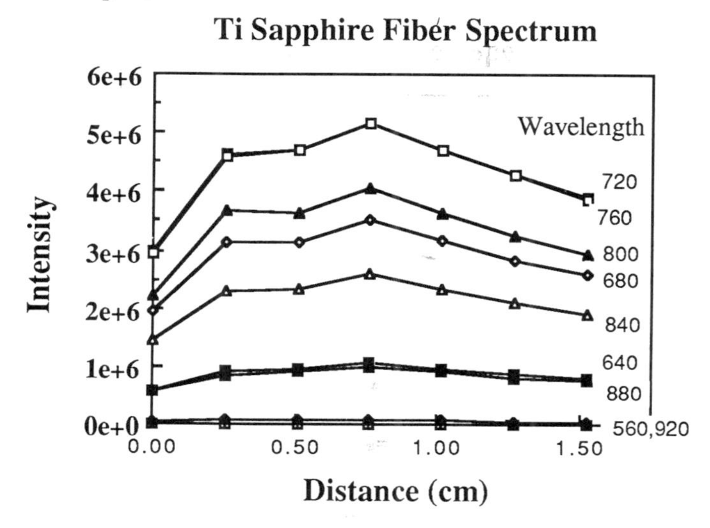

Figure 2. Measurements of the laser induced fluorescence intensity emerging from the end of the fiber for several different wavelengths as a function of excitation position.

ANALYSIS

In analyzing the loss in intensity of the laser induced fluorescence as it passes through the fiber we need to distinguish losses occurring in the fiber bulk from those which occur at the end surface of the fiber. The bulk losses result from absorption and scattering processes within the fiber and so depend on the distance traveled by the light while the losses occurring at the surface are due to the optical discontinuity and are not path dependent. Treating the bulk and surface as separate optical elements we consider the system indicated in Figure 3. The surface elements are characterized by a reflection coefficient, R_S, which depends upon the index of refraction of the fiber and is given by the Fresnel formula $R_S = [(n-1)/(n+1)]^2$. The bulk elements can be treated in terms of distance dependent reflection and transmission coefficients, R(x) and T(x), whose functional form depends on the nature of the radiation transport there. A detailed analysis of one dimensional transport with general expressions for R(x) and T(x) are presented elsewhere[4]. In the simplest case where bulk scattering processes are ignored and $T(x) = \exp(-\alpha x)$, then the steady-state flux, I, emerging from the end of a fiber of length l is

$$\Phi = (1 - R_S) \frac{e^{-\alpha x} + R_S e^{-l\alpha}}{1 - R_S e^{-l\alpha}} I_0 \qquad (1)$$

The spectral distribution of fluorescent radiation emerging from the fiber depends on the excitation position x, the intensity of the induced fluorescence I_0, and the two material parameters R_S and α. The wavelength dependence of I_0, which gives the fluorescence intensity corrected for self absorption along the fiber and the absorption spectrum $\alpha(\lambda)$ in the wavelength region where there is fluorescence can be determined using the data of Figure 2 and Equation (1). The absorption spectrum in the region where there is no fluorescence is measured directly by a transmission measurement. The absorption and fluorescence spectra obtained from an analysis of the experimental data are shown in Figure 3.

In summary we have demonstrated a simple technique for determining the optical absorption and emission spectra from crystal fibers from two direct measurements; fluorescence induced by a laser side-lighting the sample and absorption by direct transmission through the fiber. The results of the measurements on the fibers and the subsequent analysis compare well with measurements on samples grown by conventional methods[2]. This technique should prove valuable in the study of new laser materials.

REFERENCES

1. P. F. Moulton, Tunable Solid State Lasers (Springer Series in Optical Sciences, P. Hammerling, A. Budgor and A. Pinto, eds. Springer-Verlag, Berlin, 1985).
2. C. E. Byvik and A. M. Buoncristiani, IEEE J. Q. Elec., QE-21, 1619 (1985).
3. R. S. Feigelson, Growth of Fiber Crystals, Chapter 11, 1985, E. Kaldis, ed.
4. A. M. Buoncristiani and C. E. Byvik, to be submitted.

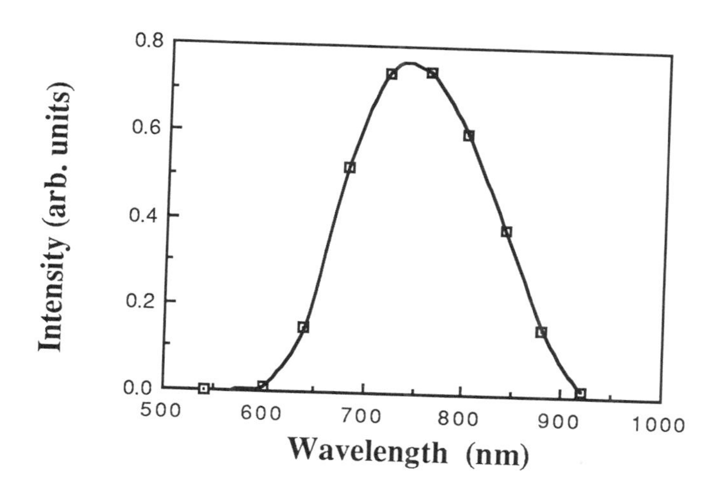

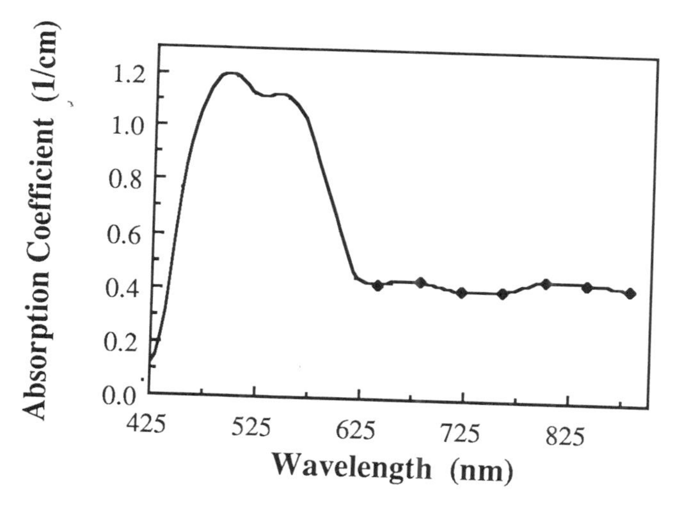

Figure 3. Emission and absorption spectra obtained by the two experiments describeds inthe text. The top graph shows the emission spectrum (corrected for self absorption) determined from laser induced fluorescence. The lower graph shows the absorption spectrum; the dotted portion is determined by a transmission measurement and the remainder by fitting to Equation 1.

OPTICAL SECOND-HARMONIC GENERATION FROM SEMICONDUCTOR SURFACES

T. F. Heinz and M. M. T. Loy
IBM Research Division, T. J. Watson Research Center,
Yorktown Heights, NY 10598

ABSTRACT

The nonlinear process of second-harmonic generation has proved to be a powerful tool for the investigation of surfaces and interfaces of centrosymmetric media. In this paper, a brief survey of the development and applications of the technique for the case of semiconductors is presented.

INTRODUCTION

For centrosymmetric materials, the nonlinear optical process of second-harmonic generation (SHG) has been found to exhibit a high degree of sensitivity to the atomic structure and composition of surfaces and interfaces. This surface-specific character of the SH radiation arises from the selection rule that prohibits a second-order nonlinear response in centrosymmetric media within the dipole approximation. Only in the region of a surface or interface, where the inversion symmetry of the bulk material is broken, can a strong nonlinear polarization be produced. In the paper, a brief review of some recent applications of the SH technique in investigations of semiconductor surfaces will be given. Particular emphasis will be placed on information concerning the symmetry and ordering of the surface region implicit in the polarization dependence of the SHG process. For a more general overview, see Ref. 1.

EARLY INVESTIGATIONS

The initial demonstration of the possibility of generating the SH of optical radiation with laser excitation was performed in 1961 by Franken and coworkers [2] using a non-centrosymmetric quartz crystal. In the following few years, the SHG process in centrosymmetric materials was the subject of considerable study. The effect was first observed in metals by Brown et al. [3] and, shortly thereafter, in centrosymmetric semiconductors by Bloembergen et al. [4,5] In these early investigations, a framework for treating the nonlinear process was developed. [5,6] An intense beam of pump radiation at frequency ω gives rise to collimated radiation at frequency 2ω propagating in the forward and reflected directions from the surface of the sample. The nonlinear process is essentially instantaneous and exhibits a quadratic relation between the intensity of the fundamental and harmonic radiation.

In regard to the origin of the nonlinearity for centrosymmetric media, it was recognized that a strong nonlinear response could arise in a very thin ($\sim$ Å) surface layer and that only a weak nonlinear response associated with magnetic-dipole and electric quadrupole terms would be present in the bulk of the material. These higher-order bulk contributions, proportional to electric-field gradients, are allowed even in a centrosymmetric medium. Despite the considerably smaller magnitude of the effective nonlinear polarization anticipated from the bulk nonlinearity compared with that of the asymmetric surface region, the magnetic-dipole and electric-quadrupole terms are present in a much larger volume and, hence, may still lead to a significant contribution to the total SH signal. [4,7] In the analytic treatment of the surface nonlinear response, the initial work placed emphasis on the strong electric-quadrupole terms associated with the abrupt change in the normal component of the electric field occurring at the surface. The influence of the altered material properties in the surface region was not considered in detail, although the effect of strong applied dc electrical fields was noted. [8] Experimental results for surfaces under ambient conditions showed little apparent sensitivity to surface treatment, in accordance with theoretical developments based primarily on the bulk material parameters. In subsequent studies performed under more controlled conditions, [9,10] the sensitivity of the SH response to the atomic properties of the surface was observed. Chen et al. [10] discovered that the SH signal from a clean Ge surface increased ten-fold upon deposition of a monolayer of Na. This effect, explained in terms of the higher nonlinear polarizability of the metallic Na surface with respect to that of a semiconductor, gave the first clear indication of the potential of the SH technique to reveal the characteristics of surfaces and interfaces on the level of a single monolayer.

SHG AS A SURFACE PROBE

The development of the SHG process as a tool for investigations of the properties of surfaces and interfaces was sparked by observation of a dramatic change in SHG efficiency from a silver surface in an electrochemical environment accompanying the adsorption of monolayers of various molecular species. [11] Although the SH signal was enhanced by the special electromagnetic properties of the roughened surface in the initial work, it was recognized that a monolayer of aligned non-centrosymmetric molecules having a typical nonlinear polarizability should be readily detectable. Subsequent studies on adsorbed molecular layers demonstrated the possibility of surface-specific spectroscopy by SHG.[12] Such measurements are accomplished by scanning the pump laser frequency and monitoring the resonant SH response when either the fundamental or harmonic frequency coincides with electronic transitions in the material system. The same principles are clearly applicable for electronic transitions in semiconductor surfaces and adsorbate-covered semiconductor surfaces, but no systematic studies have yet been performed.

The dependence of the SHG process on the polarization of the electric fields and on rotation of the sample about the surface normal has, on the other hand, been investigated for both molecular monolayers and semiconductor surfaces. From these data, the tensor character of the nonlinear susceptibility of the material can be determined. When the nonlinear response is dominated by that of aligned molecular adsorbates, the relative strength of the different tensor elements of the nonlinear susceptibility can be used to infer molecular orientational parameters, provided that at least partial information on the molecular nonlinearity is available. [13] In the case of semiconductor crystals, the symmetry properties of the SH response should reflect the long-range ordering of the crystal and of the surface of the crystal. This behavior was indeed manifest in the experimental studies of SHG from centrosymmetric semiconductors reported by Guidotti et al. [14] and Tom et al. [15] in 1983. In these works a marked dependence in the SH efficiency was seen from oxidized Si and Ge surfaces as a function of crystallographic face and of rotation of a given face about its surface normal. This strong anisotropy in the SHG process stands in contrast with usual linear optical properties, which are dominated by an isotropic bulk response. [16] Before continuing the discussion of SH studies of semiconductors, we present some the basic formalism describing the second-order nonlinear response of a centrosymmetric medium.

Symmetry Analysis

The dependence of the SH response on crystal face and orientation can be understood directly in terms of a symmetry analysis of the bulk and surface contributions to the nonlinear process.[15,17,18] The nonlinearity of the surface region (where the atomic structure of the material is perturbed from the bulk or strong electric field gradients exist) can be modeled as a sheet of polarization. The nonlinear susceptibility $\chi_s^{(2)}$ relates the electric field vector $\mathbf{E}(\omega)$ at the surface to the induced (effective) nonlinear source polarization in the surface layer:

$$[P_s^{NLS}(2\omega)]_i = \Sigma_{j,k} [\chi_s^{(2)}]_{ijk} E_j(\omega) E_k(\omega). \quad (1)$$

The third-rank tensor $\chi_s^{(2)}$, expressible formally by second-order perturbation theory as an appropriate sum of products of matrix elements of the interaction Hamiltonian and the current operator,[19] is determined by the electronic properties of the surface of the material. The symmetry properties of $\chi_s^{(2)}$ will, of course, reflect those of the surface region itself. As for the bulk contribution to the SHG process, we can write the nonlinear source polarization arising from magnetic-dipole and electric-quadrupole terms in the form

$$[P_b^{NLS}(2\omega)]_i = \Sigma_{j,k,l} [\chi_q^{(2)}]_{ijkl} E_j(\omega) \nabla_k E_l(\omega). \quad (2)$$

Here the symmetry properties of the fourth-rank tensor $\chi_q^{(2)}$ will be determined by those of the bulk material.

The forms of the third- and fourth-rank tensors $\chi_s^{(2)}$ and $\chi_q^{(2)}$ for a given symmetry class are well known.[19] For the bulk properties, the most interesting case is that of the m3m symmetry found in the diamond structure of Si and Ge crystals. Although the dominant linear optical response for a material with m3m symmetry is isotropic, the fourth-rank tensor $\chi_q^{(2)}$ exhibits anisotropic behavior. Employing the previously established notation[5] for the isotropic components of the tensor, we can express Eq. (2) as[15]

$$[P_b^{NLS}(2\omega)]_i = (\delta - \beta - 2\gamma)(\mathbf{E}\cdot\nabla)E_i + \beta E_i(\nabla\cdot\mathbf{E}) + \gamma\nabla_i(\mathbf{E}\cdot\mathbf{E}) + \zeta E_i \nabla_i E_i. \quad (3)$$

The coefficients β, γ, δ, and ζ are simply proportional to elements in $\chi_q^{(2)}$, but written in such a way as to indicate the isotropic character of the terms associated with β, γ, and δ and the anisotropic character of the term in ζ. It should be noted that for a homogeneous medium, no contribution of the form $E_i(\nabla\cdot\mathbf{E})$ is present, since $\nabla\cdot\mathbf{E}$ vanishes; Maxwell's equations also imply that the first term in Eq. (3) disappears for excitation by a single plane wave. Certain simplified expressions have been given for the bulk nonlinearity. In the low-frequency limit, it is predicted that the nonlinear response will be isotropic; within a closure approximation, the magnitude of the nonlinearity can then be related to the linear dielectric constant of the medium.[5] For finite frequencies, a formula for the bulk nonlinear susceptibility has been developed based on a bond polarizability model.[20] This model predicts comparable tensor elements for the isotropic and anisotropic response.

The properties of the surface nonlinear susceptibility $\chi_s^{(2)}$ are not governed solely by the nature of the bulk crystal, but reflect the condition and cut of the surface itself. Our only general guideline is that the symmetry group of a surface should be a subgroup of that of the bulk. Table I summarizes the tensor structure of $\chi_s^{(2)}$ for all symmetry classes consistent with long-range translational order in the surface plane and an assumed asymmetry with respect to the surface normal. It can be seen that only for surfaces with 4mm and 6mm symmetry is the SH response indistinguishable from that of an isotropic surface (with inversion symmetry in the plane). Development of a detailed microscopic theory for predicting the magnitude of the symmetry-allowed elements of $\chi_s^{(2)}$ has been limited primarily to work on molecular adsorbates[21] and to model metal surfaces.[22] Progress towards a complete theoretical formulation of the microscopic origin of the nonlinearity of semiconductor surfaces or, more generally, of surfaces exhibiting anisotropy [23] would be highly desirable.

Table I. Number of nonzero and independent elements of the surface nonlinear susceptibility tensor $\chi_s^{(2)}$ for SHG from crystal faces of differing symmetry.

Surface Symmetry	Nonzero elements of $\chi_s^{(2)}$ for SHG	Independent elements of $\chi_s^{(2)}$ for SH
1	27	18
m	14	10
2	13	8
2mm	7	5
3	19	6
3m	11	4
4	11	4
4mm	7	3
6	11	4
6mm	7	3

Oxidized Semiconductor Surfaces

The theory outlined above, coupled with appropriate choices for the symmetry of the surface, was capable of reproducing the experimental data. [15,17,18] In particular, for oxidized Si(100) surfaces, the data were reproduced assuming a surface with 4m symmetry. In this case, the anisotropic response arises only from the bulk material and the bulk coefficient ζ can be determined. For oxidized Si(111) surfaces, 3m symmetry is expected; both the surface and bulk regions then contribute to the anisotropic SH response. From measurements on both crystal faces, it was possible to compare the relative importance of the surface and bulk anisotropic terms for SHG. The Si(111) surface with a native oxide gives rise to roughly comparable surface and bulk contributions to the SH radiation [15] under pump excitation at a wavelength of 532 nm.

The strong dependence of the SHG process on crystal structure and its instantaneous response make the technique especially appealing for studies of phase transitions occurring on short time scales. This notion was beautifully illustrated in the work of Shank, Yen, and Hirlimann. [24] In their experiment, an intense subpicosecond light pulse served to induce melting of an oxidized Si(111) substrate. The SH radiation from an attenuated probe pulse was then recorded as a function of delay time. The anisotropy in the SH response was found to vanish on the time scale of 1 psec, which was taken as indicative of the time required to melt the crystal.[25] Recent advances in these studies include measurements of melting of Si(100) surfaces. [26] For this case, the anisotropic response can arise only from the bulk material, eliminating any ambiguity concerning the spatial region (surface or near-surface bulk) responsible for the dependence of the SH signal on crystal orientation.

Clean, Reconstructed Semiconductor Surfaces

The relationship between the atomic arrangement in the surface region and the polarization and orientation dependence of the SH radiation is clearly illustrated for clean, well-ordered crystal surfaces in ultrahigh vacuum. Properties of such reconstructed surfaces, in which the atomic positions have been altered in order to lower the energy associated with bonds broken in forming the surface, are naturally of central importance in surface science. For the Si(111) surface, either of two reconstructions exist: the metastable Si(111)-2x1 structure (produced by cleaving the bulk crystal) and the equilibrium Si(111)-7x7 structure. Both of these reconstructed surfaces have been probed by the SH technique [27] Very dissimilar polarization dependences for the SHG process were observed in the two cases. Since the bulk structures are identical, we can conclude that the surface contribution to the SH radiation is strong and reflects the differing electronic structure of the two reconstructions. Indeed, analysis of the polarization dependences indicated that the Si(111)-7x7 surface had the full 3m symmetry of the bulk crystal, while the symmetry of the Si(111)-2x1 surface was lowered to that of a single mirror plane (determined by the cleavage direction). The existence of the mirror symmetry planes could be verified to high accuracy by a nulling technique. The measurement is accomplished by examining the SH reponse for fundamental and harmonic electric field vectors polarized perpendicular to the possible mirror plane. For the case of complete symmetry, no SHG is allowed. The implications of these precise symmetry determinations on models of the surface structure, particularly for the Si(111)-2x1 reconstruction, are discussed in Ref. 26.

The difference in symmetry properties of the Si(111)-2x1 and 7x7 reconstructions could be exploited to investigate the Si(111)-2x1 → Si(111)-7x7 surface phase transformation. [27] The rate of this thermally driven phase transformation was determined by monitoring the disappearance of a SH signal forbidden for the 3m symmetry of the Si(111)-7x7 surface, but allowed for the lower symmetry Si(111)-2x1 reconstruction.

For the clean, reconstructed Si(111) surfaces, it was found [27,28] that the (anisotropic) surface contribution to the SHG process completely dominated that of the bulk. This behavior is to be contrasted with that for the oxidized surfaces, for which the (anisotropic) surface and bulk terms were determined to be of comparable importance. [15,18] The large nonlinearity of these surfaces can be attributed to the presence of strong transitions between well-defined surface states. For the measurements performed with 1.06μm pump radiation, the fundamental and harmonic frequencies lie much more nearly in resonance with transitions between surface electronic states than between the bulk electronic states, which are significant only for energies above the direct bandgap of the material. [27] Recent studies by Tom and Aumiller [29] have examined the nature of the nonlinear response from the clean Si(111)-7x7 surface in considerable detail. A strong variation in certain independent elements of $\chi_s^{(2)}$ with temperature and oxygen exposure have been established. In addition to measurements of the magnitude of different tensor elements of $\chi_s^{(2)}$, relative phases of some of the elements were deduced and indicated the influence of resonant transitions for the 1.06 μm excitation used in the experiment. These data, in conjunction with results obtained from other surface probes, may provide a means of establishing which surface state transitions are important for the isotropic and anisotropic parts of $\chi_s^{(2)}$.

Reactions on Semiconductor Surfaces

The high sensitivity of the SHG process to the atomic structure and symmetry of the surface layer has been utilized in several studies of modifications and reactions on reconstructed semiconductor surfaces. One class of applications involves assessing the degree of surface ordering by means of the strength of the anisotropic terms in the SH response. This simple scheme has been applied to problems in which the chemical composition of the surface is unaltered, but the crystal structure of the surface is disrupted. In this manner, it has been possible to monitor in real time the disordering of Si(111)-7x7 surfaces induced by ion bombardment [28,30] and by deposition of Si on substrates held at room temperature. [31] From these investigations, an average surface area disturbed by each incoming paticle (energetic ion or Si atom) could be inferred. Unlike electron spectroscopies, which also provide monolayer sensitivity, the purely optical SH technique is compatible with environments other than high vacuum. In practical applications, this feature can be quite significant.

For the problem of Si deposition, the SH technique has been demonstrated to be capable of distinguishing epitaxial and non-epitaxial crystal growth with submonolayer sensitivity. [32] To understand some of the fundamental parameters in epitaxial growth, the SH method has been used to monitor the annealing of amorphous Si films of monolayer thickness on Si(111)-7x7 surfaces. [32] Si layers are formed by deposition of atomic Si on the reconstructed Si surface held at room temperature. From the rate of reordering of the amorphous surface layer upon annealing the substrate at various temperatures, the activation energy for the migration of Si adatoms on a Si(111)-7x7 surface has been inferred.

The SH process has also been applied to examine the Si(111)-7x7 surface during the initial stages of oxidation. Both the isotropic [29,33] and anisotropic [28,29] elements of the surface SH response have been recorded. The very high sensitivity of the SH signal to the degree of oxidation, which arises both from the change in the degree of ordering (for the anisotropic terms) and from the large expected perturbation of the surface electronic states, has permitted the real-time studies of oxidation and of thermally induced oxygen desorption to be performed. Analysis of such data can yield important information on the reaction kinetics. In Ref. 33, values for a pre-exponential factor and an activation energy are extracted for the process of oxygen desorption from a Si(111) surface.

The somewhat more complex, but very intriguing system of Au on Si(111) has been probed with the SH technique by McGilp and Yeh.[34] Pronounced changes in the anisotropic response were observed as a function of the amount of metal deposited and the annealing conditions. The authors present evidence that even in the case where a thick metal-silicide layer has been formed, the SH response may be dominated by that of the buried interface. Thus, the SH technique can be easily applied to probe interfaces between two solids, as has previously been seen [1] for other interfaces, such as the liquid-solid interface, between two dense media.

CONCLUDING REMARKS

As this overview has illustrated, considerable progress has been made over the past few years in understanding the nature of the SHG process at the surface of centrosymmetric semiconductors. The sensitivity of the SH response to the electronic properties of the surface is known to manifest itself in a strong dependence cf the nonlinear susceptibility on the chemical nature of the surface. This behavior, which is also apparent for surfaces of other solids, [1] has been clearly observed in studies of semiconductor surfaces under controlled conditions. In addition to this effect, however, the investigations of semiconductor surfaces have also brought out the the influence of the detailed atomic structure of the surface on the SHG process. The phenomenological description of SHG in centrosymmetric media that has now been developed can account for the resulting relationship between SH polarization dependence and crystal structure. The challenge of formulating a tractable theory of a more quantitative character, which involves details of the surface electronic structure, still remains.

While the theory of surface SHG continues to advance, numerous applications of the technique to obtain new information about semiconductor surfaces have also emerged. Some of the more unique studies rely on the capability of examining surfaces under reaction conditions and interfaces between semiconductors and other dense media. Another especially promising direction is the use of the time resolution inherent in the SH technique. One example, mentioned above, involves the dynamics of melting. Many other important problems concerning the lifetime of surface excitations may also, it is hoped, be addressed by this technique. A further avenue which has not yet been fully exploited is the possibility of non-linear spectroscopy. This could take the form either of resonant SHG [12] or of three-wave mixing with two different frequencies. [35] The latter approach retains the surface specific character of the SHG process for centrosymmetric media, but permits the frequency range examined to be conveniently extended into the infra-red. As an example of the potential of such studies, resonant SHG and three-wave mixing spectroscopy have recently been applied to examine the electronic structure of the epitaxial insulator/semiconductor interface formed by CaF_2/Si(111). [36] A strong resonant response has been observed for this system, thus permitting a determination of the energy gap between the empty and filled electronic states localized at the interface.

REFERENCES

1. Y. R. Shen, J. Vac. Sci. Technol. B 3, 1464 (1985); Ann. Rev. Mat. Sci 16, 69 (1986); in Chemistry and Structure at Interfaces: New Laser and Optical Techniques, edited by R. B. Hall and A. B. Ellis (Verlag-Chemie, Weinheim, 1986), p. 151.
2. P. A. Franken, A. E. Hill, C. W. Peters, and G. Weinreich, Phys. Rev. Lett. 7, 118 (1961).
3. F. Brown, R. E. Parks, and A. M. Sleeper, Phys. Rev. Lett. 14, 1029 (1965); F. Brown and R. E. Parks, Phys. Rev. Lett. 16, 507 (1966).
4. N. Bloembergen and R. K. Chang, in Physics of Quantum Electronics, edited by P. L. Kelley, B. Lax, and P. E. Tannenwald (McGraw-Hill, New York, 1966) p. 80; N. Bloembergen, R. K. Chang, and C. H. Lee, Phys. Rev. Lett. 16, 986 (1966).
5. N. Bloembergen, R. K. Chang, S. S. Jha, and C. H. Lee, Phys. Rev. 174, 813 (1968); 178, 1528(E) (1969).
6. C. C. Wang and A. N. Duminski, Phys. Rev. Lett. 20, 668 (1968); C. C. Wang, Phys. Rev. 178, 1457 (1969).
7. P. Guyot-Sionnest, W. Chen, and Y. R. Shen, Phys. Rev. B 33, 8254 (1986).
8. C. H. Lee, R. K. Chang, and N. Bloembergen, Phys. Rev. Lett. 18, 167 (1967).
9. F. Brown and M. Matsuoka, Phys. Rev. 185, 985 (1969).
10. J. M. Chen, J. R. Bower, C. S. Wang, and C. H. Lee, Optics Commun. 9, 132 (1973); Jpn. J. Appl. Phys. Suppl. 2, Pt. 2, 711 (1974).
11. C. K. Chen, T. F. Heinz, D. Ricard, and Y. R. Shen, Phys. Rev. Lett. 46, 1010 (1981); Chem. Phys. Lett. 83, 455 (1981); Phys. Rev. B 27, 1965 (1983).
12. T. F. Heinz, C. K. Chen, D. Ricard, and Y. R. Shen, Phys. Rev. Lett. 48, 478 (1982).
13. T. F. Heinz, H. W. K. Tom, and Y. R. Shen, Phys. Rev. A 28, 1883 (1983).
14. D. Guidotti, T. A. Driscoll, and H. J. Gerritsen, Solid State Commun. 46, 337 (1983); T. A. Driscoll and D. Guidotti, Phys. Rev B 28, 1171 (1983).
15. H. W. K. Tom, T. F. Heinz, and Y. R. Shen, Phys. Rev. Lett. 51, 1983 (1983).
16. The primary linear optical response is isotropic in cubic materials, but weak anisotropic terms are associated both with higher-order bulk contributions and with the lowered symmetry of surfaces. Sensitive measurements of these anisotropic effects can yield information on the surface properties. See D. E. Aspnes and A. A. Studna, Phys. Rev. Lett. 54, 1956 (1985) and references therein.
17. J. A. Litwin, J. E. Sipe, and H. M. van Driel, Phys. Rev. B 31, 5543 (1985); J. E. Sipe, D. J. Moss, and H. M. van Driel, Phys. Rev. B 35, 1129 (1987).

18. O. A. Aktsipetrov, I. M. Baranova, and Yu. A. Il'inskii, Zh. Eskp. Teor. Fiz. 91, 287 (1986) [Sov. Phys. JETP 64, 167 (1987)].
19. Y. R. Shen, The Principles of Nonlinear Optics (Wiley, New York, 1984).
20. T. F. Heinz, H. W. K. Tom, X. D. Zhu, and Y. R. Shen, J. Opt. Soc. Amer. B 1, 446 (1984).
21. See, for example, B. Dick, Chem. Phys. 96, 199 (1985).
22. M. Weber and A. Liebsch, Phys. Rev. B 35, 7411 (1987) and references therein.
23. The symmetry considerations presented here apply to any ordered material and are not restricted to semiconductors. Indeed, single crystal metals [H. W. K. Tom and G. D. Aumiller, Phys. Rev. B 33, 8818 (1986)] and insulators [E. Matthias et al., in this volume] have been shown to exhibit a strong anisotropic response.
24. C. V. Shank, R. Yen, and C. Hirlimann, Phys. Rev. Lett. 51, 900 (1983).
25. H. W. K. Tom and G. D. Aumiller, in this volume.
26. The dynamics of melting in the non-centrosymmetric GaAs crystal has also been investigated by means of SHG. See A. M. Malvezzi, J. M. Liu, and N. Bloembergen, Appl. Phys. Lett. 45, 1019 (1984); S. A. Akhmanov et al., J. Opt. Soc. Amer. B 2, 283 (1985) and references therein.
27. T. F. Heinz, M. M. T. Loy, and W. A. Thompson, Phys. Rev. Lett. 54, 63 (1985).
28. T. F. Heinz, M. M. T. Loy, and W. A. Thompson, J. Vac Sci. Technol. B 3, 1467 (1985).
29. H. W. K. Tom and G. D. Aumiller, to be published.
30. Ion bombardment of oxidized Si samples has also been studied by means of the SH method: S. A. Akhmanov, V. I. Emel'yanov, N. I. Koroteev, and V. N. Seminogov, Usp. Fiz Nauk 147, 675 (1985) [Sov. Phys. Usp. 28, 1084 (1985)].
31. T. F. Heinz, M. M. T. Loy, and W. A. Thompson, in Laser Spectroscopy VII, edited by T. W. Hänsch and Y. R. Shen (Springer, Berlin, 1985), p. 311.
32. S. S. Iyer, T. F. Heinz, and M. M. T. Loy, J. Vac. Sci. Technol. 5, 709 (1987); T. F. Heinz, M. M. T. Loy, and S. S. Iyer, in Photon, Beam and Plasma Stimulated Chemical Processes at Surfaces, edited by V. M. Donnelly, I. P. Herman, and M. Hirose (Materials Research Society, Pittsburgh, PA, 1987) Mat. Res. Soc. Symp. Proc. 75, 697, (1987).
33. H. W. K. Tom, X. D. Zhu, Y. R. Shen, and G. A. Somorjai, in Proc. 17th International Conference on the Physics of Semiconductors, edited by J. D. Chadi and W. A. Harrison (Springer, Berlin, 1985), p. 99; Surface Sci. 167, 167 (1986).
34. J. F. McGilp and Y. Yeh, Solid State Commun. 59, 91 (1986).
35. X. D. Zhu, J. Suhr, and Y. R. Shen, Phys. Rev. B 35, 3047 (1987); J. H. Hunt, P. Guyot-Sionnest, and Y. R. Shen, Chem. Phys. Lett. 133, 189 (1987).
36. E. Palange, T. F. Heinz, and F. J. Himpsel, to be published.

PROBING SURFACE ELECTRONIC STRUCTURE OF IONIC CRYSTALS BY SECOND HARMONIC GENERATION AND LASER-INDUCED DESORPTION

E. Matthias, J. Reif, P. Tepper, H.B. Nielsen
Fachbereich Physik, Freie Universität Berlin, D-1000 Berlin 33, FRG

A. Rosén, E. Westin
Institute of Physics, Chalmers University of Technology and
University of Göteborg, S-41296 Göteborg, Sweden

ABSTRACT

Measurements of laser-induced desorption under ultrahigh vacuum conditions and second harmonic generation on surfaces in air provide evidence for surface electronic states in the band gap of ionic materials. When irradiating $BaF_2(111)$ with green laser light narrow resonances of electron and ion emission yields indicate resonance enhancement of multiphoton absorption. Using fundamental light of 532 nm, second harmonic generation behaves different for $BaF_2(111)$ and NaCl(111) on one side and $CaF_2(111)$ on the other. In the first two cases, its azimuthal dependence proves that the second harmonic yield originates from local dipole interaction at the surface. In contrast, the $CaF_2(111)$ surface appears to be more inert and the second harmonic signal originates predominantly from quadrupole interaction in the bulk.

INTRODUCTION

The problem of occupied states in the band gap at the surface of ionic materials is of importance both from the fundamental and the applied physics point of view. Photoemission studies of alkali halides[1,2] and $CaF_2(111)$[3], as well as electron energy loss spectroscopy (EELS) of alkali halides[4] and $CaF_2(111)$[5] show the existence of occupied states in the band gap. Their origin is still subject to speculation. We are interested in this problem because any electronic structure in the band gap will significantly influence the interaction between high-intensity laser radiation and surfaces. For example, why does damage for many transparent materials start at the surface, and what is the underlying photon absorption mechanism? We are trying to pursue these questions by investigating electron and ion emission following laser irradiation of surfaces cleaved in ultrahigh vacuum, and by studying the second harmonic generation (SHG) in reflection from polished surfaces in air. The results obtained by the first technique for $BaF_2(111)$ have been reported at ILS-II[6] and will be repeated here only briefly.

LASER-INDUCED DESORPTION

Assuming that the surface has been prepared in such a way that there exist occupied and empty states in the band gap, then three basic absorption processes sketched in Fig. 1 can be imagined for photons in the visible spectral range. Resonant *single photon* ab-

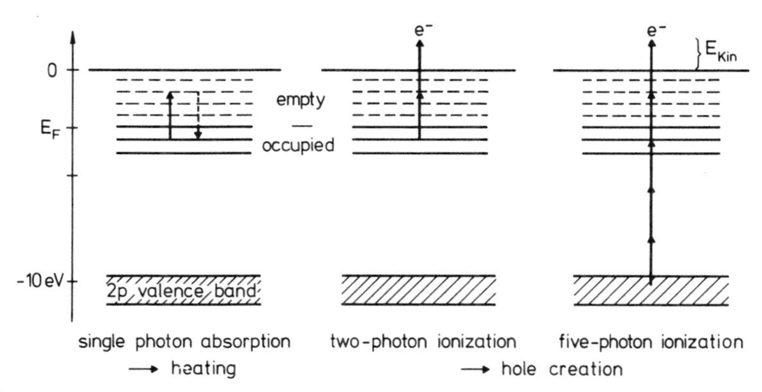

Fig. 1. Possible absorption of photons in the visible spectral range by occupied surface states of fluoride crystals.

sorption will cause local heating which eventually could lead to thermal desorption.[7] For pulse lenghts in the nanosecond range, the initial slope of the laser pulse probably generates heat by this process, which might even create more defect states thus providing for further photon absorption. *Two-photon* absorption is likely to occur at higher intensities and will proceed resonantly enhanced provided the photon energy matches resonances of the electronic structure. Although a hole is generated, two-photon ionization is in this case not expected to be a bond-breaking process. Finally, for still higher laser intensities a *five-photon* ionization will set in and will be all the more likely the more it is resonantly enhanced, i.e. it crucially depends on the surface electronic structure. Both the latter processes have been observed for BaF_2 (111) in the green spectral range and are reported among others in Refs. 6 and 8. Five-photon ionization from the 2p valence band is bond breaking. Nevertheless, we found experimental evidence that two holes in the valence band are necessary for the desorption of positive ions. First of all F^+ was observed which requires two electrons to be removed from the lattice constituent F^-. Secondly, the intensity dependence of the positive ion emission yield pointed toward a ten-photon process. For several reasons a coherent ten-photon absorption is most unlikely, and we interpreted this experimental result as evidence that *two holes* in the valence band, each generated by five-photon ionization, are necessary to remove *one positive ion*, similar to Auger-induced desorption.[9]

From the electron and ion emission results it is not possible to infer the nature of the electronic states in the band gap. Such knowledge, on the other hand, is essential for a more quantitative understanding of the photon absorption processes. It would also enable one to prepare the surface at will, either to make it more inert or more susceptible to laser absorption. Karlsson et al.[3] tried to accomplish this by irradiating their CaF_2 layer with high-energy photons. We cleaved our BaF_2 crystals in ultrahigh vacuum in order to achieve reproducibility and to avoid the influence of adsorbates. In this way we most likely produced among others structural defects in form of missing neutral fluorine atoms, reminiscent of F-centers in the bulk. However, cleaving is a microscopically uncontrolled process with the result that the density of defect states varies considerably across the surface. Correspondingly, different irradiated spots behave differently and not all lead to spectral emission yields like the ones reported in Refs. 6 and 8. We have tried various ways to improve the preparation of a cleaved surface, among others electron bombardment, but we have not yet been successful in preparing in a reproducible manner surfaces which show spectra like the ones in Refs. 6 and 8.

CLUSTER CALCULATIONS

Electron[5] and high-energy photon[3] irradiation of $CaF_2(111)$ surfaces cause fluorine deficiencies in the surface layer which will lead to defect states in the band gap. By electron energy loss spectroscopy Saiki *et al.*[5] have identified occupied states at about a few eV below the conduction band, and it is our expectation that these surface defect states generated by electron bombardment also occur when cleaving the crystal.

Surface F-centers or smaller patches of missing neutral fluorine can very well be simulated in molecular cluster calculations. Therefore, in an attempt to better understand the experimental results for multiphoton ionization and laser-induced ion emission, we have computed the total electron density for planar clusters modelling both stoichiometric and fluorine deficient surfaces. Because of the approximations and the finite cluster size the results will only be of qualitative value, nevertheless they provide a guideline for developing a better physical insight into photon absorption processes at ionic surfaces.

Details of the calculations are reported in Ref.10. The resulting density of states for a stoichiometric 13-ion cluster centered around a Ba^{2+} in C_{3v} symmetry is shown in Fig. 2a. The gap between valence and conduction band is about that for the bulk and there are *no* occupied states in the band bap. The rich structure of both bands is to a large extent an artifact due to the small cluster size. If one removes the three uppermost fluorine atoms as neutrals, leaving the three excess electrons distributed over the whole remaining 10-ion cluster, the density of states in Fig. 2b results. The main difference compared to Fig. 2a is that now occupied states occur around -3 eV binding energy. Their energy location is of course uncertain because of the approximations made including a relaxation shift of about 1 eV. In fact, ground state calculations[10] place the occupied states closer to -2 eV. Nevertheless, the calculations prove that the removal of neutral fluorine leads to dangling electrons which form occupied states in the upper half of the band gap as well as bound resonances. The observed[6,8] two-photon ionization with 2.4 eV photon energy, indicated by arrows in Fig. 2b, is considered to be proof for the existence of such fluorine deficient states. In addition, it is comprehensible that the two-photon ionization is resonantly enhanced by rather narrow excited states, as suggested by the wavelength dependence of the photoemission reported in Refs. 6 and 8. The measured five-photon ionization[6,8] and its wavelength dependence can be viewed along the same line (cf. Fig. 2b).

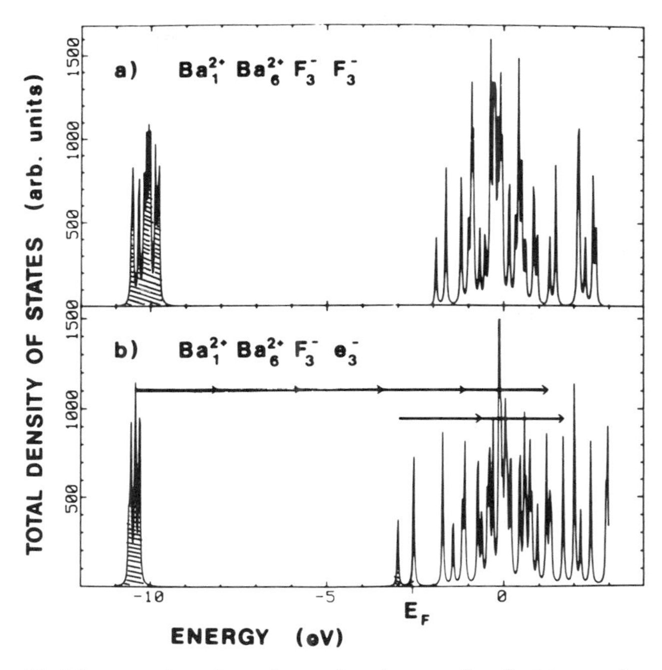

Fig. 2. Predictions of molecular cluster calculations for the surface electronic structure of $BaF_2(111)$. (a) represents the density of states for a stoichiometric surface while (b) indicates the changes when three neutral fluorine atoms are removed. The states have been broadened by a Lorentzian of 20 meV width, and relaxation shifts for ionization from the 2p band are included.

SECOND HARMONIC GENERATION (SHG)

It is of interest to know what happens to surface F-centers and other defect states when the surface is exposed to air. To what extent will the adsorbed molecules change the surface electronic structure and will a fluorine vacancy act as an active site to attract, e.g., H_2O? As opposed to photoemission, laser-induced desorption, EELS and other techniques, SHG allows to investigate surfaces in situ. Two kinds of information can be obtained:
(1) From the azimuthal dependence of the SHG yield one can find the directions along which the electrons in the surface layer are free to oscillate. (2) Resonance enhancement of the SHG yield tells us about the excited states of the system[11]. In the following we will show the for cubic transparent materials spatial and energy dependence of SHG are correlated in the sense that an azimuthal anisotropy goes together with a wavelength dependence and vice-versa.

For a comprehensive review of SHG in reflection from cubic centrosymmetric single crystals we refer to the paper of Sipe *et al.*[12] We have measured[13] the azimuthal dependence of SHG in reflection from polished surfaces of $BaF_2(111)$ and (100), $CaF_2(111)$, and NaCl(111) for three specific polarizations of the fundamental

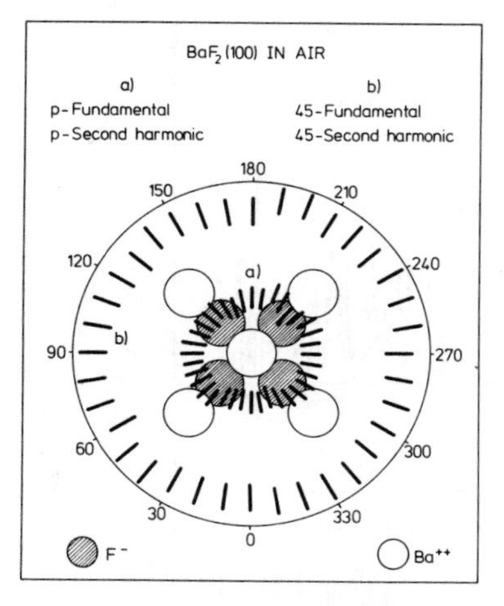

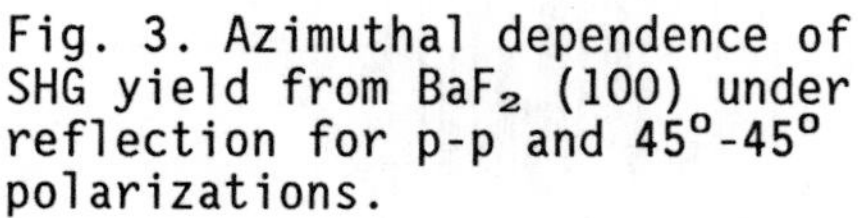

Fig. 3. Azimuthal dependence of SHG yield from BaF_2 (100) under reflection for p-p and 45°-45° polarizations.

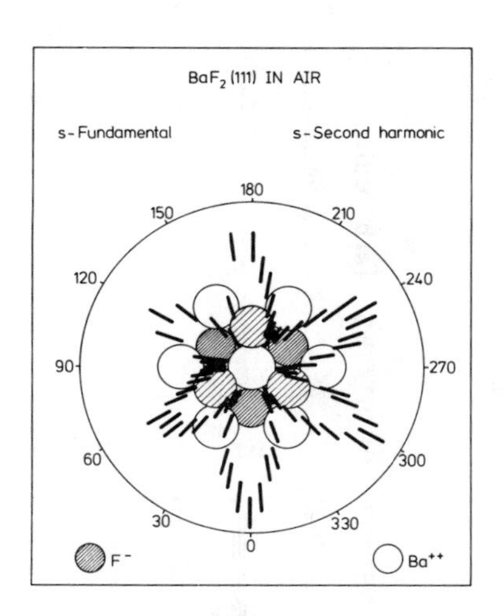

Fig. 4. Azimuthal dependence of SHG yield from BaF_2 (111) under reflection for s-s polarization.

Table I. Azimuthal dependence of second harmonic generation in reflection for selected combinations of light polarizations and surface orientations (according to Ref. 13). The angle of incidence is 45°.

Polarization		Second Harmonic Intensity	
ω	2ω	(100) (orientation)	(111)
s	s	no SHG	$\propto I_s^2(\omega)\cdot\lvert\sin 3\psi\rvert^2$
p	p	$\propto I_p^2(\omega)B_{100}^2$	$\propto I_p^2(\omega)\cdot\lvert A\cos 3\psi + B_{111}\rvert^2$
45°	45°	$\propto I_{45}^2(\omega)C_{100}^2$	$\propto I_{45}^2(\omega)\cdot\lvert(a\cos 3\psi + b\sin 3\psi) + C_{111}\rvert^2$

and second harmonic light. In Table I the expected[13] azimuthal variations of the SHG yield are listed (ψ is the angle of rotation about the surface normal). The prediction is that the (100) surface generates an *isotropic* second harmonic (SH) distribution for p-p and 45°-45° polarizations and no SH light at all for s-s polarizations. Experimentally, indeed no SHG yield was observed off the (100) surface of BaF_2 for s-s polarizations and the predicted isotropy for other polarizations was verified as shown in the polar diagram of Fig. 3. In contrast, the (111) surface of the same crystal gives rise to the sixfold anisotropy of the SHG for s-s polarization illustrated in Fig. 4, and to a threefold anisotropy for p-p and 45°-45° polarizations as shown in the right part of Fig. 5. Similar results were observed for NaCl(111) and (100) surfaces.

The derivation of the expressions in Table I shows[13] that the azimuthal anisotropy originates from the local dipole interaction at the surface. The phenomenological theory does not reveal any information about the nature of the surface dipoles. That this is of great importance, however, is demonstrated by the results for CaF_2(111) in Fig. 6, which came out contrary to the predictions in Table I and in striking contrast to the data for BaF_2(111) (Fig. 4) and NaCl(111). In fact they are rather similar to the results for a (100) surface displayed in Fig. 3. We interpret these facts as evidence that there is no surface dipole contribution to the SHG for CaF_2(111) at the fundamental wavelength of 532 nm and that the SH signal is entirely generated by quadrupole interaction. For BaF_2(111) and NaCl(111) on the other hand there apparently exists a large surface susceptibility which must be connected with a different electronic structure. To test this we have measured the wavelength dependence of the SHG yield for BaF_2(111). The result is displayed in the left part of Fig. 5. The three spectra a, b, and c were taken at different azimuthal positions. They prove the correlation between azimuthal asymmetry and wavelength dependence. The isotropic part of the SH signal shows no wavelength dependence and

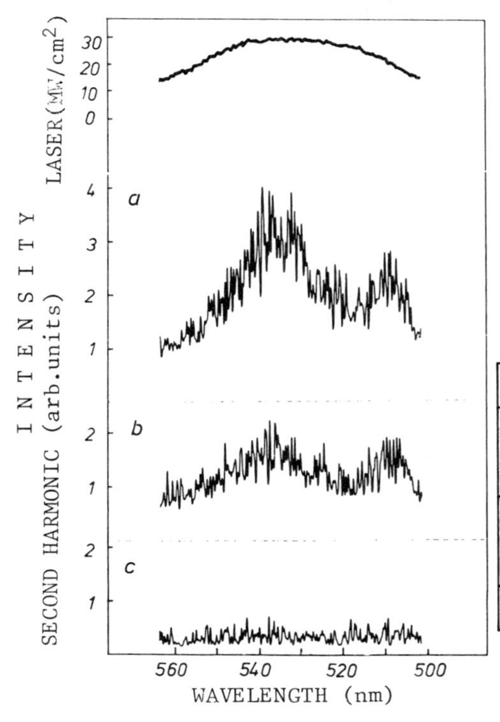

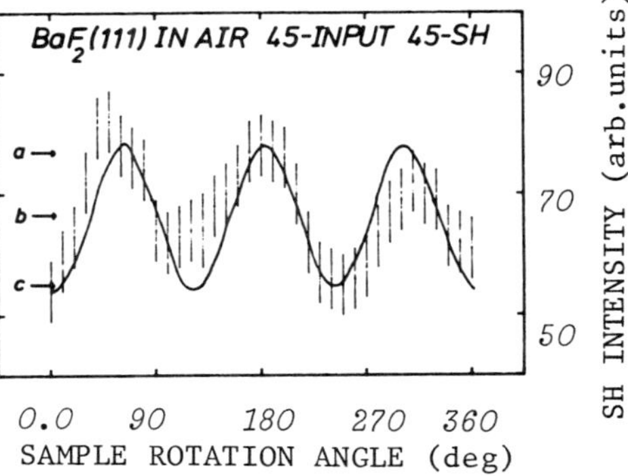

Fig. 5. Wavelength dependence of SHG for the three azimuthal positions 0°, 30°, and 60°. For comparison the laser intensity is shown on top. The expanded scale for the azimuthal dependence indicates a large isotropic background with no wavelength dependence (c).

only the anistropic contribution does. Of course these first measurements do not tell about nature and energetic position of those states either but the difference between a rather inert surface like CaF_2(111) and more hygroscopic ones like NaCl(111) and, to a lesser extent BaF_2(111) is obvious. By measuring combined azimuthal and wavelength dependencies in well-controlled atmospheres it should be possible to elucidate the nature and origin of the electronic states involved. Methodically, this type of combined measurements allows one to separate dipole and quadrupole contributions in SHG.

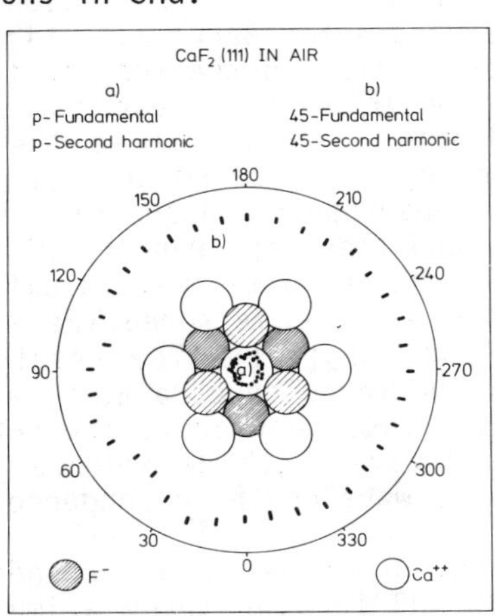

Fig. 6. Azimuthal dependence of SHG yield from CaF_2(111) under 45° reflection for p-p and 45°-45° polarizations. No SH signal was observed for s-s polarizations.

ACKNOWLEDGEMENTS

This work was supported by the Deutsche Forschungsgemeinschaft, Sfb 6, and the Swedish Natural Science Research Council. Helpful discussions with professors Y.R. Shen and G. Marowsky are gratefully acknowledged.

REFERENCES

1. L. Ernst, Surf. Sci. 176, L825 (1986)
2. W. Pong et al., J. Electr. Spectr. Rel. Phenom. 21, 261 (1980)
3. U.O. Karlsson et al., Phys. Rev. Lett. 57, 1247 (1986)
4. G. Roy et al., Surf. Sci. 152/153, 1042 (1985)
5. K. Saiki et al., Surf. Sci. 192, 1 (1987)
6. E. Matthias et al., in *Advances in Laser Science-II*, M. Lapp et al. eds. (AIP Conference Proceedings No. 160, 1987)
7. J.P. Cowin et al., Surf. Sci. 78, 545 (1987)
8. E. Matthias et al., J. Vac. Sci. Technol. B5, 1415 (1987)
9. M.L. Knotek and P.J. Feibelman, Phys.Rev.Lett.40, 964 (1978)
10. A. Rosén et al., Physica Scripta, in print
11. T.F. Heinz et al., Phys. Rev. Lett. 48, 478 (1982)
12. J.E. Sipe et al., Phys. Rev. B35, 1129 (1987)
13. J. Reif et al., submitted to Applied Physics B.

Time-Resolved Study of Laser-Induced Disorder of Si Surfaces

H. W. K. Tom, G. D. Aumiller, and C. H. Brito-Cruz
AT&T Bell Laboratories, Holmdel, NJ 07733

ABSTRACT

Optical second-harmonic studies show that the electronic structure in the top 75 - 130 Å of a crystalline Si surface loses cubic order only 150 fsec after the Si is excited by an intense 100 fsec optical pulse. This suggests that atomic disorder can be induced directly by electronic excitation, before the material becomes vibrationally excited.

INTRODUCTION

Inducing phase transitions with intense optical pulses has been a controversial topic because dynamic effects other than lattice heating can in principle drive the phase transition.[1] However, even experimental work with 90-fs time resolution has been interpreted to be consistent with lattice-heating.[2,3] Using second-harmonic generation (SHG) as a probe of cubic symmetry in Si, we find that the SH signal associated with order in the topmost atomic layers is lost <150 fs after an intense pump pulse, consistent with a nonthermal electronic origin of the disordering process. Besides being of fundamental interest, this possibility has obvious practical importance in laser-assisted semiconductor processing and surface photochemistry in general.

EXPERIMENTAL DATA

Laser pulses of 75-fsec duration (110 fsec autocorrelation FWHM) were obtained by dispersion-compensating the amplified output of a colliding pulse mode-locked laser operating at 610 nm. The time-delayed probe pulses were focussed on the sample to a 25-micron spot which was smaller than the pump's 75-micron spot diameter to insure even excitation. A different position on the sample was used for each shot of the laser. Fig. 1c shows the pump-probe cross-correlation measured by surface sum-frequency generation. It marks the arrival of the pump pulse on the surface and the effect of non-collinear excitation geometry on the time resolution. The dashed line is a fit with the autocorrelation of intensity $I(t)= \mathrm{sech}^2(t/t_p)$ where $t_p = 67$ fsec.

The top two panels of Fig. 1 show the s- and p-polarized SH intensity (separated using a beam-splitting polarizer) generated by a p-polarized probe as a function of the probe time delay. All the data in Fig. 1 were obtained from a Si(111) surface excited with enough pump fluence (0.2 J/cm^2) so the after-the-fact melt spot diameter was 75-micron. The sensitivity of SH to surface structural symmetry has been demonstrated.[5] The Si(111) surface is 3m-symmetric, and for p-polarized input, the s-SH field is $E_{s,111} = A \sin(3R)$, while the p-SH field is $E_{p,111} = B + C \cos(3R)$,

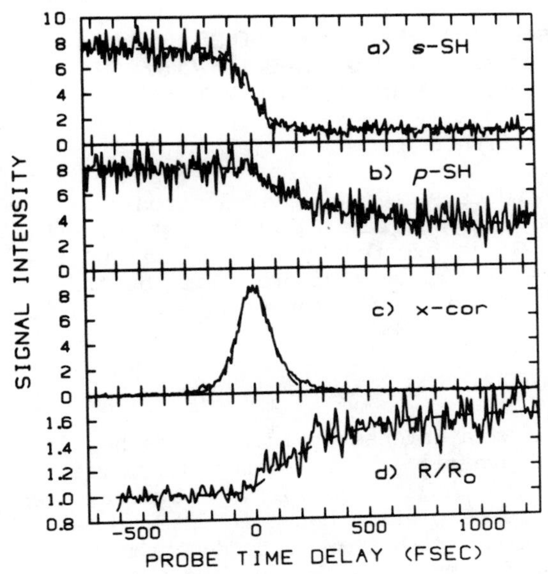

Fig. 1. Signal intensity from Si(111) surface vs. probe time delay after a 2 X threshold pump pulse. Panels from the top: a) s-polarized SH, b) p-polarized SH, c) pump-probe cross-correlation, d) p-polarized linear reflection. Solid line: data. Dashed line: fits explained in text.

where A and B are constants that depend on the excitation geometry and the SH susceptibility tensor elements, and R is the orientation angle between the Si $2\bar{1}\bar{1}$-axis and the plane of incidence. In the absence of 3m symmetry A = C = 0. The data in Fig. 1 are obtained with R=90°, eg., $E_{s,111}$ = A and $E_{p,111}$ = B. One sees that A decreases rapidly while B decreases slowly to a new saturation value.

The data are fit with the dashed curves $A(t)=A_b+(A_0-A_b)\exp((t_0-t)/t_A)$ where t_0=53.3 fsec and t_A=100 fsec for $t>t_0$. A(t) has been convolved with the sech^2 time profile of the probe pulse. An abrupt change in A(t) does not fit the data as well. The constant background A_b comes physically from the edges of the probe beam that are not excited above damage threshold. We can fit the data of Fig. 1b with $B(t)=B_f-(B_0-B_f)\exp(-t/t_B)$, where t_B=333 fs. The ratio between final and initial values of B, B_f/B_0=0.63. For lower pump intensity the SH signals change on the same timescales but to different saturation values consistent with smaller after-the-fact damage spot size.

Fig. 1d shows the p-polarized linear reflection change after excitation. The data is similar to that reported in Ref. 2 and is fit with the same model used by Shank, et al in which a thin layer[5] of molten Si expands from the surface into the bulk. The fit uses expansion velocity of 2 X 10^6 cm/sec. Our SH results are consistent with the earlier SH work by Shank, et al,[3] however they measured the time-evolution of $|B+C|^2$ and $|B-C|^2$ and were thus unable to distinguish that the order-dependent parameter C (proportional to A) decreased rapidly.

We can test whether or not the rapid decrease in A from the Si(111) surface is due to changes in order in the top 10 Å of the surface or the deeper near-surface bulk by measuring the SH from the 4m-symmetric Si(100) surface. For 4m-symmetric surfaces, rotational anisotropy is due entirely to the bulk.[4] SH signals similar to the curves in Fig. 1a and b are shown in Fig. 2a and b for Si(100). In this case, the SH fields have the forms:[4] $E_{s,100} = A' \sin(4R)$, and $E_{p,100} = B' + C' (\cos^4 R + \sin^4 R)$. The data were obtained for the angle between the 100-axis and plane of incidence, $R=22.5^o$ so that $E_{s,100} = A'$ and $E_{p,100} = B' + .75*C'$. The dashed lines in Fig. 2 are fits using the same time constants as used to fit the data in Fig. 1a and 1b. Despite the poor signal to noise of the order-dependent signal $|E_{s,100}|^2 = |A'|^2$,[6] it is clear that it has the same rapid decrease as seen in $|E_{s,111}|^2 = |A|^2$ in Fig. 1b. Thus, the rapid decrease in signal comes from changes in the electronic properties of bulk Si rather than from the $Si\text{-}SiO_2$ interface. Bulk here refers to the topmost 75 to 130 Å (using solid to liquid values for the linear dielectric constant, respectively) because that is the escape depth for the probe signal at the SH wavelength.

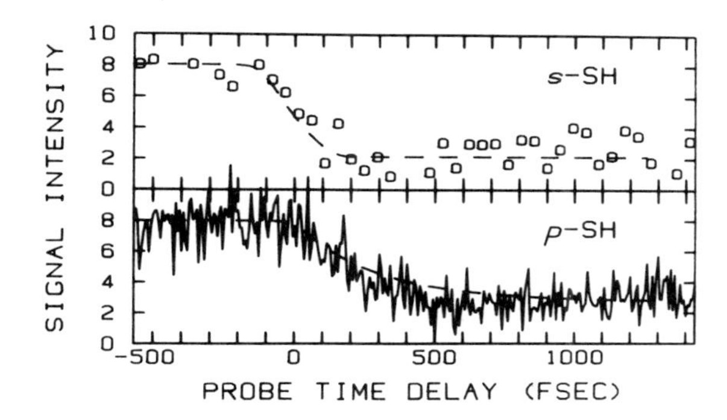

Fig. 2. SH Intensity from Si(100) surface vs. probe time delay after a 2 X threshold pump pulse. Panels from top: a) s-polarized SH, b) p-polarized SH. Solid line: data. Dashed line: fits explained in text.

DISCUSSION

The fast decrease of A and A' implies that the top 75-130 Å is disordered in < 150 fsec. Atomic disorder in less than 2 electron-phonon relaxation times requires that the atoms move out of lattice positions directly because the electronic states have been changed in analogy to the Franck-Condon effect in molecules or Jahn-Teller distortion of defects in solids. Though disorder seems surprisingly fast, it is mechanically reasonable: the atoms need acquire a velocity of only 10^5 cm/sec to move 1/4 lattice space in 100 fsec. The electronic potential driving disorder can also be very large since the laser excites 5% of the valence band[7] assuming linear absorption and more if we consider nonlinear absorption.[8] The promotion of 10% of the bonding orbitals to free carriers would already put the equivalent of a charged defects at next-nearest-neighbor sites. At 40% carrier occupation the lattice is predicted to be unstable to sheer stress.[1] At less than 40%, perhaps the lattice is unstable to the small amount of vibrational energy transfered to the lattice in the first 100 fsec. In any case, if the atoms disorder this quickly, the lattice is still vibrationally cool compared to the equilibrium melt.

The slow decrease in B and B' shows that once threshold is

exceeded, the highly excited system does not attain the equilibrium state of liquid molten (and metallic!) Si for at least 300 fsec. This conclusion is valid whether we model the process as occuring homogenously in the top 130 Å or as occuring in a thin layer of highly excited and disordered material which expands into the bulk. The slow change in p-SH may be due to the disordered Si being initially an amorphous semiconductor rather than a molten metal.

One caveat to the interpretation is that the SH data measure atomic structure indirectly through the electronic susceptibility. We can not rule out the possibility that the laser excitation bleaches the interband transitions or that the high carrier density screens the cubic ionic potential. However, severe bleaching is probably unlikely because there is no large instantaneous change in the linear reflection or the p-polarized SH. Screening[9] is also unlikely to cause a complete loss of rotational anisotropy and the s-polarized SH, since the Si conduction band and higher states have extremely non-spherical band structure.[10] While these arguments are not a proof, there is convincing reason to believe that the rapid reduction in SH signal is due to atomic disorder in < 150 fsec.

CONCLUSION

The order-dependent SH reflection from the top 75-130 Å of Si after an intense optical pulse are found to be consistent with the lattice disordering in the first 100 fsec while the lattice remains relatively cold. The order-independent SH reflection from the top 75 Å shows that the optical properties of molten Si are obtained in about 0.3 psec as expected from electron-phonon relaxation rates and previous measurements.

REFERENCES

1. J. A. Van Vechten, R. Tsu, and F. W. Saris, Physics Letters 74A, 422 (1979).
2. C. V. Shank, R. Yen, and C. Hirlimann, Phys. Rev. Lett. 50, 454 (1983).
3. C. V. Shank, R. Yen, and C. Hirlimann, Phys. Rev. Lett. 51, 900 (1983).
4. H. W. K. Tom, T. F. Heinz, and Y. R. Shen, Phys. Rev. Lett. 51, 1983 (1983).
5. We use the dielectric constants of molten Si from G. E. Jellison, Jr. and D. H. Lowndes, Appl. Phys. Lett. 47, 718 (1985); K. M. Shvarev, B. A. Baum, and P. V. Gel'd, Sov. Phys. Sol. St. 16, 2111 (1974).
6. From Eqs.in Ref. 4, one can show that $|A'|^2/|A|^2 \sim 1/30$.
7. M. C. Downer and C. V. Shank, Phys. Rev. Lett. 56, 761 (1986).
8. T. F. Bogess, Jr., A. L. Smirl, S. C. Moss, I. W. Boyd, E. W. Van Stryland, IEEE J. Quantum Electron. 21, 488 (1985).
9. H. M. van Driel, Appl. Phys. Lett. 44, 617 (1984).
10. M. L. Cohen and V. Heine, in Solid State Physics, eds. H. Ehrenreich, F. Seitz, and D. Turnbull (Academic, New York, 1970) Vol 24, p. 38.

CHEMISTRY, STRUTURES, DYNAMICS AND KINETICS OF ADSORBATES ON SURFACES BY FOURIER TRANSFORM INFRARED SPECTROSCOPY

Y. J. Chabal
AT&T Bell Laboratories
Murray Hill, New Jersey 07974

ABSTRACT

The usefulness of conventional Fourier transform infrared spectroscopy in the study of surface phenomena is assessed by reviewing several specific examples: H_2O adsorbed on Si and the structure and dynamics of H on Si, Ge and on transition metal surfaces.

INTRODUCTION

The knowledge of the chemical nature, the structure, the dynamics and kinetics of adsorbates is important for all surface processes. Among several techniques, surface infrared spectroscopy (SIRS) can provide useful information about each of these aspects. While new techniques involving the use of lasers are being developed, conventional Fourier transform ir spectroscopy is a versatile tool to identify which problems may require laser spectroscopy. We review results obtained by FT-SIRS on simple and well defined systems. Such systems include (a) H_2O on Si(100), (b) the structure of H adsorbed at Si and Ge surfaces, and (c) the dynamics of H on Si(100), W(100) and Mo(100). Based on these examples, a critical discussion of the technique can then be presented.

EXAMPLES

H_2O on Si(100)

Upon exposing a clean Si(100) surface, characterized by a (2x1) LEED pattern, to H_2O, two IR absorption bands could be clearly observed (1) in the frequency range 1000-4000cm^{-1}: a sharp mode at 2082cm^{-1} assigned to an Si-H stretch vibration and a slightly broader mode at 1660cm^{-1} assigned to an O-H stretch vibration (Fig. 1). The absence of the H_2O scissor mode at 1600cm^{-1} and the broad H-bonded H_2O stretch at 3400cm^{-1} confirmed that H_2O does not adsorb molecularly as originally suggested by UPS measurements (2) but rather dissociates into H and OH, as first inferred from EELS measurements (3).

Furthermore, by using vicinal samples with single domains of (2x1) symmetry, polarization studies showed that the Si-H modes did not correspond to H at steps but rather to H on the terraces (4). From this observation, it was concluded that water dissociation takes place on the terraces themselves.

Finally, different states of the Si oxidation upon annealing a water-exposed Si(100) surface could be identified from new Si-H stretch bands appearing at higher frequency (5). These bands are associated with a Si atom with one or two O backbonded to it. Thus, the chemical and structural arrangement of oxygen may be inferred indirectly from the Si-H vibrational spectrum.

Hydrogen at Si(100) and Ge(100) Surfaces

The problem of interest was to identify all the hydride phases that form upon exposure to H atoms. It was known that the LEED patterns changed upon exposure giving notably a very sharp (2x1) at intermediate exposure and a (1x1) on Si(100) or poor (2x1) on Ge(100) at saturation exposure. It was also known that nominally flat surfaces had a high density of steps.

The contribution of SIRS, based on the analysis of the Si-H and Ge-H *stretch bands* and on the *polarization* of various components, was to identify very specific hydride phases (e.g. monohydride, dihydride, H at steps) and to correlate these phases with exposure and annealing of the substrate. Results of ab-initio cluster calculations were also important to the positive assignment of the various bands. The results are summarized in Table 1. The schematic drawings of the monohydride and dihydride structures and their associated normal modes (involving the stretch motion only) are shown in Fig. 2.

The process of mode assignment first required the preparation of *single* (2x1) domains by using vicinal surfaces and to identify the most stable hydride phase, i.e. the only phase remaining after the highest annealing, the Si(100)-(2x1) H phase (6). The vibrational spectrum associated with this pure phase is greatly simplified with only two modes, one at 2099cm^{-1} polarized perpendicular to the surface, the other at 2087cm^{-1} polarized parallel to the surface. Polarization considerations showed that no H was adsorbed at steps and that these two modes corresponded to H adsorbed on the terraces. The final assignment to the monohydride phase (Fig. 2a) was done by comparing the polarization of each mode, the sign and value of the measured splitting between the modes and their relative intensity to the results of ab-initio cluster calculations (6).

Next, the modes associated with the dihydride phase (Fig. 2b) were identified on samples saturated with H at 100° C or below. Again, the polarization sign and value of splitting and relative intensity could be compared to ab-initio cluster calculations (7). In particular, it was found that the symmetric stretch was now at a *lower* frequency than the antisymmetric stretch and that the splitting was larger for the dideuteride than for the dihydride. These striking observations were well predicted by theory and helped assign the measured spectra which were quite complex because several phases were present (see Table 1).

For both the monohydride and dihydride, the assignments were confirmed by determining the value of the isolated frequencies obtained by exposing the surface to isotopic mixtures with mostly deuterium (90%D,

10%H) and measuring the Si-H modes. The isolated frequencies fall roughly in between the symmetric and antisymmetric stretch frequencies (6,7). Similar measurements were performed on Ge(100) (8).

The main results are that (a) the H-saturated Si(100) surface has roughly equal amounts of mono- and di-hydride, in contrast to the Ge(100) surface where only monohydride can be formed, (b) H adsorbed at steps is *less* stable than H in the monohydride configuration, and (c) nominally flat Si(100) surfaces (i.e. not vicinal) have more steps and defects than corresponding Ge(100) surfaces.

Dynamics of H on Si(100)

Once the structure of an adsorbate is known, dynamical information can be obtained by IR absorption lineshape analysis. In the favorable case of the Si(100)-(2x1)H phase for which a *single* structure (monohydride) is present and stable over a wide range of temperature (40-500K), the large variations in the linewidth ($\Delta\tilde{\nu}=1$ to $5cm^{-1}$) measured as a function of temperature could be attributed to coupling to phonons (9).

Using the values for the stretch and bend force constants derived from the ab-initio cluster calculations, molecular dynamics simulations were performed to obtain both the vibrational lifetime of the Si-H stretch and the spectrum (linewidth) of the modes. The results showed that the lifetime and dipole broadening were negligible ($\sim 10^{-3}cm^{-1}$ for H and $10^{-2}cm^{-1}$ for D) and that the dominant broadening mechanism was due to dephasing, i.e. anharmonic coupling between the Si-H bend and stretch modes (9).

This work shows that lineshape analysis of adsorbates on semiconductors are unlikely to be useful to measure vibrational *lifetimes.* Furthermore, our work on other substrates and crystallographic planes indicates that, in general, inhomogeneous broadening dominates the width of adsorbate vibrational spectra. Therefore, dynamical information may only be obtained in favorable cases.

Structure and dynamics of H on W(100) and Mo(100)

On these two substrates, Electron Energy Loss Spectroscopy (EELS) data have shown that H adsorbs at a bridge site for all coverages (10,11). At low coverage, H pinches the two surface metal atoms to form a dimer, resulting in a higher frequency of the symmetric stretch mode, ν_1. At saturation coverage, the metal surface relaxes back to a bulk-like arrangement and all bridge sites are occupied (2H per metal atom, θ=2ML). The ν_1 frequency is lowest ($\nu_1^W=1070cm^{-1}$ and $\nu_1^{Mo}=1020cm^{-1}$).

Using a single grazing-incidence reflection, the H vibrational spectrum was investigated in the frequency range, 700-4000cm^{-1}. At low coverages (θ=0.02 to θ=1.8), IR measurements of the ν_1 mode could be correlated to LEED observations of 2D phases (12,13). In particular, the manner in which new phases develop as a function of coverage (phase transitions) could be inferred from the behavior of the ν_1 mode frequency. At very low coverages (θ<0.3ML), the H adatoms were found to form islands on Mo(100) on an otherwise bare substrate, while they were distributed uniformly on W(100). These results pointed out that the effective adatom-adatom interactions can be very different on similar substrate, suggesting that substrate phonons may play an important role (14).

At saturation coverage (θ=2ML), a new feature appeared in the vibrational spectra (Fig. 3) which was not expected nor observable by EELS (15,16). This new feature at 1270cm^{-1} on W(100) and 1302cm^{-1} on Mo(100) was assigned to the overtone of the wag mode, $2\nu_2$ (17). Its strength, characteristic derivative-like line-shape and lack of temperature dependence are conclusive evidence for a strong coupling between the $2\nu_2$ mode and surface electronic states. Furthermore, the highly asymmetric lineshape is inconsistent with inhomogeneous broadening. In this particular case, therefore, the vibrational lifetime could be extracted from the line shape analysis: 0.5ps and lps for H on W(100) and Mo(100), respectively.

For adsorbates on a metallic substrate, it appears therefore that information on the vibrational energy transfer to the substrate can be obtained from SIRS. Such dynamical measurements are important both to the field of surface science (18) and to the ultimate understanding of chemical reactions at surfaces. However, just as in the case of semiconductor substrates, line-shape analysis will be hampered by inhomogeneous broadening mechanisms. Nevertheless, typical lifetimes of adsorbates on metal (~1ps) are fast enough to make their direct detection by laser spectroscopy more difficult than on semiconductor substrates, which may often leave line-shape analysis as the best probe.

CONCLUSIONS

The above examples demonstrate that quantitative chemical, structural and dynamical information can be obtained by FT-SIRS. More work is being devoted to improve the time resolution of FT-SIRS to the millisecond range so that kinetic measurements can become more routine, and to improve the sensitivity in the low frequency range (100-1000cm^{-1}) where the adsorbate-substrate modes occur.

REFERENCES

1. Y. J. Chabal and S. B. Christman, Phys. Rev. B*29,* 6974 (1984).
2. D. Schmeisser, F. J. Himpsel and G. Hollinger, Phys. Rev. *B27,* 7813 (1983).

3. H. Ibach, H. Wagner and D. Bruchmann, Solid State Commun. *42,* 457 (1982).
4. Y. J. Chabal, J. Vac. Sci. Technol. *A3,* 1448 (1985).
5. Y. J. Chabal, Phys. Rev. *B29,* 3677 (1984) and unpublished.
6. Y. J. Chabal and K. Raghavachari, Phys. Rev. Lett. *53,* 282 (1984).
7. Y. J. Chabal and K. Raghavachari, Phys. Rev. Lett. *54,* 1055 (1985).
8. Y. J. Chabal, Surf. Sci. *168,* 594 (1986).
9. J. C. Tully, Y. J. Chabal, K. Raghavachari, J. M. Bowman and R. R. Lucchese, Phys. Rev. *31,* 1184 (1985).
10. M. R. Barnes ad R. F. Willis, Phys. Rev. Lett. *41,* 1729 (1978).
11. F. Zaera, E. B. Kollin and J. L. Gland, Surf. Sci. *166,* L149 (1986).
12. J. J. Arrecis, Y. J. Chabal and S. B. Christman, Phys. Rev. *B33,* 7906 (1986).
13. J. A. Prybyla, P. J. Estrup and Y. J. Chabal, J. Vac. Sci. Technol.
14. J. A. Prybyla, P. J. Estrup, S. C. Ying, Y. J. Chabal and S. B. Christman, Phys. Rev. Lett.
15. Y. J. Chabal, Phys. Rev. Lett. *55,* 845 (1985).
16. J. E. Reutt, Y. J. Chabal, and S. B. Christman, J. Electron Spectrosc. Rel. Phenom.
17. Y. J. Chabal, J. Vac. Sci. Technol. *A4,* 1324 (1986).
18. Y. J. Chabal, J. Electron Spec. Related Phenom. *38,* 159 (1986).

TABLE 1: H-Stabilized Phases on Si(100)

Formation Temp. (K)	Desorption Temp. (K)	LEED Pattern	Structure
600	700	2x1	Monohydride (M)
450	500	2x1	M + H at steps
380	425	3x1 (ordered)	M + Dihydride (D) + H at steps
300	425	1x1 (disordered)	M + D + H at steps

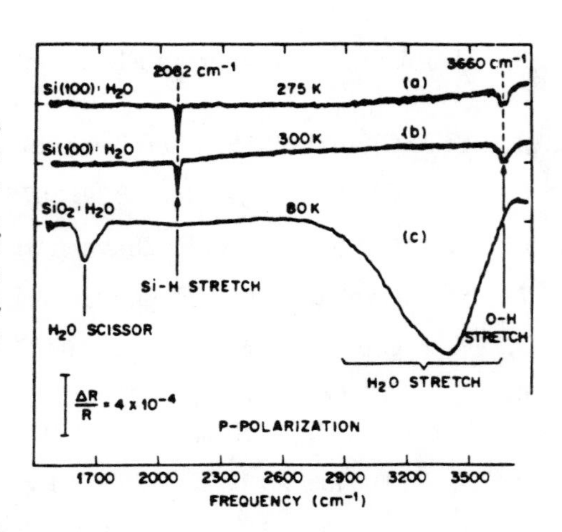

Fig. 1: Surface infrared spectra obtained upon exposure of clean Si(100)-(2x1) to (a) 0.5L water at $T_s = 275K$, and (b) 10L water at $T_s = 300K$. Curve (c) is obtained upon exposure of an oxidized Si(100) surface (native oxide) to 10L water at $T_s = 80K$. Data were taken at the exposure temperatures as indicated on each spectrum.

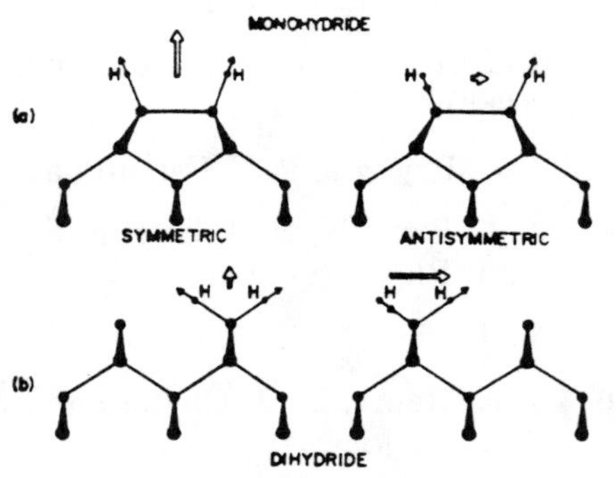

Fig. 2: Scale drawing of (a) the monohydride, and (b) the dihydride structures. The arrows (not to scale) represent the direction of H displacements (solid arrows) and the *polarization* (double arrows) of the two normal modes involving the stretching of SiH bonds. The different magnitude of the double arrows schematically indicates that the net dipole associated with each normal modes are different for the two structures, with $\mu_{\parallel} < \mu_{\perp}$ for the monohydride and $\mu_{\parallel} > \mu_{\perp}$ for the dihydride.

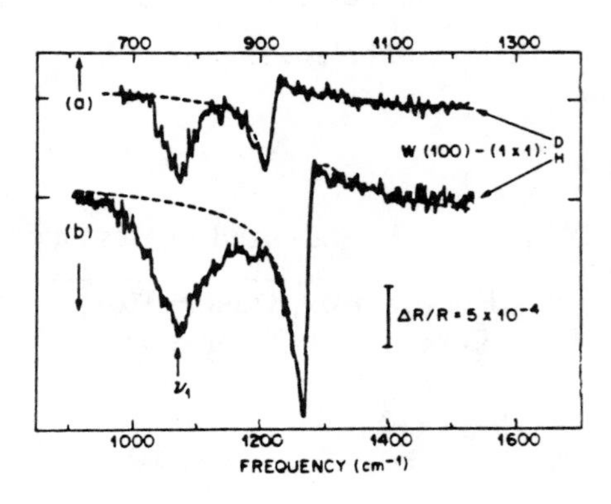

Fig. 3: Vibrational spectra associated with D- and H- saturated W(100) surfaces (10L exposure). The resolution is (a) $4cm^{-1}$ and (b) $2cm^{-1}$. A linear background subtraction was used (see refs. 15-17). The dashed lines are the fits using the line shape derived by Langreth (see ref. 15).

OPTICAL SECOND HARMONIC GENERATION STUDY ON SURFACTANT MONOLAYER: MOLECULAR ORIENTATION DEPENDENCE ON THE CHAIN LENGTH

Mahn Won Kim and Su-Nin Liu
Exxon Research and Engineering Company
Annandale, NJ 08801

ABSTRACT

Optical second harmonic generation (OSHG) is an effective surface probe to study the molecular orientation of surfactants at various interfaces including the air/water inferface. A recently formulated theory shows that the molecular orientation of surfactants depends on the interaction between the molecules and substrates, and on the surface concentration of surfactants.

We have measured the orientation of the -OH group in different carbon length alcohols (normal butanol -- normal octadecanol) by using OSHG technique. The results indicate that the molecular orientation strongly depends on the surface concentration rather than the chain length of the n-alcohols.

INTRODUCTION

The study of monolayer structure of surfactants at the air-liquid interface has attracted attention from researchers in a variety of disciplines. (1) Surfactant monolayer structures are important in the study of, for example, model biological membranes, micelles, foam and interfacial mass transfer. Unfortunately, there has been a limited experimental information of the structure at the molecular level until recently. A few studies have used internal reflection excited fluorescence (2), synchrotron x-ray radiation (3) and resonance Raman scattering (4) to determine the monolayer structure at interfaces. There are few theoretical models (5-6) which show the change of surfactant structure (orientation) by minimizing the free-energy of the configuration of the surfactants in a monolayer at the air-liquid interface. In this paper, we report the orientation of monolayers of n-alcohols as a function of the carbon length and how they compare with the recent theoretical predictions.

EXPERIMENTAL

Optical second harmonic generation technique at interfaces has been presented in previous papers[7-9]. Our experimental set-up is the same as reported on previous papers. We used the null angle technique as well as the several different components of the surface nonlinear susceptibility $\chi_s^{(2)}$ by which we confirmed that the molecular nonlinear polarizability is dominated by a single component $\alpha_{\xi\xi\xi}^{(2)}$ along the specific molecular axis ξ which is randomly oriented with respect to the surface normal. Furthermore, we have previously [11] determined that the source of $\chi_s^{(2)}$ is the alcohol group -OH, not the carbon chain. The main difference between this experiment and the previous one is the preparation of the air/liquid interfaces with surfactants. Due to the high solubility of short chain alcohols in water, the surface was prepared by saturating the alcohols in water. At the saturation point the surface tension no longer can be reduced by further addition of alcohols. The maximum reduction of surface tension is defined as the equal spreading pressure (esp). We used $C_{18}H_{37}OH$ as a reference because it is a very stable monolayer due to its insolubility in water.

RESULTS/DISCUSSION

Figure 1 shows the value of θ(•) and the value of $\sqrt{I}$ (o) as a function of the number of chain. $\sqrt{I}$ is proportional to the number of molecules at the surface, and the figure shows that the orientation of molecules seem to have linear dependence on the length of the carbon chain. $\sqrt{I}$ is also proportional to the carbon-length which indicates that the signal is either coming from the carbons of the tail or from the difference of the surface density. We found from a previous experiment [11] that $\sqrt{I}$ dependence on carbon length can be ruled out. With the stable $C_{18}H_{37}OH$ monolayer as a reference, we showed that the different chain-length gives a different concentration at the surface at equal spreading pressure (esp). Fig. 2 shows θ and $\sqrt{I}$ vs the surface concentration [11] of the stable $C_{18}H_{37}OH$ monolayer. Again, both $\sqrt{I}$ and θ are proportional to concentration. Since we do not know the surface concentration of n-alcohols, we replot Fig. 2 in terms of θ vs $\sqrt{I}$ in Fig. 3 which shows the results of θ as a function of $\sqrt{I}$ for all experimental data. We see that there is no systematic dependence of carbon number on the molecular orientation, but rather the dependence of the chain length of alcohols to the surface concentration at the esp. Therefore, the angle variations due to the change of the carbon chain length of alcohol stems from the surface concentration, and not from tail-tail interactions. This strong dependence on surface concentration was predicted by one of the theories.

CONCLUSION

We implemented OSHG technique to measure molecular orientation of surfactants, namely, different chain length alcohols, to compare recently developed theoretical predictions of monolayer structures at the air/water interface. The results show that the molecular orientation is strongly dependence on the surface concentration of surfactant. Further studies are needed to measure the chain orientation, for which theoretical calculations were done, rather than that of the -OH group.

REFERENCES

(1) G. J. Hanna and R. D. Noble, Chem. Rev. 85, 583 (1985).
(2) L. E. Morrison and G. Weber to be published.
(3) P. Datta, J. B. Peng, B. Lin, J. B. Ketterson, M. Prakash and P. Georgopoulos, Phy. Rev. Lett.
(4) T. Takenaka and T. Nakanager, Phys. Chem. 80, 475 (1976).
(5) S. Safran, M. O. Robbins and S. Garoff Phys. Rev. A 33, 2186 (1986).
(6) J. M. Carson and J. P. Sethna, to be published.
(7) Th. Rasing, Y. R. Shen, M. W. Kim and S. Grubb. Phys. Rev. Lett. 55, 2903 (1985).
(8) Th. Rasing, Y. R. Shen, M. W. Kim. P. Valint Jr., and J. Bock, Phys. Rev. A 31, 537 (1985).

(9) S. G. Grubb, M. W. Kim, Th. Rasing and Y. R., Shen, Langmuir (in print).
(10) T. Heinz, Ph.D Thesis, Univ. of California, Berkeley (1982).
(11) Th. Rasing. G. Berovic, Y. R. Shen, S. G. Grubb and M. W. Kim Chem. Phy. Rev. Lett. 130, 1, (1986).
(12) G. F. Gaines, Jr., Insoluble Monolayers at Liquid Gas Interfaces (Wiley, NY 1966).

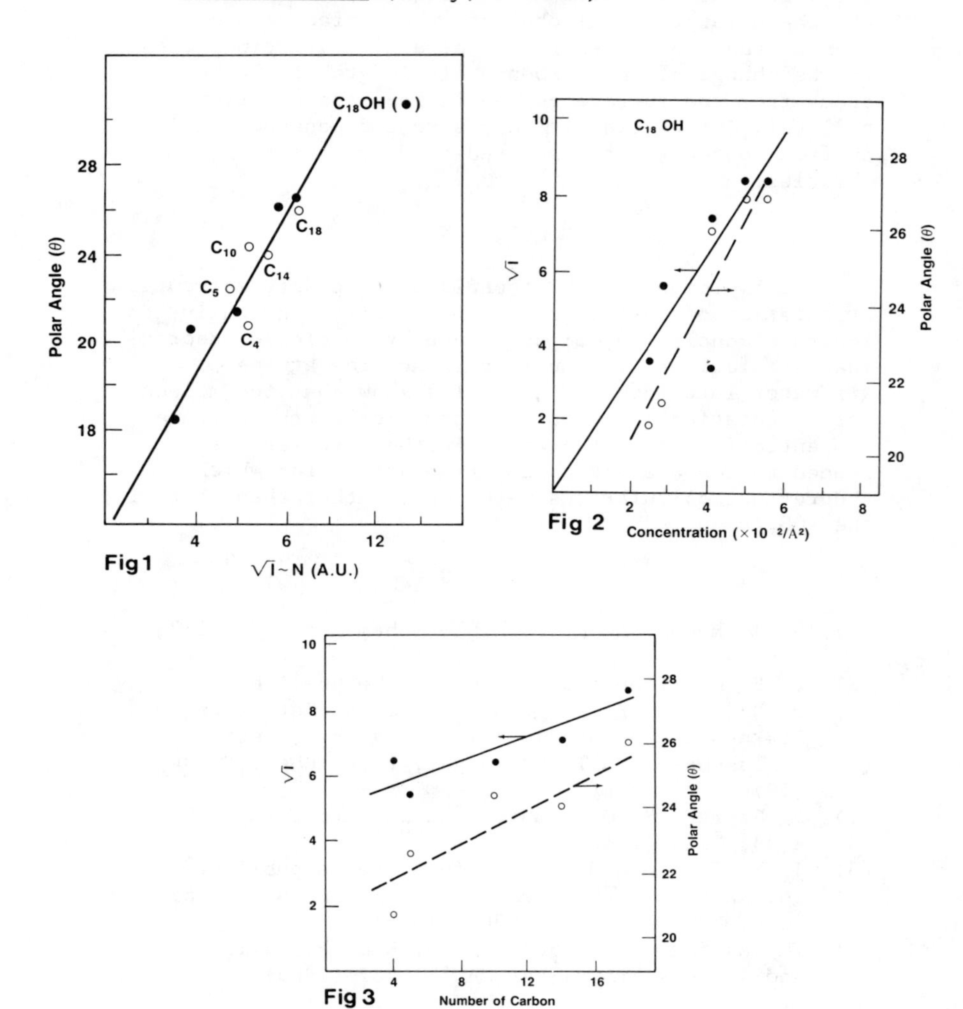

Fig 1

Fig 2

Fig 3

OPTICAL AND MICROWAVE PROPAGATION IN RANDOM DIELECTRIC AND METALLIC MEDIA

A.Z. Genack, L.A. Ferrari, J. Zhu and N. Garcia
Queens College of CUNY, Flushing, NY 11367

J.M. Drake
Exxon Research and Engineering Company, Annandale, NJ 08801

ABSTRACT

Measurements of steady state and picosecond optical transmission through a sample composed of TiO_2 microparticles embedded in polystyrene are consistent with a model of photon diffusion with a coefficient $D = 4.8 \times 10^5$ cm^2/s and photon absorption time τ_{abs} = 294 ps at 589 nm. Measurements of transmission of K band microwave radiation in a random sample of copper spheres in paraffin indicate that the diffusion model is not adequate at certain frequencies.

Scattering of electromagnetic waves by disorder dramatically reduces the average transmitted intensity[1,2], and broadens the distribution of transit times and lengthens their average[3]. This gives rise to an enhanced sensitivity of the transmitted wave to properties of inhomogeneous samples. This is demonstrated here in optical and microwave measurements of the scale dependence of transmission in random dielectric and metallic samples. We will present results of the dependence of the total transmission upon sample thickness L, T(L) and of pulsed optical transmission through a slab of thickness L, T(t;L).

We first examine propagation in the weak scattering limit in which the wavelength λ is less than the transport mean free path $\lambda \lesssim \ell$. The interference of partial waves within the medium can then be neglected in calculating characteristics of average transmission. The quantities T(t;L) and T(L) can then be determined by considering the random walk of photons with step size ℓ or equivalently by solving the diffusion equation with diffusion coefficient $D = v\ell/3$ where v is the effective velocity in the medium. The average photon transit time is proportional to L^2/D for lengths $L < (D\tau_{abs})^{\frac{1}{2}} = L_{abs}$ where τ_{abs} is the photon absorption time in the medium and L_{abs} is the exponential decay length of transmission for samples with $L > L_{abs}$. The tail of the transmitted pulse falls exponentially at a rate,

$$1/\tau = 1/\tau_{abs} + \pi^2 D/L^2, \qquad (1)$$

for $t >> L^2/D$.

Results of optical transmission are presented for a wedged sample[2] composed of a 2:3 mixture by particle volume of TiO_2 particles with diameters $d \sim 0.2 \mu m$ and polystyrene spheres. The sample is heated so that the polystyrene melts and coats the TiO_2 particles. The thickness L is determined by the position of focus of the indicent beam.

The transmitted pulse from a mode locked, cavity dumped Ar^+ laser at 514.5 nm is measured by time correlated single photon counting. The instrumental response to the incident laser pulse and to the transmitted pulse for L = 500μm are shown in Figure 1. The tail of the pulse accurately fits an exponential curve reconvoluted with the instrument response. A fit of the exponential decay rate $1/\tau$ as a function L to eq. 1 gives $D = 3.0 \times 10^5$ cm^2/s and $\tau_{abs} = 126$ ps. Similar measurements at 581 nm using a synchronously pumped dye laser gives $D = 4.5 \times 10^5$ cm^2/s and $\tau_{abs} = 243$ ps. This gives $L_{abs} =$ 105μm at 518 nm.

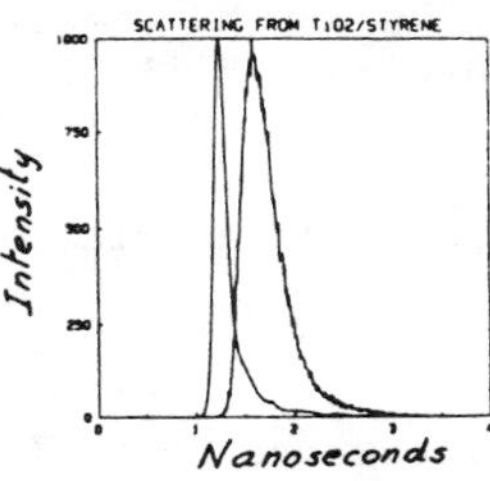

Figure 1 Incident and transmitted laser pulse at 514.5nm for L=500μm.

The transmission of a cw single frequency dye laser at 589 nm versus L for the same sample is shown in Fig.2.[2] The solid line is a fit of the results to a solution of the diffusion equation,

$$T(L) = \beta\alpha\ell/\sinh(\alpha L) \simeq \begin{cases} \beta\ell/L, & L < L_{abs} \\ 2\beta\alpha\ell \exp(-\alpha L), & L > L_{abs}, \end{cases} \quad (2)$$

where $\alpha = L_{abs}^{-1}$ and β is of order of unity. The spikes in transmission seen in Figure 2 are due to pits in the surface. A calculation based on transport theory for a normally incident beam with the diffuse intensity assumed to be zero at the boundaries gives $\beta = 5/3$. The exponential decay of T(L) with length $L_{abs} = 112$μm is in consistent with the value determined from the pulsed transmission experiment at 581 nm. The value of D given by the pulsed experiment is one half the value obtained by fitting T(L) to eq. 2 using a value of v determined from a measurement in reflection of Brewsters angle[2]. The discrepancy may indicate that the velocity at the surface is less than v, the velocity in the bulk of the sample. This may occur if TiO_2 particles are pulled out of the surface as the surface is polished.

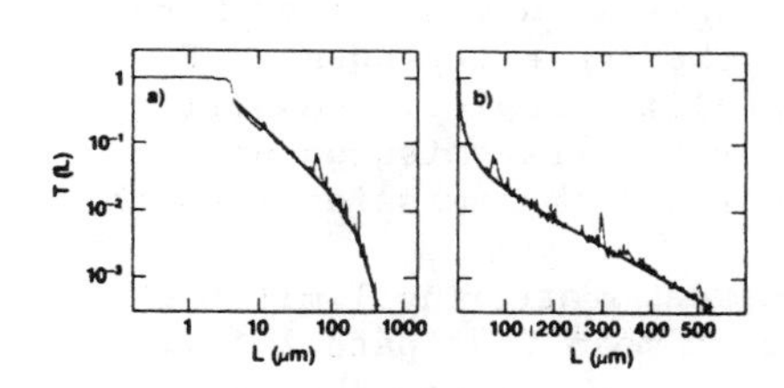

Figure 2. Transmission versus L at 589 nm.

In the presence of strong scattering $\lambda \gtrsim \ell$ wave interference not only leads to large intensity fluctuations in the medium as is the case for weakly scattering samples but gives rise to enhanced backscattering and results in reduced transmission which is the origin of localization[1,4]. It is directly observed in enhanced reflectivity within a cone about the incident laser direction[5-7].

We present evidence of reduced transport due to coherent backscattering of microwave radiation through a random sample of copper plated spheres with radius a = 0.218 cm embedded in paraffin slabs at an average packing fraction f ≈ .186. The transmitted microwave energy is sampled by a wire probe coupled to a planar-doped barrier diode detector. An average is obtained by moving the detector transversely at the output face and by shuffling the order of the slabs. Results for two microwave frequencies are shown in Figure 3. The data at 19.04 GHz indicates an initial drop of T(L) ~ $1/L^2$ instead of 1/L as is observed at 24.15 GHz or in the optical transmission results shown in Figure 2. The initial decay of T(L) as $1/L^2$ can be explained by assuming the diffusion coefficient in eq. 2 is scale dependent. This is a consequence of the enhanced opportunities for coherent backscattering as the sample thickness increases. For L < ξ, where ξ is the coherence length of the wave in the medium, D ≈ $V\ell^2/3L$ and T(L) ~ $1/L^2$. Future studies will examine samples composed of a mixture of metal and dielectric spheres which are tumbled as the transmission is measured. This will facilitate extensive configuration averaging which is required to improve the accuracy of these experiments.

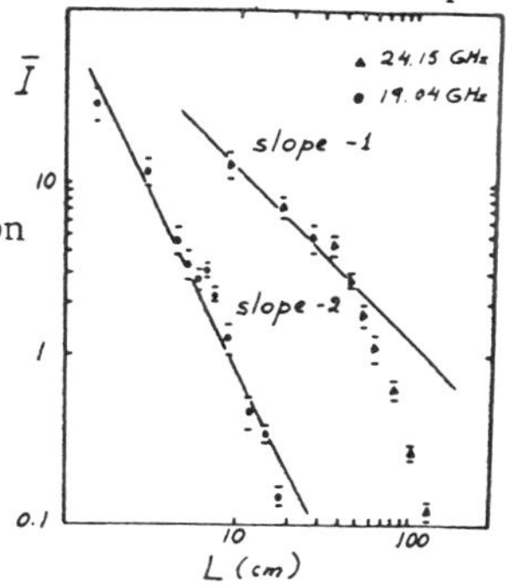

Figure 3. Microwave Transmission

In conclusion we have shown that optical and microwave propagation in random media can be characterized by a study of the scale dependence of steady state and transient transmission. A dramatic reduction and slowing down of the transmitted energy is observed.

It is a pleasure to acknowledge fruitful discussions with S. John and the technical assistance of E. Kuhner. This work is supported by grants from the Exxon Research Foundation and by a PSC-Board of Higher Education Faculty Research Award.

REFERENCES

1. P.W. Anderson, Phil. Mag., 52, 505 (1985).
2. A.Z. Genack, Phys. Rev. Lett. 58, 2043 (1987).
3. G.H. Watson Jr. P.A. Fleury and S.L. McCall, Phys. Rev. Lett. 58, 945 (1987).
4. S. John, Phys. Rev. Lett. 53, 2169 (1984).
5. Y. Kuga and A. Ishimaru, J. Opt. Soc. Am. A1, 831 (1984).
6. M.P. Van Albada and A. Lagendijk, Phys. Rev. Lett., 55, 2692 (1985).
7. P.E. Wolf and G. Maret, Phys. Rev. Lett., 55, 2696 (1985).

RAMAN SOLITONS IN HOMOGENEOUSLY AND INHOMOGENEOUSLY BROADENED MEDIA

Kai J. Druhl
Maharishi International University, Fairfield IA 52556

ABSTRACT

We discuss the effects of coherence decay on the stability of Raman solitons using analytical and numerical methods. Solitons are found to decay for detuning from Raman resonance as small a few percent of the line width. Solitons are more stable in inhomogeneously broadened media than for homogeneous broadening.

The observation of solitons in stimulated Raman scattering (SRS) was first reported in 1983[1]. Subsequent experiments showed that solitons may both narrow and decay under the influence of coherence decay. In this paper we predict that Raman solitons will show appreciable decay for detuning from Raman resonance as small as a few percent of the line width. This may have important implications for the detailed study of frequency spectra in Raman generation.

In the presence of coherence decay the equations for SRS are:

$$A_{Pz} = - Q\, A_S \quad , \quad A_{Sz} = Q^* A_P \tag{1}$$

$$Q_t = A_P A_S^* + Q^{(1)} \tag{2}$$

$$Q^{(1)} = - \gamma\, Q \tag{3}$$

$$Q^{(1)} = \int_{-\infty}^{t} dt'\, \dot{g}(t-t')\, A_P A_S^* (t') \tag{4}$$

A and Q denote the slowly varying amplitudes of fields and coherent medium excitation, while z and t are propagation distance and intrinsic time. Equations (3) and (4) hold for homogeneous and inhomogeneous broadening. γ is the homogeneous line width and $g(t)$ is the Fourier transform of the spectral distribution function for the inhomogeneous line. The dot on g denotes the derivative. For the exact one soliton solution in the absence of coherence decay we have:

$$A_P A_P^* = \rho \operatorname{sech}^2(A) \tag{5}$$

$$A = \omega_R (t - t_0) \quad , \quad \rho = \omega_R^2 / (\omega_R^2 + \Delta\omega^2) \tag{6}$$

The quantity $\Delta\omega$ is the detuning from exact resonance, while ω_R determines the temporal width. These soliton parameters are related to the real and imaginary part of the corresponding eigenvalue in the context of the inverse scattering transform. For nonzero detuning the pump pulse does not reach its maximal value: $\rho < 1$. The following equations express the balance of energy and momentum flow and dissipation in the medium:

$$(Q^*Q)_t + (A_P^*A_P)_z = -2\,(Q^*Q^{(1)}) \tag{7}$$

$$\mathrm{Im}(Q^*Q_z)_t + \mathrm{Im}(Q^*A_P^*A_S^*) = -2\,\mathrm{Im}(Q^*Q_z^{(1)}) \tag{8}$$

For pulses much shorter than the coherence time (hypertransient approximation) the optical fields will closely resemble the one soliton form. Due to coherence decay the soliton parameters will now be functions of z. We can obtain differential equations for these parameters by inserting the one soliton solutions into equations (7) and (8) and integrating over t . We find for the maximal pump intensity in the case of homogeneous broadening:

$$\mu_R = \Delta\omega/(\omega_R^2+\Delta\omega^2) = \text{const} \quad , \quad \rho(z) = \rho(0)\exp(-\alpha) \tag{9}$$

$$\alpha = -4\gamma\mu_R^2 z \approx -4\gamma(\Delta\omega^2/\gamma^4)z \approx -2\,(\Delta\omega^2/\gamma^2)\,G \tag{10}$$

The final expressions for the attenuation coefficient are valid if the detuning is much smaller than the inverse temporal width and the latter is approximately equal to the spectral line width. The coefficient $G = 2\,z/\gamma$ is the steady state gain. In this form equation (10) allows to estimate the detuning directly from experimentally observable parameters. Note that the decay rate is proportional to the square of the detuning. For a detuning of only 5% of the line width and a gain of $G = 10$ a decay of almost 40% is predicted.

Predictions for the rate of soliton narrowing can be obtained in a similar way. For exact resonance we find that the width decreases with an inverse square root law for homogeneous broadening and an inverse third root law for inhomogeneous broadening. The smaller effect of inhomogeneous broadening is due to the more gradual decrease of the temporal correlation function $g(t)$, which is of second order in t at $t = 0$. For homogeneous broadening on the other hand $G(t) = \exp(-\gamma t)$, which decreases to first order.

Numerical studies for the case where the pulse width is comparable to the coherence time confirm the expected greater stability of solitons in the inhomogeneous case. In Figures 1 and 2 we show the propagation of a soliton in a medium with homogeneous and inhomogeneous broadening for incremental gain steps of 2. The soliton width is initially equal to the coherence time, and the detuning is 5% of the line width. The rate of decay is much slower than predicted from the hypertransient approximation, which is not valid initially. The soliton amplitude decreases to about 90% and continues to decrease for homogeneous broadening, while increasing again for inhomogeneous broadening. In both cases the soliton narrows to about 20% of its original width.

These results are encouraging for possible applications in optical signal processing and transmission. Soliton decay is found to be much stronger in generation from a weak initial Stokes field, and can provide an important signature for small fluctuations in frequency[2] and for phase fluctuations.

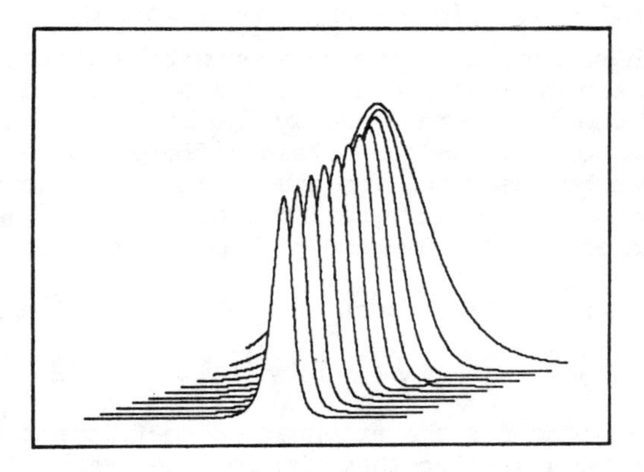

Figure 1. Soliton decay for homogeneous broadening.

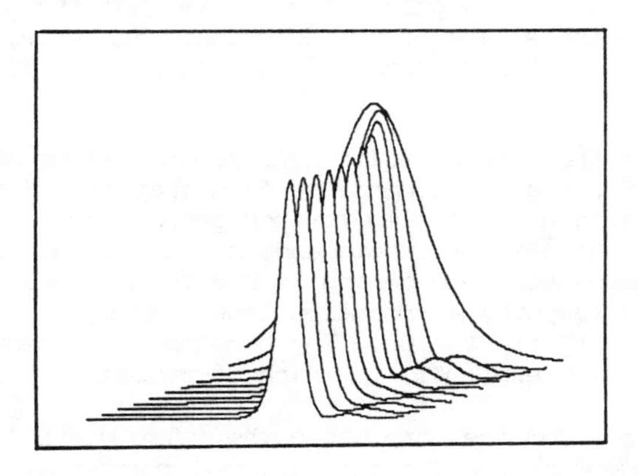

Figure 2. Soliton decay for inhomogeneous broadening.

This work was supported by the U.S.Army Office of Research.

1. K.J. Druhl, R.G. Wenzel, J.L. Carlsten, Phys. Rev. Lett. 51, 1171 (1983).
2. John Carlsten, D.C. MacPherson, "Quantum fluctuations in stimulated Raman scattering line width", these Proceedings.

EFFECTS OF MEDIUM SATURATION ON THE PROPAGATION OF RAMAN SOLITONS

Christian J. Tourenne and Kai J. Druhl
Maharishi International University, Fairfield IA 52556

ABSTRACT

Propagation of Raman solitons in a homogeneously broadened medium is studied for the case where appreciable medium saturation occurs. We use the differential conservation law for energy to obtain an equation for the temporal soliton width. In the limit of small width earlier results are reproduced, while for large width a linear decrease of width with gain is predicted. Numerical calculations give excellent agreement with our analytical results.

For sufficiently high intensity the effects of medium saturation and Stark shift become important in stimulated Raman scattering. Soliton solutions for this case have been given by Steudel[1]. We apply the method of integrating differential conservation laws[2] to obtain an equation for the change in soliton width due to homogeneous broadening. This methods gives both a simple derivation of analytical results and a physical understanding of the mechanism of narrowing.

We use the same notation as in reference 2. In the absence of Stark shift the equations for the fields are the same, while we now have an additional variable R for the medium, which is the difference in population between the lower and upper level of the Raman transition. We assume that coherence decay is much faster than population relaxation, and neglect the latter. The equations for the medium are then:

$$Q_t = A_P A_S^* R - \gamma Q \tag{1}$$

$$R_t = -4 \, \mathrm{Real}(Q^* A_P A_S^*) \tag{2}$$

For exact resonance the one soliton solution and its temporal width W are given by:

$$A_P A_P^* = \omega_R^2 / D \quad , \quad Q = \sqrt{\omega_R^2 - 1} \; \cosh(A)/ D \tag{3}$$

$$D = (\omega_R^2 - 1) \cosh^2(A) + 1$$

$$W = \int_{-\infty}^{+\infty} dt \, A_P A_P^* = \ln[(\omega_R + 1)/(\omega_R - 1)] \tag{4}$$

Note that the width increases without limit as the parameter ω_R goes to 1. In this limit the medium gets just completely inverted in the leading edge, and stays in the upper level[1].

The conservation law for energy gives expression to the balance of emission and absorbtion for the medium and gain and loss for the optical fields. Integration over time leads to an equation for the soliton width:

$$(A_P^* A_P)_z = R_t \quad (5)$$

$$dW/dz = R(+\infty) - R(-\infty) \quad (6)$$

The last term in equation (6) does not vanish since due to the coherence decay some fraction of population gets trapped in the upper level. This can be calculated from (1) and (2) as:

$$4 (Q^* Q)_t = -(R R)_t - 8 \gamma (Q^* Q) \quad (7)$$

$$R^2(+\infty) - R^2(-\infty) = -8 \gamma \int_{-\infty}^{+\infty} dt \, (Q^* Q) \quad (8)$$

In the limit where the width is much smaller than the coherence time we can evaluate the time integrals by assuming that the fields and medium amplitude have one soliton form with z-dependent parameter. This gives the final equation for the soliton width:

$$dW/dz = -\gamma (2 \omega_R + (\omega_R^2 - 1) W) / \omega_R^4 \quad (9)$$

In the limit of small width (large ω_R) we obtain the result for unsaturated media[3]:

$$W^{-2} (z) = W^{-2} (0) + 4 \gamma z \quad (10)$$

For large initial values however the width decreases linearly:

$$W(z) = W(0) - 4 \gamma z \quad (11)$$

This is due to the fact that dissipation occurs only in the regions of large Q in the leading and trailing edges, whose width is almost independent of the total soliton width in this limit.

The physical mechanism of soliton narrowing becomes transparent form the derivation above. Gain and loss for the pump field is always exactly balanced by emission and absorption from the medium (equation 5). On the other hand the change in coherent vibrational energy is equal to the rate of emission from the medium plus the rate of dissipation into noncoherent vibrational and translational modes (equation 7). Since the rate of dissipation is always positive, and the total change in coherent vibrational energy is assumed to be zero, the net amount of emission from the medium is negative (equation 8). The total amount of energy in the pump field and hence the width of the soliton will thus decrease.

In figures 1 and 2 below we compare results for the soliton width. The solid curve is the result from a numerical integration of the

full equations 1, 2 and the field equations and the circles represent the result from the approximate equation 9 .
In figure 1 the initial value of ω_R is 4.0, the line width is 0.1 and the maximal propagation distance is 62.5 units. This case comes close to the limit of narrow solitons (equation 10).
In figure 2 the initial value of ω_R is 1.1, the line width is 0.02 and the maximal propagation distance is 50 units. In this case the more linear initial decrease in width as suggested by equation (11) is clearly visible.

FIGURES

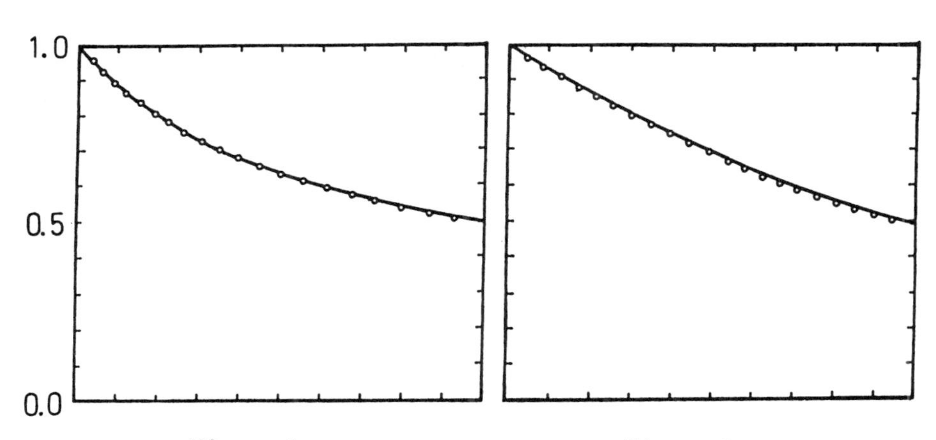

Figure 1.
Soliton width for narrow solitons as function of distance.

Figure 2.
Soliton width for broad solitons as function of distance.

This work was supported by the U.S. Army Office of Research.

1. H. Steudel, Physica D 6, 155 (1983).
2. K.J. Druhl, "Raman solitons in homogeneously and inhomogeneously broadened media", these Proceedings.
3. K.J. Druhl, J.L. Carlsten, R.G. Wenzel, J. Stat. Phys. 39, 615 (1983).

PROPAGATION OF A LASER BEAM THROUGH SELF-INDUCED TURBULENCE.

Shirish M. Chitanvis
Los Alamos National Laboratory, Theoretical Division,
Los Alamos, New Mexico 87545.

ABSTRACT.

Using a model developed by us previously for thermal fluctuations induced in clean or aerosol/obscurant-loaded air by a high-energy laser beam, we compute the mutual coherence function of this high energy laser beam as it propagates through self-induced turbulence. We use the plane wave approximation to the laser beam. Results are presented for the case of a CW beam with a flux level of $5\times10^3 W/cm^2$ propagating through air loaded with fog-oil droplets of total optical depths upto 2.5. The computed mutual coherence function indicates that propagation through induced turbulence in this case can affect the average size of a speckle spot considerably.

SELF-INDUCED TURBULENCE.

Previously,[1] we studied the effect of a high energy laser beam on the fluctuations of hydrodynamic variables. We identified three time scales in the problem, viz., $\tau_H = d/c_s$, the hydrodynamic time for a sound wave to traverse the diameter d of the beam at the speed of sound c_s ; $\tau_T = \rho_0 C d^2/\kappa$ the characteristic time for thermal conductivity effects to be established (ρ_0 is density of the atmosphere, C is the specific heat and κ is the thermal conductivity), and τ_p the pulse length of the laser. The most important regime was identified as the so-called "isobaric regime", when the pulse length τ_p is longer than τ_H , the time for hydrodynamic effects to be established, but τ_p is considerabley shorter than τ_T. In this isobaric regime, the temperature fluctuations were found to be of the form:

$$<\delta T(z,\rho)\, \delta T(z,\rho')> = <\delta T(z,\rho)\, \delta T(z',\rho')>_{eq.} \exp\left\{ \frac{(\gamma-1)}{P_0} \int_0^t dt'\, \Sigma(z,\rho,z',\rho',t') \right\} \tag{1}$$

$$\Sigma(z,\rho,z',\rho',t) = \alpha(\lambda) < I(z,\rho,t) > + \varkappa(\lambda) < I(z',\rho',t) > \tag{2}$$

and where $<\delta T\ \delta T'>_{eq}$ is the initial. equilibrium temperature fluctuation formula, given by Kolmogorov, γ is the ratio of specific heats of the atmosphere, P_0 is the ambient pressure in the isobaric regime, $<I>$ is the flux received at any given point $(z,\boldsymbol{\rho})$ at an instant of time t and $\alpha(\lambda)$ is the absorption coefficient of either clean air or aerosol-loaded air at the laser wavelength λ. Following convention, we now assume that in the isobaric regime the dielectric constant "fluctuations" are of the same form as the temperature fluctuations:

$$< \delta\varepsilon(z,\rho,t)\,\delta\varepsilon(z,\rho',t) > = < \delta\varepsilon(z,\rho,t)\,\delta\varepsilon(z,\rho',t) >_{eq.}\ \exp\left\{\frac{2\alpha(\gamma-1)}{P_0}\int_0^t dt'\,\Gamma(z,0,t')\right\} \tag{3}$$

where $<\delta\epsilon\ \delta\epsilon'>_{eq.}$ is the given by Kolmogorov's formula, and it goes as $|\boldsymbol{\rho}-\boldsymbol{\rho}'|^2$ for $|\boldsymbol{\rho}-\boldsymbol{\rho}'| \to 0$. The mutual coherence function Γ is :

$$\Gamma(z,\boldsymbol{\rho},\boldsymbol{\rho}',t) = (c/4\pi) < E(z,\boldsymbol{\rho},t)\ E(z,\boldsymbol{\rho}',t) > \tag{4}$$

In the plane wave approximation, the paraxial equation for the mutual coherence function of a scalar wave-field reduces to:[2]

$$\left[\frac{\partial}{\partial z} + \Phi(z,\rho-\rho',t)\frac{k^2}{4} + \alpha\right]\Gamma(z,\rho-\rho',t) = 0 \tag{5}$$

$$\Phi(z,\rho,t) \approx 6.56\ \mathfrak{C}_n^2(z,\rho,t)\ \rho^2/l_0^{1/3} \tag{6}$$

$$\mathfrak{C}_n^2 = C_n^2 \cdot \exp\{\ 2\alpha(\gamma-1)/P_0\ {}_0\!\int^t dt'\ \Gamma(z,0,t')\ \} \tag{7}$$

C_n^2 is the usual structure constant, and l_0 is the inner length scale of turbulence.[2] A closed form solution of Eqn. (5) is trivial to obtain. A graphical presentation is given in figs. 1-3 for the case of a "flat" laser pulse of 1s duration, delivering 5×10^3 W/cm^2 to an atmosphere loaded with fog-oil. The optical depth of fog-oil is taken to be 2.0 and 2.5. At these flux levels, fog-oil does not vaporize,[3] and the main effect of the fog-oil is to absorb energy from the beam and deposit it via conduction into the ambient air, thereby heating it to a far greater level than would normal atmospheric absorption. The curvature along the ρ -axis gives an indication of the average size of the speckle spot.

---

[1] S.M. Chitanvis, "Induced Turbulence in aerosol-loaded atmospheres", Proceedings of the 1987 CRDEC Conference on Aerosol and Obscuration Research, Aberdeen, Md.

[2] A. Ishimaru, "Wave Propagation and Scattering in Random Media" vol.II, Aca.Press, NY (1978).

[3] F.G. Gebhardt and R.E. Turner, CSL Contractor Rept. ARCSL-CR-80039 (1980)

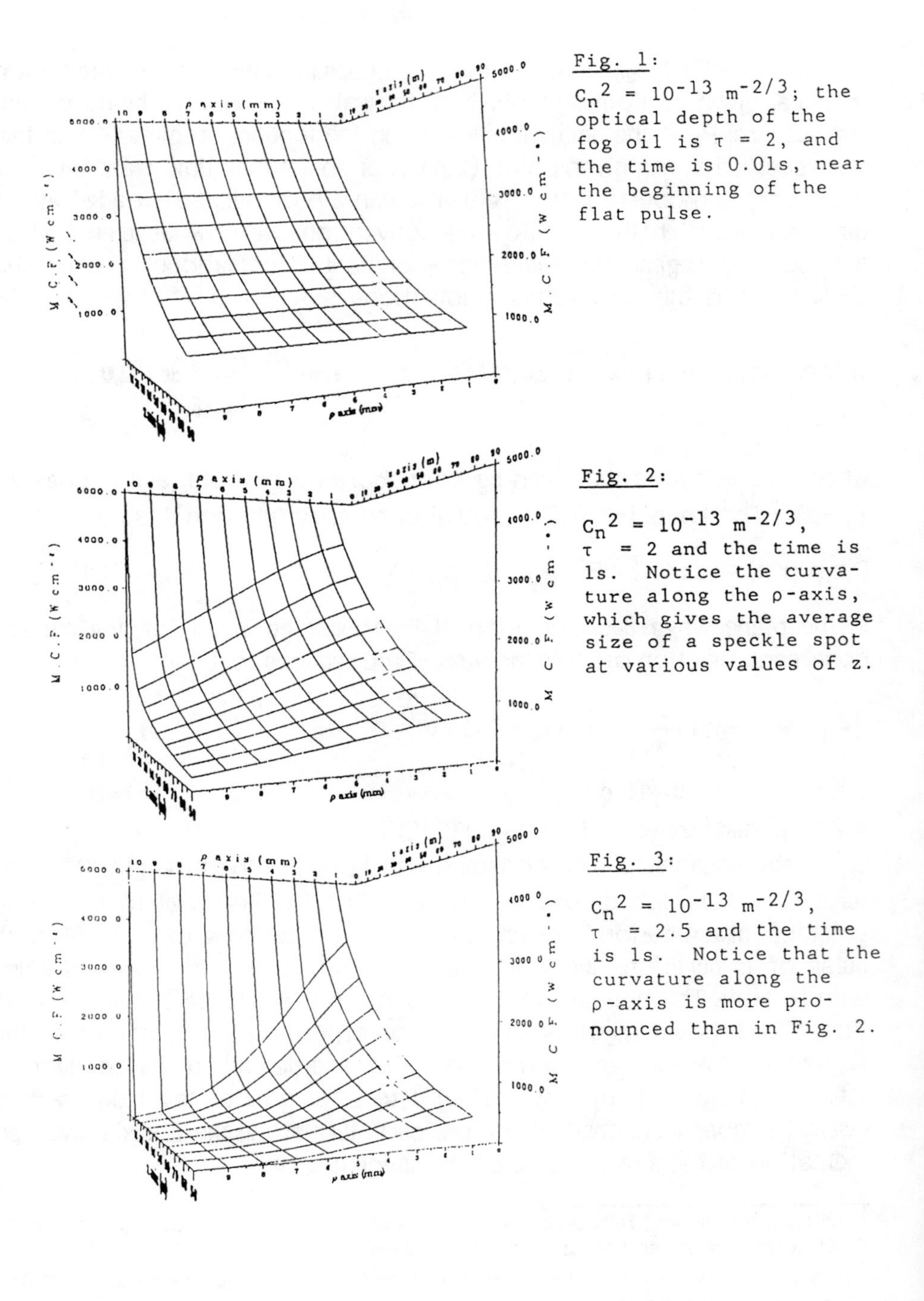

Fig. 1:
$C_n^2 = 10^{-13}\ m^{-2/3}$; the optical depth of the fog oil is $\tau = 2$, and the time is 0.01s, near the beginning of the flat pulse.

Fig. 2:
$C_n^2 = 10^{-13}\ m^{-2/3}$, $\tau = 2$ and the time is 1s. Notice the curvature along the ρ-axis, which gives the average size of a speckle spot at various values of z.

Fig. 3:
$C_n^2 = 10^{-13}\ m^{-2/3}$, $\tau = 2.5$ and the time is 1s. Notice that the curvature along the ρ-axis is more pronounced than in Fig. 2.

CLOUD AND WATER-SURFACE DIAGNOSTICS USING AN AIRBORNE POLARIZATION LIDAR

I.E.Penner, I.V.Samokhvalov, V.S.Shamanaev, V.E.Zuev
Institute of Atmospheric Optics, Siberian Branch USSR Academy of Sciences, Tomsk, 634055, USSR

Information on the cloud phase state and the sea-surface structure was extracted based on the temporal dependence of the degree of the return signal polarization.

The results of depolarization ratio measurements at one point of the cloud boundary obtained by different authors are analyzed in monograph[1]. It is shown that the errors in the cloud state phase determination and snow surface selection can appear when the spatial depolarization profile is not taken into account.

In order to eliminate the ambiguity involved in the identification of the cloud type three parameters describing the depolarization profile were used: the mean cloud depolarization gradient μ, the cloud top depolarization ratio D_0, and the relative displacement of the polarized and depolarized component maxima Δt (see Fig.1). The parameter space for water and ice clouds is given for a double variance. This multiparametric analysis minimizes the cloud phase-state diagnostics error.

The major features of the depolarization profiles observed from clouds are retained in the depth resolved measurements of the laser light attenuation by sea water. Conventionally, the lidar depolarization ratio increases with the penetration depth. In the first approximation, the depolarization ratio D also increases linearly in "pure" water. However, due to strong effect of multiple scattering, the depolarization gradient μ [km^{-2}] depends on the lidar field of view (FOV). The following approximated formula was obtained at the 440m flight altitude

$$\mu(\theta)=\mu(5mrd)+5\frac{km^{-2}}{mrd}\times[\theta-5mrd]. \qquad (1)$$

Here θ is FOV, $5 \leq \theta \leq 10$ mrad. The lidar installation is described in [1].

The depth profiles show characteristically large depolarization values at the air-water interface amounting to D_0 = 0.3 - 0.4 with standard deviation, for the majority of cases, as high as 60%. At a fixed depth under the sea surface D is found to fluctuate within 29 ± 7%. Sometimes, one can observe the sea areas with anomalous depolarization behavior, when the depolarization ratio is found to decrease with the penetration depth.

To determine the laser radiation attenuation coef-

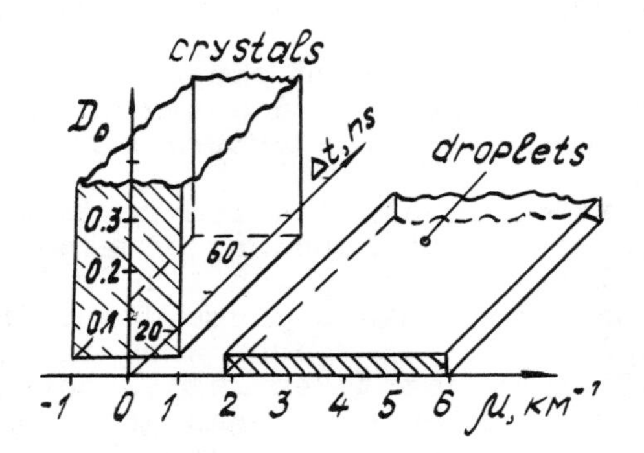

Figure 1. A 3-D parameter space for water and ice clouds

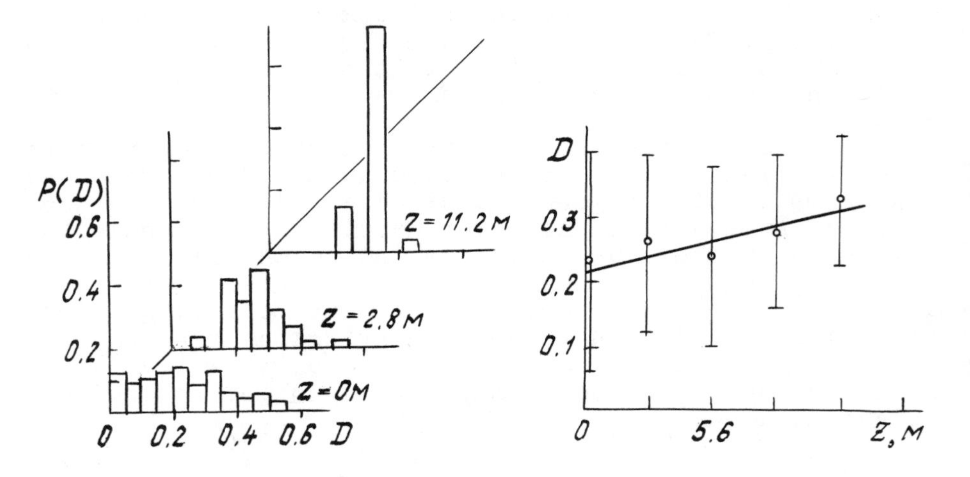

Figure 2. Experimental depth profile of the depolarization probability distribution (left). Depth profile of the depolarization ratio for this case (right).

ficient in water ε (m^{-1}), the method of taking a logarithm of lidar return trailing edge derivative was used. The ε value was also affected by multiple scattering through FOV. The approximated formula experimentally obtained is

$$\varepsilon = \varepsilon(3mrd) + 1.9\times 10^{-2}\frac{M^{-1}}{mrd}\times[\theta - 3mrd]. \qquad (2)$$

The theoretical calculations by the Monte-Carlo method give the following expression for similar conditions

$$\varepsilon = 0.2M^{-1} + \varepsilon(3mrd) + 2.2\times 10^{-2}\frac{M^{-1}}{mrd}\times[\theta - 3mrd]. \qquad (3)$$

For "pure" sea water, the standard deviation of ε is 12-25%.

The lidar depolarization variance is a useful parameter to know for the determination of the sea-surface thermodynamic state. Figure 2 shows experimental depth profiles of the depolarization probability distribution $P(D)$ down to 11 m below the water surface. The right-hand graph illustrates $D(z)$ for this case. $P(D)$ is seen to have a non-Gaussian form for all the depths. For numerical evaluation of the thermodynamic state of the sea water, the standard deviation was calculated to be $\delta D = 52 \pm 12\%$ for a depth of 2.8 m and a flight altitude of 120 m. The water had the attenuation coefficient $\varepsilon = 0.12 \pm 0.04\ m^{-1}$ that corresponded to the cold/less cold water boundary. The same characteristics were obtained for the thermodynamic equilibrium zone. The depolarization probability distribution is seen to be more narrow. For a depth of 2.8, $\delta D = 10 \pm 8\%$. At lower depths $D(z)$ is also smaller than in Fig.2. The water turbidity is found to decrease. From the physical point of view, the 3-fold reduction of the depolarization variance is easily explainable. The frontal zone is characterized by an intense mixing of the water mass. The resulting inhomogeneities of the hydrosol concentration and type lead to the enhanced depolarization fluctuations.

REFERENCES

1. Laser sounding of the troposphere and underlying surface. Novosibirsk, Nauka, 1987, 264 p., ed. by V.E.Zuev.

POTENTIALITIES OF ADAPTIVE PHASE CORRECTION OF THERMAL DISTORTIONS IN HIGH-POWER LASER BEAMS

V.E. Zuev, P.A. Konyaev, V.P. Lukin
Institute of Atmospheric Optics, USSR Academy of Sciences, Siberian Branch, Tomsk, 634055, USSR

ABSTRACT

High-power laser beam propagation through random-inhomogeneous weakly absorbing media was studied by means of the computer simulation technique. For a numerical solution of the nonstationary 4-D self-consistent problem, a modified splitting method was developed which provides high accuracy and operation rate. The beam power optimization curves were derived on the basis of the computer simulation results for the case of thermal blooming and adaptive phase correction in the turbulent atmosphere.

BASIC EQUATIONS

The nonstationary thermal blooming of a coherent laser beam propagating along Z-axis is described by a quasi-linear parabolic equation and a heat-transfer equation in dimensionless variables /1/:

$$\frac{\partial U}{\partial z} = \frac{i}{2}\left(\frac{\partial^2}{\partial x^2} + \frac{\partial^2}{\partial y^2} + \tilde{T}\right)U, \tag{1}$$

$$\frac{\partial \tilde{T}}{\partial t} + \frac{\partial \tilde{T}}{\partial x} + I = 0, \tag{2}$$

where $\tilde{T}$ is the random-inhomogeneous temperature field that, at zero time t, has a spectral density of the type

$$\Phi_{\tilde{T}}(\varkappa_x, \varkappa_y) = 0.033\, C_T^2 (\varkappa_x^2 + \varkappa_y^2 + \varkappa_0^2)^{-11/6} \tag{3}$$

Here C_T^2 is the dimensionless structural temperature characteristic. For simplicity, the transverse wind velocity is assumed to be constant, independent of Z and directed along X-axis. The beam intensity I is expressed in units of the characteristic scale of the thermal blooming intensity.

POWER OPTIMIZATION

The nonlinear nature of thermal blooming is known to lead to the nonlinear relation of the receiver-plane power density to that on the emitting aperture. Numerical solutions to the dimensionless equations (1,2) allowed us to derive universal power optimization curves for the

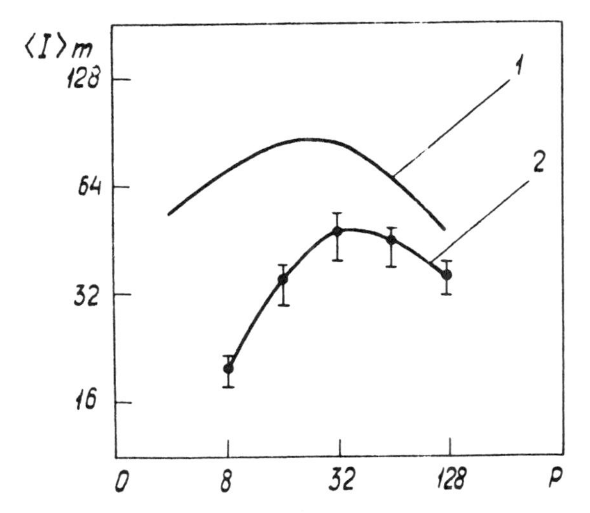

Fig.1. Power optimization for thermal blooming in homogeneous (1) and random-inhomogeneous media.

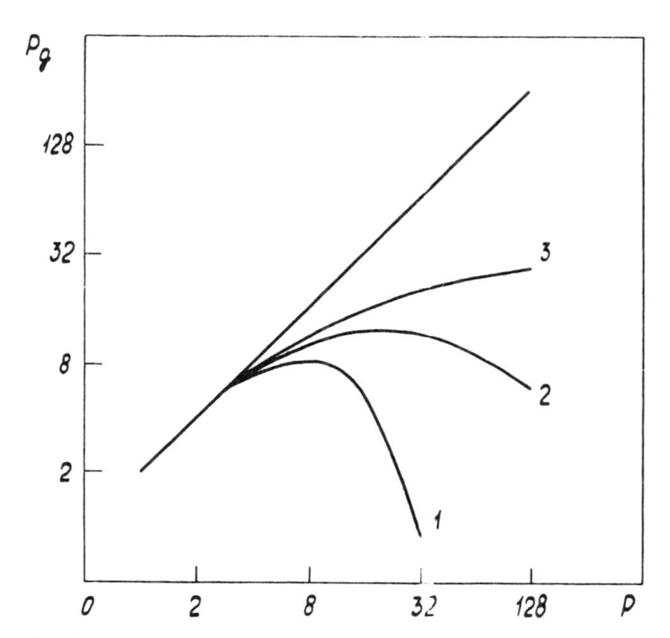

Fig.2. Power optimization for adaptive phase correction of thermal blooming: 1 - without correction; 2 - adaptive guidance; 3 - phase conjugated correction.

weakly-absorbing homogeneous and random-inhomogeneous atmosphere. Figure 1 shows the mean beam intensity in the receiver plane as a function of the emitting aperture power for a homogeneous medium (curve 1) and for a random-inhomogeneous environment with C = 2000 (curve 2). The atmospheric turbulence is seen to increase the optimal transmitter power, decreasing the highest mean receiver intensity by several fold.

ADAPTIVE PHASE CORRECTION

The proposed simulation technique was used to estimate the efficiency of the adaptive compensation for the beam thermal blooming in the turbulent atmosphere. To this end, the nonlinear equations (1,2) for a high-power beam coupled with a linear equation analogous to Eq.(1) for a reference beam counterpropagating along Z-axis were solved numerically. The phase-conjugated technique was simulated by substituting the inverted transverse phase profile for the reference beam into the complex field distribution of the high-power beam.

Figure 2 illustrates the above method by the computed power-optimization plots. Curve 1 shows the diffraction spot power P versus the emitted power P for the case of no correction. Curve 3 plots the same for the phase-conjugated correction. The compensation is seen to increase the optimal transmitter power by an order of magnitude, enhancing the mean receiver power by several fold. Shown for comparison is curve 2 depicting the efficiency of the adaptive guidance system eliminating the wind-induced nonlinear beam refraction. The straight line refers to the linear case where $T \equiv 0$ (see Eq.(1)).

Also, the power optimization curves for the combined effect of the atmospheric turbulence and thermal blooming were computed. The examination of the plots derived suggests a conclusion that the phase-conjugated correction efficiency is limited mainly by the nonlinear beam self-action and is insensitive to the atmospheric turbulence. However, the turbulence effect may influence the practical implementation of the phase-conjugated technique, which would require sophisticated flexible or segmented mirror systems.

REFERENCES

1. P.A. Konyaev, V.P. Lukin, Thermal distortions of focused laser beams in the atmosphere, Appl. Opt., Vol.24, No.3, 1985, pp.415-421.

TRANSITION FROM SUPERFLUORESCENCE (SF) TO AMPLIFIED SPONTANEOUS EMISSION (ASE): A COMPUTATIONAL EXPERIMENT

Farrès P. Mattar*
Department of Physics, New York University, New York, NY 10003
and George R. Harrison Spectroscopy Laboratory
Massachusetts Institute of Technology, Cambridge, MA 02139

ABSTRACT

A semiclassical formalism with quantum initiation and an additional Langevin fluctuation source is described. It has been successfully implemented.

I. INTRODUCTION

Boyd et al.+ recently reported an experimental observation on the evolution of SF into ASE, combining two cooperative physical processes previously studied separately. These results could **not** be quantitatively explained by current SF[1-16] and ASE[8c] theories. To describe **the intermediate regime,** Shuurmans et al.[8c] suggested a physical model which **requires not only quantum initiation as in SF but also fluctuation forces throughout the interaction.** To rigorously account for nonlinearities and reach quantitative agreement with experiment, we develop a new code by combining features of two programs developed to study SF for two and three level atoms. The first analysis provides the only quantitative agreement with the Cs SF while the second dealt with coherent pumping effects on the SF buildup (by including two-color Langevin forces in the density matrix at every time slice). The computations articulate how fluctuations evolve in conjunction with finite beam diffraction effects. The elimination of quantum fluctuations for small Fresnel number and generation of phase waves for large beam are reported.

II. SEMICLASSICAL FIELD EQUATIONS

The calculation of SF pulse evolution in the nonlinear regime is necessarily a computational problem if propagation is included explicitly. We use an algorithm presented elsewhere[17] to analyze the effects of coherent initiation, cooperative buildup propagation, density variations, diffraction effects and quantum noise sources on SF emission.

$\zeta = \mathrm{Re}\{E \exp[i(\omega t-kz)]\}$ (1a), $\quad P = \mathrm{Re}\{iP \exp[i(\omega t-kz)]\}$ (1b),

$\zeta = (2\mu E/h)\tau_R$ (2), $\quad P = P/\mu$ (3), $\quad \tau = (t-z/c)/(\tau_R)$ (4)

$\Delta\Omega = (\omega_o-\omega)\tau_R$ (5), $\quad \eta = z/\ell$ (6a), and $\xi = z\alpha_R$ (6b); with $\ell=c\tau_p$ (6c) for Self-Induced-Transparency SIT[2] propagation or $\ell=c\tau_R$(6d) for SF.

Here ζ and P are the slowly varying envelope of the electric field and polarization; τ the retarded time; η the normalized axial coordinate; μ the transition dipole moment matrix element; τ_R the cooperative radiation time; and $g(\Delta\Omega)$ the Doppler distribution. The paraxial-wave equation is equivalent to a nonlinear Schroedinger equation with the potential replaced by the polarization[8g-h,12,15]

$$-\frac{i}{4F}\nabla_T^2\zeta + \frac{\partial\zeta}{\partial\eta} = (\alpha_R\ell)d \int P(\Delta\Omega)g(\Delta\Omega)d(\Delta\Omega) \,, \qquad (7)$$

*Supported in part by ARO, AFOSR, NSF and ONR; +Private communication

$$\int_{-\infty}^{\infty} g(\Delta\Omega)d(\Delta\Omega) = \left(\frac{\pi}{\tau_2^*} \right)^{1/2} \int_{-\infty}^{\infty} \exp\{-[(\Delta\Omega)\tau_2^*]^2\}d(\Delta\Omega) = 1 \ , \quad (8)$$

$$d = \exp[-(\rho/\rho_N)^m] = N(\rho)/N(o) \quad (9)$$

where ρ_N is the (1/e) atomic density distribution radial width. For uniform density, d=1, i.e., $\rho_N \to \infty$. For m>0, the radial population density distribution is variable: m=2,4,6 respectively for a Gaussian, superGaussian and hyperGaussian density profile. The effective Beer's length is defined as

$$\alpha_R = 4\pi^2\omega_o\mu^2 N\tau_R/\hbar c \quad (10)$$

where ω_0 is the transition frequency and N is the atomic number density. The Fresnel number is $F=\pi r_p^2/\lambda L$[11] with L the cell length and $Z_d=\pi r_p^2/\lambda=\kappa^{-1}$ (12) the diffraction length $F=(\kappa L)^{-1}$ (13). Should one select $L=\alpha_R^{-1}$ one obtains F_g the gain-length Fresnel number $F_g= \pi r_p^2/\lambda\alpha_R^{-1}=Z_d\alpha_R=\alpha_R/\kappa$ (14). F_g is the effective gain to diffraction-loss ratio. Its ratio to the usual Fresnel number is the medium total gain $F_g/F=\alpha_R L$ (15); using $\xi=z\alpha_R$ (16) as the axial coordinate one obtains:

$$-\frac{i}{4F_g}\nabla_T^2\zeta + \frac{\partial\zeta}{\partial\xi} = d\int P(\Delta\Omega)g(\Delta\Omega)d(\Delta\Omega) \ . \quad (17)$$

Diffraction is taken into account by the transverse Laplacian $\nabla_T^2 = \frac{1}{\rho}\frac{\partial}{\partial\rho}\left(\rho\frac{\partial}{\partial\rho}\right)$ (18), with $\rho=r/r_p$, for cylindrical geometry. The boundary condition $\partial_\rho\zeta=0$ (19) at $\rho=\rho_{max}$ corresponds to completely absorbing walls. To insure that (1) the entire field is accurately simulated, (2) no artificial reflections are introduced at the numerical boundary $\rho_{max}>>r_p$, and (3) near-axis fine diffraction variations are resolved, the sample cross section is divided into non-uniform stretching cells [17].

The model includes the effects of both **spatial averaging,** associated with radial density variations, and **diffraction coupling.** The first calculations described a geometry with cylindrical symmetry (two spatial dimensions). It is valid for small F, i.e., with strong diffraction coupling. For large F azimuthal symmetry is absent and two transverse dimensions are required. The Cartesian model is needed to describe short-time-scale phase and amplitude fluctuations which result in multiple transverse modes initiation and lead to multi-directional output with hot spots.

III. PHYSICAL MODEL: THE MATERIAL RESPONSE

The medium response is not instantaneous. It is described by the density matrix.

$\partial P/\partial\tau+\beta P=\zeta W+fW$ (20a); $\partial W/\partial\tau+\gamma(W-W^e) = -\frac{1}{2}(\zeta^*P+\zeta P^*) - 2(f^*P+fP^*)$(20b)

$\beta = (1/\tau_2) + i(\Delta\omega)$ (21a); $\gamma = 1/\tau_1$ (21b)

$$<f(\tau)f^{\dagger}(\tau)> = \frac{1}{N_i\tau_R}\delta(\tau-\tau') \quad (22a); \qquad <f^{\dagger}(\tau)f(\tau')> = 0 \quad (22b)$$

$f = |f| \exp(i\emptyset)$ (23a); $0 \le \emptyset \le 2\pi$ (23b); $\underline{P}(f^2)=1/(\pi\sigma)\exp[-(f/\sigma)^2]$(24)

$\sigma = \tau_n^2/(\tau_R L/c)N_i = \tau_n^2/(\tau_R\tau_E N_i) = \tau_n^2/N\tau_c$ (25) with $\tau_c^2 = \tau_R\tau_E$ (26);

$\tau_R=L/c[2\pi|\mu|^2/\hbar N_i]^{-1}$(27) with τ_E & τ_c the escape and cooperative times[4] respectively while $N\rho$ is the number of radial shells.

$$N_i = N(\rho_i) = \exp[-(\rho_r/r_n)^m]\{\pi(\rho_{i+1/2}^2 - \rho_{i-1/2}^2)/B\} \quad (28a)$$

$$B=\pi\rho_{i/2}^2+\sum_{i=2}^{N\rho-1}\exp[-(\pi_i/r_N)^m]\{\pi(\rho_{i+1/2}^2-\rho_{i-1/2}^2)\}+\{\exp[-(r_{N\rho}/r)^m]\pi(\rho_{N\rho-1/2}^2)\} \quad (28b)$$

In the **on-resonance uniform-plane-wave** regime ($F=\infty$) the light-matter equations reduce to a **Sine-Gordon** equation: $\partial^2\Theta/\partial\eta\partial\tau=\sin\Theta$ (29) where $P=\sin\Theta$ (30) and $\Theta=\int_{-\infty}^{\tau}\zeta d\tau'$ (31) is the area.

Initially, the polarization is assumed to be random in phase relative to the coherent emission which eventually evolves. The probability P(u,v) that the transverse polarization has components u and v is a Gaussian distribution[9-11,12b-c,13]

$P(u,v)dudv= \frac{1}{\pi\delta^2}\exp[-(u^2+v^2)/\delta^2]du\,dv$(32) where $\delta_i=<\Theta^2>^{1/2}=2/\sqrt{N_i}$ (33)

for the quantum initiation to be properly represented. The angular brackets denote an ensemble average. Equation (33) is easily checked using $u^2+v^2-1=W^2\sim\sin^2\Theta\sim\Theta^2$(34) for small Θ as assumed here; then $P(\Theta^2)d\Theta^2\simeq[1/\delta^2]\exp[-(\Theta/\delta)^2]d\Theta^2$ (35). With R a **random** number between 0 and 1, the probability that Θ^2 is less than Θ_o^2 is

$\int_o^{\Theta_o^2}P(\Theta^2)d\Theta^2 = 1 -\exp[-(\Theta_o/\delta)^2]=1-R$(36); and $\Theta_o^i=(2/\sqrt{N_i})[\ln 1/R]^{1/2}$(37).

The random numbers used in Eq. (37) and in randomizing φ between 0 and 2π are obtained from a table of random numbers. The table starting address is changed at the beginning of each run[12b-c,13].

The Langevin force acts as a continuous temporally fluctuating noise source[8g,14,15]. The values selected for $|f|$, φ vary for each trajectory according to the statistical distributions. It can also give rise to SF initiation. The SF evolves in z, ρ, and τ due to the initiation instigated by Θ,φ while ASE is driven by $|f|$ and $\emptyset$. A complete ensemble of simulations must be constructed. **One then takes the ensemble averages of the different trajectories.** For example the delay-statistics τ_D is calculated by

$$\sigma(\tau_D)/\bar{\tau}_D \equiv [\left(\sum_{i=1}^{NR}(\tau_D^i - \bar{\tau}_D)^2/NR\right)^{1/2}/\bar{\tau}_D]\ \{\ 1 \pm 1/\sqrt{NR-1}\ \} \quad (38)$$

where $\bar{\tau}_D$ is the mean value of the pulse peak delay, σ is the standard deviation and NR is the number of trajectories in the ensemble. Occasionally, one encounters a first "peak" that is **not** the highest peak.

If one uses the absolute peak, the value of the standard deviation is dominated by that one trajectory in a set of NR trajectories. Such trajectories can be interpreted as phase waves[11].

IV. NUMERICAL RESULTS

The influence of a Langevin noise source on the output powers shown in Fig. 1 for both planar and non-planar configurations.

V. CONCLUSION

A family of SF and ASE calculations was carried out without any linearization[16] as a function of the Fresnel number and pulse length for a given atomic number density & a specific radial profile.

REFERENCES

1. R.H. Dicke, Phys. Rev. 93, 99 (1954).
2. S.L. McCall, **Self-Induced-Transparency by Pulse Coherent Light,** Ph.D. Thesis, Department of Physics, University of California at Berkeley (1968) Section II, 5-33.
3. D.C. Burnham & R.Y. Chiao, Phys. Rev. 188, 667 (1969).
4. F.T. Arecchi & E. Courtens, Phys. Rev. A2, 1730 (1970).
5. R. Bonifacio et al., Phys. Rev. A4, 302 and ibid 854 (1971); R. Bonifacio et al., Phys. Rev. A11, 1507 and ibid A12, 587 (1975).
6. R. Friedberg et al., Phys. Lett. A37, 285 (1971); A38, 227 (1972) and Opt. Commun. 10, 298 (1974).
7. N. Skribanowitz et al., Phys.Rev.Lett. 30, 309 (1973); & in **Laser Spectroscopy**, ed. R.G.Brewer & A.Mooradian (Plenum, 1975) pp.379.
8. [a] J.H.Eberly, Am. J. Phys. 40, 1374 (1972); [b] M.S.Feld et al., in **Coherent Nonlinear Optics**, ed. M.S.Feld & V.S. Letokhov, (Springer-Verlag, 1980) pp. 7-57; [c] M.F.H.Schuurmans et al., in **Advances in Atomic and Molecular Physics**, ed. D.R.Bates & B.Bederson (Academic, 1981) pp. 168-228; [d] Q.H.F.Vrehen et al., in **Dissipative Systems in Quantum Optics**, ed. R.Bonifacio, (Springer-Verlag, 1982) pp. 111-147; [e] A.V.Andreev et al., Sov. Phys. **Uspekhi** 23, 483-514 (1981); [f] M.Gross et al., **Phys. Repts.** 93, 301-396 (1982); [g] F.P.Mattar, SPIE 380, 508-542 (1983); [h] F.P.Mattar et al., in **Multi-Photon Excitation and Dissociation of Polyatomic Molecules**, ed. C.D.Cantrell, (Springer-Verlag, 1986) pp.223-282.
9. F.Haake et al., Phys. Rev. A5, 1454 (1972), ibid A13, 357 (1976); Phys. Lett. 68A, 29 (1978); and Phys. Rev. Lett. 42, 1740 (1979).
10. [a] D. Polder et al., Phys. Rev. A19, 1192 (1979); [b] Q.H.F. Vrehen et al., Phys. Rev. Lett. 43, 343 (1979).
11. [a] F.A. Hopf, Phys. Rev. A20, 2054 (1979); [b] F.A. Hopf & E.A. Overman, II, Phys. Rev. A19, 1180 (1979).
12. [a] F.P.Mattar et al., Phys.Rev.Lett. 46, 1123 (1981); [b] F.P. Mattar, SPIE 288, 353-361 (1981); [c] F.P.Mattar et al., in **Coherence & Quantum Opt. V**, ed. L.Mandel & E.Wolf (Plenum, 1984) pp.487-490. Fig.3 **displays temporal ringing in the central shell** even with diffraction and quantum fluctuation accounted for. This exhibition of ringing has been confirmed experimentally in ^{85}Rb [12d]; & [d] D.Heinzen et al., Phys. Rev. Lett. 54, 677 (1984).
13. P.D. Drummond & J.H. Eberly, Phys. Rev. A15, 3446 (1982).
14. C.M. Bowden & C.C. Sung, Phys. Rev. Lett. 50, 156 (1983).
15. [a] F.P. Mattar et al., in **Coherence and Quantum Optics V,** ed. L. Mandel & E.Wolf (Plenum, 1984) 507-514; [b] F.P.Mattar et al., in **Tech. Digest 1986 Ann. Meet. Opt. Soc. Am.**, p.68, paper #TuM6.

16. J. Mostowski & B. Sobolewska, Phys. Rev. A28, 2573 (1983).
17. F.P. Mattar, Appl. Phys. 17, 53 (1978).

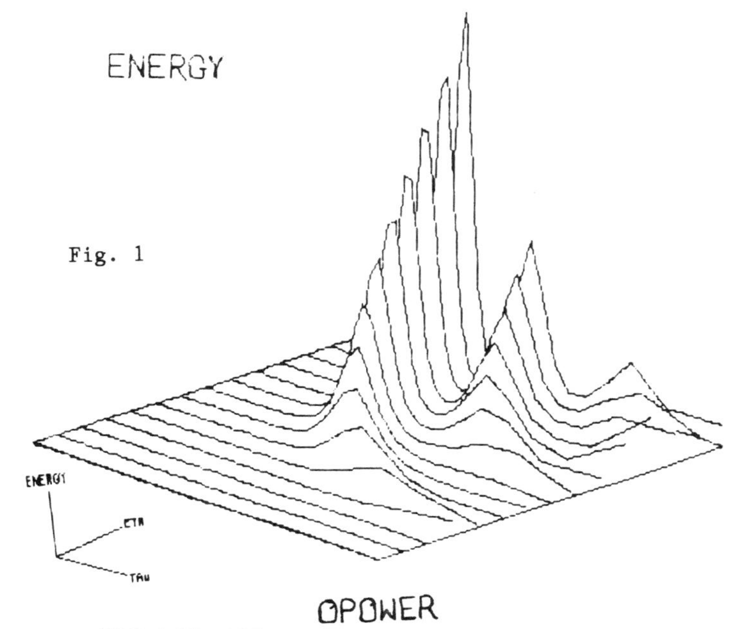

FIG. 1. The ASE on-axis energy in the uniform-plane-wave regime.

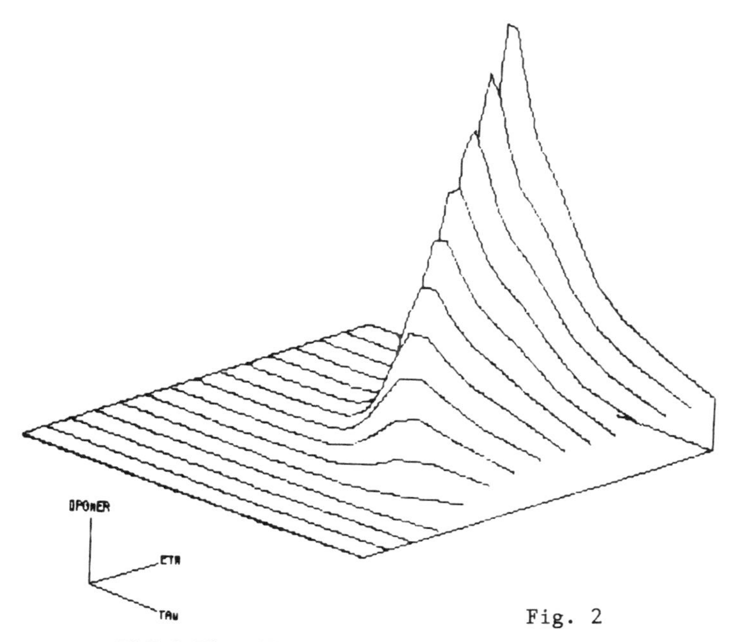

FIG. 2. The ASE output power with diffraction accounted for.

RABI-SIDE BANDS GENERATION IN OPTICALLY THICK MEDIUM†

F.P. Mattar*,** and P. Berman*
Department of Physics, New York University*, New York, NY 10003
and George R. Harrison Spectroscopy Laboratory
Massachusetts Institute of Technology**, Cambridge, MA 02139

ABSTRACT

By adopting phase-integral methods we compute how an off-resonant probe coherently extracts more energy from an on-resonance pump in an optically thick three-level atom than a resonant probe from a larger pump. In both situations the time-integrated area of the pump must exceed 5π. The population accumulated in the intermediate level exhibits a peak at a detuning equal to the pump Rabi frequency. This is a generalization of the Rabi side bands since propagation effects are accounted for. The pulse shape or pulse length affects the interaction.

I. INTRODUCTION

Analytic and numerical calculations for the propagational dependence of Rabi-splitting resulting from a strong resonant pump and a weak probe are presented. A semi-classical formalism was adopted for both the perturbational (using stationary phase and steepest-descent asymptotic techniques known as uniform approximation[1]) and the rigorous computational[2] treatments. This calculation can be considered as a Raman-like Double Self-Induced-Transparency (SIT)[3] where the strong pump (2π on-axis area) experiences coherent pulse break up and depletion while the weak probe (0.02π area) cooperatively builds up similar to a superfluorescence (SF)[4] emission. Different transition gain ratios (i.e., oscillator strength: $\mu^2\omega$) are considered to insure that the weak probe does not get (a) delayed with respect to the pump or (b) get out of synchronization and cease to overlap. The probe detuning superfluorescence must be as large as the input on-axis pump Rabi frequency. The interplay of nonlinear Raman gain action and dispersion give rise to a number of effects previously studied independently such as self-phase modulation, wave front encoding, transverse ring formation, self-focusing, quasi-trapping and asymptotic three-level solitary waves.

With the goal of understanding various aspects of Raman propagation, we have analyzed the effect of pump strength, namely (a) Rabi-splitting (side-band in $\rho_{33}(\infty)$); (b) coherent pulse break-up "à la SIT"; (c) pulse shape (d) self-focusing; and (e) pump detuning. This can lead to quasi-trapping since lensing effects[5] associated with the phase φ that evolves due to dispersion can balance the Φ originating from diffraction coupling. We have studied the effects of probe detuning to elucidate which probe detuning leads to larger probe magnification for a given pump strength. We assessed the dependence of relative delay between pump and probe as a function of: (a) pump strength; (b) probe detuning; (c) beam profile; (d) pulse shape; and (e) gain ratio (that is, the oscillator's strength ratio). We have proceeded in the understanding of the effect of optical thickness "$\alpha\ell$" in conjunction with pump depletion for: (a) resonant and off-resonant pump; in (b) the uniform-plane-wave UPW and non-UPW regimes.

† Supported by ARO, AFOSR, NSF, ONR, and NYU Research Challenge Fund

II. NUMERICAL RESULTS

The probe being a small fraction, f, of the pump one can linearize the field-matter equations and solve them. The pump is of zeroth order, the probe builds up as a first order. The energy flows from the pump to the probe only. The probe can **not** radiate energy back onto the pump.

The pump experiences an absorbing transition while the probe sees gain after the first half of the 2π pump transfer to the population from the ground state to the upper level. The pump evolves like an SIT pulse while the weak probe builds up like an SF.

In the first appendix, the analytical expression for the medium response ρ_{33}, is outlined. The numerics were validated by reproducing known analytical results. In Figs. 1 and 2 the frequency dependent evolution of ρ_{33} as one varies the pulse shape is illustrated at the input plane for different areas. In the first family of plots, the area and the peak Rabi frequency ν_R remain constant while the pulse length τ_p varies. In the second figure one keeps the pulse length constant while changing the peak Rabi frequency.

As one modifies the pulse shape the numerical integration results in unequal pulse lengths with equal Rabi-frequency peak values of ρ_{33} all occurring at the same frequency offset in the first figure. The constant pulse-length shapes lead to unequal pump Rabi frequency. This in turn gives unequal ρ_{33} maxima as well as unequal frequency sites reported in Double Resonance Studies[5].

In the second appendix, we present an analytical expression obtained perturbatively for the pump evolution in dense medium. As it propagates the pump is partially absorbed. Its Rabi frequency diminishes causing the side-band to move closer to line center. That is, as the propagational distance increases the frequency profile of ρ_{33} gets narrower.

The pump is affected by the probe only in the nonlinear regime. The probe feedback onto the pump is of a second order. As long as the analysis is linear, the pump propagation is governed by the SIT, two-level, equations.

In Figs. 3 and 4, we show the effect of propagation (two different η) in the UPW regime for a family of pulse shapes of constant areas and maximum Rabi frequencies. The reshaping of ρ_{33} versus Δ as a function of $\{\eta\}$ is shown for a group of unequal areas $\{s\}$ with a specific pulse shape. The peak of ρ_{33} associated with a larger pump area occurs at a larger detuning than the one associated with a weaker area both at the input plane and at different penetration thicknesses as expected analytically.

III. FUTURE WORK

The nonlinear feedback response of the probe onto the pump regime can be treated rigorously using previously developed algorithms[7]. Due to the dispersion one could expect a much more complex physical situation than the one in Raman-like analyses reported by Mattar[8] in these proceedings.

REFERENCES

1. (a) F.W.J. Oliver, Phil. Trans. Roy. Soc. London A247, 38 (1955); (b) C. Chester, B. Friedman & F. Ursell, Proc. Camb. Phil. Soc. 53, 599 (1957); (c) M.V. Berry, Proc. Phys. Soc. 89, 479 (1966);

(d) B.H. Bransden, **Atomic Collision Theory** (Benjamin,1970) pp. 85-95; (e) A. Erdelyi,"Uniform asymptotic expansion of integrals", pp. 149-168, in **Analytic Methods in Mathematical Physics;** (f) R.G. Newton, 'Limiting cases & approximations' in **Scattering Theory of Waves & Particles** (McGraw-Hill, 1966) pp. 72-96; (g) M.S. Child, "Semi-classical Elastic Scattering" in **Molecular Collision Theory** (Academic Press, 1974) pp. 60-73; (h) E.J. Robinson, J. Phys. B. 13, 2359 and ibid 13, 2243 (1980); (i) L.B. Felsen & N. Marcuvitz, **Radiation and Scattering of Waves,** (Prentice-Hall, 1973).

2. F.P. Mattar, SPIE 540, 588 (1985) and **Laser Spectroscopy VII,** ed. T.W. Hänsch & Y.R. Shen (Springer-Verlag, 1985) p. 218.
3. S.L. McCall & E.L. Hahn, Phys. Rev. Lett. 18, 908 (1967), Phys. Rev. 183, 457 (1969).
4. (a) M.S.Feld & J.C.MacGillivray, in **Coherent Nonlinear Optics,** ed. by M.S. Feld & V.S. Letokhov (Springer-Verlag, 1980) pp.7-57; (b) Q.H.F. Vrehen & H.M. Gibbs, in **Dissipative Systems in Quantum Optics,** ed. R. Bonifacio (Springer-Verlag, 1981) pp. 111-147; (c) M. Gross & S. Haroche, Phys. Repts. 93, 301-396 (1986); (d) F.P. Mattar et al., SPIE 380, 508-542 (1983) and in **Multiphoton Excitation and Dissociation of Polyatomic Molecules,** ed. C.D. Cantrell, (Springer-Verlag, 1986) pp. 223-282.
5. H.A. Haus, **Waves & Fields in OptoElectronics** (Prentice-Hall,1984).
6. P.T. Greendland, J. Phys. B18, 401 (1985) & Optica Acta 33, 723 (1986).
7. F.P. Mattar, Appl. Phys. 17, 53 (1978).
8. F.P. Mattar, **Physical Mechanism in On-Resonance Propagation of Two Beams in a Three Level System,** submitted Prog.Quantum Electronics.

APPENDIX I

Asymptotic Evaluation of the Population in the Intermediate Level

$$\rho_{33}(\infty,\Delta)=\left|\frac{1}{2i}\int_{-\infty}^{\infty}\zeta\exp[i(\Delta\tau\pm\Theta)]d\tau\right|^2=\left|\frac{1}{2i}\int_{-\infty}^{\infty}\zeta[\exp(i\varphi_+)-\exp(i\varphi_-)]d\tau\right|^2 \quad (I.1a)$$

$\varphi_{\pm}=\Delta\tau\pm\Theta\cdot_{\cdot}\cdot\varphi(\tau_o^{F,B})=\Delta\tau_o^{F,B}+\Theta(\tau_o^{F,B})$ where Δ is a contraction of $\Delta\omega_b$ (I.1b)

I - Two Unequal Real Roots

The stationary point of φ occurs at $\tau=\tau_o$ for which $[\partial\varphi/\partial\tau]_{\tau=\tau_0}=0$ and $[\partial^2\varphi/\partial\tau^2]_{\tau=\tau_0}\neq 0$.

$\partial\varphi_{\pm}/\partial\tau=\Delta \pm \partial\Theta/\partial\tau=\Delta \pm \zeta=$zero since $\partial\Theta/\partial\tau=\zeta$ one has $\zeta(\tau_o^F)=\zeta(\tau_o^B)$ (I.2)

for either positive or negative detuning, Δ, the phase has an extremum value when $\zeta(\tau_0)=|\Delta|$. Thus Δ is a saddle point and the phase can be written as: $\varphi_{\pm}(\tau)=\varphi_{\pm}(\tau_0)+1/2[\partial^2\varphi/\partial\tau^2]_{\tau=\tau_0}(\tau-\tau_0)^2$ (I.3); thus

$$I = \zeta(\tau_o)\ \{\exp[i\varphi(\tau_o^F)]\int_{-\infty}^{\infty}\exp[i/2\ [\partial^2\varphi/\partial\tau^2]_{\tau_o^F}(\tau-\tau_o)^2]d\tau$$
$$+ \exp[i\varphi(\tau_o^B)]\int_{-\infty}^{\infty}\exp[i/2\ [\partial^2\varphi/\partial\tau^2]_{\tau_o^B}(\tau-\tau_o)^2]d\tau\} \quad (I.4)$$

$$\int_{-\infty}^{\infty}\exp(\pm iax^2)dx = [\pi/a]^{1/2}\exp[\pm i\pi/4] \quad (I.5)$$

$$I = \zeta(\tau_o)\ \{\{\ \exp[i\varphi(\tau_o^F)]\ \{\pi/|[\partial^2\varphi/\partial\tau^2]_{\tau_o^F}|\}\ \exp[+i\pi/4]$$
$$+ \exp[i\varphi(\tau_o^B)]\ \{\pi/|[\partial^2\varphi/\partial\tau^2]_{\tau_o^B}|\}^{1/2}\exp[-i\pi/4]\ \}\} \quad (I.6a)$$

$$I = \zeta(\tau_o)\ \{\{\ \{\pi/|[\partial\zeta/\partial\tau]_{\tau_o^F}|\}^{1/2}\exp[i\varphi(\tau_o^F)]\ \exp[+i\pi/4]$$
$$+ \{\pi/|[\partial\zeta/\partial\tau]_{\tau_o^B}|\}^{1/2}\exp[i\varphi(\tau_o^B)]\ \exp[-i\pi/4]\ \}\} \quad (I.6b)$$

since $[\partial\zeta/\partial\tau]_{\tau_o^F} = -[\partial\zeta/\partial\tau]_{\tau_o^B}$; and $[\partial^2\varphi/\partial\tau^2]_{\tau_o^{F,B}}=[\partial\zeta/\partial\tau]_{\tau_o^{F,B}}$ (I.7)

$$I = \zeta(\tau_o)\{\pi/[\pm i/2|[\partial^2\varphi/\partial\tau^2]_{\tau_o^F}|]^{1/2}\{\exp[i(\tfrac{\pi}{4} + \varphi(\tau_o^F))]+\exp[i(\tfrac{\pi}{4} -\varphi(\tau_o^B))\} \quad (I.8)$$

With $\delta\varphi\equiv\varphi(\tau_o^B)-\varphi(\tau_o^F)=\Delta(\tau_o^B-\tau_o^F)+[\Theta(\tau_o^B)-\Theta(\tau_o^F)]$ and writing τ_o^F as τ_o (1.9)

$$I = 2\zeta(\tau_o)\ \{\pi/[\partial\zeta/\partial\tau]_{\tau_o}\}^{1/2}\cos[1/2(\delta\varphi)] \quad (I.10)$$

$$\rho_{33}(\infty,\Delta) = \left|2\zeta(\tau_o)\{\pi|[\partial\zeta/\partial\tau]_{\tau_o}|\}^{1/2}\cos\ [1/2(\delta\varphi)]\right|^2 \quad (I.11)$$

II. One Single Real Root

A more general situation evolves at the critical point where one has coalescence of stationary roots, $\tau_o^F = \tau_o^B \equiv \tau_R$ (I.12); and $[\partial\zeta/\partial\tau]_{\tau_R}$=zero$\rightarrow[\partial^2\Theta/\partial\tau^2]_{\tau_R}=[\partial^2\Theta/\partial\tau^2]_{\tau_R} = 0$ (I.13); the phase is expanded as follows: $\varphi=\Theta(\tau_R)+(\tau-\tau_R)[\partial\Theta/\partial\tau]_{\tau_R}+\frac{1}{6}[\partial^3\Theta/\partial\tau^3]_{\tau_R}(\tau-\tau_R)^3$ (I.14)

$$\rho_{33}(\infty,\Delta) = \zeta^2(\tau_R)\left|\int_{-\infty}^{\infty}\exp\{i(\Theta(\tau_R)+[\frac{\partial\Theta}{\partial\tau}]_{\tau_R}(\tau-\tau_R) + \frac{1}{6}[\frac{\partial^3\Theta}{\partial\tau^3}]_{\tau_R}(\tau-\tau_R)^3)\}d\tau\right|^2 \quad \text{(I.15)}$$

$u\equiv[\frac{1}{2}[\partial^3\Theta/\partial\tau^3]_{\tau_R}]^{1/3}(\tau-\tau_R)$(I.16a); $x\equiv[\partial\Theta/\partial\tau]_{\tau_R}/\{1/2[\partial^3\Theta/\partial\tau^3]_{\tau_R}\}$(I.16b)

$$\rho_{33}(\infty,\Delta) = \frac{(2\pi)^2\zeta^2(\tau_R)}{[1/2[\partial^3\Theta/\partial\tau^3]_{\tau_R}]^{2/3}}\left|\frac{1}{2\pi}\int_{-\infty}^{\infty}\exp[i(xu+u^3/3)]du\right|^2 \quad \text{(I.17)}$$

Ai(u) is the Airy rainbow integral $Ai(u)\equiv 1/2\pi)\int_{-\infty}^{\infty} du \exp[i(xu+(1/3)u^3)]$ (I.18)

$$\rho_{33}(\infty,\Delta)=\left|2\pi\Delta\{1/2[\partial^2\zeta/\partial\tau^2]_{\tau_R}\}^{-1/3}Ai(\zeta(\tau_R)\{1/2[\partial^2\zeta/\partial\tau^2]_{\tau_R}\}^{-1/3})\right|^2 \quad \text{(I.19a)}$$

$$\rho_{33}(\infty,\Delta)=\left|2\pi\Delta\{1/2[\partial^2\zeta/\partial\tau^2]_{\tau_R}\}^{-1/3}Ai(\Delta\{1/2[\partial^2\zeta/\partial\tau^2]_{\tau_R}\}^{-1/3})\right|^2 \quad \text{(I.19b)}$$

APPENDIX II
SIT PROPAGATION IN OPTICALLY-INTERMEDIATELY-THICK MEDIA

In the on-resonance UPW field-matter equation are: $\partial\zeta(\rho,\eta,\tau)/\partial\tau = gP(\rho,\eta,\tau)$ (II.1a). To treat diffraction, one must add to the RHS the Laplacian term "$-i\nabla_T^2\zeta$", $\partial^2\zeta/\partial\rho^2+(1/\rho)\partial\zeta/\partial\rho$ (II.1b); $\partial P(\rho,\eta,\tau)/\partial\tau=\zeta(\rho,\eta,\tau)W(\rho,\eta,\tau)$ (II.2a) $\partial W(\rho,\eta,\tau)/\partial\tau=-\zeta(\rho,\eta,\tau)P(\rho,\eta,\tau)$ (II.2b); Eq. (II.2) leads to: $P(\rho,\eta,\tau) = \sin\Theta(\rho,\eta,\tau)$ (II.3a)

$$\Theta(\rho,\eta,\tau) = \int_{-\infty}^{\tau}\zeta(\rho,o,\tau')d\tau'; \quad \Psi(\rho,\eta,\tau)) = \int_{-\infty}^{\tau}\sin\Theta(\rho,\eta,\tau')d\tau' \quad \text{(II.3b)}$$

By combining (II.1) and (II.3) one obtains the Sine-Gordon Equations:

$$\partial^2\Theta(\rho,\eta,\tau)/\partial\eta\partial\tau=\sin\Theta(\rho,\eta,\tau) \quad \text{(II.4)}$$

whose hyperbolic secant solution is well known[3]

$$\zeta(\rho,\eta,\tau) = \zeta(\rho,o,\tau) - g\int_{o}^{\eta}\{\sin[\int_{-\infty}^{\tau}\zeta(\rho,\eta',\tau')d\tau']\}d\eta' \quad \text{(II.5)}$$

For different pulse shapes one obtains for small η:

$$\tilde{\zeta}(\rho,\eta,\tau) \cong \zeta(\rho,o,\tau) - g\ \eta[\sin\int_o^{\tau} \zeta(\rho,o,\tau')d\tau'] = \zeta_o - g\eta\sin\Theta_o \quad (II.6)$$

$$\zeta(\rho,\eta,\tau) = \zeta(\rho,o,\tau) - g\int_o^{\eta} d\eta'\sin[\int_{\infty}^{\tau} d\tau'\ [\ \zeta(\tau') - g\eta'\sin\int_{\infty}^{\tau'} \zeta(\tau'')d\tau''] \quad (II.7)$$

$$\cong \zeta(\rho,0,\tau) - g\sin\Theta_o\int_o^{\eta} d\eta'\cos(g\psi_o\eta') = \zeta_o - g\cos\Theta_o\int_o^{\eta} d\eta'\sin(g\psi_o\eta') \quad (II.8)$$

ADDENDUM

For a hyperbolic secant square pulse, one obtains an analytical expression for the area which leads to ρ_{33} as follows:

$$\zeta = \zeta_0 \text{sech}^2\tau, \quad \Theta = \int_{-\infty}^{\tau} \zeta(\tau')dT = \tanh\tau, \quad \text{and} \quad \mu \triangleq \zeta_0^2$$

$$\rho_{33} = |(\pi\Delta/2)\exp[-i\mu]\,{}_1F_1((1+1/2i\Delta),2,2i\mu)|^2$$

with ${}_1F_1(x)$ the confluent hypergeometric function.

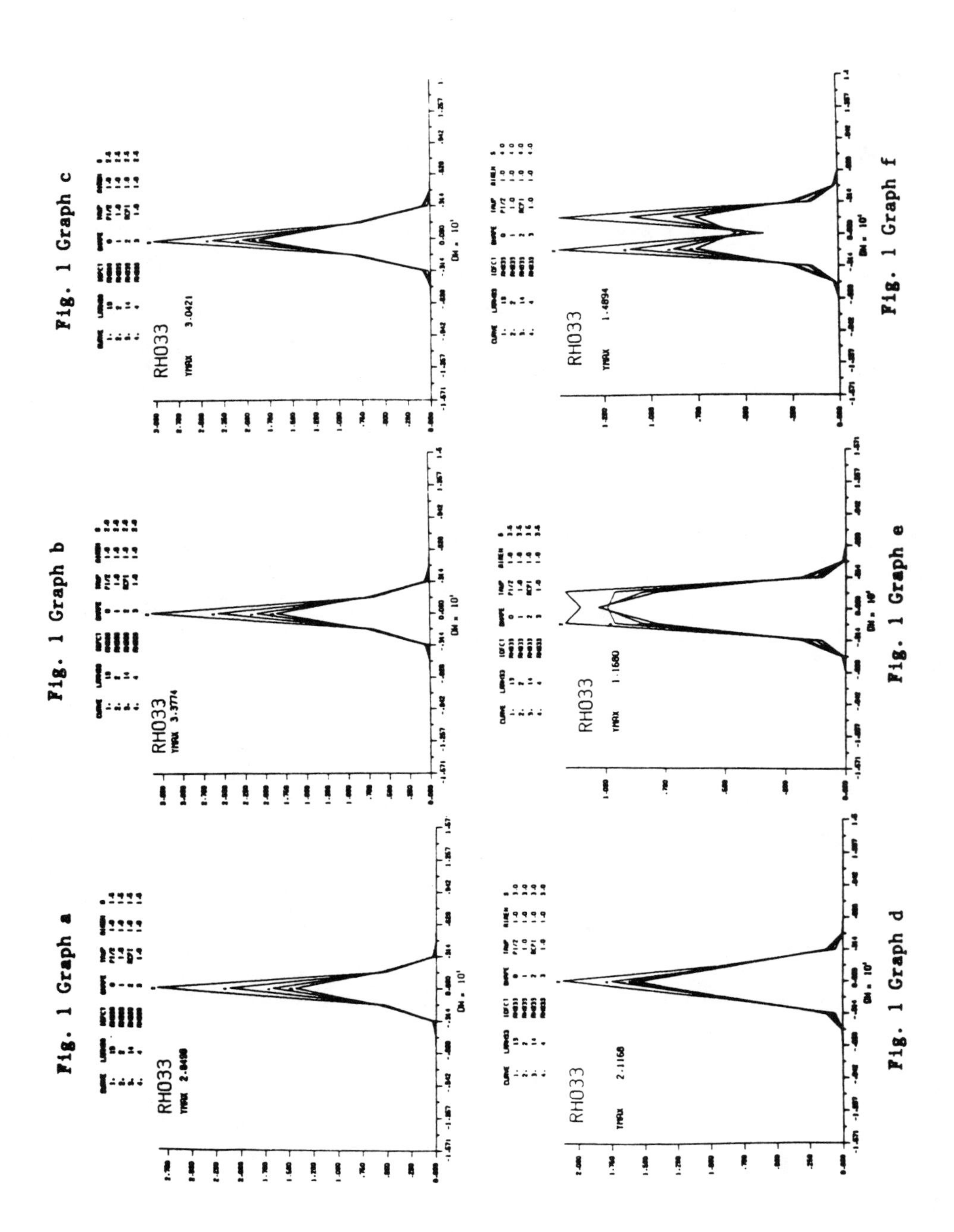

FIG. 1. ρ_{33} in arbitrary units is shown versus the probe detuning Δ for four different shapes of equal areas, s/π, equal Rabi frequency ν_R and unequal pulse length τ_p. For each graph the pump area varies as follows: (a) 1.5π; (b) 2π; (c) 2.5π; (d) 3π; (e) 3.5π; (f) 4π; (g) 4.5π; (h) 5.0π; (i) 5.5π; (j) 6.0; (k) 6.5π; (l) 8.5π; (m) 9π; (n) 9.5π; and (0) 10π. Until a threshold area, $\theta = 3.5\pi$, ρ_{33} has only one peak which always occurs on-resonance. For larger areas, ρ_{33} exhibits two off-resonance peaks. Beginning with $\theta = 8.5\pi$, a resonant peak for ρ_{33} reappears in addition to the two off-center peaks. The peak values of ρ_{33} vary with the shape, however, its location remains the same.

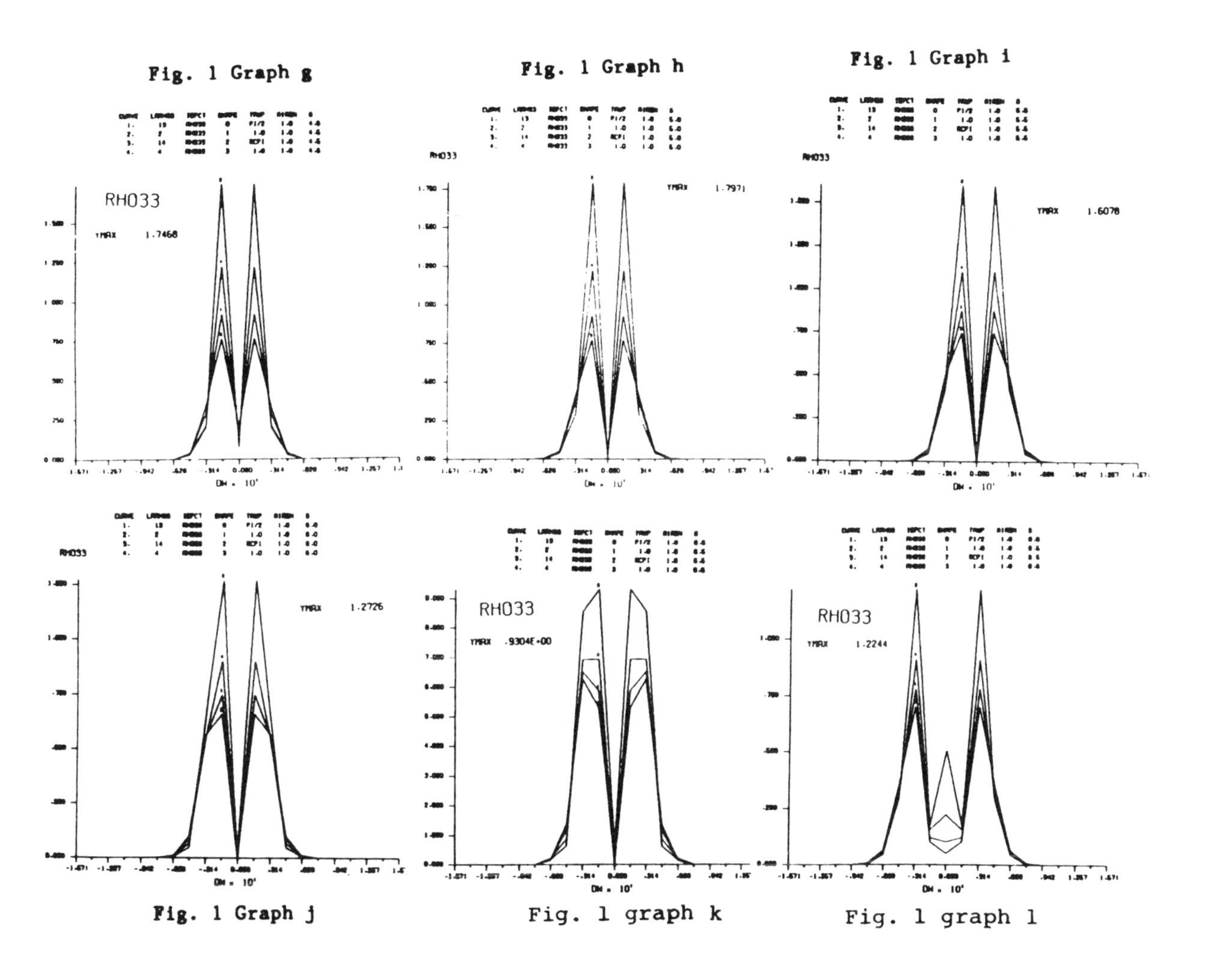

Fig. 1 Graph j

Fig. 1 graph k

Fig. 1 graph l

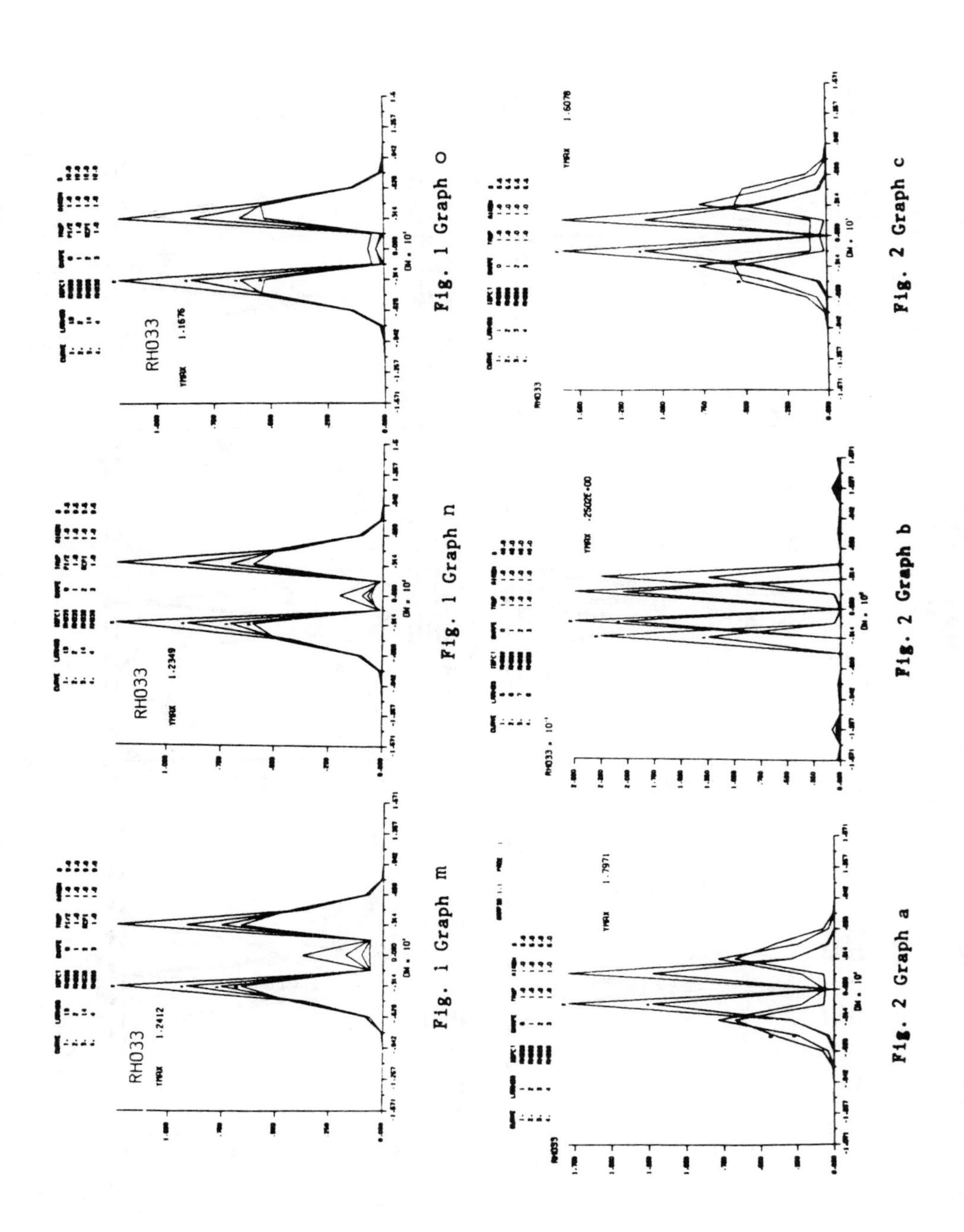

FIG. 2. ρ_{33} in arbitrary units is plotted versus the probe detuning Δ for four different shape pulses, of equal area and pulse lengths but of unequal Rabi frequency ν_R. Both the value of ρ_{33} and its site varies from one shape to the other. Graph a represents a 5.5π area while graph b corresponds to a 40π and graph c results from a 5.5π.

LRRH33 SIMULATION NO. 25

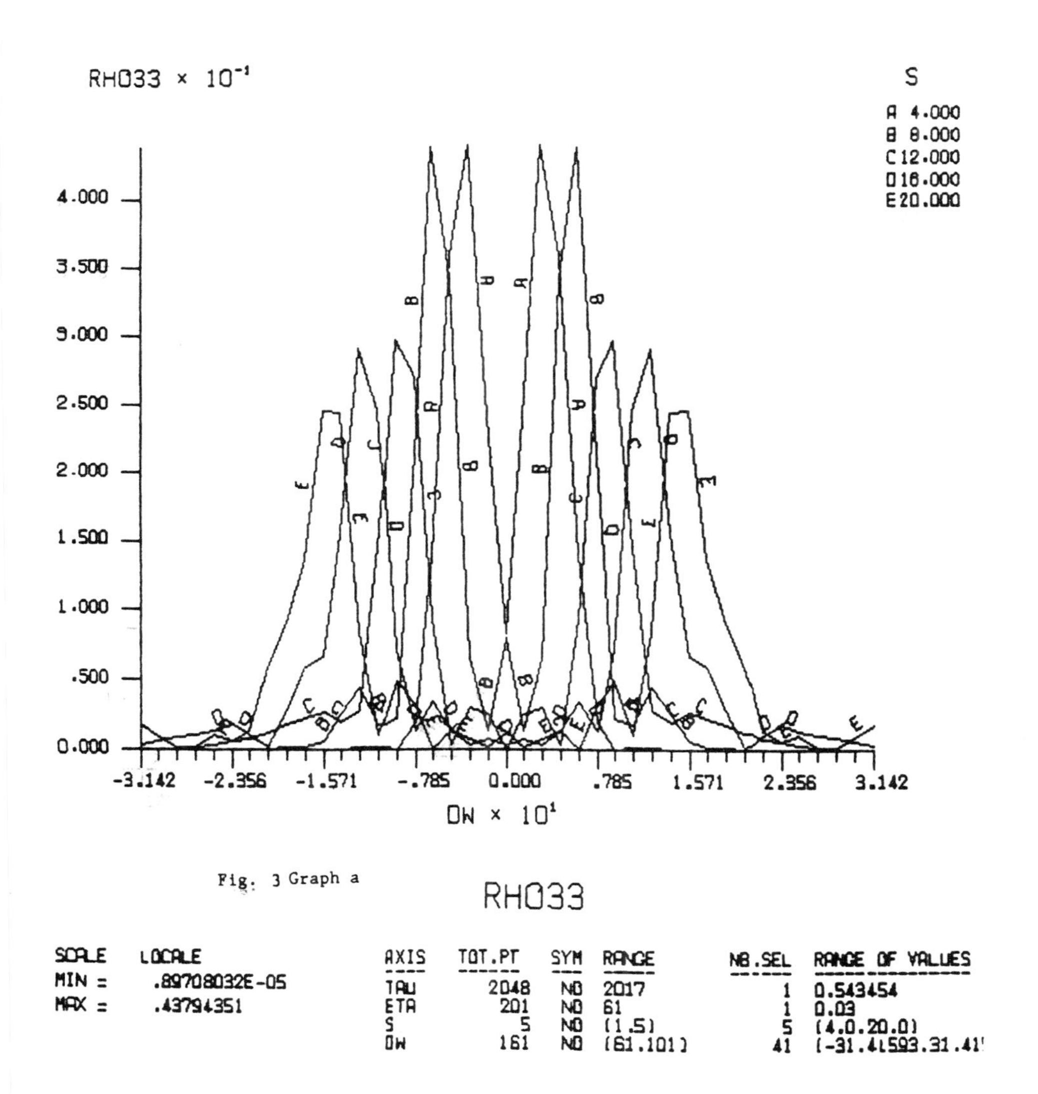

SCALE	LOCALE
MIN =	.89708032E-05
MAX =	.43794351

AXIS	TOT.PT	SYM	RANGE	NB.SEL	RANGE OF VALUES
TAU	2048	NO	2017	1	0.543454
ETA	201	NO	61	1	0.03
S	5	NO	(1.5)	5	(4.0.20.0)
DW	161	NO	(61.101)	41	(-31.41593.31.41

FIG. 3. The influence of areas on the probe buildup is shown as it is affected by propagation. In graph a $\eta = 0.15$ while in graph b $\eta_{+}0.03$. The peak of ρ_{33} associated with the larger pump area occurs at a larger detuning as expected analytically.

LRRH33 SIMULATION NO. 25

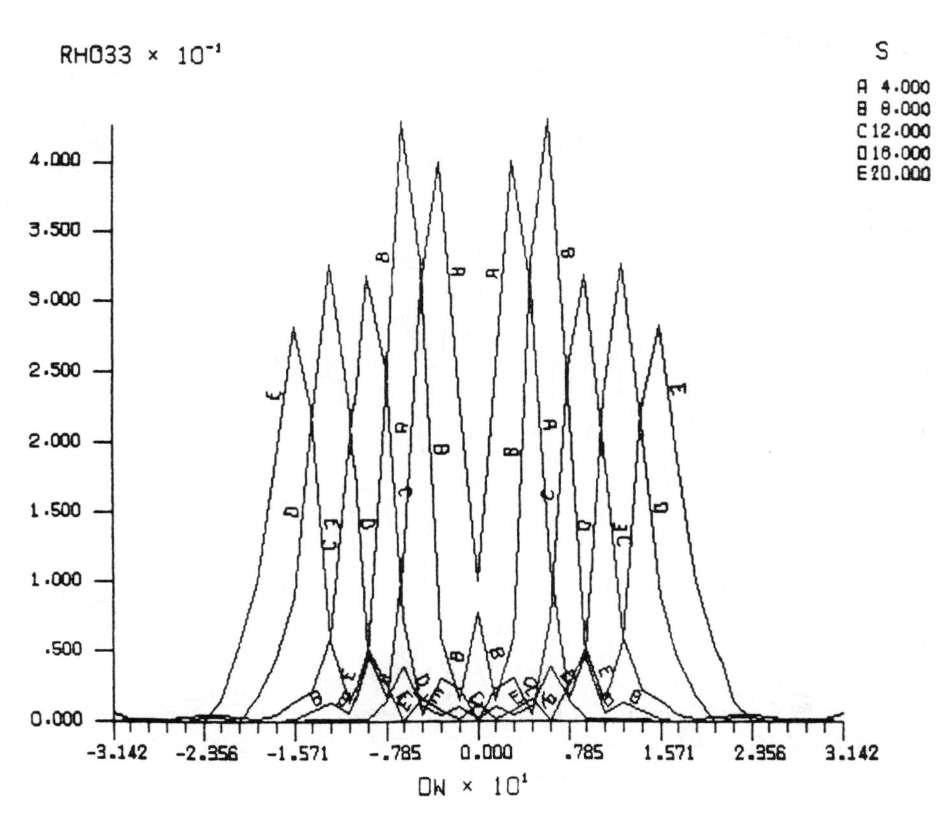

Fig. 3 Graph b

RHO33

SCALE	LOCALE
MIN =	.15500903E-04
MAX =	.42753006

AXIS	TOT.PT	SYM	RANGE	NB.SEL	RANGE OF VALUES
TAU	2048	NO	2017	1	0.543454
ETA	201	NO	31	1	0.015
S	5	NO	(1.5)	5	(4.0.20.0)
DW	161	NO	(61.101)	41	(-31.41593.31.4

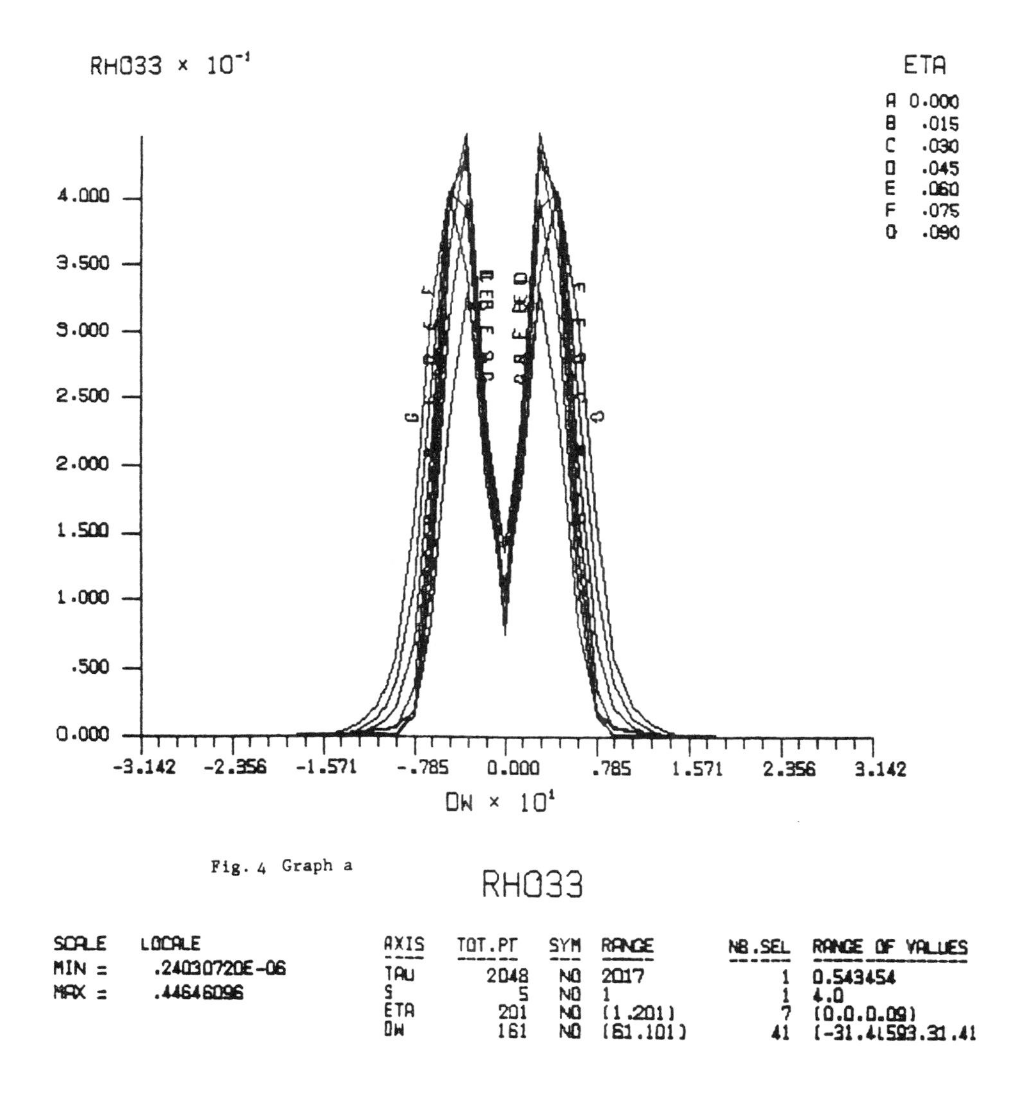

FIG. 4. The role of propagation on the medium response is shown for a given family of areas. In graph a the input area is π while in graph b its a larger one, namely 5π.

LRRH33 SIMULATION NO. 25

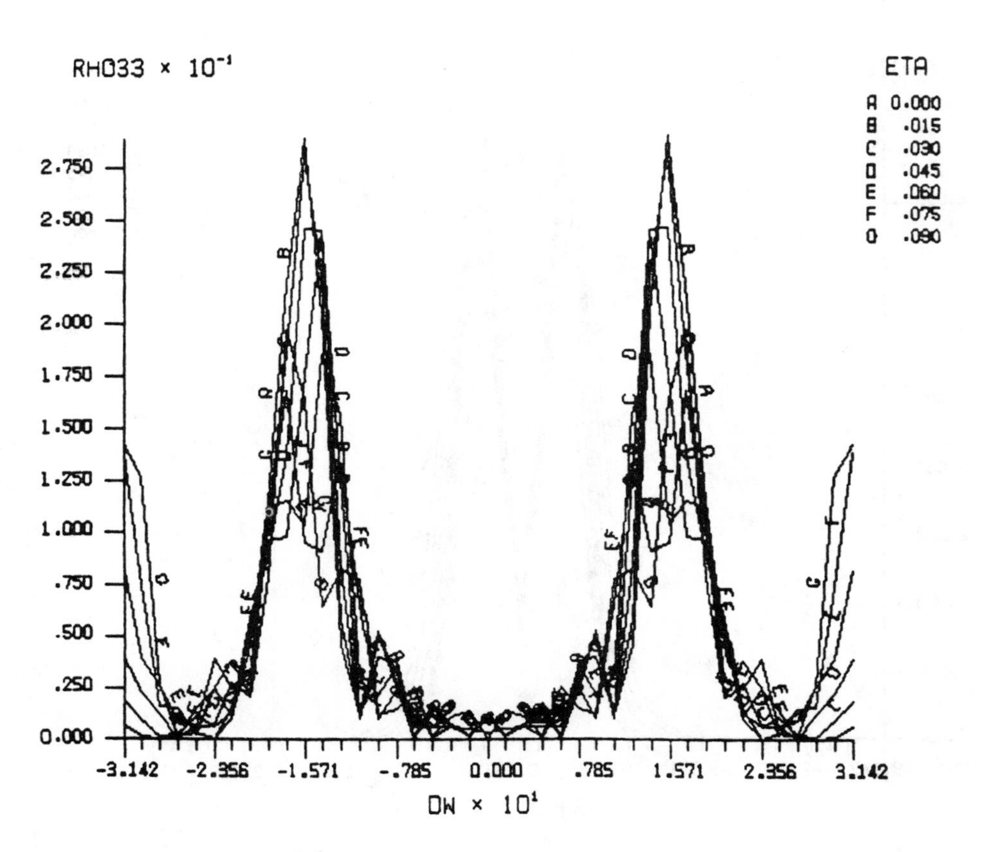

Fig. 4 Graph b

RHO33

SCALE	LOCALE
MIN =	.85441420E-04
MAX =	.28904413

AXIS	TOT.PT	SYM	RANGE	NB.SEL	RANGE OF VALUES
TAU	2048	NO	2017	1	0.543454
S	5	NO	5	1	20.0
ETA	201	NO	(1.201)	7	(0.0.0.09)
DW	161	NO	(61.101)	41	(-31.41593.31.415

DOUBLE SELF-INDUCED-TRANSPARENCY (SIT) IN THREE-LEVEL ABSORBERS†

F.P. Mattar, J.A. DeLettrez, J. Teichmann and J.P. Babuel-Peyrissac
Department of Physics, New York University, New York, NY 10003

ABSTRACT

The goal of this communication is to show that different transmission characteristics are obtained for each pulse shape when diffraction is accounted for even if the time-integrated areas are equal. For weakly overlapped two-field propagation it was argued that each pulse evolution obeys its own area theorem. In our case the pulses are strongly overlapped and they mutually influence each other. We have worked out the transverse variation beyond the physical situation where a specific self-similar Gaussian mode can be assumed. Detailed temporal and frequency reshaping of coherent pulses with finite beam extent in a three-level absorber is reported.

I. INTRODUCTION

Since the classic work of McCall & Hahn[1] in two-level atomic vapor, the subject of coherent Optical Soliton propagation has attracted the interest of many physicists[2-5]. The pulse duration is short compared to the relaxatiom times of the resonant medium. The atomic medium response exhibits inertia, it thus can **not** be characterized by an instantaneous susceptibility, it requires a density matrix formalism. The uniform-plane-wave (UPW) SIT phenomenon is described by the nonlinear pendulum equation also known as the Sine-Gordon equation. To interpret some features in the various observations Newstein & co-workers[6] accounted theoretically for phase and transverse variations. They found out a novel on-exact resonace self-focusing. Self-focusing has always been analyzed off-resonance without any contribution on-resonance. The scalar beam propagation is described by two Schröedinger equations in which the potential is replaced by the nonlinear polarizations. This SIT self-focusing was observed in Na, Ne and I[7]. More recently Konopnicki and Eberly[8] generalized the UPW two-level SIT by addressing the **simul**taneous propagation of two different-wavelength soli**tons** in a three-level absorber which they labeled **simultons**. They specified the conditions under which the pulse shape is preserved and the time-coincidence is maintained. Subsequently, Babuel-Peyrissac et al.[9] studied the effects of diffraction on the simultons evolution. We herein calculate how **time-dependent diffraction** is differently exhibited for pulses of distinct shapes and equal areas.

II. NUMERICAL EXPERIMENTS

II(a). The Physical Model

The three-level atomic configuration is cascade. To insure equal group velocity v_g, the pulse Rabi frequencies ν_R (thus, the areas) and the initial level populations (σ) are restricted; $\nu_{R12}^2+\nu_{R23}^2=4$ and $\sigma_{11,22,33}=2/3$, 1/3 & 0. Both active transitions are absorbing. The coherent exchange of energy between each pulse and its transition as well as cross-frequency to the other pulse reduces significantly the pulse group velocity to a fraction of c the velocity of light in vacuum. Only two transitions $\lambda_{12},\lambda_{23}$ are active, the third λ_{13} is forbidden due to parity consideration. The frequencies of the active transitions are unequal ($\lambda_{12}=0.6\mu,\lambda_{23}=0.3\mu,\lambda_{12}/\lambda_{23}=2$)

† Supported by ARO,AFOSR,NSF,ONR,CEA and NYU Research Challenge Fund

the dipole moments are distinct (μ_{12}=1.0 db, $\mu_{23}=1/\sqrt{2}$ db, $\mu_{12}/\mu_{23}=\sqrt{2}$) leading to $\beta=\mu_{12}^2\lambda_{12}^{-1}/\mu_{23}^2\lambda_{23}^{-1}=1$. However the effective gain lengths are equal ($g_{12}=g_{23}$=62 cm^{-1}). The atomic density N is $1.6028\times10^{12}/cm^3$. The HWHM pulse lengths and beam widths are respectively chosen as τ_p=0.2938ns and r_p=0.03816 cm for both beams. The beam diffraction which is governed by the Fresnel number, F ($F=\pi r_p^2/\lambda L$, with L the length of the cell containing the metallic vapor) is twice as large for the beam interacting with transition $|2\rangle\rightarrow|3\rangle$ than for the one interacting with transition $|1\rangle\rightarrow|2\rangle$.

The interaction is characterized by a single parameter F_g the gain-length Fresnel number. $F_g=\pi r_p^2/\lambda g^{-1}$. Their ratio $\gamma=F_{g12}/F_{g23}=(\lambda_{23}/\lambda_{12})(g_{12}/g_{23}=\beta(\lambda_{23}/\lambda_{12})\neq1$. The competition between diffraction and nonlinear action is different from one beam to the other.

II(b). Definition of Diagnostic Functions

To substantiate longitudinal and transverse reshaping in conjunction with energy cross-depletion and frequency conversion, we recorded the on-axis energy {ECLA,ECLB}, its Fourier transform (on-axis power spectrum: SPFE2); radially-integrated energy (output power: OPOWR) and corresponding Fourier Transform {SPFOP}; time-integrated area {AIRA,AIRB} and time-integrated energy (fluence: E2TORA & E2TORB) radial profiles for four distances of propagation {Z=0, 20.2, 40.4 & 60.6 cm}. The on-axis areas, fluences and effective pulse length {TAUEFF} and beam width {RHOEFF} are exhibited as a function of Z. Three pulse shapes are studied: hyperbolic secant, Gaussian and Lorentzian. The input pulse lengths and areas remain constant. Only the peak Rabi frequencies vary from one shape to the other. The beam profile is Gaussian.

II(c). Numerical Results

In Fig. 1, we contrast the pulse characteristics at four Z: variations of **on-axis Energy** and **output Power** vs. τ; profile of area and fluence vs. ρ. The pulse peaks experience an increasing delay with Z. One can distinguish the fields from each other and note their strong departure from their initial profile. The pulse breakup into multi-pulses that are readily observed as early as Z=20.2 cm. (These multi-pulses illustrate changes of phase and beam narrowing.) The group velocities cease to remain equal: Lasera and Laserb peaks clearly do not coincide at Z=60.6 cm while the acquired radial structure illustrates how much the beam profiles are distorted. **Both temporal-synchronization and shape-preserving features of the simulton are lost.**

In Fig. 2 we show for the sech pulse at the four Z the **on-axis power spectrum** both on-axis (at R=0) and off-axis (for R=0.036) and the **Fourier transform of the output power.** The two functions are very different substantiating the crucial role of diffraction. Significant distortion for a given characteristic is exhibited in the frequency domain as a function of Z. There is a marked difference between the responses of Lasera and Laserb.

In Fig. 3 we display as a function of Z for each shape four features: the on-axis time-integrated area, effective beam radii and effective pulse lengths. Self-focusing appears for each shape and for each transition but its threshold varies. For the three shapes, Lasera area (fluence) self-focuses at a shorter Z than Laserb, however, its magnification is smaller. The distance over which on-axis enhancement takes place is considerably shorter for Lorentzian than for Gaussian or hyperbolic-secant pulses. The effective beam widths are seen to narrow then expand as evidence of self-focusing & self-defocusing. A second self-focusing may arise for the Lorentzian. The effective pulse length is shown to first broaden for the three shapes then it exhibits a stabilization over Z.

In Fig. 4, we show three composite graphs of Fourier transform of the output power, at those Z, one for each shape. The rate of peak absorption and frequency narrowing over Z of the Lorentzian pulse is considerably larger than it is for hyperbolic secant and Gaussian pulses. Multi-lobe formation is shown as Z increases: the Fourier transform first develop shoulder then goes through zeros.

III. CONCLUSION

We have developed a methodology with which we examine thoroughly field and matter characteristics and analyze their reshaping and distortion in R, Z, T and ω resulting from the co-propagation of strong-overlapping beams in a three-level medium.

REFERENCES

1. S.L. McCall & E.L. Hahn, Phys. Rev. 183, 457 (1969).
2. F.A. Hopf & M.O. Scully, Phys. Rev. 179, 399 (1969).
3. A. Icsevgi & W.E. Lamb, Jr., Phys. Rev. 185, 517 (1969).
4. G. Lamb, Jr., Rev. Mod. Phys. 43, 94 (1971); J.C. Diels & E.L. Hahn, Phys. Rev. A10, 2501 (1974); E. Courtens, in **Laser Handbook**, ed., F.T.Arecchi & E.O. Shultz-DuBois, (North-Holland, 1972) pp. 493-556; R.E. Slusher, **Progress in Optics XII**, ed. E. Wolf (North-Holland, 1974) pp. 53-100.
5. [a] F.Y.F. Chu et al., Phys. Rev. A12, 2060 (1975); [b] E. Hanamura, J. Phys. Soc. Japan 37, 1598 (1974); [c] N. Tan-No et al., Phys. Rev. A12, 159 (1975); [d] J.N.Elgin, Phys. Lett. A80, 140 (1980); [e] J. Satsuma et al., Suppl. Progr. Theor. Phys. 55, 284-306 (1974); [f] V.E. Zakharov et al., Sov. Phys. JETP 34, 615 (1971); [g] M.G. Raymer et al., Phys. Rev. A24, 1980 (1981); [h] J.L.Carlsten et al., J. Stat. Phys. 39, 615 & 6 (1986); [i] D.J.Kaup, Physica D19, 125 (1986); [j] J.R.Ackerhalt et al., Phys. Rev. A33, 3185 (1986; and [k] J.C.Englund et al., Phys. Rev. Lett. 57, 2661 (1986).
6. [a] M.C. Newstein et al., **Progress Report JSEP-TAC: Quantum Electronics & Optics Program of the Microwave Research Institute,** Polytechnic Institute of New York #36 (1971) pp. 140-144; #37 (1972) pp. 117-120; [b] M.C. Newstein et al., IEEE J. Quantum Electron. QE-10, 743 (1974); and [c] M.C. Newstein et al., IEEE J. Quantum Electron. QE-13, 507 (1977).
7. H.M.Gibbs et al., Opt. Commun 18, 199 (1976); W. Krieger et al., Z. Phys. B25, 297 (1976); H.M. Gibbs et al., Phys. Rev. Lett. 37, 1743 (1976) & J.J.Bannister et al., Phys.Rev.Lett. 44, 1062 (1980).
8. M.J. Konopnicki and J.H. Eberly, Phys. Rev. A24, 2567 (1981).
9. [a] J.P. Babuel-Peyrissac et al., APS/DEAP 1984 post deadline paper#S18; [b] F.P.Mattar et al., in **Modeling & Simulation XVII**, ed. W.G. Vogt & M.H. Mickle, (Instr.Soc.Am., 1987) pp.1387-1558.

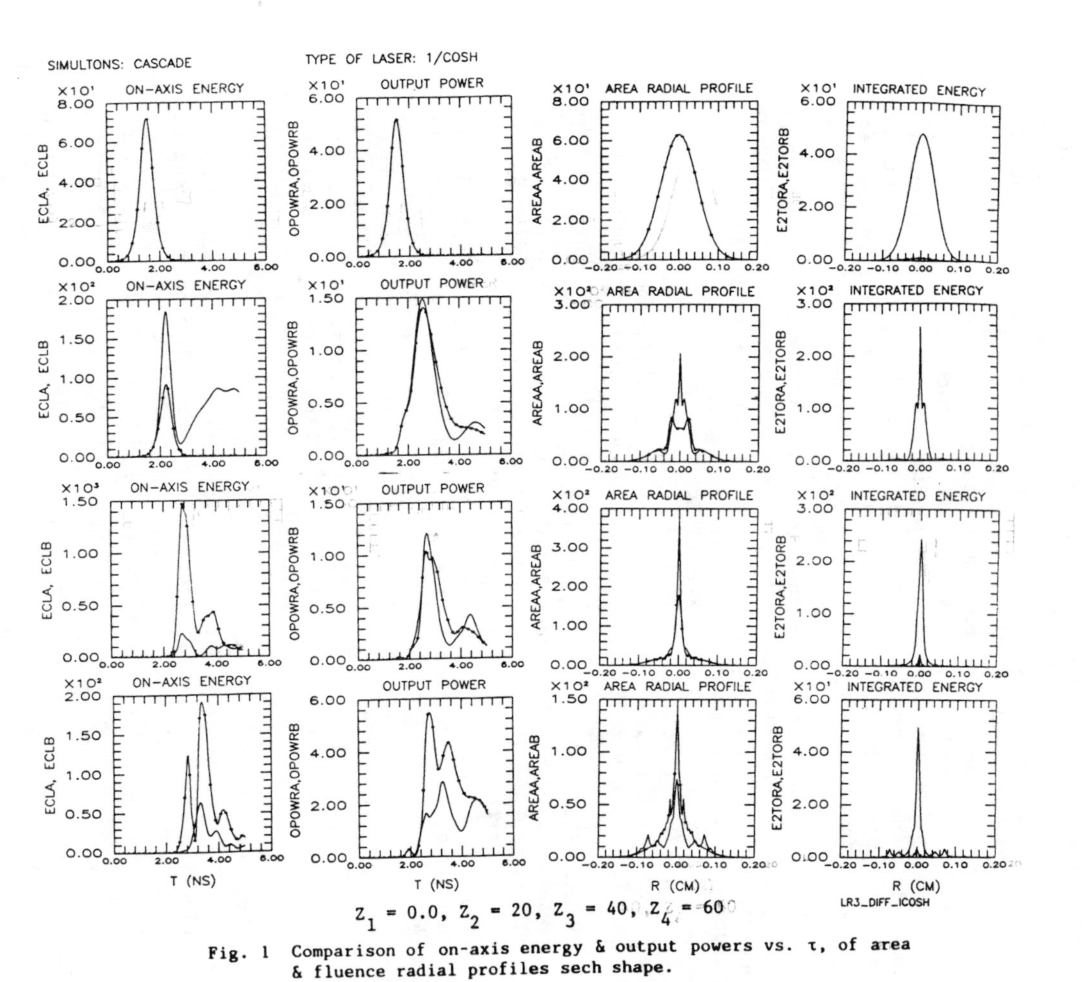

$Z_1 = 0.0,\ Z_2 = 20,\ Z_3 = 40,\ Z_4 = 60$

Fig. 1 Comparison of on-axis energy & output powers vs. τ, of area & fluence radial profiles sech shape.

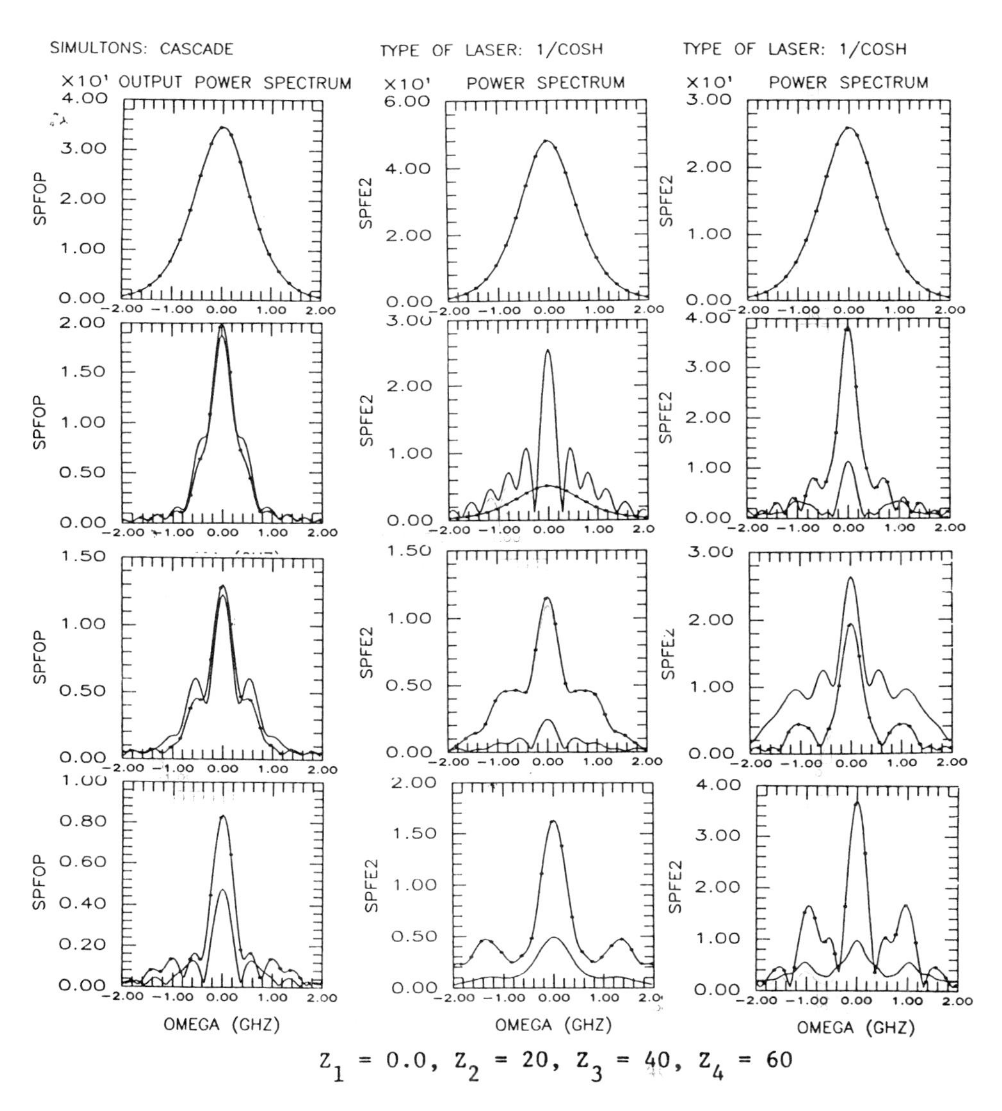

$Z_1 = 0.0,\ Z_2 = 20,\ Z_3 = 40,\ Z_4 = 60$

Fig. 2 Comparison of output power spectrum & of power spectrum for on- & off-axis (ρ=0 & .036 cm) for the sech pulse.

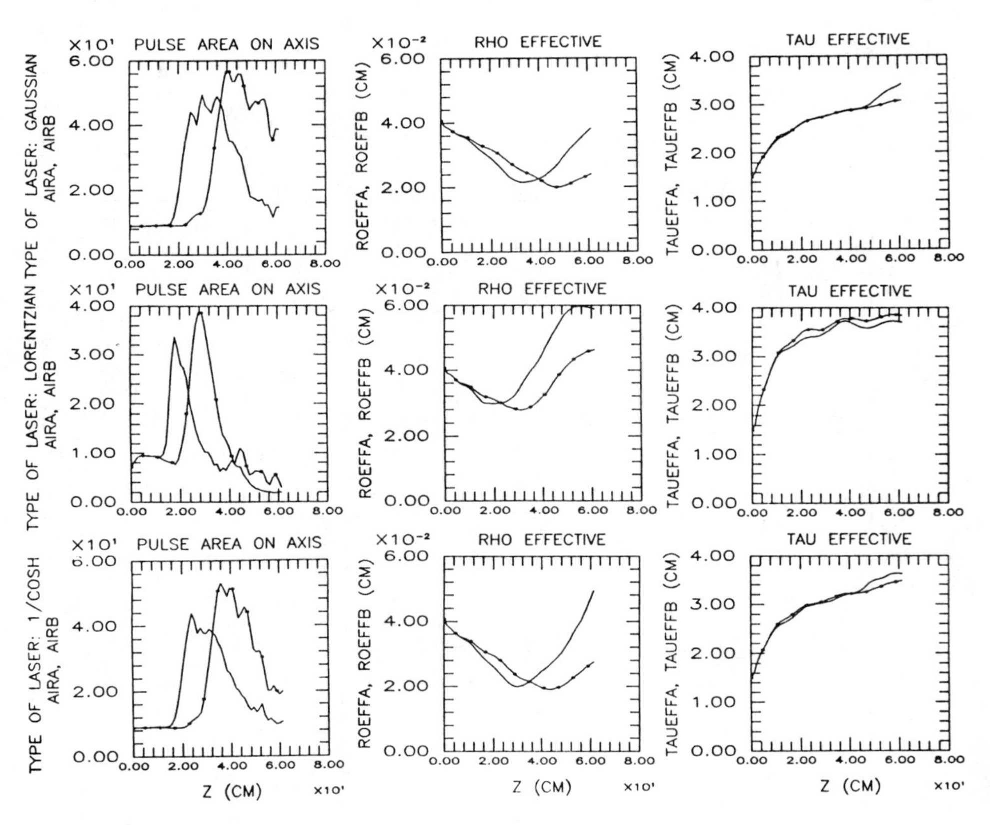

Fig. 3 Contrast vs. Z of on-axis area, effective beam radius & pulse length for Gaussian, Lorentzian & sech shapes

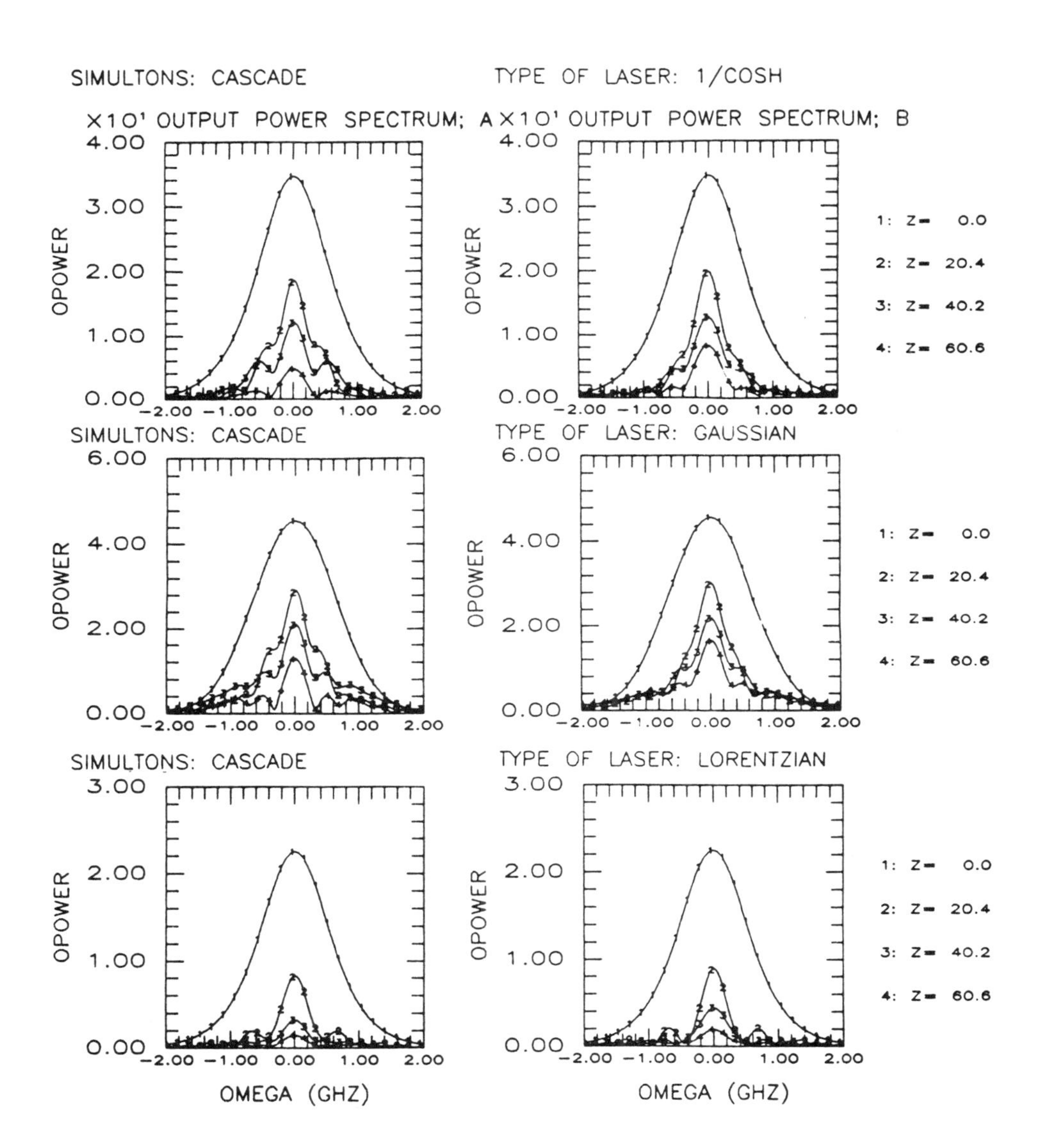

Fig. 4 Comparison of output-power spectrum and of on-axis power spectrum for hyperbolic-secant, Gaussian and Lorentzian shapes

TRANSVERSE ON-RESONANCE ASYMPTOTIC RESHAPING IN COHERENT PROBE SOLITON FORMATION IN A THREE-LEVEL SYSTEM†

Farrès P. Mattar
Department of Physics, New York University, New York, NY 10003
and George R. Harrison Spectroscopy Laboratory
Massachusetts Institute of Technology, Cambridge, MA 02139

ABSTRACT

The coherent frequency conversion between on-resonance pump and probe beams is examined in the physical situation where the nonlinear gain balances the diffraction for the probe transition. The semi-classical paraxial-Maxwell-Bloch formalism is adopted as introduced in propagational studies in two-level Self-Induced-Transparency (SIT)[1]. The probe builds up from an input deterministic seed. The asymptotic stabilization of the probe area is reported in conjunction with pump depletion and diffraction rigorously accounted for. Without diffraction the soliton does not appear. The Laplacian term (which describes transverse coupling) is included.

I. INTRODUCTION

The coherent on-resonance co-propagation of two beams in a three-level system[2] leads in the asymptotic regime to a soliton. This research was motivated by swept-gain superradiance asymptotic evolution[3,4]. Diffraction is identified as a key element in soliton formation and as the source of self-phase modulation[5] and transverse energy flux[6]. For long interaction lengths the probe (pr) and pump (p) beams ($\alpha_{pr}\ell >> \alpha_p \ell >> 1$), the calculations show a clear **sequence** of events: the **probe** experiences an initial **buildup**, gain **saturation**, and then a **stabilization** which implies the creation of a soliton. All this occurs while the pump depletes.

For a probe soliton to evolve, the accumulated diffraction loss $\kappa\ell$ has to balance out the nonlinear amplification $g\ell$. This condition requires a near-unity gain-Fresnel number $F_g=g/\kappa \geq 1$ with g the gain $g=\alpha(c\tau_p)$; α the Beer's length; τ_p the pulse length and $\kappa=\pi r_p^2/\lambda$ where r_p is the beam width; otherwise, transverse self-lensing instabilities predominate and the beam quality collapses. In addition several conditions must be satisfied: $T_2 << \tau_c$ (with τ_c being the cooperation time); $\alpha=g/T_2$; $\tau_E=(c\kappa)^{-1}$ (τ_E is an escape time associated with the diffraction length κ^{-1}); $\tau_s=\kappa T_2/g=T_2/F_g$; and $\tau_c^2=T_2/gc$ (i.e., $\tau_c^2=\tau_E\tau_s$). Both $\tau_s=\tau_R(\kappa z)$ and $F_g=F(gz)$ are independent of z unlike τ_R and F. **This asymptotic evolution into a soliton is equivalent to self-trapping.**

The soliton is defined as a pulse for which the time-integrated-area (THETA), time-integrated-energy fluence (ENER), radially-integrated energy output power (OPOWR), effective temporal length τ_{eff} (TAUEFF) and effective spatial width ρ_{eff} (RHOEFF) do **not** vary appreciably over several optical thicknesses in the asymptotic regime.

† Supported by ARO, AFOSR and NSF.

II. THREE-LEVEL ASYMPTOTIC CALCULATIONS

While achieving a unity gain-to-loss ratio (i.e., a gain-transition to diffraction-loss balance) for the probe one is confronted with a large pump Beer's length. Even though the pump sees an optical thickness ($\alpha_p \ell > 7$) which would cause an SIT self-focusing[7] with a magnification larger than seven, there is essentially **no** significant on-axis pump energy enhancement, leakage to probe. Initially, the on-axis pump energy experiences an enhancement of 1.69; subsequently it subdues and depletes, its trend to self-focusing is quenched by the probe buildup. The probe displays the same z-independent on-axis area stabilization as its two-level "swept-gain superradiance" counterpart[4]. Its area reaches a peak of about 5π and then decays down to an asymptotic stabilized value of 4π after some minor oscillations. Without diffraction there can not be a soliton. The probe pulse area in normalized Rabi frequency units is twice as large as the pump. The probe to pump fluence ratio is eight while it was initially one-tenth of a thousand.

The on-axis asymptotic stabilization of the probe area and the continuous depletion of the pump **area** is shown in Fig. 1 while their three-dimensional distribution is shown in Fig. 2. The **temporal and transverse reshaping dynamics** associated with the output power is shown in Fig. 3.

By strengthening diffraction one obtains a **different** threshold for asymptotic area stabilization. The peak probe output power in Rabi units is larger than that of the pump, while its pulse length is significantly shorter. In the asymptotic regime, the probe peak precedes the pump peak.

The probe **asymptotic** area-stabilization **generalizes** the studies of (i) a swept-gain two-level superfluorescence (SF) solitary wave formation when the gain compensates linear loss, in the UPW regime[2] or when diffraction and density variation effects are accounted for[3]; and of (ii) pump depletion and diffraction effects for a three-level SF[8]. The observation of a FIR CO_2-pumped swept-gain SF soliton in CH_3F, D_2 and NH_3 by Rosenberger[8] has enticed us to study the process as accurately as possible.

Universal **transmission characteristics** of the probe versus pump area (Fig. 4) have been compiled as (i) a function of the propagation distances η for specific radii ρ namely on-axis (graph a) or (ii) across the beam for given η well within the asymptotic regime (graph b). **These energy conversion transmission graphs exhibit a multi-value dependence** since the probe stabilizes after building up while the pump continues to deplete.

III. CONCLUSION

We have calculated the effect of diffraction and pump dynamics on the coherent probe on-resonance amplification and Raman-like frequency conversion. **The probe exhibits asymptotically a solitary wave whose area is stabilized versus the propagational distance while the pump does not cease to deplete.** The pump beam narrows and the probe experiences beam blooming and pulse compression. Neither linearization of the density matrix nor a self-similar radial profile were assumed.

REFERENCES

1. S.L. McCall & E.L. Hahn, Phys. Rev. 183, 457 (1969).
2. F.P. Mattar, **Physical Mechanism in Coherent On-Resonance Propagation of Two Beams in a Three-Level System**, sub. Proc. ILS-3.
3. R. Bonifacio et al., Phys. Rev. A12, 2568 (1975).
4. C.M. Bowden & F.P.Mattar, SPIE 288, 364 (1981), & 369, 151 (1983).
5. [a] H.A. Haus, Appl. Phys. Lett. 8, 128 (1966) & **Waves and Field in OptoElectronics** (Prentice Hall, 1984); [b] **Progress in Quantum Electronics**, Vol. 4, ed. J.H. Sanders & S. Stenholm (Pergamon Press, 1975): (i) J.H. Marburger pp. 35-110 and (ii) Y.R. Shen pp. 1-34; [c] Y.R. Shen, **The Principles of Nonlinear Optics**, (Wiley,1984); [d] M.Lax et al., J. Appl. Phys. 52, 109-125 (1985).
6. F.P.Mattar & M.C.Newstein,IEEE J. Quantum Electron.13, 507(1977).
7. F.P. Mattar, et al., in **Modeling and Simulation XVII**, ed. by W.G. Vogt & M.H. Mickle, (Instr. Soc. Am., 1987) pp. 1387-1558.
8. [a] A.T. Rosenberger and T.A. DeTemple, Phys. Rev. A24, 868 (1981); and [b] A.T. Rosenberger, H.K. Chung and T.A. DeTemple, IEEE J. Quantum Electron, QE-20, 523 (1984).

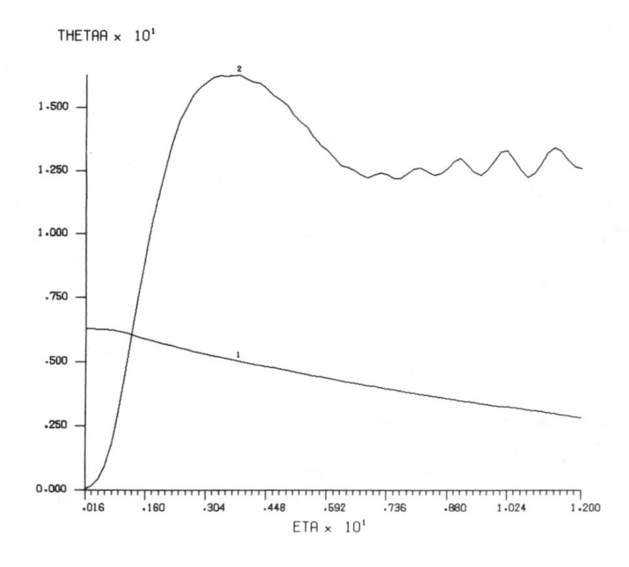

FIG. 1. Comparison vs. η of on-axis pump & probe areas.

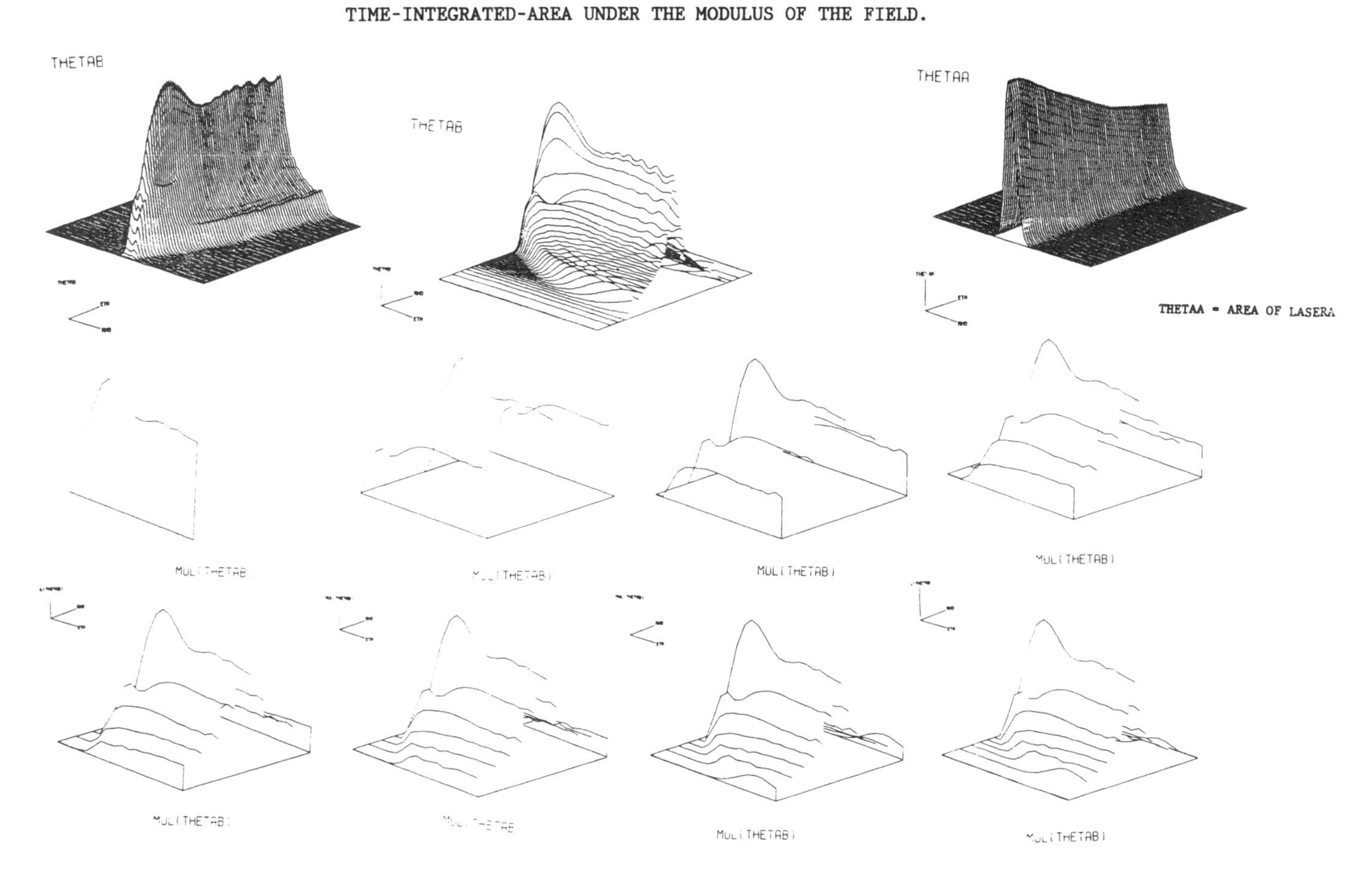

FIG. 2. Three-dimensional plots vs. ρ and η of the probe and pump areas.

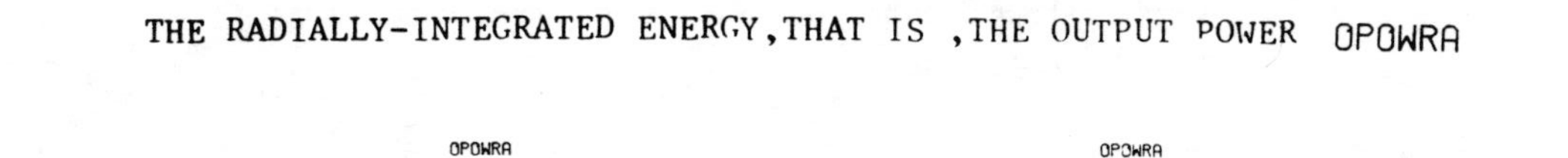

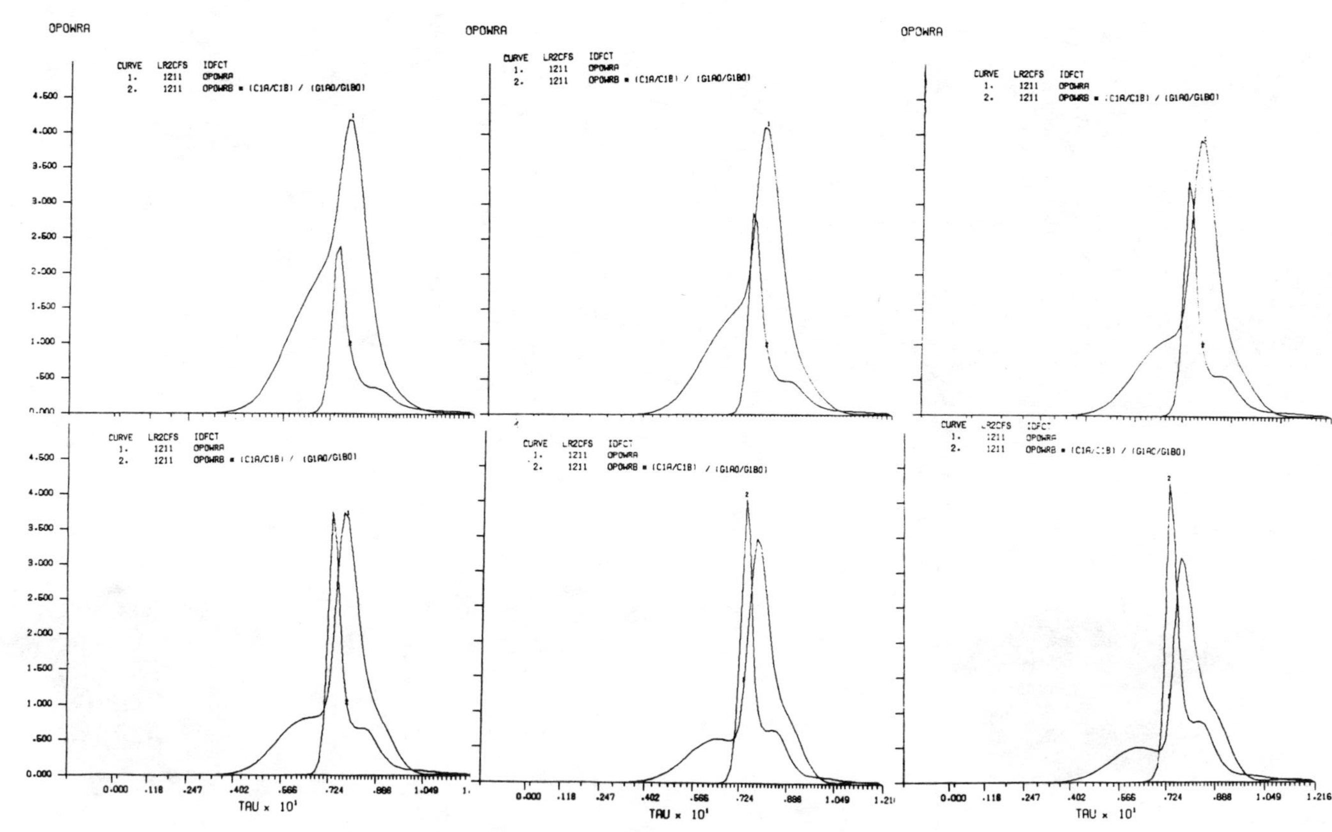

FIG. 3. Comparison of pump & probe output powers vs. τ for different η.

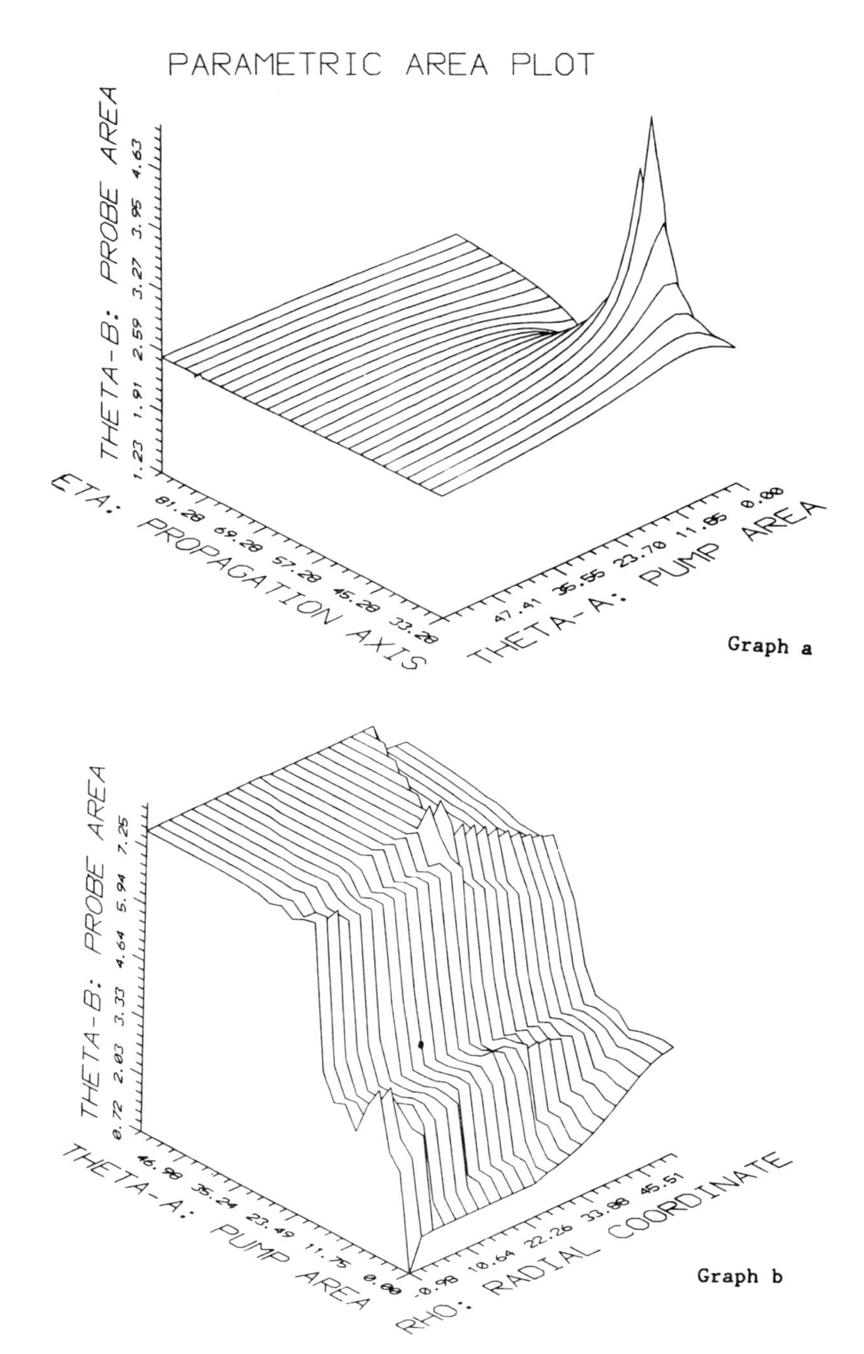

FIG. 4. Phase diagrams of pump & probe areas (a) on-axis; (b) across the beam in the asymptotic regime.

TRANSVERSE ON-RESONANCE RESHAPING IN COHERENT PUMP SOLITON ASYMPTOTIC EVOLUTION IN A THREE-LEVEL SYSTEM†

Farrès P. Mattar
Department of Physics, New York University, New York, NY 10003
and George R. Harrison Spectroscopy Laboratory
Massachusetts Institute of Technology, Cambridge, MA 02139

ABSTRACT

The coherent frequency conversion between on-resonance pump and probe beams is examined in the physical situation where the medium nonlinearity balances diffraction for the pump transition. The probe soliton is destroyed and a **pump soliton** is formed. This **pump depletion reversal** is reported in a self-consistent semi-classical calculation in which diffraction is rigorously accounted for.

I. INTRODUCTION

The observation of a **pump soliton**[1] in Stimulated Raman Scattering (SRS), is obtained when an optical phase shift imposed on the Stokes beam is initiated. Understanding the result has motivated a number of theoretical uniform-plane-wave (UPW) analyses based on either the solution of the SRS equations formulated to simulate collisional coherence damping[2] or through Inverse-Scattering-Theory (IST)[3] in conjunction with a linear loss. The question of Stokes initiation by four-wave mixing[4] and through quantum fluctuations[5] was treated in the UPW regime.

Using self-consistent computational methods published previously [6], we studied the **on-resonance** propagation of two pulses in a three-level atom[7], one strong pump and one initially weak probe. We have established that **probe solitary wave[8] behavior can occur here provided the medium nonlinear gain it experiences balances the linear diffraction.**

We examine the conditions whether or not the pump can display solitary behavior in this resonant study. Diffraction is critical in soliton formation. It is the source of self-action phenomena[9] such as self-phase modulation, transverse energy flux and self-lensing processes.

The probe (pr) and pump (p) interaction lengths ($\alpha_{pr}\ell >> \alpha_{p}\ell >> 1$), are long. The calculations show a clear **sequence** of events: the **probe** experiences an initial **buildup**, gain **saturation**, and then a **stabilization** which implies the creation of a soliton. During this time the pump depletes continuously. Further propagation results in **probe decay** and **destruction** of its soliton concurrent with a **pump recovery** similar to the anomalous Raman **pump-depletion reversal**. This is followed by **pump stabilization**, or **'pump soliton'** assuming a sufficient optical thickness for the pump. For further optical thickness **the pump soliton is destroyed.** The pump area experiences on-axis enhancement (small self-focusing). **The pump soliton formation is repeated.** However, the probe soliton does not reproduce itself. We stress the **orderly** evolution of these events. The probe experiences pulse compression and beam blooming. The pump beam narrows.

The soliton is defined as a pulse for which the time-integrated-area (THETA), time-integrated-energy fluence (ENER), radially-integrated energy output power (OPOWR), effective temporal length τ_{eff}

† Supported by ARO, AFOSR and NSF.

(TAUEFF) and effective spatial width ρ_{eff} (RHOEFF) do **not** vary appreciably over several optical thicknesses in the asymptotic regime.

II. THREE-LEVEL ASYMPTOTIC CALCULATIONS

In this regime the **probe soliton**[8] is destroyed. A pump soliton formation occurs at the pump frequency. **The accumulated nonlinear action experienced by the pump compensates its diffraction.** The probe incurs depletion while the pump shows a depletion reversal then displays an asymptotic area-stabilization. This calculation is the on-resonance equivalent of the Raman soliton[1,2].

When appropriate phase variations take place, the pre-stage of an anomalous pump recovery is set up. The probe stabilization ceases. The probe depletes its energy back to the pump, but the pump begins to recover while the probe depletes. The pump is renewed until its area, fluence and beam width ρ_{eff} asymptotically stabilize. The probe beam continues to be broader than the pump. The on-axis pump soliton area is smaller than its initial '2π' value which is in direct contrast to what evolves for the probe: its input area was 0.02π while its maximum 3.31π. This represents a significant gain factor.

The pump first undergoes stripping, a gradual narrowing and depletion then its rate of absorption diminishes until it vanishes and then reverses itself.

The conditions which support a pump soliton destroy those for the probe wave and vice versa; we cannot have both at the same time. In Fig. 1 the pump and probe **areas** are plotted versus ρ and η side by side. The pump depletes while the probe grows. After reaching an absolute peak the probe saturates, ceases to grow, decreases and oscillates about a quasi-steady-state value until it stabilizes. When appropriate phase variations evolve the probe **departs** significantly **from its asymptotic** value, depletes and experiences radial distortion. **The probe and pump area** profiles experience **different rates of change (growth or decay)** along η. In the left-hand graph, the probe area displays a well-defined profile on- and near-center stabilization as a solitary wave. The pump, in the right-hand graph, is going through reductions in the depletion rate. Afterwards, the probe behavior becomes erratic while the pump depletion is interrupted. The pump recovers, stabilizes and forms a channel of essentially constant width. Both beams have departed from their input profiles.

In Fig. 2, a comparison of the **on-axis** probe and pump areas or fluence exhibits a **non-coincidence** of the stabilizations. **The two solitary waves, first probe then pump, occur in sequence.** The succession of probe amplification to absorption leading to probe area oscillations are drastic. This leads to **only** a pump soliton while this **on-resonance theory** leads to **both** probe and pump solitons.

In Fig. 3, we establish for **even longer propagation** distances that the **pump refocuses** then its area is **re-stablized** as expected. However, the **probe soliton does not reproduce** itself. The probe area continues to deplete and goes through wild oscillation.

Figure 4 shows that τ_{eff}, defined in terms of its output power, and ρ_{eff}, defined in terms of its fluence, saturate and stabilize over an extended optical thickness.

Should one plot the on-axis **transmission characteristics** of the probe versus pump area as a function of the propagation distances η

one obtains a multi-value dependence since the pump recovers and stabilizes while the probe depletes reversing its traditional role.

III. CONCLUSION

We have calculated the effect of diffraction and pump dynamics on the coherent probe on-resonance amplification and Raman-like frequency conversion. **In addition to the probe soliton one obtains a pump soliton.** The pump soliton occurs exclusively. The soliton formation can only occur whenever the diffraction loss equals the medium nonlinear medium action.

REFERENCES

1. J.L. Carlsten et al., SPIE Vol. 380, 201-207 (1983).
2. K.J. Drühl et al., Phys. Rev. Lett. 51, 1171 (1983); and R.G. Wenzel et al., J. Stat. Phys. 39, 615; and 621 (1985).
3. D.J. Kaup, Physica 19D, 621 (1986).
4. J.R. Ackerhalt & P.W. Milonni, Phys. Rev. A33, 3185 (1986).
5. J.C. Englund & C.M. Bowden, Phys. Rev. Lett. 57, 266 (1986).
6. F.P. Mattar, Appl. Phys. 17, 53 (1978); F.P. Mattar & J.H. Eberly in Laser-Induced Processes in Molecules: Physics & Chemistry, ed. K.L. Kompa & S.D. Smith (Springer-Verlag, 1970) pp.61-65; and F.P. Mattar & M.C. Newstein, Comp. Phys. Commun. 20, 139 (1980).
7. F.P.Mattar, **Physical Mechanism in Coherent On-Resonance Propagation of Two Beams in a Three-Level System**, sub. Prog. Quantum Electron.
8. [a] F.P. Mattar, **Transverse Resonant Asymptotic Reshaping in Coherent Probe Soliton Formation in a Three-Level System**; and [b] F.P.Mattar, et al., in **Modeling and Simulation XVII**, ed. by W.G. Vogt & M.H. Mickle, (Instr.Soc.Am., 1987) pp.1387-1558.
9. [a] H.A. Haus, Appl. Phys. Lett. 8, 128 (1966) & **Waves and Field in OptoElectronics** (Prentice Hall, 1984); [b] **Progress in Quantum Electronics**, Vol. 4, ed. J.H. Sanders & S. Stenholm (Pergamon Press, 1975): (i) J.H. Marburger pp.35-110 and (ii) Y.R. Shen pp.1-34; [c] Y.R. Shen, **The Principles of Nonlinear Optics**, (Wiley, 1984); [d] M.Lax et al., J. Appl. Phys. 52, 109-125 (1985); and [e] F.P. Mattar & M.C. Newstein,IEEE J. Quantum Electron.13, 507(1977).

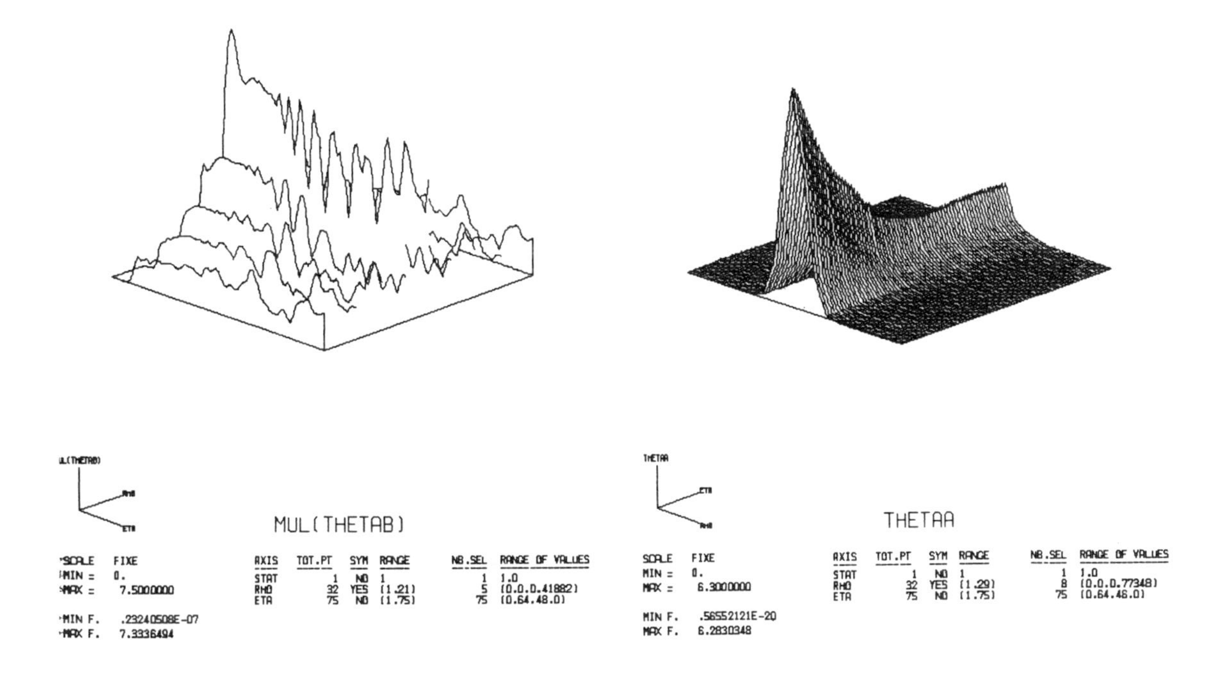

FIG. 1. Isometric plots of pump and probe areas vs. ρ and η.

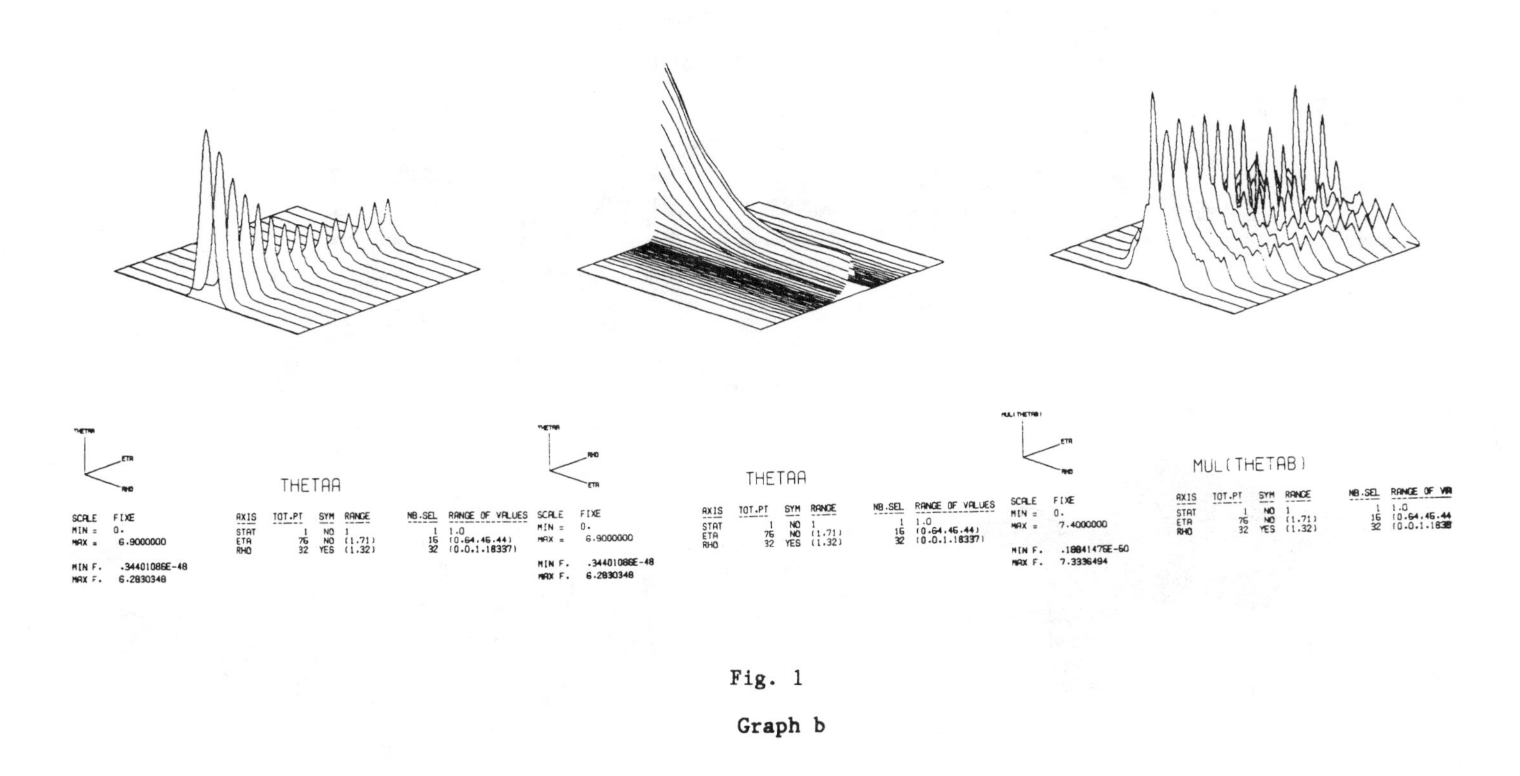
THETAA
ETA
RHO
SCALE FIXE
MIN = 0.
MAX = 6.9000000
MIN F. .34401086E-48
MAX F. 6.2830348
THETAA
AXIS TOT.PT SYM RANGE
STAT 1 NO 1
ETA 76 NO (1.71)
RHO 32 YES (1.32)
NB.SEL RANGE OF VALUES
1 1.0
16 (0.64.46.44)
32 (0.0.1.18337)
THETAA
RHO
ETA
SCALE FIXE
MIN = 0.
MAX = 6.9000000
MIN F. .34401086E-48
MAX F. 6.2830348
THETAA
AXIS TOT.PT SYM RANGE
STAT 1 NO 1
ETA 76 NO (1.71)
RHO 32 YES (1.32)
NB.SEL RANGE OF VALUES
1 1.0
16 (0.64.46.44)
32 (0.0.1.18337)
MUL(THETAB)
ETA
RHO
SCALE FIXE
MIN = 0.
MAX = 7.4000000
MIN F. .18841479E-60
MAX F. 7.3336494
MUL(THETAB)
AXIS TOT.PT SYM RANGE
STAT 1 NO 1
ETA 76 NO (1.71)
RHO 32 YES (1.32)
NB.SEL RANGE OF VA
1 1.0
16 (0.64.46.44
32 (0.0.1.1838

Fig. 1

Graph b

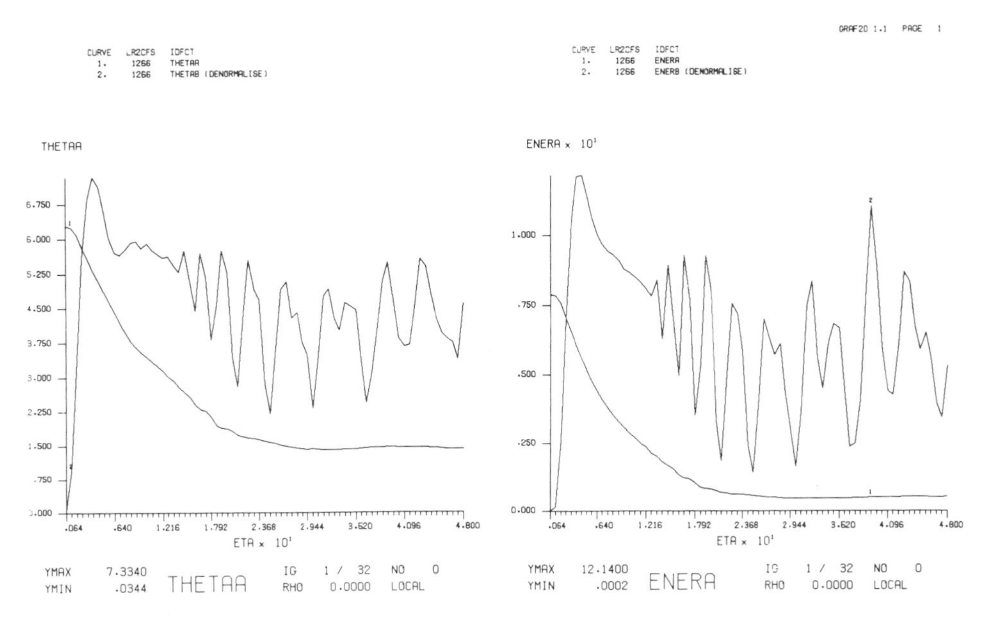

FIG. 2. On-axis pump and probe pulse areas.

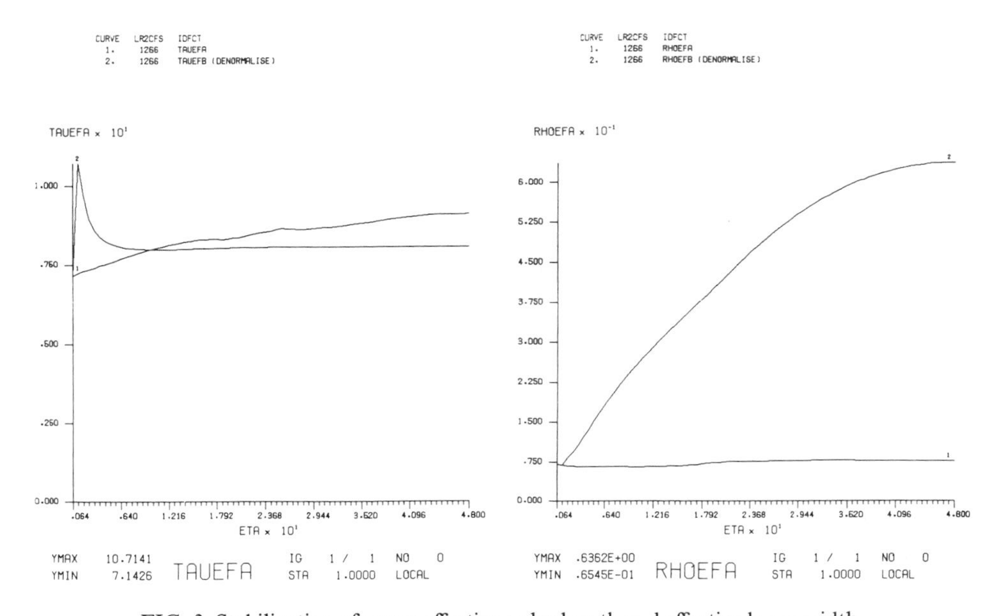

FIG. 3. Stabilization of pump effective pulse length and effective beam width.

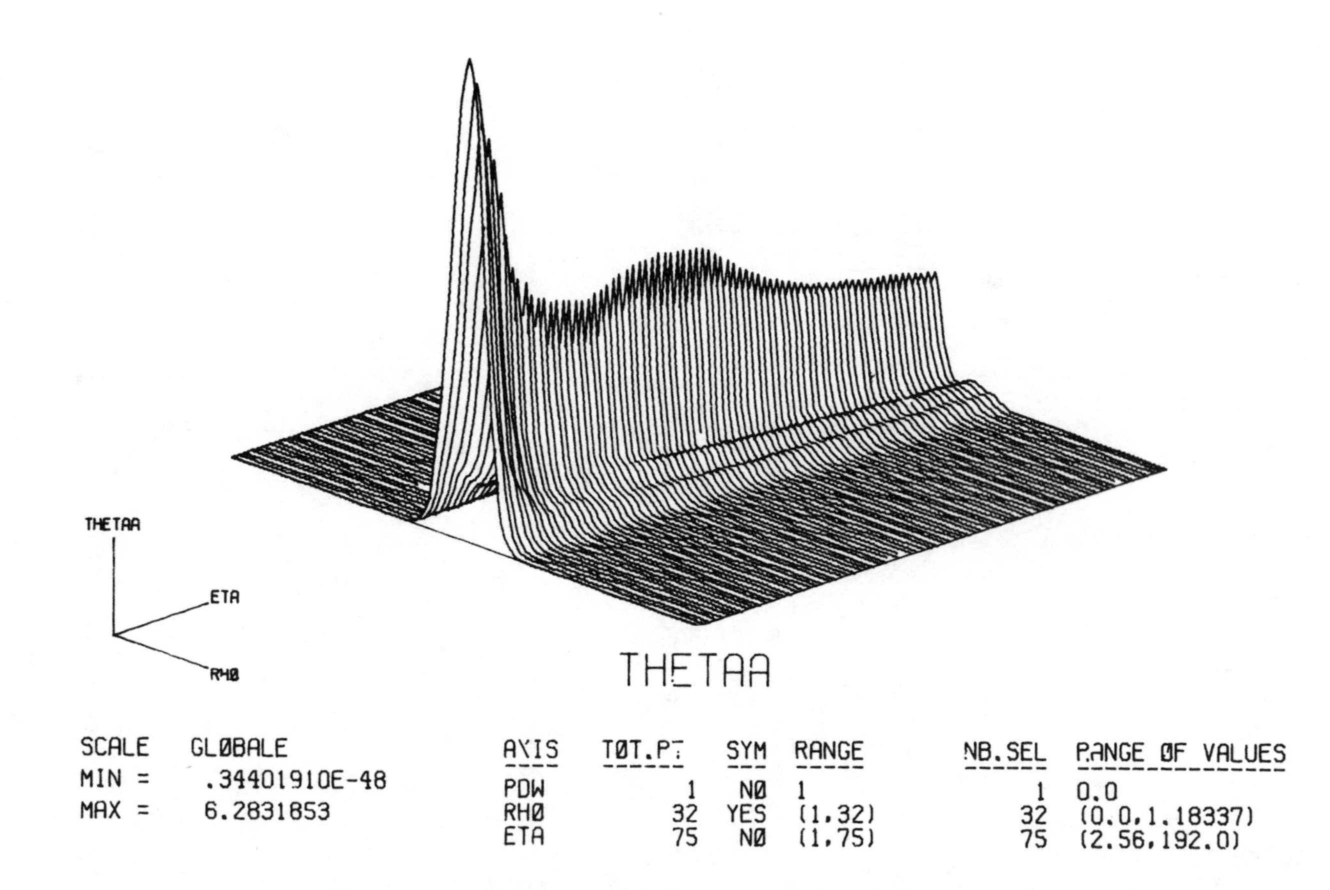

FIG. 4. Pump soliton formation, destruction and renewal.

PHYSICAL MECHANISMS IN COHERENT ON-RESONANCE PROPAGATION OF TWO DIFFERENT-WAVELENGTH BEAMS IN A THREE-LEVEL SYSTEM*

Farrès P. Mattar
Department of Physics, New York University, New York, NY 10003
and George R. Harrison Spectroscopy Laboratory
Massachusetts Institute of Technology, Cambridge, MA 02139

ABSTRACT

The coherent energy transfer from a strong pump beam at a given frequency to an initially weak probe beam injected at another frequency is reported. This exchange is achieved optimally by matching velocities to maintain maximum pulse overlap throughout the motion. The central role of pump depletion, transverse effects and time-dependent phase variations in the co-propagation in a medium is revealed.

I. INTRODUCTION

Previous analyses of coherent on-resonance propagation have included either two-field three-level in the uniform-plane-wave (UPW) [1] or transverse effects in one-field two-level[2] but never both. This study reports simultaneous treatment of both effects. The calculation strives to achieve rigorous analysis of this nonlinear interaction with maximum accuracy and minimum computational effort by introducing adaptive stretching and rezoning transformations[3].

The primary objective of this paper is to show how the transition nonlinearity (absorptive for the pump and amplifying for the probe) changes the transverse profile of the beams and how the diffraction acts on the distorted shape. We have derived, rather than assumed, the transverse field variation (TFV). We have determined the TFV such as would develop for arbitrary input pulses into an initially grounded three-level system. The longitudinal profile of the pulses gets distorted by interaction with the medium, while the atomic transient dynamics is altered by the two different wavelength pulses. The modified medium then re-affects the field profile.

In this study we examine diffraction in conjunction with the propagation of two **strongly-overlapped** pulses of distinct areas in a three-level system with a λ configuration initially at ground state.

II. THE PHYSICAL MODEL

The semiclassical formalism (which amounts to representing the field classically and describing the atomic system quantum-mechanically) is adopted. The pulse length is considerably shorter than any depopulation or dephasing relaxation times. **This type of interaction originated with the discovery of Self-Induced-Transparency (SIT)**[4].

The equations of motion consist of two paraxial-Maxwell equations coupled through a density matrix. The medium response is not instantaneous. The polarization P_a in the pump transition is driven by both the pump E_a and probe E_b fields. With Q a pseudo-quadrupole, the product E_bQ corresponds to an oscillation at the pump frequency. With Q* the complex conjugate of Q, the product E_aQ^* contributes to the temporal rate of change of P_b the polarization associated with the probe transition. Frequency conversion from pump to probe is studied: the pump loses energy and the probe acquires it and grows.

* Supported by ARO, AFOSR and NSF.

The pump evolves as a SIT pulse to zeroth order in δ. The probe is similar to superfluorescence SF emission predicted by Dicke[5] and observed by Feld et al.[6]. The probe builds up as first order in δ, whereas the feedback of the probe onto the pump is of second order δ^2. As soon as the probe grows, the pulses lose their independence.

For optimum probe build-up to occur, conditions must favor an efficient exchange of energy between pump and probe. The two pulses must overlap as much as possible throughout their motion. Because the delay experienced by a pulse in a resonant two-level system is proportional to the reciprocal area Θ, the pulse length τ_p, and the Beer's length (α^{-1}), a strong pump propagates with the same group velocity of a weaker probe **only** when $\alpha_{pr} >> \alpha_p$. This is achieved in a metallic vapor with different oscillator strengths $\mu^2_{pr}\omega_{pr} >> \mu^2_p\omega_p$. Moreover, group velocity depends upon the character of the medium. The initial populations dictate that the pump meet an absorbing transition, while the probe an amplifying one. As soon as a π-part of the pump area is absorbed and establishes a population in the upper level, each pulse encounters a different medium: the pump absorptive; the probe emissive. The pump propagates with a velocity smaller than the velocity of light while the probe travels with a larger velocity. This probe peak acceleration and pump peak retardation effectively reduces the required ratio of Beer's lengths.

III. THREE-LEVEL CALCULATIONS

A look at propagation of the field in a two-level medium provides insight into the physics of two-pulse propagation in a three-level medium in conjunction with diffraction. The pump is analogous to a SIT pulse, while the counterpart of the probe pulse is an SF emission. The SF emission can be initiated by a tipping angle or equivalently by a small seed whose area equals the tipping angle.

The medium provides amplification for the probe while the transverse communication across the beam imposes a diffractive loss[7]. The beam reshapes longitudinally and transversely. A measure of the gain/loss condition would be a Fresnel number defined in terms of the Beer's length transition associated with a specific radial profile of N, the pre-excited atomic density. F_g, $F_{gp,pr} = \pi r^2_p / \lambda_{p,pr}\alpha^{-1}_{p,pr}$, is the ratio of the nonlinear effect (Beer's length) to the Rayleigh diffraction length (linear loss $\kappa^{-1}_{p,pr} = \pi r^2_p / \lambda_{p,pr}$): $F_{gp,pr} = \alpha_{p,pr} / \kappa_{p,pr}$.

In the three-level case, the pump experiences **greater** absorption than a SIT pulse in a two-level medium because of its leakage to the probe. Conversely, the probe buildup will be **weaker** than that of a two-level SF. To illustrate the pump depletion we compare in Fig. 1a the pump area (THETA) as a function of the propagation distance with a corresponding SIT pulse: The SIT pulse (Curve 2) is absorbed considerably less than the pump (Curve 1).

An optically driven three-level (OD3) probe with a finite depleting pump (labeled Curve 2) and a two-level probe with infinite pumping (labeled Curve 1) are compared in Fig. 1b. In the two-level analysis, the probe-seed continuously experiences a population inversion and thus an initial gain at small τ. The OD3 probe does **not** experience a gain **until** a π-portion of the pump depletes the ground state and fills the upper level of the λ-scheme. The two-level probe threshold comes earlier. The density and gain necessary for the two-level probe buildup are smaller; the delay of the two-level probe is smaller. The maximum energy and temporal width are both larger in the two-level case. The OD3 probe must have a gain larger than the two-level case to insure equal delays and probe peak energies.

In a rigorous two-field three-level atom interaction, the strong pump depletes its energy (ENRGYa) both temporally and spatially so that the initially weak probe energy (ENRGYb) builds up. At the input plane, the on-axis pump area is 2π, while the on-axis input seed which represents the probe is only 0.02π. The probe emerges with appreciable amplitude after a certain propagation distance. The pump depletion which accompanies the probe growth is shown. In graph a of Fig. 2 the **output power** (OPOWRa,b) of both pump and probe exhibits a strong temporal distortion that departs from the input Gaussian profile. The probe pulse is temporally narrower than the pump pulse. The probe peak Rabi frequency is larger than that of the pump. The probe pulse front edge occurs at about the temporal location of the pump peak. The temporal and spatial thresholds for the probe buildup are observable. However, for sufficiently large $\alpha_b \ell$, the probe buildup slows down, tapers, or even changes sign.

The interplay of pump depletion and diffraction regime on the temporal dynamics is presented in Fig. 3 by displaying a family of comparison plots of the pump and probe **areas** (THETAa,b) for a family of propagation distances, Θ_a (THETAa) for the pump and Θ_b (THETAb) for the probe. The pump area is seen at first to experience **stripping** around the radius corresponding to an area $\Theta_a = \pi$. This agrees with McCall & Hahn's SIT theory. Subsequently, the pump area profile develops a radial structure similar to that of a conical emission and steadily shrinks. At the same time, the profile of the probe pulse area does not develop any ring and blooms beyond the free space diffraction spreading. Once the Rabi frequency exceeds that of the pump, its area ceases to build up, regresses and acts as a pump by radiating out energy to the pump instead of receiving it. First Θ_a depletes while Θ_b builds up and eventually saturates, then Θ_b decreases and Θ_a stabilizes. This oscillatory interchange of roles takes place when relative phase variations between pump and probe occur. This predominately appears when diffraction is important.

When diffraction plays a role the pump experiences, at small $z(0<\alpha_a z<2)$ before any probe feedback arises, an on-axis energy enhancement (see Fig. 4) due to a flux redistribution from the peripheral shells toward the beam center. The impact of this light diffracting from the peripheral shells toward the beam center is significant due to its continuous amplification by atoms in the intermediate shells. These atoms have been excited by previous passages of higher intensity light. This self-focusing trend slows down the rate of pump depletion. The pump energy experiences the same spatial redistribution displayed in SIT self-focusing[2]. Beyond a certain optical thickness, the feedback of the probe buildup affects the pump. The pump loses some of its energy to the growing probe seed, the pump's enhancement is interrupted, and its trend to self-focusing instability is aborted. Even the pump beam-center ceases to grow. Meanwhile the probe peak advances in time from one η (ETA) frame to the other. The presence of the probe quenches the pump's trend to self-focus. However, the pump peak output power decreases right from the beginning. The pump on-axis energy enhancement is not reflected in the output power.

In graph a of Fig. 5 the probe energy (ENRGYb) is presented as an isometric plot versus ρ and τ at different η accentuating relative motion among neighboring shells and temporal distortion, such as pulse compression. Graph b of Fig. 5 shows the pump energy (ENRGYa) plotted versus τ and ρ for different η to emphasize spatial reshaping, i.e., beam narrowing. This illustrates the nonlinear buildup of the probe at the expense of the pump. The probe growth rate and the pump depletion rate **vary** as a function of the propagation distance.

This calculation represents the on-resonance equivalence to Raymer et al.'s[8] UPW Raman analysis. It generalizes Herman et al.'s[9] UPW study of weakly overlapped two-pulse propagation in a three-level system. In addition diffraction coupling is included.

CONCLUSION

Rigorous calculations accounting for the interplay of diffraction coupling with the medium inertial response in conjunction with pump depletion exhibit the **inevitable** spatial and temporal redistribution of pump and probe energy.

REFERENCES

1. M.J. Konopnicki and J.H. Eberly, Phys. Rev. A24, 2567 (1981).
2. F.P. Mattar and M.C. Newstein, IEEE J. Quantum Electron. QE-13, 507 (1977).
3. F.P. Mattar, Appl. Phys. 17, 53 (1978).
4. S.L. McCall and E.L. Hahn, Phys. Rev. 183, 457 (1969). SIT is a solution to the Sine-Gordon equation.
5. R.H. Dicke, Phys. Rev. 93, 99 (1954).
6. N. Skribanowitz, I.P. Herman, J.C. MacGillivray and M.S. Feld, Phys. Rev. Lett. 30, 309 (1973).
7. F.P. Mattar, H.M. Gibbs, S.L. McCall and M.S. Feld, Phys. Rev. Lett. 46, 1121 (1981).
8. M.G. Raymer & J. Mostowski, Phys. Rev. A24, 1980 (1981).
9. B.J. Herman, P.D. Drummond, J.H. Eberly & B. Sobolewska, Phys. Rev. A30, 2462 (1984).

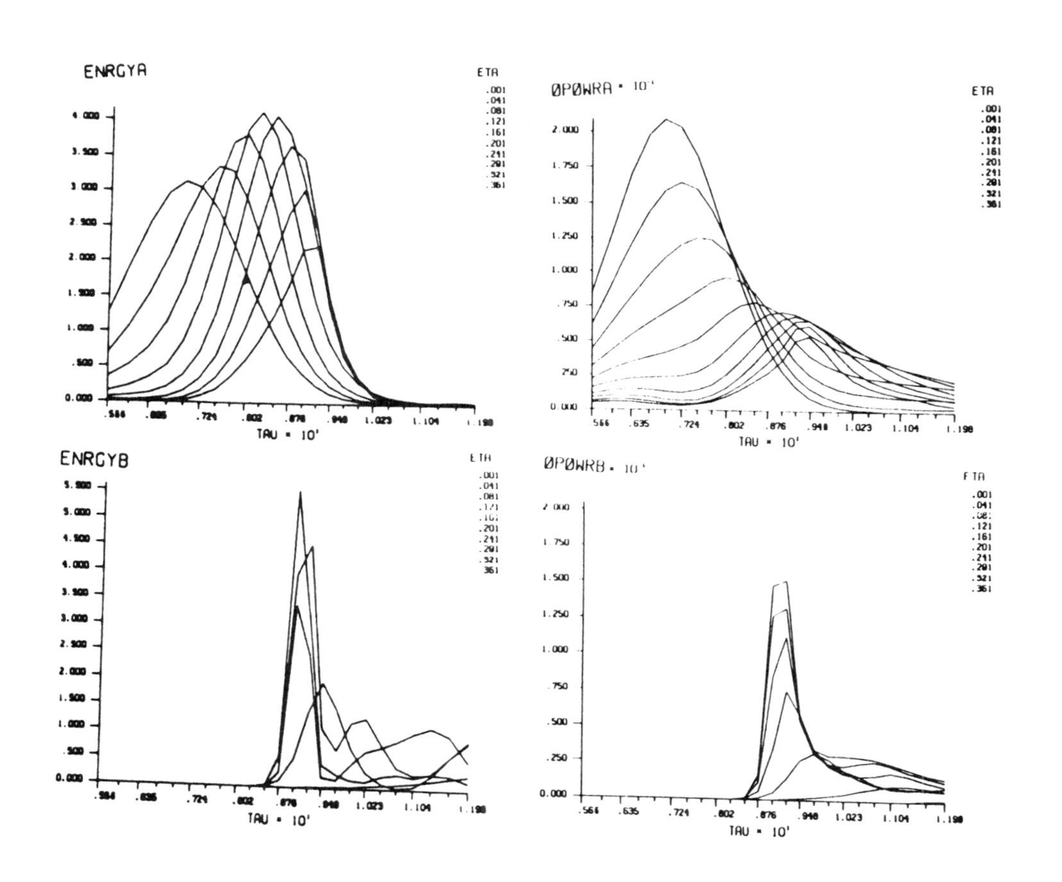

FIG. 4. Pump pulse distortion and probe pulse compression.

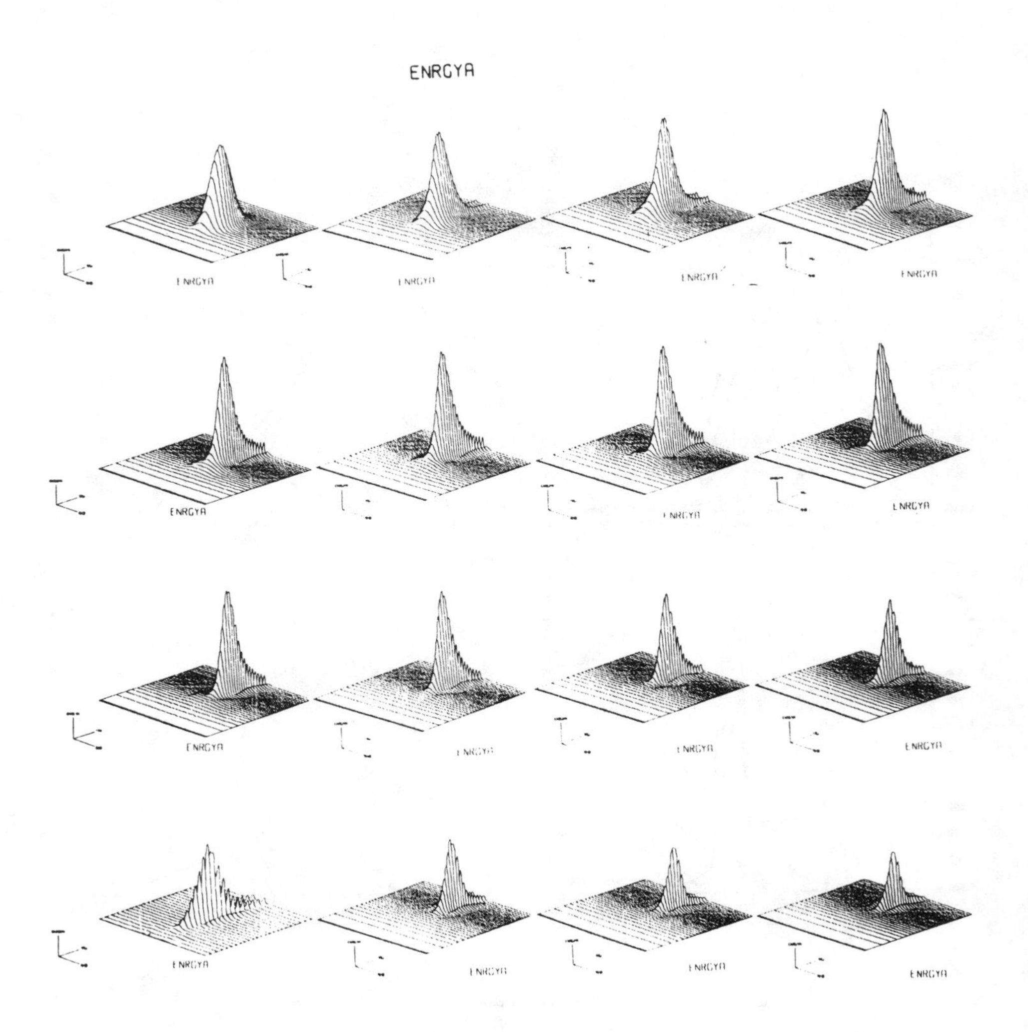

FIG. 5. Three-dimensional distribution of the pump energy versus ρ and τ in graph a; and probe energy versus ρ and τ in graph b for different η.

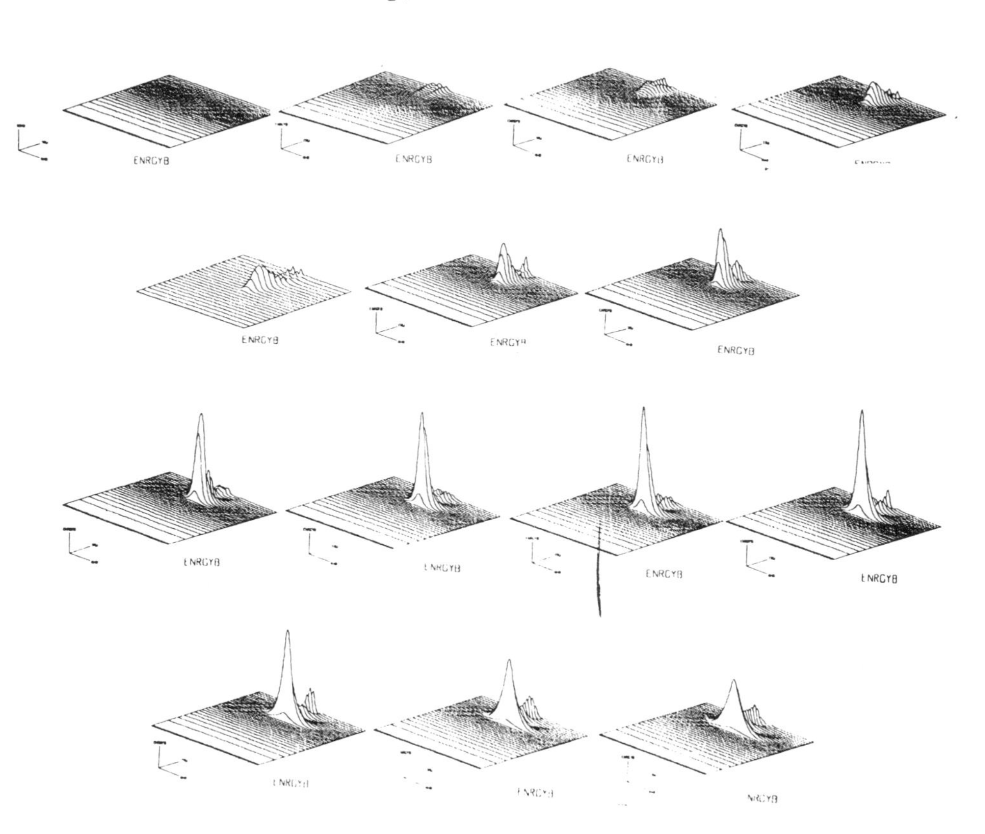

Fig. 5 Graph b

ON-RESONANCE DIFFRACTION-INDUCED PHASE IN TWO-FIELDS THREE-LEVEL COHERENT PROPAGATION†

J. Teichmann# and F.P. Mattar*
Department de Physique, Université de Montreal#, Montreal
Department of Physics, New York University*, New York, NY 10003

ABSTRACT

Coherent pulse propagation began with McCall's & Hahn's discovery of Self-Induced-Transparency (SIT)[1] theoretically and experimentally. Even for collimated beams of finite extent a diffraction-induced phase develops. Using a perturbational treatment[2-5] the time- and radially-dependent phase φ, its wavefront curvature $\partial\varphi/\partial\rho$, and the associated transverse energy flux current $J_T=A^2\partial\varphi/\partial\rho$ are calculated. The energy is redistributed across the beams. The evolving pulse departs from the uniform-plane-wave (UPW) counterparts. The onset of the radial reshaping and dynamic self-action is demonstrated.

I. CONCEPTUAL ROLE OF DIFFRACTION

All previous two-field three-level[6,7] analyses are carried out in the UPW regime. However, the experimental observations are made with finite beams and therefore often depart from these predictions (Fig. 1). The observed on-axis energy is larger than the one predicted by the UPW theory due to the competition between coherent transient and self-focusing effects. If one wants the experimental observations to agree with the UPW calculations, one must select a detector narrower than the beam widths.

To illustrate the effect of finite beam we review the analogous result when a one-field beam enters a two-level medium. The total transient field is made up of individual UPW pulses of different strength for different radii. Each shell obeys its own time-integrated area theorem. Pulse areas smaller than π gets totally absorbed while those between π and 2π evolve asymptotically towards a 2π. The beam wings are stripped. However in a higher order theory with diffraction accounted for the various shells are cross-coupled and inter-communicate. Partial stripping takes place. **Distinct resonant interactions occurs for different beam radii.** Absorptive and dispersive parts of the polarization vary across the beam making the field mattter energy transfer radially dependent. The more intense a coherent pulse is, the faster it travels in an absorbing medium.

The group velocity of the pulse peak at the center shell exceeds the corresponding off-axis group velocity. The peak of the pulse propagating in an absorbing medium is delayed with respect to a frame moving with the velocity of light. A relative motion between neighboring radii occurs (Fig. 2).

Associated with this relative motion between adjacent pencils, a sign variation occurs for the transverse Laplacian [*L*] term. At the input plane its contribution is negative. As the pulse propagates along η, at a later instant of time τ_0, the *L* eventually vanishes. Still further away, its contribution for lagging times τ_0 becomes positive. Thus, the *L* sign is a function of time differing at the pulse leading portions from its value at the trailing edges, the phase (φ) changes along with these amplitude (A) changes. These phase variations induce a dynamic radial energy current ($J_T=\zeta\nabla_T\zeta^*-\zeta^*\nabla_T\zeta$) which flows inwardly at some times and outwardly at others.

† Supported by ARO, AFOSR, NSF and ONR.

The weak diffraction coupling characterized by a large Fresnel number F can be **treated perturbatively** for short propagation distances. By expanding the fields and material variables in powers of the reciprocal of F Eq.(3) and by considering the UPW solution as the zeroth order Eq.(4) for pulses whose areas change across the beam, one **can** calculate the evolving phase due to diffraction coupling. Validity is limited to the propagation distances where the UPW solutions do **not** differ significantly from their three-dimensional counterparts Eq. (5). Critical prediction of the perturbation theory is the development of substantial focusing wavefront curvature in the pulse tail in an absorber. The fact that the phase is smaller on the beam wings than on axis is indicative of inward radial energy flow.

The **phase velocity is larger than the group velocity** associated with the peak intensity. As the pulse propagates into the absorber, its peak experiences a temporal lagging which brings it towards the same temporal site of the phase variation. This leads to a greater overlap A and $\partial\varphi/\partial\rho$. When this transverse reshaping reaches a critical value the beam-rim energy flows towards the center causing a self-focusing of the on-axis fluence.

II. MATHEMATICAL FORMULISM

For simplicity the relaxation times are assumed infinite.

$$- i F^{-1}_{a,b} \nabla^2_T\zeta_{a,b} + \partial\zeta_{a,b}/\partial\eta = g_{a,b}P_{a,b} \tag{1}$$

$$\partial_\tau P_a = \zeta_a W_a - \frac{1}{2}\zeta_b Q \tag{2a}$$

$$\partial_\tau P_b = \zeta_b W_b - \frac{1}{2}\zeta_a Q^* \tag{2b}$$

$$\partial_\tau Q = \frac{1}{2}(\zeta_a P^*_b + \zeta^*_b P_a) \tag{2c}$$

$$\partial_\tau W_a = -\frac{1}{2}(\zeta^*_a P_a + \zeta_a P^*_a) - \frac{1}{4}(\zeta^*_b P_b + \zeta_b P^*_b) \tag{2d}$$

$$\partial_\tau W_b = -\frac{1}{4}(\zeta^*_a P_a + \zeta_a P^*_a) - \frac{1}{2}(\zeta^*_b P_b + \zeta_b P^*_b) \tag{2e}$$

with fields and material variables developed in a power series

$$\zeta_{a,b} = \zeta_{0a,0b} + F^{-1}_{a,b}\zeta_{1a,1b} = \zeta_{0a,0b} + F^{-1}_{a,b}(\zeta_{1ra,1rb} + i\zeta_{1ia,1ib}) \tag{3a}$$

$$P_{a,b} = P_{0a,0b} + F^{-1}_{a,b}P_{1a,1b} = \zeta_{0a,0b} + F^{-1}_{a,b}(P_{1ra,1rb} + iP_{1ia,1ib}) \tag{3b}$$

$$Q = Q_0 + F^{-1}Q_1 = Q_0 + F^{-1}(Q_{1r} + iQ_{1i}), \text{ with } F = F_a >> F_b \tag{3c}$$

$$W_{a,b} = W_{0a,0b} + F^{-1}W_{1a,1b} = W_{0a,0b} + F^{-1}W_{1a,1b}, \tag{3d}$$

The pump Fresnel number is larger than that of the probe while the probe gain g_b is larger than the pump gain g_a ($g_b >> g_a$). The UPW equations are recuperated: $\partial_\eta\zeta_{0a,0b} = g_{a,b}P_{0z,0b}$ (4)

The diffraction-induced phase is given by

$\partial_\eta\zeta_{1a,1b} = \nabla^2_T\zeta_{0a,0b} + g_{a,b}P_{1a,1b}$ (5a) with $g_{a,b}P_{1a,1b} << i\nabla^2_T\zeta_{0a,0b}$ (5b) here $\zeta_{0a,0b}$, $P_{0a,0b}$ and Q_0 are known analytical solutions. At the input plane $\zeta_{1a,1b}=0$, $P_{1a,1b}=0$, $W_{1a,1b}=0$ and $Q_1=0$; as the pulse propagates, the imaginary part of the fields grow and self-phase modulation results. The rate of growth of $\zeta_{1ia,1ib}$ and its sign depend on the radial variation of $\zeta_{0a,0b}$. The quantities $\zeta_{1ra,1rb}$; $P_{1ra,1rb}$ and $W_{1ra,1rb}$ and Q_{1r} **may** remain zero for all η. From this, one can calculate: $\zeta_{1ia}=\int_0^\eta d\eta' \nabla_T^2\zeta_{0a,0b}=(\nabla_T^2\zeta_{0a,0b})\eta$ (6a). For Gaussian beams, $\zeta_{a0,b0}\exp[-\rho^2]$, one has $\nabla_T^2\zeta_{a0,b0}=4\zeta_{a0,b0}(\rho^2-1)$ (6b) and hence obtain the phase $\varphi_a=\arctan(F_{a,b}\zeta_{1ia,1ib}/\zeta_{0a,0b})=\arctan(4F_{a,b}(\rho^2-1)\eta)$ (7). By iteration, one obtains a better phase estimator $\varphi^c_{a,b}$ **without** assuming that diffraction predominates the nonlinear interaction.

$$\partial_\eta\zeta^c_{1a,1b} = \nabla_T^2\zeta_{0a,0b} + g_{a,b}P^c_{1a,1b}((\zeta_{1ia,1ib})) \tag{8}$$

with the corrected $P^c_{1ia,1ib}$ evaluated in terms of the predicted fields $\zeta_{1ia,1ib}$.

IV. NUMERICAL RESULTS

This theory is applied to the **simul**taneous propagation of soli**tons**[7]. One observes a self-phase development, and the growth of an energy current[8]. These results have been validated by rigorous diffraction calculations[9]. The evolving phase and the energy current are respectively shown versus τ and ρ after a propagation distance Z in Fig. 3-4. From this one sees that the energy propagating moves into the region where a focusing phase variation has emerged.

V. CONCLUSION

We have obtained an analytical expression for the phase variations that originate from the Laplacian term in the scalar-wave equations. Incident beams with sufficient energy are subjected to significant transverse reshaping as they propagate. This occurs **even** in the near field of the effective input aperture.

REFERENCES

1. S.L. McCall & E.L. Hahn, Phys. Rev. 183, 457 (1969).
2. M.C.Newstein et al., **Progress Rept.JSEP-TAC:Quantum Electron.Prog Microwave Res. Inst.**,Polytech. Inst. NY 36,140(1971);37,117(1972)
3. M.C.Newstein & N.Wright, IEEE J.Quantum Electron.QE-10,743 (1974)
4. M.C.Newstein & F.P.Mattar,IEEE J.Quantum Electron.QE-13,507(1977)
5. [a] M.LeBerre et al., Phys.Rev. A25,1604(1982), & A29,2669 (1984); [b] M.LeBerre et al.,SPIE 369,269(1983) & J.Opt.Soc.Am. B2, 956(1985);[c] J.Teichmann et al., **Abst. Dig. Ann. Meet. Canadian Assoc. Phy. & Astronomical Soc.**, Victoria, B.C., Canada 1983; & Opt.Commun. 54, 33 (1985).
6. [a] F.Y.F.Chu et al., Phys.Rev. A12, 2060 (1975); [b] E.Hanamura, Opt.Commun. 9,396(1973) [c] N.Tan-No et al., Phys. Rev. A12, 159 (1975); [d] J.N.Elgin et al., Opt.Commun. 18,250(1977); [e] D.Kaup, Physica D6,143 (1983).
7. [a] M.J. Konopnicki & J.H. Eberly, Phys. Rev. A24, 2567 (1981); [b] C.R.Stroud Jr. & D.A.Cardimona, Opt. Commun. 37, 221 (1981).
8. [a] J. Teichmann, Technical report, Departement de Physique, Universite de Montreal, (1987).

9. F.P. Mattar et al., in **Advances in Laser Science-I**, ed. W.C. Stwalley & M. Lapp, (Am. Inst. Phys., 1986) Vol. 146, pp.**324-328**.

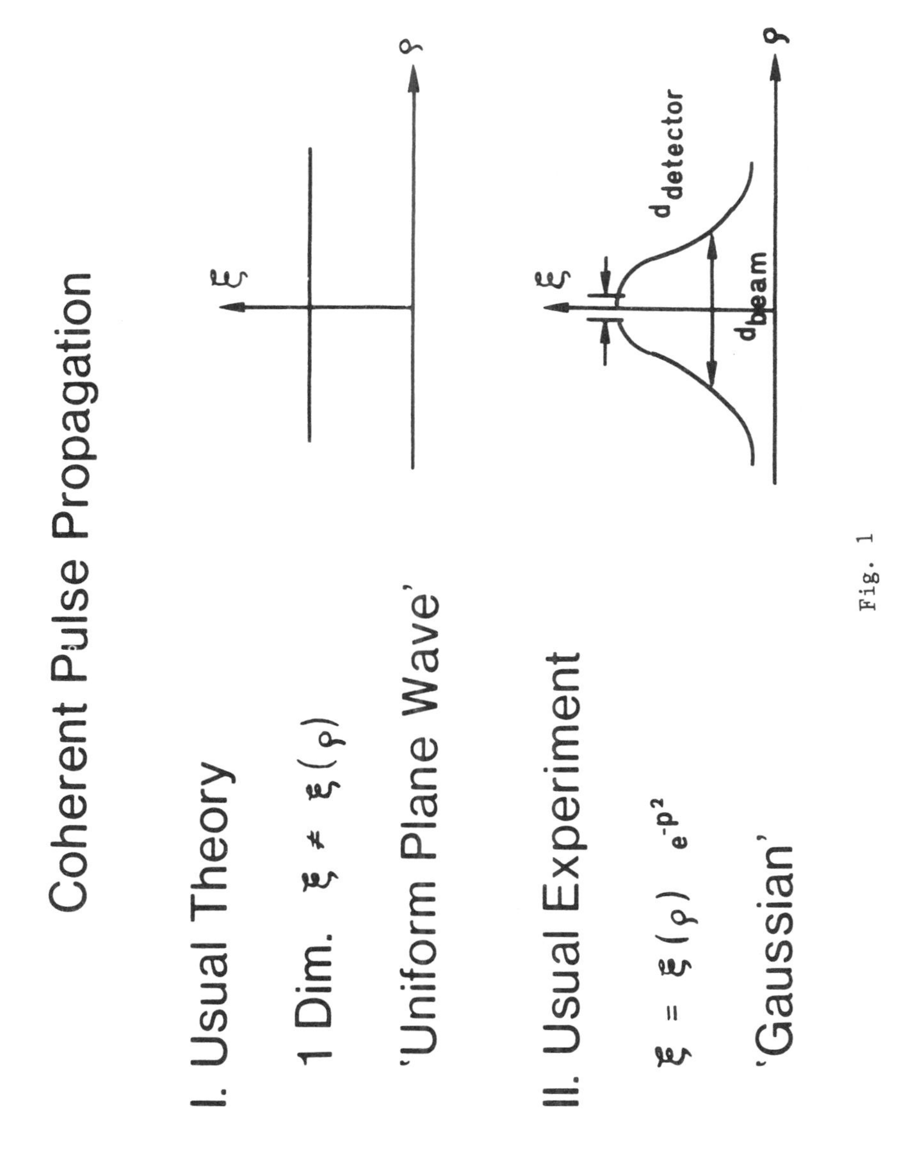

Fig. 1

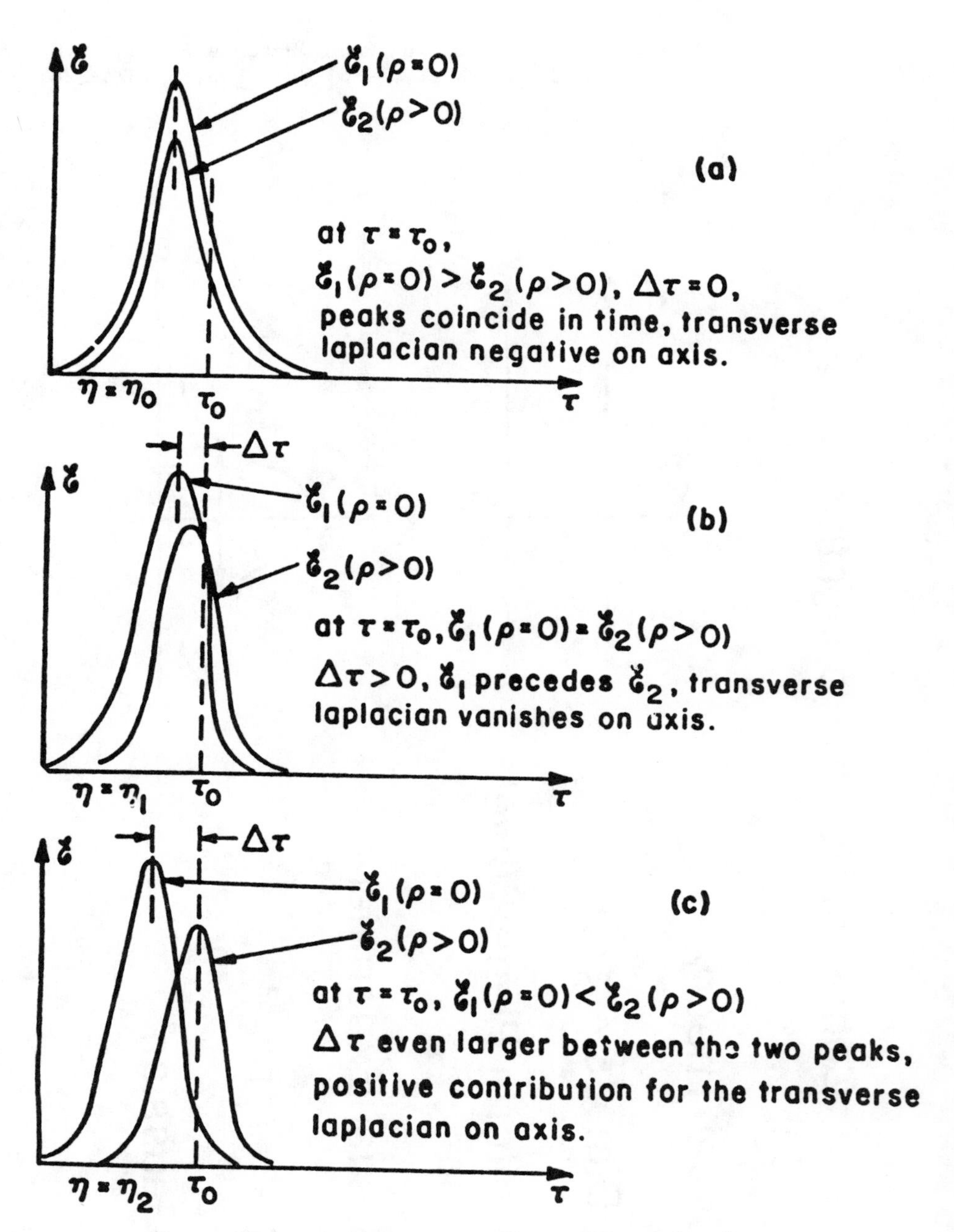

Fig. 2 Relative displacement between unequal pulses.

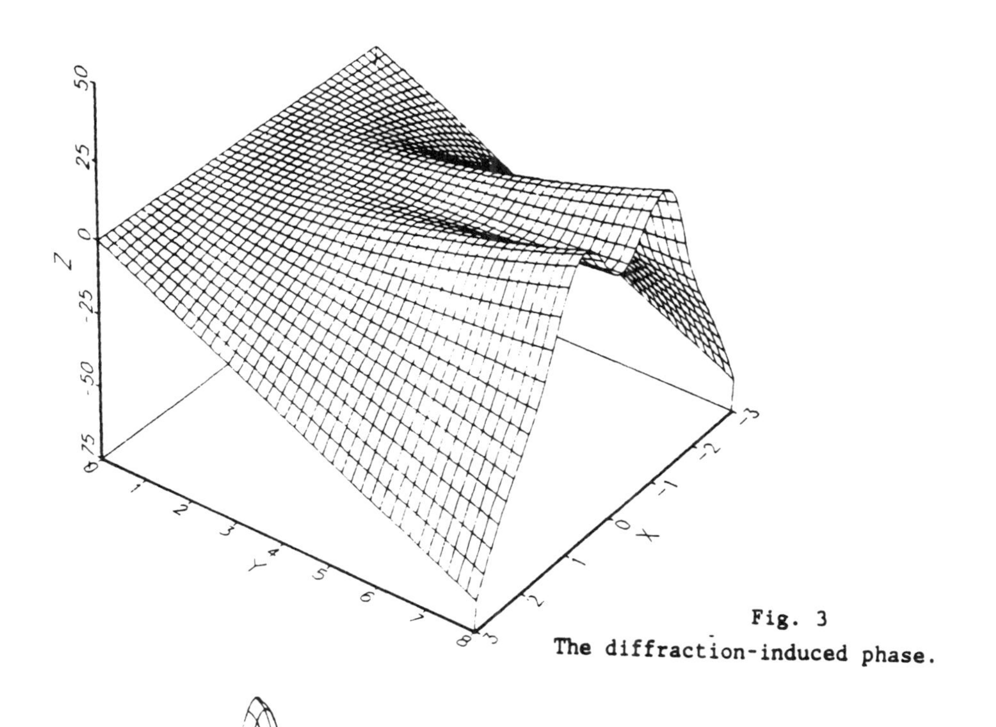

Fig. 3
The diffraction-induced phase.

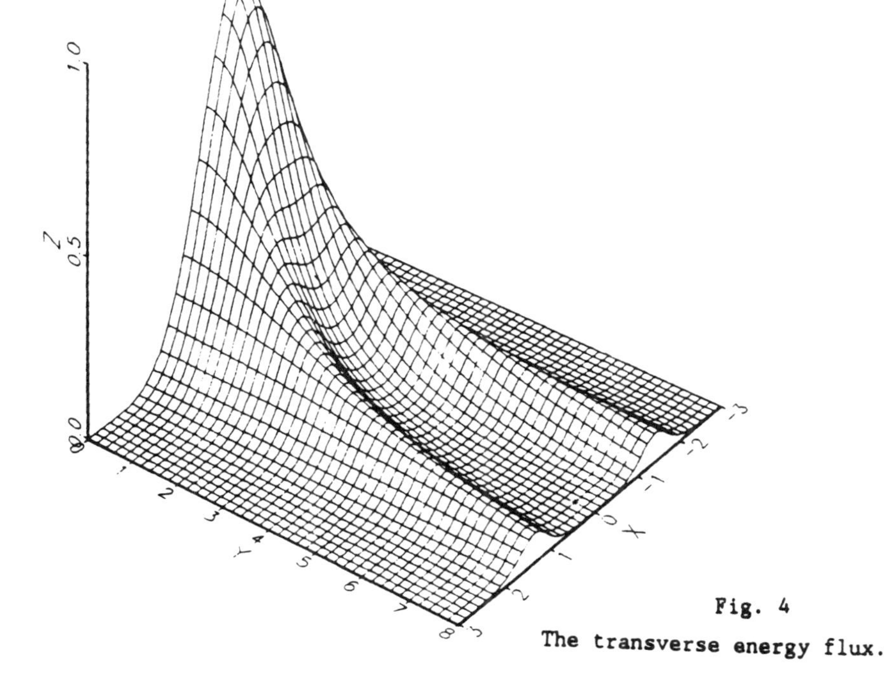

Fig. 4
The transverse energy flux.

PROPAGATION OF CO_2-LASER RADIATION PULSES IN A CLOUD MEDIUM

R.Kh. Almaev, L.P. Semenov, A.G.Slesarev, O.A.Volkovitsky
Institute of Experimental Meteorology, Obninsk, USSR, 249020

The paper gives a theoretical study of propagating laser pulses series in a cloud medium with the intensity sufficient for explosive drop shattering.

Let a cloud medium in a semispace $z \geqslant 0$ moving along the X-axis with a constant velocity V towards the z-axis be subjected the incidence of laser radiation pulses with a duration t_p and a pulse recurrence interval t_r. We shall consider a situation when the cloud medium shift caused by wind transport at an individual pulse action time is small as compared with the typical cross-section of a radiation beam. Let us consider also that between the pulses spatial variations of the drop distribution function and of the temperature profile within the clearing zone occur only as a result of wind transport.

At the conditions postulated the propagation process of an arbitrary sequence of j-th pulse is described with the equation system /I,2/.

The set of equations has been solved numerically for radiation pulses with different intensity distributions and various parameters of the medium.

Characteristic changes of the optical depth and temperature distributions in the area cleared are shown in Figs. I-3. The figures give the optical depths profiles of the cleared zone at wavelengths of probing and acting radiation(Figs.I and 3) and temperature distributions (Fig.2) for $y = 0$, $z = 0$ as well as beam trajectories by the end of an N-th pulse action. Variants a), b) of Fig.I refer to different intervals of t_r.

The calculations were performed for intensity distributions within a pulse of one maximum(Figs.I,2) and with two maxima (Fig.3).

The analysis of the results obtained (a part of which is given in curves of Figs.I-3 has shown that at irradiating a cloud medium with a series of laser beams beginning with some moment t_s a regime of the medium and a laser beam interaction is established when changes of their characteristics with time in the clearing zone is periodically repeated. This means that at $t > t_s$ the self-action dynamics of any pair of

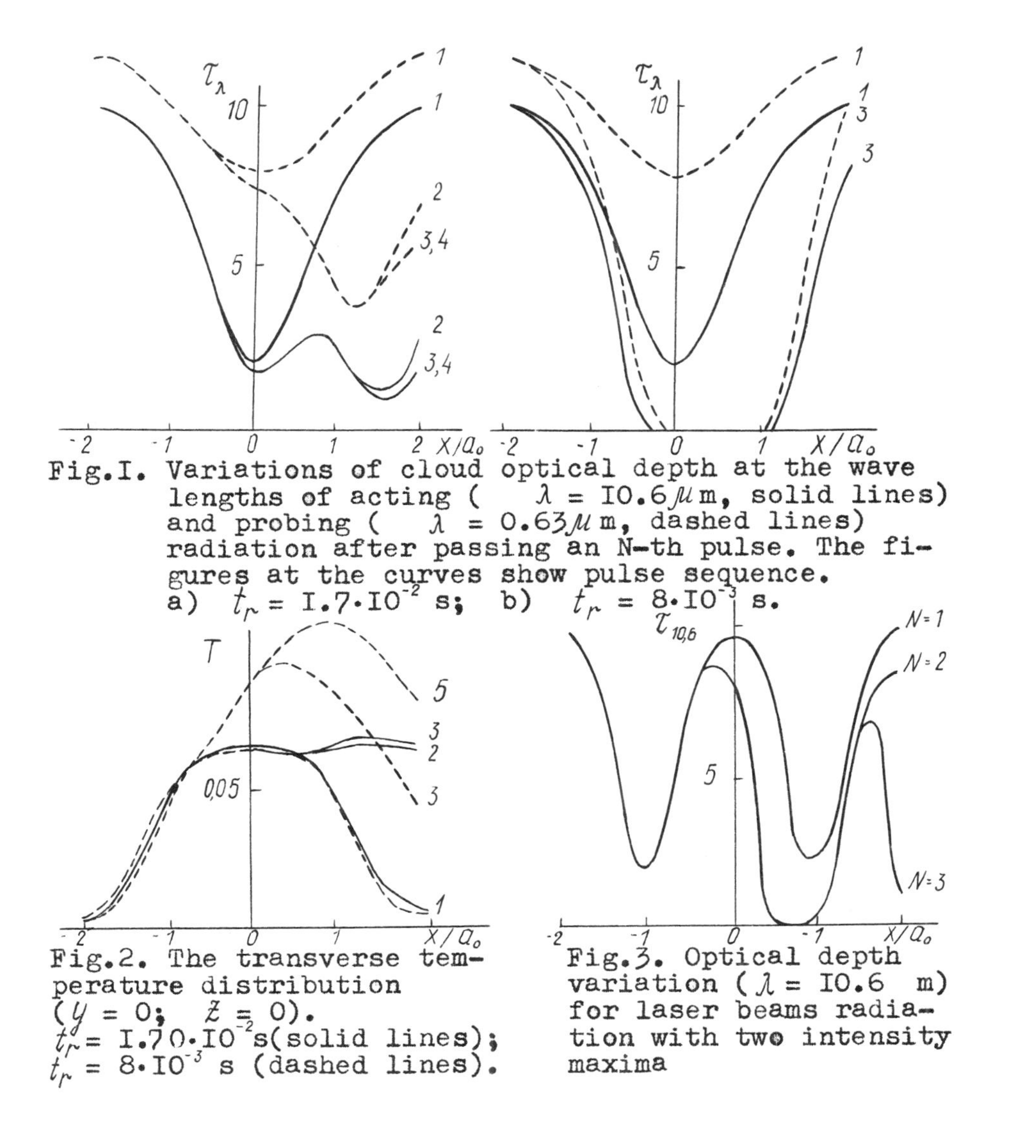

Fig.1. Variations of cloud optical depth at the wave lengths of acting (λ = 10.6 μm, solid lines) and probing (λ = 0.63 μm, dashed lines) radiation after passing an N-th pulse. The fi-gures at the curves show pulse sequence. a) $t_r = 1.7\cdot10^{-2}$ s; b) $t_r = 8\cdot10^{-3}$ s.

Fig.2. The transverse temperature distribution (y = 0; z = 0). $t_r = 1.70\cdot10^{-2}$ s (solid lines); $t_r = 8\cdot10^{-3}$ s (dashed lines).

Fig.3. Optical depth variation (λ = 10.6 m) for laser beams radiation with two intensity maxima

pulses is analogous. The time when this regime is established depends mainly on frequency and characteristic time of (cloud) medium transverse wind shifts. Thus for a sequence with $t_r = 1.7\cdot10^{-2}$ s (Figs.1a,2,3) the periodic regime is established after the third pulse; with $t_r = 8\cdot10^{-3}$ s (Figs 1b, 2) it is attained after the fifth pulse. From the figures it is seen also that with increasing the pulse frequency clearing medium transparency within $t_s < t < t_a$ (t_a is the series duration) is improved at the wavelengths of both acting and probing radiation. It is noted that in contrast to stationary clearing of a moving cloud medium with CO_2 laser radiation, when overheating and transparency of a clearing zone increase monotonously at moving along the X axis to the downwind edge of a beam, the given distributions of temperature and transparency of the medium at $y = 0$, $z =$ const. are nonmonotonous. Thus maximum transparency zone attained at clearing with a series of pulses can be not at the downwind edge of the clearance zone but it can be shifted towards the beam axis. In its turn, the nonmonotonous character of temperature distributions along the X -axis may lead to the formation of zones with radiation focusing (Fig.2).

The comparison of the results obtained in irradiation of a cloud medium with a series of laser pulses with different intensity profiles has shown that it is possible to increase maximum level of clearing by special setting of intensity distributions in a beam at the same energy consumption. The above said is illustrated with Figs.Ia and 3 from which it follows that when clearing a medium with pulses of the intensity with two maxima there is formed a zone of complete transparency. At the same time when clearing a medium with pulses of the intensity with one maximum some residual optical depth is sustained even in zones of maximum transparency.

REFERENCES

I. R.Kh. Almaev, L.P. Semenov, A.G. Slesarev, Proc. of IEM, issue 40(I23), IO (I986).
2. R.Kh.Almaev, Yu.S. Sedunov, L.P. Semenov, A.G. Slesarev, O.A. Volkovitsky, Infrared Physics, vol.85, No.I/2, 475 (I985)

PROPAGATION OF A PULSE TRAIN OF CO_2 LASER RADIATION WITH DIVERGENCE THROUGH A CLOUD

O. A. Volkovitsky
Institute of Experimental Meteorology,
Obninsk 249020, USSR

A possibility to obtain approximate formulas for describing the main peculiarities of CO_2 laser diverging radiation pulses propagation in a moving cloud is considered. It is shown that determination of the dependence of a change in optical depth of a cloud over the time of pulse action on its thermal action function can be reduced to solution of an algebraic cubic equation.

Propagation of a pulse series of intensive laser radiation in a moving droplet aerosol can be presented as an interchange of two processes. During the action of a pulse (t_p), droplets evaporate, aerosol clearing process takes place. During the time (t_c) between the subsequent pulses the cleared zone is replenished due to wind shifts. The character of radiation propagation in an aerosol medium depends on the relation between t_p and t_c . A consideration of a case when $t_c > t_p$ and $t_c < d_0/V$ is likely to be most interesting. Schematically such a process for a CO_2 laser diverging beam with a uniform transverse intensity distribution is shown in Fig. 1.

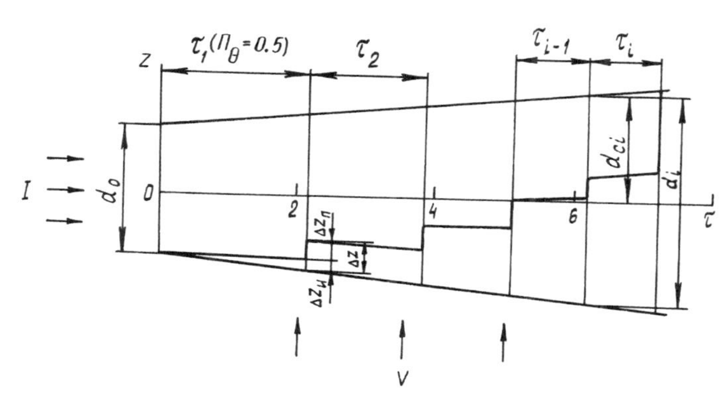

Figure 1. A scheme of CO_2 laser diverging pulse series propagation in a cloud.

During the first pulse t_{p1} action a displacement of a present aerosol transparency level $\Pi_\theta = (I(\tau)/I_0)(1+\theta\tau)^2$ will occur by the value τ_1 from the beginning of the path. During this time the cleared zone will be replenished by $\Delta z_p = Vt_p$. Between the pulses further replenishment of the zone will occur by the value of $\Delta z_c = Vt_c$. At a sequential action of a series of i-th pulses we shall have a cleared zone length $\sum_{j=1}^{i} \tau_j$. The zone cross section dimensions will be

$$d_{ci} = d_i - i\Delta z = d_o(1+\theta\sum_{j=1}^{i} \tau_j) - iV(t_p + t_c) \quad . \tag{1}$$

The number of pulses N causing $d_{ci}=N=0$ can be found from the condition that

$$N = d_o(1+\theta\sum_{i=1}^{N} \tau_i)\, V^{-1}(t_p + t_c)^{-1} \quad . \tag{2}$$

To find the dependence of τ_i on the pulse thermal action function $q_{\theta i}$ one can use an approximated description of continuous radiation propagation of a CO_2 laser diverging beam in a droplet aerosol [1].

Under a fine-droplet approximation and of a given radiation field (at limited values of the function $\psi = \theta\tau < 10$ $(\theta = \varphi / d_o \alpha_o)$, where φ is the angle of radiation divergence, α is the extinction coefficient of CO_2 laser radiation) as well as under other approximation, accepted in [1], it is possible to write for a clearance zone produced with a series of i-th pulses:

$$\sum_{j=1}^{i} \tau_j = \ln\left[(1-\Pi_\theta)\Pi_\theta^{-1} \exp(\sum_{j=1}^{i} q_{\theta j}) + 1\right] \quad . \tag{3}$$

During t_p the aerosol medium can be accepted motionless and the expression for the thermal action function may be written as $q_{\theta j} = C\bar{I}_j t_p$, where $\bar{I}_j$ is the radiation effective intensity, $C = 3\Lambda_o\beta/4\rho L$, ρ , and L are the density and evaporation heat for water, β is the coefficient of heat losses. Let us find $\bar{I}_j$ from the condition of equality of pulses radiation energies from pulses within the cylindrical and divergence clearing zones with optical depth τ_j and the base area S_j :

$$I_j S_j^o \tilde{\tau}_j = \bar{I}_j S_j^o \int_0^{\tilde{\tau}_j} (1+\theta\tau)^2 d\tau \; .$$

$$\bar{I}_j = I_j \left(1 + \theta\tilde{\tau}_j + \frac{\theta^2}{3}\tilde{\tau}_j^2\right)^{-1} . \tag{4}$$

The integration limit $\tilde{\tau}_j = \sum_{K=1}^{j} \tau_K(\Pi_{\theta 2}) - \sum_{K=1}^{j-1} \tau_K(\Pi_{\theta 1})$ depends on how the transient zone of radiation and aerosol interaction is defined. (Here $\Pi_{\theta 2}$ is the transparency level close to 0; $\Pi_{\theta 1}$ is the level close to unity). Using (3) we find

$$\tilde{\tau}_j = \ln\left\{\left[(1-\Pi_{\theta 2})\Pi_{\theta 2}^{-1} \exp(\sum_{K=1}^{j} q_{\theta K}) + 1\right]\left[(1-\Pi_{\theta 1})\Pi_{\theta 1}^{-1} \exp(\sum_{K=1}^{j-1} q_{\theta K}) + 1\right]^{-1}\right\} \tag{5}$$

For I_j, considering that at the optical depth $\sum_{K=1}^{j-1} \tau_K(\Pi_{\theta 1})$ the radiation energy consumption in a well-cleared zone ($\Pi_{\theta 1}$ is close to unity) is small to evaporate droplets one can write

$$I_j = I_o\left[1+\theta\sum_{K=1}^{j-1}\tau_K(\Pi_{\theta 1})\right]^{-2} . \tag{6}$$

Using (4) and (6) for $q_{\theta j}$ we obtain

$$q_{\theta j} = q_o\left[1+\theta\sum_{K=1}^{j-1}\tau_K(\Pi_{\theta 1})\right]^{-2}\left[1+\theta\tilde{\tau}_j+\frac{\theta^2}{3}\tilde{\tau}_j^2\right]^{-1} . \tag{7}$$

The solution of equation system (3),(5),(7) can be made by the iterative method.

The calculation procedure of $\sum_{j=1}^{i}\tau_j$ allows for simplifications in practically important cases when $\sum_{j=1}^{i} q_{\theta j} \gg 1$ at $1-\Pi_\theta/\Pi_\theta\, e^{\sum_{j=1}^{i} q_{\theta j}} \gg 1$.

In this case equation systems (3),(5) and (7) are reduced to an algebraic equation of the 3-rd power:

$$y^3+3m_K y+2n_K=0 , \tag{8}$$

where

$$m_K=\frac{a_K}{3}\left(\frac{1}{\theta}-\frac{a_K}{3}\right), \quad n_K=\left(\frac{a_K}{3}\right)^3-\frac{a_K^2}{6\theta}-\frac{a_a-3a_K}{2\theta^2}-\frac{1}{2\theta^3}$$

At $K=1$ $\quad a_1=\ln\left[\Pi_\theta(1-\Pi_{\theta 1})\,\Pi_{\theta 1}^{-1}(1-\Pi_\theta)^{-1}\right]$,

At $K\geqslant 2$ $\quad y=\tau_1+\left[\frac{1}{3}a_1+\frac{1}{\theta}+\ln(1-\Pi_{\theta 2})\,\Pi_{\theta 2}^{-1}\right]$.

$$a_K=\ln\left[\Pi_{\theta 2}(1-\Pi_{\theta 1})\,\Pi_{\theta 1}^{-1}(1-\Pi_{\theta 2})^{-1}\right] ; \quad y_K=\tau_K+\frac{1}{\theta}-\frac{2}{3}a_K .$$

This equation has only one real solution corresponding to the formulation of the problem, so that $D=n_K^2+m_K^3>0$,

$$y=\sqrt[3]{-n_K+\sqrt{n_K^2+m_K^3}}+\sqrt[3]{-n_K-\sqrt{n_K^2+m_K^3}} .$$

Further simplification is possible if $a_1=0$ and $\tilde{\tau}_j=\tau_j$. The solution is reduced to a formula

$$\tau_K=\frac{1}{\theta}\left[\sqrt[3]{1+3\theta q_{\theta K}}-1\right], \quad q_{\theta K}=q_o\left(1+\theta\sum_{K=0}^{K-1}\tau_K\right)^{-2}, \quad \tau_0=\ln 2 . \tag{9}$$

To illustrate this we shall give an example of the calculation results with the data of Table 1.

Table 1

Parameter	$C, \frac{cm}{W\cdot s}$	$I_o, \frac{W}{cm^2}$	t_p, s	q_o	$\Pi_{\theta 1}=1-\Pi_{\theta 2}$
Value	0.3	$2\cdot 10^3$	$5\cdot 10^{-3}$	3	0.9
Parameter	φ, radn	d_o, cm	α_0, cm^{-1}	θ	V, cm/s
Value	$1.2\cdot 10^{-3}$	12	10^{-3}	0.1	200

The calculations were carried out by the iterative method with egs.(3) and (7). The calculation results of the distribution Π_θ depending on τ in the cleared part of the action zone for four pulses are given in Fig.2. The number of pulses, the propagation of which in the example considered gives a "stationary" clearing ($d_{ci} \approx 0$) (found on the base of (2)), appears to be $N = 6$. At estimating it is accepted that $t_c = 2t$.

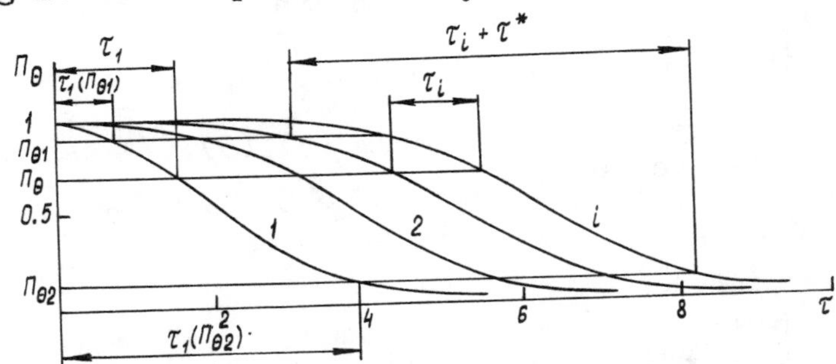

Figure 2. The distribution of transparency level Π_θ in the clearing zone.

The calculation results of the dependence of τ_i on q_i with (9) is shown in Table 2.

Table 2

Pulse	1	2	3	4
τ_i, (25)	2.26	1.53	1.26	1.09
τ_i, (3),(10)	2.13	1.66	1.33	1.09

The calculations have given, as it could be expected, somewhat different results as compared to the calculations with the iterative method, but this discrepancy is insignificant.

Reference

1. Volkovitsky O.A., Sedunov Yu.S., Semenov L.P. Propagation of Intensive Laser Radiation in Clouds. - Leningrad, Gidrometeoizdat, 1982, 312 p.

INFLUENCE OF RAMAN SELF-PUMPING ON FUNDAMENTAL SOLITONS IN FIBERS

Lin-jie Qu   Shu-sheng Xie   Shi-chen Li
Department of Precision Instrument, Tianjin University, China

ABSTRACT

With computer simulation we study energy spectra evolution of the fundamental solitons, propagating down fibers, caused by stimulated Raman process. We find that the energy spectra will distort together with the self-frequency shift effect. Hence the pulses whose pulsewidth are lower than subpicosecond cannot propagate stably in fibers.

INTRODUCTION

In 1986, F.M.Mitschke and L.F.Mollenauer first experimentally discovered self-frequency effect in a fiber. For a soliton with wavelength $\lambda=1.5\,\mu m$ and pulsewidth $\tau=120$ fsec, they have observed a 20THz (10%) frequency shift in 52m of the test fiber. Moreover they pointed out that the effect is caused by self-pumping of the soliton. Later, J. P. Gorden introduced the Raman effect into the nonlinear Schrödinger equation. He started from Fourier transformation of the modified nonlinear Schrödinger equation and obtained the relationship between self-frequency rate and pulse-width.

In this paper, we simulate and analyse energy spectrum evolution of a fundamental soliton propagating down a fiber by means of basic theory of stimlated Raman scattering. In addition to confirming the essential result of Ref. 1, 2, we have obtaind some new conclusions.

BASIC THEORY

It's known that the frequency spectrum of fundamental soliton in dimentionless form is

$$u(\Omega)=1/2\,\mathrm{sech}(\pi\Omega/2) \tag{1}$$

where $\Omega=(\omega-\omega_0)t_c$, ω_0 is the mean frequency of the soliton, $t_c=\tau/1.763$, and τ is the pulsewidth of the soliton.

Because the peak power of a narrow soliton is very high, this may make stimulated Raman scattering effect stronger. The pulsewidth is inversely proportional to the spectral width. Therefore the spectral width of a soliton with narrow pulsewidth is very broad. For example, if the pulsewidth of a fundamental soliton is 100 fsec, then the spectral width is 3.0 THz, which is less than the Raman bandwidth and Raman frequency shift (13.2 THz) corresponding to the gain peak. Thus, for a soliton with narrow pulsewidth, there will be strong interaction among all fourier components of its spectrum.

On account of Raman interaction, for Fourier component with frequency ω, all components whose frequency are higher than ω will yield gain to it, so the gain factor should be $\exp[\xi\int_0^\infty G'(\omega')u^2(\omega+\omega',\xi)d\omega']$, on the other hand, as the Fourier component with frequency ω yields gain to all Fourier components whose frequency are lower than ω, it will be incured a loss factor being $\exp[-\xi\int_0^\infty G'(\omega')u^2(\omega-\omega',\xi)d\omega']$ also, where $G'(\omega')$ is the Raman gain coefficient in dimentionless form, ξ is the transmitting distans in dimentionless form. Moreover, we introduce the fiber linear loss factor $\exp(-\alpha'\xi)$, where α' is the fiber loss rate in dimentionless form. Finally, we obtain the energy spectrum of a fundamental soliton at ξ in a fiber:

$$W(\omega,\xi)=\pi/2\ \mathrm{sech}^2[\pi(\omega-\omega_0)t\ /2\ \exp\{\xi[\int_0^\infty G'(\omega')u^2(\omega+\omega',\xi)d\omega'-\int_0^\infty G'(\omega')u^2(\omega-\omega',\xi)d\omega'-\alpha']\}\ . \qquad (2)$$

A CALCULATION RESULT AND ANALYSIS

We suppose that, the mean wavelength of fundamental soliton $\lambda=1.5\mu m$, the fiber dipersion parameter D=15 psec/nm/km, the optical Kerr coefficient $n_2=3.2\times10^{-16}$ cm^2/w, the Raman gain peak value is 6.6×10^{-12} cm/w (corresponding value of G' is 0.492), the fiber loss $\alpha=0.042$/km (corresponding value of α' is $7.554\times10^{-6}\tau^2$). For convenience we regard the Raman gate curve as a Lorentzian line shape, and set the line width to be 250cm^{-1}. Afterwards, we simulate the energy spectrum evolution of a 100 fsec fundamental soliton transmitting down the fiber.

Fig.1 shows the energy spectra of a 100 fsec fundamental soliton at z=0, 5.66m, 7.92m, 9.05m, respectively. Though it does not show the whole evolution process of the energy spectrum, the main characters of the evolution is presented clearly.

First, the mean frequency of the soliton continuously downshifts as it propagates in the fiber. For example, the mean frequency is 200THz at z=0, 199.3THz at z=7.92m, 198.8THz at z=7.92m. This is known as the soliton self-frequency shift effect.

Second, the energy spectrum gradually broadens as it propagates in the fiber. For example, the spectral width $\Delta\nu=3.0$THz at z=0, 3.3THz at z=7.92m, 3.9THz at z=7.92m. At the same time, the energy spectrum gradually becomes asymmetric and deviates from the shape of square of hyperbolic secant function. The change of the spectrum certainly leads to the change of the pulse shape. In such a case, the pulse cannot preserve the characters of fundamental soliton.

Third, after the 100 fsec fundamental soliton has propagated through roughly 6.28m of the fiber, a Stokes peak appears in its spectrum. And the Stokes spectrum cannot be seperated from the soliton spectrum. Henceforth, energy transfers from the soliton to the Stokes wave quickly. After transmission of 12.56m roughly, the $1.5\mu m$ soliton disappears, because it has changed into the

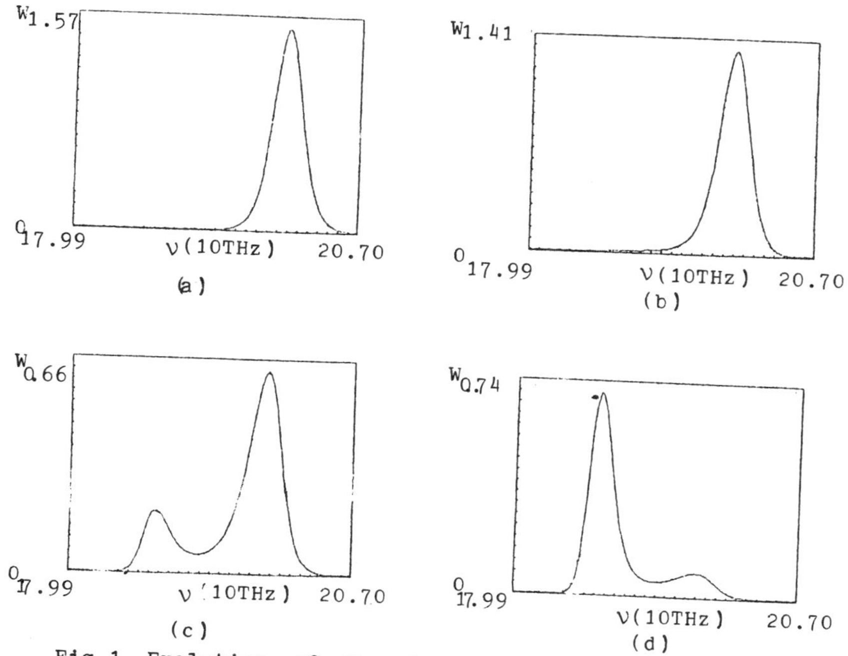

Fig.1 Evolution of the fundamental soliton energy spectrum during transmission, for wavelength $\lambda=1.5\mu$ m, pulsewidth $\tau=100$fsec, dispersion parameter D=15psec/nm/km, optical Kerr coefficient $n_2=3.2\times10^{-16}$ cm^2/w. (a) z=0, (b) z=5.66m, (c) z=7.92m, (d) z=9.05m

Stokes pulse completely.

CONCLUSION

Stimulated Raman process affects strongly transmission of the solitons, whose pulsewidth are narrower than subpicosecond, in a fiber. This is shown not only on obvious self-frequency shift effect, but also on the strong distortion of the energy spectrum. So the fundamental soliton pulse, whose pulsewidth are narrower than subpicosecond, cannot be formed in certain fibers.

References

1 F.M.Mitschke and L.F.Mollenauer, Opt. Lett. 11, 659 (1986).
2 J.P.Gordon, Opt. Lett. 11, 662 (1986).

V. PHOTOCHEMISTRY, PHOTOPHYSICS, AND PHOTOBIOLOGY

PROBING GAS-SURFACE SCATTERING BY LASER MULTIPHOTON IONIZATION*

G. O. Sitz, A. C. Kummel and R. N. Zare
Stanford University
Department of Chemistry
Stanford, California 94305

When an atom or molecule scatters from a surface, the forces acting during the encounter are not isotropic. Instead they possess directionality, and this directional character alters the form of the velocity and angular distribution of the scattered particles as well as the spatial distribution of the angular momentum vectors. These features are conveniently probed via laser multiphoton ionization. New results will be presented for N_2 scattering from Ag(111) in which the N_2 molecules are detected in a quantum-state-specific manner using 2+2 resonance enhanced multiphoton ionization. A well-characterized Ag(111) surface in an ultra high vacuum system is bombarded with a beam of supersonically cooled N_2 seeded in He or H_2. The scattered molecules are rotationally excited and exhibit a "rotational rainbow", that is, an excess population at large ΔJ. By varying the polarization of the laser beam, it is also possible to determine the even and odd moments of the J vector spatial distribution. The even moments correspond to alignment, the odd to orientation.

* This work has been supported by the Office of Naval Research under N00014-78-C-0403.

PHOTOFRAGMENTATION DYNAMICS OF ICl-RARE GAS VAN DER WAALS COMPLEXES[a)]

Janet C. Drobits, John M. Skene and Marsha I. Lester
Department of Chemistry, University of Pennsylvania
Philadelphia, Pennsylvania 19104-6323

ABSTRACT

The vibrational predissociation dynamics of ICl-rare gas van der Waals (vdW) complexes in the A $^3\Pi_1$ and B $^3\Pi_{0+}$ states have been examined at the state-to-state level of detail through optical-optical double resonance methods. After state selective preparation of the vdW complex, the nascent ICl product distribution is probed via excitation of the ICl fragments to an ion pair state. By populating different vibrational levels of the vdW complex, the energetics and lifetime of the unimolecular reaction can be selectively altered. A significant fraction of the available energy is converted to rotational excitation of the ICl product in a highly nonstatistical manner.

INTRODUCTION

Weakly bound van der Waals (vdW) complexes represent prototype systems for examining photochemical change in small molecules. The complexes can be selectively prepared in metastable states having an internal energy content which exceeds the vdW bond strength. Energy is transferred from the initial vibrational state to the vdW bond dissociation coordinate, resulting in the rupture of the weak vdW bond. The rate of vibrational predissociation is a measure of the coupling between the initial state and the continuum of states in the separated molecule limit. The dynamics of the photodissociation event are reflected by the manner in which excess energy is distributed over final product quantum states. In particular, the rotational distribution of the photofragments is expected to be a sensitive probe of the potential energy surface along the reaction coordinate.

The vibrational predissociation dynamics of vdW complexes are being examined with increasingly more detail. Initial work[1] had been directed towards obtaining high resolution fluorescence excitation spectra of the complexes. When spectral linebroadening is detected, the homogeneous component of the linewidth is used to infer vibrational predissociation lifetimes. By dispersing the induced emission, the primary predissociation channels could be identified. For halogen-rare gas complexes, a propensity has been found to populate the highest energetically allowed vibrational level of the halogen fragment. Little information was known, however, about the rotational excitation imparted to the diatomic fragment.

a) This work is supported in part by the National Science Foundation (CHE-85-08552) and the Petroleum Research Fund.

More recently, state-to-state methods have been used to explore the vibrational predissociation dynamics of vdW complexes. A double resonance technique has been devised in this laboratory to prepare ICl-rare gas (ICl-Rg) vdW complexes in a specific vibrational level in either the A $^3\Pi_1$ or B $^3\Pi_{0+}$ valence electronic states and then probe the resultant photofragments.[2-5] A partial potential energy diagram for ICl in the X ($^1\Sigma+$) and A ($^3\Pi_1$) states and corresponding levels for ICl-Rg is shown in Fig. 1. The pump laser prepares the vdW complex in the A state with v_A quanta of ICl stretch. Following vibrational predissociation of the complex, the ICl photofragments remain in the electronically excited A state but typically lose one quantum of ICl vibrational excitation ($\Delta v=-1$):

$$\mathbf{ICl\text{-}Rg\ A(v_A) \longrightarrow ICl\ A(v_A-1,\ J_A) + Rg.}$$

The product state distribution of the halogen fragment is probed with a second tunable dye laser which excites ICl from the A state to the β ($\Omega=1$) ion pair state. The total fluoresence emanating from the ion state is monitored. In addition, as the probe laser is scanned, vdW complexes in sufficiently long-lived intermediate levels of the A (or B) state can be promoted to an ion pair state via a sequential two-photon excitation process.[2,4] Through temporal delay between the pump and probe lasers, the rate of vibrational predissociation can be measured in real-time.[2]

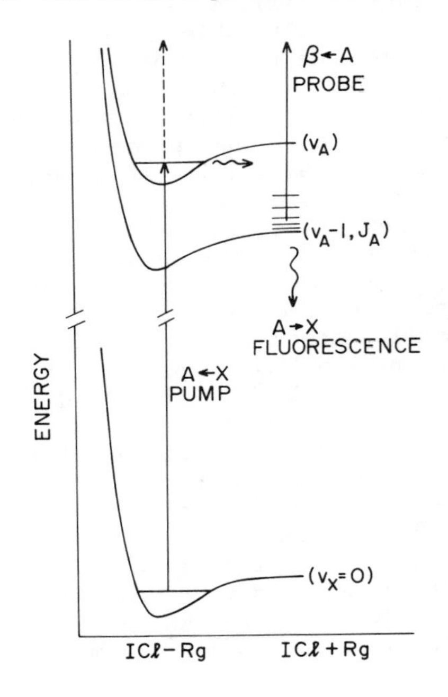

Fig. 1: Potential energy diagram depicts the optical-optical double resonance excitation scheme used to monitor the vibrational predissociation dynamics of ICl-He and ICl-Ne vdW complexes in the A and B electronic states.

The energy available to fragments is the difference between the energy liberated by the loss of one quantum of ICl stretch and the vdW bond dissociation energy. The excess energy is converted into rotational and translational energy of the fragments, subject to energy and angular momentum constraints. The balance of energy had previously been assumed to go primarily into the translational motion of the separating fragments. Our results indicate that a significant fraction of the available energy is deposited into ICl rotational degrees of freedom in a highly nonstatistical manner.

EXPERIMENTAL

ICl-Ne and ICl-He vdW complexes are formed in a continuous supersonic free jet expansion of ICl seeded in various He/Ne carrier gas mixtures.[4] A single excimer laser pumps two narrow bandwidth tunable dye lasers which operate in the spectral region about ICl A-X (B-X) and β-A (E-B) transitions. The two laser beams are spatially overlapped but temporally delayed from 0-20 ns when copropagated through the expansion region. Induced fluorescence from either single or sequential two-photon optical transitions are detected with an appropriate photomultiplier tube/filter combination and processed with a boxcar integrator interfaced to a laboratory computer.

NASCENT PRODUCT STATE DISTRIBUTIONS

The ICl-Ne vibrational predissociation dynamics can be significantly altered through initial state preparation. By populating different halogen vibrational levels in the complex, the energetics of the unimolecular reaction can be dramatically changed. The energy released per quantum changes with vibrational level due to anharmonicity in the halogen potential. Our results indicate that the vibrational predissociation lifetime and the distribution of available energy over rotational and translational degrees of freedom are strong functions of the initial I-Cl vibrational excitation in the vdW complex.

ICl-Ne A state complexes containing 10-25 quanta of halogen stretch (v_A) constitute a particularly rich system for study since the vibrational predissociation lifetime varies over three orders of magnitude, from nanoseconds to a few picoseconds. At low vibrational levels the complexes are exceeding long-lived, permitting direct time-domain measurements of the ensuing dynamics.[2] In contrast, at high vibrational levels vibrational predissociation is rapid as evidenced by homogeneous broadening of the A-X vdW excitation features.[5] The observed trend of decreasing lifetime with increasing halogen vibrational level in ICl-Ne A state complexes is qualitatively consistent with "energy gap"[6,7] models. Since the vibrational predissociation lifetime has been correlated with the amount of momentum to be disposed into products,[6,7] the exact partitioning of excess energy over rotations and translations is also of fundamental importance.

Extensive rotational excitation of ICl is observed in the

photofragmentation of ICl-Ne $A(v_A=14)$ complexes.[2] The extent of ICl rotational excitation via the $\Delta v=-1$ channel changes with the initial halogen vibrational level. As the complex is excited to higher vibrational levels, the maximum ICl rotational level populated, J_{max}, decreases. The highest rotational state formed when the $v_A=14$ level is prepared is $J_{max}=29$. This decreases to $J_{max}=22$ and 12 for the $v_A=19$ and 23 levels, respectively. The fraction of available energy that flows into ICl rotations also decreases from nearly 0.70 at $v_A=14$ to less than 0.33 at $v_A=23$, suggesting a change in vibration to rotation coupling.

At $v_A=23$, vibrational predissociation proceeds with the loss of one or two quanta of I-Cl excitation at comparable rates. The rotational state distributions for the two channels, however, differ enormously. Low rotational levels of ICl $A(v_A=22)$ are primarily populated, $J_A=1-8$, accessing only a small portion of the allowed final states. ICl $A(v_A=21)$ fragments are produced in a broad distribution which extends nearly to the energetic limit. The fraction of available energy deposited into rotations at the maximum populated rotational level is 0.33 and 0.69 for the $\Delta v=-1$ and $\Delta v=-2$ processes, respectively.

The general trends of the rotational product distributions can be qualitatively understood if the vibrational predissociation process is viewed as the second half of a rotationally inelastic collision event. Both the initial I-Cl vibrational excitation of the complex and the number of vibrational quanta lost change the amount of energy released to fragments. This is equivalent to changing the relative translational energy in a full collision. Thus, by analogy to inelastic atom-diatom scattering,[8] the more energetic "half-collision" can be expected to yield a larger average amount of rotational excitation and a broader range of significantly populated final rotational states. These trends are readily observed in our experimental results.[5]

ROTATIONAL RAINBOW SCATTERING

The primary photodissociation channel for ICl-He and ICl-Ne complexes in the B $^3\Pi_{0+}$ state also occurs with the loss of one quantum of halogen stretch. The predissociation dynamics evolve on a nanosecond timescale,[9] yet result in sharply peaked rotational product state distributions.[3] Vibrational predissociation of ICl-He $B(v_B=2)$ vdW complexes generates the ICl $B(v_B=1)$ rotational distribution displayed in Fig. 2. The distribution rises to primary maximum at approximately $J_B=7$ and falls off to about half of the peak population by $J_B=12$. Above $J_B=12$ the rotational distribution is no longer symmetric about the primary peak and goes through a secondary maximum or shoulder at $J_B=16$. The bimodal form of this distribution is discussed below. The ICl photofragments detected upon vibrational predissociation of ICl-He $B(v_B=3)$ complexes exhibit a completely analogous rotational distribution, where the bimodal form of the distribution is also clear. Rapid electronic predissociation of ICl or poor Franck-Condon overlap preclude examination of other B state levels.

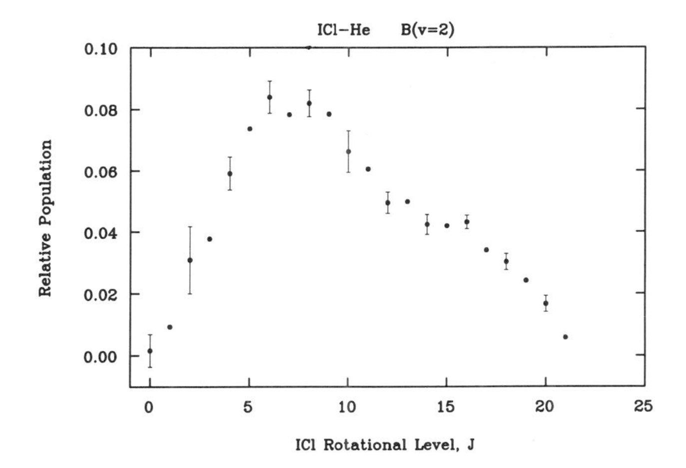

Fig. 2: Nascent rotational distribution of ICl B(v_B=1) state photofragments following vibrational predissociation of ICl-He complexes along the primary Δv=-1 channel.

In contrast, the ICl products from the photofragmentation of ICl-Ne B(v_B=2) vdW complexes follow a distinctly different distribution which is shown in Fig. 3. There is little population in low rotational levels, specifically J_B<10. The distribution sharply rises to a maximum for J_B=15-18 and falls off greatly by J_B=25. Rotational levels as high as J_B=33 are observed, but with little population in each rotational state. In comparing Figs. 2 and 3, we note that the primary peak of the rotational distribution directly scales with the square root of the reduced mass of the vdW complex.

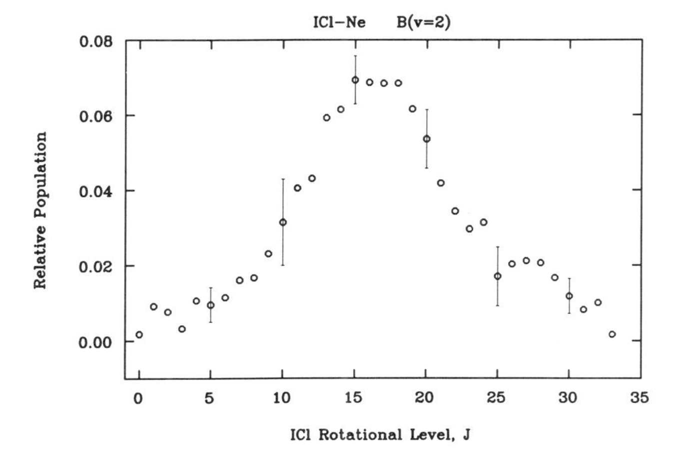

Fig. 3: Rotational excitation of ICl B(v_B= 1) products accompanying vibrational predissociation of ICl-Ne B(v_B=2).

We assert that the dominant contribution to the rotational distribution results from final state interactions occuring upon the sudden separation of the fragments. The excess energy released upon vibrational deactivation of the halogen molecule induces a rotationally inelastic scattering event. Many aspects of the experimental product distributions point towards a scattering explanation. The ICl products formed upon ICl-He photofragmentation exhibit a bimodal rotational distribution. The predominant peak in the rotational distribution scales with the reduced mass of the complex. Furthermore, a larger amount of rotational excitation occurs for more energetic half-collisions.

The distinct structures in the rotational product distributions are attributed to rotational rainbows. Rotational rainbows have been detected for a variety of inelastic scattering processes when there is a strong variation of the interaction potential over angular coordinates.[10,11] The rainbow appears under the conditions of maximum rotational angular momentum transfer. Rotational rainbows are a dynamical effect which scale with collision energy, reduced mass and the initial vibrational state of the molecule. The asymmetry of the ICl-He interaction potential arising from the heteronuclear ICl molecule can lead to a double rainbow in the rotational product distribution. The experimental observation of a bimodal ICl rotational distribution from ICl-He vdW complexes, suggests that the initial wavefunction for the metastate vdW complex accesses a broad range of scattering angles. From these dynamical observations, we hope to obtain an estimate of the "rigidity" or "floppiness" of the vdW complexes which ultimately undergo the vibrational predissociation process.

REFERENCES

1. D.H. Levy, Adv. Chem. Phys. 47, 323 (1981).
2. J.C. Drobits, J.M. Skene, and M.I. Lester, J. Chem. Phys. 84, 2896 (1986).
3. J.M. Skene, J.C. Drobits, and M.I. Lester, J. Chem. Phys. 85, 2329 (1986).
4. J.C. Drobits and M.I. Lester, J. Chem. Phys. 86, 1662 (1987).
5. J.C. Drobits and M.I. Lester, J. Chem. Phys. 88, xxxx (1988).
6. J.A. Beswick and J. Jortner, Adv. Chem. Phys. 47, 363 (1981).
7. G.E. Ewing, Faraday Discuss. Chem. Soc. 73, 325 (1982); G.E. Ewing, J. Phys. Chem. 91, 4662 (1987).
8. R. Schinke and V. Engel, J. Chem. Phys. 83, 5068 (1985); R. Schinke, J. Chem. Phys. 85, 5049 (1986).
9. J.M. Skene and M.I. Lester, Chem. Phys. Lett. 116, 93 (1985).
10. A.W. Kleyn, Comments At. Mol. Phys. 19, 133 (1987).
11. R. Schinke, J. Phys. Chem. 90, 1742 (1986).

INVESTIGATION OF THE TRANSITION STATE REGION OF NEUTRAL BIMOLECULAR REACTIONS BY NEGATIVE ION PHOTODETACHMENT

T. Kitsopoulos, R. B. Metz, A. Weaver, and D. M. Neumark
Department of Chemistry, University of California, Berkeley, CA

It has long been recognized that one of the most important regions of the potential energy surface for a chemical reaction is in the vicinity of the transition state. This is where reactants are transformed to products, and a fundamental understanding of the microscopic, interatomic forces that govern this transformation has been a central goal in the field of reaction dynamics. Considerable effort has been devoted to the development of sophisticated state-to-state reactive scattering experiments[1] in which one attempts to extract information on the transition state region from the dependence of product final state distributions on reactant attributes. An alternative approach has been to develop a spectroscopic probe of the transition state region.[2] Such an experiment should, in principle, be easier to interpret in terms of the detailed features of the potential energy surface than a reactive scattering experiment. We have developed a novel application of negative ion photoelectron spectroscopy in pursuit of the second approach. Specifically, we have observed vibrational and electronic structure in collision complexes of the Cl + HCl and I + HI reactions via photoelectron spectroscopy of $ClHCl^-$ and IHI^-.

The principle of the experiment is as follows. $ClHCl^-$ is a linear, symmetric ion bound by 1 eV with respect to dissociation.[3,4] *Ab initio* calculations on the Cl + HCl reaction predict a barrier to reaction of 6.3 kcal/mole and a linear or nearly transition state.[5] In addition, the recently determined Cl-Cl equilibrium distance in the ion of 3.112 Å[6] is only slightly longer than the corresponding length at the saddle point on the reactive surface. Therefore, since photodetachment is a vertical process, we expect to access the transition state region of the Cl + HCl surface by photodetaching $ClHCl^-$. The experiment was motivated by the possibility that the photoelectron spectrum of $ClHCl^-$ would exhibit vibrational structure characteristic of the unstable ClHCl complex, and that this information would provide a detailed spectroscopic probe of the transition state region for the Cl + HCl reaction.

A pulsed negative ion photoelectron spectrometer[7,8] is used in these studies. $ClHCl^-$ is generated by expanding a mixture of CF_2Cl_2, HCl, and He through a pulsed molecular beam valve and crossing the neutral beam with a 1 keV electron beam just outside the valve orifice. The ions are mass-selected with a time-of-flight mass spectrometer. The mass-selected ions are irradiated by an excimer laser pulse, and a small fraction of the ejected photoelectrons is energy-analyzed with a second time-of-flight system.

Figure 1 shows the photoelectron time-of-flight and kinetic energy spectra for $ClHCl^-$ and $ClDCl^-$. The spectra were taken using the ArF laser line at 6.42 eV. The electron energy resolution at

1.5 eV was 35 meV. Both spectra were averaged for 200,000 laser shots.

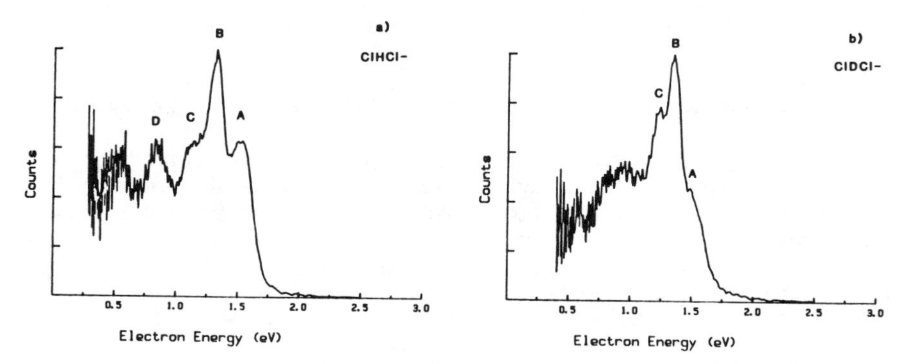

Figure 1. a) Photoelectron spectrum of $ClHCl^-$. Peak positions are (A) 1.512 eV, (B) 1.316 eV, (C) 1.138 eV, (D) 0.850 eV; b) Photoelectron spectrum of $ClDCl^-$. Peak positions are (A) 1.482 eV, (B) 1.345 eV, (C) 1.235 eV.

The estimated peak center positions are indicated in the figure. A comparison of the spectra shows that peak A is at nearly the same energy in both spectra, but the spacing between the peaks in the $ClDCl^-$ spectrum is less than in $ClHCl^-$. This suggests we are observing a progression in a vibrational mode of the ClHCl complex involving H atom motion, and that peak A at 1.512 eV is from the 0-0 transition. The nature of the complex must be that the H atom is interacting strongly with both Cl atoms; if we were accessing the reactant or product valley of the Cl + HCl surface, a considerably higher frequency progression in the HCl stretch(0.357 eV) would be expected. Since the ion and predicted minimum energy configuration near the neutral transition state are both linear(or nearly linear), it is reasonable to assign the observed vibrational progression to the asymmetric stretch of the ClHCl complex. From the electron affinity of Cl(3.617 eV)[9] and the dissociation energy of $ClHCl^-$(1.02 eV),[4] any structure at electron kinetic energies below 1.78 eV in an ArF photoelectron spectrum corresponds to neutral states that lie <u>above</u> the Cl + HCl(v=0) asymptote. Thus, all the peaks represent transitions to states that can dissociate to Cl + HCl(v=0).

The origin of the peaks in the photoelectron spectra can be understood by considering vibrational motion of the ClHCl complex on a collinear Cl+HCl potential energy surface. Figure 2 shows a model LEPS surface[10] for this reaction plotted in hyperspherical coordinates. The shaded region indicates the part of the surface that has reasonable Franck-Condon overlap with the ion. The equilibrium geometry of the ion lies near the center of the shaded region, and the extent of the region represents the 90% probability amplitude boundaries for the vibrational ground state of $ClHCl^-$. Because of the large Cl/H mass ratio, the asymmetric stretch normal mode for the ClHCl complex is basically along a line of constant r (see figure 2).[11] This motion is nearly perpendicular to the entrance and exit valleys on the surface. We therefore expect weak coupling between the asymmetric stretch vibration and dissociation of the complex.

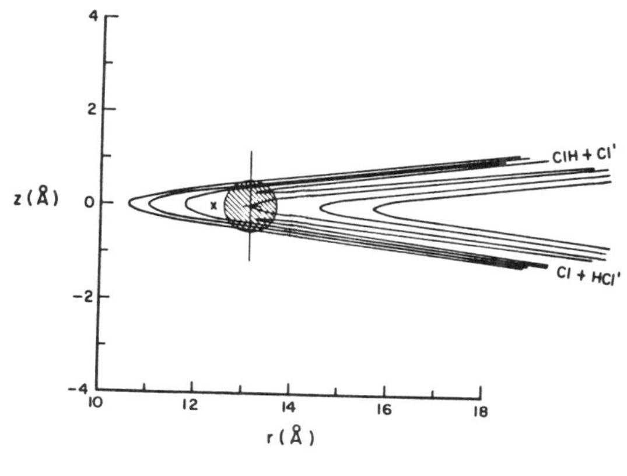

Figure 2. Collinear LEPS surface for Cl+HCl'→Cl'+HCl from reference 11 plotted in hyperspherical coordinates(r,z). Saddle point is marked with x. Franck-Condon accessible region from $ClHCl^-$ (v=0) is shaded (see text). The vertical line through the shaded region illustrates the asymmetric stretch motion of the ClHCl complex.

$$r \cong \sqrt{\frac{\mu_{Cl,HCl}}{\mu_{HCl}}}\, r_{Cl'-Cl}, \quad z=1/2(r_{Cl'-H}-r_{H-Cl}).$$

This allows us, in a first approximation, to compare our experimental results to this or any other model surface by treating the asymmetric stretch of the unstable complex like the analogous motion in a stable molecule. The effective potential for the asymmetric stretch of the complex at the geometry accessed by photodetachment can be found by taking a cut through the LEPS surface at constant r passing through the center of the shaded region. This is a double-minimum potential. One can solve for the eigenvalues and compare their spacing with the peaks in the photoelectron spectrum. Assuming the transitions originate from the v=0 state of the ion, only transitions to even asymmetric stretch states of the neutral are allowed. In addition, we can simulate the expected intensity distribution in the photoelectron spectrum by calculating the Franck-Condon factors between the v=0 state of $ClHCl^-$ and the neutral asymmetric stretch states. Our preliminary calculations indicate that while the model LEPS surface gives eigenvalues that are within 25% of the experimental values, the intensities are quite far off. We are currently devising new surfaces which should improve the fit to the data.

A feature of note in the spectra is that the peaks are considerably broader than the instrumental resolution of 30-35 meV. One possibility is that the complex dissociates rapidly after photodetachment. However, another intriguing possibility is suggested by reactive scattering calculations[10] on Cl + HCl which show sharp resonance structure attributed to long-lived, vibrationally excited ClHCl states. If this picture is correct, then at least some of the peaks in our spectra are actually composed of several closely spaced transitions. These transitions involve excitation of symmetric stretch levels in the ClHCl complex; the predicted spacing between them is comparable to our current resolution. Higher resolution exepriments are currently underway to determine if there is in fact any further underlying structure in the photoelectron spectra of $ClHCl^-$ and $ClDCl^-$.

Finally, the photoelectron spectrum of IHI^- is shown in Figure 3. The fastest peaks in the spectrum(marked A) appear analogous to the $ClHCl^-$ spectrum and are probably a progression in the asymmetric stretch of the IHI complex. In addition, two prominent bands appear at lower electron kinetic energy. We spectulate that these result from transitions to excited electronic states of the I + HI system. Possible candidates are the two $^2\Pi$ spin-orbit excited states[12] that result from the interaction of $I(^2P_{3/2,1/2})$ with HI.

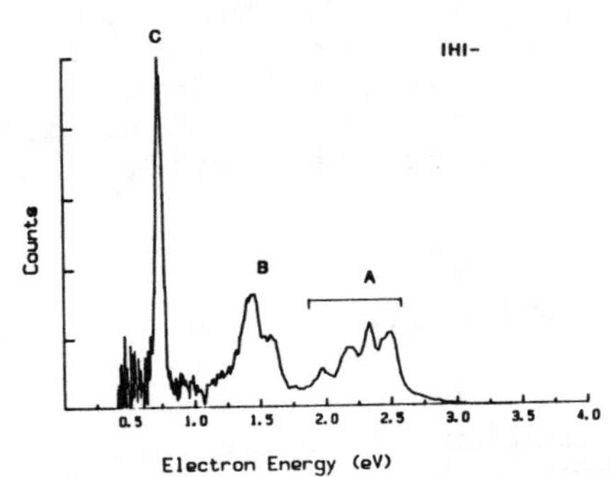

Figure 3. IHI^- photoelectron spectrum. A, B, and C are attributed to different electronic states of I+HI.

REFERENCES

1. S. R. Leone, Ann. Rev. Phys. Chem. 35, 109 (1984).
2. J.C. Polanyi, M. G. Prisant, and J. S. Wright, J. Phys. Chem. 91, 4727 (1987); P. R. Brooks, Chem. Reviews (to be published).
3. C. A. Wight, B. S. Ault, and L. Andrews, J. Chem. Phys. 65, 1244 (1976); D. E. Milligan and M. E. Jacox, J. Chem. Phys. 53, 2034 (1970).
4. G. Caldwell and P. Kebarle, Can. J. Chem. 63, 1399 (1985).
5. B. C. Garrett, D. G. Truhlar, A. F. Wagner, and T. H. Dunning, Jr., J. Chem. Phys. 78, 4400 (1983).
6. K. Kawaguchi, J. Chem. Phys. (submitted).
7. L. A. Posey, M. J. Deluca, and M. A. Johnson, Chem. Phys. Lett. 131, 170 (1986).
8. R. B. Metz, T. Kitsopoulos, A. Weaver, and D. M. Neumark, J. Chem. Phys. 88, xxxx (1988).
9. H. Hotop and W. C. Lineberger, J. Phys. Chem. Ref. Data 14, 731 (1985).
10. D. K. Bondi, J. N. L. Connor, J. Manz, and J. Romelt, Mol. Phys. 50, 467 (1983).
11. D. C. Clary and J. N. L. Connor, J. Phys. Chem. 88, 2758 (1984).
12. N. C. Firth and R. Grice, J. Chem. Soc. Faraday Trans. 2 83, 1023 (1987).

LASER-ASSISTED CHEMISTRY IN THE REACTION OF $Mg(^1S)$ WITH CO_2 TO YIELD $MgO(B^1\Sigma^+)$

Joseph J. BelBruno and George A. Raiche
Dartmouth College, Department of Chemistry, Hanover, NH 03755

ABSTRACT

A photon of energy approximately equal to that of the endoergicity is shown to promote the title reaction. The photon is not resonant with any state of the reactants or products, but is absorbed by the quasimolecule formed during the collision. The process is carried out in an optical heat pipe and the reaction is observed by means of emission spectroscopy. A kinetic model yielding a bimolecular-type rate expression is shown to qualitatively predict the experiment results.

INTRODUCTION

A description of the crucial events and forces involved when atoms and molecules interact is fundamental to the detailed understanding of chemical reactivity. Both reaction rates and product distributions are determined during the critical moments in which the atoms belong to neither reactants nor products, but rather some transient quasimolecule, intermediate between the two states. The goal of this research is an improvement of the understanding of the reaction event and an investigation into the feasibility of influencing the reactivity through manipulation of the transient complex.

THE KINETIC MODEL

We have developed a model[1] which does not depend upon the availability of high quality potential surfaces, but still conveys a qualitative description of the experiment. The model is also useful in choosing reactions which may exhibit laser-assisted effects.

The problem is approached as the unimolecular decay of an excited transient complex

$$A + B \rightarrow [A\text{-}B] + h\nu \rightarrow [A\text{-}B]^* \rightarrow \text{Products} \tag{1}$$

The concentration of these species is determined using the radial distribution function and first-order perturbation theory. Statistical decay into products is assumed and the following expression is obtained

$$\text{Rate} = \frac{16\pi^4 \mu^2 I}{c\, h^2 (\Delta\omega)^2} N_A N_B k_d \int r^2 \exp[-V(r)/kT]\, dr \tag{2}$$

Many of the quantities required in equation 2 are unknown. Therefore, using the known transition moment of one the reactants, the appropriate detuning, a $1/r^6$ potential and known molecular constants, the final rate constant expression is

$$k_d = \frac{1.38 \times 10^{23} \mu_A^2 P_d \tau^{-1} I r_m^2}{(\Delta\omega_A)^2} \tag{3}$$

A comparison of the model with early experiments may be found in Ref.1.

The title reaction is endoergic[2] by ~36,000. cm^{-1}. This energy is greater than that required to pump the Mg(^{1}P) resoanance level. Moreover, other experiments have indicated that direct pumping of the resonance level is ineffective in promotion of the title reaction and will not be a competitive process in any event. Application of the model as described above[3] predicts a cross section of 0.004Å for the laser-assisted process.

EXPERIMENTAL

The apparatus and technique have been previously described[3]. A frequency doubled dye laser (10μJ, collimated) is the pump source and is scanned ~1000 cm^{-1} toward the wings from the Mg resonance line. The reactants are prepared in a heat pipe and the emission is dispersed by a monochromator before averaging with a gated integrator. The typical point represents ~1500 laser shots.

RESULTS AND DISCUSSION

Typical experimental results (excitation profiles) are shown in Fig.1. The triangles show the normalized emission (signal x detuning^2) at 500nm as a function of detuning from the Mg resonance for a mixture of 50mtorr of CO_2/2 torr of He at 350°C. The squares show the signal in the absence of CO_2. The additional signal in the presence of CO_2 is obvious. The signal only occurs in the blue wing, in agreement with the energy

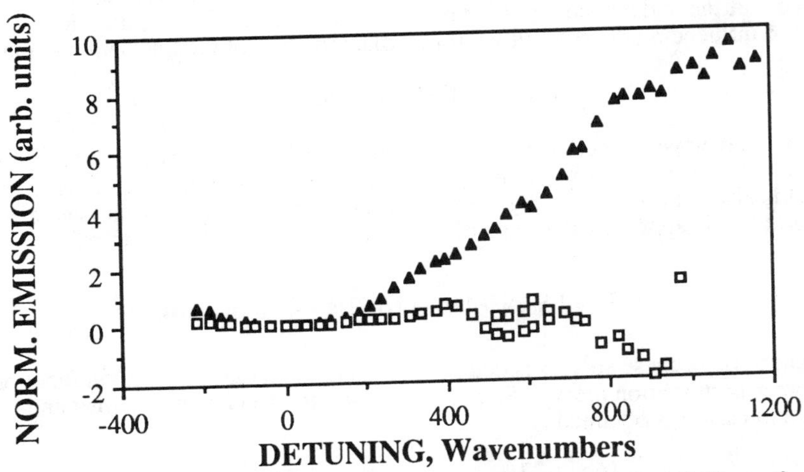

Figure 1. Excitation Profile for Laser-Assisted Reaction

requirements of the process. Fig. 2 presents a portion of the emission spectrum for a detuning of ~590cm^{-1} in the presence (triangles) and absence (squares) of CO_2. A significant emission spectrum was recorded even at such a large detuning. The emission is linear in Mg concentration, CO_2 pressure (below ~100mtorr) and laser energy. Collisional line broadening has been eliminated as a source of the emission by substituting Ar for CO_2 and comparing the excitation profiles. ASE and multiphoton

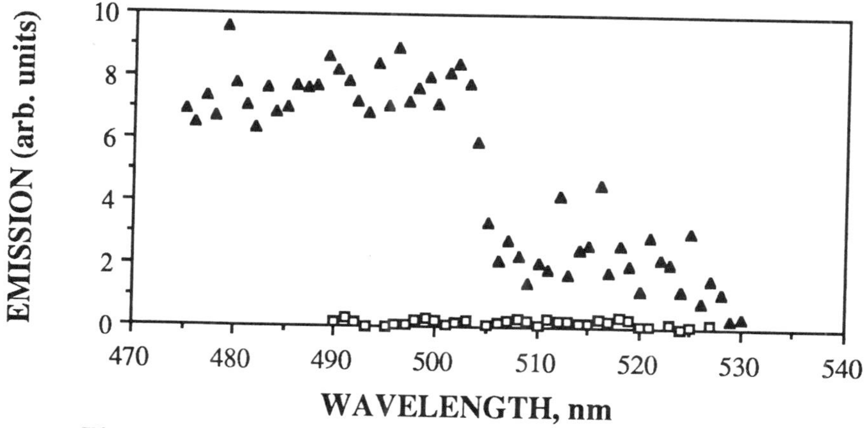

Figure 2. Emission Spectrum for Laser-Assisted Reaction

processes have also been investigated and not found to be related to the signal. In addition, the signal is coincident with the pump laser, eliminating a long chain of kinetic processes as the source.

The absence of any signal which could be attributed to non-reactive scattering into a $Mg(^1P)+CO_2(X)$ channel differentiates this work from a previous study[4] involving H_2 rather than CO_2. Since the energetics are similar to the title reaction, the lack of a significant signal for this channel may indicate the existence of dynamical restrictions in the collision process. This is an obvious topic for future research.

In summary, we have observed an emission signal attributable to $MgO(B^1\Sigma^+)$ in the far-blue wing of the Mg resonance spectrum. This signal results from the laser-assisted reaction of Mg with carbon dioxide by transfer of the reactants to a potential surface correlating with the observed products.This research also represents an extension of the laser-switched process to slightly larger molecular systems. Calculations along the lines of Kleiber and Stwalley[4] are now in progress.

REFERENCES

1. J.J. BelBruno, Chem. Phys. Lett. **117**, 592 (1985).
2. K.P. Huber and G. Herzberg, *Molecular Spectra and Structure Volume 4* (Van Nostrand Reinhold Co., NY 1979).
3. G.A. Raiche and J.J. BelBruno, Chem. Phys. Lett., submitted.
4. P.D. Kleiber, A.M. Lyyra, K.M. Sando, V. Zafiropulos and W.C. Stwalley, J. Chem. Phys. **85**, 5493 (1986).

PHOTOFRAGMENTATION OF SEMICONDUCTOR POSITIVE CLUSTER IONS

Q. Zhang, Y. Liu, R. F. Curl, F. K. Tittel, and R. E. Smalley
Rice Quantum Institute, Rice University, Houston, TX 77251

ABSTRACT

Si, Ge, and GaAs positive cluster ions containing up to sixty atoms have been prepared by laser vaporization and supersonic beam expansion and their laser photofragmentation studied using tandem time-of-flight mass spectrometry. The fragmentation pattern observed for GaAs is very similar to that observed for metal clusters and is consistent with the sequential loss of atoms. Si and Ge clusters appear to fragment by a fission process, Si_n^+ fragments primarily into positive ions in the 6-11 size range with a subsidiary channel corresponding to loss of a single atom. Ge_n^+ also gives clusters in the 6-11 size range at relatively high fluence and has a channel corresponding to loss of a single atom at low fluence. At intermediate fluences, channels corresponding to sequential loss of Ge_{10} and Ge_7 are observed.

INTRODUCTION

Semiconductors are generally covalent solids with sp^3 hybridization having the diamond structure. However, the electronic and geometric structures of semiconductor clusters are expected to be quite different from the bulk because of extensive surface restructuring. The aim of this work is to investigate the properties of semiconductor clusters with the hope of gaining insight into the chemical forces driving surface restructuring.

EXPERIMENTAL

The apparatus has been described elsewhere[1,2]. Briefly, laser vaporization of semiconductor material from a rotating disc takes place in the center of a helium carrier gas pulse. The hot plasma is entrained into a flow tube where clustering and thermalization takes place, and is then cooled to a few K by supersonic expansion into a vacuum. The supersonic jet is skimmed into a molecular beam containing neutral and ionic clusters. Positive clusters are extracted into a field-free flight tube, mass resolved by time-of-flight, and detected by an in-line detector at the end of the tube. In these studies, the maximum cluster size produced was n=80 for Si_n^+, n=50 for Ge_n^+, and x+y=40 for $Ga_xAs_y^+$. Clusters can be mass selected by a timed gate, and fragmented by a probe laser. After interaction with the probe laser, the fragmentation products are accelerated into a perpendicular flight tube, mass-analyzed by time-of-flight, and detected by two microchannel plates.

RESULTS AND DISCUSSION

The mass distributions of the Si, Ge, and GaAs positive cluster ions are smooth for n>20 with no indication of magic numbers. For the discrete near-UV probe laser wavelengths (third and fourth harmonic of the Nd:YAG and ArF and KrF excimer lines) used in these studies there is no qualitative dependence on probe laser wavelength, but the fragmentation patterns do depend strongly upon laser fluence.

Fig. 1 shows a typical fragmentation pattern of a GaAs cluster ion. This pattern strongly resembles those found for the fragmentation of metal cluster ions[3]. It suggests the sequential loss of atoms. However, there is a clear even/odd intensity alternation in the product distribution with the odd fragments being more intense than their even neighbors. A similar intensity alternation has been observed[4] for the fragmentation of GaAs negative cluster ions. These observations suggest that the cluster ions containing an even number of electrons contain no dangling bonds.

The fragmentation patterns of Si and Ge positive cluster ions are very different from GaAs or metal clusters and appear to be dominated by fission processes. Thus for Si_n^+ and Ge_n^+ with n=12 to 29 the observed daughter ions are about half the original size. Similar behavior has been observed[4] for the corresponding negative ions and the patterns are roughly the same for the ions of the same size in the two charge states with a few exceptions. Si_n^+ with n>30 fragment either by loss a single atom or by "explosion" into ions in the 6-11 size range. Intermediate products are not observed. We believe that Si_6^+ - Si_{11}^+ are the largest fragments as the ionization potential is expected to decrease with increasing size and therefore the charge should settle on the largest fragment as particles separate. On the other hand, this fragmentation pattern is observed for Ge_n^+ with n 30-50 only at higher laser fluences. At low fluences Ge_n^+ appears to lose Ge_{10}, and sometimes Ge_7, in a stepwise manner as shown in Fig. 2.

It appears unlikely that these clusters are loosely bound aggregates as the fragmentation of Si_{60}^+ to Si_{10}^+ is at least quadratic in ArF fluence as shown in Fig. 3 indicating that this fragmentation is a two-photon process. Thus the loss of Ge_{10} and Ge_7 seems to indicate that these neutral clusters are exceptionally stable or there is a natural cleavage process giving rise to these fragments.

ACKNOWLEDGMENT

This work was supported by the Army Office of Scientific Research and the Robert A. Welch Foundation.

REFERENCES

1. Q. Zhang, Y. Liu, R. F. Curl, F. K. Tittel, and R. E. Smalley, J. Chem. Phys. (accepted).
2. S. C. O'Brien, J. R. Heath, R. F. Curl, and R. E. Smalley, J. Chem. Phys. 88 (2) (1988).
3. P. J. Brucat, L. -S. Zheng, C. L. Pettiette, S. Yang, and R. E. Smalley, J. Chem. Phys. 84, 3078 (1986).
4. Y. Liu, Q. Zhang, F. K. Tittel, R. F. Curl, and R. E. Smalley, J. Chem. Phys. 85, 7434 (1986).

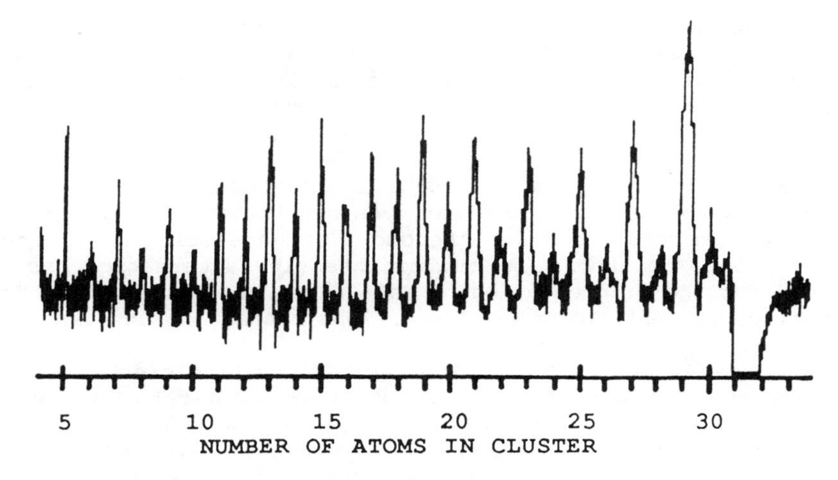

Fig. 1. Fragmentation of GaxAsy+ with x+y=31 with 27 mJ/cm^2 355 nm laser. The results were obtained by subtracting two data sets: one with laser on, and the other with laser off.

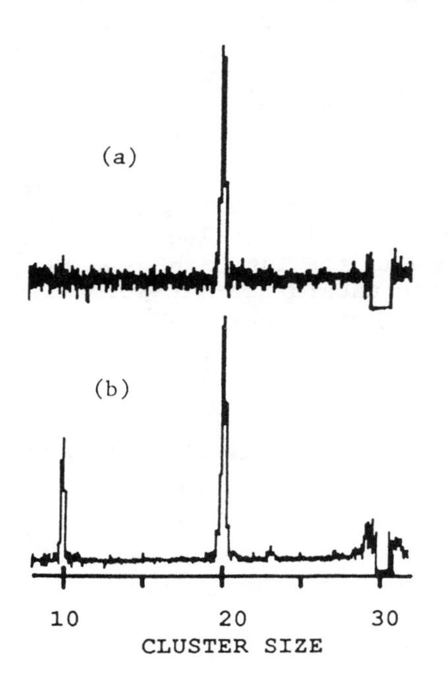

Fig. 2. Fragmentation of Ge_{30}^+ with (a): 1 mJ, and (b): 5 mJ 532 nm laser.

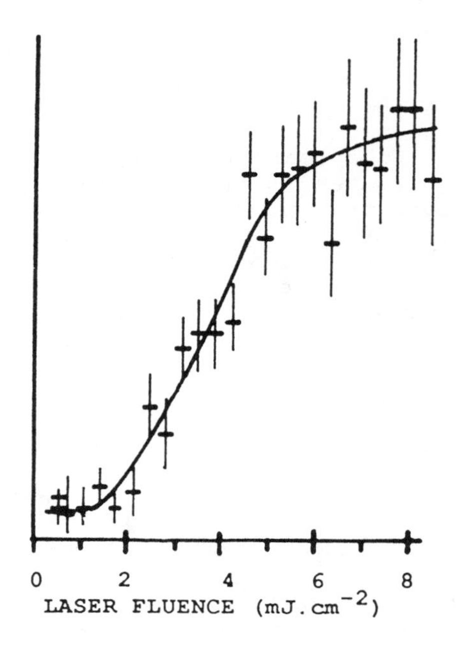

Fig. 3. Fluence dependence of Si_{10}^+ signal intensities produced from Si_{60}^+ with 193 nm laser.

COOPERATIVE FLUORESCENCE AS A PROBE OF COLLISION AND FRAGMENTATION DYNAMICS

G. Kurizki and G. Hose
Chemical Physics Department, Weizmann Institute of Science
Rehovot, Israel 76100

A. Ben Reuven
School of Chemistry, Tel-Aviv University
Ramat-Aviv, Israel 69978

In an experiment performed by Grangier, Aspect and Vigué[1], following Diebold's suggestion[2], interference effects were observed in the time-resolved intensity of Ca($^1P \to {}^1S$) fluorescence subsequent to the dissociation of Ca_2 via the $^1\Pi_u^*$ state. This interference was recognized[1] to be the consequence of molecular ungerade (u) parity, which requires the atomic fragments A,B to form a symmetric superposition $2^{-1/2}\{|^1P\rangle_A|^1S\rangle_B + |^1P\rangle_B|^1S\rangle_A\}$. Hence, the emitting atoms are correlated dipoles whose emissions interfere, and the interference pattern changes with the difference in optical paths from the two receding atoms to the detector, $\vec{k}\cdot\vec{R}(t)$, where $\vec{k}$ is the emission wavevector and $\vec{R}(t)$ the changing interatomic separation.

A detailed analysis of this phenomenon, referred to as time-resolved cooperative fluorescence (TRCF), was given[3] based on Agarwal's[4] master-equation approach. When this theory is applied to <u>spinless</u> homonuclear diatoms, dissociated by one photon (via a u state) or two photons (via a g state) and emitting a single photon, the following total emission rate $\dot{P}$ (integrated over emission angles) is obtained[3]:

$$\dot{P}(t)=\sum_{\pm} \rho_{\pm\pm}(0)[\gamma\pm\Gamma_\Lambda(R(t))]\exp[-\int_0^t (\gamma\pm\Gamma_\Lambda)dt'] \qquad (1)$$

Here +(-), pertain to u(g) parity and Λ denotes $^1\Pi^*$ or $^1\Sigma^*$ excitations. The $\Lambda_{u(g)}$ adiabatic states contribute to the emission according to their relative populations $\rho_{\pm\pm}(0)$, where R(0) is chosen to be a separation of few nm, at which only dipole-dipole interactions are significant. The instantaneous rate of decay $\gamma\pm\Gamma_\Lambda$ of each such state is the inverse lifetime of the $^1P \to {}^1S$ atomic transition γ modified by $\pm\Gamma_\Lambda(R(t))$, representing the effect of interference integrated over all emission angles. This rate is periodic in the argument kR(t), and exhibits oscillations (ringing) at the frequency $k\dot{R}=kv$. The nonexponential decay factor in Eq. (1) arises from the cumulative population decay up to time t.

The diagnostic significance of TRCF patterns describable by (1) lies, firstly, in their sensitivity to Λ: because for $kR \gtrsim 2\pi$ $\Gamma_\Pi/\gamma \sim \sin kR/kR$, while $\Gamma_\Sigma/\gamma \sim (kR)^{-3}$, Π^*-state dissociation produces much more pronounced ringing than Σ^*-state dissociation. Secondly, the initial emission rate $\dot{P}(0)$ reflects the parity of the state: At small separations $kR \ll 1$ we have $\Gamma_\Pi \simeq \Gamma_\Sigma \simeq \gamma$, so that superradiance occurs for a u state (constructive interference), whereas subradiance ($\dot{P}(0)/\gamma \simeq 0$) occurs for a g state (destructive interference), in keeping with the molecular dipole selection rules which are valid for $kR \ll 1$. It is therefore possible to infer from the emission pattern the predominant Λ-state and parity for a dissociation channel with unknown characteristics, e.g. in dissociative recombination of a diatomic ion with an electron.

In search of further diagnostic applications of TRCF we have studied dissociating homonuclear alkali diatoms. In what follows the theory of an experiment suggested for Li_2 is outlined[5]:

(a) Excitation: The continuum of the $1^1\Sigma_u^+$ state is excited selectively by absorbing a photon whose wavelength varies between 4306Å ($1^1\Sigma_u^+$ dissociation threshold) and 3960Å (below which the $^1\Pi_u$ dissociation channel is accessible). The excitation should be preceded by vibrotational cooling of the Li_2 ground state in order to avoid dynamical Coriolis couplings.

(b) Dynamical evolution: Following the above excitation, the diatom evolves into a superposition of two adiabatic states while traversing the region $1nm \lesssim R \lesssim 20nm$. These two states have $\Omega = |\Lambda + \mathcal{S}| = 0$ projections of the total electronic angular momentum on the internuclear axis (Λ and $\mathcal{S}$ being the orbital and spin projections) because the value $\Omega=0$ of the excitation state $^1\Sigma_u^+$ is conserved for negligible Coriolis couplings. Spin-orbit coupling $\delta = 0{,}3\ cm^{-1}$ mixes the eigenstates of the dipole-dipole interaction $V(R)$, $^1\Sigma_u^+(\Omega=0)$ and $^3\Pi_u(\Omega=0)$, creating two adiabatic states labelled by $|\pm\rangle$ with potential energies $\varepsilon_\pm(R)$. At $R \sim 1nm$ ($\delta \ll |V(R)|$) $^1\Sigma_u^+ \to |-\rangle$. At $R \gtrsim 20nm$ ε_+ and ε_- coincide with the energetic limits of $^2P_{3/2} + {}^2S_{1/2}$ and $^2P_{1/2} + {}^2S_{1/2}$, respectively ($\varepsilon_+ - \varepsilon_- \simeq \delta$). The nonadiabatic wavefunction $\psi = b_+|+\rangle + b_-|-\rangle$ stops evolving at $R \gtrsim 20nm$, as found by integrating the two semiclassical coupled equations for $\dot{b}_\pm(t)$ driven by the radial kinetic energy of dissociation . As the dissociation velocity v grows, the population of the excitation state $\rho_{--} = |b_-|^2$ is increasingly transferred to that of the other state $\rho_{++} = |b_+|^2$, while the amplitude $A = 2|b_+ b_-|$ and phase $\Phi = \arg(b_+/b_-)$ of the coherence between the states vary non-monotonically.

(c) Emission: Following the delay time t_D required to cross the region of dynamical evolution, the temporal variation of $b_\pm$ is mainly due to radiative coupling. Our general master equation[3] yields for the emission rate of $Li(^2P \to {}^2S)$, without resolving the doublet lines:

$$\dot{P}(\tilde{t} = t - t_D) \simeq \sum_\pm \rho_{\pm\pm}(t_D)(\gamma + \Gamma_\pm)\exp[-\int_0^{\tilde{t}} (\gamma + \Gamma_\pm) d\tilde{t}'] + A(t_D)\Gamma_c \cos(\delta\tilde{t} + \Phi)\exp[-\int_0^{\tilde{t}} (\gamma + \Gamma_+ + \Gamma_-)\, d\tilde{t}'] \qquad (2)$$

where $\Gamma_\pm \cong \pm(2\Gamma_\Pi/3 - \Gamma_\Sigma/3)$; $\Gamma_c \simeq 2^{1/2}(\Gamma_\Pi + \Gamma_\Sigma)/3$. In Fig. 1 we have plotted two normalized TRCF patterns (solid curves) $\dot{P}(\tilde{t})/\gamma$ obtained from (2) with $\rho_{\pm\pm}$, A and Φ calculated semiclasically for the corresponding v. The coherence term oscillates at two combination ("superbeat") frequencies $\delta \pm kv$. The population contribution (dashed curves) oscillates with the ringing frequency kv and has the initial value $1+(\rho_{++}-\rho_{--})/3$. Hence the population difference can be inferred from the initial asymmetry of the fast superbeats about the normal rate $\dot{P}/\gamma=1$. The overall initial rate $\dot{P}/\gamma \simeq 1+A\cos\Phi$ can serve to deduce Φ, which determines along with the ungerade parity the initial superradiant or subradiant character of the emission. Thus, whereas conventional spectroscopy can determine only the relative populations of adiabatic states, TRCF patterns are sensitive to their coherence too, and therefore can provide more complete information on nonadiabatic dissociation.

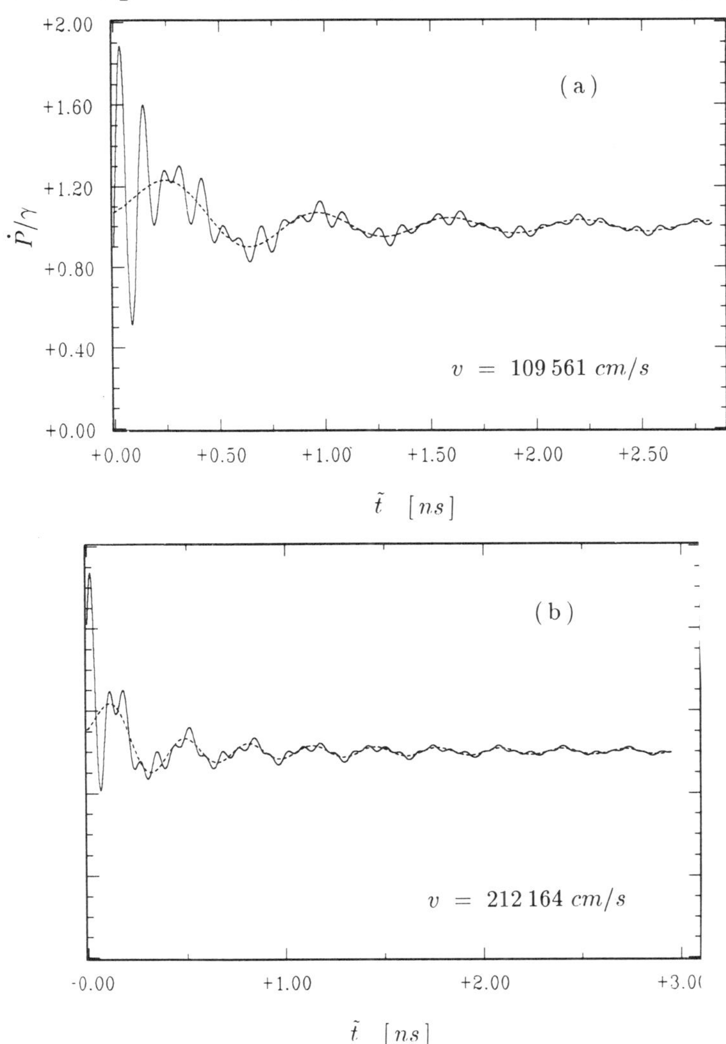

Figure 1. TRCF pattern for Li_2 dissociated via $1^1\Sigma_u^+$ at two dissociation velocities corresponding to (a) $\cos\Phi>0$ and (b) $\cos\Phi<0$.

References

1. P. Grangier, A. Aspect and J. Vigué, Phys.Rev.Lett. 54, 418 (1985).
2. G.J. Diebold, Phys.Rev.Lett. 51, 1344 (1983); Phys.Rev. A32, 1458 (1985).
3. G. Kurizki and A. Ben-Reuven, (a) Phys.Rev. A32, 2560 (1985); (b) Phys.Rev. A36, 90 (1987).
4. G.S. Agarwal, Springer Tracts 70, 1974.
5. G. Kurizki, G. Hose and A. Ben-Reuven, Phys.Rev.Lett. (submitted).

AN ADVANCED AIRBORNE PHOTOFRAGMENTATION/TWO-PHOTON LASER-INDUCED FLUORESCENCE INSTRUMENT FOR THE INSITU DETECTION OF THE ATMOSPHERIC TRACE GAS SPECIES OF NO, NO_2, NO_y, NH_3. AND SO_2*

J. Bradshaw, S. Sandholm, C. Van Dijk, M. Rodgers, and D. Davis
Georgia Institute of Technology, School of Geophysical Sciences, Atlanta, Georgia 30332

ABSTRACT

The third generation Georgia Tech. airborne laser-induced fluorescence (LIF) system currently being developed will offer a significant enhancement in overall system versatility. This newly configured system utilizes a common set of excitation lasers for the detection of the atmospheric trace gas species NO, NO_2, NO_y, NH_3, and SO_2. All of these systems have demonstrated performance in the parts-per-trillion by volume (pptv) range (10^7-10^8 molecules/cm^3) under ambient field sampling conditions. In addition to these now field proven detection systems, several other trace gas detection systems are currently undergoing laboratory development. These include systems for detecting the species OH, HONO, H_2S, CS_2, and CH_2O.

INTRODUCTION

Earlier laser efforts at detecting a variety of trace gas species under atmospheric conditions of composition and pressure have involved single-photon excitation of the species with the monitoring of the resulting fluorescence occurring at wavelengths that are red shifted relative to the excitation wavelength. Under most atmospheric sampling conditions, these single-photon LIF (SP-LIF) techniques exhibit detection limits which are controlled by the noise associated with the background fluorescence occurring from other gas phase species, aerosols, and the cell walls of the sampling chamber. If this non-resonant fluorescence background is spectrally and temporally stable, conventional weak signal extraction techniques work well. Unfortunately, the background is typically found to vary significantly as a result of inhomogenities in the chemical composition of the atmosphere. The resulting impact on instrument performance is that of significantly degrading the achievable detection limit. While the near-simultaneous differential SP-LIF methods (2λ-SP-LIF) compensate to some degree for the temporal variabilty of the background fluorescence, these systems still exhibit detection limits which are still controlled by non-resonant background fluorescence noise. The attributes and limitations of the 2λ-SP-LIF techniques for atmospheric monitoring of the trace species OH, NO, and SO_2 have now been well documented[1].

* Supported by NSF grant ATM-861026 and by NASA grant NAG-1-50

Several atmospheric trace gas species have been demonstrated to avail themselves to the method of sequential two-photon LIF (STP-LIF). Among these are the species NO, OH, Hg, and I_2[2]. In this technique, the molecule is stepwise excited first into an upper excited state (either vibronic or electronic) at a wavelength, λ_1, and then further excited into a still higher energy electronic state at a second wavelength, λ_2. The resultant fluorescence is then observed at a wavelength, λ_3, which is spectrally shifted toward the blue from either the λ_1 or λ_2 excitation wavelengths. In these systems the laser associated resonant and non-resonant background signals can be reduced to negligible levels by the proper choice of solar blind PMT's in conjunction with long-wavelength cutoff filters. The detection limits of these STP-LIF systems have been proven to be one to two orders of magnitude lower than their 2λ-SP-LIF counterparts under ambient sampling conditions.

Although the 2λ-SP-LIF and the STP-LIF techniques can be expected to provide an effective means of detecting numerous trace gas species at natural tropospheric concentration levels, a great many others are not detectable by these methods due to the absence of bonding excited states that fluorescence strongly. In the latter case, the absence of fluorescence can be overcome if the species can be made to photodecompose yielding a photofragment species which can be probed via LIF (PF-LIF)[3]. However, it should be pointed out that monitoring the luminescence emitted by photofragments which are born directly into excited states will, in general, not provide a sensitive atmopsheric sensor. In this latter case, the resulting luminescence signal will be accompanied by a large non-resonant background fluorescence signal produced from the photolysis laser. The differential LIF method used in the 2λ-SP-LIF approach can not usually be applied due to the broadband absorption characteristics normally associated with this class of species.

However, if the photofragmentation species produced by the photolysis laser at wavelength λ_1 is born in a long lived metastable or excited vibronic state, a type of LIF excitation can be used in which the resultant fluorescence wavelength, is blue shifted relative to the second excitation laser wavelength λ_2. The laser generated noise characteristics of this system will be nearly identical to those demonstrated by the STP-LIF techniques. In the photofragmentation case, an appropriate time delay is utilized to allow the background signal produced by the photolysis laser to decay to near negligible levels.

The systems to be deployed in our third generation airborne LIF instrument incorporates combinations of the 2λ-SP-LIF, the STP-LIF, and the PF-LIF methodologies.

CURRENT FIELD TESTED SYSTEMS

The newly configured Ga. Tech. airborne LIF system utilizes a common set of excitation lasers for the detection of NO, NO_2, NO_y,

LASER-INDUCED FRAGMENTATION OF VAN DER WAALS CLUSTERS INVOLVING LARGE AROMATIC MOLECULES

R. Jefferson Babbitt, Andrew J. Kaziska, Andrea L. Motyka, Stacey A. Wittmeyer and Michael R. Topp

University of Pennsylvania, Philadelphia, Pennsylvania 19104-6323

FLUORESCENCE SPECTROSCOPY OF PERYLENE 1:1 COMPLEXES WITH ARGON AND METHANE

Studies of the dissipation of excess molecular vibrational energy represent an important part of current research in clusters, as well as in the condensed phase. In fact, there are strong indications from condensed phase work that, at high energies, intramolecular coupling effects outweigh vibrational energy relaxation to the matrix, so that very similar results begin to be obtained for condensed phase and isolated molecule studies in this regime.(1)

The presence of low-frequency modes in van der Waals clusters provides new channels for the coupling of vibrational energy out of optically prepared "zero-order" states of an aromatic molecule. This is readily demonstrated for the case of perylene complexation with Ar and CH_4, shown in Fig. 1 (left) and (right), respectively.

Vibrational predissociation is indicated by the appearance of emission characteristic of the "cold" parent species, P, following the injection of 705 cm^{-1} and 900 cm^{-1}, respectively into the Ar and CH_4 clusters.(2) Thus, the emission spectrum for 705 cm^{-1} excitation of perylene/Ar_1 is a simple sum of "cold" bare molecule emission and "unrelaxed" cluster emission, indicating that about 40% of the clusters have vibrationally predissociated. A result for 900 cm^{-1} excitation is also given, showing a slightly broadened bare molecule emission with just a trace of residual "unrelaxed" cluster emission. This shows vibrational predissociation to be much faster than the fluorescence decay rate for this excitation energy. Data have also been obtained for perylene/Ar_1 at 780 cm^{-1} and 796 cm^{-1} vibrational energy, indicating an intermediate predissociation efficiency (70-80%). Using the fluorescence decay time as an approximate reference, this indicates that the predissociation lifetime varies over the range 5 ns to <100 ps for a 200 cm^{-1} increase in internal energy.

The binding energy of perylene/$(CH_4)_1$ is reliably calculated to be about 30% higher than that of the Ar_1 complex.(3) The dispersed fluorescence spectra show that, at 353 cm^{-1}, the energy is 70-80% randomized. At 705 cm^{-1} excitation, the unrelaxed emission now corresponds only to ≈4% resonance emission, equivalent to a randomization time <100 ps. There is no evidence, at 705 cm^{-1} excitation, for predissociation. At 900 cm^{-1}, the spectrum is almost indistinguishable from that of bare perylene excited into the electronic origin. This indicates that very little excess energy is available for the perylene fragment. Moreover, even at this energy, which is only about 100 cm^{-1} more than the estimated binding energy, virtually no component of the cluster fluorescence remains, indicating a quite rapid predissociation event.

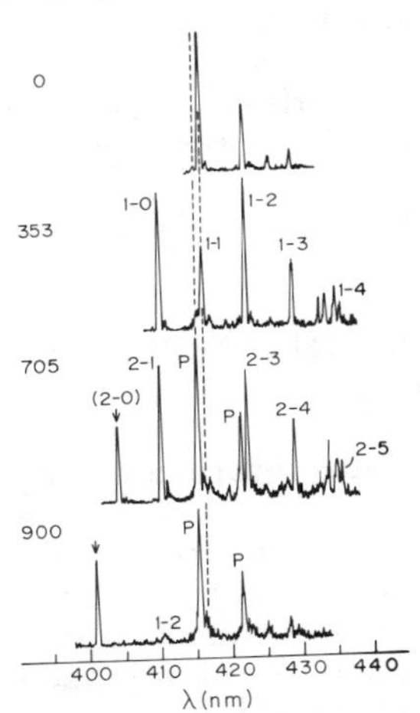

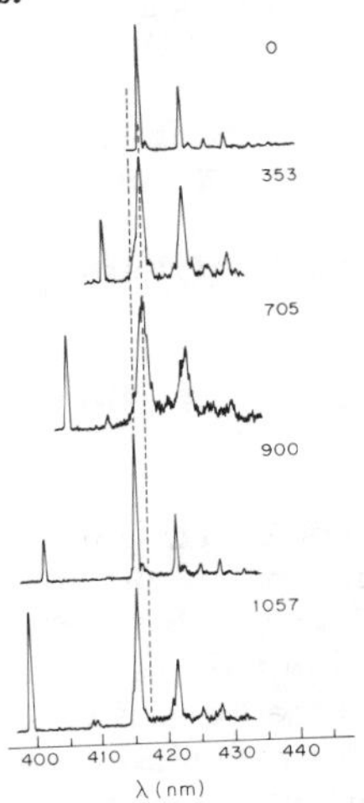

This sudden onset indicates significantly different behaviour from the argon case, associated with the extent to which the energy is randomized. Therefore, the systems are dynamically different in the sense that the energy is rather imperfectly randomized in the Ar complex, even above the predissociation threshold, so that a full RRKM approach cannot reasonably be applied.[4] On the other hand, the presence of extra degrees of freedom in the CH_4 complex readily randomizes the energy below the predissociation threshold, as may be seen from the 705 cm^{-1} fluorescence data, so that more favourable conditions are achieved for rapid coupling to the exit coordinate, and a very rapid turn-on of the predissociation event is observed.

MULTIPHOTON IONIZATION STUDIES OF PPF (2,5-DIPHENYL FURAN) COMPLEXES WITH ARGON AND METHANE

We have constructed an apparatus which uses pulses from two synchronized picosecond laser sources for two-photon ionization. The present apparatus has 50 ps resolution, and uses a 10 Hz pulsed jet: a 1 mm skimmer ≈20 mm downstream provides a clean molecular beam for the ionization experiments. Pressure variation of subatmospheric argon behind the nozzle was used to generate different Ar complexes, while CH_4 could be admitted through a side arm to generate PPF/CH_4 complexes. A time-of-flight mass spectrometer having a 30 cm field-free drift tube was used, in conjunction with a transient recorder, coupled to a computer.

Complex ion signals due to PPF/Ar (mass 260) were observed on transitions corresponding to 0^0_0, 256^1_0 and 256^2_0 of PPF, consistently with stably bound complexes. Figure 2 (left) shows a sequence of time-of-flight traces for 256^1_0 excitation of PPF, PPF/Ar and PPF/Ar_2. The binding energy of PPF complexes with Ar and CH_4 is estimated to be about 20% lower than for the corresponding perylene complexes. Therefore, excitation into the next higher strong transition of the PPF complexes, 867^1_0,(5) may be expected to exhibit faster vibrational predissociation than for perylene complexes excited

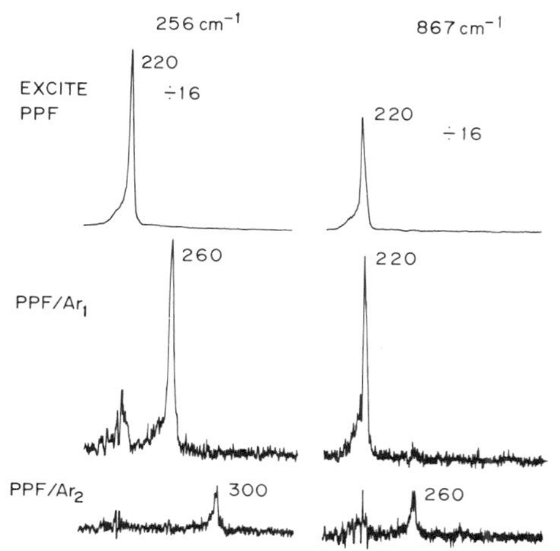

into 900 cm^{-1}. This was in fact found to be the case since, although the 867^1_0 transition of PPF is about as strong as 256^1_0, the corresponding resonance due to the Ar_1 complex did not give rise to a distinguishable mass peak at 260. Instead, as Fig. 2 (right) shows, a significant signal was observed in the parent mass channel (i.e. PPF^+; 220). Similarly, excitation of PPF/Ar_2 at 867 cm^{-1} gave a signal due to $(PPF/Ar)^+$ at mass 260. This indicates that photofragmentation has taken place at 867 cm^{-1} internal energy, on a time scale substantially less than 50 ps. Similar data have also been obtained for the $(CH_4)_1$ complexes of PPF, where 867^1_0 excitation was only observed to give a signal in the parent mass channel.

These studies are currently being extended to a shorter time scale, to carry out a systematic survey of the predissociation dynamics of perylene complexes. Improvement of the time resolution to ≈1 ps will allow us to study S_2 cases, involving pyrene and coronene,(6) where ultrafast predissociation of molecular complexes will be studied for internal energies in excess of 4000 cm^{-1}.

References:

1. B.P. Boczar and M.R. Topp, **Chem. Phys. Lett.**, 108, 490 (1984).
2. M.M. Doxtader and M.R. Topp, **J. Phys. Chem.**, 89, 4291 (1985).
3. M.J. Ondrechen, Z. Berkovitch-Yellin and J. Jortner, **J. Am. Chem. Soc.**, 103, 6586 (1981).
4. D.F. Kelley and E.R. Bernstein, **J. Phys. Chem.**, 90, 5164 (1986); L.R. Khundkar, R.A. Marcus and A.H. Zewail, **J. Phys. Chem.** 87, 2473 (1983).
5. E.A. Mangle, P.R. Salvi, R.J. Babbitt, A.L. Motyka and M.R. Topp, 133, 214 (1987).
6. E.A. Mangle and M.R. Topp, **J. Phys. Chem.**, 90, 802 (1986); E.A. Mangle and M.R. Topp, **Chem. Phys.**, 112, 427 (1987); R.J. Babbitt, C-J. Ho and M.R. Topp, **J. Phys. Chem.** 91, 5599 (1987).

TIME-RESOLVED FTIR EMISSION STUDIES OF MOLECULAR PHOTOFRAGMENTATION INITIATED BY A HIGH REPETITION RATE EXCIMER LASER

T. Rick Fletcher and Stephen R. Leone*
Joint Institute for Laboratory Astrophysics
University of Colorado and National Bureau of Standards
and Department of Chemistry and Biochemistry, University of Colorado
Boulder, Colorado 80309-0440

ABSTRACT

The availability of high repetition rate (>300 Hz) excimer lasers provides new opportunities for studies of molecular processes by time-resolved FTIR spectroscopy. An overview of the technique is given and state resolved infrared emission results are presented for the triatomic radical, C_2H, generated by photolysis of C_2H_2 at 193 nm. Electronic emission from the low lying $\tilde{A}^2\Pi$ state of C_2H is observed, along with high vibrational levels of the ground state which gain intensity by coupling with the vibrationless level of $\tilde{A}^2\Pi$.

The study of polyatomic photodissociation has been limited by the inability to detect with quantum state resolution those fragments which contain three or more atoms. We have applied time-resolved Fourier transform infrared (FTIR) emission spectroscopy to the study of photodissociation and report here the observation of a fully resolved spectrum of a triatomic radical (CCH) formed by photofragmentation of acetylene.

A schematic of the apparatus is given in Fig. 1. The experimental apparatus[1,2] consists of a high repetition rate (>300 Hz) excimer laser, a commercial 0.02 cm^{-1} resolution FTIR and a photolysis chamber equipped with multipass mirrors for the laser beam and collection optics for the infrared (IR) fluorescence. The IR emission results from dissociation of precursor molecules by pulses of light from the excimer laser. The resulting emission is collected by spherical mirrors and imaged into the emission port of the FTIR. The interferograms are collected at set delay times after the laser pulses.

The timing is controlled as follows. A He:Ne laser beam is used in the interferometer for measuring the mirror displacement, producing a sine wave at the photodetector as the moving mirror is swept. Each time a positive zero crossing occurs in the He:Ne sine wave, a signal is sent to a computer which waits a delay time, ΔT_1, and then samples the IR detector. The same zero crossing signal is also sent to an external delay generator, which triggers the excimer laser following a shorter delay, ΔT_2. Since $\Delta T_2 < \Delta T_1$, the laser is pulsed before the detector is sampled, allowing interferograms to be collected at a selectable time ($\Delta T_1 - \Delta T_2$) after the laser pulse. This scheme allows the entire interferogram to be collected at a constant delay time following laser photofragmentation. Even though fluctuations in the velocity of the moving mirror occur, no errors in the temporal resolution are introduced, since the laser trigger and detector sampling are coupled to the actual sine wave zero crossing.

*Staff Member, Quantum Physics Division, National Bureau of Standards.

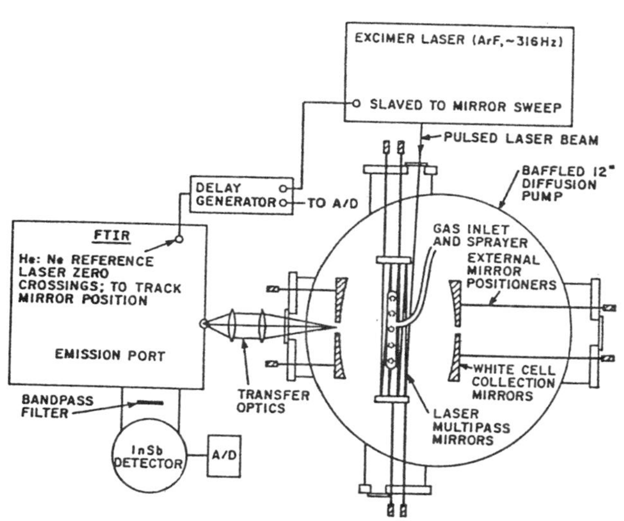

Fig. 1. Schematic of the experimental apparatus.

The temporal resolution is presently limited by the speed of the IR detector. For the data presented here, a delay time of 7 μs is used. Typical precursor pressures are ~0.15 Pa. The photolysis cell is pumped by a 12 inch diffusion pump and flow rates are ~10^{16} molecules s^{-1}. Under these conditions the photofragments have suffered on average ~0.1 hard sphere collisions prior to observation by the FTIR. When observing the C_2H radical considered below, ~10^8 excited molecules cm^{-3} per quantum state produces a S/N of unity in a single sweep of the mirror. Between 16 and 50 sweeps are typically averaged to provide sufficient S/N for population measurements.

An interesting application of the technique is to examine the photodissociation of acetylene using an ArF laser as the photolysis source. Dissociation at 193 nm produces a hydrogen atom and the triatomic radical, C_2H. This radical has a low-lying electronic state ($\tilde{A}^2\Pi$, T_0 ~ 4000 cm^{-1}) which mixes with high vibrational levels in the ground state ($\tilde{X}^2\Sigma^+$). An emission spectrum recorded 7 μs after photolysis of C_2H_2 is shown in Fig. 2. At present, five transitions belonging to C_2H have been assigned in this spectrum. Three of the bands are high-lying vibrational levels in the ground state which couple with the $\tilde{A}^2\Pi$ (0,0,0) level. These are the (0,1,2) level of $\tilde{X}$ at 4106 cm^{-1}, the (1,1,0) level of $\tilde{X}$ at 4011 cm^{-1} and the (0,5,1) level of $\tilde{X}$ at 3785 cm^{-1}. A fourth weaker band is the residual of the $\tilde{A}^2\Pi$ (0,0,0) state, observed at 3692 cm^{-1}. The strongest band, occurring at 3772 cm^{-1}, is the relatively unmixed state, $\tilde{A}(0,1,0)^2\Sigma^-$. The assignments given here are taken from Curl and co-workers, who observed C_2H by absorption in a gas discharge using a color center laser.[3]

The spectrum shown in Fig. 2 is taken at 0.6 cm^{-1} resolution. This is sufficient resolution to provide information on which vibronic states are produced by photofragmentation of C_2H_2 at 193 nm. An advantage of the FTIR emission method is that the instrumental resolu-

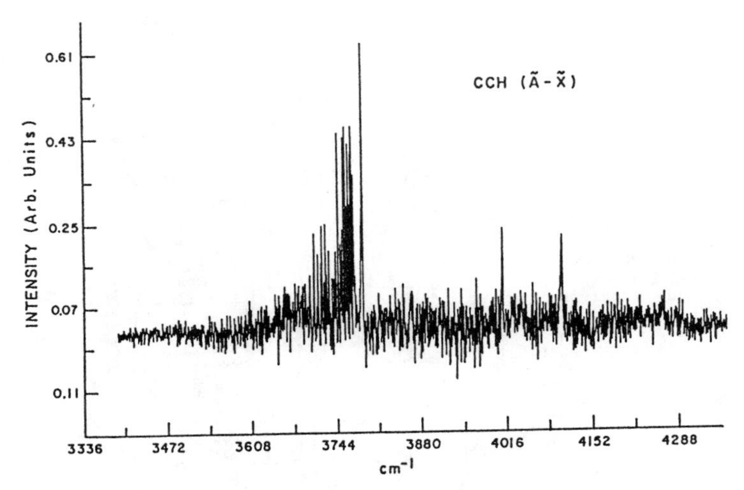

Fig. 2. Emission spectrum of C_2H. C_2H_2 pressure is 0.13 Pa and delay time is 7 μs.

tion can be high enough to investigate complex polyatomic spectra, although higher resolutions require longer data collection times. This can be a severe limitation when running the excimer laser at 300 Hz. We have found that without correcting for fluctuations in the excimer laser power, the useful lifetime is on the order of 10^7 pulses or ~10 hours of continuous use.

The spectra of C_2H at 0.2 cm^{-1} resolution has also been recorded, which is sufficient to provide details on the rotational state distribution for the least perturbed state, $\tilde{A}(0,1,0)^2\Sigma^-$. Since the other states contain significant ground state character, much higher resolution is necessary in order to separate the individual rotational levels spectroscopically. The results for $\tilde{A}(0,1,0)^2\Sigma^-$ have been analyzed and will be presented in detail elsewhere.[4] Briefly, the rotational energy of the $\tilde{A}(0,1,0)^2\Sigma^-$ state is observed to be less than that of the precursor C_2H_2. The most likely explanation of this observation is a constraint due to conservation of angular momentum and total energy. Fragmentation is allowed to occur from bent geometries only when the initial $\vec{J}$ of C_2H_2 and the orbital angular momentum, $\vec{L}$, are antiparallel. In this way, the orbital angular momentum of the departing fragments effectively cancels the initial rotation of C_2H_2, thereby lowering the rotational energy of product C_2H. The time-resolved FTIR method offers the potential for many new investigations of polyatomic fragmentation.

1. T. R. Fletcher and S. R. Leone, J. Chem. Phys., in press.
2. E. L. Woodbridge, T. R. Fletcher, and S. R. Leone, J. Phys. Chem., submitted.
3. R. F. Curl, P. G. Carrick, and A. J. Merer, J. Chem. Phys. 82, 3479 (1985).
4. T. R. Fletcher and S. R. Leone, J. Chem. Phys., submitted.

SUBPICOSECOND EXCIMER-FORMATION DYNAMICS IN ORGANIC MOLECULAR CRYSTALS

Leah Ruby Williams* and Keith A. Nelson†
Department of Chemistry
Massachusetts Institute of Technology
Cambridge, MA 02139

ABSTRACT

Subpicosecond excited-state relaxation times in the organic molecular crystals pyrene and α-perylene are observed via femtosecond time-resolved transient absorption spectroscopy. Forward impulsive stimulated scattering with a single ultrashort excitation pulse is demonstrated in a crystal.

INTRODUCTION

Excimer formation in organic molecular crystals has attracted special attention because it provides examples of oriented bimolecular reactions for which the starting configuration is extremely well defined. Pyrene and α-perylene are classic examples of excimer forming crystals.[1] The crystal structure consists of closely spaced pairs of the planar aromatic molecules. Upon photoexcitation, the molecules in a pair form an electronically stabilized excited state dimer in which the intermolecular distance is contracted. The "reaction coordinate" is primarily a 1-dimensional motion of the two molecules in a pair toward each other. In pyrene, excimer formation occurs at all temperatures, indicating that it occurs in a barrierless potential. Perylene shows temperature dependent excimer formation, with excimer fluorescence not observed below about 50K.

EXCIMER-FORMATION DYNAMICS

In order to investigate excimer formation dynamics, we performed femtosecond time-resolved transient absorption experiments on single crystals of pyrene and α-perylene. The excitation pulse was a 65-fs, 1-μjoule pulse centered at 615 nm, generated by a synchronoulsy pumped and amplified femtosecond dye laser system which has been described previously.[2] Part of the dye laser output was focussed into a 1-cm cell of D_2O to generate a white-light continuum. A portion of the continuum was selected with an interference filter (10-nm FWHM) and used as the probe. Signal was recorded as the change in probe transmission $(I - I_0)$ versus delay of the probe.

Figure 1 shows pump-probe data at a probe wavelength of 560 nm in pyrene (solid line) and α-perylene (dashed line). An autocorrelation of the excitation pulse is shown with crosses for comparison. Consider first the data in pyrene. Before $t = 0$ the probe is not absorbed by the sample. The probe transmission decreases sharply at $t = 0$, following the temporal profile of the excitation pulse, and then recovers roughly exponentially to reach a constant level. Scans taken out to much longer delays (not included here) show that the constant level is unchanging from 0.5 ps to several hundred picoseconds.[3] The interpretation of the data is straightforward. The 615-nm excitation pulse produces excited states via two-photon absorption, as demonstrated earlier[4] and confirmed by intense blue fluorescence visible to the eye. The initially excited state absorbs the 560-nm probe whose transmission therefore decreases at $t = 0$. Relaxation into the excimer state, which absorbs the probe pulse less strongly, occurs with an excimer

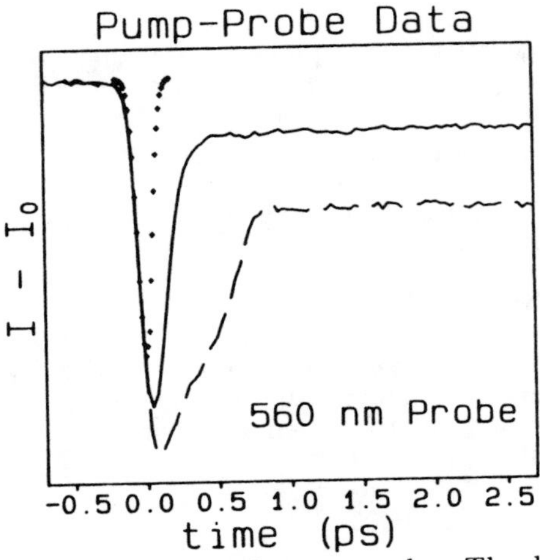

Figure 1. Time-resolved pump-probe data in pyrene (solid line) and α-perylene (dashed line) single crystals at 300K recorded with 560 nm probe pulses. Autocorrelation of the 65 fs excitation pulse is shown with crosses. Excited-state relaxation occurs in less than one picosecond in both materials.

formation time of 140 femtoseconds. The lack of further spectral change out to several hundred picoseconds confirms that the state formed is the excimer. The only spectral evolution which occurs at times longer than several hundred picoseconds is due to excimer relaxation to the ground state, with a 110 ns excimer lifetime. The data in perylene has similar features to that in pyrene. However, the signal in perylene takes approximately twice as long to rise to the long-lived excimer level and has a distinctly different shape than the signal in pyrene.

The difference in the data for pyrene and perylene may arise from differences in the excited state potentials. An excimer formation time of 140 fs in pyrene is consistent with a model of a bimolecular reaction in a barrierless potential.[3,4] The excimer formation time may be "mechanically" limited by molecular inertia, i.e., by the amount of time required for two neighboring molecules to move along the reactive potential toward each other and into the excimer configuration. Perylene excimer formation, on the other hand, occurs in a potential with a barrier. It is possible that the several slopes apparent at short times in the perylene data represent the approach to and crossing of the potential barrier. Further analysis of the shape of the short-time perylene data is currently being carried out.

OSCILLATIONS IN PROBE PULSE TRANSMISSION

In addition to creating electronic excited states through two-photon absorption, the single ultrashort excitation pulse creates a coherent population of optic phonons via impulsive stimulated Raman scattering (ISRS).[5] The excitation pulse (which must be short in duration compared to a single vibrational period) exerts a temporally impulsive driving force on the vibrational mode through forward Raman scattering which is stimulated because the Stokes frequency is contained within the bandwidth of the pulse. The probe pulse then encounters a sample which is undergoing coherent optic phonon oscillations. When the probe pulse is at a wavelength which is absorbed by the sample, the probe transmission oscillates at the phonon frequency as shown in Fig. 2. Initially the oscillations were interpreted in terms of phonon-induced shifts in the crystal's ground state absorption spectrum.[3,5] However, the oscillations are in fact due to coherent scattering of the probe pulse, which like the excitation pulse exerts an impulsive force on the vibrational mode. If the probe arrives at the sample at a time such that its force is in phase with the coherent vibrational oscillations already under way, then the

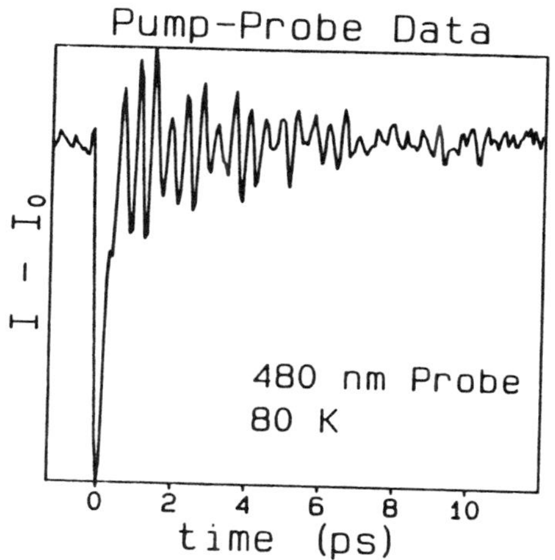

Figure 2. Pump-probe data from α-perylene crystal at 80K recorded with 480 nm probe pulses which are absorbed by the perylene ground state. Oscillations in signal are due to coherent optic phonons excited via ISRS. The phonon oscillations give rise to coherent scattering of the probe pulse, which is alternately red- and blue-shifted and whose transmission therefore oscillates at the phonon frequency.

vibrational motion is amplified and the probe pulse, having given up energy to the vibrational excitation, emerges red-shifted. If the probe pulse arrives out of phase, it opposes the vibrational motion already under way, takes energy from the vibrational excitation, and emerges blue-shifted. In the experiment whose results are shown in Fig. 2, the probe wavelength overlaps the sharply rising red edge of the perylene absorption spectrum. As the probe is alternately blue and red-shifted by the vibrational wave, more or less probe light is absorbed by the sample and thus the transmitted intensity oscillates. This observation confirms the earlier prediction of coherent optic phonon generation in a solid by a single ultrashort pulse.[5] The alternate red- and blue-shifting described has been observed directly in ISRS experiments on molecular vibrations.[2,6] The current results further establish the general occurrence of ISRS, through which a sufficiently short light pulse initiates coherent vibrational oscillations whenever it passes through a Raman-active medium.[5]

ACKNOWLEDGEMENTS

This work was supported in part by U.S. ARO No. DAAL03-86-K-0002 and a grant from the donors of the Petroleum Research Fund, administered by the ACS.

REFERENCES

*AT&T Bell Laboratories PhD Scholar
†Presidential Young Investigator Awardee and Alfred P. Sloan Fellow

1. J. B. Birks, A. A. Kazzaz and T. A. King, Proc. Roy. Soc. A291, 556 (1966).
2. S. Ruhman, A. G. Joly, B. Kohler, L. R. Williams and K. A. Nelson, Rev. de Phys. Appl. 22, 1717 (1987).
3. L. R. Williams and K. A. Nelson, J. Chem. Phys. 87, 7346 (1987).
4. L. R. Williams, E. B. Gamble, Jr., K. A. Nelson, S. De Silvestri, A. M. Weiner and E. P. Ippen, Chem. Phys. Lett. 139, 244 (1987).
5. Y.-X. Yan, E. B. Gamble, Jr. and K. A. Nelson, J. Chem. Phys. 83, 5391 (1985).
6. S. Ruhman, A. G. Joly and K. A. Nelson, J. Chem. Phys. 86, 6563 (1987).

DIRECTIONAL PHOTODISSOCIATION DYNAMICS OF ALKYL NITRITES AND NITROSAMINES AT 250 AND 350 NM

E.Radian, D.Schwarz-Lavi, R.Lavi, I.Bar and S. Rosenwaks
Ben-Gurion University of the Negev, Beer-Sheva 84105, Israel

ABSTRACT

The geometry and symmetry of the first two excited singlet states, S_1 and S_2, of tert-butyl nitrite (TBN) and dimethylnitrosamine (DMN) were studied using laser induced polarization spectroscopy at 250 and 350 nm. The scalar and directional properties of the NO fragment illustrate the differences between S_1 and S_2 and between TBN and DMN dissociation dynamics.

INTRODUCTION

This paper presents a short summary of the work done in our laboratory on the characterization of the nascent NO following excitation of tert-butyl nitrite (TBN) and dimethylnitrosamine (DMN) to two different potential surfaces. TBN, $(CH_3)_3CONO$, was irradiated at 365.8 and 351.8nm [S1←S0 (π^*←n)] and at 250nm [S2←S0 (π^*←π)]. DMN, $(CH_3)_2NNO$, was irradiated at 363.5nm [S_1←S_0(π^*←n)] and 250nm [S_2←S_0(π^*←π)]. The nascent NO was monitored using single photon laser induced fluorescence.

RESULTS

The rotational alignment factors $A_0^{(2)}$ for the NO fragment following TBN photolysis at 365.8, 351.8 and 250nm are:

$A_0^{(2)}$ = 0.38±0.19 for 365.8 nm, S_1←S_0 transition;
$A_0^{(2)}$ = 0.42±0.10 for 351.8 nm, S_1←S_0 transition;
$A_0^{(2)}$ = -0.30±0.06 for 250.0 nm, S_2←S_0 transition.

The Λ-doublet population ratios $\Pi(A'')/\Pi(A')$ for NO from TBN photolysis at 365.8, 351.8 and 250nm are:

$\Pi(A'')/\Pi(A')$ = 1.73±0.14 for 365.8 nm, S_1←S_0 transition;
$\Pi(A'')/\Pi(A')$ = 2.02±0.10 for 351.8 nm, S_1←S_0 transition;
$\Pi(A'')/\Pi(A')$ = 2.70±0.15 for 250.0 nm, S_2←S_0 transition.

The spin-orbit population ratio for TBN

fragmentation at 250nm is almost statistical. At 365.8 and 351.8nm average population ratios $n(^2\Pi_{1/2})/n(^2\Pi_{3/2})$ of 1.52±0.11 and 1.59±0.10 respectively, are obtained.

Our analysis for nascent NO indicates a high rotational non-Boltzmann distribution peaking at high rotational levels (J"~31.5 for the S_1 state, and J"~53.5 for the S_2 state).

The relative vibrational population of nascent NO after TBN photodissociation at 365.8nm are: 0.05, 0.47, 0.43 and 0.05 for v"= 0, 1, 2, 3, respectively. At 351.8nm 0.02, 0.35, 0.42, 0.21 for v"= 0, 1, 2, 3. At 250nm 0.92, 0.07 and ~0.01 for v"= 0, 1, 2.

The $A_0^{(2)}$ values for the NO fragment following DMN photolysis at 363.5 and 250nm are the following :

$A_0^{(2)}$ = +0.18±0.08 for 363.5 nm, $S_1 \leftarrow S_0$ transition;

$A_0^{(2)}$ = -0.19±0.06 for 250 nm, $S_2 \leftarrow S_0$ transition.

The average Λ-doublet population ratios for NO $(X^2\Pi_{3/2}, v''=1)$, for J">24.5 are:

$\Pi(A'')/\Pi(A')$ = 1.59±0.23 for 363.5nm, $S_1 \leftarrow S_0$ transition;

$\Pi(A'')/\Pi(A')$ = 0.76±0.05 for 250nm, $S_2 \leftarrow S_0$ transition.

Our analysis for nascent NO indicates a high rotational non-Boltzmann distribution peaking at high rotational levels (J"~26.5 for v"=1 after irradiation at 363.5 nm, and J"~31.5 for v"=1 after irradiation at 250nm). For 363.5nm the average rotational energy is 1510, 1230 and 1070 cm^{-1} for v"= 0, 1, 2, respectively, and for 250nm it is 2000 and 1640 cm^{-1} for v"= 0, 1.

The relative vibrational population of nascent NO after DMN photodissociation at 363.5nm are: 0.92, 0.07 and ~0.01 for v"= 0, 1, 2, respectively. For 250nm it is: 0.95, 0.045 and ~0.005 for v"= 0, 1, 2.

DISCUSSION

<u>TBN</u>. The $S_1 \leftarrow S_0$ $(\pi \leftarrow n)$ transition moment is perpendicular to the molecule plane. The directional properties indicate that the dissociation occurs from a largely planar transition state creating NO fragment which rotates with its **J** vector parallel to **μ**. The population preference of the $\Pi(A'')$ antisymmetric component reveals the A" symmetry of the excited state. The preference of the $\Pi_{1/2}$ spin-orbit state might point on exit-channel

interactions leading to electronic energy transfer between the RO and NO fragments. The derived scalar quantities point on a non-Boltzmann rotational distribution and on wavelength dependent vibrational distribution, controlled by a vibrational predissociation decay mechanism. The $S_2 \leftarrow S_0$ ($\pi^* \leftarrow \pi$) transition moment is in the molecule plane. As shown by the vectorial properties, the dissociation from this state also occurs from a largely planar transition state creating NO fragment with **J** perpendicular to **μ** and to the molecule plane. The population preference of the Π(A") antisymmetric component reveals the planarity of the process and predict an approximate A" symmetry for the second fragment. The scalar quantities show an increase in the rotational and translational energies deposited in the NO fragment, but a decrease in vibrational energy as compared to the S_1 fragmentation. The dynamics described by the impulsive model in its rigid limit is in good agreement with the experimental results.
DMN. The photofragmentation is characteristic of a direct dissociation mechanism on a repulsive potential surface for both dissociation wavelengths. The impulsive model in its rigid limit gives a resonable fit to the average rotational and vibrational energy obtained both for S_1 and S_2 states. The main forces act along the bond which breaks causing the angular momentum of the NO fragment to be perpendicular to the plane of the parent molecule. Two main differences between the two potential surfaces can be seen. One is the direction of the transition dipole moment for excitation and the other the symmetry of the two excited states with respect to reflection of the spatial coordinates of all the electrons in the plane of rotation of the parent molecule. The results of the rotational alignment give experimental evidence that in the first singlet excited state the transition dipole moment is perpendicular to the plane of the parent molecule and in the second excited state it lies parallel to this plane along the bond which breaks. The Λ-doublet population ratio obtained for the two processes implies an A" symmetry for the S_1 state and an A' symmetry for the S_2 state. Full account of this work will be the subject of future publications.

Table I. ESCA results for CVD films grown by pyrolysis of various Si-source materials in H_2

Sample[a]	H_2	Pressure	T°C	(C/Si)[b]
SiH_4	30%	6 Torr	825	0.08
ES	14%	21 Torr	630	0.26
t-BS	3%	20 Torr	610	0.24
PS	15%	14 Torr	743(c)	2.43

a SiH_4 = silane; ES = ethylsilane; t-BS = t-butylsilane; PS = phenylsilane

b Values after argon ion sputter etching cycle.

c SiH_2 absorption intensity similar to that of ES and t-BS at 620 °C.

ACKNOWLEDGMENT

The authors wish to gratefully acknowledge the contributions of Dr. Arthur Barry in the preparation of t-butylsilane. This work was supported partially by a grant from the IBM Corporation and the Materials Characterization Program in the Department of Chemistry, University of Arizona.

REFERENCES

1. G. H. Atkinson in Advances in Chemical Reaction Dynamics, pp. 207-228, P. M. Rentzepis and C. Capellos, eds. (Reidel, Dordrecht, 1986) and references therein.
2. J. J. O'Brien and G. H. Atkinson, Chem. Phys. Lett. 130, 321 (1986).
3. J. M. Jaskinski, B. S. Meyerson and B. A. Scott, Ann. Rev. Phys. Chem. 38, 109 (1987) and references therein.
4. J. J. O'Brien, N. Goldstein and G. H. Atkinson in Laser Applications to Chemical Dynamics, M. A. El-Sayed, ed., Proc. SPIE pp. 87 - 96 (1987).
5. J. J. O'Brien and G. H. Atkinson, unpublished results.

PHOTODECOMPOSITION OF CO_2 BY 193 nm RADIATION

W. T. HILL, III and B. P. TURNER
Institute for Physical Science and Technology
University of Maryland, College Park, Maryland 20742

ABSTRACT

Dispersed fluorescence has been employed to study photo-fragmentation of CO_2 at 193 nm. Our measurements show a prominent photolysis channel producing excited atomic carbon; the decomposition pathway is associated with multiphoton dissociation of CO_2. Furthermore, we observe that multiphoton dissociation at this wavelength is preferred over multiphoton ionization.

DISCUSSION

Our apparatus was specifically designed to allow the dispersed fluorescence of the photofragments to be measured down to 200 nm as a function of the excitation wavelength. The CO_2 (99.99% purity) was continuously flowed through a cell while the pressure was maintained at about 500 mTorr. The fragmentation was initiated by a line-narrowed (1 cm^{-1}), tunable (over 150 cm^{-1}) ArF^* excimer laser (193 nm, 51800 cm^{-1}) focused to yield a power density of 5 GW/cm^2. The fluorescence was monitored by a photomultiplier after being filtered by either a 10 nm band-pass filter, centered at 250 nm, or a 3/4 m monochromator (0.1 nm resolution). Data acquisition and wavelength control of both the laser and the monochromator were handled by an IBM PC.

Figure 1 shows the fluorescence excitation spectrum of CO_2 obtained with the 250 nm filter. This filter will transmit several CO_2^+ fluorescence bands.[1] These bands have been observed following single photon excitation of CO_2 in the 65 nm region;[2] three 193 nm photons could excite these same bands as well. Our dispersed fluorescence measurements over the 10 nm wavelength spread of the filter, however, show virtually no CO_2^+ fluorescence. A portion of this spectrum is displayed in Fig. 2. The fluorescence signal in Figs. 1 and 2 originates from the 2p3s $^1P^o$ -> $2p^2$ ^1S transition in atomic carbon.[3] The spectrum is complicated by the facts that the field intensities are

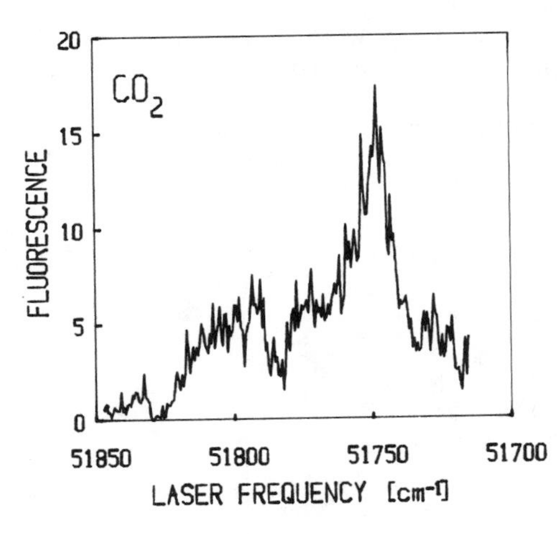

Figure 1. Excitation Spectrum.

high and a near coincidence in the C spectrum at 193 nm involving the $2p^2\ ^1D \rightarrow 2p3s\ ^1P^o$ transition exists.[3] As a result, the ^1D and $^1P^o$ states experience AC stark splitting,[4] which results in the two-component fluorescence spectrum.[5]

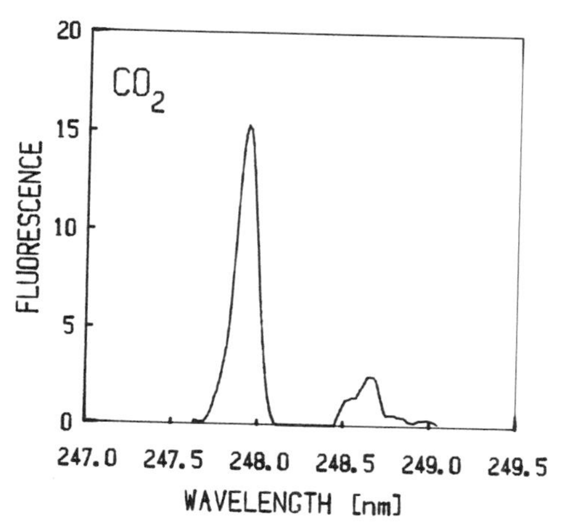

Figure 2. Dispersed Fluorescence.

There are several possible fragmentation pathways which could produce C in the $^1P^o$ state. Space does not permit a detailed discussion of these here. It is instructive, however, to highlight the decomposition channels associated with a few specific dissociation thresholds. The lowest relevant dissociation threshold (4.40×10^4 cm^{-1}),[6] requiring only one photon, relies on the excitation of CO_2 to the A state followed by predissociation into $CO(X^1) + O(^3P)$. If such a predissociation is to take place, there would have to be significant singlet-triplet mixing -- the A state of CO_2 is a singlet,[6] while the molecular state leading asymptotically to $CO(X^1) + O(^3P)$ is a triplet. The sharp contrast between the fluorescence excitation spectra of CO_2, Fig. 1, and CO, Fig. 3, shows that this evidently does not occur. Figure 3 was obtained in the same way as Fig. 1. The resonant structure is due to the rotational levels of the a^3 state of CO; these states act as intermediate states for multiphoton dissociation of $CO(X^1)$ into $C(^1D$ or $^1P^o) + O(^3P)$.[5,7]

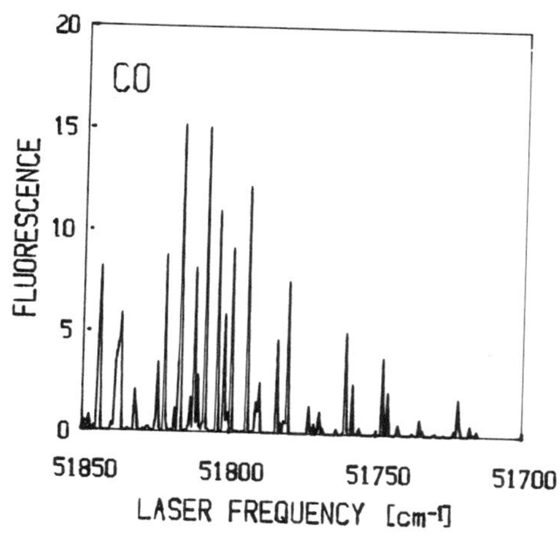

Figure 3. Excitation Spectrum.

If CO_2 absorbs a second photon, its energy would lie above the dissociation threshold for producing $C(^1D) + O_2(X^3)$ (1.03×10^5 cm^{-1}).[3,8] Once in the ^1D state, the strong laser field would equilibrate the population in the ^1D and $^1P^o$ states even at large (50 cm^{-1}) detunings.[5] A dissociation of this kind would require the CO_2 excited state to be very bent. Since the A state is bent (and could provide the necessary intermediate states for multiphoton excitation),[6] it is quite reasonable that some fraction of the molecules will be excited to higher bent states which are either predissociated or unbound. Each would ultimately decompose into C + O_2. After absorbing two photons, CO_2 could also dissociate into excited states of CO, such as, the a^3 state (9.24 x 10^4 cm^{-1}).[6,9] Additional photons would then dissociate the CO to produce $C(^1D$ or $^1P^o)$. Finally, if CO_2 absorbs three photons,[5] it is possible to completely breakup the molecule into $C(^1D) + 2O(^3P)$ (1.44 x 10^5 cm^{-1}).[3,8]

The fluorescence exhibits a nonlinear (less than quadratic) dependence on both the laser intensity and the CO_2 pressure. This implies that the fragmentation process depends on multiphoton absorption and collisions. The fact that the signal does not scale with a higher power of the intensity suggests that there is saturation of some of the intermediate states and/or that there is a competing process interfering with the flurorescence, such as ionization of the $^1P^o$ C atoms.

In sum, we have shown that the primary contribution to atomic carbon production is not due to single photon dissociation of CO_2. We have not, however, determined which of the multiphoton channels is most important nor do we understand the origin of the resonance feature in Fig. 1. With the additional capability of vacuum ultraviolet fluorescence detection and direct ion detection, we can study other final states and, hence, gain more insight into the fragmentation dynamics of CO_2 and other molecules. This work is currently in progress.

ACKNOWLEDGMENTS

We thank P. S. Julienne for helpful discussions. This work was supported by the National Science Foundation (grant Nos. PHY-84-06192 and PHY-84-51284), the Research Corporation and Baltimore Gas and Electric Company.

REFERENCES

1. R.W.B. Pearse and A.G. Gaydon, The Identification of Molecular Spectra, p 117 (Chapman and Hall, London, 1976).
2. J.A. Guest, M.A. O'Halloran and R.N. Zare, J. Chem. Phys. 81 2689 (1984).
3. C.E. Moore, Atomic Energy Levels, Vol. 1, Natl. Stand. Ref. Data Ser. Natl. Bur. Stand. 35 (1949).
4. S. Feneuille, Rep. Prog. Phys. 40, 1257 (1977).
5. B.P. Turner and W.T. Hill, III, To be published.
6. G. Herzberg, Electronic Spectra of Polyatomic Molecules, (Van Nostrand Reinhold Co. New York, New York, 1966).
7. J. Bokor, J. Zavelovich and C.K. Rhodes, J. Chem. Phys. 72, 965 (1980).
8. E.P. Gentieu and J.E. Mentall, J. Chem. Phys. 58, 4803 (1973).
9. P.H. Krupenie, The Band Structure of Carbon Monoxide, Natl. Stand. Ref. Data Ser. Natl. Bur. Stand., 5 (1966).

MATCHING AND OPTIMIZING OF THE EXPERIMENTAL PARAMETERS FOR SYNTHESIS OF Si_3N_4 BY LASER

Li Daohuo
Anhui Institute of Optics and Fine Mechanics, Academia Sinica

INTRODUCTION

In experimental research on the laser-induced vapor phase synthesis of Ceramic powder Si_3N_4, a close dependence of powder characteristics on parameters such as laser intensity, mixing ratio and flow speed of gaseous mixtures, and gas pressure in reaction cell has been found[1][2]. This paper focuses its discussion on the matching law among parameters of laser intensity (w/cm^2), flow speed of gaseous mixture V(cm/s), and pressure in reaction cell P (atm), and the relationship between reaction temperature T(°C) and I-V-P, and a requirement for stability degree of system parameters has been given and the theoretical basis and data provided for controlling of synthesis system and selection of its optimum parameters.

RATIO OF I/V

Temperature of reaction gases rises quickly to 10^5–10^7 °C/s, while laser heating of the flowing gaseous mixture. In a certain threshold of temperature, nucleation and its growth appear[3], in which temperature change of the reacting gaseous mixture is expressed as[4].

$$(T_2-T_1)/(T_2+T_1)=R1Id/2VC_p \tag{1}$$

Let $R1d/2C_p=A$ then from (1) we have:

$$T_2=T_1\ (1+I/V\ A)/(1-I/V\ A) \tag{2}$$

Equation (2) is a very important formula, giving the relationship between reaction temperature T (i.e.T_2) and ratio of I/V. By utilizing (2), the ratio of I/V in the temperature threshold of synthesis reaction will be readily ratio of I/V in the temperature threshold of synthesis reaction will be readily given. It is a very interesting result, since the basis for selection and matching of I-V is provided, so long as the I/V ratio is known.

Under our experimental condition, A=0.07163. in general, the reaction temperature of laser synthesis of Si_3N_4 is in the range of 800 C - 1200 C, within which I/V ratio is: $13<I/V<14$. The ratio of I/V, limited to a very narrow range value, indicates that there exists a critical requirement for stability of I and V. Therefore, the rational ratio of I/V should be selected at the corresponding points which satisfy the optimum temperature of sysnthesis reaction. In our case, the ratio of I/V is only limited to the corresponding value around 13.3

VALUE OF VP

Gas pressure P in reaction cell is an important parameter.

$$T_2=T_1\ (1+1/VP\cdot B)/(1-1/VP\cdot B) \tag{3}$$

where, $B=-(lgI/I_o)\cdot RI/2C_p$ (B depends on I). Under a certain laser intensity, a certain value of B could be measured. From Equation (3), VP>B. If the measured I/I_o is 89% (satisfying single pass absorption), then B_1=31.6586 for I =3000W/cm^2, B_2=52.7643 for I = 5000W/cm^2. The corresponding value of $(VP)_{B1}$ is 33,2311 ($(VP)_{B2}$= 55.3852), while reaction temperature if 1031°C. Around this value, rate of temperature change is so large that temperature changes sharply with decrease of VP value. This observation tells us that a stable control of VP in experiment should be preferably taken at the side of larger value of VP. From Equation (3) and according to the experimentally measured values, the curves of T-VP are plotted in Fig. 1 and Fig. 2, where the quantitative relationships of T-VP

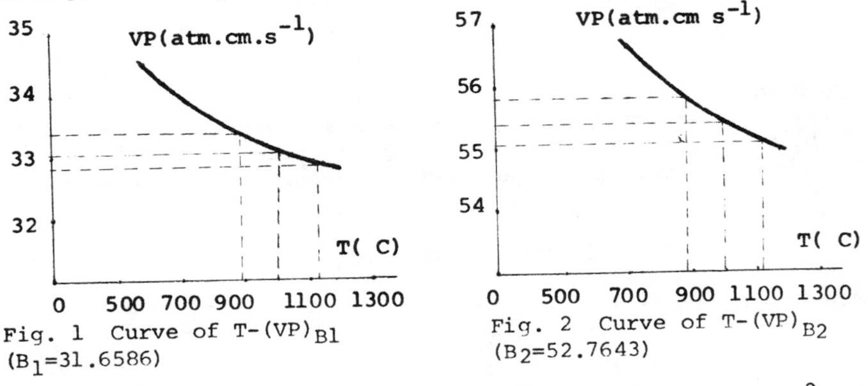

Fig. 1 Curve of T-$(VP)_{B1}$ (B_1=31.6586)

Fig. 2 Curve of T-$(VP)_{B2}$ (B_2=52.7643)

are shown respectively for I=3000 W/cm^2 and for I=5000 W/cm^2. The value of B is larger, the change of T with VP is smaller, therefore the laser synthesis system operated under the high value of B is more stable.

RELATIONSHIP OF I-VP

To generalize the results, the relationship of I-VP will be further discussed.

From Equation (3)

$$I_o=IEXP\left\{1/I\cdot 2C_p/R\cdot\left[(T_2-T_1)/(T_2+T_1)\right]\cdot VP\right\} \tag{4}$$

When reaction temperature is 1000°C, we have $\alpha=2C_p/R\cdot[(T_2-T_1)/$

$(T_2+T_1)]$ =9.3487 and Equation (4) may be written as

$$I_o = I\,EXP(dVP/I) \tag{5}$$

which is a semi-emperical equation, suitable to laser induced vapor phase synthesis of Si_3N_4. For other systems of laser vapor phase synthesis, value in Equation (5) has to be modified according to values of T and I given by experiment. From Equation (5), a curve of I -VP may be plotted (Fig. 3). It is an extremely important curve. Moreover, other experimental parameters such as I, V, P, could be selected simultanuously at every time of experiment. Based on the results above, we are able to take optimum control in the experiment, making the estimation of experimental results possible.

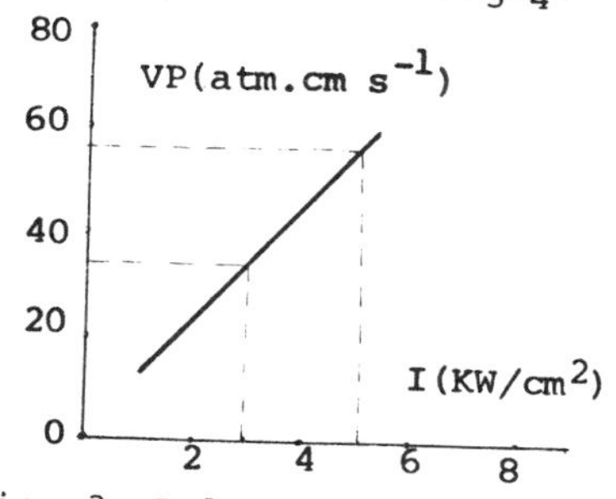

Fig. 3 Relationship of I-VP

STABILITY DEGREES OF I. V. P

Analysis in above section indicate that parameters I. V. P have a narrow range of allowed value in the threshold range of reaction temperature. Reaction can not go on properly, once the change of a certain parameter exceeds the allowed value range. For stable operation, parameters I. V. and P must have a certain degree of stability. From Fig. 2, for I_1=3000W/cm^2 and I_2=5000W/cm^2 we have V_1=225.6±1.7 (cm/s) V_2=375.95±2.85 (cm/s) the stability degrees of flow speed of reaction gases are repectively: δ_{v1} = 0.735%, δ_{v2} = 0.758%, under the above condition, the stability of laser intensity for the stable operation of system are δ_{I1} = 0.752%, δ_{I2} = 0.752%, and the relative stability of pressure in reaction cell are: δ_{P1}= 0.826%, δ_{P2} = 0.794%. That is, under higher pressure, stable operation has a more critical requirement of more stable pressure.

REFERENCES

1. R. A. Marra and J. S. Haggerty, Synthesis and Characteristics of Ceramic Powders made from laser Heated Gases. Ceramic Engineering and Science Proceedings, 3, 1-2, 3-19 (1982)
2. Li Daohuo et al. Chinese Journal of Lasers, Vol. 13, No. 9 P. 523, (1986)
3. R. A. Marra, J. H. Flint and J. S. Haggerty, Homogeneous Nucleation and Growth of silicon powders from laser Heated Si_3N_4, ADA/44439, (1984)
4. Jeffery I. Steinfeld, Laser-Induced Chemical Processes, P. 211 (1981)

Photodissociation of $Fe(CO)_5$ at 193 nm

U.Ray, G.Bandukwalla, B.K.Venkataraman, S.L.Brandow, Z.Zhang, M.F.Vernon

Department of Chemistry, Columbia University, New York, NY 10027

ABSTRACT

The photodissociation of a supersonically cooled beam of iron pentacarbonyl has been studied. Our results show that the photodissociation pathway at 193nm involves a three step, sequential, statistical cleavage of the three M-CO bonds.

INTRODUCTION

Previous photodissociation studies of $Fe(CO)_5$ at 193nm are consistent with a mechanism which involves sequential elimination of CO ligands[1-3]. However, the metal containing product has never been detected under collision free conditions. Here we report the results of a crossed laser-molecular beam study of the photodissociation of $Fe(CO)_5$ by ArF radiation, where the metal containing photofragments are detected mass spectroscopically.

EXPERIMENTAL APPARATUS

All the experiments were carried out in the crossed laser-molecular beam apparatus, equipped with a rotatable mass spectrometer (electron bombardment ionizer, rf quadrupole mass filter, Daly ion detector[4]). A molecular beam of $Fe(CO)_5$ is formed by expanding a 1.5% $Fe(CO)_5$/Ar mixture through a .004" nozzle at 25°C at a total pressure of 400 torr. These conditons produced a beam comprising of >95% of $Fe(CO)_5$ monomers. The collimated beam of molecules interacts with the laser photons, and the products, scattered to various laboratory angles, travel a fixed distance (~21cm) to the ionizer. At high laser energies (>2mj/cm^2), we observe the appearance of a fast peak in the photofragment TOF spectra, when the mass spectrometer is set at the mass of Fe^+, which is from a two-photon process. To measure only the single photon dissociation products, the photofragment TOF spectra were recorded at laser energies below the level of the onset of the two photon channel.

RESULTS AND ANALYSIS

At the laboratory detector angles of 10°, 15°, 20° and 25°, TOF signals are detected only for ion masses corresponding to Fe^+, $Fe(CO)^+$, $Fe(CO)_2^+$ The normalized TOF spectra recorded for any of these ions arealways identical at all four laboratory detector angles. Also, the ion count rate ratios at the four angles is the same for the three detected ions (Fe^+: $Fe(CO)^+$: $Fe(CO)_2^+$ = .714:1.00:.572). These two observations are consistent with a single neutral photoproduct, cotaining at least two COs which can fragment in the electron bombardment ionizer to produce lighter daughter ions. We observe the parent and all the daughter fragments $Fe(CO)_x^+$, (x=0-4) when the detector looks directly at the $Fe(CO)_5$ molecular beam, with relative intensities within one order of magnitude. Unless the fragmentation pattern changes considerably after photoexcitation, we should have been able to detect signals at the parent ions $Fe(CO)_n^+$(n>3), if such neutral species were formed in the photodissociation process. Hence we conclude that $Fe(CO)_2$ is the only primary product at 193nm.

We have analyzed the complete time-of-flight data to obtain the product velocity distribution $P(v^2)$ by the IQ method[5]. The plot of $\ln(P(v^2))$ vs v^2 gives a straight line fit. Our experimental data is consistent with a dissociation mechanism which involves sequential, statistical elimination of three COs. To compare the measured $Fe(CO)_2$ velocity distribution with a stepwise dissociation model, the data has been convoluted over the velocity distributions of the intermediate steps. If each intermediate step is assumed to be statistical, it can be shown that the final velocity distribution of $Fe(CO)_2$ will have the following form:

$$P(v^2) \sim \exp(-\gamma v^2) \qquad (1)$$

where γ is related to the masses and temperatures of the intermediate steps[6]. From our experimental meausurements, we determine γ to be .097 ± .01, which agrees well with the value of .101 calculated using the microcanonical sequential model used by Waller et al.[3] to fit the CO(v,j) distributions. We cannot obtain direct information on the number of dissociation steps in our experiments. However, a one-step, concerted mechanism, using the metal-ligand bond dissociation energies of Engelking et.al[7] would produce $Fe(CO)_2$ with a high average internal energy content (~30 Kcm^{-1}). Such a hot $Fe(CO)_2$ fragment should be unstable and lose an additional CO. Our angular scan data shows that the experimentally detected ratio of $Fe(CO)^+ : Fe(CO)_2^+$ is produced only by the fragmentation of neutral $Fe(CO)_2$ by electron bombardment ionization and cannot originate from ionization of two neutral photofragments, $Fe(CO)_2$ and FeCO.

CONCLUSION

We conclude that under *collision-free conditions*, $Fe(CO)_2$ is the only photoproduct formed by the single photon absorption of $Fe(CO)_5$ at 193nm. The recoil velocity distribution of the detected photoproduct agrees quantitatively with the CO(v,j) measurements of Waller et al[3], and show that at 193nm the photodissociation process involves a three step, statistical elimination of CO ligands.

REFERENCES

1. J.T.Yardley, Barbara Gitlin, Gilbert Nathanson, Alan M.Rosen, J.Chem.Phys.,74, 370, (1981).
2. Eric Weitz, J.Phys.Chem.,91, 3945, (1987).
3. I.M.Waller, H.F.Davis, J.W.Hepburn, J.Phys.Chem., 91, 506, (1987).
4. N.R.Daly, Rev.Sci.Instrum.,**31**, 264, (1960)
5. G.Bandukwalla, M.F.Vernon, to be published.
6. U.Ray, G.Bandukwalla, B.K.Venkataraman, S.L.Brandow, Z.Zhang, M.F.Vernon, submitted to J.Chem.Phys.
7. P.C.Engelking, W.C.Lineberger, J.Am.Chem.Soc.,101,5569, (1979).

MULTIPHOTON IONIZATION AND FRAGMENTATION STUDY OF ACETONE USING 308nm LASER RADIATION

Liu Houxiang, Li Shutao, Han Jingcheng,
Zhu Rong, Guan Yifu and Wu Cunkai

Laboratory of Laser Spectroscopy, Anhui Institute of Optics & Fine Mechanics, Academia Sinica, Hefei, China

ABSTRACT

Multiphoton ionization and fragmentation (MPI-F) of acetone molecules using 308nm laser radiation was studied by using a molecular beam and quadrupole mass spectrometer. The ion peaks of acetone molecule appear at m/e=15 and 43, corresponding to the two fragments CH_3^+ and CH_3CO^+. It is considered that these two ions are, respectively, formed by direct (2+1) and 2-photon ionization of methyl and acetyl radicals, generated by photodissociation of acetone molecule.

INTRODUCTION

Since the acetone molecule is a good donor of methyl and acetyl radicals in organic chemistry, one often uses acetone as the precursor of methyl and acetyl radicals to study their spectroscopy and photochemistry[1,2]. Massaki et al[3] have recently studied MPI spectroscopy of the acetone molecule at 193nm and 248nm. Their work has revealed the character of the photochemistry and MPI spectroscopy of acetone . In particular the $^1(n,\pi^*)$ valence state, and the $^1(n,3s)$ Rydberg state, serve as resonant intermediate states for 2-photon ionization.

The acetone molecule has a weak $^3(n,\pi^*)$ valence state at 308nm, which is heavily predissociative, and a $^1(n,4s)$ Rydberg state (153nm), close to the two-photon resonant absorption wavelength of the XeCl excimer laser (154nm). Furthermore, acetyl and methyl do not have one-photon absorption bands at 308nm. Thus we expect that MPI spectroscopy using 308nm laser radiation will be able to provide some new information. For example, it might reveal the characters of spectroscopy and photochemistry of acetone occuring via $^3(n,\pi^*)$ and $^1(n,4s)$ resonant intermediate states as well as the effect of predissociation of the $^3(n,\pi^*)$ state. It might also provive a useful method for preparing methyl and acetyl radicals, for use in studies of reaction kinetics.

EXPERIMENTAL

The experimental system has been described in detail previously[4]. In the present experiment the molecular beam is crossed at right angles with the focused laser beam in a horizontal plane. The ion signals are detected with a mass spectrometer above the crossing point. The ion signals are amplified and then transmitted through a Boxcar into a X-Y recorder.

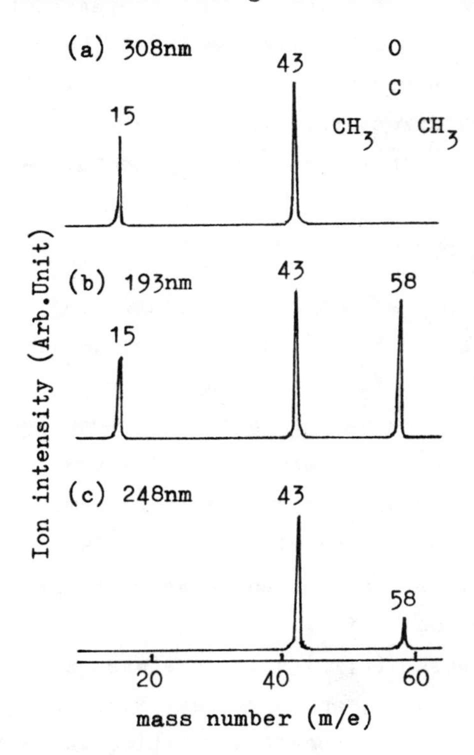

Fig.1 The MPI mass spectrum of acetone. (a) the present work; (b) and (c) from ref.3.

The MPI mass spectrum of acetone molecule obtained at 308nm is shown in Fig.1 along with BaBa[3] results. our experimental results, as shown in Fig.1 (a), differ apparently from others, ie. Fig.1 (b) and (c). It suggests that the MPI mechanism for acetone at 308nm is different from those at 193nm and 248nm.

The laser power dependence of the ion signals is measured under a variety of pressures. The slopes of typical log(ion signal) versus log(laser intensity) lines fitted to data for m/e=15 and 43 are 3.6 and 3.0, respectively.

The intensities of ion signals increase linearly with pressure except for high pressures where the saturation occurs. This fact excludes the influence of intermolecular collisions.

KINETIC ANALYSIS

For 308nm laser radiation two resonant intermediate states[5], the first excited triplet state $\tilde{A}\ ^3A_2$(4.04ev) and the Rydberg state $\tilde{D}\ ^1B_2$(8.06ev), are involved. The two-photon absorption of acetone, $\tilde{D}\ ^1B_2 \leftarrow \tilde{X}\ ^1A_1$ must pass through a real state, $\tilde{A}\ ^3A_2$. In this $\tilde{A}\ ^3A_2$ state, the absorption is very weak and the predissociation is rapid, so that the two-photon transition often fails to occur. That is, the second step of the absorption ladder does not compete with the rapid dissociation into two neutral fragments CH_3 and CH_3CO. The above analysis is consistent with the fact that no parent ion via the (2+1)-photon ionization process is observed in the present experiment. The decrease in the parent ion intensity (relative to that at 193nm) at 248nm is mainly due to the population leakage from the resonant intermediate state $\tilde{A}\ ^1A_2$ via rapid intersystem crossing into the strongly dissociating state $\tilde{A}\ ^3A_2$.

The methyl ion CH_3^+ is generated by (2+1)-photon ionization via resonant intermediate state $\tilde{C}\ ^2E''$ in methyl radical. The acetyl radical reachs up to ionization continuum directly through two-photon resonant absorption. These analyses are in correspondence with our experimental results.

REFERENCES

1. D.A.Parkes, Chem.Phys.Lett. 77, 527 (1981)
2. H.Adachi, N.Basco and D.G.L.James, Intl.J.Chem.Kinet. 13, 1251 (1981)
3. M.BaBa, H.Shinohara and N.Nishi, Chem.Phys. 83, 221 (1984)
4. J.C.Han, Y.F.Guan, H.X.Liu, C.K.Wu, R.Zhu and S.T.Li, Acta Opt.Sinica (in Chinese) 7(3), 216 (1987)
5. H.E.O'Neal and C.W.Larson, J.Phys.Chem. 73, 1011 (1969)

ENERGY TRANSFER STUDIES WITH LASERS

Ralph E. Weston, Jr.
Chemistry Department, Brookhaven National Laboratory
Upton, New York 11973

ABSTRACT

Using pulsed lasers to produce energetic species, we have observed their interaction with other molecules in real time by absorption spectroscopy with a cw infrared diode laser. This technique has been used to observe the vibrational excitation of carbon dioxide by energy transfer from azulene and azulene-d_8 excited to energies of 30,000 and 40,000 cm^{-1}, and the relative populations of the bending and antisymmetric stretching modes of CO_2 are found to be very nonstatistical. Diode laser spectroscopy of CO has also been utilized to determine the efficiency of energy transfer from $O(^1D)$ atoms to vibrations of CO excited to v=1 through 4.

INTRODUCTION

Tunable infrared diode lasers have been widely used in spectroscopic studies of stable molecules, radicals, and molecular ions. However, there are relatively few examples[1,2] of their use in time-resolved experiments, either in classical kinetic experiments or in state-to-state studies of reaction dynamics. Perhaps because diode lasers operate in a cw mode, there is a mind-set against using them to do transient spectroscopy in real time. However, by using a pulsed source to initiate the process being monitored, a time scale is readily established. We first used diode lasers in time-dependent spectroscopy in a determination of the vibrational populations of carbon dioxide excited in collisions with translationally energetic H atoms.[3] In these time-resolved studies, diode laser absorption spectroscopy has the advantages of spectral resolution high enough for the observation of individual vibrational-rotational transitions, sensitivity high enough to permit measurements under nearly single-collision conditions, and simplicity of intensity calibration compared with fluorescence methods. Disadvantages are the relatively slow IR detector response ($\gtrsim$ 100 ns), the short tuning range of an individual diode (15-50 cm^{-1}), and the tediousness of wavelength calibration.

A typical experimental arrangement of the sort used in our current work is shown in Fig. 1.

ENERGY TRANSFER FROM VIBRATIONALLY EXCITED AZULENE TO CO_2

The vibrational excitation and de-excitation of molecules by collisions with the "bath gas" plays a very important role in the mechanism of a unimolecular reaction. In the past, some idea of the amount of vibrational energy transferred per collision and the relative effectiveness of various bath gases has been inferred from

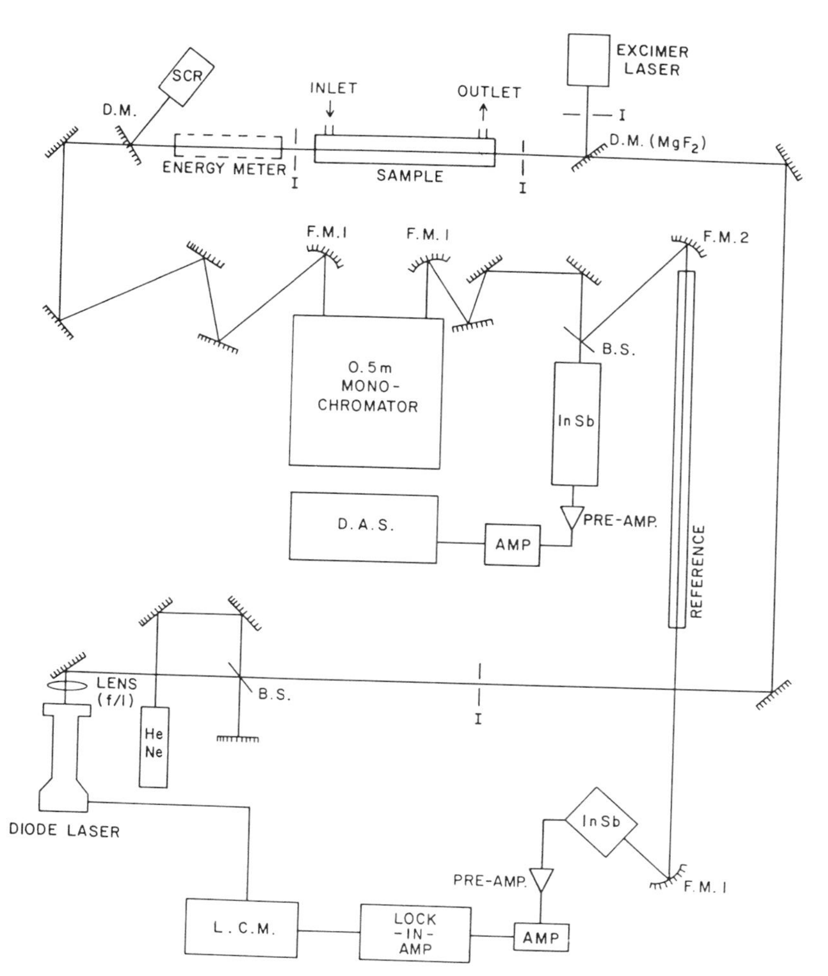

Figure 1. Experimental arrangement for time-resolved diode laser absorption spectroscopy.

measurements of the reaction rate as a function of bath gas pressure. More recently, experiments have been devised in which the population of highly vibrationally excited polyatomic molecules is directly determined.[4-7] This approach is beginning to give us a detailed picture of what happens to the "reacting" molecule as a result of collisions, but it provides no information about the ultimate disposition of the energy lost from the reactant. It occurred to us that the diode laser provided a probe to determine whether any of this energy is ultimately converted to vibrational energy of the bath gas, and if so, how much, and in what specific vibrational modes.[8]

The experimental approach was based on the same unusual photophysics of the azulene molecule that Barker has utilized in his IR fluorescence experiments.[4] Excitation of azulene in the visible spectral region to the S_1 state or in the UV to the S_2 state produces vibrationally excited S_0 molecules because of very rapid internal conversion. Thus, on the time scale of subsequent collisions, molecules are instantaneously produced with a well-defined amount of vibrational energy. In our experiments, the azulene molecules were prepared in the S_2 state by excitation with a nitrogen laser at 337 nm or a KrF excimer laser at 248 nm. The bath gas used in our experiments was carbon dioxide, because of our previous experience in the diode laser spectroscopy of this particular species. Populations can be determined in each of the three vibrational modes: ν_1(symmetric stretch, 1388 cm^{-1}), ν_2(degenerate bend, 667 cm^{-1}), and ν_3(antisymmetric stretch, 2349 cm^{-1}). In spite of the fact that the symmetric stretching mode is IR inactive, transitions such as $10^01 \leftarrow 10^00$ have the high intensity associated with the large transition dipole of the antisymmetric stretching mode. In addition, all transitions that involve a one-quantum change in ν_3 are at nearly the same wavelength, since the energy levels differ only by small amounts due to anharmonicity corrections. With luck, it becomes possible to probe several such transitions with a single diode laser. Relative populations in the ν_1 and ν_2 modes are partially mixed, because of Fermi resonance between ν_1 and $2\nu_2$. However, population exchange between the combined ν_1-ν_2 modes and the antisymmetric stretching mode is slow enough so that relative populations $[2N(\nu_1) + N(\nu_2)]/N(\nu_3)$ can be determined.

The results of the azulene-carbon dioxide experiments indicate that approximately 25% of the initial energy in the azulene molecule is converted to vibrational energy of the CO_2 molecule as a result of collisional energy transfer. Whether this is in qualitative agreement or disagreement with a theory that assumes no vibrational excitation will take place[9] is like deciding whether the glass is half full or half empty. Perhaps even more striking is the distribution of energy among the vibrational modes of carbon dioxide shown by the data in Table I, where the experimental values for population ratios are compared with those calculated from a statistical model. It can be seen that the relative excitation of the bending mode is much greater than the statistical model predicts. Some modification of this type of model to favor energy transfer when there is near resonance between the energies of the

Table I. Relative populations of CO_2 vibrational modes excited by V-V transfer from azulene.

Molecule, energy in cm^{-1}	$N(\nu_2)/N(\nu_3)$ Theory	$N(\nu_1)/N(\nu_3)$ Theory	$[2N(\nu_1) + N(\nu_2)]/N(\nu_3)$ Theory	$[2N(\nu_1) + N(\nu_2)]/N(\nu_3)$ Expt.
Azulene (30600)	17.1	2.9	22.9	352(66)
Azulene (40980)	15.2	2.7	20.6	309(24)
Azulene-d_8 (31480)	19.2	3.1	25.4	178(60)

Standard deviations in parentheses.

donor and acceptor molecules may be applicable. This hand-waving argument is based on the fact that azulene has many vibrational frequencies in the range of 500 to 1000 cm^{-1}, fairly close to the bending mode frequency of carbon dioxide, but no frequencies near that of the antisymmetric stretch. However, deuterium substitution in azulene shifts the C-H stretching modes down to ~ 2300 cm^{-1}, closer to the CO_2 stretching mode. This may account for the decreased population of the bending mode shown in Table I for azulene-d_8.

It should be emphasized that all of these measurements were made under conditions such that rotational relaxation of the CO_2 takes place. If we were able to achieve greater sensitivity and to work at lower pressures, it should be possible to determine nascent rotational populations of the carbon dioxide. In this way, additional information about the ultimate disposal of the azulene energy would be provided. This type of experiment can, and should, be extended to other donor and acceptor molecules before any generalities are made about vibrational energy transfer from highly excited molecules. We have done some preliminary experiments with perfluorobenzene, which has photophysical properties similar to those of azulene, as the donor.

VIBRATIONAL EXCITATION OF CO IN THE QUENCHING OF $O(^1D)$

The temperature of the upper atmosphere is largely determined by the electronic-to-vibrational energy conversion process

$$O(^1D) + N_2(v=o) \rightarrow O(^3P) + N_2(v > 0).$$

The excited O atoms are produced by the photolysis of ozone, or at higher altitudes, by photodissociation of molecular oxygen. It is difficult to measure the efficiency of the above process, but the efficiency of N_2 relative to that of CO has been determined,[10] and the absolute efficiency of transfer to CO has also been measured.[10,11] Using classical photochemical techniques, Slanger and Black[10] obtained a value of 40% for this efficiency. Shortridge and Lin[11] used a CO laser to determine vibrational populations of collisionally excited CO, and found that levels up to v=7 were populated, with a nearly Boltzmann distribution and a vibrational temperature of 8000 K. The efficiency determined in their work was 21%; we hoped to resolve this discrepancy of nearly a factor of two.

In our experiments, $O(^1D)$ atoms were produced by excimer laser photolysis of a suitable precursor, usually O_3 at 248 nm. The number of excited atoms generated was determined by measuring the laser power behind the cell with the cell empty and with an O_3/CO mixture flowing through it. Knowing the extinction coefficient then made it possible to calculate the number of atoms. The measurement of vibrational energy in CO was carried out with a diode laser, with which levels from v=0 to v=4 could be probed. We made use of the fact that CO(v=1) is a metastable state because ladder-climbing processes such as

$$CO(v) + CO(v=0) \rightarrow CO(v-1) + CO(v=1)$$

occur rapidly (~0.1 times the gas-kinetic collision rate), whereas vibrational relaxation from the CO(v=1) level

$$CO(v=1) + CO(v=0) \rightarrow CO(v=0) + CO(v=0)$$

is extremely slow. Therefore, with CO pressures of a few torr, it is possible to count the total number of vibrational quanta by measuring the absorption of the diode laser beam due to CO(v=1). In this way a value of 25+8, -5% for the efficiency of E-V energy transfer was obtained. Using $C^{18}O$, we also corroborated the finding of Shortridge and Lin[11] that quenching involves the formation of a CO_2 collision complex in which either C-O bond can break, leading to isotopic scrambling.

ACKNOWLEDGEMENT

This reasearh was carried out at Brookhaven National Laboratory under contract DE-AC02-76CH00016 with the U.S. Department of Energy and supported by the Division of Chemical Sciences, Office of Basic Energy Sciences.

REFERENCES

1. R. J. Balla and L. Pasternack, J. Phys. Chem. 91, 73 (1987).
2. P. H. Beckwith, C. E. Brown, D. J. Danagher, D. R. Smith, and J. Reid, Appl. Opt. 26, 2643 (1987), and references cited therein.

3. J. O. Chu, C. F. Wood, G. W. Flynn, and R. E. Weston, Jr. J. Chem. Phys. 80, 1703 (1984); 81, 5533 (1984); J. A. O'Neill, J. Y. Cai, G. W. Flynn, and R. E. Weston, Jr., J. Chem. Phys. 84, 50 (1986); S. A. Hewitt, J. F. Herschberger, G. W. Flynn, and R. E. Weston, Jr., J. Chem. Phys. 87, 1894 (1987).
4. J. R. Barker, J. Phys. Chem. 88, 11 (1984), and earlier references cited therein.
5. H. Hippler, L. Lindemann, and J. Troe, J. Chem. Phys. 83, 3906 (1985), and earlier references cited therein.
6. T. J. Wallington, M. D. Scheer, and W. Braun, Chem. Phys. Lett. 138, 538 (1987).
7. R. J. Gordon, Comm. At. Mol. Phys., in press.
8. W. Jalenak, R. E. Weston, Jr., T. J. Sears, and G. W. Flynn, J. Chem. Phys. 83, 6049 (1985).
9. Y. N. Lin and B. S. Rabinovitch, J. Phys. Chem. 74, 3151 (1970).
10. T. G. Slanger and G. Black, J. Chem. Phys. 60, 468 (1974).
11. R. G. Shortridge and M. C. Lin, J. Chem. Phys. 64, 4076 (1976).

HIGH RESOLUTION SUB-DOPPLER EXPERIMENTS ON BENZENE

E. W. Schlag, H. J. Neusser, E. Riedle
Institut für Physikalische und Theoretische Chemie
Technische Universität München, D-8046 Garching, West-Germany

ABSTRACT

It is shown that sub-Doppler spectroscopy enables one to resolve individual rotational states in the S_1 manifold of polyatomic molecules. This is an essential to the understanding of the primary photophysics within the molecule. Spectra of benzene are found to undergo substantial changes as the vibrational energy is raised within S_1. Due to the increased density of vibrational states, Coriolis coupling, which is already seen at low energies, can lead to effective IVR above 3000 cm^{-1} excess energy. This onset of IVR may be responsible for the onset of "Channel Three" in benzene and probably produces gross changes in the photophysical behavior of any polyatomic molecule.

INTRODUCTION

Contrary to small molecules, which can only decay by radiative processes or at most dissociation, excited large polyatomic molecules can also undergo intramolecular nonradiative decay. One of the prime examples that showed the existence of such processes is the benzene molecule. This was concluded from the distinct deviation of the fluorescence quantum yield from unity after excitation of the molecule into the S_1 electronic state /1/. The reason for this deviation is the relaxation of the molecule into the triplet manifold, i.e. Inter-System-Crossing (ISC). Due to the high density of triplet states at the energy of the S_1 state, ISC is believed to be in the statistical limit and leads to the irreversible decay observed. A fairly smooth increase in the decay rate with the vibrational excess energy in S_1 would be expected from this model. However, a sudden nearly complete disappearance of fluorescence was observed starting at about 3000 cm^{-1} excess energy /2/. This drop in quantum yield could not be explained with an ISC process alone. In addition, from experiments aimed at estimating linewidths of states above 3000 cm^{-1} in S_1 an even more drastic increase of the nonradiative rate at around this energy was concluded /3/. Recent decay time measurements confirmed the strong increase of the decay rate, but the increase was not found to be as large /4/. Since none of the known relaxation processes of large molecules could explain these observations, a mystic "Channel Three" was invoked /3/. The nature of this "Channel Three" could not be determined from low resolution experiments with vibronic state resolution, and especially the discrepancy between decay rate measurements and "linewidth" measurements of /3/ remains a puzzle. To clarify the situation, decay time measurements and measurements of the homogeneous linewidth of single defined quantum states at rotational resolution are needed. Due to the Doppler broadening of about 1.7 GHz such experiments have to be performed with sub-Doppler resolution. To gain insight into the exact nature of the nonradiative process, states at the onset of "Channel Three" have to be observed.

EXPERIMENTAL TECHNIQUES

The experimental set up used for the recording of Doppler-free two-photon spectra and the investigation of the decay of individual levels has been described in detail previously /5,6,7/. Doppler-free excitation takes place if a molecule simultaneously absorbs two photons of equal energy and opposite direction of propagation from a standing wave light field. The light source in our experiments is a cw frequency stabilized single mode ring dye laser. For extremely high spectral resolution and very high sensitivity a cw Doppler-free experiment is performed within a concentric external cavity whose length is variable and locked to the laser frequency /5/. The UV-fluorescence following the two-photon-excitation of the molecules is monitored. The use of the external cavity increases the sensitivity by two orders of magnitude as compared to the simpler arrangement of a backreflected laser beam. A resolution of better than 10 MHz can be obtained with this set up /5/.

For the investigation of the decay of individual levels pulsed excitation of the molecules has to be used. Light pulses of 500 KW peak power and nearly Fourier transform-limited bandwidth are produced by pulsed amplification of the cw light /6/. With an additional parasitic cavity around the second amplifier the pulse length can be varied between 2.5 ns and 10 ns /7/. The decay time measurements are not performed within an external resonator like in the cw set up. Rather after passing through the sample cell the laser beam is reflected back into itself to allow the Doppler-free absorption. The resulting UV-fluorescence signal is either integrated for the recording of spectra or its time behavior is recorded with a transient digitizer. In the Doppler-free two-photon absorption process all molecules regardless of their velocity contribute to the observed signal. Molecules are only excited to one single level if the laser frequency is set to a resolved rotational line in the spectrum regardless of the number of lines within the Doppler width. This allows the observation of the decay of an individual level /6,7/.

RESULTS AT LOW EXCESS ENERGY

To understand the rotational structure of vibronic bands, that is observable at sub-Doppler resolution, vibronic bands with very low excess energy in S_1 have to be analyzed at first. The deviations observed in the "Channel Three" regime can then be attributed to the unknown nonradiative process.

The lowest S_1 state reached by a strong two-photon transition is the 14^1 state at 1570 cm^{-1} vibrational excess energy. Both with the cw set up and the pulsed one the observed sub-Doppler spectrum of the $14^1{}_0$ band is a line spectrum, with most of the lines corresponding to single rovibronic transition /8,9/. Part of the cw recording of the blue part of the Q-branch of this band is shown in figure 1. Each rotational line shown has been assigned within the model of a semirigid symmetric top. For the whole band more than 5000 lines were identified and a fit to the observed line positions yields extremely accurate values for the rotational constants and the quartic centrifugal distortion constants /10/.

In contrast to the situation described so far, in some places of

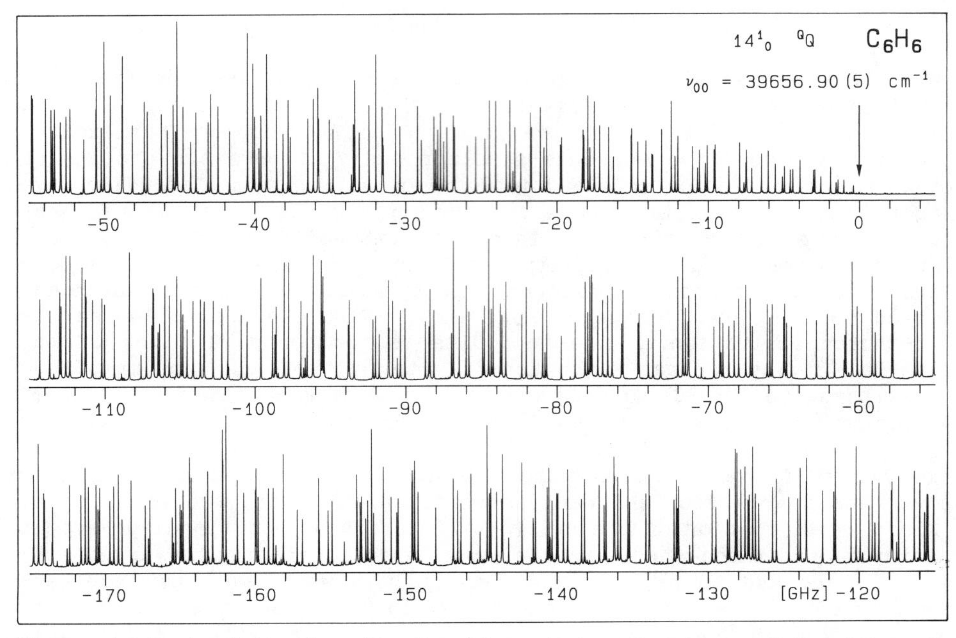

Figure 1. Part of the Doppler-free two-photon spectrum of the Q-branch of the $14^1{}_0$ band of C_6H_6 /10/.

the spectrum, a line predicted by the symmetric top model is not found in the experimental spectrum. Instead two smaller lines are observed, one each to both sides of the expected position. These two lines can be explained as transitions to the two quasi-eigenstates, that originate from the coupling of the "light" zero-order rotational state of the 14^1 manifold to a "dark" zero-order state /11/. Due to the isolated nature of the coupling it can be concluded that the dark state belongs also to the S_1 manifold /11/. A detailed analysis shows that for fixed K values all lines within a range of J values show this kind of perturbation. The distance of the two quasi-eigenstates from the expected position of the light state shows the typical form of an avoided crossing /11/. Since both eigenstates belonging to one light state can be observed in our spectra, we are able to determine the size of the coupling matrix element for each pair. The coupling element on the order of 1 GHz is found to depend strongly on both J and K. Its exact dependence on the rotational quantum numbers agrees very well with that expected for perpendicular Coriolis coupling. We therefore conclude that some of the rotational states of the 14^1 vibronic state are coupled to rotational states of another vibrational state within S_1 by this mechanism. Since the two vibronic states have slightly differing rotational constants, only some of the rotational states will be in resonance and this explains the isolated nature of the perturbation.

With the pulsed set up the decay time of selected single quasi-eigenstates in 14^1 can be measured. We observed, that all states found to be unperturbed possess the same decay time /6/. This shows that the nonradiative decay of S_1 benzene does indeed happen in the statistical limit and does not depend on the rotation of the molecule. However, for all perturbed states a significantly shorter decay time was found /6/. This can be explained by the fact that the dark vibronic state coupled to the 14^1 state dacays much faster. The observed decay is then dependent on the degree of admixture of the dark state. Combining our knowledge from the spectroscopic analysis and from the decay time measurements we can even determine the decay time of the pure dark state. In this way the perturbations observed in the spectrum allow us to investigate the decay behavior of vibronic states that cannot be directly excited by either one- or two-photon absorption. The fact that the dark states have a faster nonradiative decay is important for the understanding of the observations within the "Channel Three" regime.

RESULTS AT HIGH EXCESS ENERGY

The 14^1 state of benzene discussed so far has a vibrational excess energy well below the onset of "Channel Three" at around 3000 cm^{-1}. On the contrary, the energy of the $14^1 1^2$ state is 3412 cm^{-1} and just slightly above this threshold. The rotational structure of the $14^1{}_0 1^2{}_0$ two-photon transition leading to this state should be nearly identical with that of the $14^1{}_0$ band. However, a strong deviation is observed in the experiment /12/. This is demonstrated in figure 2, where the blue edge of the Q-branch of the $14^1{}_0 1^2{}_0$ is compared with the same part of

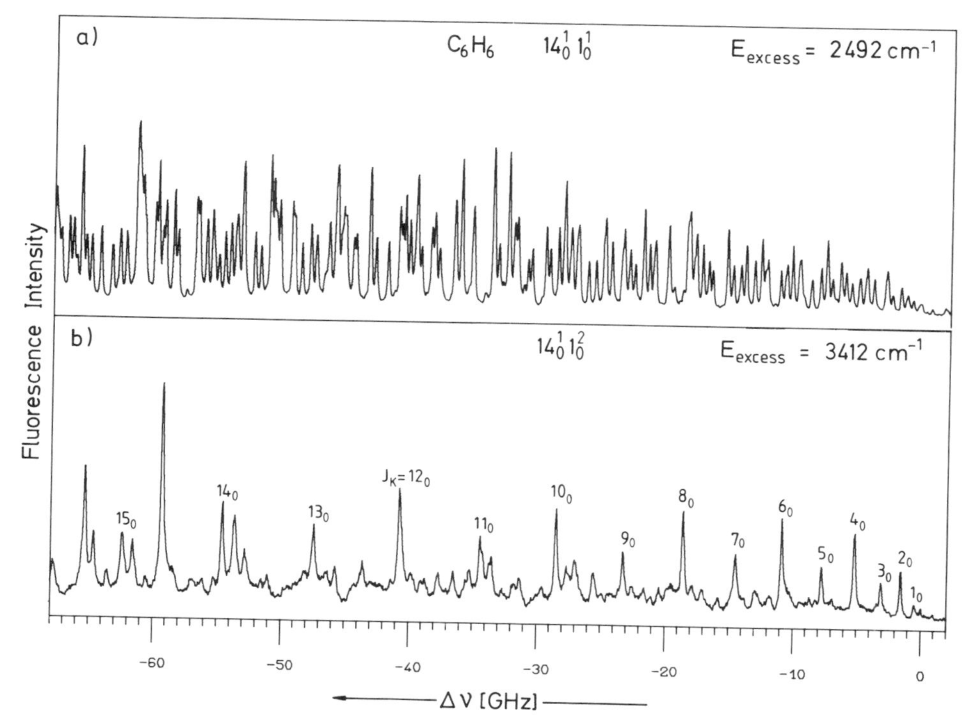

Figure 2. Comparison of identical parts of the Doppler-free two-photon spectrum of the $14^1{}_0 1^1{}_0$ band and $14^1{}_0 1^2{}_0$ band of C_6H_6 /12/.

the $14^1{}_0 1^1{}_0$ band /12/. The structure of the latter band was found to agree quite well with a symmetric top model /9/. In the $14^1{}_0 1^2{}_0$ band a strongly decreased number of lines is seen. In an absorption spectrum all rotational lines would be present, but due to a stronlgy rotational dependent nonradiative decay most rotational states do not fluoresce and are therefore not observed in our fluorescence-excitation spectrum. From the positions of the remaining lines and their alternating intensities it was found, that in the part up to about -60 GHz only lines with K = 0 are seen /5,12/. This can be understood by a strong parallel Coriolis coupling to a fast decaying dark state, since parallel Coriolis coupling does affect all rotational states but the K = 0 states. The excistence of such a state with a very fast nonradiative decay is consistent with the observation made at low excess energy.

With our pulsed set up the decay times of the K = 0 states (seen as sharp lines in the spectrum of figure 2) were measured. Typical results are shown in figure 3 /7/. For all states single exponential decays were found within the experimental resolution and the decay rate increases monotonically with J. The values of the decay rate agree very well with the values found from the measurement of the homogeneous linewidth of the same lines /7/. The J-dependence of the decay fits nicely the dependence expected for an additional perpendicular Coriolis coupling to a broadened dark state. The magnitude of the coupling matrix element is identical with the one found from the analysis of the perturbations in the $14^1{}_0$ band. However, since the dark state is strongly broadened, no splitting of the lines results, but rather the light state is broadened and superimposed on the dark state. The decay behavior of the sharp one of the two lines, i.e. the observed increase in the nonradiative decay rate, is determined by the degree of mixture and the decay behavior of the dark state. The broad dark state is believed to be a combinational state containing quanta of low frequency out-of-plane modes, which are known to enhance the nonradiative decay /13/.

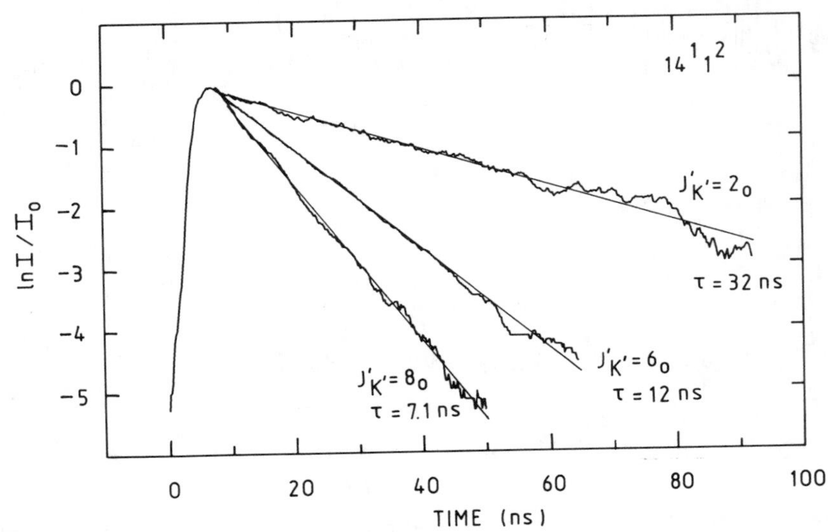

Figure 3. Decay curves for K = 0 states of the 14^1 state of C_6H_6 /7/.

A NEW METHOD FOR PROBING HIGHLY VIBRATIONALLY EXCITED MOLECULES

Constantine Douketis and James P. Reilly
Indiana University, Bloomington, Indiana, 47405

ABSTRACT

In this paper the $\Delta v=4$ vibrational overtone spectrum of room temperature hydrogen peroxide vapor is reported as recorded under low (1 cm^{-1}) and high (Doppler limited) resolution conditions. The complexity of the observed spectra motivated a study of this same transition under supersonic beam conditions. Preliminary results obtained in this study are reported.

INTRODUCTION

One of the important remaining frontiers of modern spectroscopy involves the study of molecules in very highly excited vibrational states. These states can be accessed for study in a few different ways. One of the cleanest and certainly the most direct involves exciting overtone transitions. Although molecular overtone spectra have been under investigation for over one hundred years, recent developments in techniques that employ laser light sources have facilitated these studies and led to renewed interest in this field during the last decade. Molecular overtone spectra can be broadly classified into two types. Those of small molecules such as diatomics, water, acetylene, methane, ammonia etc. consist of bands with cleanly resolved rotational structure. In some of these cases, the rotational structure is easily assignable, in other cases it has never been analyzed. Overtone bands of larger molecules like methanol, ethane, benzene, pyrrole consist of unresolved rotational contours some of which exhibit hints of discrete structure. One of our research group's principal goals in recent years has been to reduce the complexity of sharp-lined but unassignable overtone bands and to look for sharp structure in the diffuse overtone bands of larger molecules by investigating these transitions under supersonic beam conditions. In a previous study, the $\Delta v=4$ overtone band of water vapor was successfully recorded under supersonic beam conditions[1]. In the present study, a somewhat larger "intermediate size" molecule, hydrogen peroxide, is investigated under both room temperature and supersonic beam conditions.

EXPERIMENTAL

The apparatus that we used to record gas phase overtone spectra has been previously described[1]. Briefly, a Spectra Physics Model 380D ring laser, operating under either multimode or single mode conditions, is our source of near-infrared radiation. A glass photoacoustic cell with a Knowles Model 1785 microphone is mounted in the cavity of this laser and is used to record bulk gas overtone spectra. For cold molecule studies, a continuous supersonic beam is orthogonally crossed with light from the dye laser. Molecules convey their vibrational excitation to a liquid helium-cooled bolometer, and the wavelength dependence of the vibrational excitation spectrum is recorded. Because of the geometry of the apparatus, Doppler broadening is drastically reduced and typical linewidths that we observe for our overtone transitions are approximately 6 Mhz.

RESULTS AND DISCUSSION

A low resolution photoacoustic spectrum of room temperature H_2O_2 vapor in 100 torr of krypton gas is displayed in Figure 1. This spectrum shows a remarkable

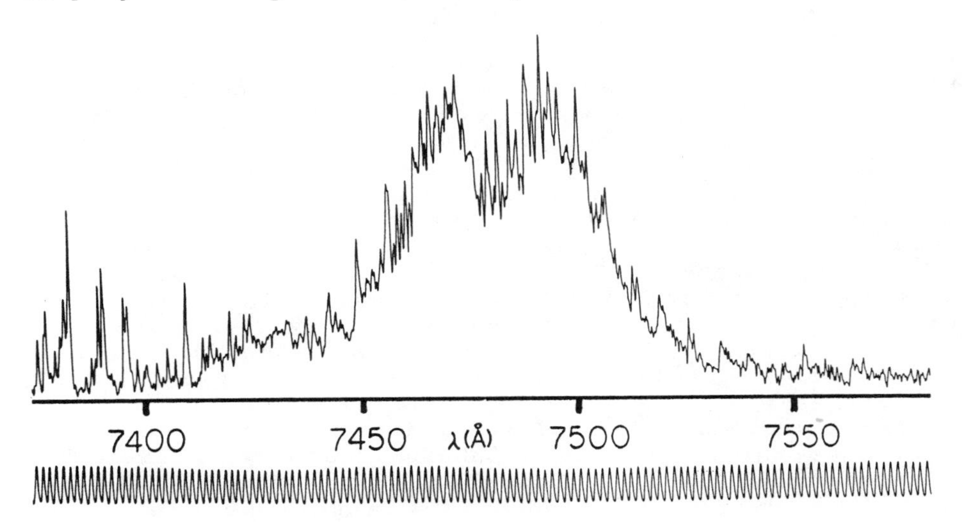

Figure 1. Room temperature photoacoustic spectrum of H_2O_2 at 1 cm^{-1} resolution. An etalon transmission spectrum (free spectral range = 2.863 cm^{-1}) is recorded at the bottom. The lines on the short wavelength end are due to water.

resemblance to that recently recorded by Crim and coworkers[2]. This is quite significant, considering that they were observing a photodissociation/OH excitation spectrum, and it would have been quite conceivable for this to exhibit a somewhat different rotational state dependence than the absorption spectrum itself. Although our low resolution spectrum does exhibit evidence of sharp-line structure, the rotational contour is poorly resolved. In order to investigate whether the structure in this spectrum could be better resolved with improved experimental conditions, or whether it reflected the fundamental character of H_2O_2's overtone absorption spectrum, we began recording photoacoustic spectra with our ring dye laser operating under single mode conditions. Demonstrating the resolution obtainable under these conditions, Figure 2 displays photoacoustic absorption spectra of the R(7) single rovibrational line in the (10300) ← (00000) overtone spectrum of acetylene. In Figure 2A the acetylene pressure was 0.01 torr, and the dominant line broadening mechanism is Doppler broadening. The observed linewidth of 200 Mhz is about a factor of four smaller than one would expect. This narrowing results from the limited acceptance solid angle of the microphone in our photoacoustic cell. In Figure 2B, the pressure was 15 torr, and both Doppler and

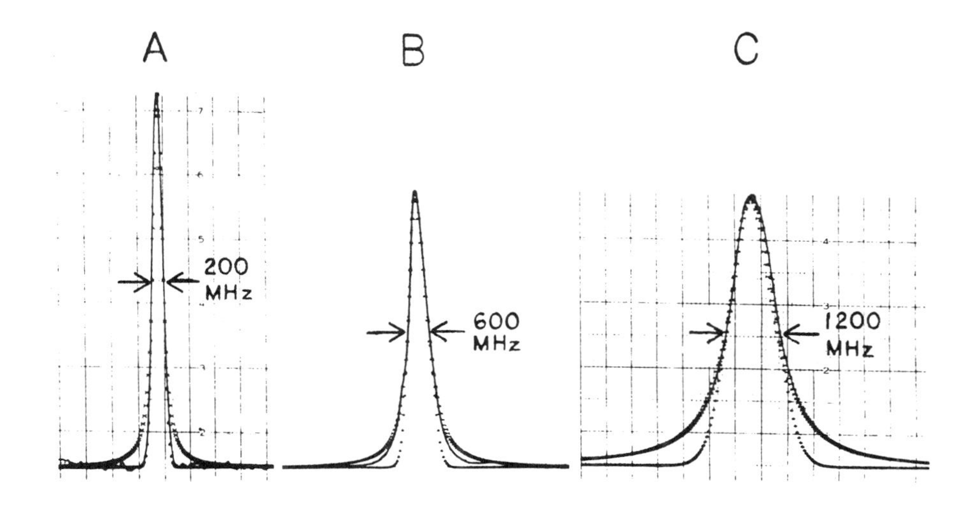

Figure 2. R(7) in the (10300) ← (00000) band of acetylene at 7876.401 Å (12692.67 cm^{-1}). A) 0.01 torr, B) 15 torr, C) 300 torr. The solid line is experimental data, the dotted curves are calculated Gaussian and Lorenzian contours.

pressure broadening contribute significantly to the lineshape. Finally, at 300 torr in Figure 2C a predominantly pressure broadened lineshape was observed. Based on estimated pressure broadening coefficients for acetylene,one would expect a linewidth of 1.2 Ghz, in good agreement with what is observed. Therefore we conclude that with our photoacoustic spectrometer and single mode laser, our spectral linewidths are fully characterized.

Figure 3 displays a photoacoustic spectrum of hydrogen peroxide recorded under high resolution laser operating conditions. The observed linewidths are about 30% larger than the Doppler width due to the introduction of buffer gas. It is quite evident that most of the rotational structure of the overtone band has been resolved. In contrast with Figure 1, we would expect that a straightforward spectroscopic analysis of all of the lines in this overtone band may be feasible. However, at least two complications are expected. First, H_2O_2 is an asymmetric top molecule, so the spectrum will not appear as an orderly pattern of easily assignable lines. Second, hydrogen peroxide can undergo low frequency torsional motions that should lead to at least one hot band in the vibrational overtone spectrum. As Hirota has previously shown for the $\Delta v=2$ and $\Delta v=3$ overtone spectra, this essentially doubles the number of lines observed, and causes them to fit not one, but two asymmetric top contours[3]. In order to aid in the assignment of the rovibrational lines appearing in Figure 3, we have initiated a series of studies of hydrogen peroxide's $\Delta v=4$ overtone band under supersonic beam conditions. As a guide to aid us in searching for

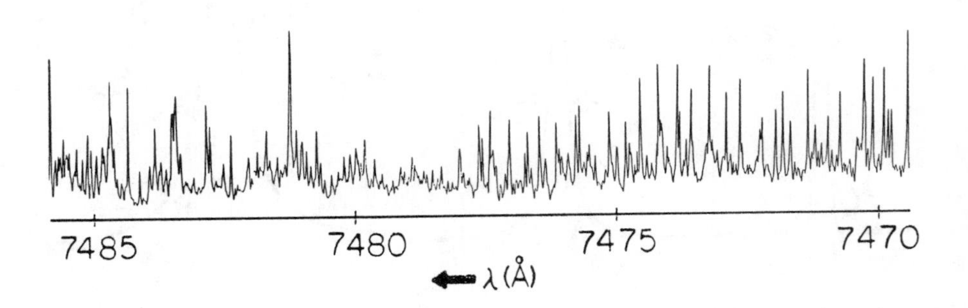

Figure 3. High resolution photoacoustic spectrum of H_2O_2 in 100 torr Kr at 295 K near the band origin. Several continuous scans of 50 Ghz width were overlapped to generate this spectrum.

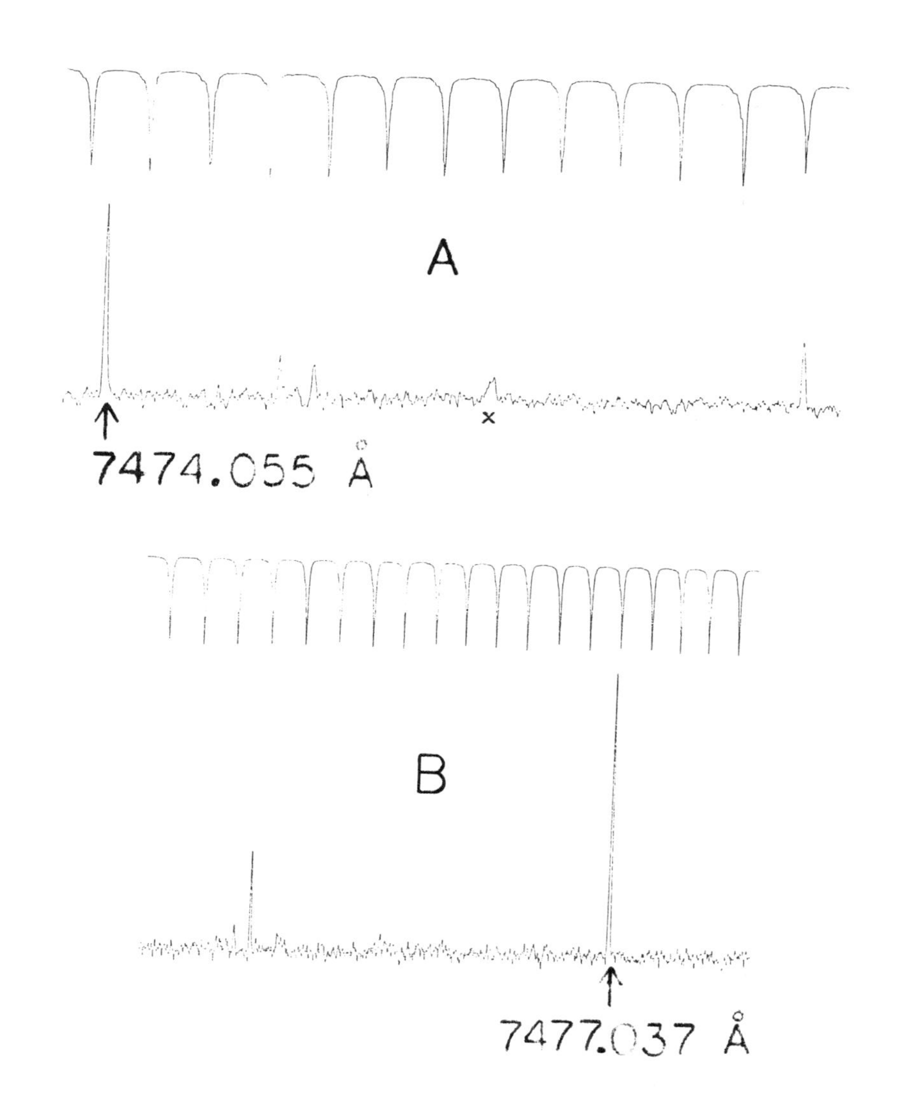

Figure 4. Two regions in the overtone spectrum of H_2O_2 as recorded on the molecular beam. The signal marked x in A) is spurious.

spectral lines, we used the pulsed laser, pulsed free jet work of Butler et al[2]. In their OH photodissociation-excitation work on a cold beam, they observed a handful of lines having somewhat different linewidths. We have scanned over these lines under ultrahigh resolution, Doppler-free beam conditions and see considerable evidence of line splittings. Examples of two such scans, along with a 150 Mhz etalon marker, are displayed in Figure 4. We find that all of the lines in the H_2O_2 $\Delta v=4$ overtone spectrum are as sharp as our limiting spectral resolution (6 Mhz). Our present work is focussing on fitting the high resolution data into patterns that can be described by an asymmetric top formula, and then applying this information to assigning the room temperature photoacoustic spectrum displayed in Figure 3.

In the course of performing the bulk gas photoacoustic investigations described above, anomalous results were sometimes obtained. For example, the

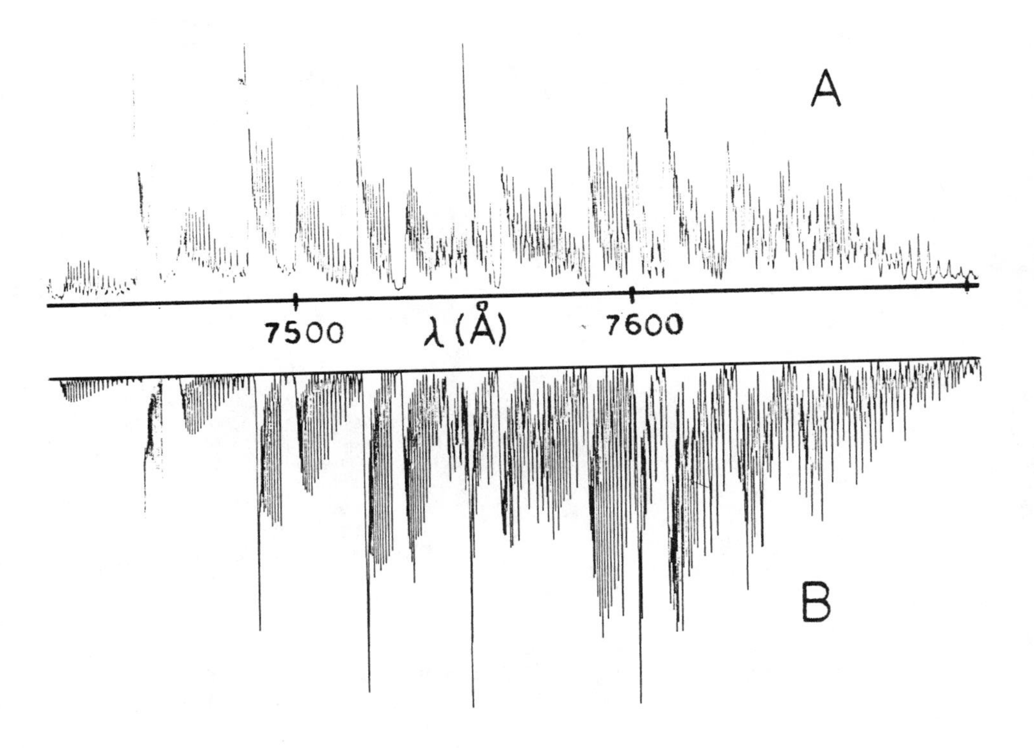

Figure 5. A) Low resolution (1 cm^{-1}) photoacoustic spectrum of HOO free radical. B) An asymmetric rigid rotor simulation of an HOO vibrational band.

photoacoustic spectrum in Figure 1 was not obtained when the peroxide vapor was first introduced into the absorption cell. Rather, the spectrum displayed in Figure 5A was recorded. The intensity of this spectrum decayed on a timescale of approximately 30 minutes, suggesting that it might be due to some transient intermediate. We hypothesized that this spectrum might involve an electronic transition of the HOO free radical[4]. Using a computer program that simulates the rotational structure of an asymmetric top, in conjunction with the known rotational constants for the ground state of HOO and estimated constants for the excited state, we synthesized a rovibrational band for this species. It is displayed in Figure 5B. It is quite evident that the calculated spectrum matches that which was experimentally observed. The significance of this observation is two-fold. First, an electronic absorption spectrum of this free radical in the visible spectral region has not previously been recorded. Considering the interest in this species due to its role in combustion processes[5] and in atmospheric chemistry[6] this new absorption may be useful in kinetic studies. Second, since no deliberate effort was made to generate this species in our photoacoustic cell, the mechanism by which it was produced deserves consideration. It must involve some sort of catalytic process on the surface of our glass sample cell. Whether heavy metal ions or silicate structures in the glass are leading to the breakdown of hydrogen peroxide is not presently understood, but future efforts will be dedicated to probing this.

This work has been supported by the National Science Foundation. JR is a Camille and Henry Dreyfus Foundation Teacher-Scholar.

REFERENCES

1. C. Douketis, D. Anex, G.E. Ewing, J.P. Reilly, J. Phys. Chem. 89, 4173 (1985).
2. L.J. Butler, T.M. Ticich, M.D. Likar, F.F. Crim J. Chem. Phys. 85, 2331 (1986).
3. E. Hirota, J. Chem. Phys. 28, 839 (1958).
4. C. Douketis, M. Scotoni, J.P. Reilly, J. Chem. Phys. (submitted).
5. J. Warantz, in Combustion Chemistry, W.C. Gardiner ed., Springer, New York, (1984).
6. B.A. Thrush, Acct. Chem. Rev. 14, 116 (1982).

DYNAMICS OF THE IR LASER INDUCED GEOMETRICAL REARRANGEMENT OF 2-FLUOROETHANOL IN SOLID ARGON AT 11 K

Zakya H. Kafafi,* Charles L. Marquardt and James S. Shirk
Naval Research Laboratory, Washington, D.C. 20375
*Sachs/Freeman Associates, Landover, MD 20785

ABSTRACT

IR laser excitation of either the C-H or O-H fundamental vibrational modes of 2-fluoroethanol results in photoorientation of the Gg′ isomer as well as conversion to its less stable Tt form. Vibrational mode assignments were made to the absorption bands of the Gg′ isomer of 2-fluoroethanol based on its IR dichroism. Analysis of the IR spectral dichroism of the Tt photoproduct suggests that it is formed by either simultaneous or stepwise rotation of the CH_2F group about the C-C bond and the hydroxyl group about the C-O bond, with the C_2H_2O skeleton remaining almost stationary.

INTRODUCTION

Certain classes of molecules isolated in cryogenic matrices undergo photoorientation and/or photoisomerization following selective IR laser excitation of vibrational modes.[1] An example of such a system is the IR laser photolysis of 2-fluoroethanol in which both photoisomerization and photoorientation were observed.[2]

In the present paper the IR dichroic spectra of the Gg′ and Tt isomers of 2-fluoroethanol are reported. Vibrational mode assignments of the Gg′ isomer and the mechanism of its photoisomerization to the Tt form are inferred from these experimental results.

EXPERIMENTAL SECTION

A 0.1% mixture of 2-fluoroethanol in argon was condensed onto a polished copper surface at 23 K and cooled to 11 K. The matrix isolated molecule was vibrationally excited by irradiating into a specific vibrational mode with polarized IR light from a color center laser. The IR dichroism was measured by recording the Z- and Y-polarized infrared spectra of the matrix isolated 2-fluoroethanol before and after photolysis. The Z and Y axes were defined as the laboratory axes parallel and perpendicular to the electric vector of the IR laser beam.

RESULTS AND DISCUSSION

Figure 1 shows typical dichroic spectra obtained after irradiation of the matrix isolated sample with Z polarized infrared laser light. These spectra show that both the reactant and product isomers are photooriented in solid argon. The dichroism exhibited by the reactant isomer Gg′ is used to assign the vibrational modes of the molecule. The dichroism of the product isomer Tt demonstrates that the photoisomerization leaves the product in a specific

orientation relative to the reactant.

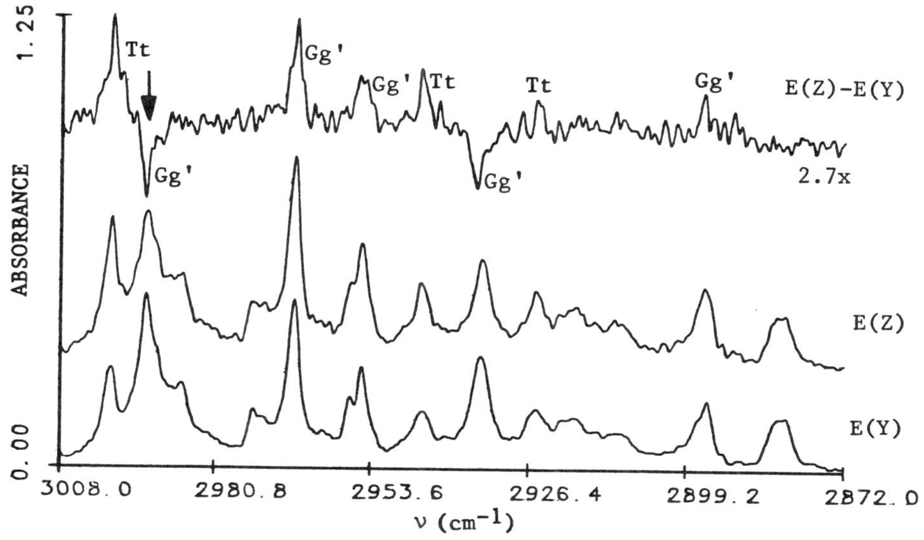

Figure 1. Polarized matrix isolation infrared spectra of the C-H stretching region of the Gg′ and Tt isomers of 2-fluoroethanol after IR laser irradiation at ν=2991.7 cm^{-1}. In E(Z,Y), the electric vector of the observing FTIR beam is parallel, perpendicular to the electric vector of the IR laser beam.

Vibrational mode assignments were deduced from a qualitative analysis of the IR dichroism. For example by examining the dichroism of the absorption bands of the Gg′ isomer of 2-fluoroethanol in Fig.1, it is possible to make vibrational mode assignments to all of the C-H stretching modes of this molecule. The ν=2991.7 cm^{-1}, photolysis frequency, was resonant with that of the C-H asymmetric stretching mode of the CH_2OH group of 2-fluoroethanol.[3] The 2966.3 cm^{-1} peak showed dichroism opposite in direction to the 2991.7 cm^{-1} peak. Since the asymmetric and symmetric stretching modes have mutually perpendicular transition dipole moments, the 2966.3 cm^{-1} peak was assigned to the C-H symmetric stretching mode of the CH_2OH group. Relative to the 2991.7 cm^{-1} peak, the absorption bands at ν=2954.7 and 2933.9 cm^{-1} showed dichroism in the opposite and in the same directions, respectively. The transition dipoles of the C-H asymmetric and symmetric stretching modes of the CH_2F group form angles close to 60° and 45° respectively with the transition dipole of the C-H asymmetric stretching mode of the CH_2OH group. Qualitatively, the angles larger and smaller than the magic angle~54° should show dichroism in the positive and negative directions, respectively.[4] Since the 2954.7 cm^{-1} band shows dichroism in the positive direction it is assigned to the asymmetric stretching mode. Similarly, since the 2933.9 cm^{-1} absorption is dichroic in the negative direction it is assigned to the symmetric stretching mode. Similar arguments were made for the vibrational mode assignments in other regions of the infrared absorption spectra of the Gg′ and Tt isomers of 2-fluoroethanol. A detailed discussion of the vibrational mode assignments is the subject of a future publication.

The dynamics of the photoisomerization may be deduced from an analysis similar to the one given in the previous section. In the top trace of Fig. 1, the difference band at 2997.8 cm^{-1} of the Tt product[5] is opposite in direction but equal in value to that at 2991.7 cm^{-1} of the Gg′ reactant. This large positive dichroism exhibited by the Tt isomer suggests that the transition dipoles of the C-H asymmetric stretching modes of the Gg′ and Tt isomers are parallel. This is most likely to occur if the CH_2O group remained almost stationary during the isomerization process. A similar dichroism was observed in the C-O stretching region of the reactant and product isomers confirming that the CH_2O group did not move much during the geometrical rearrangement. The peak at 2943.5 cm^{-1} shows a dichroism similar to that exhibited by the 2997.8 cm^{-1} peak relative to the 2991.7 cm^{-1} peak. This dichroism suggests that the transition dipole moment of this product vibrational mode and that of the C-H asymmetric stretching mode of the CH_2OH group of the Gg′ reactant isomer are parallel to each other. Since the transition dipole moments due to the 2991.7 cm^{-1} and 2997.8 cm^{-1} peaks are parallel to one another and the latter is also parallel to that of the 2943.5 cm^{-1} peak, it follows that the transition dipole moment of the unassigned mode is parallel to that of the C-H asymmetric stretching mode of the CH_2OH group of the Tt isomer. Hence this peak is assigned to the C-H asymmetric stretching mode of the CH_2F group of the Tt isomer. One may also notice that the dichroism exhibited by this 2943.7 cm^{-1} absorption band is in the same direction as that due to the C-H asymmetric stretching mode of the CH_2F group of the Gg′ reactant isomer indicating that rotation of the CH_2F group about the C-C bond has taken place during the photoisomerization process. Similar observations were made for the internal rotation about the C-O bond.

Summarizing, the observed dichroism on the Tt isomer is consistent with the CH_2O group remaining stationary while the CH_2F and OH groups rotate about the C-C and the C-O bonds, respectively. A detailed analysis of the mechanism of the photoisomerization will be published elsewhere.

REFERENCES

1. H. Frei and G. C. Pimentel, Ann. Rev. Phys. Chem. 36, 491 (1985).
2. W.F. Hoffman and J.S. Shirk, J. Phys. Chem. 89, 1715 (1985).
3. The mode assignment of this absorption peak was deduced from its dichroism relative to the O-H absorption band when the molecule was excited at the O-H stretching frequency.
4. E.W. Thulstrup and J. Michl, Spectroscopy with Polarized Light; VCH Publishers: Deerfield Beach, FL, 1986.
5. This absorption is assumed to be due to the C-H asymmetric stretching mode of the CH_2OH group of the Tt isomer.

EXCHANGE REACTIONS OF ATOMIC HYDROGEN WITH ACETYLENE AND ETHYLENE

S. Satyapal, G. Johnston, J. Park and R. Bersohn*
Columbia University, New York, NY 10027

ABSTRACT

Total reaction cross sections for the exchange reactions of atomic hydrogen with C_2D_4, C_2D_2 and CH_3CCD were obtained by the technique of laser induced fluorescence. The translational energies of nascent D atom products were determined from their Doppler-broadened LIF profiles. In all three cases it is shown that the fraction of available energy resulting in translation of the D product does not agree with a statistical model of energy equilibration.

INTRODUCTION

The reactions of atomic hydrogen with acetylene and ethylene have been of interest for a number of years. Absolute rate parameters have been determined by several investigators using various techniques such as pulse radiolysis-resonance absorption and flash photolysis resonance fluorescence.[1,2] The detailed dynamics of these reactions however, is still not completely understood. We have probed H and D atoms using LIF in order to measure rate constants and total reaction cross sections for the exchange reactions of H with deuterated ethylene, acetylene and methyl acetylene(d_1). Furthermore, measuring the translational energy of the D atom product may provide conclusive qualitative information on the lifetime of the activated complex and on the number of modes involved in the equilibration of available energy.

EXPERIMENTAL

The experimental set-up consists of a Lumonics HyperEx-440 excimer laser which photolyzes H_2S at 248 nm (KrF), producing nearly monoenergetic H atoms with a translational energy of ~25.1 kcal/mol. H and D atoms are probed in the 121.6 nm region by frequency tripling the output of a dye laser (Lambda Physik FL2002E) in krypton. The resolution of the dye laser is 0.04 cm^{-1} with an intracavity etalon before tripling. The LIF intensity is monitored by a photomultiplier and sent to a boxcar averager. The output is then received by a microcomputer and plotted on a chart recorder.

RESULTS AND DISCUSSION

For the general reaction H+AD→HA+D, the rate equation may be integrated at short times to give k=[D]/([H][AD]t). The total reaction cross section σ, is given by k/<v>rel. Typical delay times used in our experiment were around 100 nsec. Also note that typical pressures insured pseudo-first order conditions. The calculated values of k and σ are shown below.

Table I. Rate constants and reaction cross sections

	k(cm^3/molec-sec)	σ(A^2/molec)
$H+C_2D_4 \rightarrow D+C_2D_3H$	3.03(±0.44)x10^{-10}	2.12(±0.30)
$H+C_2D_2 \rightarrow D+C_2DH$	2.17(±0.21)x10-10	1.51(±0.14)
$H+CH_3CCD \rightarrow D+CH_3CCH$	8.63(±0.67)x10^{-11}	0.60(±0.19)

According to ab initio calculations[3,4], the reaction of H with either acetylene or ethylene involves substantial C-H bond deformation and C-C lengthening. As a hydrogen atom approaches an acetylene molecule the minimum barrier to reaction is 6.2 kcal/mol, followed by a deep 40.1 kcal/mol potential well. Since the barrier for abstraction is 32.4 kcal/mol, the addition reaction is the only predominant process at the energies we use in our experiments. If the hot vinyl radical is not stabilized by collisions, it will dissociate back into the reactants. The energetics for H addition to ethylene is similar, with a 2.5 kcal/mol barrier and a 38.0 kcal/mol potential well. Thus from geometric considerations as well as due to the barrier difference, we would expect a slightly larger cross section for $H+C_2D_4$ than for $H+C_2D_2$, which is indeed the result we obtained.

With regard to the translational energy of the D product, it should be noted that in all three cases the Doppler profiles were not Gaussian. The translational energy may be calculated using:

$$\langle v_z^2 \rangle = \Sigma f(v_z) v_z^2 / (\Sigma f(v_z)).$$

If all possible final states have an equal probability of being occupied, the kinetic energy distribution will be given by

$$f(E_T) = A E_T^{1/2} (E_{AVL} - E_T)^n,$$

resulting in:

$$\langle E_T\rangle = \int[E_T f(E_T)dE_T]/\int f(E_T)dE_T = 3E_{AVL}/(2n+5),$$

where n is the number of vibrational modes. Note that this assumes a rigid rotor-harmonic oscillator approximation for the product molecule. One can see that the experimental results are quite different.

Table II. $\langle E_T\rangle/E_{AVL}$

	n	$f_T = \langle E_T\rangle/E_{AVL} = 3/(2n+5)$	f_T(expt'l)
$H+C_2D_4 \rightarrow D+C_2D_3H$	12	~10%	~35-40%
$H+C_2D_2 \rightarrow D+C_2DH$	7	~16%	~45-50%
$H+CH_3CCD \rightarrow D+CH_3CCH$	15	~9%	~45-50%

Because the experimental values of f_T are larger, we may already infer that the reaction takes on a more local character, involving fewer modes in the equilibration of the available energy than predicted by a statistical model. Furthermore, the fraction of E_{AVL} converted into kinetic energy of D is about the same for both the acetylene and methylacetylene reactions indicating that the extra modes provided by the methyl group are not involved in the equilibration of the available energy.

REFERENCES

1. K. Sugawara, K. Okazaki, S. Sato, Bull. Chem. Soc. Jpn. 54, 2872 (1981).
2. J. H. Lee, J. V. Michael, W. A. Payne, L. J. Stief, J. Chem. Phys. 68, 1817 (1978).
3. W. L. Hase, H. B. Schlegel, J. Chem. Phys. 86, 3901 (1982).
4. S. Nagase, C. W. Kern, J. Am. Chem. Soc. 101, 2544 (1979).

ISOTOPE EFFECTS ON RAPID VIBRATIONAL DEACTIVATION OF LASER-EXCITED NITROUS OXIDE BY SULFUR HEXAFLUORIDE

K. L. McNesby, M. C. Longuemare, and R. D. Bates, Jr.
Georgetown University, Dept. of Chemistry, Washington, D.C. 20057

ABSTRACT

Single exponential 4.5 micron fluorescence decay times are observed from samples containing SF_6 and $^{14}N^{14}NO$ or $^{14}N^{15}NO$. The nitrous oxides are excited directly to the 00^01 level by a Q-switch CO_2 laser. Decay times are dominated by the partial pressure of SF_6 in the sample. At room temperature the rate constant for $^{14}N^{14}NO$-SF_6 is 1200 $ms^{-1}torr(SF_6)^{-1}$, and for $^{14}N^{15}NO$-SF_6 is 200 $ms^{-1}torr(SF_6)^{-1}$. The cross sections for each isotopically differentiated N_2O decreases as the temperature increases. The excited N_2O level for $^{14}N^{15}NO$ is ~50 cm^{-1} lower than that for $^{14}N^{14}NO$. This increase in the energy defect can account for the decrease in efficiency by long range coulombic forces.

INTRODUCTION

In experiments involving CO_2 laser-excited SF_6 and the probe molecule N_2O, rapid sensitization of vibrations in N_2O has been observed.[1] This vibrational excitation through collisions with a second species was determined to take place on a time scale much shorter than that on which the majority of the energy deposited in the system by the CO_2 laser pulse is converted into heat. In fact, it also occurs more rapidly than the time required for collisions to redistribute energy among the SF_6 vibrational modes, at least for large N_2O mole fractions. After sensitization, the energy deposited in the N_2O drains back out very rapidly, resulting in a pulse of excitation reaching the lowest level in the 2224 cm^{-1} stretch in N_2O.[2] If $^{14}N^{15}NO$ is used instead, the excitation is too slow to provide the rapid pulse of excitation, as shown in Fig. 1.

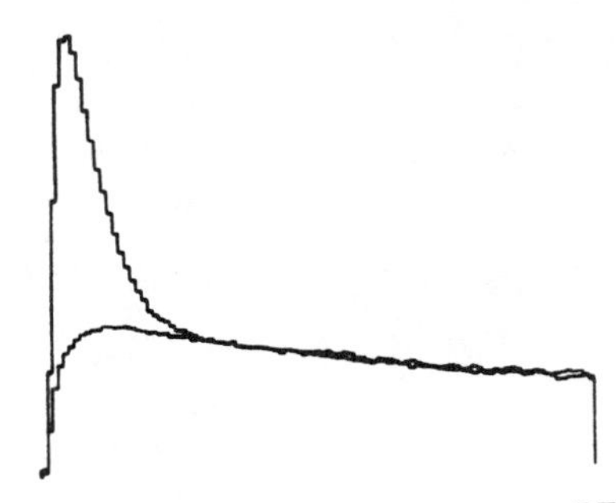

Figure 1. Observed 4.5 μm fluorescence signals after averaging for samples with identical pressures of 0.05 torr SF_6 and 0.4 torr $^{14}N^{14}NO$ (upper trace) or $^{14}N^{15}NO$ (lower trace). Other experimental parameters are identical. A CO_2 laser pulse excited the SF_6 in the samples.

RESULTS

These observations prompt a closer look at SF_6-N_2O mixtures. In this study, the experiment described above is done in reverse:

the laser pulse excites N_2O directly. Fluorescence is monitored from the 2224 cm^{-1} stretch in N_2O. SF_6 acts as a quencher. Specific CO_2 laser lines pumped each N_2O species directly.[3]

Experiments in which either $^{14}N^{14}NO$ or $^{14}N^{15}NO$ was pumped produce single exponential fluorescence decays. Decay constants are the sum of several first order processes. Subtracting N_2O-N_2O and radiative terms and working at pressures where diffusion is negligible gives linear plots of k(corrected) versus SF_6 partial pressure with slopes equal to $k_{N_2O-SF_6}$ as shown in Table I.

TABLE I. Rate constants and probabilities for deactivation of $^{14}N^{14}NO$ and $^{14}N^{15}NO$ by SF_6 as a function of temperature.

T(K)	$k_{N_2O-SF_6}$ ($ms^{-1}torr^{-1}$)	Probability
	for $^{14}N^{14}NO$	
238	1751 ±178	0.142
255	1422 ± 83	0.120
267	1276 ± 52	0.110
283	1129 ± 53	0.100
296	1018 ± 70	0.092
306	992 ± 18	0.091
314	967 ± 17	0.090
326	860 ± 20	0.082
343	823 ± 49	0.080
366	720 ± 69	0.073
	for $^{14}N^{15}NO$	
223	266 ± 22	0.021
250	175 ± 45	0.015
273	141 ± 55	0.015
301	209 ± 4	0.019
318	145 ± 10	0.014

Figure 2 shows the dominant downward trend in k for N_2O-SF_6 with rising temperature. This characteristic decreasing efficiency as temperature increases indicates the importance of long range forces in a near resonant V-V process.[4] The $^{14}N^{15}NO$ probabilites are lower and show a milder decrease with increasing temperature.

DISCUSSION

These results show that deactivation of vibrationally excited N_2O (00^01) by SF_6 proceeds by a very rapid process governed by the long-range coulombic forces. The distinctive 1/T dependence obtained indicates that energy differences between initial and final vibration-rotation states must be a small fraction of kT. The decrease in probability by a factor of more than 6 and apparent flattening out of the cross section versus temperature for the $^{14}N^{15}NO$ indicate the coupling is slower and governed by a larger

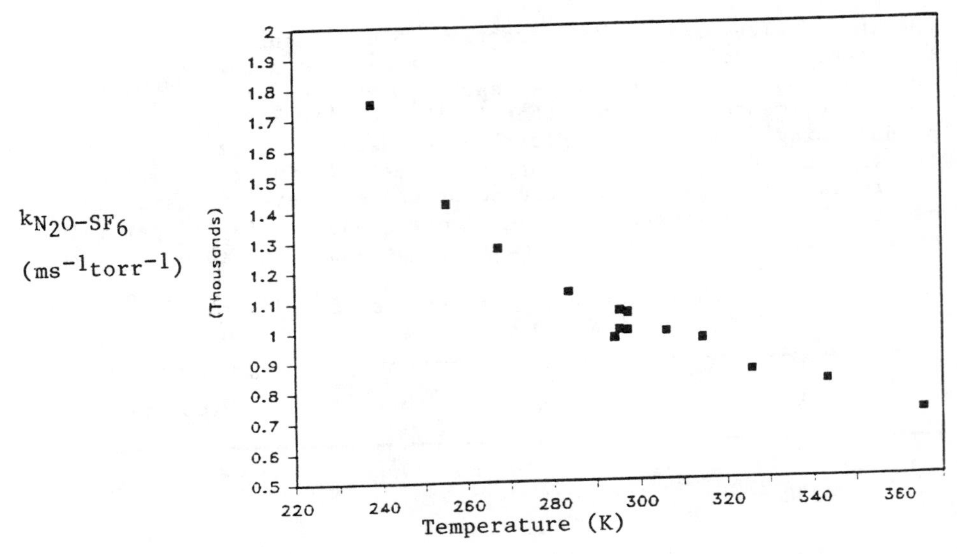

Figure 2. Corrected $^{14}N^{14}NO-SF_6$ rate constant versus temperature.

energy gap. The deactivation of vibrationally excited CO_2 (00^01) by SF_6 probably takes CO_2 from (00^01) to (02^00) at 961 cm^{-1}, which can be referred to as a 10 μm mechanism.[5] For N_2O a 4 μm mechanism involving deactivation of N_2O (00^01) to the ground state has been hypothesized.[1] These data do not discriminate between the 4 μm and 10 μm pathways for the deactivation of N_2O by SF_6. The mechanism by which energy is exchanged between excited SF_6 and N_2O in experiments where the SF_6 receives several quanta of excitation may or may not be related to the efficient V-V processes that occur during collisions under low excitation conditions.[6] Clearly, the deactivation of N_2O by SF_6 is extremely fast with a cross section 1/10th gas kinetic; it proceeds by near resonant intermolecular processes; and the larger energy gap in $^{14}N^{15}NO$ results in a decrease in the rate constant and in sensitivity to temperature.

REFERENCES

1. A. Fahr and R.D. Bates, Jr., Chem. Phys. Letters 71, 381 (1980).
2. A. Fahr and R.D. Bates, Jr., Chem. Phys. 105, 449 (1986).
3. K. McNesby, Ph.D. thesis, Georgetown University, 1987.
4. R.D. Sharma and C.A. Brau, Phys. Rev. Letters 19, 1273 (1967); W.G. Tam, Can. J. Phys. 50, 2691 (1972); R.J. Cross and R.G. Gordon, J. Chem. Phys. 45, 3571 (1966); R.T. Pack, J. Chem. Phys. 72, 6140 (1980); D.E. Godar and R.D. Bates, Jr., unpublished results.
5. J.C. Stephenson and C.B. Moore, J. Chem. Phys. 52, 2333 (1970).
6. Y. P. Vlahoyannis, N. Presser, and R.J. Gordon, Chem. Phys. Letters 106, 157 (1984); M. Koshi, Y.P. Vlahoyannis, and R.J. Gordon, J. Chem. Phys. 86, 1311 (1987).

REACTIONS OF He_2^+ AND $He(2^3S)$ WITH SELECTED ATOMIC AND MOLECULAR SPECIES*

J. M. Pouvesle, A. Khacef, and J. Stevefelt
GREMI, University of Orleans
B.P. 6759, 45067 Orleans Cedex 2, France
and
H. Jahani, V. T. Gylys, and C. B. Collins
The University of Texas at Dallas
Center for Quantum Electronics
P.O. Box 830688, Richardson, TX 75083-0688

ABSTRACT

Super-Langevin rates have been demonstrated for the reactions of He_2^+ with molecular gases and no evidence of saturation at 6 atm pressure of helium diluent was found. For the reactions of $He(2^3S)$, three-body channels have been well characterized, and they occur at rates of the order of 10^{-31} cm^6s^{-1}.

INTRODUCTION

This work concerns the general applicability of three-body effects to a wide set of reactant species which could teach reaction principles important to high pressure lasers undergoing continued development. Previous papers[1,2] have focused upon the demonstration of true super-Langevin rates for the reactions of He_2^+ with a few species over a range of pressures to 6 atm. On the other hand, it was shown[3] that the anomalously large values reported[4] for the pressure-dependent part of the reactions of $He(2^3S)$ with at least one species, N_2, was due to a more reactive component, $He(^3\Sigma, \nu^*)$. When examined independently, the termolecular components of the neutral particle reactions were shown[2,3] to be more than one order of magnitude smaller than values previously reported,[4] while for the ions the large values of termolecular rates were confirmed.

EXPERIMENTAL AND ANALYTICAL METHOD

As in the previous work,[1-3] measurements were made in the afterglows of preionized discharges into the gas samples flowing in geometries resembling those encountered in high pressure laser systems. A constant and small component of N_2 was added to the reacting mixture as a convenient monitor of the energy storing species, He_2^+ and $He(2^3S)$. Two experimental systems were employed, using substantially different discharge energies to enhance separate afterglow periods corresponding to the reactions of He_2^+ and $He(2^3S)$, respectively, and each of exponential character. The relevant reaction frequencies, ν, could then be determined from the time-resolved fluorescence from the product ion, $N_2^+(B)$, $\nu = I^{-1}dI/dt$.

From the basic definition of a reaction rate coefficient, k, it follows that $k = \partial\nu/\partial[X]$, where X is the reacting species and where it may be expected that $k = k_2 + k_3[He]$ from independent bimolecular and termolecular reaction channels. The rates k_2 and k_3 for the reactions of He_2^+ with O_2 and CO_2 were measured with three different amounts of N_2 additive. The close agreement obtained serves to confirm the experimental procedures.

RESULTS

Results of this work are summarized in Table I. Reactant pressures ranged from 0 to 6 Torr, while diluent pressures of He ranged from 0.4 to 6.2 atm.

Table I. Summary of the biomolecular and termolecular rate coefficients, in units of 10^{-10} cm^3s^{-1} and 10^{-30} cm^6s^{-1}, respectively, for the reactions of He_2^+ and $He(2^3S)$ measured in this work.

Reactant	He_2^+		$He(2^3S)$	
	k_2	k_3	k_2	k_3
Ne			0.036 ± 0.010	0.020 ± 0.018
Ar	2.0 ± 0.7	7.2 ± 1.2	0.76 ± 0.01	0.20 ± 0.04
Kr			1.24 ± 0.02	0.164 ± 0.045
Xe	2.4 ± 0.5	9.5 ± 1.5	1.8 ± 0.2	0.22 ± 0.36
H_2			0.37 ± 0.05	0.09 ± 0.10
N_2	11.5 ± 1.2	13.1 ± 0.2	0.68 ± 0.03	0.15 ± 0.04
O_2	9.3 ± 0.8	24.3 ± 1.5	2.54 ± 0.03	0.16 ± 0.06
CO_2	13.6 ± 0.8	51 ± 2	7.4 ± 0.4	0.20 ± 0.70
N_2O	27.7 ± 5.2	16.0 ± 4.9	6.4 ± 0.5	0.67 ± 0.95

CONCLUSIONS

The drastic effect of the termolecular channel upon effective rates for the reactions of He_2^+ is most readily appreciated from Fig. 1(a). It can be seen that the Langevin value proved to be no limit to the total rate of reaction, and no evidence of saturation was found, even at 6 atm pressure of diluent. Moreover, the results for this extensive examination served to confirm generally the earliest e-beam measurements of ion-molecule reactions at atmospheric pressures[5] conducted upon 38 reactions of the ions He_2^+, Ne_2^+, and Ar_2^+. In almost 95 percent of the cases studied in those experiments, the three-body reaction channel would dominate at pressures only slightly above 1 atm of diluent. The present reexamination implies that the most appropriate a priori assumption is that the three-body ion-molecule reactions in atmospheric pressures of inert gases will strongly accelerate the total rate of reaction, unless the system is one of the ~ 5 percent for which termolecular kinetics is inhibited.

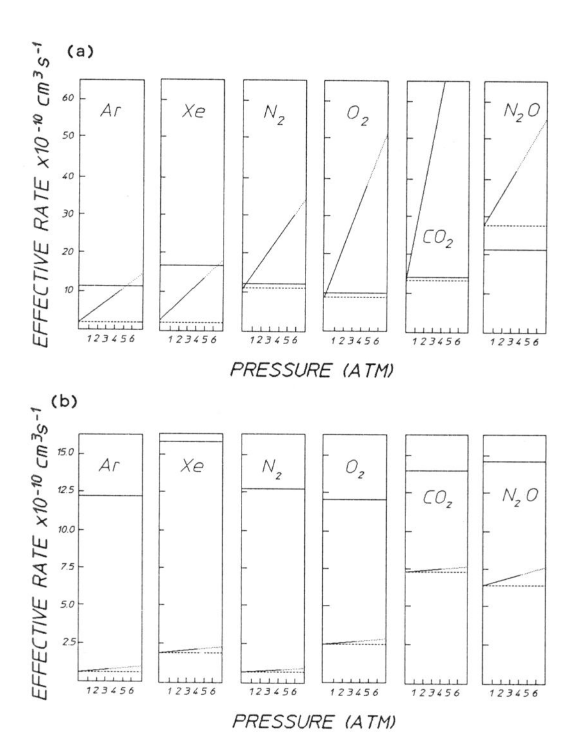

Summary plot of the effective rate coefficients as functions of pressure of helium for various gases. (a) Reactions of He_2^+. (b) Reactions of $He(2^3S)$. Solid horizontal lines indicate the classical Langevin rate (a) and the analogous rate for an attractive polarization potential (b), and the dashed horizontal lines show the bimolecular rates. Shown by the diagonals is the linear dependence of effective rates on helium pressure. The solid portions of the lines indicate the helium pressure range over which each gas mixture was studied.

The bimolecular rate coefficients for the reactions of $He(2^3S)$ with various atoms and molecules obtained in this work are in general agreement with the literature. Comparison with the theoretical rate coefficients[6] for classical capture by an attractive polarization potential shows the reaction probabilities to range from 0.004 (Ne) to 0.52 (CO_2), as seen in Fig. 1(b). For all reactants under study in this work, three-body channels have been identified. However, these channels for the reactions involving neutral triplet metastables are, as seen in Fig. 1(b), relatively unimportant in commonly used plasmas up to a few atmospheres pressure.

*Supported in part by NSF Grant ECS 8314633 and in part by NATO Grant 655/84.

REFERENCES

1. C. B. Collins, Z. Chen, V. T. Gylys, H. R. Jahani, J. M. Pouvesle, and J. Stevefelt, IEEE J. Quantum Elec., QE-22, 38 (1986).
2. V. T. Gylys, H. R. Jahani, Z. Chen, C. B. Collins, J. M. Pouvesle, and J. Stevefelt, Advances in Laser Science - I, edited by W. C. Stwalley and M. Lapp (AIP Conference Proceedings No. 146, New York, 1986) p. 560.
3. J. M. Pouvesle, J. Stevefelt, and C. B. Collins, J. Chem. Phys. 82, 2274 (1985).
4. C. B. Collins and F. W. Lee, J. Chem. Phys. 70, 1275 (1979).
5. C. B. Collins and F. W. Lee, J. Chem. Phys. 68, 1391 (1978).
6. K. L. Bell, A. Dalgarno, and A. E. Kingston, J. Phys. B 1, 18 (1968).

INTRAMOLECULAR PROTON TRANSFER IN ASYMMETRIC DOUBLE MINIMUM POTENTIAL FUNCTIONS

G. D. Gillispie
North Dakota State University, Fargo, North Dakota 58105

ABSTRACT

Excited singlet state proton transfer has been demonstrated in hydroxy-substituted anthraquinones via the technique of laser induced fluorescence. Both low temperature Shpol'skii matrix and free jet sample environments have been used. A comparison of vibronic intensity patterns shows the proton transfer to be largely invariant to the sample environment. Well defined rotational contours are observed in the free jet excitation spectra. Phosphorescence spectra have also been measured for some of these molecules and these indicate that the S_1 and T_1 states probably have very different potential functions for the hydrogen bond.

INTRODUCTION

Over the past decade there has been considerable interest in the excited state proton transfer process. Many studies of vibronically resolved spectra and time evolution of the emission have been published. Most of the latter work has not led to observation of a risetime for the Stokes shifted emission characteristic of the ESPT process. This has been taken by some as evidence that the proton transfer takes place along a single minimum potential function. However, an alternative is that the potential function is a double minimum with the proton transfer occuring rapidly via tunneling even at liquid helium temperature. In principle, the vibronically resolved spectra can answer the single vs. double minimum potential function question, but in practice the issue remains undecided.

Little is known about excited state proton transfer or the properties of intramolecular hydrogen bonds in triplet states. Usually the presence of a strong intramolecular hydrogen bond is associated with rapid internal conversion in the singlet manifold, thereby circumventing the triplet states almost altogether.

Our work on 1,5-dihydroxyanthraquinone provides perhaps the best experimental evidence in favor of an S_1 double minimum potential function along the proton transfer coordinate. The previously published work emphasized Shpol'skii matrix work but we now have completed a fairly extensive free jet excitation spectrum study; that and related work is summarized here. We have also found an additional example of unusual triplet state behavior for an intramolecularly hydrogen bonded species, 6-aminobenzanthrone. It thus joins 6-hydroxybenzanthrone and 9-hydroxyphenalenone as behaving in this fashion.

The Shpol'skii matrix fluorescence and fluorescence excitation spectra of 1,5-DHAQ each show a two region character.[1] We have assigned the two regions as arising from transitions to one or the other of the two wells in the double minimum potential function. The lower energy region in the excitation spectrum is of much lower intensity than in the higher energy portion, which begins approximately 1100 cm^{-1} above the origin band. Typical rotational contours for the low and high energy regions are shown below:

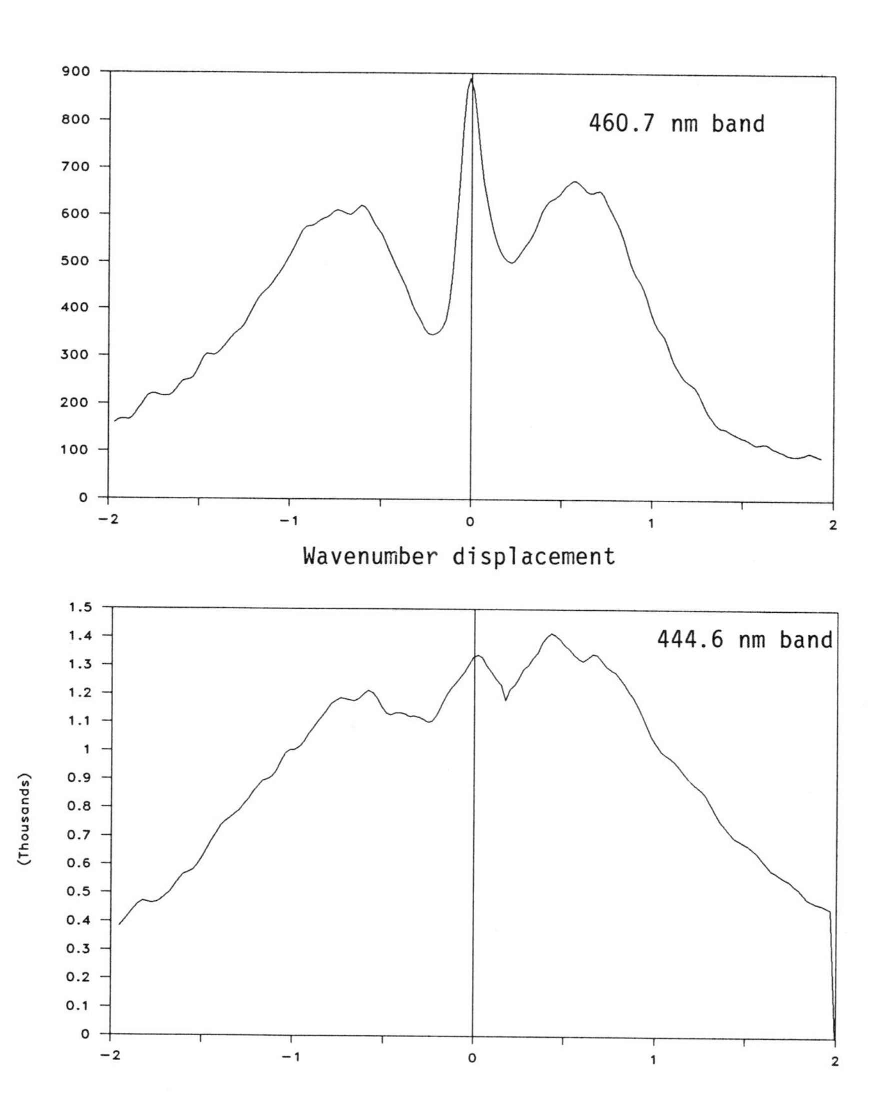

The width of the Q-branch in the 460.7 nm band rotational contour is laser limited (0.09 cm^{-1} resolution). All transitions with less than 1100 cm^{-1} exceess energy in S_1 have similar rotational contours. Likewise in the high energy region, most band contours resemble that for the 444.6 nm band. However, we have found at least one weak band in this region with the same contour as the low energy bands.

Considerable work is necessary before these results can be fully interpreted. However, we tentatively propose that the high energy region, characterized by many bands of higher intensity but much less spectral regularity, represents transitions to levels with excess vibrational energy greater than the barrier height to proton transfer. These levels are strongly coupled to levels corresponding to motion along the proton transfer coordinate. The one band in the high energy region, but having a contour more typical of the low energy region, represents a transition terminating on an S_1 level not coupled to the proton transfer motion.

Work is currently in progress on the jet spectra of several other molecules with intramolecular hyrogen bonds, including 1,8-dihydroxyanthraquinone, 1-aminoanthraquinone, and 9-hydroxyphenalenone. The other two anthraquinones have rotational contours similar to that of 1,5-DHAQ. Also, in 1,8-DHAQ, the change in rotational contour occurs lower in the S_1 manifold than in 1,5-DHAQ. In our previous reports of vibronically well-resolved phosphorescence spectra of internally hydrogen bonded molecules[2,3], we commented on the differences in the fluorescence and phosphorescence vibronic patterns. Now we have added spectra for 6-aminobenzanthrone and find these discrepancies to be even more pronounced. Both the fluorescence and phosphorescence spectra show reasonably strong origin bands indicative of an allowed transition. It is therefore surprising that we are unable to assign with confidence a single mode active in both the fluorescence and phosphorescence. The 6-aminobenzanthrone molecule also resembles 6-hydroxybenzanthrone and 9-hydroxyphenalenone in that the triplet state lifetime is greatly increased when the hydrogen bonded proton is deuterium substituted. Further work in this area will attempt to identify more hydrogen bonded molecules with detectable phosphorescence and to directly probe the T_1 vibrational manifold.

REFERENCES

1. M.H. Van Benthem and G.D. Gillispie, J. Phys. Chem. 88, 2954 (1984).
2. G.D. Gillispie, M.H. van Benthem, and M. Vangsness, J. Phys. Chem. 90, 2596 (1986).
3. G.D. Gillispie, J. Chem. Phys. 85, 4825 (1986).

ELECTRONIC QUENCHING AND FLUORESCENCE LIFETIME OF THE TRIATOMIC RADICAL NCN ($A^3\Pi_u$)

Gregory P. Smith, Richard A. Copeland, and David R. Crosley
Chemical Physics Laboratory
SRI International
Menlo Park, California 94025

ABSTRACT

Laser-induced fluorescence of NCN was observed downstream of a microwave discharge of N_2 and CF_4 in He via the $A^3\Pi_u$-$X^3\Sigma_g^-$ transition near 329 nm. The zero-pressure fluorescence lifetime of the 000 vibrational level of the A-state was measured from the time-resolved laser-induced fluorescence to be 183 ± 6 ns. Thermally averaged total removal cross sections were determined at 300 K for Ar, Kr, Xe, O_2, CO, CO_2, N_2O, SF_6, NO, and NO_2. Upper limits were obtained for the inefficient quenchers He, N_2 and CF_4. The large variations in cross sections among the colliders do not correlate with the predictions of simple electronic quenching theories.

INTRODUCTION

Efforts in our laboratory have been directed toward developing laser-induced fluorescence (LIF) diagnostics for flame species and other reactive intermediates.[1] The necessary data base includes spectroscopic, transition strength, and quenching information. Lately, we have extended the studies from simple diatomic radicals to triatomics such as NH_2 and NCO.[2] This study commences similar efforts for the LIF detection of the NCN radical. Given its ~100 kcal/mole stability and the large extent of carbon-nitrogen bonding found in many nitramine propellant systems, this radical could be of chemical significance in propellant pyrolysis and combustion. Basic spectroscopic data are available from the absorption study of the A-X transition at 329 nm by Herzberg and Travis.[3]

EXPERIMENTAL METHOD

The NCN radicals were produced in a low-pressure microwave-discharge (20W) flow tube containing 1 Torr He, 10 mTorr CF_4, and 2-5 mTorr N_2. Some 30 cm downstream from the discharge, the NCN radicals were excited by the frequency-doubled output of a Nd-YAG pumped dye laser. A two-lens system focussed the fluorescence onto a 0.35 m monochromator (~4 nm bandpass), tuned to the band origin at 329 nm. Gases were added to the flow 14 cm upstream from the laser beam. Under these conditions the added quencher and helium carrier gas were well mixed. Total pressure was measured with a capacitance manometer, and quencher partial pressures were determined using calibrated mass flowmeters. For both quenching and lifetime studies, the fluorescence signals from the photomultiplier (9558) were processed by a transient digitizer (10 ns/channel), averaged by a laboratory computer, and stored for later analysis. Single exponential least-squares fits were made to the decays from 90% to 10% of the peak signal amplitude. A linear least-squares fit of the decay constant versus quencher pressure furnished the quenching rate constants. Excitation scans between 325 and 330 nm of the 000-000 and 010-010 vibrational bands were used to compare with the absorption study of Herzberg and Travis[3] and thus verify that we were exciting the NCN radical. No other interfering species were detected in this wavelength region.

FLUORESCENCE LIFETIMES

A series of fluorescence lifetime measurements has been made for the most prominent features of our NCN excitation spectrum. Since a pressure increase up to 6 Torr of He had no effect on the observed lifetime, the reported lifetimes measured at 1 Torr He represent purely radiative values. They show no significant variation for the various 000-000 and 010-010 band heads as assigned in Ref. 3 (and corrected in Ref. 5), and give an average, weighted value of 183 ± 6 ns. This stated 2-σ error is estimated from the variations in the measured values for different rotational features in the $\Delta v = 0$ region; measurements for an individual feature typically show 4 ns average deviations and 2 ns precision. This consistency and the lack of spectral perturbations strongly suggest that nonradiative processes do not contribute significantly to this lifetime. Thus LIF detection of NCN is facile, and the measured lifetimes represent the radiative lifetime of NCN(A).

QUENCHING CROSS SECTIONS

Quenching rate constants for the 329 nm NCN bandhead LIF were measured for ten colliders and upper limits were obtained for three more. The linear fit of these decay constants as a function of collider pressure as shown in Figure 1 gives the quenching rate constant k. Data for O_2, CO_2, N_2O, and Xe are illustrated. The mass dependent relative collision velocity is then divided out, giving the thermally-averaged quenching cross section $\sigma = k/v$. The results are given in Table I, with 2-σ errors. No measurements could be made for hydrogen, methane, water, or ammonia due to reactive destruction of the NCN.

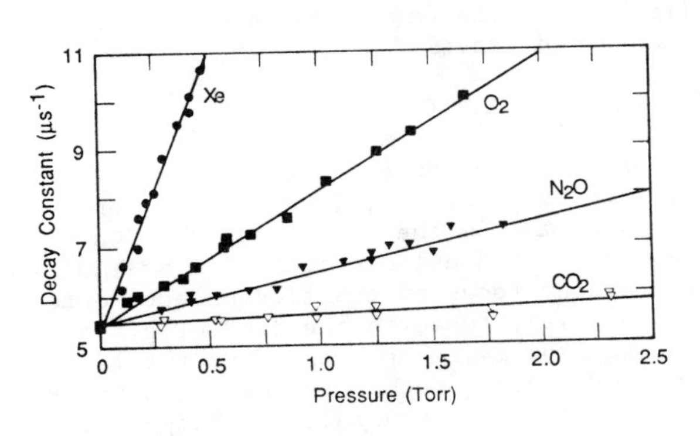

Figure 1

Table I. $NCN(A^3\Pi_u)$ Thermally Averaged Total Removal Cross Sections.

Collider	He	Ar	Kr	Xe	N_2	CF_4	SF_6
$\sigma_Q(\text{Å}^2)$	<1.8	1.2±1.1	5.9±1.1	345±18	<0.7	<3.0	7.3±2.7
Collider	O_2	CO	CO_2	N_2O	NO	NO_2	
$\sigma_Q(\text{Å}^2)$	83±4	125±13	4.6±2.1	32±3	305±37	164±16	

The results display several interesting features. A dramatic increase in quenching is evident among the rare gases as they become heavier. The molecules with unpaired electrons (O_2, NO, NO_2) are good quenchers. The huge variation observed among the quenchers, however, is not easily accommodated by existing theories relating collisional probabilities to molecular properties. For example, CO is a very good quencher, in marked contrast to the slow quencher nitrogen. No clear explanation exists for why N_2O is a seven-fold better quencher than the isoelectronic CO_2. We therefore conclude that a simple, attractive forces, orbiting collision model such as we used with some success to explain OH quenching[5] will not fit, correlate, or predict the NCN A-state quenching results. Quantitative calculations confirm this. This model also did poorly in predicting the quenching of several diatomics recently examined, such as CH, NH, and NS. We do note the interesting fact that the quenching behavior of NCN appears quite similar to that of $NH(A^3\Pi)$.[6]

In summary, the collisional dynamics of NCN is unique among triatomic species studied thus far. It has a well-isolated excited electronic state, in contrast to other triatomics previously studied, such as NO_2 and SO_2 which are highly perturbed. The collisional quenching of NCN shows a collider specificity reminiscent of diatomic species, in contrast to the universally large removal cross sections of other triatomics.

ACKNOWLEDGEMENTS

This work was supported by the U.S. Army Research Office.

REFERENCES

1. D. R. Crosley, High Temp. Mat. Proc. 7, 41 (1986); D. R. Crosley and G. P. Smith, Opt. Eng. 22, 545 (1983).
2. R. A. Copeland, D. R. Crosley, and G. P. Smith, Twentieth Symposium (International) on Combustion, p. 1195 (1984).
3. G. Herzberg and D. N. Travis, Can. J. Phys. 42, 1658 (1964).
4. C. Devillers and D. A. Ramsay, Can. J. Phys. 49. 2839 (1971).
5. P. W. Fairchild, G. P. Smith, and D. R. Crosley, J. Chem. Phys. 79, 1795 (1983).
6. N. L. Garland, J. B. Jeffries, D. R. Crosley, G. P. Smith and R. A. Copeland, J. Chem. Phys. 84, 4970 (1986).

EXTRACTION OF THE CENTRE-OF-MASS DIFFERENTIAL CROSS SECTION FROM CROSSED LASER MOLECULAR BEAM TIME OF FLIGHT DATA USING THE INVERSE QUADRATURE METHOD

G. Bandukwalla, M. Vernon
Columbia University, Dept.of Chemistry, NYC, NY. 10027

ABSTRACT

We present a novel and very general method for extracting the centre-of-mass differential cross section (DFCS) from crossed laser molecular beam time of flight data. Our method employs the fact that for the counts in each time bin the DFCS is detected only locally in the dependent variables (such as angular and energy dependence in the centre-of-mass frame). Combined with the technique of Gaussian Integration, this leads to a powerful scheme for extracting the DFCS to an arbitrary precision subject to the experimental error.

INTRODUCTION

In all crossed laser molecular beam photodissociation experiments one is faced with the difficulty of how to extract the centre-of-mass differential cross section (DFCS) from the time of flight (TOF) data

$$d\sigma / d\Omega \; dE = DFCS(E,\theta) \tag{1}$$

The DFCS is expressed in the centre-of-mass variables of translational energy E and θ the angle between the laser polarization vector and the centre-of-mass recoil velocity.

In general the problem consists of solving a multidimensional integral equation which takes into account the exprimental resolution determined by the finite width of the time channels as well as the finite interaction volume, detection volume and the spread in the molecular beam velocity.

The experimentally obtained counts in the ith time channel are related to the DFCS via the following expression:

$$Ri = C \int_{\Delta t} dt \int_{\Delta I} dr \int_{\Delta D} dR \int dv_B \; P(v_B) J(t, v_B, r, R) DFCS(E,\theta) \tag{2}$$

where ΔI is the interaction volume, ΔD is the detection volume, Δt is the width of the time channel $P(v_B)$ is the molecular beam velocity distribution and $J(t,v_B,r,R)$ is the Jacobian[1] for the laboratory to the centre-of-mass reference frame and C is the normalization constant.

THE INVERSE QUADRATURE (IQ) METHOD

Using the principle of Gaussian Integration[2] we observe that Eq.2 can be approximated in terms of the formal expansion

$$R_i \approx \sum_{\mu=1}^{M} \sum_{\nu=1}^{N} C_i^{\mu\nu}\, \mathrm{DFCS}(E_i^{\mu},\theta_i^{\nu}) \qquad (3)$$

where $C_i^{\mu\nu}$ are the Gaussian Quadrature weights and the E_i^{μ}, θ_i^{ν} are the Gaussian Quadrature points. M and N are related to the degree of the polynomial expression for $\mathrm{DFCS}(E,\theta)$. In our case however we are faced with the problem that the $\mathrm{DFCS}(E,\theta)$ is an unknown function.

The IQ Method is based on the fact that Eq.3 becomes exact if the $\mathrm{DFCS}(E,\theta)$ were a polynomial of order (2M-1,2N-1) ie

$$\mathrm{DFCS}(E,\theta) = \sum_{\mu=1}^{2M-1} \sum_{\nu=1}^{2N-1} a_{\alpha\beta}\, (E)^{\alpha}\, (\theta)^{\beta} \qquad (4)$$

Due to the experimental resolution, the domain of translational energies E and angle θ probed for each time channel will be sufficiently small. A truncated Taylor series expansion should then be an accurate way of representing the DFCS. This however is equivalent to expressing the DFCS as shown in Eq.4 above.

The quadrature weights $C_i^{\mu\nu}$ and the quadrature points E_i^{μ}, θ_i^{ν} in Eq. 3 can now be computed by requiring that Eq.3 and Eq.2 be independent of the coefficients $a_{\alpha\beta}$.

The "Linear Approximation" (M=N=1) plays the most fundamental role in the IQ method. This can be easily worked out by inserting Eq.4 (with M=N=1) into Eq.3 and Eq.2. The weights C_i^{11} and the points E_i^{1} and θ_i^{1} are given by

$$\langle C_i^{11}\rangle = \langle 1\rangle_i \ , \ E_i^{1} = \langle E_i\rangle/\langle 1\rangle_i \ , \ \theta_i^{1} = \langle\theta\rangle_i/\langle 1\rangle_i \qquad (5)$$

where the expectation value $\langle X\rangle_i$ is identical to the right hand side of Eq.2 with the DFCS(E,θ) replaced by X. The expressions in Eq.5 can be evaluated numerically. The DFCS(E,θ) can now be obtained from the set of counts $\{R_i\}$ as shown below

$$\mathrm{DFCS}(E_i^{1},\theta_i^{1}) = R_i \ / \ C_i^{11} \ \text{(Linear Approximation)} \qquad (6)$$

Using Eq.6 we can directly relate the error in the extracted DFCS to the error in the experimentally obtained counts.

This scheme of linearization is the most important one for practical applications. A detailed description of analyzing isotropic and anisotropic photodissociation processes as well as higher order approximations (Quadratic Approximation) usng the IQ Method are given in reference 3.

CONCLUSION

a) The IQ Method can be used for extracting the DFCS from TOF data obtained from crossed laser molecular beam experiments.The method does not make any a priori assumptions about the functional form of the DFCS.

b) The IQ Method provides a systematic way of testing and improving the accuracy of the extracted DFCS.

c) The IQ Method readily incorporates experimental errors and transforms these into errors in the extracted DFCS in an unbiased fashion.

d) The IQ Method is very efficient and involves only elementary numerical computations.

REFERENCES

1. G. L. Cachen, J. Hussain, R. N. Zare, J. Chem. Phys. 69, 1737 (1978)
2. J. Mathews and R. L. Walker, Mathematical Methods of Physics (W. A. Benjamin Inc., Second Edition), p. 351-353.
3. G. Bandukwalla, M. Vernon, to be published

LASER SPECTRA FROM HUMAN NORMAL AND TUMOR LUNG AND BREAST TISSUES

R. R. Alfano, G. C. Tang, Asima Pradhan and Michael Bleich
Institute for Ultrafast Spectroscopy and Lasers
Photonics Application Laboratory
The City College of New York
New York, NY 10031

Daniel S. J. Choy
Laboratory for Investigative Cardiology
St. Luke's-Roosevelt Hospital Center
Columbia University College of Physics & Surgeons
New York, NY 10025

S. J. Wahl
Department of Pathology
Lenox Hill Hospital
New York, NY 10021

ABSTRACT

Using different excitation wavelengths we measured and analyzed the steady state and time-resolved laser spectra emitted from normal and tumor human lung and breast tissues. Differences measured in the spectra and lifetimes of normal and tumor tissues are attributed to environmental transformations and charge buildup in the cancerous tissues. These results may offer new diagnostic methods to determine cancer.

INTRODUCTION

Recently, there have been major breakthroughs in the use of laser spectroscopy as a unique and sensitive approach to reveal changes in the physical and chemical properties that occur in healthy and abnormal cells in tissues. Alfano and coworkers[1,2] have established that the fluorescence spectroscopy and relaxation times from malignant and normal rat tissues were different.The differences were attributed to the transformations of local environment surrounding the fluorophors assigned to be flavins and porphyrins in the normal and cancerous rat tissues.

Most recently, we have extended our laser spectroscopy research from animal to human tissues[3]. In the previous reports,only one laser wavelength at 488 nm was used to the excite the samples. In the present paper,three different wavelengths were used to photo-excite the spectra from normal and tumor lung tissues. The fluorescence kinetics from lung tissues is also reported.

METHODS AND MATERIALS

An Argon ion laser beam at 514.5 nm, 488 nm and 457.9 nm was focused on the front surface of the tissue to a spot size of 200 µm. The fluorescence from the front surface was collected and focused into the entrance slit of a double 1/2m grating scanning spectrometer blazed at 500 nm. A photomultiplier tube located at the exit slit of the spectrometer measured the intensity at different wavelengths. The output of the PMT was connected to a lock-in-amplifier and an X-Y recorder combination to display each spectrum. Picosecond time resolved fluorescence measurements were performed using a streak camera system.

The tissue samples were solid chunks, not cut to any particular specificity and were a few millimeters thick. Each tissue sample was placed in a borosilicate glass test-tube with a cap for spectroscopic studies. The samples were run within two to four hours after removal on the same day of extraction. The measurements were made on at least three locations on each sample and at least three samples to assure reproducibility. No chemicals or dyes were added.

RESULTS

Typical fluorescence spectra from normal (N) and tumor (T) human lung tissues are shown in Fig.1 to Fig.3. The principal spectral peaks excited at 457.9 nm ,488 nm and 514.5 nm are located at 496 nm, 509 nm and 531 nm for normal tissues, respectively,and at 503 nm,515 nm and 537 nm for tumor tissues,respectively. There is one subsidiary maximum located at about 605 nm in the normal lung tissue spectra. The tumor tissue spectra only showed a monotonic decrease with less structure.

For comparison , we have included normal and tumor breast tissue spectra excited at 457.9 nm in Fig.4. The principal spectral peaks are located at 505 nm and 515 nm for tumor and normal tissues,respectively.

The fluorescence profiles in time for spectral band at wavelength center at 600 nm excited by a picosecond 530 nm laser pulse are displayed in Fig.5. The profiles were found to be nonexponential in time. To fit these curves, a double exponential $I(t)=A_f e^{-(t/\tau_f)}+A_s e^{-(t/\tau_s)}$ was used. The A_f and A_s are the intensity amplitudes of the fast and slow components at t=0 and τ_f and τ_s are the decay times for the fast and slow components. The decay times are accurate within 25%. In Fig.5a,the fluorescence decay times of the fast and slow components from the normal of the lung were about 220 ps and 2650 ps,respectively. In Fig.5b,the fluorescence decay time of the fast and slow components for the tumor tissue of the lung were about 120 and 2600 ps,respectively.

DISCUSSION

The the experimental results (Fig. 1) using 457.9 nm excitation for lung tissues indicate about 7 nm spectral red shift from normal to tumor lung tissue spectrum. A red shift of 6 nm was also observed using 488 nm and 514.5 nm excitation (Fig. 2 and Fig.3). However,a blue shift was observed in rat kidney tissue.[1,2] The tumor lung tissue spectra are much smoother than normal lung tissue spectra. The lack of structure at 605 nm in tumor tissue may be associated with intensity and bandwidth increase of second band at wavelength center at 555 nm or decrease in the native porphyrins.

The observed 6 nm red shifts of the main peak and the smoothing of spectra for tumor lung tissues are attributed to the physiological and biochemical transformation of the lung cells from the normal tissues. Previous fluorescence studies have shown that when protein containing fluorophors gain positive (negative) charge ions, their spectral maxima have been blue (red) shifted[4]. Hence, the observed red shifts in the main maxima of the human lung tumor spectra suggest an accumulation of negative charge ions in the malignant cells' intracellular environment. One may conjecture that the most likely fluorophors giving rise to the observed spectral signature at about 520 nm are the flavins in the mitochondria. The spectral feature at 605 nm may be associated with porphyrins. The original of the 555 nm feature is being investigated but may be tentatively assigned to melanins[5] or keratins.[1]

The flavin band observed for the normal breast tissue near the 515 nm peak exhibits a blue shift of about 10 nm in the tumor tissue (see Fig.4) spectra. This is opposite to the red shift observed for lung tissues. This blue shift may suggest a buildup of positive ions in the breast tumors. The 555 nm and 600 nm subsidiary bands are similar to lung tissue bands discussed in the previous paragraph and most likely arise from the same type of molecules. The lack of structure in the porphyrin spectral region of the cancerous breast tissues is consistent with spectra measured from lung cancer tissues

The time-resolved fluorescence shows that the decay time of tumor tissues is faster than that of normal tissues. This suggests an enhancement of the nonradiative processes in tumors.

Much more research will be required to characterize the spectral features of human malignant and benign tumors and normal tissues. If neoplastic tissues have clearly-defined and reproducible fluorescence properties, this spectroscopy approach may provide an immediate and "light-biopsy" diagnosis of cancer in vivo and could even be considered as a substitute for needle biopsy or some similar cytologic examination[6].

This research is supported by ONR and Hamamatsu photonics.

REFERENCES

1. R. R. Alfano, D.B. Tata, J. J. Cordero, P. Tomashefsky, F. W. Longo, and M. A. Alfano, Laser induced fluorescence spectroscopy from native cancerous and normal tissues. IEEE J. Quantum Electron.vol.QE-20, 1507-1511 (1984).

2. D. B. Tata, M. Foresti, J. Cordero, P. Tomashefsky, M. A. Alfano, R. R. Alfano, Fluorescence Polarization Spectroscopy and Time-Resolved Fluorescence Kinetics of Native Cancerous and Normal Rat Kidney Tissues. Biophys. J. 50, 463-469 (1986).

3. R. R. Alfano, G. C. Tang, W. Lam, D. S. J. Choy, E. Opher, Fluorescence spectra from cancerous and normal human breast and lung tissues. IEEE J. Quantum Electron.,vol. QE-20,1806-1811 (1987).

4. B. Honig, Theoretical Aspects of Photoisomerization. In Biological Events Probed by Ultrafast Laser Spectroscopy, R. R. Alfano, editor, Academic Press Inc., New York, 285 (1982).

5. S. D. Kozikowski, L. J. Wolfram, R. R. Alfano, Fluorescence

Spectroscopy of Eumelanins. IEEE Quant. Electron, vol.QE-20, 1379-1382 (1984).
6. R. R. Alfano, M. A. Alfano, Medical Diagnostics: A New Optical Frontier. Photonics Spectra 19, 55-60 (Dec. 1985) and US patent pending.

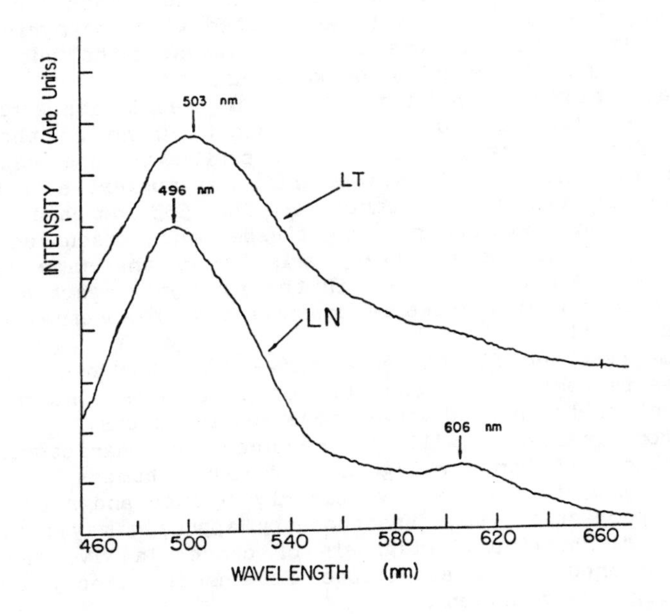

Figure 1: Steady state laser fluorescence spectra of human lung tissues: labeled (a) LN=normal (17mw) and (b) LT=tumor (10mw) excited by 457.9 nm,2 mv sensitivity.

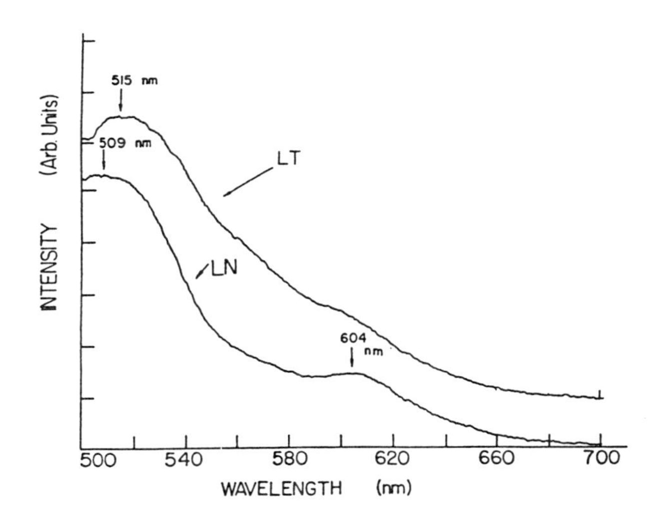

Figure 2 : Steady state laser fluorescence spectra of human lung tissues:
labeled (a) LN=normal (70mw) and (b) LT=tumor (25mw) excited by 488 nm,2 mv sensitivity.

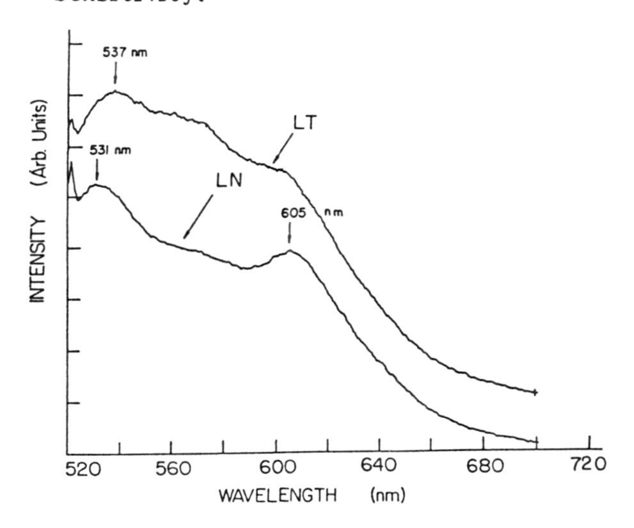

Figure 3: Steady state laser fluorescence spectra of human lung tissues: labeled (a) LN=normal (100mw) and (b) LT=tumor (100mw) excited by 514.5 nm,2mv sensitivity.

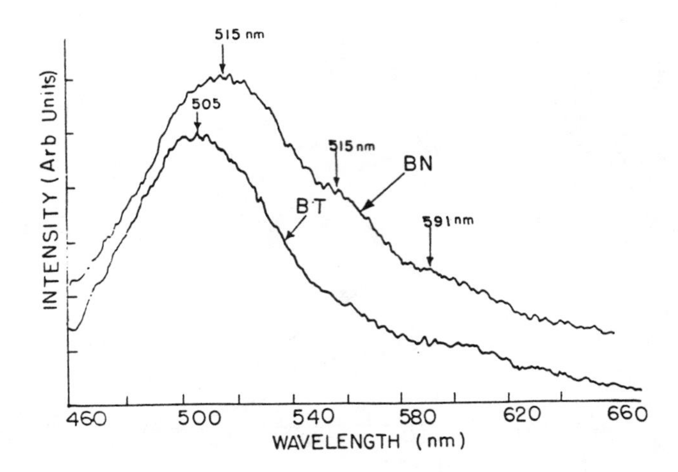

Figure 4: Fluorescence spectra of human breast tissue : labeled (a) BN=normal (20 mw) and (b) BT=tumor (1.6 mw) excited by 457.9 nm, 1 mv sensitivity.

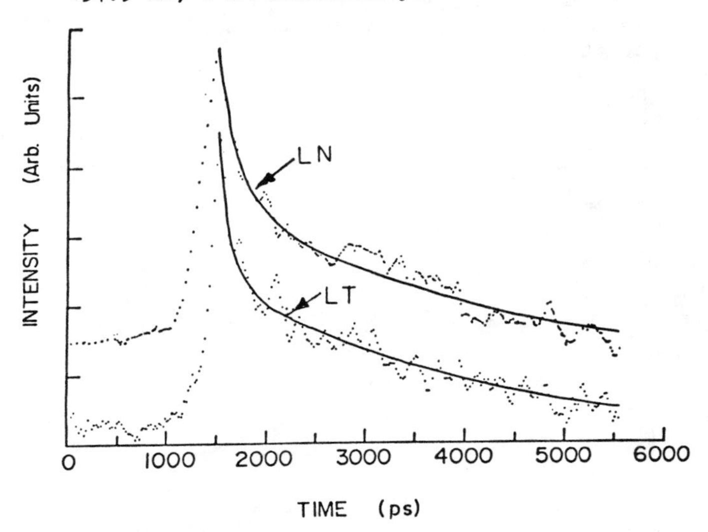

Figure 5: kinetic fluorescence profiles from human lung tissues at the wavelength center at 600 nm (bandwidth 40 nm) excited by 530 nm picosecond pulse. (a) From normal lung tissue τ_f=220 ps, τ_s=2650 ps. (b) From tumor lung tissue τ_f=120 ps, τ_s=2600 ps.

VIBRATIONAL ENERGY RELAXATION PROCESSES IN HEME PROTEINS*

L. Genberg, F. Heisel, G. McLendon and R. J. Dwayne Miller
Dept. of Chemistry University of Rochester Rochester, NY 14627

ABSTRACT

Vibrational energy relaxation pathways from optically excited heme proteins are studied using a transient thermal phase grating technique which monitors the solvent lattice temperature. Vibrational energy transfer from the porphyrin ring to the protein backbone leads to extensive delocalization of the energy in the protein helix which is efficiently transferred to the water interface in less than 20 psec. A slower relaxation process on the nanosecond time scale is also observed. The slow relaxation component is attributed to slow conformational relaxation processes of high potential energy states of the heme proteins accessed during the high internal energy conditions of the optically excited molecule.

INTRODUCTION

Heme proteins provide ideal model systems for studying vibrational energy relaxation, and are of intrinsic interest from the standpoint of understanding phonon modes and energy flow in protein matrices. Optical excitation at 355 nm can selectively deposit 28,000 cm^{-1} into the heme porphyrin ring on a sub-picosecond to picosecond timescale. The heme porphyrin ring is covalently bonded to the protein through one or two coordination sites about the iron. Since these bonds are orthogonal to the porphyrin ring, the vibrational motion along this axis will only be weakly coupled to the ring vibrational modes. Vibrational energy transfer of this large amount of excess energy must channel through the van der Waals interactions between the porphyrin ring and the surrounding protein amino acid residues. In order to understand large amplitude biopolymer motions with large correlation lengths in detail, the vibrational energy relaxation processes driving these motions and the dispersion of the various modes involved needs to be addressed experimentally. A new method based on the transient grating technique has been developed which enables direct monitoring of the flow of vibrational energy into the intermolecular bath. The heme proteins used in this study were cyanomethemoglobin, metcytochrome c, metmyoglobin and methemoglobin. These systems provide a variety of different coordination ligands and van der Waals contact points of the heme to the protein backbone to study the effects of site geometry on vibrational energy transfer.

*This work was supported, by the Department of Energy, Grant #81-049

RESULTS AND DISCUSSION

The experimental setup has been described elsewhere.[1] When excited states are extremely short lived, as in the present case for optically excited heme groups, the excited state grating contribution is generally absent in liquids. The diffraction efficiency in transient grating spectroscopy is dominated by phase grating contributions caused by the non-radiative deposition of thermal energy into the solution lattice. A detailed theoretical description is needed for the thermal phase grating formation in order to obtain quantitative information on energy relaxation processes. Previous treatments of this problem have assumed an instantaneous change in the lattice temperature to explain acoustic oscillations characteristically observed in this experiment.[2] Our theoretical analysis is a completely generalized approach which accounts for both the finite timescale for energy relaxation processes and the effects of thermal conductivity on the grating amplitude. A detailed discussion of the theoretical work will be given elsewhere.[3] The salient details are that the grating formation timescale is limited by speed of sound restrictions which can be made shorter by using smaller fringe spacing; and the acoustic oscillations can be cancelled by using Gaussian excitation pulse durations equal to one acoustic cycle at the given fringe spacing. Counterpropagating excitation beam geometry provides the shortest fringe spacing for a given excitation wavelength and eliminates acoustic oscillations in the thermal grating signal. For counterpropagating 366 nm excitation the intrinsic temporal resolution from speed of sound limitations is 22 psec in water.

Representative results are shown in figure 1 for cytochrome c using narrow (4.6°) angle beam geometries. The signal shows a slowly rising component of the thermal grating (800 psec) which is seen as a decrease in the depth of acoustic modulation of the signal. This result demonstrates there is both a fast and slow component to the relaxation of excited heme proteins. Similar decays were observed for the other proteins but with varying slow relaxation components. With counterpropagating excitation geometry, the acoustic modulation seen in figure 1 is eliminated and the dynamics of the thermal grating can be resolved. The thermal grating formation dynamics are shown in figure 2. The upper curve is the grating signal rise time for disodium fluorescein in ethanol. This grating signal is due to excited state formation and gives the instrument response for an instantaneous rise time. The next lowest curve depicts the data for cyanomethemoglobin in an aqueous buffer. The grating rise time is 17 ± 4 psec and is clearly discernible in comparison to the fluorescein data. Nearly identical rise times were observed for cytochrome c and metmyoglobin. The lowest curve is the thermal grating rise time of Malachite Green in ethanol (2 psec lifetime). The observed rise is 28 ± 4 psec. The slower rise time corresponds to the 30% slower speed of sound in ethanol than in water.The observed rise times are all in the 20 psec range which demonstrates that the fast vibrational energy relaxation component is faster than the temporal resolution of the technique. Therefore, the vibrational relaxation process from the heme porphyrin ring to the protein backbone must be faster than 20 psec. In addition, for the energy to be transferred to the aqueous bath on this timescale, the vibrational energy must be highly dispersed over the protein structure for it to occur so rapidly on the protein's exterior. These results are in excellent agreement with predictions of a recent molecular dynamics simulation.[4]

REFERENCES

1. L. Genberg, F. Heisel, G. McLendon and R.J.D. Miller, J. Phys. Chem. 91, 5521 (1987).
2. K.A. Nelson and M.D. Fayer, J. Chem. Phys. 72, 5205 (1980).
3. L. Genberg, Q. Bao, S. Gracewski and R.J.D. Miller, to be submitted to J. Appl. Phys.
4 E.R. Henry, W.A. Eaton, and R.M. Hochstrasser, P.N.AS. 83. 9092 (1986).

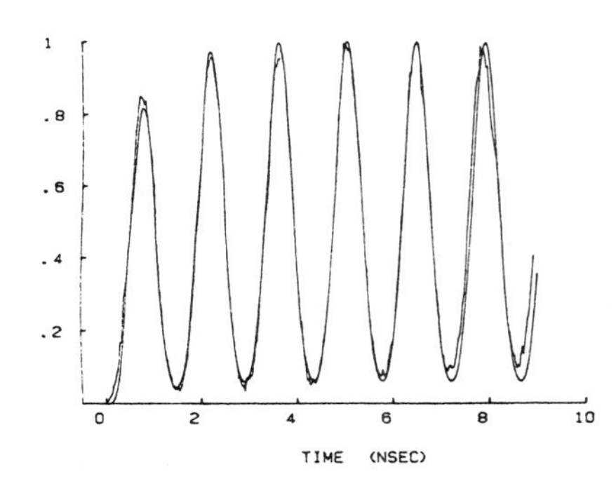

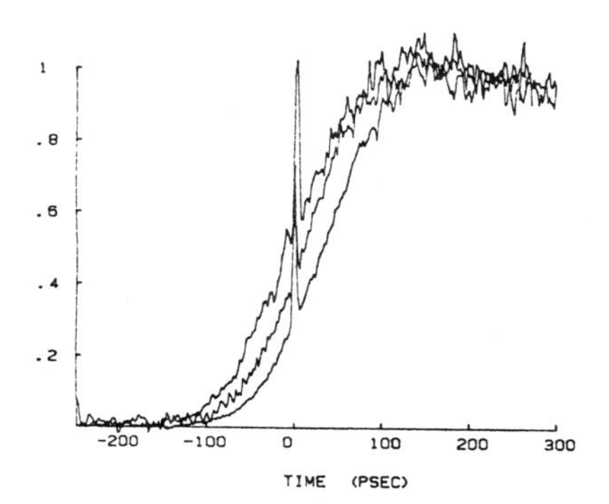

Figure 1.
Narrow angle studies of cytochrome c with 355 nm excitation and 532 nm probe. The theoretical fit to the data is the solid line with parameters τ_1 < 20 psec, τ_2 = 800 psec weighted 55/45 respectively. The mismatch in the fit at long time delays demonstrates that the overall protein relaxation is occuring over a range of timescales.

Figure 2.
The rise time of the thermal phase grating with counterpropagating 366 nm excitation (curves described in text).

THE STRUCTURE AND DYNAMICS OF MOLECULAR LIQUIDS PROBED WITH FEMTOSECOND OPTICAL KERR EFFECT SPECTROSCOPY

Dale McMorrow, William T. Lotshaw and Geraldine A. Kenney-Wallace
Lash Miller Laboratories, Department of Chemistry
University of Toronto, Toronto, Canada M5S 1A1

ABSTRACT

The dynamical behavior of simple molecular liquids is investigated using the time-resolved optical Kerr effect. These studies provide direct information on the distribution of molecular environments in the various liquids, and give insight into the local intermolecular potentials. The effects of variations in the intermolecular potential on the various relaxation components of CS_2 are investigated through dilutions in effectively inert alkane solvents. The interaction-induced contributions to the induced birefringence are observed to increase with decreasing CS_2 concentration.

INTRODUCTION

The dynamical behavior of molecular and atomic liquids has been the subject of extensive experimental and theoretical investigation. With recent developments in the generation of ultrashort laser pulses, it is now possible to observe directly dynamical events which occur on the femtosecond timescale, providing information which is complementary to that obtained from the various frequency domain techniques. In this paper we discuss recent time domain investigations of molecular dynamics in numerous molecular liquids utilizing optically heterodyned optical Kerr effect spectroscopy (OHD-OKE). These studies provide new and detailed insight into the structure and sub-picosecond dynamics of simple liquids. In particular, they reveal directly the existence of a distribution of locally ordered liquid structures, the consequences of dynamical fluctuations which modulate these structures, and the contribution of "interaction-induced" effects to the induced birefringence.

EXPERIMENTAL

The OHD-OKE experiments were performed at 295 K using 65-75 fs pulses (100-110 fs autocorrelation FWHM) centered at 633 nm generated with synchronously pumped anti-resonant ring dye laser. Details of the laser system and optical heterodyne detection techniques are given elsewhere[1,2]. All experimental data presented here was obtained using an out-of-phase local oscillator and, as such, represent the real part of $\chi^{(3)}$, the third order nonlinear susceptibility[2]. The transmission of the Kerr cell using optical heterodyne detection[2] is given by,

$$T(\tau) = \int_{-\infty}^{\infty} G_0^{(2)}(t)\, R(t-\tau)\, dt \tag{1}$$

Optical heterodyne techniques[3,4] render the detected signal linear in both the experimentally measured laser-pulse autocorrelation, $G_0^{(2)}(t)$, and the impulse response of the sample, R(t), permitting the straight forward analysis of the entire temporal profile of the data. In what follows, the sample response R(t) is assumed to be of the form,

$$R(t) = \sigma(t) + \sum_i r_i(t) + \sum_j q_j(t) \qquad (2)$$

where $\sigma(t)$ is the ensemble averaged purely electronic hyperpolarizability, the $r_i(t)$ are the *intermolecular* nuclear contributions to the signal, and the $q_i(t)$ represent *intramolecular* Raman resonances.

ORIGIN OF THE SUBPICOSECOND RESPONSES

Recent investigations[1,2] of the OKE in simple molecular liquids have revealed a complex temporal behavior at times less than 500 fs. The OKE of all liquids thus far investigated composed of molecules possessing a permanent polarizability anisotropy can be resolved into four dynamically distinct responses. Of these, one is instantaneous on the time scale of the laser pulse used and is identified with the purely electronic hyperpolarizability; the remaining three are decidedly noninstantaneous, exhibiting pronounced inertial effects at short times. The four responses have been assigned[2] to: 1) the purely electronic hyperpolarizability; 2) a coherently excited, rapidly dephased ensemble of intermolecular librators ($\tau_{1/e}$<170 fs, ν =25-60 cm^{-1}); 3) an exponentially decaying local translational anisotropy ($\tau_{1/e}$~400-700 fs); and 4) an exponentially decaying orientational anisotropy ($\tau_{1/e}$>1 ps).

Responses 2 and 4 originate in orientational motions and, as such, are absent in the OKE of liquids composed of atoms or molecular species possessing an isotropic polarizability. The molecular librational motion $r_2(t)$ is a consequence of *local structure* in the liquid, with the functional form of this response giving direct information on the distribution of local intermolecular potentials (*vide infra*). The translational anisotropy $r_3(t)$ is assigned by analogy with the exponentially decaying response present in the OKE of carbontetrachloride[2] and cryogenic rare-gas liquids[5]. The decay of this response is causally related to *translational density fluctuations* in the liquid, with the signal intensity arising from the intermolecular "interaction-induced" distortion of the molecular polarizabilities. In what follows the two exponentially decaying nuclear responses, $r_3(t)$ and $r_4(t)$, are treated phenomenologically with the responses modeled using an impulse response of the form[2],

$$r_i(t) \propto \sinh(t/\beta_i)\, \exp(-t/\tau_i) \qquad (3)$$

This functional form accounts for the observation that the nuclear responses can not instantaneously follow the intensity profile of the ultrashort laser pulse[1,2,6,7], while maintaining the experimentally observed exponential decays. In addition to the responses noted above, low frequency intramolecular vibrational modes may be resonantly and coherently driven by the different Fourier components of the ultrashort laser pulse. These responses, represented by $q_i(t)$ in equation 2, are observed superimposed on the intrinsic Kerr responses 1-4, and are not considered explicitly here.

THE DISTRIBUTION OF MOLECULAR ENVIRONMENTS

The study of molecular librational motion provides the opportunity to study directly the distribution of molecular environments in condensed phase systems. It is well established that a distribution of environments gives rise to inhomogeneous broadening of electronic and intramolecular vibrational bands. In these cases the environmental perturbation is generally small compared to the *intramolecular* potentials involved, resulting in small frequency shifts about a central transition frequency, which itself is generally solvent dependent. In the case of *intermolecular* librational motion, however, the vibrational potential is defined by the local intermolecular potential, and the distribution of oscillation frequencies gives direct information on the distribution of these intermolecular potentials. For a particle in a locally harmonic potential in the limit of zero damping this relationship is

$$\omega_{oi} = (\mu_i/I)^{1/2} \tag{4}$$

where ω_{oi} is the undamped characteristic frequency of the oscillator, μ_i is the restoring force of the harmonic potential, and I is the moment of inertia. When damping is considered equation 4 becomes

$$\omega_i = \frac{1}{2}\{(\xi_i/I)^2 - 4(\mu_i/I)\}^{1/2} \tag{5}$$

where ξ_i is an effective frictional coefficient which, in the current treatment, contains contributions from both energy relaxation and dephasing of the coherently excited ensemble of oscillators and is responsible for homogeneous broadening of the transition. When a distribution of molecular environments is considered the coherent vibrational amplitude is given by a superposition of the different homogeneously broadened frequency components,

$$r_2(t) = \sum_i B_i(\omega_i)\, r_{2i}(\omega_i, t) \tag{6}$$

where $B_i(\omega_i)$ is the amplitude of the ith frequency component, the time dependence of which, in the harmonic approximation, is given by,

$$r_{2i}(\omega_i, t) \propto \sin(\omega_i t)\, \exp(-t/\tau_{2i}) \tag{7}$$

In general, for a given distribution of oscillators, equation 6 must be evaluated numerically. However, for a continuous, Gaussian distribution of frequencies, $B(\omega)=\exp[-(\omega-\omega_o)^2/2\alpha^2]$, the response $r_2(t)$ may be approximately represented by,

$$r_2(t) \sim \sin(\omega_o t)\, \exp(-t/\tau_2)\, \exp(-\alpha^2 t^2/2) \tag{8}$$

In this expression α is a measure of the width of the inhomogeneous frequency distribution and $\omega_o/2\pi$ is its center frequency. The time constant τ_2 contains contributions from the dephasing and population decay and is related to the homogeneous line width in the frequency domain spectrum. Interference from the different frequency components of the coherently excited ensemble of oscillators results in a rapid dephasing (inhomogeneous) of the coherent signal. For the systems investigated to date, the librational response exhibits an average oscillation period on the order of 1 ps with a signal decay on the timescale of 150 fs.

Figures 1 and 2 give comparisons of the response generated with equation 1 for neat CS_2 and benzene, respectively, with and without the consideration of inhomogeneous broadening effects. The upper panel of each gives the raw OHD-OKE data. In the middle panels this data (dots) is analyzed (solid curve) considering only exponential (homogeneous) dephasing of the librational response (equation 6). Gaussian (inhomogeneous) dephasing (equation 8) is

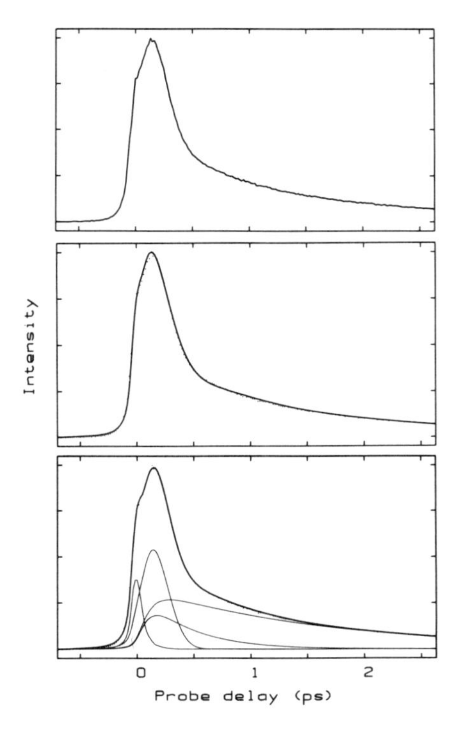

Figure 1. *OHD-OKE data for neat CS_2: (upper) raw data; (middle) raw data (dots) with fit (solid) considering only exponential dephasing of the librational response; (lower) same with consideration of Gaussian dephasing. Parameters used in obtaining lower curve given in table 1.*

Figure 2. *OHD-OKE data for neat benzene as in figure 1. Parameters used in obtaining the lower curve:* $\tau_4 = 2.05$ *ps;* $\tau_2 = \tau_3 = 385$ *fs;* $\beta = 1/\omega_o = 110$ *fs* ($\bar{\nu}_o = 48.2$ cm^{-1}); $\alpha = 7.7$ ps^{-1} ($\Delta\bar{\nu} = 68$ cm^{-1}); $A_1:A_2:A_3:A_4 = 0.36:0.31:0.20:0.13$.

considered in the lower panels. Also shown in the lower panels are the individual component responses generated with equations 3 and 8, with the electronic response represented by a scalar multiple of the laser pulse autocorrelation function. While the consideration of only homogeneous dephasing mechanisms reproduces the essential qualitative features of the data, the consideration of inhomogeneous dephasing produces a significant (but sometimes subtle) improvement in the quantitative description of the data. In particular, the Gaussian dephasing term of equation 8 broadens the librational response function while maintaining the rapid decay following the signal maximum. For CS_2, neither the shoulder on the leading edge nor the position of the signal maximum are adequately accounted for with exponential dephasing, while both are well reproduced in the lower panel of figure 1. For benzene the shortcomings of purely exponential dephasing are more evident. It is not possible to simultaneously reproduce the rapid signal decay and the sharp "break" in the data near 500 fs with an exponentially damped librational response. In the lower panel of figure 2 the entire temporal profile of the benzene data is well described. We note that the presence of a pronounced shoulder in the depolarized Rayleigh spectrum of benzene has been discussed in terms of molecular librational motion[9]. A complete analysis of the OKE data for benzene and several benzene derivatives will be presented elsewhere[10].

INTERACTION-INDUCED EFFECTS

The intermediate lifetime, exponentially decaying component $r_3(t)$ is assigned[2] to the imposition and decay of a local anisotropic distribution of the molecular centers of mass. The intensity of this response is believed to arise from the *intermolecular "interaction-induced" distortion of the molecular polarizabilities*. Such "interaction-induced" signal contributions have been discussed extensively with regard to light scattering studies on atomic and molecular liquids[11,12]. The current time domain data permits a clear identification of the orientational contributions to the signal, allowing us to separate out and study independently the dynamics and intensity of the signal component associated with translational motion.

The "interaction-induced" distortion of the molecular polarizability is inherently a many-body effect. The amplitude of this signal contribution is expected to exhibit a pronounced sensitivity to the symmetry in the local liquid structure through cancellation effects between two, three, and four body contributions to the effective molecular polarizability[11,12]. The dynamics of the interaction-induced signal intensity are those of the intermolecular distances and orientations and, in light-scattering experiments, are causally related to *local density fluctuations* in the liquid[12]. Within the Born-Oppenheimer approximation, intermolecular interaction-induced contributions to the signal will adiabatically follow the nuclear translational and orientational motions of the anisotropic molecular species. The response $r_3(t)$ is ascribed to the former, while interaction-induced contributions to the librational signal intensity are expected but are not considered explicitly. In the current experiment the anisotropy is induced by the intense optical field, and will decay through the random, thermal fluctuations of the liquid. This decay thus exhibits the characteristic time of local translational density fluctuations.

Cancellation between the two, three, and four body interaction terms in the effective molecular polarizability can be probed by changing the density of a pure liquid, or by dilution in a non-interacting solvent. We have investigated the molecular dynamics of the probe CS_2 in dilutions with n-pentane, isoheptane, and n-tetradecane. The essential results of this investigation are presented in Table 1. Figure 3 gives the experimental data for 0.20 volume fraction CS_2 in isoheptane, together with the best fit and individual component curves generated using equations 1, 2, 3 and 8.

Over the limited range of concentrations which we report, the relative amplitude A_3 increases monotonically on dilution in each case, as is expected for an "interaction-induced" signal amplitude as three and four body contributions become less significant. On higher dilution the amplitude A_3 is expected to turn over and decrease with decreasing CS_2 concentration, if we have correctly identified the origin of this response. Lowering the temperature of the pure liquid will decrease the interaction-induced signal intensity by promoting cancellation among many-body terms due to a more symmetrical local structure at higher density.

Table 1

volume fraction CS_2	η^a (cp)	Electronic Response	Librational Response				Translational Anisotropy			Orientational Anisotropy		
		A_1^b	A_2	$\bar{\nu}_0{}^c$ (cm^{-1})	$\tau_2{}^e$ (fs)	$\Delta\bar{\nu}^f$ (cm^{-1})	A_3	τ_3 (fs)	$\beta_3{}^g$ (fs)	A_4	$\tau_4{}^h$ (ps)	$\beta_4{}^g$ (fs)
1.0	0.366	0.27	0.39	35.4	400	39.1	0.14	400	150	0.20	1.61	150
Tetradecane												
0.75	0.476	0.28	0.38	31.2	400	36.8	0.17	400	170	0.17	1.78	170
0.50	0.656	0.27	0.36	28.7	440	32.1	0.25	440	185	0.12	2.01	185
0.20	1.154	0.33	0.35	25.9	475	31.0	0.25	475	205	0.07	2.71	205
0.00	2.215	-	-	-	-	-	-	-	-	-	-	-
Isoheptane												
0.75	0.353	0.26	0.40	33.2	410	36.1	0.17	410	160	0.17	1.62	160
0.50	0.353	0.21	0.39	27 9	450	34.7	0.24	450	190	0.16	1.61	190
0.20	0.369	0.23	0.35	24.7	415	29.5	0.27	415	215	0.15	1.63	215
0.0	0.370	-	-	-	-	-	-	-	-	-	-	-
n-pentane												
0.75	0.325	0.23	0.40	33.2	435	34.0	0.19	435	169	0.18	1.54	160
0.50	0.292	0.20	0.36	30.3	425	34.0	0.23	425	175	0.21	1.43	175
0.25	0.259	0.24	0.31	27.9	430	30.0	0.29	430	190	0.16	1.35	190
0.0	0.221	-	-	-	-	-	-	-	-	-	-	-

[a] measured at 295±0.1 K; error ± 0.002 for each

[b] error in all amplitudes ±5%; $A_1+A_2+A_3+A_4=1$

[c] $\bar{\nu}_0 = \omega_o/2\pi c$; error in ω_o +5%, -40% [14]

[e] τ_2 fixed at value of τ_3, error ±50 fs

[f] $\Delta\bar{\nu} = \alpha(\ln 2)^{1/2}/\pi c$, FWHM of inhomogeneous distribution; error, +15%, -3%

[g] $\beta_3 = \beta_4 = 1/\omega_o$

[h] errors in orientational decays: neat CS_2, ±0.05 ps; 75%, ±0.1 ps; 50%, ±0.15 ps; 25%, 20%, ±0.2 ps

THE INTERMOLECULAR POTENTIAL

In addition to the behavior of the amplitude A_3 on dilution, several other trends are evident in the data of Table 1: i) the time constant for the orientational diffusion process τ_4 varies directly with the shear solution viscosity and is invariant in the isoheptane series; ii) the decay of the translational anisotropy τ_3 remains essentially constant in all solutions investigated, independent of viscosity, iii) the inhomogeneous dephasing rate α for the librational motion decreases upon dilution in all solutions reflecting a decrease in the distribution of molecular environments; and iv) the frequency of libration ν_o increases on dilution, apparently independent of the solution viscosity. The qualitative manifestations of these trends are evident in a comparison of the data of figures 1 and 3. In particular, a significant broadening of the response and a shift in the signal maximum is observed on dilution with each alkane solvent.

The dynamical form of the librational response contains specific information about both the curvature of intermolecular potential (in the oscillation frequency) and the equilibrium distribution of molecular environments (through

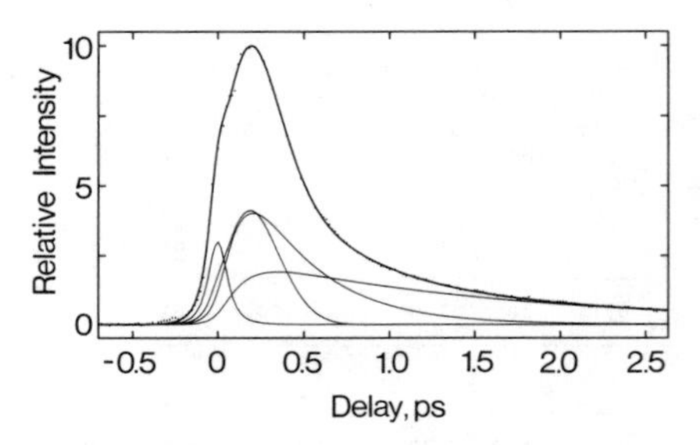

Figure 3. *OHD-OKE data (dots) for 0.20 volume fraction CS_2 in isoheptane with fit (solid) and individual component curves. Parameters are given in table 1.*

the inhomogeneous damping coefficient α). The observed variations in the librational frequency on dilution are consistent with the changes in the intermolecular potential expected when CS_2 molecules are replaced by the weakly interacting alkanes. As these variations scale with the number density of CS_2 molecules in the same fashion for each dilution series, it appears that the intermolecular potential is determined primarily by CS_2-CS_2 interactions.

It is informative to note that while the librational response gives direct information about the nature of the local intermolecular potential, the diffusive decay of the orientational anisotropy varies with the solution viscosity in a manner that fails to reflect the variations of the microscopic potential with changing intermolecular interactions. The isoheptane series is particularly enlightening in this respect because the decay of the orientational anisotropy remains constant on dilution. A more detailed discussion of these observations will be presented elsewhere[8].

We gratefully acknowledge the support of the U.S. Office of Naval research and the Natural Sciences and Engineering Research Council of Canada, and thank Kathy Carpenter for providing density and viscosity measurements on the solutions investigated.

REFERENCES

1. C. Kalpouzos, W.T. Lotshaw, D. McMorrow, and G.A. Kenney-Wallace, J. Phys. Chem. 91, 2028 (1987); Chem. Phys. Lett. 136, 323 (1987); in Laser Applications to Chemical Dynamics. Proceedings of the SPIE, vol. 742, 1987, pp 47-53.
2. D. McMorrow, W.T. Lotshaw, and G.A. Kenney-Wallace, IEEE J. Quant. Elect., February, 1988.
3. M.D. Levenson and G.L. Eesley, Appl. Phys. Lett. 19 1 (1979).
4. E.P. Ippen and C.V. Shank, Appl. Phys. Lett. 26, 92 (1975).
5. B.I. Green, P.A. Fleury, H.L. Carter, and R.C. Farrow, Phys. Rev. A 29 271 (1984).
6. S. Ruhman, L.R. Williams, and K.A. Nelson, J. Phys. Chem. 91, 2237 (1987).
7. J. Etchpare, G. Grillon, J.P. Chambaret, G. Hamoniaux, and A. Orszag, Opt. Comm. 63, 329 (1987).
8. D. McMorrow, W. T. Lotshaw, and G.A. Kenney-Wallace, Optical Society of America, Annual Meeting, Rochester, NY, October 18-23, 1987; C. Kalpouzos, D. McMorrow, W.T. Lotshaw, and G.A. Kenney-Wallace, Chem. Phys. Lett., submitted.
9. H. Versmold, in Molecular Liquids. Dynamics and Interactions, A.J. Barnes, W.J. Orville-Thomas, and J. Yarwood, eds. (NATO ASI Series, Reidel, 1984) pp 275-308.

10. W.T. Lotshaw, D. McMorrow, and G.A. Kenney-Wallace, in preparation.
11. P.A. Madden and D.. Tildesley, Mol. Phys. 49 193 (1983); Mol. Phys. 55 969 (1985); T.I. Cox and P.A. Madden, Mol. Phys. 39 1487 (1980).
12. P.A. Madden, in Molecular Liquids. Dynamics and Interactions, A.J. Barnes, W.J. Orville-Thomas, and J. Yarwood, eds. (NATO ASI Series, Reidel, 1984) pp 431-474.

MOLECULAR DYNAMICS IN PURE AND MIXED LIQUIDS PROBED BY FEMTOSECOND TIME-RESOLVED IMPULSIVE STIMULATED SCATTERING

S.Ruhman, Bern Kohler, Alan G. Joly, and Keith A. Nelson
Department of Chemistry, Massachusetts Institute of Technology, Cambridge, Massachusetts 02139

ABSTRACT

Recent time-domain observations of molecular dynamics in simple liquids are discussed. Molecular orientational motion is shown to be vibrational (i.e. librational) in character at short times. In carbon disulphide (CS_2), librational dephasing is due primarily to inhomogeneity in the liquid. Temperature-dependent frequencies and inhomogeneous dephasing rates have been determined, providing information about configuration-averaged intermolecular potentials and about the extent of inhomogeneity in the potentials. Experiments carried out under various conditions on CS_2, CS_2/hydrocarbon mixtures, and other liquids are discussed.

INTRODUCTION

An understanding of subpicosecond molecular dynamics in liquids has long been sought.[1,2] On longer time scales, molecular motion appears diffusional in character, and both translational and rotational diffusion rates have been reported for many liquids.[3,4] However, the elementary center-of-mass translational and orientational motions may be vibrational in character, as suggested theoretically and from computer simulations.[5,6] Crudely, one may envision molecules occupying transient local potential minima inside "cages" formed by their immediate neighbors. Motion about these multidimensional potential minima should appear vibrational on time scales short compared to cage rearrangement times.

Characterization of intermolecular vibrational motion can permit elucidation of net intermolecular forces in liquids. Conventional vibrational spectroscopy, e.g. light scattering spectroscopy, of these low-frequency motions, which are expected to undergo rapid dissipation and dephasing, is often frustrated since they give rise to broad spectral features centered at or near zero frequency shift.[1,2] Impulsive stimulated light scattering (ISS), an entirely time-domain light scattering technique, is well suited for study of such motions.[7,8] ISS data from carbon disulphide (CS_2) liquid at room temperature, recorded with 70-fs time resolution, were reported at ILS-II and elsewhere in 1986.[9,10] The time-delayed four-wave mixing or "transient grating" experimental geometry is shown schematically in Fig. 1, and typical V-H ISS data (analogous to depolarized or V-H light scattering data[7,8]) from CS_2 at 300K are shown in Fig. 2. In these experiments, the excitation pulses exert impulsive torques on molecules through impulsive stimulated rotational Raman scattering (essentially the impulsive limit of the optical Kerr effect), acting on single-molecule polarizability anisotropies. Coherent scattering or "diffraction" of the variably delayed probe pulse is used to measure the extent of orientational

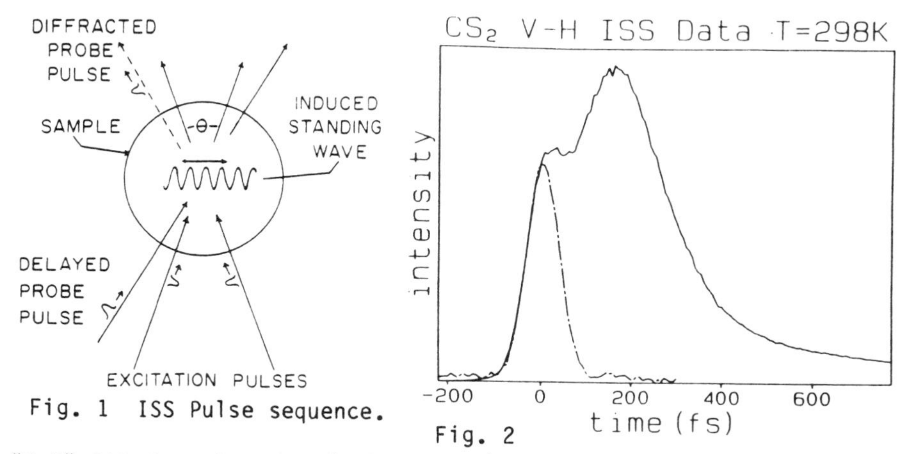

Fig. 1 ISS Pulse sequence.

Fig. 2
"V-H" ISS data from CS_2 (solid curve) at 298 K. The polarizations of the excitation pulses were V and H, the probe pulse was V, and the signal was H. The continued rise in signal intensity following the excitation driving force (broken curve) is apparent.

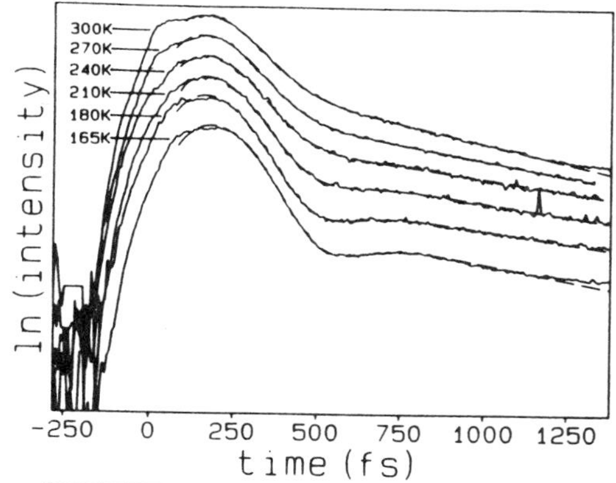

Fig. 3
T-dependent V-H ISS data from CS_2. A weak second oscillation appears at lower T, demonstrating the vibrational (i.e., librational) nature of the orientational motion under observation. Fits (dashed curves) are based on Eq. (2).

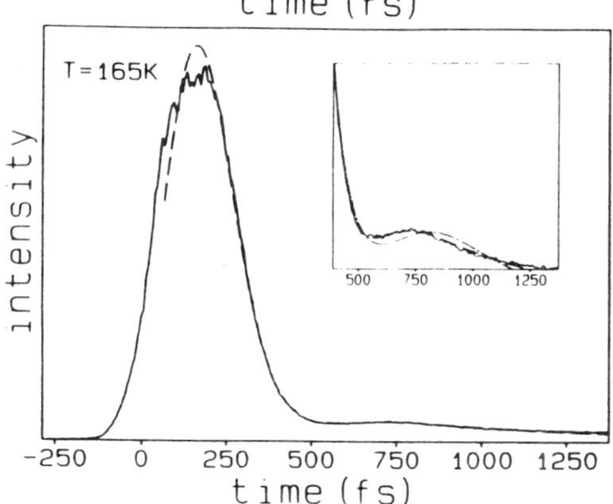

Fig. 4
V-H ISS data from CS_2 at 165K. The inset shows fits based on Eq. (1) (broken curve) which assumes purely homogeneous dephasing, and Eq. (2) (dashed curve, barely visible under the data) which assumes purely inhomogeneous dephasing.

alignment. The data clearly show that molecular motion away from equilibrium continues for ~100 fs after the driving force has ended. The rise in signal intensity is followed by a rapid decline, and then by a slower decay toward zero. The initial rise and fall were interpreted in terms of overdamped molecular librational motion.[8-11]

Several research groups have since reported similar results in various molecular liquids.[11-20] Here we discuss results which (1) confirm unambiguously the interpretation of time-domain ISS data in terms of intermolecular librational motion; (2) indicate the predominant mechanism for librational dephasing; (3) yield quantitative values for the temperature-dependent librational frequency and dephasing rate in CS_2; and (4) discuss additional experiments which may provide more information.

TEMPERATURE-DEPENDENT CS_2 RESULTS

Figure 3 shows V-H ISS data, plotted on a logarithmic scale, from CS_2 liquid at several temperatures between 165K and 298K. The main feature of note is the weak second oscillation which grows in as the temperature is reduced. This shows beyond any doubt that intermolecular vibrational motion, which is underdamped at low temperatures, is being observed.

Figure 4 shows the data from CS_2 at 165K, with the weak oscillatory feature enlarged in the inset. Two fits to the data are also shown in the inset. The broken curve is based on the form,

$$G^{\varepsilon\varepsilon}_{1212}(t) = Ae^{-\gamma t} \sin \omega t + Be^{-\Gamma t} , \tag{1}$$

where $G^{\varepsilon\varepsilon}$, the dielectric tensor impulse response function (also called the nonlinear susceptibility tensor, χ), is the quantity whose square is measured in ISS data.[7,8] Different light polarizations probe different projections of $G^{\varepsilon\varepsilon}$. V-H ISS measures $G^{\varepsilon\varepsilon}_{1212}(t)$, whose Fourier transform is probed in V-H light scattering spectroscopy. The first term in Eq. (1) describes the underdamped oscillatory motion observed at short times, characterized by frequency ω and homogeneous dephasing rate γ. This term is associated with local molecular librational motion. The second term describes the exponential decay, with relaxation rate Γ, observed in signal at long times. This is associated with collective rotational relaxation (i.e., rotational diffusion) through which the fluid returns to the isotropic state. The dashed curve is based on the functional form

$$G^{\varepsilon\varepsilon}_{1212}(t) = Ae^{-\Delta^2 t^2/2} \sin \omega_A t + Be^{-\Gamma t}, \tag{2}$$

where homogeneous dephasing has been neglected and the vibrational motion is now characterized by an inhomogeneous distribution of local frequencies whose average is ω_A and whose spread is Δ.[12-14] The far superior fit of the dashed curve indicates that librational dephasing is predominantly inhomogeneous. The data therefore yield information about the configuration-averaged intermolecular potential and the extent of inhomogeneity in the potential. Quantitative temperature dependent values of ω_A and Δ, and approximate values for the librational force constant and its inhomogeneity, have been reported.[12-14] The frequency increases at lower temperatures because the density increases and the molecular "cages" are tighter. The inhomogeneity, i.e. Δ, decreases at lower T. These effects combine to make oscillatory behavior more apparent as the temperature is reduced.

CS_2/CYCLOHEXANE SOLUTIONS

Figure 5 shows V-H ISS data from several mixtures of CS_2 and cyclohexane at room temperature,[13] with the CS_2 volume fraction ranging from 1% to 80%. The most striking feature is the similarity among all the data. Other than the relative intensities of the electronic responses at $t \approx 0$, the short-time (i.e. $t < 1$ ps) data are nearly identical. This suggests, though does not prove, two conclusions: (1) the same type of molecular motion, i.e. orientational (as opposed to translational) motion, is observed in all the data; and (2) the net intermolecular potentials which mediate librational dynamics are similar in all the solutions. Conclusion (2) should certainly depend on the choice of solvent, and we note that Kenney-Wallace _et al._ have examined CS_2 mixtures with other hydrocarbons[20] and seen concentration-dependent short-time dynamics. Conclusion (1) is an issue because the excitation pulses, in addition to exerting torques on molecules through single-molecule polarizability anisotropies, can also act on interacting molecule pairs, triplets, etc. through higher-order "interaction-induced" (I-I) polarizabilities.[1,2] For example, the excitation pulses may exert an impulse force on two neighboring molecules which drives them toward each other, and the resulting correlated motion away from equilibrium may be detected by the probe pulse. (Both excitation and probe pulses act on the interaction-induced polarizability of the molecule pair). There is uncertainty about the extent of I-I contributions to light-scattering from CS_2 and other liquids.[21] These contributions should be negligible in ISS data from a 1% solution of CS_2 in cyclohexane (note that the cyclohexane polarizability is comparatively small). Therefore, the 1% solution data reflect only orientational motion. More specifically, they can be related to single-molecule and pair orientational correlations,[12-14] and in a 1% solution the latter can be neglected. The results in Fig. 5 suggest that orientational motions (and mainly single-molecule orientational correlations) make the dominant contribution to data even in pure CS_2.

OTHER RESULTS

ISS data from molecular liquids have been recorded with various polarization combinations. With V-polarized excitation, probe, and signal light, $G^{\varepsilon\varepsilon}_{1111}(t)$ is probed.[11,13] Figure 6 shows data recorded with V-polarized excitation pulses and H-polarized probe and signal light. In this case, $G^{\varepsilon\varepsilon}_{1122}(t)$ is probed. The electronic and nuclear contributions have different signs, leading to a sharp "dip" in the data at $t \gtrsim 0$ where they cancel.[13] Etchepare _et al._[18,19] demonstrated polarization combinations which measure linear combinations of $G^{\varepsilon\varepsilon}_{1111}$ and $G^{\varepsilon\varepsilon}_{1212}$. With some combinations, either the electronic contribution to the data or the nuclear contribution due to single-molecule polarizabilities can be eliminated. Elimination of the electronic response permits analysis of data even near t=0. Elimination of the single-molecule polarizability contribution may permit scattering due to higher-order I-I polarizabilities, and in particular center-of-mass translational motions, to be elucidated.

We have discussed "short" and "long" time responses due to local and collective orientational motions, respectively (see Eq. (2)). Careful analysis of data from several liquids including CS_2 indicates an additional "intermediate-time" decay (i.e. decay times ~ 0.5 ps). Kenney-Wallace et al. have fit this with an additional exponentially decaying term,[16,17] and argued plausibly that translational motions are being observed through I-I polarizabilities. Here we suggest another possible interpretation indicated by theoretical work of Lynden-Bell and Steele[6] and currently being tested. The intermediate decay may reflect loss of vibrational coherence due to "cage" fluctuations which cause local librational frequencies to fluctuate. This may be accounted for by a lineshape theory due to Kubo[6] which describes the Guassian decay of vibrational correlations at short times and the exponential decay at longer times.

Finally, other ISS experiments have been carried out in which intramolecular vibrational oscillations are observed in real time.[8,13,17,22] These experiments may permit unique determination of single-molecule orientational correlations, whose effects on the time-dependent oscillatory signal intensity are analogous to their effects on the wings of intramolecular vibrational Raman spectra.[1,3]

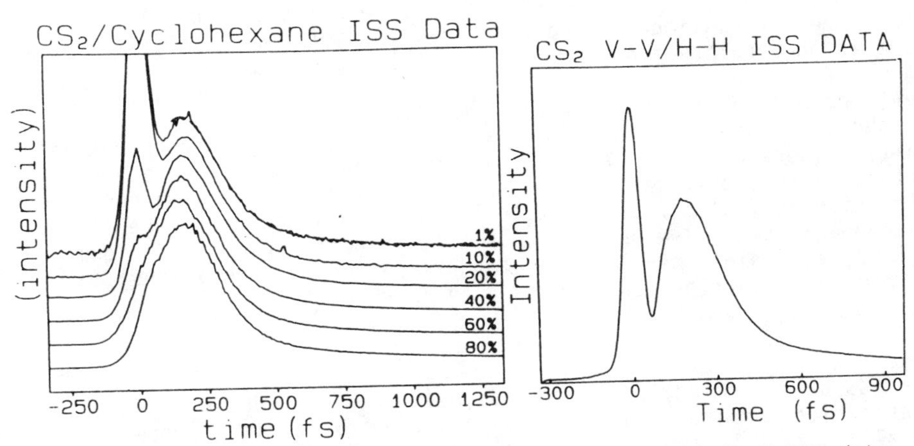

Fig. 5 V-H ISS data from CS_2/cyclohexane solutions at 298K.

Fig. 6 V-V/H-H ISS data from CS_2 at 298K.

CONCLUSIONS

In the intervening year between ILS-II and ILS-III, much progress has been made in time-domain vibrational spectroscopy of molecular liquids. In the near future, pressure-dependent studies should be important for comparison to theory and computer simulation and because the vibrational character of intermolecular motions should become more pronounced in liquids at high densities. Experiments with appropriate polarization combinations or on molecules of appropriate symmetry (e.g. tetrahedral molecules[17]) should allow the vibrational nature of center-of-mass translations to be characterized. We anticipate continued rapid progress in this area.

REFERENCES

1. W.G. Rothschild, Dynamics of Molecular Liquids (Wiley, 1984).
2. D. Kivelson and P.A. Madden, Ann. Rev. Phys. Chem. 31, 523 (1980).
3. B.J. Berne and R. Pecora, Dynamic Light Scattering (Wiley, 1976).
4. G.R. Fleming, Chemical Applications of Ultrafast Spectroscopy (Oxford University Press, New York, 1986).
5. B. Guillot, S. Bratos, and G. Birnbaum, Phys. Rev. A 22, 2230 (1980).
6. R.J. Lynden-Bell and W.A. Steele, J. Phys. Chem. 88, 6514 (1984).
7. Y.-X. Yan and K.A. Nelson, J. Chem. Phys. 87, 6240; 6257 (1987).
8. Y.-X. Yan, L.-T. Cheng, and K.A. Nelson in Advances in Nonlinear Spectroscopy, vol. 15, ed. by R.J.H. Clark and R.E. Hester (Wiley, Chichester, 1987), p. 299.
9. K.A. Nelson, Invited paper at ILS-II, Seattle, October 1986; Plenary Lecture at the Quasielastic Light Scattering Spectroscopy-II (QELSS-II) Conference, Worcester, June 1986.
10. L.R. Williams, S. Ruhman, A.G. Joly, B. Kohler, and K.A. Nelson, in Advances in Laser Science - II, ed. by M. Lapp, W.C. Stwalley, and G.A. Kenney-Wallace (AIP, 1987), p. 408.
11. S. Ruhman, L.R. Williams, A.G. Joly, B. Kohler, and K.A. Nelson, J. Phys. Chem. 91, 2237 (1987).
12. S. Ruhman, B. Kohler, A.G. Joly, and K.A. Nelson, Chem. Phys. Lett. 141, 16 (1987).
13. S. Ruhman et al., Rev. de Phys. Appl. 22, 1717 (1987).
14. S. Ruhman, B. Kohler, A.G. Joly, and K.A. Nelson, IEEE J. Quantum Electron., in press.
15. C. Kalpouzos, W.T. Lotshaw, D. McMorrow, and G.A. Kenney-Wallace, J. Phys. Chem. 91, 2028 (1987).
16. W.T. Lotshaw, D. McMorrow, C. Kalpouzos, and G.A. Kenney-Wallace, Chem. Phys. Lett. 136, 323 (1987).
17. D. McMorrow, W.T. Lotshaw, and G.A. Kenney-Wallace, IEEE J. Quantum Electron., in press.
18. J. Etchepare et al. Opt. Commun. 63, 329 (1987).
19. J. Etchepare et al. Rev. de Phys. Appl. 22, 1749 (1987).
20. C. Kalpouzos, D. McMorrow, W.T. Lotshaw, and G.A. Kenney-Wallace, Chem. Phys. Lett., in press.
21. L.C. Geiger and B.M. Ladanyi, J. Chem. Phys. 87, 191 (1987).
22. S. Ruhman, A.G. Joly, and K.A. Nelson, J. Chem. Phys. 86 6563 (1987); IEEE J. Quantum Electron., in press.

DIRECT STRUCTURAL CHARACTERIZATION OF CHARGE LOCALIZATION IN METAL TO LIGAND CHARGE TRANSFER COMPLEXES

L. K. Orman, D. R. Anderson, and J. B. Hopkins.
Louisiana State University, Baton Rouge, LA 70803

ABSTRACT

The metal to ligand charge transfer states of Ru(bipyridine)$_3^{2+}$, Ru(bipyrimidine)$_3^{2+}$, and Ru(dimethylbipyridine)$_3^{2+}$ have been studied using picosecond Raman spectroscopy. The spectrum associated with the charge localized species appears within the 25ps pulsewidth of the laser. This is true even for viscous solutions where solvent reorganization times are much slower than the experimental observation time. In fact, it is found that the charge localized species is generated instantly in glycerol solutions below the glass point.

INTRODUCTION

The metal to ligand charge transfer states (MLCT) of Ru(bipyridine)$_3^{2+}$ and similar complexes have been intensely investigated. In a classic experiment using resonance Raman spectroscopy Woodruff et. al.[1] determined that the electron is localized on a single bipyridine ligand on a time scale of $\leq$ 10 ns. Additional experiments have confirmed this initial result.[2] Even so, there have been several reports of evidence for configurations consisting of electrons delocalized over all ligands. This latter result comes from low temperature experiments where solvent motions and reorganization can be frozen out. As such, it has been postulated that the MLCT state is delocalized upon initial photoexcitation, and that following this that rapid solvent reorganization traps the electron in one ligand potential well.

We have investigated the behavior of the MLCT states on time scales fast compared to the rate of solvent reorganization. This is achieved by using picosecond Raman spectroscopy to detect the MLCT states of the previously mentioned complexes in viscous or frozen solutions.

EXPERIMENTAL

All ruthenium complexes used were synthesized using techniques found in the literature.[3,4] Ru(bpy)$_3^{2+}$ was purchased from Aldrich and used as received. All products were analyzed using mass spectroscopic methods.

The laser system utilizes the amplification of modelocked Nd:YAG laser pulses by regenerative amplification in tandem with temporal pulse compression.[5,6] The third harmonic of the NdYAG (3547Å) at pulse energies of 25μJ/pulse was used to produce the MLCT state and detect it using Raman spectroscopy. The sample cell was rotated at speeds to ensure that each laser shot interrogated a fresh region of sample.

RESULTS

The picosecond Raman spectrum of $Ru(bpy)_3^{2+}$ in glycerol is shown in Figure 1. The top view shows the spectrum of the complex taken at a temperature of -10° which is well below the glass point. The bottom view shows the spectrum of the comply above the glass point at a temperature of 35°C. The large peak at 1460 cm^{-1} is due to the solvent. Four bands at 1211 cm^{-1}, 1283 cm^{-1}, 1425 cm^{-1} and 1545 cm^{-1} are known to correspond to the localized MLCT excited state.[1] Ground state bands[1] are observed at 1318 cm^{-1}, 1562 cm^{-1}, and 1607 cm^{-1}. From the figure it is very clear that the ratio of the excited state peaks to ground state peaks is the same in both spectra above and below the glass point in glycerol.

We have performed similar experiments in glycerol for the $Ru(dimethyldipyridine)_3^{2+}$ and $Ru(bipyrimidine)_3^{2+}$ complexes with similar results. From this data we conclude that the electron localized configuration is born within the pulsewidth of the laser (≈25 ps). In addition, it would appear that solvent reorganization processes have little effect on producing the trapped electron configuration.

The implications of these experimental results can be viewed in two ways. First, that the electron localized configuration occurs because the ligand-ligand coupling is too small to permit fast electron communication between ligands. Secondly, that electron transfer occurs from vibrational levels well below the barrier. Although the data is not overwhelming, we favor the latter mechanism or a combination of the two mechanisms because it is consistent with the results of a second paper published in these proceedings.

REFERENCES

1. P.G. Bradley, N. Kress, B.A. Hornberger, R.F. Dallinger, and W.H. Woodruff, J. Am. Chem. Soc. 103, 7441 (1981).
2. J.V. Caspar, T.D. Westmoreland, G.H. Allen, P.G. Bradley, T.J. Meyer, and W.H. Woodruff, J. Am. Chem. Soc. 106, 3492 (1984).
3. P.A. Mabrouk and M.S. Wrighton, Inorg. Chem. 25, 526 (1986).
4. M. Hunziker and Andreas, Ludo., J. Am. Chem. Soc. 99, 7370 (1977).
5. Y.J. Chang, C. Veas, and J.B. Hopkins, Appl. Phys. Lett., 49, 1758 (1986).
6. Y.J. Chang, D.R. Anderson, and J.B. Hopkins, International Conference on Lasers 1986, R.W. McMillam, p. 169 STS Press 1987.

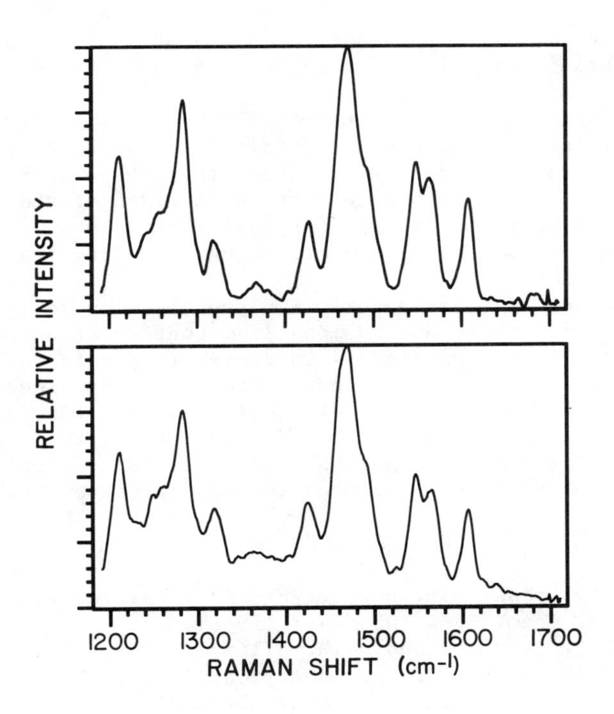
RELATIVE INTENSITY
1200
1300
1400
1500
1600
1700
RAMAN SHIFT (cm-l)

Figure 1

ROTATIONAL REORIENTATION AND ISOMERIZATION OF TRANS-STILBENE

S. K. Kim, S. H. Courtney[‡], and G. R. Fleming
The University of Chicago, Chicago, Illinois 60637

ABSTRACT

We have tested the validity of using solvent shear viscosity as a measure of friction involved in the dynamics of t-stilbene. The rotational reorientation and the isomerization in n-alkanes are considered.

INTRODUCTION

Ultrafast techniques have made it possible to investigate hydrodynamic theories in the picosecond regime. The photoisomerization of t-stilbene has been widely used to test the Kramers theory for barrier crossing[1]. But the small size of the probe suggests that a macroscopic solvent viscosity (η) may not be an adequate measure of the friction. In contrast the rotational reorientation time (τ_{OR}) is expected to provide a more accurate measure of the microscopic friction. We applied this concept to fit the isomerization data of t-stilbene in n-alkanes to the Kramers equation. We also discuss the isomerization in n-alcohols where the dynamic nature of the solvent possibly changes the activation barrier.

LASER SYSTEM

The measurements of τ_{OR} were performed with a CPM ring laser amplified through Nd:YAG pumped dye cells. The 615nm output pulse was frequency doubled and used to induce an excited state anisotropy in the sample. The decay of the anisotropy is monitored via $S_1 - S_n$ absorption by the time delayed fundamental through crossed polarizers. All isomerization data were obtained with a single photon counting system or were taken from other sources[2].

RESULTS AND DISCUSSION

Figure 1 shows the macroscopic friction ($\propto \eta$) is not parallel to the microscopic friction ($\propto \tau_{OR} T$) when the solvent is varied. The longer alkane solvent creates a larger space where t-stilbene can rotate more freely, as reflected in the trend of the slope changes in Figure 1. However, rather surprisingly, Stokes law ($\tau_{OR} \propto \eta / T$) appears to hold within a given alkane.

‡ Present address: AT & T Bell Laboratories, Murray Hill, NJ 07974

The data agree qualitatively with a free space model[3]. But the model predicts a curvature in the plot when applied to a single solvent[4]. This suggests that the effect of density changes within a given solvent is less important than the theory predicts.

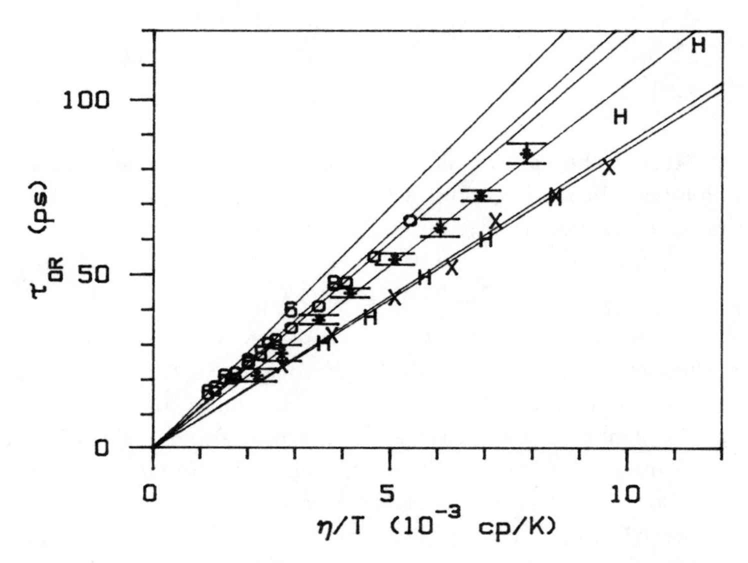

Fig.1 τ_{OR} for t-stilbene in hexane (6), octane (8),decane (o), dodecane(*), tetradecane(X), hexadecane (H). The solid lines are the least square fits to data for each alkane.

In earlier studies[5,6], using τ_{OR} as a friction measure instead of η seemed to improve the Kramers expression when applied to t-stilbene in n-alkanes at room temperature. However, this method fails when the study is extended over a range of temperature (Figure 2). The Kramers equation with the microscopic friction ς_{OR} $(=6kT\tau_{OR}T/I_{OR})$ still lacks the necessary curvature. This failure is also seen when the fit is made for a single solvent (Figure 2*c*) where according to Figure 1 η is a suitable parameter. Therefore other forms of the friction (e.g. frequency dependent friction[7]) should be considered. We also cannot exclude the possibility that the potential energy surface is multidimensional or density dependent.

While isoviscosity plots provide a fixed barrier height (~3.5 Kcal/mol) for the stilbene isomerization in alkanes, they give higher barriers for low viscosity alcohols than high viscosity alcohols (e.g. ~3.5 Kcal/mol at 1 cp and ~2 Kcal/mol at 10 cp). This trend cannot be explained by a static polarity alone.

Fig.2 Reduced isomerization rate $F(\varsigma_{ISO})$ for t-stilbene plotted versus η (*a*) and ς_{OR} (*b*) with best Kramers fits (solid curves). The plot symbols are given in Figure 1. (*c*) shows a simple Arrhenius expression (dotted curve) fits the data for a single solvent (decane) better than Kramers equation.

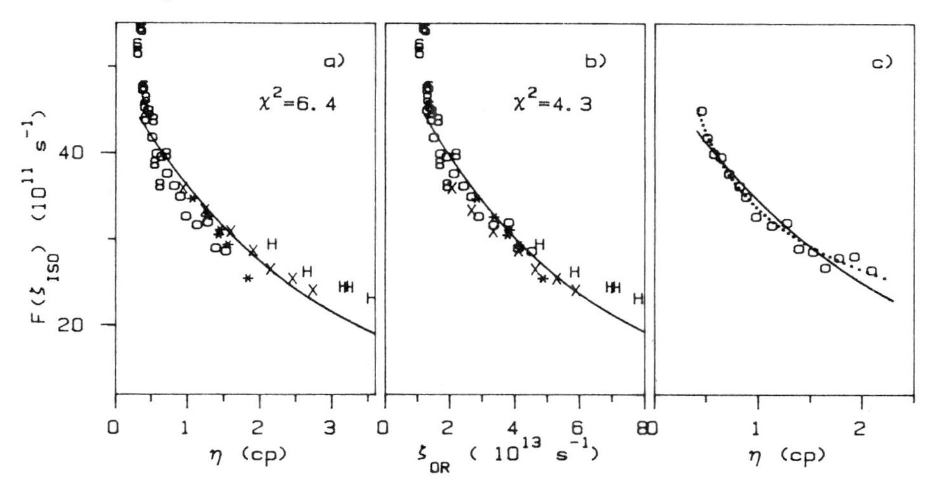

The longitudinal relaxation times (τ_L) of the alcohols are distributed over much wider range than the isomerization lifetimes (τ_{ISO}) in the corresponding alcohols. For example on going from methanol (τ_L ~9 ps at 20°C) to decanol (τ_L ~690 ps at 20°C) τ_{ISO} changes from 40 ps to 190 ps. This suggests that only the short alcohols solvate effectively the excited state during the isomerization. This effect, coupled with the static polarity effect, may change the barrier height in a way consistent with the data[8].

We conclude that in any of the dynamics considered above the macroscopic shear viscosity is not the proper measure of the friction.

REFERENCE

1. H.A. Kramers, Physica 7, 284 (1940).
2. S.H. Courtney and G.R. Fleming, J. Chem. Phys. 83, 215 (1985).
3. J.L. Dote, D. Kivelson, and R.N. Schwartz, J. Phys. Chem. 78, 249 (1983).
4. S.K. Kim and G.R. Fleming, J. Phys. Chem. in press.
5. S.H. Courtney, S.K. Kim, S. Canonica, and G.R. Fleming, J. Chem. Soc. Faraday Trans II 82, 2065 (1986).
6. M. Lee, A.J. Bain, P.J. McCarthy, J.N. Haseltine, A.B. Smith III, and R.M. Hochstrasser, J. Chem. Phys. 85, 4341 (1986).
7. B. Bagchi and D.W. Oxtoby, J. Chem. Phys. 78, 2735 (1983).
8. D.M. Zeglinski and D.H. Waldeck, J. Phys. Chem. in press.

Enhancement of Ultrafast Supercontinuum Generation in Water by Cations

P. P. Ho, T. Jimbo, Q. X. Li, Q. Z. Wang, V. Caplan and R. R. Alfano
Institute for Ultrafast Spectroscopy and Lasers
Photonics Application Laboratory
Departments of Electrical Engineering and Physics
The City College of New York, New York, NY 10031

ABSTRACT

The ultrafast Kerr gate and self-phase-modulation were found to be several times larger in saline water than in pure water. The optical Kerr effect from saturated aqueous solutions of $ZnCl_2$ was about 35 times greater. The self phase modulation effect from saturated aqueous solutions of K_2ZnCl_4 was about 10 times greater.

INTRODUCTION

One common goal in nonlinear optics research is to find a material with large optical nonlinearity and ultrafast ($<10^{-13}$ sec) response time. In this paper, we report on a ten fold enhancement of the supercontinuum generation and the optical Kerr effect in water by the addition of cations. The increase in the nonlinearity in aqueous solutions arises from the increase of the molecular density through hydration.

EXPERIMENTAL

The experimental setup has been described in reference 1. 8ps-laser pulse at 530-nm was obtained from a mode-locked glass laser system. Three different two-component salt solutions of various concentrations were tested. The solutes were KCl, $ZnCl_2$, and K_2ZnCl_4. All measurements were performed at room temperature. Typical spectra of supercontinuum pulses[2] exhibit both self-phase-modulation (SPM) and Four-Photon-Parametric-Genration (FPPG). The collinear profile arising from SPM has nearly the same spatial distribution as the incident 530 nm laser pulse. Two conical emissions correspond to FPPG pulse propagation. The angle of these wings are determined by the phase-matching condition of the generated FPPG pulse and the incident laser beam. FPPG spectra sometimes appear shaped like coaxial cones and, like the SPM spectra, frequently show modulated patterns.

Stokes side supercontinuum spectra of different aqueous solutions and neat water are shown in Fig. 1. The salient feature is a wide band SPM spectrum together with the Stimulated Raman Scattering (SRS) of OH stretching vibration around 645 nm. The addition of salts cause the SRS signal to shift towards the longer wavelength region and sometimes weaken the SRS (Fig. 1a). The SRS signals of pure water and dilute solution appear in the hydrogen-bonded OH stretching region (~ 3400 cm^{-1}). In a high concentration solution, it appears in the non-hydrogen-bonded OH stretching region (~ 3600 cm^{-1}).

To quantitatively evaluate the effect of cations on SPM generation, the SPM signal intensity of various samples at a fixed wavelength were measured and compared. The SPM signal intensity generated in K_2ZnCl_4, $ZnCl_2$, and KCl solutions at 570 nm and 500 nm were found to be montonically increased as a function on salt concentration from 0 to 10M. This measurement indicated that the SPM pulse intensity was highly dependent on salt concentration, and both the Stokes and anti-Stokes sides of the SPM signals from a saturated K_2ZnCl_4 solution were about 10 times larger than from neat water. After plotting these data as a function of K^+-ion concentration, the solutions of KCl and K_2ZnCl_4 generated almost equal amounts of SPM at the same K^+ cation concentration even though they contained different amounts of Cl^- anions.

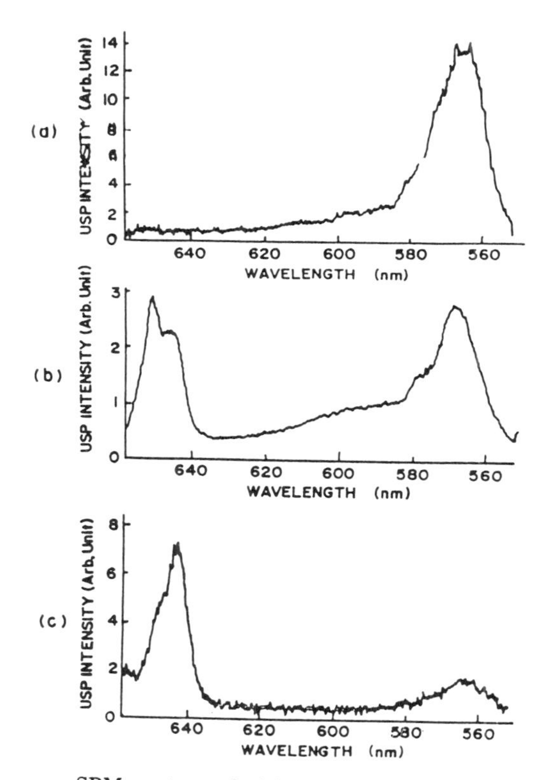

SPM spectrum of a (a) saturated K_2ZnCl_4 solution, (b) 0.6-*M* K_2ZnCl_4, and (c) pure water. The SRS signal (645 nm) is stronger in pure water, and it disappears in high-concentration solution.

This indicates that Cl^- anions are insignificant on the enhancement of SPM. Measurements of the optical Kerr effect[3] and supercontinuum in the saturated salt solutions are summarized in Table 1. The value $G_{SPM}(\lambda)$ represents the ratio of the SPM signal intensity from a particular salt solution to that from neat water at wavelength λ.

SOLUTION / SIGNAL	K_2ZnCl_4 1.9 M	KCl 4.0 M	$ZnCl_2$ 10.6 M
G_{SPM}(570nm)	11 ± 1	5.6 ± 0.9	6.6 ± 0.4
G_{SPM}(500nm)	9.5 ± 2.5	4.9 ± 0.2	4.3 ± 0.5
G_{Kerr}	16 ± 1	6.1 ± 1.4	35 ± 9

Table 1. Enhancement of the SPM and the optical Kerr effect signals in saturated aqueous solutions at 20°C.

DISCUSSION

The enhancement of the optical nonlinearity of water by the addition of cations can be explained by the cation disruption of the tetrahedral hydrogen bonded water structures and their formation of hydrated units[3]. Hydration increases N and thereby increases n_2. The ratio of the hydration numbers between Zn^{2+} and K^+ cations in aqueous solutions is ~ 11.2/7.0=1.6. This is in good agreement with our nonlinear opitcal experimental measurements : I(signal) ~ n_2^2 ~ $(N)^2$ ~ 2.5. The large discrepancy of the gain measured by the Kerr effect and SPM in $ZnCl_2$ solution can be account for the response time of the saline solution. The transmitted signal of the OKG depends on Δn while the USL signal is determined by $\partial n/\partial t$, i.e., the SPM also depends on the response time of the hydrated units. Since the Zn^{2+} hydrated units are larger than those of K^+, the response time will likewise be longer.

Additional factors may contribute to the disparate values between G_{SPM} and G_{Kerr} for $ZnCl_2$. The first is the nonlinear distortion from the salt ions and the salt-water molecules interactions. The total optical nonlinearity of a mixture depends by the coupled interactions of solute-solute, solute-solvent, and solvent-solvent molecules. The molar ratio of K_2ZnCl_4 to water was ~ 0.2. The nonlinearity contribution from the salt ion itself is about 10 time smaller than water. Therefore, the solute-solute interaction can be neglected. However, salt solute to water solvent effect may contribute additional optical nonlinearity to the water. The second is related to the mechanism of δn generation. The nonlinear susceptibility χ_{1111} is involved in the generation of SPM while the difference $\chi_{1111}-\chi_{1122}$ is responsible for the optical Kerr effect. The third is the possible dispersion of n_2 due to the difference in wavelength between the exciting beams of SPM and the optical Kerr effects.

ACKNOWLEDGMENT

This research was supported by NSF and Hamamatsu Photonics K.K.

References

1. T. Jimbo, V. Caplan, Q. Li, Q. Wang, P. Ho and R. Alfano, Opt. Lett., 12 477 (1987)
2. R. R. Alfano and S. L. Shapiro, Phys. Rev. Lett. **24**, 584, 592, and 1217 (1970).
3. P. P. Ho and R. R. Alfano, Phys. Rev. **A20**, 2170 (1979).
4. G. E. Walrafen, Advan. Mol. Relaxation Processes **3**, 43 (1972).

PICOSECOND PUMP-PROBE STUDIES OF ENERGY RELAXATION IN AGGREGATES OF PSEUDOISOCYANINE ADSORBED ON COLLOIDAL SILICA

Edward L. Quitevis, Miin-Liang Horng, and Sun-Yung Chen
Department of Chemistry and Biochemistry
Texas Tech University
Lubbock, Texas 79409

ABSTRACT

Using picosecond pump-probe techniques, we have measured transient photobleaching of J-aggregates of pseudoisocyanine adsorbed on colloidal silica. The signals decayed exponentially at low colloid concentrations, but biexponentially at higher colloid concentrations. The decay times were consistent with the lifetimes predicted by a polariton model.

INTRODUCTION

Cyanine dyes are of interest because of their special aggregation properties in solution and on surfaces.[1] Cyanine dye aggregates are spectral sensitizers of semiconductor electrodes and photographic film and are model systems for complex biological systems. We previously reported preliminary measurements of electronic energy relaxation of aggregates of pseudoisocyanine (PIC) adsorbed on colloidal silica.[2] In this paper we describe the effects of changing the colloid concentration on the observed transient bleaching signals.

The colloidal silica samples consisted of 40 Å diameter particles in an aqueous pH 10.4 solution. At this pH, silanol groups at the surface of the colloid are ionized, giving rise to a negatively charged surface. Cationic molecules such PIC are readily adsorbed on the surface of the colloid. By using colloidal silica we eliminate electron transfer processes between adsorbed molecules and the substrate.

The formation of aggregates on colloidal surface drastically changed the spectrum of a 5 x 10^{-5}M PIC solution. This is shown in Figure 1. The monomer absorption spectra had a maximum at 523 nm. Addition of colloidal silica lead to a decrease in the intensity of the monomer band and the appearance of a sharp band known as the J-band at 569 nm. The intensity of the J-band grew with increasing colloid concentration. A single sharp band at 572 nm was observed in the fluorescence spectra. The aggregates that gave rise to this band are known as J-aggregates.[1] The J-band is the result of a strongly coupled exciton state.[3] PIC molecules in these aggregates are lined up on their long edge in a brickwork arrangement.[1b] Molecular orbital calculations have shown that at least four monomer units are required to form the J-aggregate.[4]

EXPERIMENTAL DETAILS

By measuring transient photobleaching after excitation at the peak of the J-band, where there is negligible absorption due to the monomer, the electronic energy relaxation in J-aggregates can be isolated from that in the monomer. The picosecond pump-probe apparatus was previously described.[1] Briefly, a synchronously-pumped dye laser (≈ 76 MHz repetition rate) at 570 nm with 10-12 ps pulses was the excitation source. Colinear, copropagating, orthogonally polarized pump and probe beams were focused into a spinning sample. A difference-frequency modulation technique[5] was used to detect the the change in transmission of the probe beam. To eliminate the coherent coupling artifact, the signals were antisymmetrized.[6] This procedure makes use of the fact that the coherent coupling artifact is symmetric about zero delay. The antisymmetrized signal $S_a(\tau)$ is given by the convolution of the pulse autocorrelation with the antisymmetrized response function.

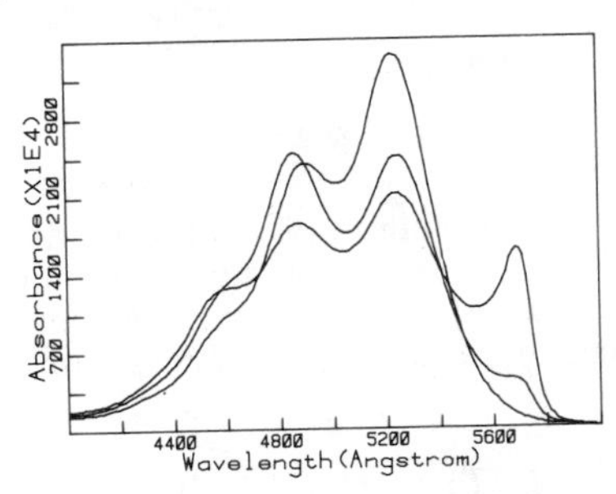

Figure 1. Spectra of 5 x 10^{-5}M PIC in colloidal silica at concentrations of 0%, 1.41%, and 7.6%. Note the increase in the intensity of the J-band with colloid concentration.

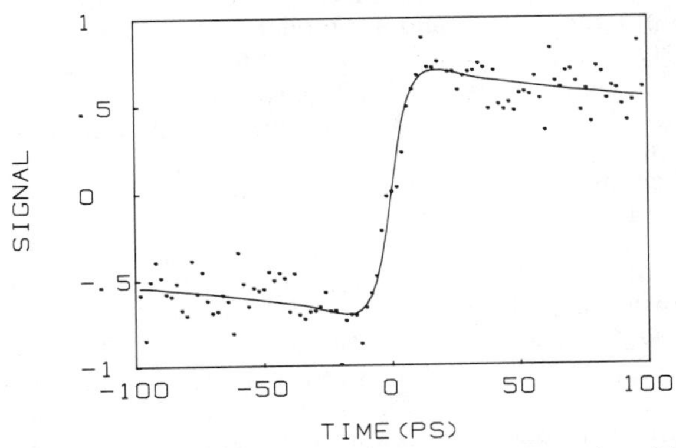

Figure 2. A typical antisymmetrized transient bleaching signal for 5 x 10^{-5} M PIC in 1.41% colloidal silica. The solid curve is a fit to the data. See text for details.

RESULTS AND DISCUSSION

Figure 2 illustrates a typical antisymmetrized signal for PIC in 1.41% colloidal silica. The solid line through the signal is a best fit to the convolution of the pulse autocorrelation with an antisymmetrized exponential response function obtained with a decay time of 185 ps and a $\chi^2 = 1.5 \times 10^{-2}$. The average of several runs was 167 ± 32 ps. In contrast, the decay was biexponential in a sample containing 7.6% colloidal silica.[2] The biexponential had a short decay component of 20 ± 2 ps and a long decay component of 250 ± 32 ps.

We suggested that a distribution of aggregates exists on the surface of the colloids.[2] If different sizes of aggregates have different lifetimes, then the decay will be nonexponential and will depend on the relative concentrations of PIC and colloid, which is consistent with our results. The decay times can be explained in terms of a polariton model for one- and two-dimensional aggregates.[7] The model couples the radiation field to the exciton state. One of the states that results from the coupling is a superradiant state. The lifetime of this superradiant state τ_N, is given by $\tau_N = \tau_0/N$, where N is the size of the aggregate and τ_0 is the radiative lifetime of the monomer. In the case of cyanine dyes τ_0 is typically 2 - 3 ns.[8] It is known that there are 60-65 ionized silanol groups at the surface of our 40 Å diameter colloids at pH 10.4.[9] If we assume PIC molecules adsorb onto the surface of the colloid by binding to ionized silanol groups, then the largest J-aggregate will consist of 60-65 monomer units. Therefore, we predict a lifetime of 30-50 ps for these large aggregates. Similarly, for J-aggregates having the minimum number of 4 monomer units, the model predicts a lifetime of 500-700 ps. The model predicts that effect of the dielectric substrate will reduce these lifetimes by at most a factor of 1/2. This would yield a lifetime of 15-25 ps for the largest J-aggregate and a lifetime 250-350 ps for the smallest J-aggregate. These lifetimes agree remarkably well with our preliminary results. Future work on these systems will focus on determining the effects of dye concentration, colloid size, and different cyanine dyes on electronic energy relaxation in J-aggregates of pseudoisocyanine.

ACKNOWLEDGEMENT

This work was supported by the Robert A. Welch Foundation, the Petroleum Research Fund as administered by the ACS, and the Research Corporation. Use of equipment and facilities at the Picosecond and Quantum Radiation Laboratory at Texas Tech University is greatly appreciated.

REFERENCES

1. (a) A. H. Herz, R. P. Danner, and G. A. Janusonis, in *Adsorption From Aqueous Solution*; edited by W. J. Weber Jr. and E. Matijevic (Rheinhold, New York, 1968) ACS Monograph Ser. No. 79, pp. 173-197. (b) A. Herz, in *The Theory of the Photographic Process, 4th ed.*; edited by T. H. James (MacMillan, New York, 1977), pp 235-250.
2. E. L. Quitevis, M.-L. Horng, and S.-Y. Chen, J. Phys. Chem., in press.
3. (a) V. Czikklely, H. D. Forsterling, H. Kuhn, Chem. Phys. Lett. **6**, 207 (1970). (b) E. G. McRae, Australian J. Chem. **14**, 354 (1961).
4. R. Ballard and B. Gardner, J. Chem. Soc. B, 736 (1971); F. Dietz, Tetrahedron, **28**, 1403 (1972).
5. E. L. Quitevis, E. F. Gudgin Templeton, and G. A. Kenney-Wallace, Appl. Opt. **24**, 318 (1985).
6. R. A. Engh, J. W. Petrich, and G. R. Fleming, J. Phys. Chem. **89**, 618 (1985).
7. R. Dicke, Phys. Rev. **93**, 99 (1954); M. R. Philpott, and P. G. Shermann, Phys. Rev. **B12**, 5381 (1975).
8. N. J. L. Roth, and A. C. Craig, J. Phys. Chem. **78**, 1154 (1974).
9. C. Laane, I. Willner, J. W. Otvos, and M. Calvin, Pro. Natl. Acad. Sci. **78**, 5928 (1981).

SPECTRAL AND TEMPORAL INVESTIGATION OF CROSS-PHASE MODULATION EFFECTS ON PICOSECOND PULSES IN SINGLEMODE OPTICAL FIBERS

P.L. Baldeck, F. Raccah, R. Garuthara, and R.R. Alfano
I.U.S.L., E.E. Department, The City College of New York, N.Y. 10031.

ABSTRACT

Cross-phase modulation (XPM) effects on co-propagating picosecond pulses have been investigated. In a pump-probe configuration, the probe pulse undergoes a substantial induced-frequency shift which can be tuned by varying the time delay between pump and probe pulses. During the Raman (SRS) generation process, the XPM induced by the pump and self phase modulation (SPM) strongly affect the spectra of Raman pulses.

INTRODUCTION

Spectral effects induced by XPM[1] were predicted in 1980 for SRS[2]. They were not observed until late 1985 when it was reported that intense infrared pulses could be used to enhance the spectral broadening of weaker 527-nm pulses co-propagating in a BK-7 glass sample[3-4]. Since then, several groups have been studying XPM effects on stimulated Raman scattering pulses[5-9], second harmonic pulses[10], and stimulated four photon mixing pulses[11].

In this presentation, we report on measurements of XPM effects on picosecond pulses for the pump-probe configuration, and for the SRS generation process.

INDUCED-FREQUENCY SHIFT OF CO-PROPAGATING OPTICAL PULSES

The first part of our research was to investigate spectral effects induced by a strong infrared pulse on a weak green pulse co-propagating in a nonlinear and dispersive optical fiber. These effects were studied for varying time delay between pulses at the optical fiber input, and increasing infrared pulse energy. A mode-locked Nd:YAG laser with a second harmonic crystal was used to produce 33-ps time-duration pulses at 1064-nm and 25-ps time-duration pulses at 532-nm . The optical path of each pulse was controlled using variable optical delays. The pulse energy of infrared pulses was adjusted in the range 1 nJ to 100 nJ, and the energy of green pulses was set to about 1 nJ.

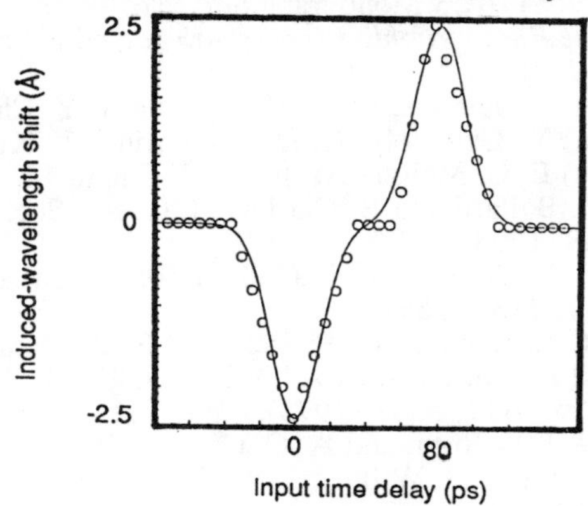

Fig. 1 Induced frequency shift versus the time delay between pump and probe pulses at the input of a 1-m long optical fiber.

The main effect of the XPM nonlinear interaction between the strong infrared pulse and the weak green pulse was to shift the carrier frequency of the green pulse. The induced-frequency shift versus the input delay between the infrared and green pulses is plotted in Fig. 1 (dots are for data, and the solid line is computed from a theory including XPM and walk-off[12]). As shown, the carrier frequency of the green pulse could be tuned both toward the shortest and longest wavelengths by varying the time delay between the infrared and green pulses at the optical fiber input. The maximum induced-frequency shift was found to increase linearly with the infrared pulse peak power. Induced-frequency shifts up to 4 Å have been observed. The blue and red induced-frequency shifts resulted from the combined effect of the cross phase modulation generated by the infrared pulse and pulse walk-off[12]. **This is a conclusive observation of XPM.**

CROSS-PHASE MODULATION & WALK-OFF EFFECTS ON RAMAN PULSES

The second part of our research was to evaluate XPM effects induced by 25-ps pump pulses at 532-nm on Raman pulses generated by SRS at 514.5-nm. Spectra of Raman pulses, which were measured for increasing pump energy at the output of a 1-m long optical fiber and a 6-m long optical fiber, are plotted in Figs. 2-a & 2-c and Figs. 2-b & 2-d, respectively. For pump intensities just above the SRS threshold, spectra of Raman pulses appeared broad, modulated, and symmetrical for both fibers (2-a & 2-b). For these low pump intensities the pulse walk off (6m corresponded to 2 walk-off lengths) did not lead to asymmetrical spectral broadenings (2-b). For higher pump intensities, Raman spectra became much wider (2-c & 2-d). In addition, spectra of Raman pulses generated in the long fiber were highly asymmetrical (2-d). The general features observed in Figs. 2 are characteristic of spectral broadenings arising from nonlinear phase modulations such as SPM and XPM. At the lowest intensities XPM could dominate, while at the highest intensities the SPM generated by the Raman pulses itself could be the most important.

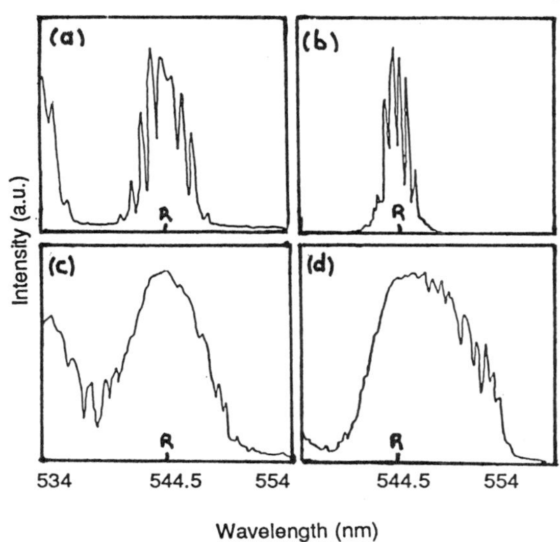

Fig. 2 Spectra of Raman pulses generated from the noise in optical fibers. Spectral features are characteristic of XPM and SPM broadenings.

Temporal measurements of the Raman generation process were perfomed to test whether the spectral asymmetry originated from the pump depletion re-shaping as in the case of longer pulses (> 500 ps)[5]. Pump and Raman profiles were measured at the output of a 17-m optical fiber using a 2-ps streak camera (Fig. 3). The dotted line is for a pump intensity at the SRS threshold, and the solid line for a higher pump intensity. The reduction of the leading edge (negative times) profile of the pump pulse in the high intensity case could be a pump depletion effect. Thus, in the case of short pulses and large GVD, the Raman pulse might not have time to completely deplete the pump edge. The effect of pump depletion re-shaping on spectral asymmetry could be less important with 25 ps pulses than with longer pulses.

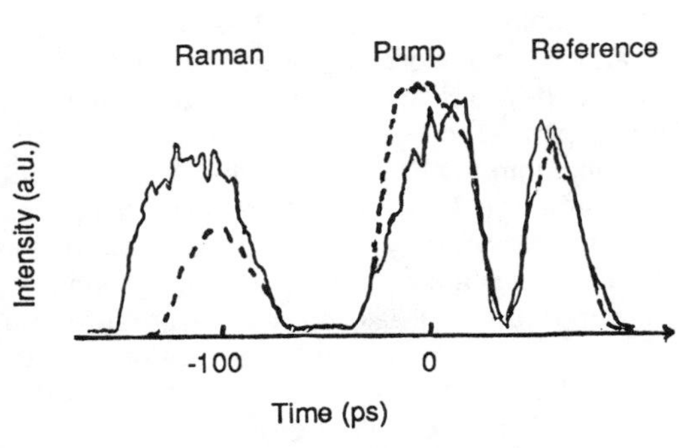

Fig. 3 Streak camera traces of the Raman process at the output of a 17-m long optical fiber.

CONCLUSIONS

XPM effects on spectra of picosecond pulses have been investigated for the cases of pump-probe and pump-Raman interactions. In the pump-probe configuration, the probe pulse can undergo a substantial induced-frequency shift which could be tuned by varying the time delay between pump and probe pulses at the optical fiber input. During the Raman generation by SRS, XPM and SPM significantly broadened Raman spectra. XPM appeared to dominate for pump intensities just above the Raman threshold. However, for larger pump intensities, the SPM generated by the Raman itself seemed to be the most important.

ACKNOWLEDGMENTS

This research is supported in part by Hamamatsu Photonics K.K..

REFERENCES

1. Cross-phase modulation has also been termed induced-phase modulation in earlier papers for the pump-probe configuration .
2. J. Gersten, R. Alfano and M. Belic, Phys. Rev. **A21**, 1222-1224 (1980).
3. J. Manassah, M. Mustafa, R. Alfano and P. Ho, Phys. Lett. **113A**, 242-247 (1985).
4. R. Alfano, Q. Li, T. Jimbo, J. Manassah and P. Ho, Opt. Lett. **11**, 626-628 (1986).
5. D. Schadt, B. Jaskorzynska, and U. Osterberg, J. Opt. Soc. Amer. **B3**,1257 (1986).
6. R. R. Alfano, P. L. Baldeck, F. Raccah and P. P. Ho, Appl. Opt. **26**, 3491-3492 (1987).
7. P.L. Baldeck, P.P. Ho, R.R. Alfano, Rev. Phys. Appl. **22**, december 1987.
8. M.N. Islam, L.F. Mollenauer, R.H. Stolen, J.R. Simpson, and H.T. Shang, Opt. Lett. **12**, 625-627 (1987).
9. J. T. Manassah, Appl. Opt. **26**, 3747-3749 (1987).
10. R.R. Alfano, Q.Z. Wang, T. Jimbo, and P.P. Ho, Phys. Rev. **A 35**, 459-462 (1987).
11. P.L. Baldeck and R.R. Alfano, J. Lightwave Technol. **L.T-5**, december (1987).
12. P.L. Baldeck, R.R. Alfano and G. Agrawal, submitted for publication.

PICOSECOND DYNAMICS OF ELECTRON TRANSFER AT SEMICONDUCTOR LIQUID JUNCTIONS*

J. J. Kasinski, L. Gomez-Jahn, L. Min, Q. Bao, and R. J. Dwayne Miller

Dept, of Chemistry, University of Rochester, Rochester, New York 14627

ABSTRACT

A surface restricted transient grating technique has been demonstrated as a sensitive probe of the ultrafast dynamics of surface reactions. Studies of interfacial electron transfer processes at single crystal n-TiO_2 (001) surfaces have shown that electron transfer is occurring to thermalized hole vacancies at the surface valence band edge. The main competing pathway is shown to be surface trap recombination and not Auger recombination, even at high carrier concentrations. Evidence is given for OH^- in the Helmholtz layer as the initial hole carrier acceptor.

INTRODUCTION

The abrupt phase discontinuity defined by a surface changes significantly the physics of numerous chemical processes. One of the most fundamental steps is that of electron transfer. The most successful treatments of this problem have assumed electron transfer processes occur directly from lattice sites at the first atomic surface layer involving electron or hole vacancies which are thermally equilibrated with the lattice i.e. thermalized surface band edge electron transfer processes. However, in the context of electron transfer processes at semiconductor surfaces it has recently been pointed out that unthermalized electron transfer processes (hot carrier effects) to below surface carrier sites could dominate under controlled junction conditions.[1] The two different mechanisms for interfacial electron transfer differ in their dynamic signature by over two orders of magnitude. Basically, hot carrier effects are expected to occur on a time scale faster than 10 picoseconds relative to thermalized carrier processes which occur on timescales slower than 100 picoseconds.

Recently, the surface restricted transient grating technique has been demonstrated as a sensitive probe of the ultrafast dynamics of the fundamental electron transfer process and competing solid state surface trapping at semiconductor liquid junctions.[2] In addition, this technique has been demonstrated as a unique probe of back electron transfer processes on a slower timescale.[3]

RESULTS AND DISCUSSION

The experimental setup is shown in figure 1. The above band gap excitation at 355 nm is used to write an ultrathin grating on the surface of single crystal (001) n-TiO_2 surfaces. This grating image is stored in the form of free carriers. Interfacial electron transfer processes and solid state competing processes which deplete the carrier density diminish the grating image. Monitoring the grating amplitude with a below band gap probe enables direct determination of interfacial electron transfer processes and surface state recombination velocities. Grating studies of n-TiO_2 in air have found non-exponential decays with 1/e lifetimes of 5.0 nsec[2] The decay is well described by diffusion controlled surface state trapping with a slow bulk recombination of 100 nsec. Controlled studies of surface adsorbed monolayers of hydroxide produced no effect on the observed trapping dynámics. The formation of a chemisorbed OH_s^- layer significantly increases the density of midgap surface states. The lack of an effect demonstrates that the site of high surface trapping is not the atomic surface layer but the surface deformation layer which extends 100Å from the surface. This result is in agreement with recent theoretical and experimental results which have shown that surface sites of titanium and oxygen do not produce mid gap energy states required for trapping.[4] The major surface state traps are oxygen defects which are more prevalent in the deformation layer than at the surface.

Direct, in-situ, studies of the same crystal, but forming a n-TiO_2/H_2O liquid junction with a 1 eV space charge field (500Å wide), are shown in figure 2. The most significant feature is the initial 460 psec decay component (~40%) which corresponds to the minority carrier lifetime within the space charge region. The quantum yield for interfacial electron transfer to hole vacancies generated within the space charge region is essentially unity. Thus, the hole depletion corresponds to the interfacial electron transfer dynamics. Theoretical estimates, within the thermalized surface band edge model for this electron transfer process, are in the sub-nanosecond timescale which is in agreement with the experimental results. This work indicates that OH^- in the Helmholtz layer is the initial hole acceptor and demonstrates that hot carrier effects are absent in the n-TiO_2/H_2O junction in contradiction to previous predictions of the effect for this system.[1] However, these results must be qualified by the amount of band bending occurring under the excitation conditions employed. Studies of lower doped samples have shown that the band bending in the data shown in figure 2 must be 50% or greater. The unbending of the bands would be expected to alter the space charge quantization effects. Higher sensitivity studies with resolution less than 100 fsec are currently in progress to determine if any hot carrier processes are occurring at the surface.

These are the first direct measurements of fundamental reaction processes at semiconductor liquid junctions. Similar studies of GaAs in aqueous and non-aqueous redox couples are in progress. On a preliminary basis the dynamics are a factor of ten faster than TiO_2 due to the higher carrier mobility. Such studies will provide a detailed description of electron transfer at surfaces, the role of surface states in this process, and rigorously test various aspects of electron transfer theory.

*This work was supported by the Department of Energy Office of Basic Sciences Gran#81-049.

REFERENCES

1. D.S. Boudreaux, F. Williams and A.J. Nozik, J. Appl. Phys. 51, 2158 (1980).
2. a) J.J. Kasinski, L. Gomez-Jahn, and R.J. Dwayne Miller, Denver, USA, ACS National Meeting 1987.
   b) J.J. Kasinski, L. Gomez-Jahn and R.J. Dwayne Miller, Phys. Rev. Lett., to be published.
3. S. Nakabayashi, S. Komuro, Y. Aoyagi and A. Kira, J. Phys. Chem. Lett. 91, 1697 (1987).
4. S. Mannix and M. Schmeits, Phys. Rev. B 30, 2202 (1984).

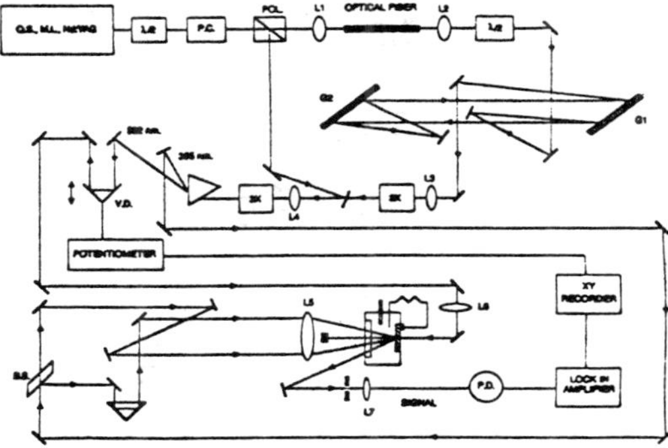

Figure 1 Experimental set up. The n-TiO_2 crystals are housed in a 3 electrode liquid junction photocell (PH=13.5) shown at the three beam crossing point. The pulses are 3psec in duration with excitation conditions less than .06 mj/cm^2. P.C.=pockel cell, pol=polarizer, L=lens, $\lambda/2$=half wave plate. V.D.=delay line, P.D.=Photodiode and G=grating.

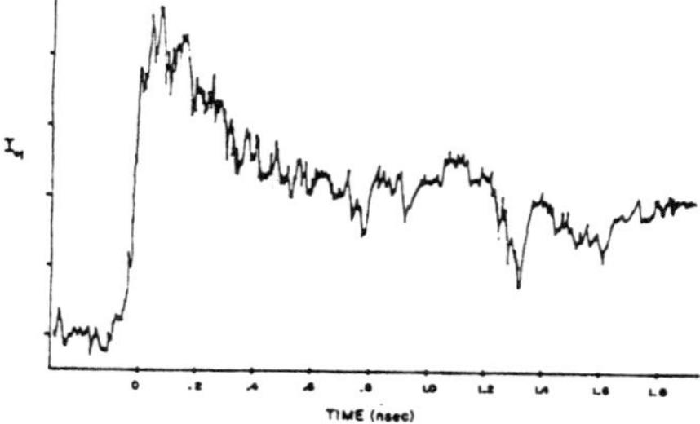

Figure 2 Transient grating studies of the n-TiO_2/H_2O liquid junction (donor concentration $1 \times 10^{19}/cm^3$).

TIME-RESOLVED STUDIES OF SOLVATION

John D. Simon† and Shyh-Gang Su
Department of Chemistry and Institute for Nonlinear Studies
University of California at San Diego, La Jolla, CA 92093

ABSTRACT

Solvent dynamics in alcohols are measured by examining the time dependent Stokes shift of the emission from the twisted intramolecular charge transfer state of bis(4-(dimethylamino)phenyl) sulphone (DMAPS). The solvation times deviate from the dielectric continuum prediction of τ_L, the longitudinal relaxation time of the solvent, but generally fall between τ_L and the Debye relaxation time, τ_D. The differences between the observed rates of solvation and those predicted by models based on a dielectric continuum description of the solvent reveal the importance of the underlying molecular aspects of solvation.

INTRODUCTION

Solvent fluctuations play an important role in chemical dynamics in solution. Interest in understanding the details of how solvent fluctuations affect chemical reactivity has prompted an effort to understand the details of nonequilibrium solvation. Experimentally these studies have largely focussed on measuring the time dependent Stokes shift of a probe molecule in polar solvents [1-5]. Due to the different charge distributions in the ground and excited electronic states, photoexcitation of a molecule generally results in a change in the magnitude and/or direction of the permanent dipole moment. As a result, the excited state charge distribution is created in the equilibrium solvation for the ground electronic state. With increasing time, the solvent restructures in response to the demands of the new charge distribution. This results in a lowering of the energy of the excited state and is revealed by a time dependent red shift (Stokes shift) in the emission spectrum.

In order to quantify the time dependent propertied of the emission spectrum, several workers have introduced the following correlation function [6,7].

$$C(t) = \frac{\nu(t) - \nu(\infty)}{\nu(0) - \nu(\infty)} \qquad 1$$

In the above equation, $\nu(t)$, $\nu(0)$, and $\nu(\infty)$, are the emission maxima at time t, t=0, and t=∞, respectively. The time dependent behavior of C(t) provides a means of examining the relaxation of the surrounding environment on the microscopic level.

In the present paper, the time dependent Stokes shift of DMAPS is measured in MeOH, EtOH, and ethylene glycol at several temperatures.

The simplest model for understanding the underlying solvent dynamics is to consider a dipole in a spherical cavity embedded in a dielectric continuum. Generally, the frequency dependent dielectric constant is expressed in a Debye form. In this case, it has been shown that C(t) should decay exponentially with a time constant of τ_L [6,7]. τ_L is

†NSF Presidential Young Investigator

related to the Debye relaxation time τ_D by $\tau_L = (\varepsilon_\infty/\varepsilon_o)\tau_D$ [8]. More detailed continuum theories show that the relaxation should be biexponential with time constants of τ_L and τ_D [9]. These models attribute the relaxation to rotational motion. Relaxation by translational motion has also been discussed and is predicted to dominate when $D\tau_D/R^2 > 1$ (D: self diffusion constant, R: cavity radius). In this case, C(t) is predicted to be nonexponential and relax to 1/e of its initial value on a time scale much faster than τ_L [10,11]. Finally, recent analytic models of charge solvation which try to include molecular details of the nearby solvent suggest that the solvation time will fall between τ_L and τ_D [12,13].

RESULTS AND DISCUSSION

In Figure 1, the time dependent emission spectrum of the TICT state of DMAPS is shown several times after excitation in ethanol at -20 C [2]. The emission spectrum shifts to decreasing energy with increasing time after excitation. In addition, the shape of the spectrum also changes.

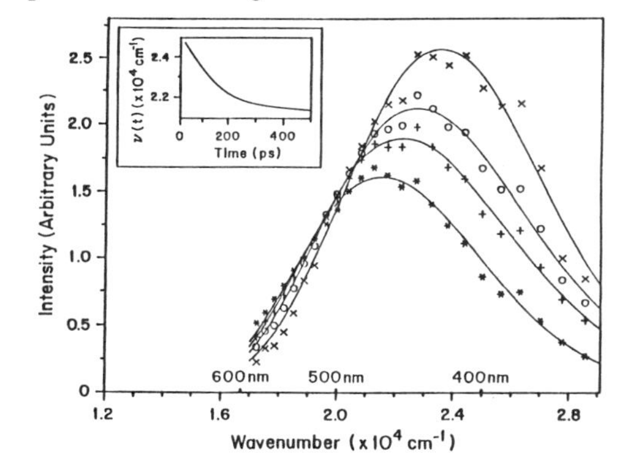

Figure 1: The emission spectrum of the TICT state of DMAPS at -20 C is plotted as a function of time: (x) 100 ps, (o) 150 ps, (+) 200 ps and (*) 400 ps. The solid lines are the best fit of the log normal line shape function to the experimental data. The insert is a plot of the emission maximum as a function of time.

From this data, C(t) can be generated. In Figure 2, ln[C(t)] is plotted for DMAPS in ethanol at -20 C. For comparison, $-t/\tau_L$ (the continuum prediction) and $-t/\tau_D$ are plotted. It is clear that the function C(t) is not well fit by t/τ_L but falls between the curves corresponding to the two limiting relaxation times.

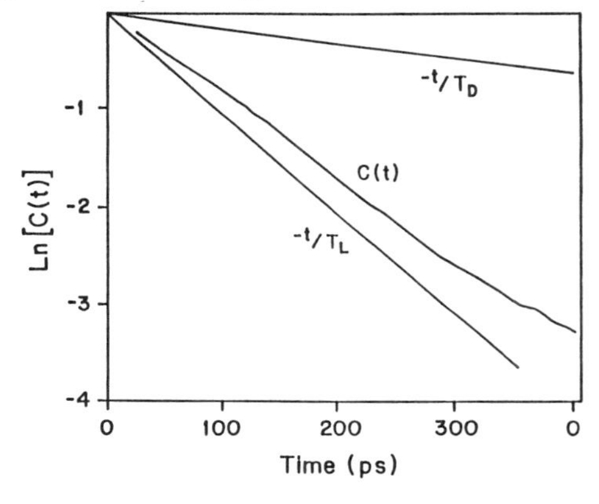

Figure 2: The log C(t) is plotted as a function of time for the TICT emission of DMAPS in ethanol at -20 C. For comparison, $-t/\tau_L$ and $-t/\tau_D$ are also plotted.

Temperature (K)	τ_s (psec)	τ_L (psec)	τ_D (psec)	ε_s	ε_∞
Ethanol					
273	43	60	337	26.0	4.65
263	64	72	456	29.8	4.74
253	113	96	632	31.4	4.85
243	200	132	899	33.8	4.97
233	356	185	1319	36.2	5.10
223	420	270	2001	38.7	5.24
Methanol					
233	67	41	290	49.2	6.91
223	106	56	414	52.7	7.19
213	197	81	609	56.4	7.51
Ethylene Glycol					
298	100	90	778	34.6	4

Table 1: Solvation times and dielectric data for methanol, ethanol, and ethylene glycol.

In Table 1, solvation data and the corresponding dielectric data are given for methanol, ethanol, and ethylene glycol. From this data, four general conclusion can be drawn. First, the time scale for solvation generally falls between τ_L and τ_D. Second, the relaxation dynamics as determined by C(t) are nonexponential. Third, with increasing value of the static dielectric constant, ε_o, the solvation time and τ_L become more dissimilar. Finally, in all the solvent studied, $D\tau_D/R^2 > 1$, yet the solvation time is longer than τ_L, in contrast to the predictions of polarization diffusion theory.

ACKNOWLEDGEMENT

This work is supported by the National Science Foundation, the Petroleum Research Foundation administered by the American Chemical Society, and the Office of Naval Research.

REFERENCES

[1] J. D. Simon , Acc. Chem. Res., submitted for publication
[2] J. D. Simon, S-G. Su, J. Phys. Chem. 91, 2693 (1987)
[3] V. Nagaragan, A. M. Brearley, T. J. Kang, P. F. Barbara, J. Chem. Phys. 86, 3183 (1987)
[4] E. W. Castner, M. Maroncelli, G. R. Fleming, J. Chem. Phys. 86, 1090 (1987)
[5] M. Maroncelli, G. R. Fleming, J. Chem. Phys. 86, 6221 (1987).
[6] B. Bagchi, D. W. Oxtoby, G. R. Fleming, Chem. Phys. 257 (1984)
[7] G. van der Zwan, J. T. Hynes, J. Phys. Chem. 89, 4181 (1985).
[8] H. Friedman, Trans. Farad. Soc. 79, 1465 (1983).
[9] P. Madden, D. J. Kivelson, J. Phys. Chem. 86, 4244 (1982).
[10] G. van der Zwan, J. T. Hynes, Physica A 121, 227 (1983).
[11] G. van der Zwan, J. T. Hynes, Chem. Phys. Lett. 101, 367 (1983).
[12] D. F. Calef, P. G. Wolynes, J. Chem. Phys. 78 4145 (1983)
[13] R. F. Loring, S. Mukamel, J. Chem. Phys. 87, 1272 (1987).

LINEAR AND NONLINEAR RESONANCE RAMAN STUDIES OF SUBPICOSECOND PHOTODISSOCIATION DYNAMICS

L. D. Ziegler and Y. C. Chung
Northeastern University, Boston MA 02115

ABSTRACT

Resonance Raman excitation profiles of rotationally resolved transitions are observed in the linear and nonlinear (hyper-) scattering of ammonia. V and J-specific dephasing times in the range from 300 to 40 femtoseconds reveal how rotational and vibrational degrees of freedom are coupled on the resonant photodissociative A state surface on this short time scale. The rates of photodissociation determined by this technique map out "dark" portions of the photoactive surface which are important to the mechanism of photodissociation on this surface.

INTRODUCTION

The analysis of the excitation frequency dependence of resonance rotational Raman transitions observed both in linear and nonlinear (hyper) Raman scattering is a sensitive measure of subpicosecond dynamical processes in isolated molecules. This technique is used to measure vibrational and rotational specific lifetimes of photodissociative levels of the lowest lying electronic state of ammonia (A state). The X → A absorption spectrum extends from 217 nm to 185 nm and exhibits a long progression in the ν_2 out-of-plane bending coordinate. Ammonia dissociates to NH_2 and H with unity quantum yield upon excitation to this quasibound surface.

Resonance Raman (RR) scattering intensity may be written as a sum of molecular transition polarizability elements C^k_q (k = 0,1,2, q = −k, ... k). When molecular states are described as Born-Oppenheimer rigid rotors the polarizability elements may be described in the Kramers-Heisenberg sum over states approach by:

$$\left| C^k_q \right|^2 = (2J+1)(2J''+1) \left| \sum_{J'} (-1)^{J'} (2J+1) \right.$$

$$\times \begin{Bmatrix} k & J'' & J \\ J' & 1 & 1 \end{Bmatrix} \begin{pmatrix} J'' & 1 & J' \\ -k & 0 & k \end{pmatrix} \begin{pmatrix} J' & 1 & J \\ -k & 0 & k \end{pmatrix} \tag{1}$$

$$\left. \times \frac{(M_z)^{ev'}_{gv''} \; (M_z)^{gv}_{ev'}}{(\nu_{gv,ev'} + \nu_{JK,J''K} - \nu_0) + i\Gamma^{J'K}_{ev}/2} \right|^2$$

In eq. (1) the resonance contribution of just one vibronic band of a parallel polarized electronic transition to the rovibrational scattering transition from vJK to v"J"K" is considered. J′ are the resonant intermediate rotational levels. For short-lived photodissociative molecular states, such as the A state of ammonia, the damping constant, $\Gamma_{ev'}^{J'K}$, is determined by the photodissociation rate and is thus simply related to the excited state lifetime. The technique described here accurately determines these quantum specific damping constants. For excitation in ammonia, only 1 vibrational level needs to be included in the sum over all states. By conservation of angular momentum considerations only 1 (S and O Raman transitions) or 2 (R and P Raman transitions) resonant rotational levels are required in the sum over rotational levels. Thus fits of eq. (1) to observed rotationally resolved scattering transitions as a function of excitation frequency (REPs) reveal v,J quantum specific damping constants i.e. lifetimes.

RESULTS AND DISCUSSION

The results of fitting the S transition REPs resonant with the second vibronic band of the X → A absorption system of NH_3 determine the following J-specific lifetimes:

J′	2	3	4	5	6	7	8
τ(fs)	141	100	91	78	75	70	68

The rate of photodissociation approximately doubles as J′ increases from 2 to 8 in the $v'_2 = 2$ band. These lifetimes are fractions of rotational periods.

We have also observed rotationally resolved resonance hyper-Raman (RHR) spectra of ammonia. Excitation frequencies in the blue are two-photon resonant with the A state. RHR intensities can be treated analogously to the description of linear RR scattering. In fact the analysis of A state resonant rotational RHR scattering intensities is easier than for the linear scattering spectra due to the simpler rotational structure of the higher order process. Only P, Q and R transitions are seen in the RHR spectra of ammonia.

RHR Excitation profile analysis of the second vibronic band reveals the same J-dependence of the photodissociation rates (within experimental error) as found in the linear analysis. This J-dependence is ascribed to centrifugal effects on the photoactive surface. The rates of photodissociation are controlled by the tunneling rate of the departing H atom through a barrier on the A state surface. Centrifugal forces contribute to a J-dependent lowering of this barrier. Consequently, the tunneling rates increase as J increases. J-dependence to the photodissociation rates indicates that the optically excited bending level is below the top of the barrier along the N-H reaction coordinate. The J-dependence of the corresponding vibronic band of ND_3 is one-half as great as that of NH_3 in accordance with this centrifugal model.

The results of linear and nonlinear resonance rotational Raman analyses may be combined to probe the subpicosecond dynamical behavior at other vibronic levels. The J-dependence of the photodissociation rates of the $v_2' = 0$ and 1 levels have been determined by linear RR analysis. $v_2' = 3$ rotationally dependent lifetimes have been determined via an RHR analysis. All of these vibronic bands exhibit photodissociation rates which increase as J increases. Furthermore, the $v_2' = 0$ and $v_2' = 1$ lifetimes are nearly identical (both range from 250-100 fs as a function of J) and then decrease for $v_2' = 2$ (140-70 fs) and $v_2' = 3$ (70-45 fs). We attribute these lifetime effects to the dependence of the barrier to photodissociation on the out of plane bending angle. With increasing number of bending quanta in the optically accessible level, the average deviation from planarity increases. The photodissociation barrier increases as the molecule deforms from planarity thus acounting for the observed shot-time rate dependence.

Support of the National Science Foundation and the Petroleum Research Fund is gratefully acknowledged.

REFERENCES

1. L. D. Ziegler, J. Chem. Phys. 84, 6013 (1986).
2. L. D. Ziegler, J. Chem. Phys. 86, 1703 (1987).
3. L. D. Ziegler and J. L. Roebber, CHem. Phys. Lett. 136, 377 (1987).
4. L. D. Ziegler, Y. C. Chung and Y. P. Zhang, J. Chem. Phys. 87, 4498 (1987).

ULTRAFAST ELECTRON TRANSFER: THE ROLE OF SOLVENT MOTION

P.F. Barbara, M.A. Kahlow, and W. Jarzęba*
Department of Chemistry, University of Minnesota, Minneapolis, MN 55455

ABSTRACT

This paper shows that the electron transfer (ET) time τ_{ET} of the intramolecular ET reaction of electronically excited bianthryl (BA) is not equal to the longitudinal relaxation time τ_l of the solvent in various polar aprotic solvents. It has been observed that microscopic solvation time τ_s is very similar to τ_{ET} for a broad range of polar aprotics.

INTRODUCTION

Electron transfer and solvation are fundamental processes which play an important role in various chemical and biological phenomena. These two processes are closely related, as the rate of electron transfer generally depends on the polar solute/solvent interactions which allow the solvent molecules to retard the progress of the reaction. In the case of small activation barrier ET reactions ($\Delta G << k_bT$) theory[1] shows that the ET time τ_{ET} ($1/k_{ET}$) should be equal to τ_s, the microscopic relaxation time of the solvent. Furthermore, according to the dielectric continuum model (DCM) of solvation τ_s should be equal to the longitudinal relaxation time $\tau_l=\tau_D\epsilon_\infty/\epsilon_s$ of the solvent (ϵ_s and ϵ_∞ are the static and infinite frequency dielectric constants, respectively, and τ_D is the dielectric relaxation time).

In this paper we attempt to test the DCM for aprotic solvents studying the kinetics of the ET of BA, which is an excellent prototype for ET reactions[2,3].

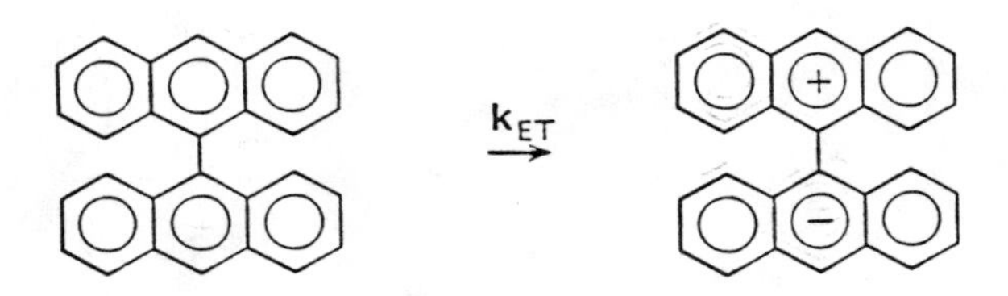

Fig.1 Electron transfer of bianthryl (BA).

EXPERIMENTAL

The fluorescence dynamics was observed using fluorescence upconversion technique apparatus which is discribed in detail in another article of this

*On leave from the Faculty of Chemistry, Jagiellonian University, Krakow, Poland.

proceedings by the same authors. Briefly, the laser source is a dye laser (Pyridine 2 dye, HITCI saturable absorber, 725 nm, 0.75 – 1.2 ps pulses) synchronously pumped by the second harmonic of actively modelocked Nd:YAG laser. The dye laser is amplified, at a 8.2 kHz repetition rate in copper vapor laser pumped amplifier. The doubled output of the amplified dye laser is focused in the sample. Time resolution is achieved by mixing the fluorescence with the 725 nm light in a KDP crystal to generate light at sum frequency (upconversion).

RESULTS AND DISCUSSION

The molecule bianthryl BA has two fluorescence bands in polar solvents. One of the bands is due to the nonpolar LE isomer, while the other is due to the

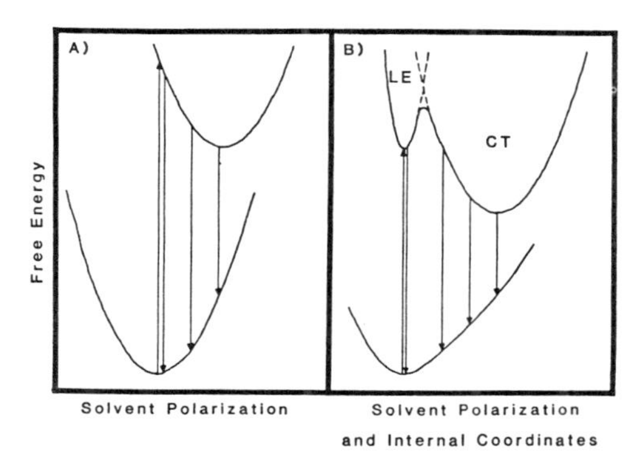

Fig. 2 A schematic representation of the nonequilibrium free energy as a function of instantaneous solvent polarization for (A) an ideal probe,(B) for bianthryl.

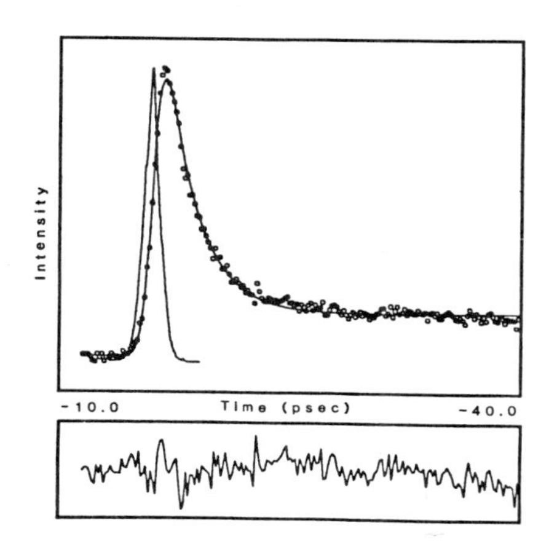

Fig. 3 Fluorescence transient for BA in propylene carbonate monitored at 420 nm.

highly polar CT isomer (Fig.2). The dynamic of the LE emission give a direct measure of the ET time τ_{ET} of electronically excited BA. Fig.3 shows emission dynamics of BA at 420 nm which should be identified with ET dynamics[2,3]. The long time–scale tail in the emission intensity is due to the equilibrium concentration of LE in S_1. We have measured τ_{ET} in various polar aprotic solvents as summarized in Table I. Polar aprotic solvents were studied as opposed to alcohols, because the solut/solvent interaction in polar aprotics is more similar to assumed in the DCM i.e. specific solvatation effects like hydrogen bonding are less important in aprotic solvents. Our data (Tab. I) on ET times of BA in polar aprotic solvents show that τ_{ET} is

Table I. A comparison of τ_{ET}, τ_s and τ_l in various solvents.

Solvent	τ_{ET}(ps)	τ_l(ps)	τ_s(ps)
acetonitrile	0.7	0.2	0.4–0.9[d]
propionitrile	1.2	0.3	1.1–1.5
butyronitrile	2.0	0.5	1.5–2.1
pentanitrile	4.4	0.7	3.6
PC[a]	3.4	5.1	4.9
triglyme	11.7	–	25
GTA[b]	750	–	820

a. Propylene carbonate. b. Glycerol triacetate.

c. The GTA numbers are from Nagarajan et al[5].t=11.5 °C. The rest of the values in the table were recorded at ambient temperature.

d. These values represent the range of observed τ_s values with different probes, see Kahlow et al[4].

not equal τ_l in general[2]. We have also measured the microscopic solvation times τ_s of the solvent by recording the excited state solvation times of various polar aromatic molecules that are not capable of ET[4]. Agreement between τ_{ET} and τ_s is fairly good, supporting the notion that the ET dynamics of BA are indeed controlled by transient solvation. The lack of agreement between τ_{ET} and τ_l apparently reflects the failure of the DCM to accurately describe transient solvation.

ACKNOWLEDGEMENT

Acknowledgement is made to the National Science Fundation (Grant No. CHE–8251158), to the National Institutes of Health (shared Instrument Grant No. RRO1439), Rohm and Haas, and Unisys.

REFERENCES

1. H. Sumi and R.A. Marcus, J. Chem. Phys., 84, 4894 (1986); D.F. Calef and P.G. Wolynes, J. Chem. Phys., 78, 470 (1983); E.M. Kosower and D. Huppert, Ann. Rev. Phys. Chem., 37, 127 (1986).
2. "Electron Transfer Times are not Equal to Longitudinal Relaxation Times in Polar Aprotic Solvents", M.A. Kahlow, T–J. Kang, and P.F. Barbara, J. Phys. Chem., in press.
3. D.W Anthon and J.H. Clark, J. Phys. Chem., 91, 3530 (1987).
4. "Transient Solvation of Polar Dye Molecules in Polar Aprotic Solvents", M.A. Kahlow, T–J. Kang, and P.F. Barbara, J. Chem. Phys., in press.
5. V. Nagarajan, A.M. Brearley, T–J. Kang, and P.F. Barbara, J. Chem. Phys., 86, 3183 (1987).

PICOSECOND RAMAN MEASUREMENTS OF INTRALIGAND ELECTRON TRANSFER IN SUBSTITUTED Ru(II) COMPLEXES

D. R. Anderson and J. B. Hopkins
Louisiana State University, Baton Rouge, LA 70803

ABSTRACT

The excited state metal to ligand charge transfer states of bipyridine and dimethyl-bipyridine have been studied using picosecond resonance Raman spectroscopy. It has been found that the rate of electron transfer between ligands in mixed ligand complexes is controlled by a slow electron transfer process probably due to tunneling.

INTRODUCTION

Recent electron transfer theory has determined two limiting cases for rates of electron transfer[1]. When formulated in the language of radiationless transition theory the limits can be viewed as arising from: (1) fast rates above the barrier to electron transfer where the rates are determined by coupling and vibrational mode dependent, density weighted Franck-Condon factors, (2) slow rates below the barrier where electron transfer occurs primarily through tunneling processes. In this paper these ideas are investigated with respect to their application in describing the rates of electron transfer between ligands in mixed ligand complexes of substituted Ruthenium (II) molecules. To this end, Raman spectroscopy has been used to measure the rates of intraligand electron transfer with picosecond time resolution.

EXPERIMENTIAL

All ruthenium complexes used were synthesized using techniques found in the literature[2]. All products were analyzed using mass spectroscopy to ensure that mixtures of the compounds were not being used.

The laser system used in these experiments employs the amplification of a pulse from a 100 MHz mode-locked Nd:YAG laser by a regenerative amplifier,[3,4] as shown in Figure 1. An optical fiber is used in this system to allow temporal pulse compression down to ≈25 ps without the use of a grating pair. In our experiments, the third harmonic (3547 Å) of the Nd:YAG laser was used at a 2 kHz repetition rate to produce the metal to ligand charge transfer state and obtain the resonance Raman spectra of the transient thus produced.

RESULTS AND DISCUSSION

The picosecond Raman spectra of the complexes $Ru(bpy)_3^{2+}$,

$Ru(bpy)_2(dimethylbpy)^{2+}$, $Ru(bpy)(dimethylbpy)_2^{2+}$, and $(Ru(dimethylbpy)_3^{2+}$ are shown in Figure 2 from top to bottom respectively The assignments of the bands are given in the figure with respect to ligand parentage B = bipyridine, M = dimethylbipyridine, and electronic state E = excited state, G = ground state. The excited state assignments were made by measuring the laser power dependence of the Raman intensities. The ligand parentage was assigned by comparing the frequencies observed in the mixed ligand complexes to those of the tris substituted complexes $Ru(bpy)_3^{2+}$ and $Ru(dimethylbpy)_3^{2+}$ shown in Figure 2. For the mixed complexes in the two center plots of Figure 2, it is clearly evident that there are excited state bands corresponding to the MLCT states of both bipyridine and dimethylbipyridine ligands. In the lower frequency region of the spectrum (not shown) there are additional peaks at 726 cm^{-1} and 739 cm^{-1} which correspond to the dimethylbipyridine and bipyridine excited states respectively.

We have measured the rates of intraligand electron transfer by varying the observation timescale of the Raman experiment. This is achieved by reducing the frequency chirp created by the optical fiber. In this way, the pulsewidth of the laser can be changed from 25 ps to 150 ps. Over this timescale no change is observed for the relative intensities of the bipyridine and dimethylbipyridine excited state bands. In fact, recent nanosecond[2] Raman experiments on the same complexes find approximately the same relative intensities for these bands as the picosecond result. The rate of electron transfer therefore must be on the order $\leq 10^9\ s^{-1}$. This relatively slow electron transfer rate is taken to indicate that the coupling between ligands is small and/or that the reaction is occuring below the barrier to electron transfer probably by way of a tunneling process.

CONCLUSION

Rates for intraligand electron transfer between bipyridine and dimethylbipyridine complexes of Ruthenium (II) have been found to be $\leq 10^9 s^{-1}$. These results have been interpreted to indicate that electron transfer between ligands is probably occuring in the slower limit by way of tunneling. These results can also be taken to indicate that the metal to ligand change transfer states of similar complexes rarely, if ever, exhibit delocalized electron behavior since the intraligand coupling appears to be too small to facilitate fast electron communication.

REFERENCES

1. M. Bixon, J. Jortner, Far. Disc. Chem. Soc., 74, 17 (1982).

2. Patricia A. Mabrouk and Mark S. Wrighton, Inorg. Chem., 25, 526 (1986).

3. Y. J. Chang, C. Veas, and J. B. Hopkins, Appl. Phys. Lett., 49, 1758 (1986).

4. Y. J. Chang, D. R. Anderson, and J. B. Hopkins, International Conference on Lasers 1986, R. W. McMillan, p.169, STS Press (1987).

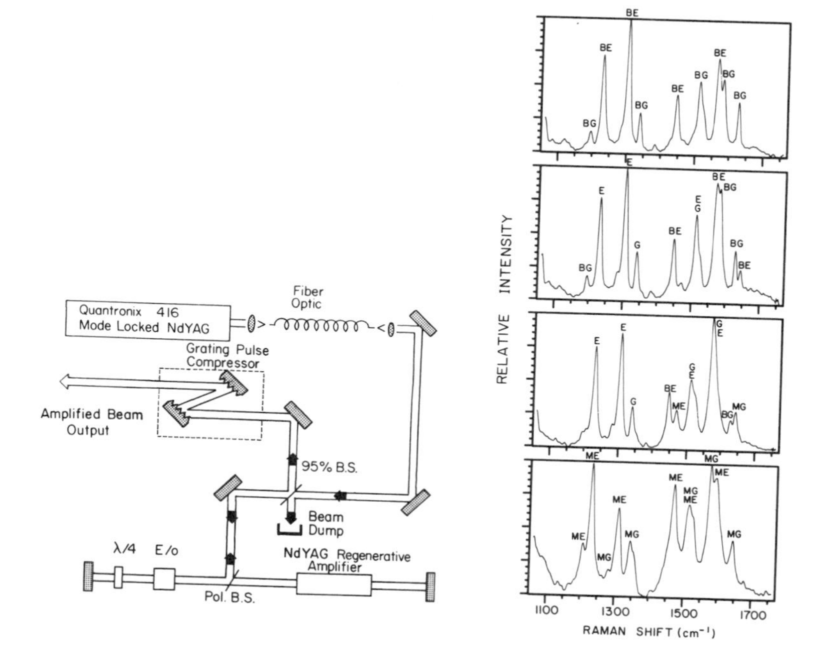

Figure 1

Figure 2

FLUORESCENCE SPECTROSCOPY OF ULTRASHORT-LIVED STATES OF AROMATIC MOLECULES

R. Jefferson Babbitt, Co-Jen Ho, Andrew J. Kaziska,
Maria I. Shchuka and Michael R. Topp

University of Pennsylvania, Philadelphia, Pennsylvania 19104-6323

INTRODUCTION

Reports in the literature have placed the relevant time regime for vibrational energy relaxation in fluid solution at room temperature in the range 0.1-1000 ps, depending on the system.(1) To examine this phenomenon for large aromatic molecules, it is necessary to determine the spectra of molecular excited states on a picosecond time scale. The present approach involves the isolation of fluorescence signals from states with intrinsically short decay times, which then act as probes of the local environment.

CORONENE

We have recently calibrated the $S_0(A_{1g}) \longleftrightarrow S_2(B_{1u})$ transition of coronene, both in absorption and emission in the condensed phase.(2) The $S_0 \rightarrow S_2$ excitation spectrum for $S_1 \rightarrow S_0$ emission has been recorded for different molecular sites in *n*-heptane at 12 K, and also under supersonic jet conditions.(3) The linewidths and lineshapes in excitation indicate a classic case of statistical limit coupling, having linewidths (2.5-5 cm^{-1}) consistent with a minimum electronic relaxation time of 1-2 ps. Direct fluorescence spectroscopy on the $S_2 \longrightarrow S_0$ transition following excitation of different features of the $S_0 \longrightarrow S_2$ absorption spectrum shows for all temperatures studied (12-300 K) that vibrational energy relaxation is incomplete on the time scale of the S_2 lifetime. The lower temperature data, including 77 K, showed fluorescence line "narrowing" consistent with the selection of homogeneous linewidths from the inhomogeneous profile, which did not undergo diffusive site relaxation on the time scale of the lifetime of the S_2 state. Thus, as Fig. 1 shows, excitation into higher vibrational levels of S_2 generates substantial amounts of "hot band" structure, and the dephasing-limited bandwidth of 125 cm^{-1} at 77K is increased by vibrational sequence congestion. In this respect the molecule, while in S_2, is in an adiabatic limit, very similarly to numerous cases which have been demonstrated under supersonic jet conditions. A key point about these spectra is that, in the region to longer wavelengths than ≈360 nm, the only differences are due to relatively subtle

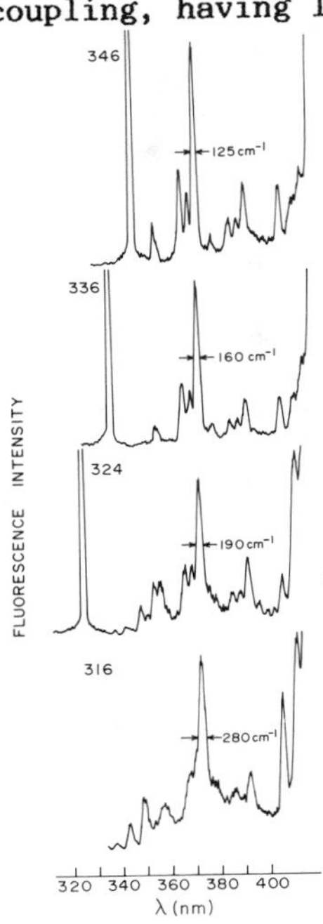

bandwidth changes. Hence only the shorter wavelength region carries the unambiguous signature of vibrational-level dependent emission.

METAL-FREE TETRAPHENYL PORPHINE (H_2TPP)

The more usual approach to the study of fluorescence from short-lived electronic states of large molecules uses consecutive two-photon excitation. This eliminates scattered laser radiation from the detection region, and the entire emission spectrum can be scanned at high sensitivity.

The Soret band emission spectra of H_2TPP and similar species have long been known from two-photon excitation experiments.[4] However, since these spectra exhibit nearly symmetrical emission profiles of ≈1000 cm^{-1} FWHM, there is always the question of the relative contributions from homogeneous and inhomogeneous broadening. A typical result for the Soret band emission spectrum (B_x; $S_3 \longrightarrow S_0$), obtained by excitation of H_2TPP in fluid solution by two photons at 532 nm, is shown in Fig. 2. The total excitation energy in this case was 34,100 cm^{-1} (≡293 nm), corresponding to ≈10,000 cm^{-1} excess energy in S_1. Like many other cases we have studied,[5] the emission spectrum for such high internal energies has no resolvable vibrational structure. Also, since the admixture of homogeneous and inhomogeneous broadening in such a spectrum, which persists for >1 ps, has not been determined, the emission in principle could correspond to S_2 states in any intermediate stage of vibrational relaxation.

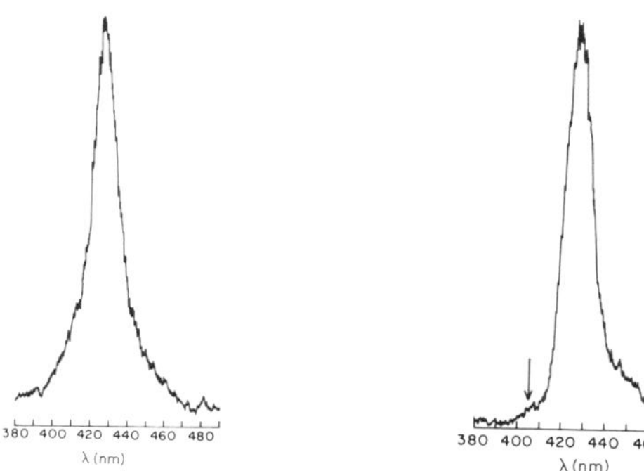

The second spectrum results from irradiation of H_2TPP with 532 and 1064 nm pulses, in such a way that any contribution from double 532 nm excitation was suppressed. The effective maximum excitation energy, corresponding to S_1(300K) + 9400 cm^{-1}, is 24,700 cm^{-1} (≡405 nm), equivalent to ≈1000 cm^{-1} excess energy in S_3. The spectrum is now seen to be unsymmetrical, with a moderately sharp cut-off on the short-wavelength edge, to slightly longer wavelengths than the arrow indicating the maximum available energy. This spectrum indicates a substantially narrower homogeneous linewidth (i.e. ≈400 cm^{-1}). The lack of emission precisely at the position of the arrow is a consequence of the symmetry of this molecule (i.e. D_{2h}), so that the $S_1(Q_x) \longrightarrow S_3(B_x)$

transition is vibronically induced through ungerade vibrational modes. Since these modes have zero Franck-Condon activity in the $S_2 \longrightarrow S_0$ emission spectrum, no prominent vibrational "hot" bands are observed, and the spectrum is diffuse. The short-wavelength edge of the spectrum therefore still represents only an upper limit on the intrinsic bandwidth. Finally, although the lifetime of the S_3 state is not known exactly, the difference between the two spectra again clearly shows that vibrational relaxation is substantially incomplete on a picosecond time scale.

3,4,9,10-DIBENZPYRENE

By contrast, the case of a non-centrosymmetric molecule, 3,4,9,10-dibenzpyrene (C_{2v}) is shown in Fig. 3.[5] This particular sample was irradiated by nanosecond pulses at 355 nm and 1064 nm. The effective total energy was equivalent to the energy of thermalized S_1 (432 nm) and a 1064 nm photon (i.e. 32,550 cm^{-1} ≡ 307.3 nm). This corresponds to a region on the low-energy edge of the molecular S_4 state, to that a mixture of S_3 and S_4 states is excited out of the inhomogeneous bandwidth. In this case, since the various transitions are all symmetry allowed, the spectrum clearly shows a quasi-resonant feature near 307.3 nm, with attendant vibronic structure, due to a mixture of S_4 and S_3 states. The fluorescence is vibrationally unrelaxed, and line-narrowed, indicating that the 180 cm^{-1} bandwidth of this spectrum, recorded in EPA glass at 77 K, is due to a convolution of linewidths due to non-diffusive dephasing and the relaxation time of the subpicosecond excited states. This sets a minimum possible lifetime on the states emitting at 307.3 nm of 60 fs. Similar data obtained at 300 K show an increase in the linewidth to about 320 cm^{-1}, which mostly represents an increase in the dephasing rate, although the extent of the inhomogeneous contribution resulting from molecular site relaxation on the time scale of the electronic relaxation time has not yet been determined.

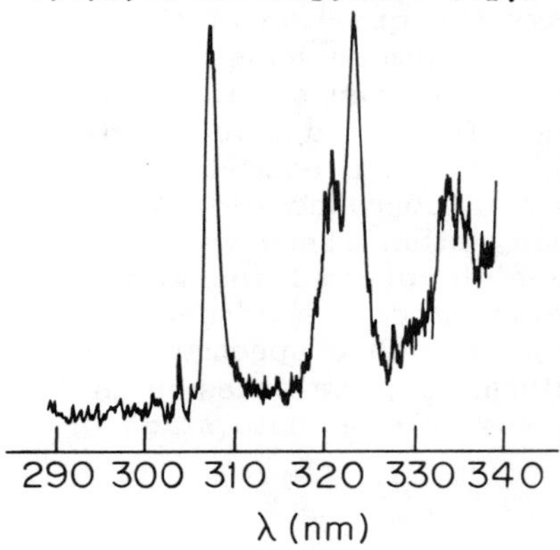

References:
1. V.E. Bondybey, **Ann. Rev. Phys. Chem.,** 35, 591 (1984); E.J. Heilweil, M.P. Cassassa, R.R. Cavanaugh and J.L. Stephenson, **J. Chem. Phys.,** 82, 5216 (1985).
2. C.J. Ho, R.J. Babbitt and M.R. Topp, **J. Phys. Chem.,** 91, 5599 (1987).
3. R.J. Babbitt, C-J. Ho and M.R. Topp, **J. Phys. Chem.,** (in press).
4. S. Tobita, K. Kayii and I. Tanaka, A.C.S. Symposium Series No. 321, Porphyrins, Excited States and Dynamics, Ch. 15, p. 219 (1986).
5. H.B. Lin and M.R. Topp, **Chem. Phys. Lett.,** 48, 251 (1977); M.R. Topp, H.B. Lin and K.J. Choi, **Chem. Phys.,** 60, 47 (1981).

Photodissociation of $Cr(CO)_6$ in Solution: Solvation Dynamics of $Cr(CO)_5$

John D. Simon† and Xiaoliang Xie
Department of Chemistry and Institute for Nonlinear Studies
University of California at San Diego, La Jolla, CA 92093

ABSTRACT

Picosecond absorption spectroscopy is used to examine the dynamics of formation of $Cr(CO)_5$(solvent) generated by photolysis of $Cr(CO)_6$ in neat alcohol and hydrocarbon solvents. In several solvents, the time resolved absorption data show that the photodissociation results in the formation of a distribution of solvated $Cr(CO)_5$ intermediates.

INTRODUCTION

The dynamics of solvation has been the topic of several recent experimental [1] and theoretical [2] works. These studies clearly show that the role of the solvent in both relaxation processes and chemical reactions cannot be completely described in terms of bulk properties of the fluid. A complete understanding requires a molecular picture of the solvating medium. In this paper we demonstrate that molecules which undergo rapid photodissociation followed by solvent coordination can reveal details of solvation dynamics on the molecular level. The molecule studied is $Cr(CO)_6$. This molecule was chosen for four major reasons.

(1) The photodissociation quantum yield is very high, ≈ 0.7 [3]. As a result complications resulting from geminate recombination are reduced.

(2) The absorption spectrum of the intermediate $Cr(CO)_5S$ is sensitive to the solvent ligand [4].

(3) The transient intermediate $Cr(CO)_5S$ is stable onto the nanosecond time scale [5].

(4) From matrix isolation studies and theoretical calculations, $Cr(CO)_5$ retains a C_{4v} geometry [6], suggesting that isomerization between C_{4v} and D_{3h} structures is not rate limiting in the solvation process.

EXPERIMENTAL

A frequency doubled mode locked CW Nd^{+3}:YAG laser is used to synchronously pump a R6G/DQOCI dye laser producing visible laser pulses with a FWHM of ≈ 0.8 ps. These pulses are amplified by a three stage longitudinally pumped pulsed dye amplifier. After amplification, the pulses are <1 ps (FWHM), ≈ 1-2 mJ/pulse, 600 nm, and have a repetition rate of 20 Hz.

The light is frequency doubled to 300 nm (0.1 mJ), directed down a variable delay line and focussed on the sample. The remaining red light is used to generate a picosecond continuum. The continuum light is used to record transient electronic spectra in a standard double beam configuration. Narrow band interference filters are used to select the desired

†NSF Presidential Young Investigator

probe wavelength.

The instrument response is determined by measuring the rise time of the $S_1 \longrightarrow S_N$ absorption of t-stilbene.

RESULTS AND DISCUSSION

In Figure 1, the transient absorption spectra recorded 50 ps after the photolysis of $Cr(CO)_6$ in room temperature methanol and cyclohexane solutions are shown. In methanol and cyclohexane, an absorption maximum of 460 nm and 505 nm, respectively is observed. These spectra correspond to the solvated complexes, $Cr(CO)_5(MeOH)$ and $Cr(CO)_5$(cyclohexane). No change in band position is observed from 50 ps to several nanoseconds after photolysis. The position of the absorption band of the alcohol complex is similar in the series of primary alcohols (methanol-hexanol).

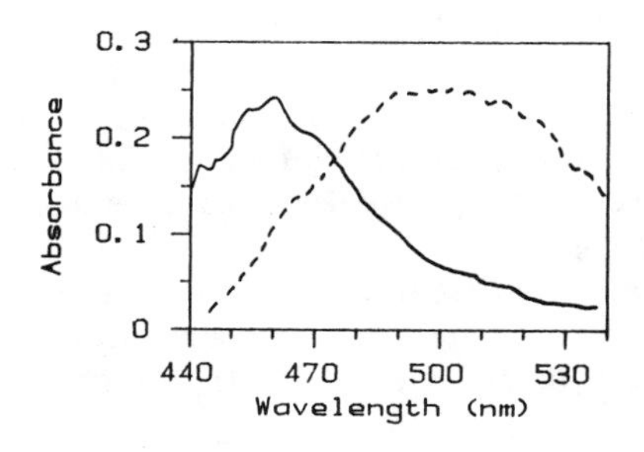

Figure 1: Continuum absorption spectra of $Cr(CO)_5$(cyclohexane) (----) and $Cr(CO)_5(MeOH)$ (——) recorded 50 ps after the photolysis of $Cr(CO)_6$ in cyclohexane and methanol, respectively.

In order to monitor the dynamics of solvation, the kinetics were monitored at 460 nm and 505 nm for the photolysis in methanol and cyclohexane, respectively [5]. The kinetics observed at 460 nm in methanol are shown in Figure 2.

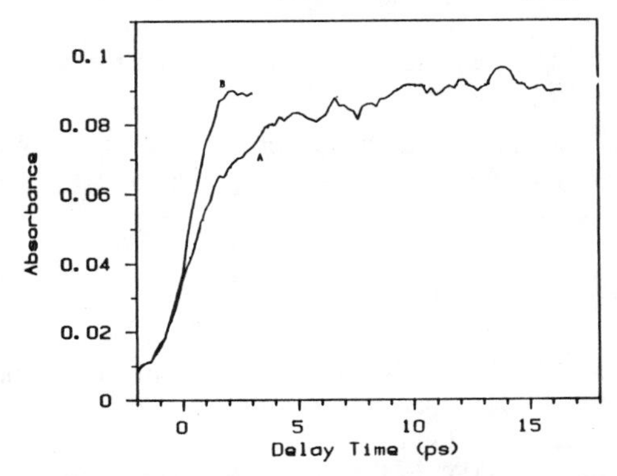

Figure 2: The time dependence of the transient absorption spectrum of $Cr(CO)_5(MeOH)$ observed after photolysis of $Cr(CO)_6$ in methanol. (A) The experimentally measured absorbance as a function of delay time between the excitation and 460 nm probe pulse. (B) Instrument response function.

In cyclohexane, the transient absorption kinetics at 505 nm track the instrument response. We are unable to resolve the formation dynamics of this solvated intermediate. From this result, we conclude that the rise time observed for the $Cr(CO)_5(MeOH)$ complex reflects the time required for the local solvent to reorganize and coordinate to the ground state $Cr(CO)_5$ fragment.

In longer chain alcohols, photodissociation can result in initial coordination of either the hydroxyl group or an alkyl group. Studies in propanol, butanol, and pentanol clearly show that photodissociation of $Cr(CO)_6$ results in formation of a distribution of hydroxyl and alkyl coordinated $Cr(CO)_5S$ intermediates [7,8]. The thermodynamically most stable complex involves coordination of the hydroxyl end of the solvent molecule. Thus, the initially formed $Cr(CO)_5$(alkyl) will restructure to form the more stable $Cr(CO)_5$(hydroxyl) product. The decay of the initially formed alkyl complex can be monitored by recording transient absorption kinetics at a wavelength where $Cr(CO)_5$(alkyl) has a stronger absorbance than $Cr(CO)_5$(hydroxyl). From Figure 1, the region around 520 nm satisfies this criterion. In Figure 3, we present the kinetics at 520 nm for the photolysis of $Cr(CO)_6$ in pentanol.

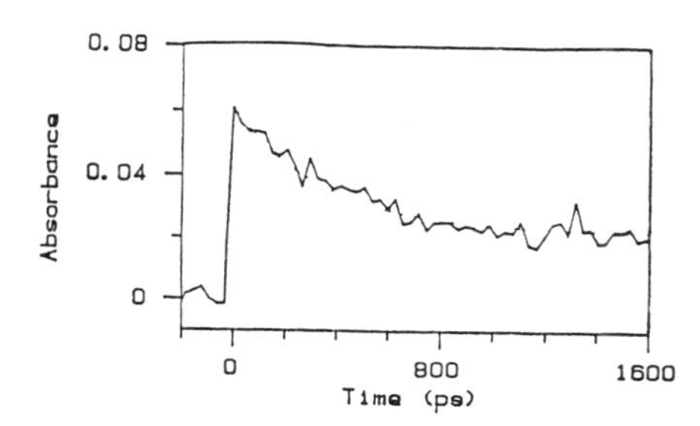

Figure 3: Transient absorption kinetics for the photolysis of $Cr(CO)_6$ in pentanol at 520 nm. The absorption rises with the instrument response and slowly decays to a constant value. For delay times longer than 800 ps, no change in the magnitude of the transient absorbance is observed.

The transient absorption at 520 nm rises with the instrument response, similar to that observed in neat hydrocarbon solvent. The decay of the signal reflects the restructuring of the alkyl complex to the more stable $Cr(CO)_5$(hydroxyl) intermediate. Examination of the kinetics at 460 nm (the absorption maximum of the hydroxyl coordinated species) confirms this conclusion. The rate of structural reorganization is dependent on the length of the alkyl chain.

ACKNOWLEDGEMENT

This work is supported by the National Science Foundation and the Office of Naval Rseearch. We also thank Newport Corporation and Klinger Scientific for donating equipment used in these experiments.

REFERENCES

[1] J. D. Simon , Acc. Chem. Res., submitted for publication

[2] J. T. Hynes, Theory of Chemical Reactions, Volume IV, CRC Press, Boca Raton (1986).

[3] J. Nasielski, A. Colas. Inorg. Chem. 17, 237 (1978).

[4] J. M. Kelley, D. V. Bent, H. Hermann, D. Schulte-Frohlinde, E. K. Gustorf, J. Organomet. Chem. 69, 259 (1974).

[5] J. D. Simon, X. Xie, J. Phys. Chem. 90, 6751 (1986).

[6] J. J. Turner, J. K. Burdett, R. N. Perutz, M. Poliakoff, Pure and App. Chem. 49, 271 (1977).

[7] J. D. Simon, X. Xie, J. Phys. Chem., in press.

[8] J. D. Simon, X. Xie, J. Amer. Chem. Soc., in preparation.

SUBPICOSECOND EMISSION SPECTROSCOPY IN THE ULTRAVIOLET BY TIME GATED UPCONVERSION.

Paul F. Barbara and Michael A. Kahlow
Dept. of Chemistry, University of Minnesota
Minneapolis, Minnesota 55455

ABSTRACT

We have built an apparatus to measure emission transients with ultrafast time resolution, following ultraviolet excitation. This system uses fluorescence upconversion to attain nearly pulsewidth limited time resolution. Laser pulses from a synchronously pumped dye laser are amplified at 8.2 kHz to ≥ 1 μJ/pulse to increase the efficiency of the frequency doubling and sum frequency generation processes. Results are shown for the same system with two dye lasers. The first is a commercially available laser with a pulsewidth of 1 picosecond; the other is a home built laser with a pulsewidth of 70 femtoseconds.

INTRODUCTION

Time gated fluorescence upconversion is a time resolved emission technique which allows one to measure emission intensity v. time with a time resolution comparable to the laser pulse width[1-3]. A sample is excited by either a fraction of the ultrafast laser fundamental or a harmonic. The resulting sample emission intensity is measured by mixing the emission photons (frequency ω_{fl}) with the laser fundamental gate pulse (frequency ω_l) to generate light at the sum frequency $\omega_{uc} = \omega_{fl} + \omega_l$. The intensity of the sum frequency light is proportional to the fluorescence intensity at a time determined by the relative delay between excitation and gate pulses. Thus, by scanning the time delay between the two pulses with an optical delay line, we can determine the decay of the fluorescence intensity with time.

In this paper, we discuss the design of the instrument; the dye laser, dye laser amplifier, and the upconversion optics. Following this are examples of how this apparatus can be used in the study of ultrafast chemistry in solution.

DYE LASER

This system has been used with two different dye lasers. The first is a commercially available dual jet dye laser (Coherent 702). Used with Pyridine 2 as the gain dye and HITCI as the saturable absorber, we obtain pulses of roughly 1 picosecond full width half maximum (FWHM) at 725 nm, with a total output of 60 mW. The second laser is based on the design of Dawson et. al.[4]. With Rhodamine 6G and DQOCI, this design has produced 70 femtosecond pulses, using intracavity prisms for compensation of group velocity dispersion. We have found that the dye combination of Styryl 8 and HITCI produces 70 femtosecond pulses with no intracavity prisms[5]. The laser output is 20 mW total power at 795 nm, and is very stable.

AMPLIFIER AND UPCONVERSION OPTICS

Pulses from the dye laser are amplified by a copper vapor laser pumped dye laser amplifier, similar to the design of Knox et. al.[5]. This amplifier, pictured in **Figure 1**, uses a 1.2 mm jet of Pyridine 2 (for 725 nm operation) or Styryl 8 (for 795 nm) as the gain medium, produces pulses of 1.3 – 2.4 μJ at 8.2 kHz, for a total single pulse amplification of 2000 – 6000 fold.

The fluorescence upconversion optics are shown in **Figure 2**. Pulses from the amplifier are frequency doubled. The resulting UV light is passed through a delay stage, and focused onto the sample. The resulting fluorescence is collected and combined collinearly with the residual fundamental (which has travelled through a fixed delay). Both fluorescence and fundamental are focused onto the KDP upconversion crystal. The resulting sum frequency light is selected with a polarizer, prism and monochromator, and detected with a solar blind photomultiplier tube.

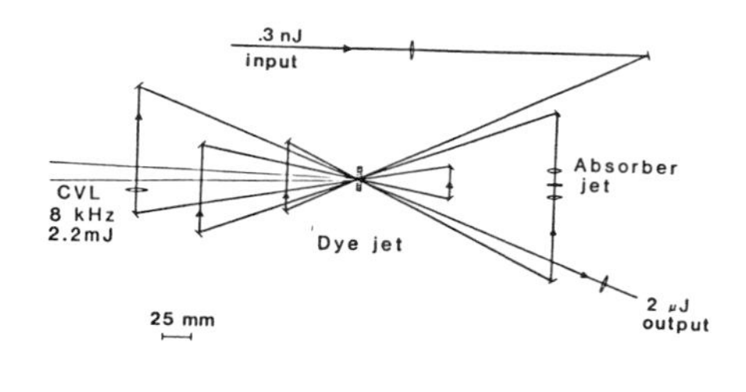

Figure 1. Scale design of the dye laser amplifier. Single ultrafast laser pulses are amplified roughly 6000X by six passes through a thick dye jet pumped by a copper vapor laser. The absorber is not needed for amplification of the 790 nm femtosecond pulses.

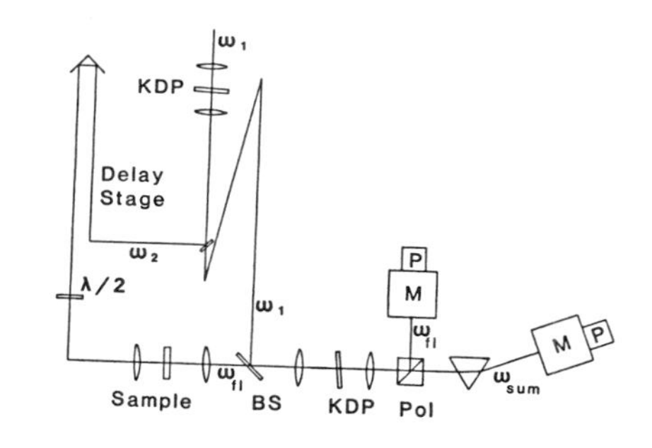

Figure 2. Design of the upconversion system. Elements are: $\lambda/2$, halfwave plate; BS, dichroic beamsplitter; Pol, Glan–Thompson polarizer; M, monochromator; and P, photomultiplier tube.

EXAMPLES

The fluorescence spectra of molecules in polar solution often show a fast red shift immediately after excitation[6]. This is due to the relaxation of the solvent polarization about the changed dipole moment of the excited molecule. Thus, one can monitor the evolution of microscopic solvent polarization by time resolved fluorescence.

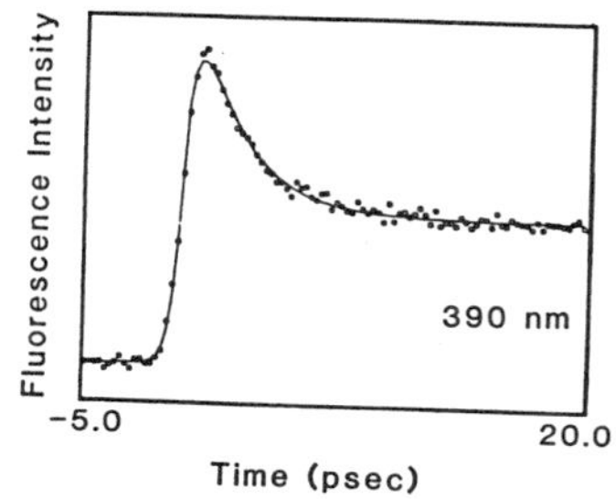

Figure 3. Emission intensity v. time transient for coumarin 102 in ethyl acetate, monitored at 390 nm. The picosecond dye laser was used for this experiment. The dots are the data points; the solid line is a best fit to the data.

Figure 3 shows the fluorescence v. time transient for coumarin 102 in ethyl acetate, monitored at 390 nm. This is on the blue edge of the fluorescence spectrum. Results at this and other wavelengths are consistent with a rapid red shift of the fluorescence.

WORKING ON THE FEMTOSECOND TIME SCALE

Attempting this experiment with an amplified femtosecond dye laser adds a number of complications. Currently, group velocity mismatch (in the nonlinear optical processes) and group velocity dispersion broaden our instrument response function to 280 femtoseconds, starting with a 70 fsec pulse. We are currently analyzing and correcting for the sources of time broadening in this system[2].

In the meantime, we have begun to use the apparatus with this faster time resolution. **Figure 4** shows the fluroescence decay of coumarin 102 in acetonitrile, measured with the 70 fsec laser. The instrument response is 350 fsec. We hope to improve on both the time resolution and signal to noise ratio of this apparatus in the near future.

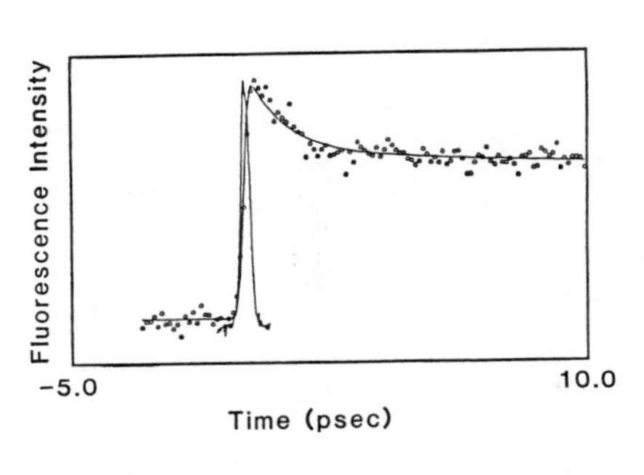

Figure 4. Emission intensity v. time transient for coumarin 102 in acetonitrile, monitored at 420 nm. This experiment was done using the femtosecond laser. The spike at zero time is the cross correlation between the excitation and probe beams, taken as the instrument response function. The fast decay is single exponential within the signal to noise.

CONCLUSION

We have built an apparatus to time resolve emission on the subpicosecond time scale with ultraviolet excitation, using a kilohertz repetition rate amplified dye laser. We have used this instrument to observe the decay of solvent polarization in polar solvents on the 1 psec time scale, and hope to extend this to the femtosecond regime in the near future.

ACKNOWLEDGEMENTS

Acknowledgement is made to the National Science Foundation (Grant No. CHE–8251158) and the Office of Naval Research.

REFERENCES

1. P.F. Barbara and A.J.G. Strandjord, ACS Symp. Ser. 236, 183 (1983).
2. M.A. Kahlow, W. Jarzeba and P.F. Barbara, "Ultrafast Emission Spectroscopy in the Ultraviolet by Timegated Upconversion", in preparation, to be submitted to Rev. Sci. Inst.
3. E.W. Castner, Jr., M. Maroncelli and G.R Fleming, J. Chem. Phys. 86, 1090 (1987).
4. M.D. Dawson, T.F. Boggess and A.L. Smirl, Opt. Lett. 12, 254 (1987).
5. W.H. Knox, J. Opt. Soc. Am. B 4, 1771 (1987).
6. M.A. Kahlow, T.J. Kang and P.F. Barbara, "Transient Solvation of Polar Dye Molecules in Polar Aprotic Solvents", J. Chem. Phys., in press.

FRINGE RESOLVED THIRD-ORDER AUTOCORRELATION FUNCTIONS

R. Fischer
Zentralinstitut für Optik und Spektroskopie, Akademie der Wissenschaften der DDR, 1199 Berlin, GDR

J. Gauger, J. Tilgner
Zentrum für Wissenschaftlichen Geraetebau, Akademie der Wissenschaften der DDR, 1199 Berlin, GDR

ABSTRACT

It is shown by numerical analysis that fringe resolved third-order autocorrelation functions are more sensitive to the pulse shape and time variations of the phase than fringe resolved second-order autocorrelation functions.

INTRODUCTION

The shortest optical pulses generated to date have durations of 6 fs[1]. Till now the experimental determination of the parameters of ultrashort optical pulses in the subpicosecond region is only possible using nonlinear optical effects[2]. Of special interest are interferometric methods which allow the determination of the pulse chirp [1,2,3]. For this reason two essential methods have been developed: the fringe resolved second-order autocorrelation [2,3]. (i.e. using second harmonic generation) and the linear autocorrelation[3].

In order to determine uniquely the characteristics of a pulse higher order correlation functions must be used [4,5]. Therefore we are interested in calculations of third-order interferometric autocorrelations which can be measured by third harmonic generation.

RESULTS

Using a fast Fourier transform algorithm we have calculated the fringe resolved third-order autocorrelation function $G^{(3)}(\tau_1,\tau_2)$,

$$G^{(3)}(\tau_1,\tau_2) \sim \int_{-\infty}^{\infty} dt \left| \left(V(t) + V(t-\tau_1) + V(t-\tau_2) \right)^3 \right|^2 \quad ,(1)$$

for different pulse shapes and time dependent phases. Here we demonstrate the results for an electric field strength of the pulse given by

$$V(t) = A(t) \exp\left\{ i \left[D\pi t - B A^2(t) \right] \right\} \quad ,(2)$$

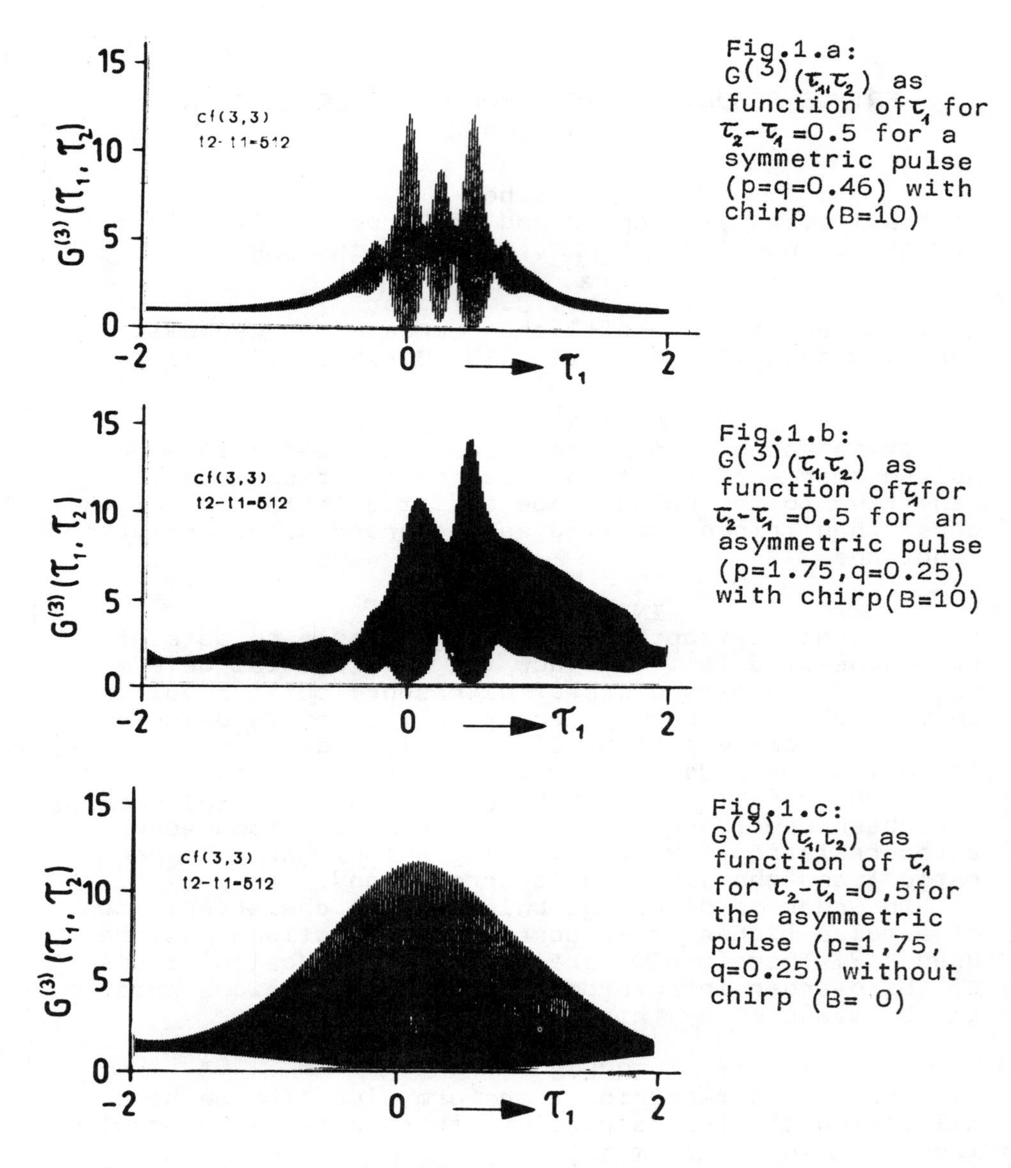

Fig.1.a: $G^{(3)}(\tau_1,\tau_2)$ as function of τ_1 for $\tau_2-\tau_1=0.5$ for a symmetric pulse ($p=q=0.46$) with chirp ($B=10$)

Fig.1.b: $G^{(3)}(\tau_1,\tau_2)$ as function of τ_1 for $\tau_2-\tau_1=0.5$ for an asymmetric pulse ($p=1.75, q=0.25$) with chirp($B=10$)

Fig.1.c: $G^{(3)}(\tau_1,\tau_2)$ as function of τ_1 for $\tau_2-\tau_1=0,5$ for the asymmetric pulse ($p=1,75$, $q=0.25$) without chirp ($B=0$)

i.e.for a pulse where the time dependence of the phase is caused by self-phase modulation,for instance in an optical fiber.For the envelop function has been used

$$A(t)=\left(\frac{p}{q}\right)^{\frac{q}{p+q}}\left(1+\frac{q}{p}\right)\left[e^{t/p}+e^{-t/q}\right]^{-1}$$

In these equations is B the chirp parameter and p,q describe the symmetry of the pulse.In the following all

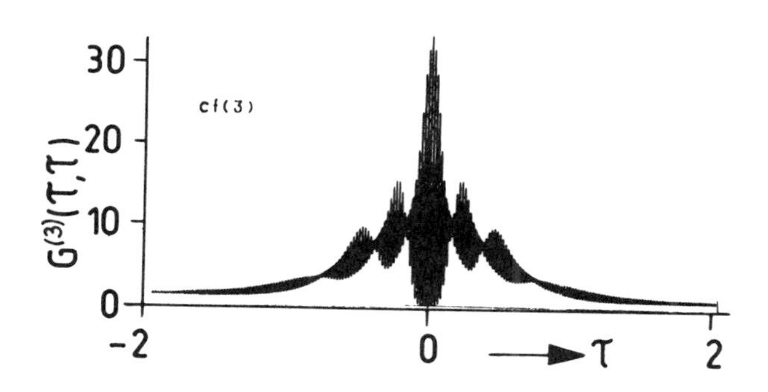

Fig.2.a: $G^{(3)}(\tau_1,\tau_2)$ for $\tau_1=\tau_2=\tau$ as function of τ for a symmetric pulse (p=q=0.46) with chirp (B=10)

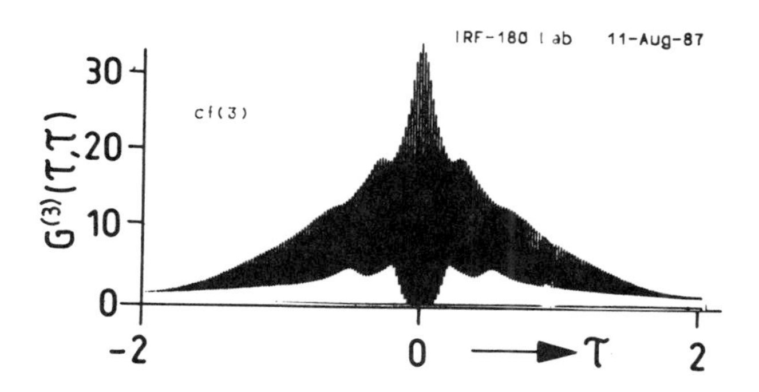

Fig.2,b: $G^{(3)}(\tau_1,\tau_2)$ for $\tau_1=\tau_2=\tau$ as function of τ for the asymmetric pulse (p=1.75, q=0.25) with chirp (B=10)

times are given in units of the pulse duration. For the parameter D in equ.(2) we used D=100;at a wavelength of 600 nm this corresponds to a pulse duration of about 100 fs.

In Figs.1 and 2 results for different parameters are shown.A comparison of the curves given in Fig.1 shows the influence of the pulse shape (symmetric or asymmetric) as well as the phase modulation on $G^{(3)}(\tau_1,\tau_2)$.The results for equal delay times $\tau_1=\tau_2=\tau$ (Fig.2)are similar with our former results for $G^{(2)}(\tau)$ (see[3],Fig.2b), for $G^{(3)}(\tau,\tau)$, however, the influence of the pulse symmetry is more distinct.

REFERENCES

1. R.L.Fork et al.,Optics Lett. 12, 483 (1987)
2. J.-C.Diels et al.,Appl.Optics 24, 1270 (1985)
3. R.Fischer et al.,American Institute of Physics Conference Proceedings NO.160,New York 1987, Advances in Laser Science-II, p.232
4. S.L.Shapiro (Edit.),Ultrashort Light Pulses, (Topics in Appl.Phys.,Vol. 18, Springer-Verlag 1977)
5. R.Trebino et al.,J.Opt.Soc.Am. B3, 1295 (1986)

A MODULAR SYSTEM FOR PICOSECOND SPECTROSCOPY

N. Kempe
Centre of Scientific Instruments of the Academy of
Sciences of the GDR, Berlin-GDR 1199, Rudower Chaussee 6

A modular system for picosecond spectroscopy developed at the Centre of Scientific Instruments of the Academy of Sciences of the GDR is described.

Picosecond lasers are playing an increasing role in biology, medicine and technology. Special equipment for such modern instrumentation consisting of different modules for picosecond light production, beam manipulation, detection of low level optical signals with picosecond time resolution, etc. is being investigated in our Centre. Some components of our system have control electronics supported by microprocessors[1]. The main properties of this system are:

- its modular character, allowing various combinations according to specific applications,
- separate operation of single units for other purposes (e.g. dye laser as tunable laser source for cw-operation, application in medicine, etc.),
- uniform electronic structure in all variants of spectrometers for control and data processing (interactive operation, use of standard interfaces, etc.),
- real-time software system in separately controllable modular sections, with possible connection with customers' computers and specialized software,
- strictly hierarchical microprocessor coupling in the system, allowing flexible control as an element of the measurement system and total automation of measurement and experiment,
- possibility of building an expert system for picosecond laser light pulse applications.

Original modular devices allow customers to upgrade the simplest system configuration to a highly sophisticated one.

One of many possible examples is the automatically controlled laser pulse source (Figure 1), which consists of the Ar-Ion Laser ILA-120 with modelocking units AOM (K)-100 (acousto-optical modulator) and AOM (E)-100 (electronic generator) pumping the dye laser FSL-101 with dye circulation system LVE-100. There are two special control units in the system: modelocking control unit MKE-100 for controlling the quality of modelocking of the Ar-laser by means of measurement of its SHG-energy and scanning correlator KRL-101 for continuous control of the output ps pulses. The special attenuator unit LRE-100 allows the control of the desired output energy. The heart of this combination is the electronically controlled cw dye laser FSL-101[1]. It has problem-oriented microprocessor control electronics built into the laser module, enabling the automatic control of optimal laser adjustment within a hierarchic data processing system with serial standard interface by means of a closed digital control loop: sensor-microprocessor-electronics-piezotranslator. The same control for the wavelength is possible by means of a stepping or linear motor for the movement of a Lyot filter and a special wavelength measuring sensor.

Many interesting applications are possible with broadband cw dye laser FSL-102. Here we have three different jets simultaneously operating in the same laser head, allowing the extension of the wavelength to 200 nm. All other parameters here are the same as in the FSL-101. There are a great number of applications of this laser, but the use in photo-acoustic spectroscopy is one of the most convincing. Another interesting example for applying the elements of our system is a microfluorometer[2] for the measurement of time resolved intercellular fluorescence with simultaneous observation of this cell (or any similar object) by means of a modern imaging processing system.

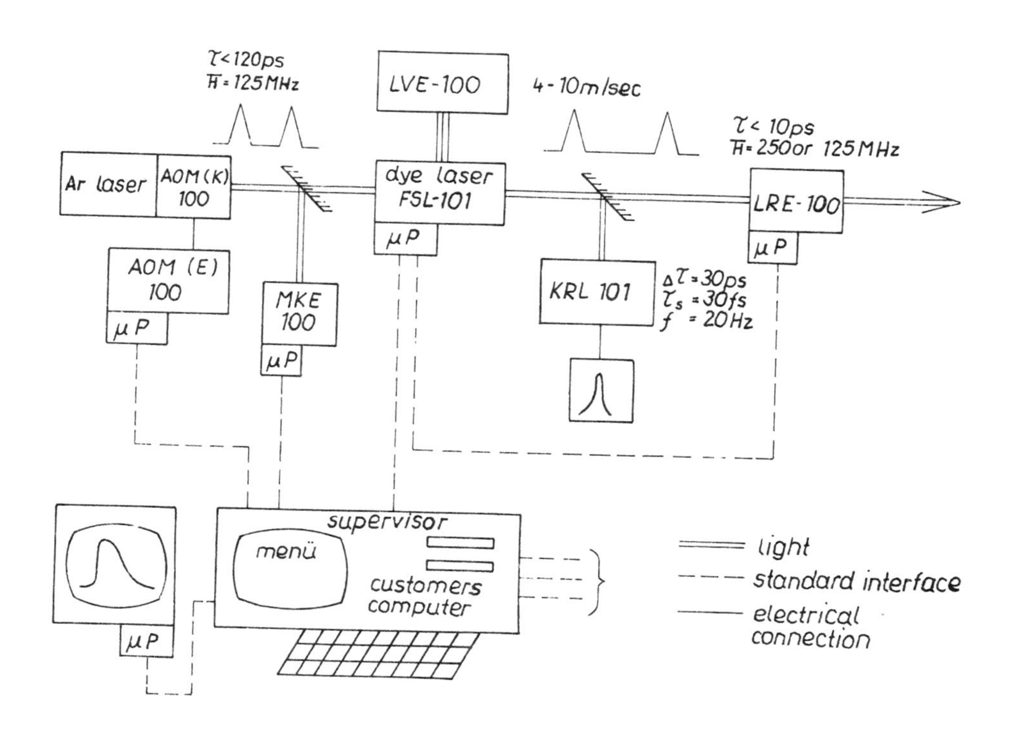

Figure 1. Automatic controlled laser pulse source.

Currently we are developing other system units such as receivers and detectors of light signals with ps-resolution, etc. For further information about our modular system for picosecond laser pulse techniques, we are ready to send more material.

1. N. Kempe, N. Langhoff and H. Schlawatzky, Scientific Instrumentation 1 (3), 53-62 (1986).

2. N. Kempe, N. Langhoff, G. Molgedey, C. Peschel, Forschungsgeräte für die Biotechnologie AdW der DDR (Heiligenstadt) 351-365 (1986).

PICOSECOND PHASE-COHERENT OPTICAL PULSES

William L. Wilson, Amy E. Frost, John T. Fourkas, G. Wäckerle, and M. D. Fayer

Chemistry Department
Stanford University, Stanford, CA 94305

ABSTRACT

A technique for performing optical coherence studies using picosecond phase-coherent pulses is being developed. A well-defined phase relationship between pulses has been demonstrated and applied in experiments performed on a sodium vapor sample, with the laser tuned to the D line. Scanning the relative phase between two temporally separate pulses results in sinusoidal modulation of the sodium fluorescence. A plot of the depth of modulation as a function of the pulse delay yields a Fourier transform spectrum of the Doppler-broadened line. This work is part of an ongoing effort to develop optical analogues to powerful and elegant pulsed NMR techniques.

Although the application of optical coherence techniques to the investigation of the properties of systems of atoms and molecules is becoming increasingly sophisticated, many of the methods common in magnetic resonance have not yet been employed in the optical regime. Sequences of optical pulses with well-defined durations, amplitudes, and temporal separations have been employed. In general, however, the optical experiments do not make use of phase relationships among the pulses, although phase relationships are of fundamental importance in all but the simplest magnetic resonance techniques.[1]

In the optical regime, photon echo, stimulated photon echo, and other phase-independent techniques have been used to probe optical dephasing and population dynamics. Phase-coherent pulse sequences in the microsecond and submicrosecond ranges, produced using acousto-optic phase shifting techniques to modulate CW lasers, have been employed.[2,3] These and similar techniques have been used to study collisional dephasing in gaseous systems.[4,5] Warren and coworkers have employed a technique based on a pulsed laser injection-locked to a single-mode dye laser to generate phase-related pulses in the nanosecond and subnanosecond regimes.[6]

Here we report a new approach for producing phase-related pulses on a time scale of tens of picoseconds. The method is general and can be applied on a subpicosecond time scale as well. It involves active stabilization of phase relationships between picosecond pulses.

A single tunable pulse is obtained by cavity-dumping a dye laser synchronously pumped by the doubled output of an acousto-optically mode-locked and Q-switched Nd:YAG laser. The dye pulses, occurring at 1 kHz repetition rate, are 20 psec long and 5μJ in energy. A single pulse is beam-split to form two pulses, one of which traverses an optical delay line having a maximum delay of 9 nsec. In addition to this long delay, there is a piezo-electric translator (PZT) which provides delays on an angstrom distance scale. This "phase scan PZT" permits the phase of one pulse to be scanned relative to the other. The two pulses are recombined with a 50% beam splitter and directed into the sample cell, which contains a low pressure of sodium (Na) vapor at 500 K. The laser is tuned to one component of the Na D line, and the fluorescence following the excitation is detected at a right angle to the beam path by a phototube linked to a lock-in amplifier.

To stabilize the phase relationship between the two pulses, small picked-off pieces of each are combined with a beam splitter, so that they are temporally and spatially coincident, forming an optical interference pattern. This pattern is observed by a photodiode. The output of this photodiode, when normalized to remove shot-to-shot laser intensity fluctuations, reflects shifts in relative phase of the pulses. This phase error signal is used in a feedback circuit to drive a second PZT delay in one of the main beam paths, locking the phase relationship between the two pulses.

The phase scan PZT is located after the pickoffs used in phase locking. Ramping the voltage to the phase scan PZT scans the phase relationship between the two pulses. Figure 1 shows the Na fluorescence as the phase is scanned. The delay between pulses is sufficient that the pulses do not overlap in time, but it is shorter than the Na free-induction decay time associated with the Doppler width of the transition. The first pulse generates a coherent superposition of the ground and excited electronic states. If the second pulse arrives in phase with the first, it will tip the vector which represents the coherent superposition further toward the excited state, and fluorescence will be at its maximum. If the second pulse is brought in 180° out of phase with the first, the vector will be tipped toward the ground state, and less fluorescence will occur. Figure 1 shows a scan through several cycles of 360° phase shift. Some free-induction decay occurs in the delay between the pulses, so that the minima in the fluorescence do not go to zero. The data clearly demonstrate that the atomic system is experiencing optical pulses with well-defined phase relationships.

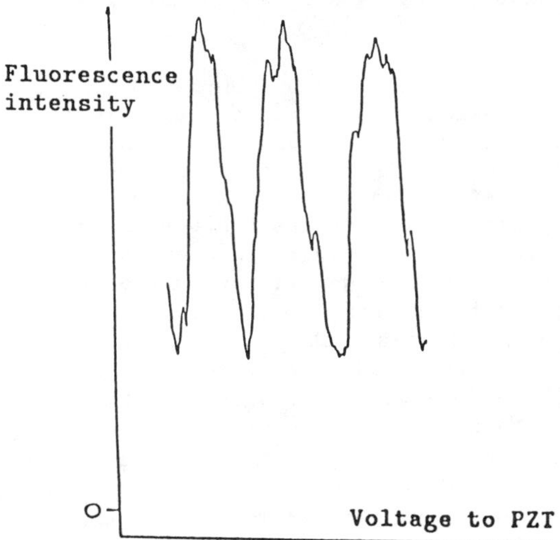

Figure 1. Fluorescence intensity as a function of voltage.

As the interval between pulses is increased, free-induction decay causes greater degradation of the phase information preserved in the excited system of atoms, so that the depth of modulation in a phase scan is reduced. A plot of the normalized modulation depth vs. the time delay between the pulses is shown in figure 2. The squares are the data points and the solid curve is the Fourier transform of the Doppler lineshape of Na at 500 K. The excellent agreement shows that measurement of

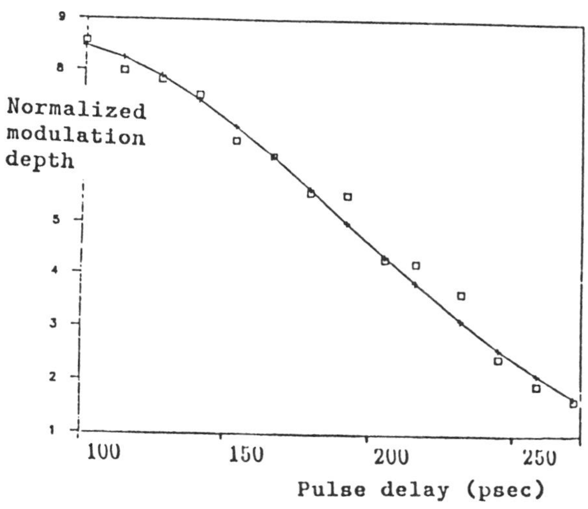

Figure 2. Normalized modulation depth vs. pulse delay

incoherent fluorescence following excitation with phase-related pulses can be used to obtain free-induction decay information and further demonstrates the phase-dependent interaction of the medium with the optical pulses. Active locking of the phase relationship between the pair of picosecond pulses makes this phase-dependent interaction possible.

ACKNOWLEDGEMENTS

This work was supported by the National Science Foundation, Division of Materials Research (# DMR 84-16343) and by the Office of Naval Research (# n00014-85-K-0409). AEF and WLW thank AT&T Bell Laboratories for graduate fellowships, JTF thanks the National Science Foundation for a graduate fellowship, and GW thanks the Deutsche Forschungsgemeinschaft for partial support.

REFERENCES

1. Abragam, A. *The Principles of Nuclear Magnetism*, Oxford University Press, London, 1961; Mehring, M., *Principles of High-Resolution NMR in Solids*, Springer-Verlag, Berlin, 1983.
2. M. J. Burns, W. K. Liu, and A. H. Zewail, in *Spectroscopy and Excitation Dynamics of Condensed Molecular Systems*, ed: V. M. Agranovich and R. M. Hochstrasser, North Holland, Amsterdam, 1983.
3. E. T. Silva, I. M. Xavier, and A. H. Zewail, J. Opt. Soc. Am. B **3**, 483 (1986).
4. A. G. Yodh, J. Golub, N. W. Carlson, and T. Mossberg, Phys. Rev. Lett., **53**, 659 (1984).
5. M. A. Banash and W. S. Warren, Laser Chemistry, **6**, 36 (1986).
6. F. Spano, M. Haner, and W. S. Warren, Chem. Phys. Lett., to appear (1987).

VI. DIAGNOSTIC AND ANALYTICAL APPLICATIONS

PULSED LASER PHOTOTHERMAL SPECTROSCOPY

Stephen E. Bialkowski
Department of Chemistry and Biochemistry
Utah State University, Logan UT 84322-0300

ABSTRACT

The use of photothermal spectroscopy (PTS) offers a wide range of flexibility and potential application to the spectroscopist. In this paper, basic theories of both pulsed and continuous excited PTS are compared. It is shown that techniques using pulsed excitation sources have the potential for greater sensitivity, less dependence on sample dynamics, faster analysis times, and smaller sampled volumes. On the other hand, continuous excitation may be used to avoid non-linear signal effects due to acoustic energy loss and optical saturation.

INTRODUCTION

The simplest way to compare pulsed to continuous excitation in PTS is through the theoretical enhancement factor. This factor is the ratio of the thermal lens spectrophotometry (TLS) signal to that of conventional absorption spectrophotometry. In the past, these enhancement factors have been based on the premise that the sample cell path length is the same for both the TLS and spectrophotometry cases. But this is not a valid assumption. Indeed, since the enhancement factors are inversely proportional to the excitation beam waist radius, and the beam waist radius will increase with path length for any real focused beam, strictly valid enhancement factor definitions must account for the finite path length effects.

PATH LENGTH EFFECTS DUE TO BEAM DIVERGENCE

Calculation the effects of beam divergence on the effective path length can be accomplished by first, assuming that the total lens strength is a sum of individual thin lens elements spaced along the path length of the sample cell, and second, integrating the individual lens elements along the total path length.[1] Using this procedure, the theoretical enhancement of continuous laser excited TLS, E_{cw}, over conventional spectrophotometry is,

$$E_{cw} = (dn/dT)(2P/k\lambda)(z/Z_r)(\mathbf{tan}^{-1}(X)/X) \qquad (1)$$

where n is refractive index; T, temperature; P, excitation power; k, thermal conductivity; λ, the wavelength of the pump laser; z is the distance between the pump and probe laser focus positions; Z_r is the Rayleigh range given by $\pi w^2/\lambda$; X, the relative path length $X=l/2Z_r$, where; l, is the actual path length; and w, minimum laser beam waist radius. The time of measurement was taken to be that where the signal is a maximum: at infinite time. Similarly, the path integrated enhancement for pulsed laser excited TLS is,

$$E_p = (dn/dT)(8E/\lambda\rho C_p w^2)(z/Z_r)[(1/(1+X^2)+\mathbf{tan}^{-1}(X)/X] \qquad (2)$$

where E is the pulse energy and, ρC_p, the heat capacity. The time is again taken to be that for the maximum signal, t=0 in this case.

Both theoretical enhancements were derived assuming a pump-probe laser setup with the probe laser beam focus located z distance in front of the pump laser focus.[2,3] This experimental feature is not critical for a comparison between the two enhancement factors. Continuous and pulsed laser excited TLS enhancement factors as a function of relative sample length are compared in Figure 1.

It is apparent from this figure that both TLS experiments favor short path length, and therefore small volume samples. It could be said that the continuous excitation experiment is more suited to long path length samples since in addition to the inverse Z_r dependence, the pulsed excited TLS enhancement is inversely proportional to the squared minimum beam waist radius. To determine just how sensitive to sample path length the two experiments are, it is useful to compare the enhancement factors at a fixed relative sample length while increasing the sample length. This can be done by noting that at fixed relative sample length, both Z_r and w^2 are proportional to **l**. Subsequently, at any fixed relative sample length, continuous excited TLS enhancements will decrease as $1/l$ while the pulsed as $1/l^2$.

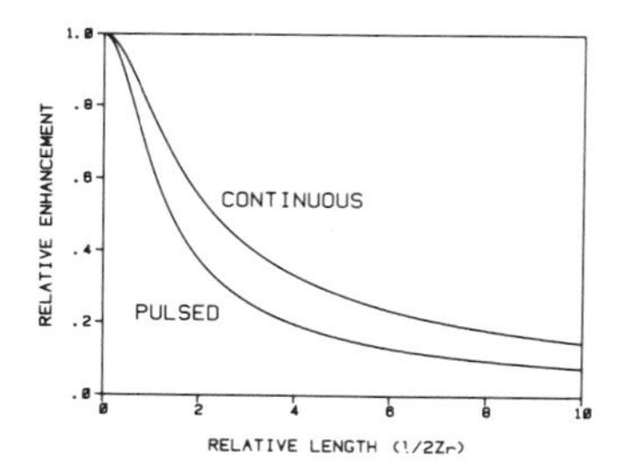

Figure 1. Comparison of continuous and pulsed excited TLS enhancement factors.

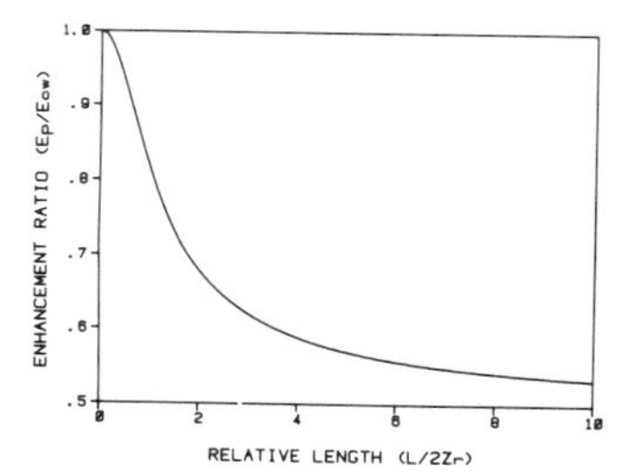

Figure 2. E_p/E_{cw} as a function of relative path length.

Having now established that TLS is a small volume analysis technique, we determine which laser setup is better for the small volume sample. The ratio of the enhancement factors may be examined.

$$E_p/E_{cw} = (2E/Pt_c)\{1 + [\tan^{-1}(X) + X^2\tan^{-1}(X)]^{-1}\} \quad (3)$$

where t_c is the thermal time constant, $t_c=w^2/4K$, K being the thermal diffusion coefficient. A plot of this ratio as a function of X is shown in Figure 2. Clearly, this relative ratio favors the pulsed laser excitation source for small volume samples. This relative ratio does not indicate which is most sensitive overall since the factors not dependent on sample length are not apparent in the sample length trend.

The path length independent ratio for the best experimental situation, *i.e.*, small sample lengths, is found by taking the limit of the ratio as X approaches zero. In this limit,

$$E_p/E_{cw} = 2E/Pt_c \tag{4}$$

A comparison of pulsed to continuous excited TLS for different materials using typical lasers sources is not a straight forward task. The ratio in Eq. 4 is perhaps more tied to technology than to physics since the ratio of pulse laser energy to continuous laser power is technically limited. But as a general rule, the pulsed laser source is favorable for samples with a high thermal diffusion coefficient. And since the thermal diffusion coefficient of gas is much greater than that of liquid phase, the pulsed laser excitation source may be best suited for the gas phase, and the continuous source in liquids.

In many real situations, the characteristic time constant is not limited by the beam waist radius to thermal diffusion ratio, but rather by the dynamics of the sample. Mass transfer and turbulence in flowing samples can limit the effective characteristic time constant to values much shorter than that predicted based only on thermal diffusion. Thus, Eq. 4 predicts only the limiting case where the sample is static. If the time constant is decreased by effects other than that of thermal diffusion, then pulsed excited TLS may be favored. One factor not accounted for in this simple prediction is that the thermal lens may retain spatial integrity and be translated while the sample is flowing in a dynamic system such found in liquid chromatography. It has been shown that by translating the probe laser beam in the direction of sample flow, the maximum signal can be found. The main drawback to this technique of optimization is that probe beam location for maximum signal is a function of sample flow rate. Thus there is no static experimental set up for this situation.

Table I. Thermal diffusion coefficients for gas and liquid.

Sample	Condition	$K(cm^2 \cdot s^{-1})$
CO_2 (gas)	300K, 1 ATM	0.110
H_2O (gas)	400K, 1 ATM	0.112
CS_2 (liq)	300K, 1 ATM	0.074
H_2O (liq)	300K, 1 ATM	0.001

ACOUSTIC LIMITATIONS OF PULSED LASER EXCITED PTS

There are two non-linear energy dependent factors that will ultimately affect pulsed excited TLS, and more generally, PTS. First is the photoacoustic effect. When the pulse energy is rapidly deposited by analyte absorption, the absorbed energy will be partitioned between the acoustic wave generated by the rapidly expanding matrix, and the heating of the matrix material local to the absorption. Although both processes result in a change in matrix density and thereby change the refractive index, pulsed laser excited PTS theories have for the most part assumed that all absorbed energy is available for matrix heating.

The acoustic power, P_a, coming from a thin line source at angular frequency, ω, and generated in an optically thin sample is[4]

$$P_a(\omega) = (l/8\rho)(\alpha\beta/C_P)^2\omega P_o(\omega)^2 \tag{5}$$

where β is the linear thermal expansion coefficient and $P_o(\omega)$ is the frequency dependent excitation power. A Gaussian shaped pulsed excitation power is given by,

$$P_o(t) = (E/\tau/\pi)e^{-(t/\tau)^2} \tag{6}$$

where τ is the pulse duration. The acoustic power spectrum obtained from the Fourier transform of Eq. 6 in Eq. 5 is,

$$P_a(\omega) = (l/8\rho)(\alpha\beta E/C_P)^2\omega e^{-(\omega\tau)^2/2} \tag{7}$$

Since the energy is an infinitesimal band of frequencies is,

$$E_a(\omega) = (1/2\pi)P_a(\omega)d\omega \tag{8}$$

integration over all frequencies yields the total acoustic energy,

$$E_a = (l/16\pi\rho)(\alpha\beta E/C_P\tau)^2 \tag{9}$$

The time constant is not necessarily that of the pulsed excitation source. The time constant can be expressed as a sum of excitation source or excited state relaxation times and acoustic propagation terms,[5]

$$\tau^2 = \tau_e{}^2 + \tau_a{}^2 \tag{10}$$

where τ_e is the combined pulse duration/excited state relaxation time, and

$$\tau_a{}^2 = w^2/2c^2 \tag{11}$$

c being the sound velocity, is the acoustic propagation time. The acoustic energy is thus,

$$E_a = (l/8\pi\rho)(\alpha\beta cE/C_P)^2/(w^2 + 2c^2\tau_e{}^2) \tag{12}$$

According to this expression, the acoustic energy is a maximum for small volume samples and for short excited state excitation/ relaxation times. Further, since the total energy absorbed by an optically thin sample is αEl, the fraction of acoustic energy absorbed by the sample is,

$$E_a/E_{abs} = (\alpha E/16\pi\rho)(\beta/C_P\tau)^2 \tag{13}$$

The troublesome feature of this last expression is that the fractional energy in the acoustic wave will increase with absorbed energy. Thus, the PTS signal is theoretically non-linear. However, the degree to which the PTS signal is non-linear is not very significant for typical experiments. For gases, β is equal to 1/T. The $(\beta/C_P)^2$ term in Eq. 13 is rather small. In fact, noting that the temperature change local to laser heating is,

$$\delta T \approx 2\alpha E/(\pi w^2 \rho C_P) \tag{14}$$

and that for ideal gases,

$$\beta = -1/V(dV/dT) \tag{15}$$

$$c^2 = PC_P/\rho C_V \tag{16}$$

the fraction of absorbed energy in the acoustic wave in the "fat" source approximation, where $\tau^2 = w^2/c^2$, is,

$$E_a/E_{abs} = 0.021\delta T/T \tag{17}$$

Typical δT in PTS are on the order of milli-Kelvin or less at room temperature. Thus the amount of energy lost in the acoustic wave is typically 7 orders of magnitude below that available for the PTS signal.

The second non-linear effect which limits pulsed excitation source PTS is due to optical saturation. Eq. 2 predicts that the pulsed excitation TLS enhancement can be arbitrarily increases by either increasing the excitation energy or by decreasing the beam waist using shorter focal length lens'. However, both of these measures increase the intensity of light in the sample. Short of catastrophic effects such as dielectric breakdown due to the large electric fields, with high intensity illumination the analyte may not relax at a rate competitive with excitation. In this event, stimulated emission is a significant excited state depopulation mechanism. This optical saturation not only limits the amount of energy able to be absorbed by the analyte, but also changes the spatial characteristics of the resulting temperature perturbation. Further, the decreased analyte adsorption may result in higher detection limits since weaker matrix absorptions would become more prevalent.

SUMMARY

To summarize, it has been shown that TLS and so therefore PTS in general, is a technique that favors small volume samples. There does appear to be ways to predict which particular excitation is best for PTS. This prediction should be based primarily on knowledge of the dynamics of the system. Dynamic systems should use a pulsed excitation source. For static systems, the path length of the particular sample should be examined. Short sample path lengths favor the pulsed excitation. And while longer sample path lengths may favor the continuous excitation source PTS experiment, the enhancement factors should be examined to ascertain that conventional spectrophotometry is not in fact better. Finally, the thermodynamics and photo-physics need to be considered. While pulsed excitation is favored for samples with a high thermal diffusion, *i.e.* for less dense sample matrices, low density often means low relaxation rates and therefore more chance for optical saturation.

REFERENCES

1. S. E. Bialkowski, Anal. Chem., 58, 1706 (1986).
2. S. E. Bialkowski, Spectroscopy, 1, 26 (1986).
3. H. L. Fang and R. L. Swofford, Ultrasensitive Laser Spectroscopy (D. S. Kliger, Editor, Academic Press, NY, 1983) Chapter 3.
4. S. Temkin, Elements of Acoustics (John Wiley, NY, 1981), p.348.
5. D. A. Hutchins and A. C. Tam, IEEE Trans, UFFC-33, 429 (1986).

PULSED LASER PHOTOTHERMAL TECHNIQUES FOR MATERIALS CHARACTERIZATION

A. C. Tam
IBM Research Div.,
Almaden Research Center
650 Harry Road,
San Jose, CA 95120-6099

ABSTRACT

Extensive progress has been made over the past several years in various laser photothermal techniques, using a pulsed laser for non-destructive photothermal generation in a sample and suitably detecting a resultant photothermal effect (e.g., acoustic, refractive index, and thermal radiation transients) by a transducer, "probe" laser beam, or an infrared sensor for various purposes of materials characterizations. Two specific examples will be described, namely, the use of photothermal probe-beam deflection technique for characterizing thermal relaxations in a gas, and the use of flash radiometry to characterize thermal contact resistance between an opaque thin-film and a substrate.

Photothermal (PT) generation refers to the heating a sample and its surroundings due to the absorption of electromagnetic radiation, and is a type of energy conversion. Using a laser beam for PT generation, many PT effects can be measured, including refractive-index gradients, acoustic emission, surface deformation, reflectivity changes, desorption, and "grey-body" infrared emission, providing useful techniques for materials characterization.[1] These PT material probing or characterization techniques generally rely on the use of high-sensitivity detection methods, involving the use of "probe" laser beams, transducers, or infrared detectors to monitor the effects caused by PT heating of a sample. Many of these PT effects occur simultaneously, e.g., PT heating of a sample in air will produce temperature rise, photoacoustic waves and refractive-index changes in the sample and in the adjacent air, infrared thermal radiation increase, etc., all at the same time. Thus, the choice of a suitable PT effect for detection will depend on the nature of the sample and its environment, the light source used, and the measurement desired.

We have previously reviewed[1, 2, 3] the various experiment arrangements and detection schemes for the different PT effects, and given a summary of possible applications. We have shown that it is possible to obtain high PT generation efficiencies using a short-pulsed laser beam, and sensitive detection schemes can be obtained using a "probe" laser beam to monitor the transient reflection or refraction from the sample or its vicinity after the pulsed laser excitation which causes temperature or pressure transients in the sample or at its surface.

One important example is the use of a probe-beam refraction (i.e. deflection) scheme to detect PT refractive-index gradients.[3] The probe deflection monitoring scheme has several advantages, namely, high detection sensitivity and spatial resolution, ability to measure prompt as well as delayed heat generation, and the quantitative nature of the detected signal so that it can be related to the photothermal generation mechanisms. These advantages are useful in many applications, including the monitoring of photochemical particulate production[4] in a gas, and the measurement of moisture adsorption and desorption on a surface in atmospheric conditions.[5] Approximate analytical solutions for typical configurations have been derived[3] to show how the observed beam deflection signal depends on various parameters, including the case of delayed heat release due to a long thermal de-excitation time constant.[4]

Another example is the use of pulsed PT radiometry (whereby transient surface cooling after pulsed heating is monitored by the infrared emission signal shape) for quantifying subsurface air-gaps[6] and thermal contact resistance between layers.[7] This method relies on the heating of the film surface by a short light pulse and detecting the subsequent infrared thermal radiation from the surface. An analytical solution to the heat diffusion equation[7] shows that in a suitable delayed time interval, the infrared signal decays exponentially with a time constant related to the air gap thickness or the thermal contact resistance at the film/substrate interface. In the latter case of contact-resistance measurement[7], changing gases in the interface at constant pressures allows us to separate the thermal conductance into two components: that due to solid contacts and that due to gas conduction.

This work is supported in part by the U. S. Office of Naval Research.

REFERENCES

1. A. C. Tam, "Application of Photoacoustic Sensing Techniques", Rev. Mod. Phys. 58 , 381 (1986).
2. A. C. Tam, "Pulsed Photothermal Radiometry for Non-Contact Spectroscopy, Material Testing and Inspection Measurement, Infrared Physics", 25 , 305 (1985).
3. A. C. Tam, "Overview of Photothermal Spectroscopy" in Photothermal Investigations of Solids and Fluids , Edited by Jeff A. Sell, Academic Press, NY, (in press, 1988).
4. H. Sontag, A. C. Tam and P. Hess, "Energy Relaxations in CS_2 and in NO_2 - N_2O_4 Vapors Following Pulsed Laser Excitation at 337.1 nm by Probe Beam Deflection Measurement", J. Chem. Phys. 86 , 3950 (1987).
5. A. C. Tam and H. Schroeder, "Laser-Induced Thermal Desorption of Moisture from a Surface in Atmospheric Conditions", J. Appl. Physics (in press, 1988).
6. A. C. Tam and H. Sontag, "Measurement of Air-Gap Thickness Underneath an Opaque Film by Pulsed Photothermal Radiometry", Appl. Phys. Lett. 49 , 1761, (1986).
7. W. P. Leung and A. C. Tam, "Thermal Conduction at a Contact Interface Measured by Pulsed Photothermal Radiometry", J. Appl. Phys., 63 , 4505 (1988).

Numerical Calculation of the Photoacoustic Signal Generated by a Droplet

Gerald J. Diebold
Department of Chemistry, Brown University
Providence, R.I., 02912

ABSTRACT

Absorption of light by a spherical droplet immersed in a fluid causes heating and a subsequent expansion of the droplet which, in turn, launches a sound wave. This paper gives numerical results for the pressure generated by a delta function light pulse using the fast Fourier transform technique.

INTRODUCTION

The problem of sound emission by a heated spherical region of a fluid has been discussed by several authors[1-5]. Closed form solutions have been obtained for sound generation by thermal expansion of the heated medium as well as for the case of heat conduction from the sphere followed by expansion of the surrounding fluid. When the heated region and the surrounding medium have different acoustic properties, reflection and dispersion at the interface between the two media complicate the calculation of the acoustic signal amplitude. Recently, Diebold and Westervelt[6] have found the solution to the problem of sound generation by an instantaneous heating of a droplet surrounded by a fluid medium. Unfortunately, the time domain solution for the acoustic signal is written in integral form, which can be evaluated explicitely only in a few limiting cases. Here, a numerical calculation of the integral is given for several values of the acoustic parameters for which evaluation of the integral is not possible.

NUMERICAL RESULTS

According to Ref. 6, the acoustic signal produced by an amplitude modulated, continuous laser beam is found by superposition of a solution to the inhomogeneous wave equation for pressure

$$\nabla^2 p - \frac{1}{c_s^2}\frac{\partial^2 p}{\partial t^2} = \frac{-\beta}{C_p}\frac{\partial H}{\partial t}, \qquad (1)$$

with a solution to the homogeneous equation (finite at the origin) to give the pressure inside the sphere. Here, c_s is the sound speed in the droplet, β is the isobaric volume expansion coefficient, C_p is the heat capacity, H is the heat per unit volume deposited, and t is the time. Outside the sphere the same equation holds but with c_f, the sound speed in the fluid, replacing c_s, and with $H = 0$ since there are no sources. The solution to this equation is a spherical wave that propagates outward from the sphere. The relative amplitudes of the solutions are found by equating the pressures and velocities of the solutions at the radius of the sphere giving a frequency domain signal that can be Fourier transformed to give the time domain signal as

$$p_f(t) = \frac{i\kappa}{2\pi}\int_{-\infty}^{\infty} \frac{(\sin\, q - q\, \cos\, q)\, e^{-iq\hat{\tau}}}{q^2\left[(1-\hat{\rho})\frac{\sin\, q}{q} - \cos\, q + i\,\hat{\rho}\hat{c}\sin\, q\right]}\, dq\,, \qquad (2)$$

where $\hat{\tau} = \frac{c_s}{a}(t - \frac{r-a}{c_f})$ is the dimensionless retarded time, $\hat{\rho}$ and $\hat{c}$ are the ratios of the densities and sound speeds of the sphere to its surrounding medium, κ is an amplitude parameter given by $\kappa = \alpha\beta E_0 c_s^2 / C_p \hat{r}$ where $\hat{r}$ is the dimensionless radial distance from the center of the sphere, *i.e.* the distance from the center of the sphere divided by the radius of the sphere, α is the optical absorption coefficient, and E_0 is the laser fluence (energy per area).

Evaluation of Eq. 2 in closed form has been possible only in the limit of equal densities ($\hat{\rho} = 1$), when the pulse length of the exciting radiation is long compared with the acoustic transit time across the droplet, and for small values of $\hat{\tau}$ when $\hat{\rho} > 1$. However, since the pressure is given as an integral, which is an explicit algebraic formula, Eq. 2 can be looked on as the complete solution to the droplet problem with the evaluation of the integral a computational detail, as is shown below.

The pressure signal was evaluated using the fast Fourier transform technique for several values of the acoustic parameters where solutions could not be found in Ref. 6. In particular, the case where $\hat{c} = 1$, but $\hat{\rho}$ differs from one was found to be difficult to evaluate. Plots of the dimensionless pressure, Eq. 2 evaluated with $\kappa = 1$, versus the dimensionless retarded time $\hat{\tau}$ are given in Figs 1 – 2. The decay of the signal is similar to the sawtooth curves found for $\hat{\rho} = 1, \hat{c} = 1$ discussed in Ref. 6; however the curvature in the decay, characteristic of frequency dependent transmission and reflection coefficients, easily distinguishes the response from the equal density curves. Of further note is the increase in curvature of the reflected waves as $\hat{\tau}$ increases. This follows from the multiple internal reflections, each reflection being frequency dependent.

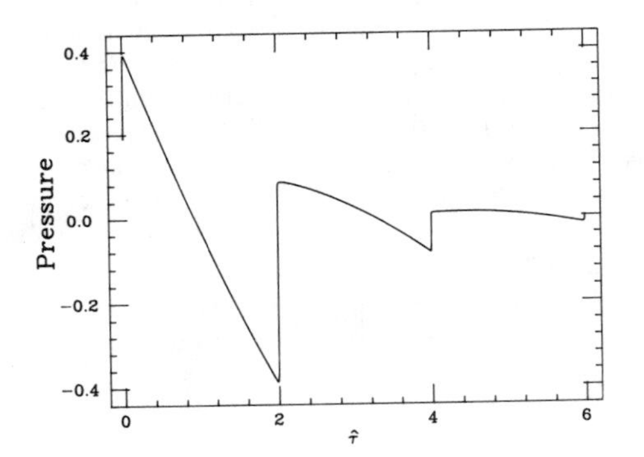

Fig. 1 *Pressure (dimensionless units) versus retarded time (dimensionless units) for* $\hat{c} = 1$, $\hat{\rho} = 1.5$

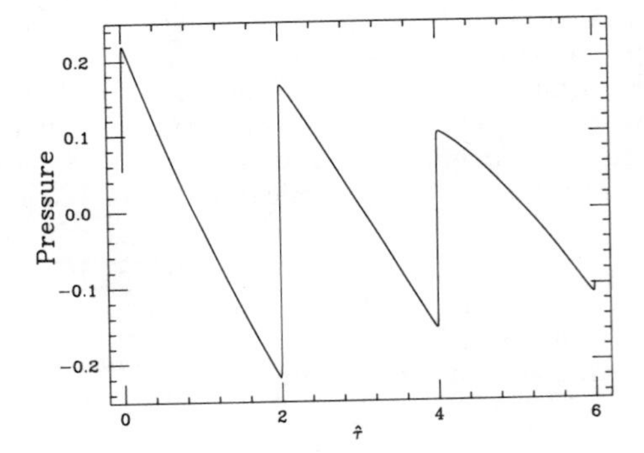

Fig. 2 *Pressure versus retarded time as in Fig. 1 but with* $\hat{\rho} = 5$

For values of $\hat{\rho}$ less than, but nearly equal to one and $\hat{c} = 1$, the poles of the integrand lie in the upper half complex q plane and there is no closed form solution for the pressure. Numerical evaluation of Eq. 2 gives the results plotted in Figs. 3 – 6. For $\hat{\rho} = 0.9$, the signal is nearly a single N shaped wave as occurs when $\hat{\rho} = \hat{c} = 1$. As $\hat{\rho}$ is made smaller, the inverted N shaped waves are seen for odd reflections but with significant curvature. Of particular note is the asymmetry of the pressure wave about zero.

This is evidence of the new periodicity in the wave that can easily be seen for $\hat{\rho} = 0.05$, where, each reflection adds in a coherent manner to give a low frequency wave that has the impulse response of a damped simple harmonic oscillator. Note that small discontinuities at multiples of $2\hat{\tau}$ are still present even at $\hat{\rho} = 0.01$. Their amplitudes diminish as $\hat{\rho}$ decreases as is evident by comparison of Figs 5 and 6. For very small values of $\hat{\rho}$ the resonance frequency ω_0 is given by $\omega_0 = (c_s/a)(3\hat{\rho})^{\frac{1}{2}}$ and the damping rate γ by $\gamma = (c_s/a)(\frac{3}{2}\hat{\rho}\hat{c})$ in accord with previous results for a pulsating bubble.

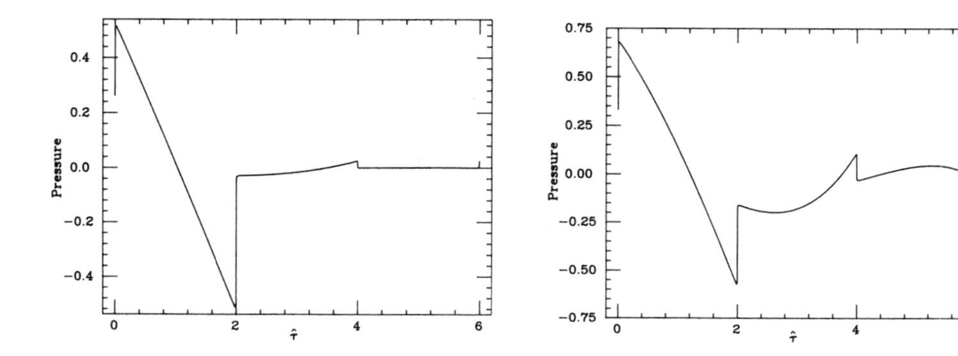

Fig.3 *Pressure versus* $\hat{\tau}$ *for* $\hat{\rho} = 0.9, \hat{c} = 1$ **Fig.4** *Pressure versus* $\hat{\tau}$ *for* $\hat{\rho} = 0.5, \hat{c} = 1$

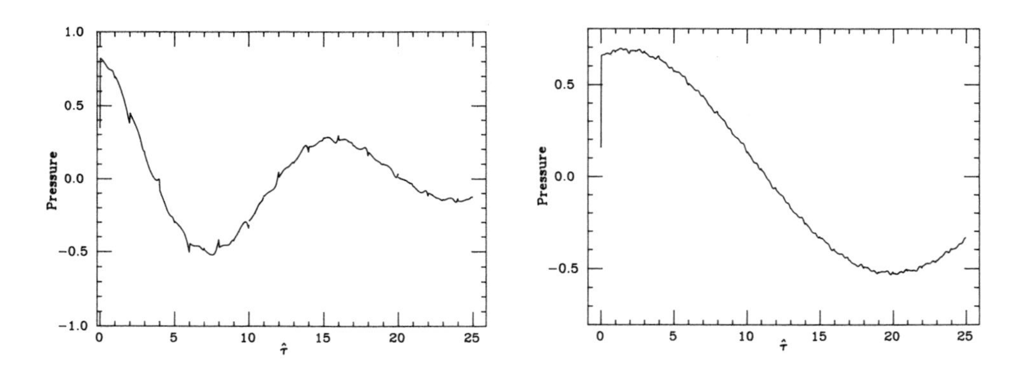

Fig.5 *Pressure versus* $\hat{\tau}$ *for* $\rho = 0.05, \hat{c} = 1$ **Fig.6** *Pressure versus* $\hat{\tau}$ *for* $\rho = 0.01, \hat{c} = 1$

These numerical results show that a droplet has a characteristic signature that is determined by its acoustic parameters relative to the surrounding medium. It should be possible to obtain a reasonable match between an experimental photoacoustic signal and a numerical calculation of Eq 2 to determine the density and the sound velocity of an unknown sphere.

This research was supported by the U. S. Department of Health and Human Services.

REFERENCES

1) T. Kitamori, M. Fugii, T. Sawada and Y. Goshi, J. Appl. Phys. **55**, 4005 (1984); **58**, 268 (1985); **58**, 1456 (1985).
2) C. Hu, J. Acoust. Soc. Am. **46**, 728 (1969).
3) F. V. Bunkin and V. M. Komisarov, Sov. Phys. Acoust. **19** , 203 (1973).
4) M. W. Sigrist and F. K. Kneubuhl, J. Acoust. Soc. Am. **64**, 1652 (1978).
5) C. H. Chan, Appl. Phys. Lett. **26**, 628 (1975).
6) G. J. Diebold and P. J. Westervelt "The Photoacoustic Effect Generated by a Spherical Droplet in a Fluid".

TIME-RESOLVED LASER-INDUCED FLUORESCENCE IN LOW-PRESSURE FLAMES

Karen J. Rensberger, Mark J. Dyer, Michael L. Wise,
and Richard A. Copeland
Molecular Physics Department, SRI International
Menlo Park, California 94025

ABSTRACT

Time-resolved laser-induced fluorescence measurements of the total removal rate constants of electronically excited states of the NH and CH radicals have been obtained in a variety of low-pressure (5 to 20 Torr) flames. The $NH(A^3\Pi)$ removal rate constants have been examined for flames containing N_2O as an oxidizer and source of nitrogen. The NH fluorescence quantum yields decrease significantly in the early portion of the flame where the temperature is lower, and become constant farther away from the burner. For CH, the $B^2\Sigma^-$ removal rate constant is about 70% faster than that of the $A^2\Delta$ in several hydrocarbon/oxygen stoichiometric flames; both rates do not vary significantly from flame to flame. The CH removal rate constants for both electronic states are constant or show a slight increase from the burner surface into the burnt gases for all the flames. Temperature profiles are obtained using excitation scans over several rotational levels for CH and NH.

INTRODUCTION

Laser-induced fluorescence (LIF) is a sensitive, selective and non-intrusive spectroscopic diagnostic technique for the detection of minor species in flames.[1] Radical intermediates such as CH and NH are ideal candidates for LIF detection because of their relatively low concentration (ppm) and easily accessible electronic states. Quantitative concentration measurements using LIF require knowledge of both the radiative and the collisional processes that occur following electronic excitation.[2]

In this work, we examine the collisional energy transfer of the $A^2\Delta$ and $B^2\Sigma^-$ states of CH and the $A^3\Pi$ state of NH by direct measurement of the time dependence of the LIF in several low-pressure premixed flat flames. At atmospheric pressure, the LIF of these radicals follows the time profile of the excitation laser because of the rapid collisional quenching; in the low-pressure flame, the quenching rate is reduced so that the fluorescence decays are significantly slower than that of the laser light and can be directly observed.

EXPERIMENTAL APPROACH

For these LIF experiments, the pulsed output of an excimer-pumped dye laser excites the radicals from the ground state to a specific level (v, J) in an excited electronic state. The radicals are generated in a premixed low-pressure laminar flame burning on a

McKenna Products flat porous plug burner of 6 cm diameter. Flames of many different fuels, oxidizers, and fuel equivalence ratios are burned in the study. The burner can be scanned vertically to examine different regions of the flame. The LIF is imaged onto the entrance slit of a monochromator acting as a wide bandpass tunable filter. A photomultiplier tube monitors the dispersed light and its output is amplified and captured using either a 100 megasample/sec transient digitizer or a boxcar integrator. The time dependent fluorescence signal is fit from 90% to 10% of its peak value to a single exponential decay constant.

$NH(A^3\Pi)$ QUENCHING

Until this study, the collisional removal rate constants of the $A^3\Pi$ state of NH in flames were unknown; however, we have now directly measured fluorescence decay constants in a number of low-pressure flames containing N_2O. Figure 1 is a typical summary plot of the data for a C_2H_4/N_2O flame at 14 Torr with a fuel equivalence ratio of ϕ = 1.06. The bottom solid line is the NH LIF signal from excitation of the $P_3(7)$ transition in the (0,0) vibrational band of the $A^3\Pi$-$X^3\Sigma^-$ electronic transition as a function of height above the burner. At this pressure the signal peaks near 5 mm and persists out past 12 mm. The boxes above the LIF signal correspond to the right-hand vertical axis and give the temperature profile extracted from NH excitation spectra. Details will be given in a future publication. The diamonds show the decay constants extracted from the temporal evolution as a function of height above the burner. They change considerably from early times in combustion (near the burner) to the burnt gases (> 10 mm). Each decay constant is composed of a pressure independent term due to the NH radiative lifetime of 2.3 μs^{-1} and a pressure dependent term due to collisional quenching. The decrease in the decay constants shown in Figure 1 for NH is typical for all the flames studied. This behavior can be explained by the change in density and relative collision velocity that results from the differences in the temperature. This accounts for most of the change in the decay constant. We therefore conclude that the effective cross section for collisional removal is roughly the same over the entire flame.

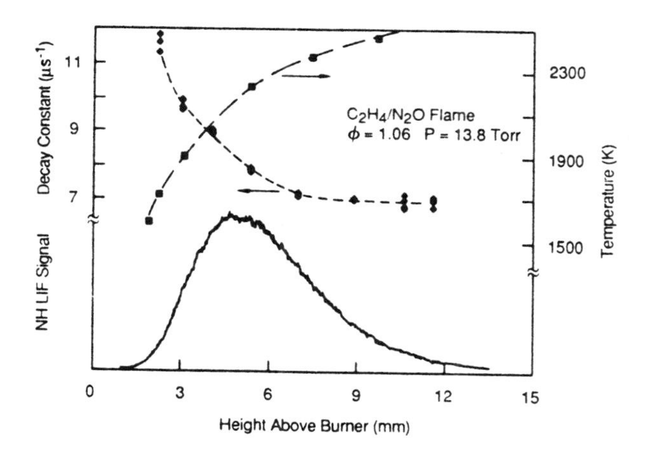

Figure 1

$CH(A^2\Delta)$ AND $CH(B^2\Sigma^-)$ QUENCHING

The electronic quenching of the $A^2\Delta$ state of CH has been studied previously for a 20 Torr methane/oxygen flame by Cattolica et al.[3]; we have now made temporally resolved measurements in several different hydrocarbon/oxygen flames for both the $A^2\Delta$ and the $B^2\Sigma^-$ electronic states of CH. We find these notable aspects of the data. The CH fluorescence quantum yields for both the B- and A-states extracted from the decay constants remain constant or decrease slightly with distance from the burner surface, in marked contrast to NH. As CH is found primarily in the region of the flame front where the temperature is increasing, a quenching cross section that also increases with distance from the burner is suggested from the data. We observe that the B-state is removed about 70% faster than the lower lying A-state for all the flames, even though the position dependences are similar. This difference in quenching was anticipated from indirect measurements in an atmospheric pressure flame by Garland and Crosley.[4] Both electronic states show a slight decrease of 10 to 20% in quenching with increasing rotational level N′ = 2 to 14.

CH B→A ELECTRONIC-TO-ELECTRONIC ENERGY TRANSFER

In addition to the total removal rates described above, we obtain information on the pathways of the removal of the CH B-state by wavelength resolution of the fluorescence. We find that a significant fraction of the molecules initially excited to the B-state fluoresce at a later time from the A-state of CH. This effect, previously observed in atmospheric pressure flames,[4] can be used advantageously to eliminate scattered laser light from the detection of CH. Exciting the B-state and observing the A-state about 50 nm away could eliminate many detection problems in systems with significant particulate scattering. The signal profiles for the B-state LIF and the B→A transferred emission show that the relative amount of electronic energy transfer is constant throughout the changing collision environment across the flame. All flames studied to date show this energy transfer effect.

ACKNOWLEDGEMENTS

We thank Dave Crosley for many helpful discussions on both the experiment and interpretation. This research is supported by the Aero Propulsion Laboratory of the Air Force Wright Aeronautical Laboratories and the National Science Foundation.

REFERENCES

1. D. R. Crosley, High Temp. Mat. and Proc. 7, 41 (1986).
2. N. L. Garland and D. R. Crosley, Twenty-First Symposium (International) on Combustion, The Combustion Institute, Pittsburgh, in press (1987).
3. R. J. Cattolica, D. Stepowski, D. Puechberty, and M. Cottereau, J. Quant. Spectrosc. Radiat. Transfer 32, 363 (1984).
4. N. L. Garland and D. R. Crosley, Appl. Opt. 24, 4229 (1985).

APPLICATION OF LASER-INDUCED FLUORESCENCE IN AN ATMOSPHERIC-PRESSURE BORON-SEEDED FLAME*

Greg R. Schneider and Won B. Roh
Air Force Institute of Technology, Wright-Patterson AFB, Ohio 45433

ABSTRACT

Laser-induced fluorescence (LIF) has been applied to the detection of a variety of combustion species as well as to BO_2 in low-pressure reacting flows. This paper reports the first known demonstration of LIF of BO_2 in an atmospheric-pressure flame. BO_2 is believed to be one of the key species in boron combustion chemistry. A CH_4/O_2/air premixed flame seeded with BCl_3 was probed with a pulsed dye laser and the resulting fluorescence detected with an optical multichannel analyzer. Non-resonant LIF was produced in the A-X system of BO_2 principally by pumping the 00^00-00^00 transition and observing the 00^00-10^00 transition although other combinations were also used. The spectra show a considerable redistribution of energy due to collisions in the flame. A two-dimensional relative fluorescence profile of the BO_2 in the flame was also obtained.

INTRODUCTION

The goal of this effort was to demonstrate the feasibility of using laser-induced fluorescence (LIF) of boron oxide radicals in atmospheric-pressure flames for probing the combustion chemistry of boron fuels. The species of interest was BO_2, which is believed to play a critical role in the boron combustion process.

LIF is a sensitive, species-selective, in-situ diagnostic that has been demonstrated with species such as OH in atmospheric-pressure flames.[1] Indeed, LIF of BO_2 has also been performed, but at low pressure (a few torr) in controlled reacting flows.[2,3,4,5] At the relatively high pressure of this experiment, collisional quenching and redistribution of energy from the excited state were expected to be much more serious effects than in the previous tests.

EXPERIMENTAL PROCEDURE

A premixed CH_4/air/O_2 flame was seeded with BCl_3 producing the green emission characteristic of boron flames. Unfortunately, it also created aerosols of solid boric acid from the reaction of the BCl_3 with water vapor in the air supply. The 1 cm diameter burner was composed of 61 capillary tubes. The burner could be remotely translated in two directions permitting investigation of different regions of the flame. Gas flows were adjusted to produce a stable conical flame approximately 6 cm tall with an inner cone 2 cm tall.

*Partially supported by the Air Force Office of Scientific Research and the Aero Propulsion Laboratory, Air Force Wright Aeronautical Laboratories, Aeronautical Systems Division

Flame emission was recorded using a 0.64 m spectrometer, a 512 diode linear array detector (PAR Model 1420), and an optical multichannel analyzer (PAR Model 1460). The system resolution (approximately 1.2 A) was sufficient to resolve vibrational bands, but not their rotational structure. For the LIF measurements, a dye laser (Lambda Physik Model FL3002E) pumped by a Nd:YAG laser (Quantel YG481) produced 7 mJ of energy in 10-15 nsec pulses at 5 Hz using Fluorscein 548 dye at 5471 A. The detector was gated on when the laser fired and then again halfway between the laser pulses. The second signal was subtracted from the first to yield a net LIF signal corrected for flame chemiluminescence. Usually 500 signal pairs were summed to produce a reasonable display.

RESULTS

The flame emission exhibited the "fluctuation bands" from BO_2, composed of heavily overlapped ro-vibrational bands. The strongest of these were centered around the 00^00-00^00 and 00^00-10^00 transitions of the A-X system at 5456 and 5790 A, respectively. The 0-0 band of CH at 4250 A was also observed.

An LIF signal was produced by pumping either the R_1 or R_2 branch of the 00^00-00^00 transition at 5456 or 5471 A, respectively. The effect was monitored by observing the entire band region (about 120 A wide) around the 00^00-10^00 transition. Figure 1 shows the signal from pumping the R_1 branch. As expected the R_1 branch of the of the 00^00-10^00 transition at 5790 A is the dominant band. The relatively strong signal from the R_2 branch at 5813 A as well as the presence of other bands is indicative of a significant amount of collisional energy redistribution. Figure 2 shows that pumping the R_2 branch of the 00^00-00^00 transition yields the expected inversion of the intensities of the R_1 and R_2 branches of the 00^00-10^00 transition.

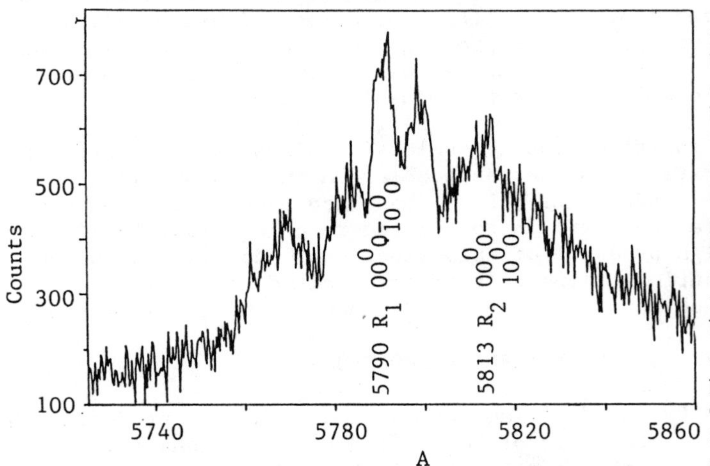

Figure 1. LIF Signal, Laser at 5456.4 A

A two-dimensional relative intensity map of the flame was obtained by pumping the R_2 band and probing horizontally across the flame in 1 mm steps and vertically in 1 cm steps. The collected signal was integrated over nearly the entire spectral region shown

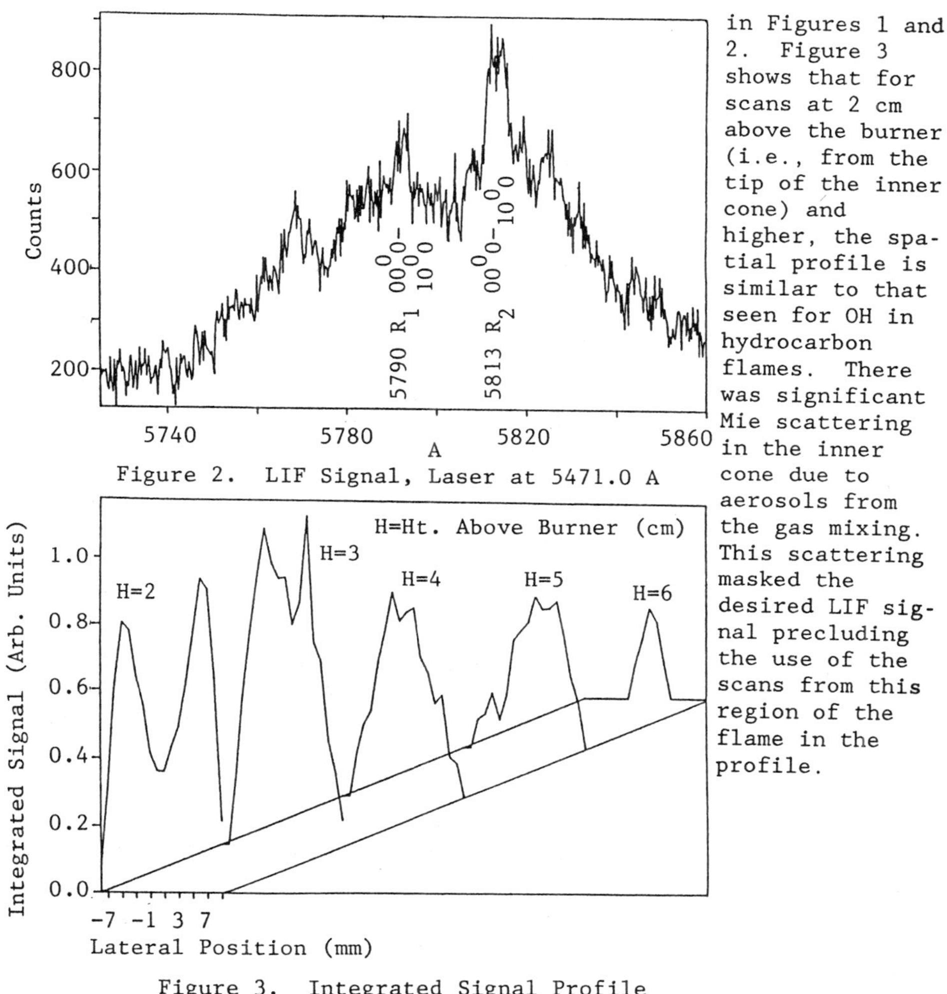

Figure 2. LIF Signal, Laser at 5471.0 A

Figure 3. Integrated Signal Profile

in Figures 1 and 2. Figure 3 shows that for scans at 2 cm above the burner (i.e., from the tip of the inner cone) and higher, the spatial profile is similar to that seen for OH in hydrocarbon flames. There was significant Mie scattering in the inner cone due to aerosols from the gas mixing. This scattering masked the desired LIF signal precluding the use of the scans from this region of the flame in the profile.

CONCLUSIONS

LIF of BO_2 has been demonstrated in an atmospheric-pressure flame. The spatial profile of the LIF signal is similar to that of OH. Significant rates of collisional quenching and energy redistribution pose a challenge to using this technique for quantitative measurements.

REFERENCES

1. D.R. Crosley and G.P. Smith, Opt. Eng. 22, 547 (1983).
2. D.K. Russell, et al, JCP 66, 1999 (1977).
3. A. Fried and C.W. Mathews, Chem. Phys. Lett. 52, 363 (1977).
4. M.A.A. Clyne and M.C. Heaven, Chem. Phys. 51, 299 (1980).
5. A. Hodgson, J. Chem. Soc., Faraday Trans. 2 81, 1445 (1985).

MOLECULAR DYNAMICS OF LASER ABLATION OF A POLYMER COATING

D. W. Noid*
Chemistry Division
Oak Ridge National Laboratory, Oak Ridge, Tennessee 37831

S. K. Gray
Department of Chemistry
Northern Illinois University, DeKalb, Illinois 60115

ABSTRACT

Large-scale molecular dynamics calculations have been carried out to simulate laser ablation on a polymer surface. The model consisted of a polyethylene chain with 400 CH_2 units placed on a polymer surface. A laser excitation was simulated by giving a small portion of the polymer chain a nonequilibrium momentum distribution. The excitation energy threshold for fragmentation of the polymers was found to be quite large, owing to the tendency of the system to transfer large amounts of vibrational energy along the chain.

I. INTRODUCTION

The microscopic details of polymer molecules exposed to intense radiation is of both scientific and practical interest. Fundamentally, there are important questions with respect to energy transfer involving collective modes and the nature of equipartitioning of energy [1]. An understanding of laser ablation and damage is of importance for a variety of medical and industrial applications [2], as well as for use in optical components [3]. Although there is a large amount of experimental work in this area, very few theoretical simulations have been reported [4].

In this paper, we explore theoretically the behavior of a polyethylene (PE) chain on a PE surface after portions of it have been excited by, say, a strong IR laser. In this molecular dynamics (MD) simulation, the details of the excitation mechanism are ignored and PE surface is idealized as a corrugated surface. Specifically, we find that a threshold pulse energy is required for laser ablation and that this threshold is raised if the excitation is spread out randomly throughout the molecule, as opposed to being localized in a group of neighboring atoms. This latter situation mimics crudely how a coiled polymer might behave. For pulse energies below threshold, we find fast equipartitioning of energy, whereas above threshold this is not the case.

Section II discusses our theoretical methods. In Section III the results of our study are presented, with a brief discussion in Section IV.

---

*Research sponsored by the Division of Materials Sciences, Office of Basic Energy Sciences, U. S. Department of Energy, under contract DE-AC05-84OR21400 with Martin Marietta Energy Systems, Inc.

II. PROCEDURE

The MD method involves specification of a classical Hamiltonian to describe the system and the numerical solution of the corresponding classical equations of motion given appropriate initial conditions [4]. Our model for polyethylene consists of 400 carbon atoms interacting via appropriate bonded and non-bonded interactions. The surface-molecule interaction was modeled by including non-bonded terms in the potential to represent the interaction of the atoms in the molecule with a corrugated surface of CH_2 groups. The Hamiltonian for our model is

$$H = \sum_{i=1}^{400} \frac{\tilde{p}_i}{2m} + \sum_{i=1}^{399} V_{bond}(r_{i,i+1}) + \sum_{i=1}^{398} V_{bend}(\theta_{i,i+1,i+2})$$
$$+ \sum_{i=1}^{400} \sum_{j=1}^{400} V_{non\text{-}bond}(r_{i,j}) + \sum_{i=1}^{400} \sum_{\substack{\text{surface} \\ \text{atoms, } j}} V_{non\text{-}bond}(r_{i,j}) , \tag{1}$$

where $\tilde{p}_i$ are cartesian momenta. $\theta_{i,i+1,i+2}$ and r_{iq} are internal bending and stretching coordinates. The potential energy functions in Eq. (1) were developed by Weber [6] to simulate liquid hydrocarbons. In our simulation, a More function $V(r_{i,i+1}) = D(1 - e^{-\alpha(r_{i,i+1} - r_e)})$ was fit to the harmonic oscillator function in this study with a dissociation energy D equal to .1324 hartrees. More details of the potential functions used can be found in Ref. 6. A limitation of this model is that no explicit H atom interactions are included-each CH_2 unit is treated as a mass 14.5 (m_i) amu atom interacting via C-C type interactions.

The initial conditions were obtained by randomly selecting all atomic momenta to generate a temperature of ∿50°K and placing the atoms at their equilibrium distances in an extended chain configuration. In one series of calculations, a localized excitation was created in 20 consecutive atoms (atoms 210 to 229) by setting their cartesian momenta equal to $P_{MAX} \times \xi$, where ξ is a random number between -1 and 1. Roughly, this corresponds to a nonequilibrium temperature for these atoms of $T \sim P_{MAX}{}^2/.25$. See Figs. 1a-c, for example, where the initial kinetic energy for each atom is displayed as the lower curves. The vertical distance of each atom above the surface is plotted on the upper curve. This is the reaction coordinate for the ablation process. The pulse energy is varied by scaling P_{MAX} and, naturally, scales as P^2_{MAX}. Another type of initial condition was a delocalized one where the excess kinetic energy was distributed to 20 atoms chosen randomly along the chain. Such an initial condition is displayed in Fig. 1c.

The analysis of equipartitioning was accomplished both visually and with the aid of the following correlation function:

$$\delta K^2(t) = \Sigma_{i=1}^{400} (\langle K \rangle - K_i(t))^2/400 \tag{2}$$

where <K> represents the average kinetic energy of the 400 atoms and $K_i(t)$ represents an individual atom (actually CH_2 group) kinetic energy.

Standard numerical techniques discussed more fully in Ref. 4 were used to solve Hamilton's equations using a CRAY XMP computer.

III. RESULTS

Figures in rows a and b display results for P_{MAX} = 30 and 37.5 au for the localized excitation initial conditions. In row c, results for P_{MAX} = 37.5 for the delocalized excitation are shown. Figures 2b-2c and 3b-3c demonstrate the resulting energy flow in the lower curve and ablation in the upper curve after 3 and 6 ps, respectively. When the atoms have completely left the surface, i.e., Z_i > 14.8 Å, they are plotted as a solid line on the upper curve at Z^i= 14.8 Å. The particular trajectory in the a row, despite a nonequilibrium temperature T ∿ 4000°K for the 20 excited atoms, does not lead to ablation (i.e., fragmentation of polymer accompanied by atoms leaving the corrugated surface). Notice that the correlation function $\delta K^2(t)$, shown in Fig. 4a, quickly decays to zero, indicating rapid equipartitioning. Figures in row b reflect the situation for a stronger pulse with P_{MAX} = 37.5, corresponding to a nonequilibrium temperature of T ∿ 6000°K. By 6 ps (Fig. 3b), ablation has occurred. The correlation function $\delta K^2(t)$, Fig. 4b, decays at a slower rate and with more noise compared to Fig. 4a, indicating a general lack of equipartitioning.

Figures in row c display the MD results for a delocalized excitation with P_{MAX} = 37.5. In this case, no ablation is observed, unlike an equivalent localized excitation in column b. Since one might expect a group of surface atoms in a coiled polymer to correspond to random atoms along the corresponding elongated chain, this type of initial delocalized state roughly mimics amorphous coil excitation; however, it should be pointed out that the lack of a regular crystalline environment would probably reduce the amount of reattachment after a bond fission. This effect would tend to lower the damage threshold.

IV. CONCLUDING REMARKS

The MD results in this paper have demonstrated the existence of a high energy threshold for laser ablation--typically, the excitation energy of the pulse is much larger than a typical C-C bond energy. The behavior of the kinetic energy correlation function serves to characterize the rate of energy flow out of the excited atoms to the rest of the chain.

A more extensive study of this problem, including estimates of the actual energy transfer rates which correspond to optical (i.e., high frequency) to acoustic in collective mode energy transfer would be useful. The kinetic energy distribution of ablated atoms is also of interest. This study will also include further analysis of the random coiled polymers to verify the prediction, based on the random excitation of the elongated chain, that the threshold for damage is even higher in this case.

Finally, we note that our study involves vibrational excitation of the ground electronic state of PE and is therefore most directly related to IR experiments as in ref. 7. It also applies to UV photoablation if a thermal mechanisms (4a) is dominant. We plan further calculations to investigate the thermal versus photochemical processes using our model.

ACKNOWLEDGMENT

We wish to acknowledge helpful discussions with Professors B. Wunderlich and G. Pfeffer and Dr. J. McKeever during the course of this work. SKG aknowledges partial support for this research from the Petroleum Research Fund.

REFERENCES

1. M. C. Carotta, C. Farrario, G. L. Vecchio, and L. Galgani, Phys. Rev. A 17, 787 (1978); G. Benettin and A. Tenenbaum, Phys. Rev. A 28, 3020 (1983); G. Benettin, L. Galgani, and A. Giorgilli, Phys. Lett. 120A, 23 (1987).
2. For a recent review, see T. A. Znotins, D. Polins, and J. Reid, Laser Focus, p. 54, May 1987.
3. R. M. O'Connell and T. T. Saito, Opt. Eng. 22, 393 (1983).
4. (a) B. J. Garrison and R. Srinivasan, J. Appl. Phys. 57, 2909 (1985).
   (b) M. D. Kluge, J. R. Ray, and A. Rahman, J. Chem. Phys. 87, 2336 (1987).
5. D. W. Noid, G. A. Pfeffer, S. Z. D. Cheng, and B. Wunderlich, J. Phys. Chem. (submitted).
6. T. A. Weber, J. Chem. Phys. 69, 2347 (1978).
7. N. Graener and A. Lauberbau, Chem. Phys. Lett. 133, 378 (1987).

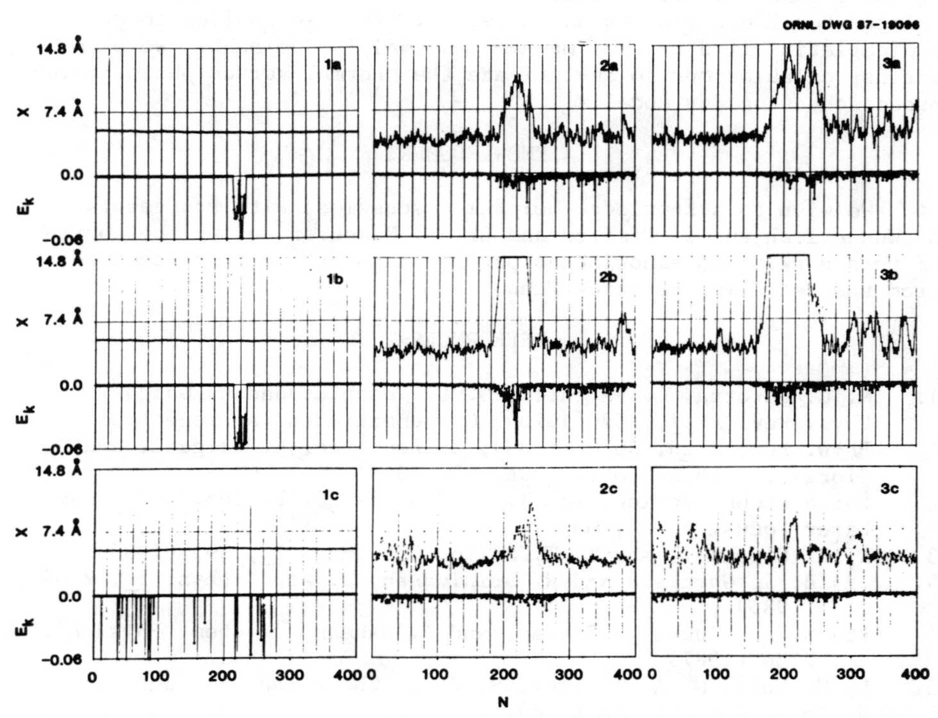

FIG. 1. Initial plot (t = 0) of Z_i vs N and E_{Ki} vs N for excitation. (a) Localized pulse of 4000° K; (b) Localized pulse of 6000° K; and (c) Delocalized pulse of 6000 °K.

FIG. 2. Propagation of initial state in Figs. 1a, 1b, and 1c to 3 ps.

FIG. 3. Further propagation of initial state in 1a, 1b, and 1c to 6 ps.

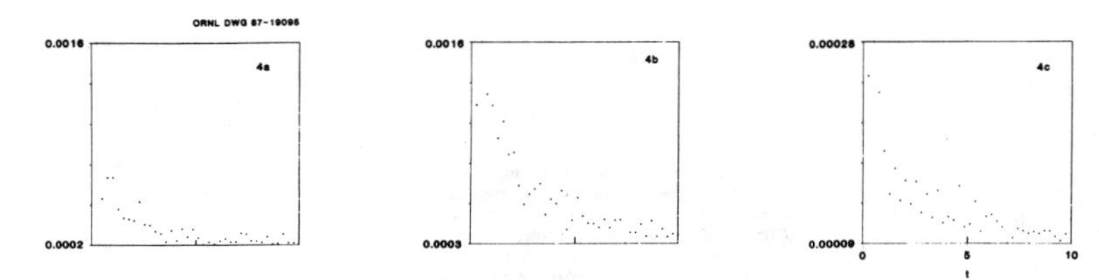

FIG. 4. Kinetic energy equilibration correlation function defined in Eq. (1) vs time for initial states in Figs. 1a, 1b, and 1c.

LASER INDUCED FLUORESCENCE STUDIES IN FLAMES VIA PREDISSOCIATING STATES: THE O_2 SCHUMANN-RUNGE SYSTEM

A. M. Wodtke and P. Andresen

Max-Planck-Institut für Strömungsforschung, D3400 Göttingen, West-Germany

L. Hüwel

Wesleyan University, Physics Department, Middletown, CT 06457

ABSTRACT

A narrowband, tunable ArF excimer laser was used to obtain dispersed laser induced predissociation fluorescence (LIPF) spectra of Schumann-Runge bands of oxygen in an atmospheric flame. Emission from individual, laser prepared rovibronic states high in the $B^3\Sigma^-$ state to vibrational levels in the $X^3\Sigma^-$ ground state as high as v" = 35 was detected from which relative emission probabilities were derived. Fluorescence excitation spectra were measured yielding direct information on spin-state selective predissociation of the B-state. Furthermore, two dimensional fluorescence field measurements of the flame were accomplished with the help of a gated image intensifier under single laser pulse conditions.

1. INTRODUCTION

In part because of it's prime importance as an atmospheric gas the spectroscopy of oxygen in general[1a] and of the Schumann-Runge system in particular[1b] have attracted wide attention over the years. The outstanding characteristic of the Schumann-Runge bands is the picosecond predissociation of the B-state into two ground state $O(^3P)$ atoms. From experimental[2] as well as theoretical[3] work the detailed mechanisms of this predissociation are thought to be well understood. In addition, high resolution absorption and emission studies have provided spectroscopic data for the B-state up to v' = 21 and for the X-state up to v" = 28. These results have been inverted to electronic potential energy curves for the two states and Franck-Condon factors for transitions involving v' = 0 - 20 and v" = 0 - 23.[4]

In the experiments reported here we used a modified commercial tunable ArF excimer laser to extract new spectroscopic information about the Schumann-Runge system. Specifically, a clearer and more detailed picture emerges for the predissociation of levels above v' = 10 with the role of $^3\Sigma^+$ and $^3\Pi$ states being emphasized. Secondly, relative emission probabilities were obtained for transitions into X-state vibrational levels within about 0.3eV of the dissociation limit. And finally, in the course of these experiments, we were able to demonstrate the feasibility of single shot, two dimensional fluorescence field measurements of O_2 under essentially quench-free conditions in an open atmosphere flame.

2. EXPERIMENTAL SET UP

A commercial tunable excimer laser operating with argon fluoride was directed into the flame of a standard butane soldering torch. A portion of the laser induced fluorescence was imaged onto the entrance slit of a 0.3m monochromator which was used with 0.2 mm slits and a 1200 lines/mm grating in 1st order yielding a 0.5 nm resolution. The dispersed fluorescence was detected by a photomultiplier and fed into a gated boxcar integrator. The signal was typically integrated over a 35 nsec wide gate centered around the laser pulse, averaged over 3 laser pulses, and then recorded. Two dimensional fluorescence images were obtained with an image intensifier equipped with 1:1 imaging optics, a UV filter to suppress scattered laser light, and a 250 nsec high voltage gate pulse applied to the photocathode so that only laser induced fluorescence but no background flame light was intensified. Further details of the experiment will be published elsewhere.

3. RESULTS AND ANALYSIS

The experimentally observed data is of two forms: laser excitation spectra, in which fluorescence is observed at a well defined narrow emission band determined by the monochromator while the laser is scanned and dispersed fluorescence spectra, where the monochromator is scanned to collect the emission of a single rovibronic level of O_2 prepared by the laser. The O_2 transitions that we observed in excitation in a flame arise from $v''=2$ and 3. At higher vibrational levels of the B-state the O_2 spin triplet becomes first partially ($v'=14$) and then completely resolved ($v'=16$) under our experimental conditions. The observed intensity patterns of these spin-triplets reflect the efficiency with which the individual spin components are being predissociated which in turn is governed by the symmetry of the predissociating state. In accordance with theory [3a] we found conclusive evidence that for $v'=14,16,17$ predissociation is predominantly mediated by a $^3\Sigma^+$ state. This is in contrast to the lower vibrational levels of the B-state whose predissociation is mainly via a $^5\Pi$ state.[3b] In contradiction to theoretical expectation we found strong evidence from the intensity ratios of partially resolved triplets in $v' = 11$ that a $^3\Pi_u$ must be responsible for at least 80% of the predissociation linewidth of this level.

Dispersed emission spectra of laser prepared individual rovibronic levels in the B-state yield sharp emission lines because the rapid predissociation prevents any significant population transfer to neighboring levels even at atmospheric pressure. After factoring out a ν^3 scaling for spontaneous emission and the detection efficiency of the experiment the intensity of these emission lines is then directly proportional to the relative emission probabilities for transitions out of the prepared level into any of the vibrational levels of the ground state. Relative emission probabilities for the $v'=14 \rightarrow v''$ transitions are listed in table 1. These numbers where obtained from dispersed emission after excitation of the P23(17) line in the $v'=14$, $v''=3$ band at 51684.4 cm^{-1}.

TABLE 1 Relative emission probabilities for $v'=14 \rightarrow v''$ transitions

v"	A(v'=14,v")	v"	A(v'=14,v")	v"	A(v'=14, v")
12	0.194	20	0.344	28	0.451
13	0.594	21	1.000	29	0.306
14	small	22	small	30	0.127
15	0.529	23	0.636	31	0.387
16	0.656	24	0.298	32	small
17	small	25	0.176	33	0.399
18	0.747	26	0.728	34	small
19	0.302	27	small	35	0.272

Two dimensional pictures of O_2 primarly in vibrational levels $v''=2$ or $v''=3$ were also obtained in these studies resulting from single laser pulse excitation. When the laser is blocked or off resonance no signal is observed. When the laser is tuned to a $v''=2/v''=3$ transition distinct regions in the flame "light up" indicating the presence of the particular state that is being excited. For example, $v''=2$ is found to be present both in the center and the wings of the flame whereas O_2 in $v''=3$ is only found in the central core.

A complete presentation of results from this investigation will be presented elsewhere.

REFERENCES

1.a) P.H. Krupenie, J. Phys. Chem. Ref. Data 1, 423 (1972)
b) K. Yoshino, D. E. Freeman, and W. H. Parkinson, J. Phys. Chem. Ref. Data 13, 207 (1984) is latest review

2.a) R. D. Hudson and S. H. Mahle, J. Geophys. Res. 77, 2902 (1972)
b) M. Ackerman and F. Biaume, J. Mol. Spectrosc. 35, 73 (1970)

3.a) P.S. Julienne, J. Mol. Spectrosc. 63, 60 (1976)
b) P.S. Julienne and M. Krauss, J. Mol. Spectrosc. 56, 270 (1975)

4. R. Harris, M. Blackledge, and J. Generosa, J. Mol. Spectrosc. 30, 506 (1969)

POLARIZATION SPECTROSCOPY IN ANALYTICAL MEASUREMENTS

Edward S. Yeung
Department of Chemistry and Ames Laboratory, Iowa State University, Ames, Iowa 50011

ABSTRACT

A new generation of instruments has been developed for using the polarization properties of light in sensitive and selective chemical measurements. The detection of rotation of the plane of polarization of light is at the micro degree level. The detection of the intensity differences between left and right circularly polarized light is at the parts-per-million level. These are orders of magnitude better than commercial instruments can provide.

INTRODUCTION

Several years ago, we reported on improved sensitivity in the measurement of rotation of the plane of polarization of light in a laser-based polarimeter[1]. This is possible because the high degree of collimation of the laser beam allows excellent rejection of stray light, and the small size of the laser beam allows one to select local regions with high extinction ratios in standard polarizers. We have already reported on applications from the determination of the various forms of cholesterol in human serum to the study of protein conformations. These biological applications are obvious since optical activity is often associated with biological activity. In this article, we report the adaptation of this technology to the detection of polarization in atoms and to Raman-induced Kerr effect spectroscopy (RIKES).

One of the main drawbacks in using lasers in analytical spectroscopy is that lasers are usually much less stable in intensity than conventional light sources. However, high frequency modulation techniques have recently become available. It is possible to stabilize intensities to the sub-parts-per-million range with the appropriate double-beam optics. As a result, circular dichroism and fluorescence-detected circular dichroism can be monitored at extremely low concentrations.

INSTRUMENTATION

Polarization Spectroscopy. The principles of this have been discussed earlier[2]. One basically saturates one of the Zeeman subcomponents of an atomic transition with, e.g. a left circularly polarized pump beam. This causes a rotation in the plane of polarization of a probe beam monitored through crossed polarizers. The optical arrangment in Figure 1 allows measurement in a conventional air/acetylene flame, into which the liquid sample is introduced. The dye laser allows tuning around the atomic resonance to confirm that a true polarization signal is obtained. The counter-propagating beams provide Doppler-free spectra. However, Doppler broadening only accounts for half the atomic linewidth here because of atmospheric operation of the system. The key is the polarization modulation on the pump beam (between left and right circular polarization). This cleanly sorts out the signal from the highly luminous and highly scattering flame source.

Raman Induced Kerr Effect. The original experiments[3] were performed in an optical arrangement similar to that for polarization spectroscopy, except that pulsed lasers were used to enhance the signal. Essentially, one can invoke the differences in Raman cross-sections for same sense versus opposite sense circular polarizations, so that a circular pump beam causes rotation in the plane of polarization of a probe beam at a Raman resonance. In liquids, the solvent gives a background signal from its Raman scattering, even far away from resonance. This is because the solvent is present at much higher concentrations compared to other species of interest in the sample. In our studies, we use a double-beam, double-cell arrangement shown in Figure 2. A pulsed pump beam and a cw probe beam is used. The pump beam enters the two cells at orthogonal linear polarizations, but both at 45° relative to the probe beam polarization. RIKES is still observed because one can invoke the differences in parallel versus perpendicular polarizations for Raman transitions. The probe beam then becomes elliptically polarized, but still passes through the analyzer (crossed polarizer) as a signal. If the two pump beams are identical in every way except for polarization, there will be a cancellation effect in the two cells when identical solutions are present. The extent of depolarization in the first cell will be exactly compensated by an opposite amount of depolarization in the second cell. This way, any signal (resonant or otherwise) from the solvent can be suppressed at the detector. When the species of interest passes through one of the cells, its RIKES signal will be recorded free from the effects of the solvent. Since in many applications in trace detection, the signal from the solvent is limiting, there is the potential to achieve lower detection limits.

Circular Dichroism. It has been known[4] that noise in cw lasers often follows a 1/f dependence. By operating at 1 MHz, one can suppress intensity noise to the one part-per-million level. This is, however, very difficult to implement in analytical spectroscopy where amplitude modulation is involved. For example, in absorption measurements, one needs to split the beam into a reference and a sample path. The two paths unfortunately are very difficult to match in every respect. In fact, the presence of a reference beam may cause deterioration in signal-to-noise if improper matching exists. A fortuitous situation exists in circular dichroism measurements. There, the difference in absorption between left and right circularly polarized light is monitored. The effect arises from optical activity of molecules, which in turn influences the absorption chromophore. It has found use in many biological applications. The two circular polarizations can simply be encoded in the same beam by polarization modulation. Since the light path is not changed, modulation and demodulation can lead to stabilities of one part per million. The instrumentation to achieve this is shown in Figure 3. Pockels cell modulation and conversion to circular polarization by a rhomb are standard techniques. The light beam simply passes through a flow cell designed to allow liquid samples to be introduced, e.g. in liquid chromatography.

Fluorescence Detected Circular Dichroism. If fluorescence can enhance detection of absorption, one can likewise enhance CD detection by monitoring fluorescence rather than the transmitted intensity.[5]. The optical arrangement to do this is shown in Figure 4. The only difference in the modulation scheme is the elimination of the rhomb, so that the production of circularly polarized light is directly from the Pockels cell. The geometry for fluorescence is also conventional.

RESULTS AND DISCUSSION

Polarization Spectroscopy. First we need to confirm that a true polarization signal is observed. As we rotate the analyzer away from null, the lineshape for Na (589 nm) changes from a Lorentzian shape to a derivative shape. This is due to interference with the local field supplied by the probe beam component that is not nulled.[1] The signal should depend linearly on the pump power, since this depletes the atomic level involved. The signal should depend linearly on the probe power, since this is the observed intensity. Experimentally, it was found that the signal depends on the square of the laser power which supplies both the pump and the probe beam, as expected. The signal is proportional to the concentration of the atomic species aspirated into the flame, making it analytically easy to apply. Finally, the signal depends on the modulation voltage on the Pockels cell. This changes the degree of modulation and thus the magnitude of the anisotropy introduced. So, clearly we are observing the correct phenomenon.

The whole purpose of the study is to see how much the sensitivity of the technique can be improved because of the better detection of polarization rotation. So, a series of Na solutions were aspirated into the flame and the signal-to-noise ratio was measured in each case. From these we determined that the detection limit is 0.03 ppb of Na in the original solution. This is comparable to laser-excited atomic fluorescence in flames, which is one of the most sensitive elemental methods, and much better than atomic absorption spectroscopy. While laser-excited atomic fluorescence will ultimately provide the best detection limits (e.g. single atom detection has been reported), there are good reasons why polarization spectroscopy is important. In highly luminous or highly scattering environments, laser-excited atomic fluorescence will suffer serious interferences. On the other hand, polarization spectroscopy has good background rejection features (collimated signal beam versus isotropic stray light), and should be able to preserve its good detection powers even in these unfavorable environments. Applications in remote sensing and in-situ measurements, e.g. in combustion chambers, are therefore likely.

Raman Induced Kerr Effect. To confirm that we are observing the selected phenomenon, we studied the signal for various experimental parameters. The signal depends linearly on the probe laser intensity. This is identical to the case of polarization spectroscopy. The signal is dependent on the square of the pump laser power. The difference is that a direct intensity measurement was used and not lock-in detection. The transmitted intensity is dependent on $\sin^2\alpha$, where α is the angle rotated off null. α is proportional to the pump power, so a quadratic signal dependence is expected. The signal was found to be quadratically dependent on concentration (acetonitrite dissolved in water). Again, α is proportional to number density, causing the quadratic dependence. We found that the detection limit in this study is roughly an order of magnitude better than previous results, when laser power and Raman cross-sections are factored in.

A goal of this study was to provide background correction for the solvent which is present at substantially higher concentrations. This is demonstrated by successively blocking the pump beams to one cell, then the other, and comparing these with the case for both beams present. When both pump beams are present and only pure solvent is present in the two cells, we find that the transmitted intensity falls to 20% of the level when either beam alone is used. This shows cancellation of the background in one cell by an opposite effect in the other cell. The reason cancellation is not 100%, as it should be, is due to the poor beam quality in the pump beam. Local hot spots made background rotation non-uniform over the probe beam cross-section. This cannot be compensated with the arrangement in Figure 2 since the propagation directions are not matched. The low repetition rate of the pump laser also makes signal averaging difficult. So, there is good probability that further work in this area can bring even higher levels of suppression and even better detection powers.

Circular Dichroism. The best way to confirm the observed phenomenon is to successively look at an absorbing but optically inactive compound and then one with a known CD effect. This is best accomplished in a liquid chromatogram. A mixture of three cobalt amine complexes were used in the sample, only one of which is optically active. Using standard absorption detection, it can be shown that the three are well separated in time. In the CD mode, only the optically active component showed a large peak, as expected. It turns out that there is the possibility of a false signal when the intensities of the left and right circular components are not identical. An absorbing species will cause a difference in attenuation, i.e. a false signal. By proper tuning of the modulator, this artifact can be eliminated. This in fact brings to question some of the previously published CD spectra, that there may be present artifacts due to the instrument.

The level of detection is improved to about 5 x 10^{-7} absorbance units. This is a factor of 110 better than the standard commercial instrument. It turns out one can also perform these measurements at small volumes, e.g. 2.6 µL in a 1-cm path. This is compatible with microbore chromatography. Noise is degraded somewhat due to focusing and the associated aperture effect in the small volume, but a detection limit of 5.6 ng is still obtained.

Fluorescence Detected Circular Dichroism. The optical arrangement for FDCD detection is shown in Figure 4. As a test case, riboflavin, a known CD active and fluorescent compound, was chosen. Using standard column chromatography, as little as 170 pg of riboflavin can be detected. This is an improvement over the transmission CD technique discussed above. However, because the solvent is normally non-fluorescing, there is no convenient signal to lock onto in setting up the instrument. So, it is much more difficult to balance the intensities of the left versus right circularly polarized light. False signals are more common. It may be possible in the future to include a small amount of fluorescing material in the solvent just to achieve a good null.

The FDCD results here confirm the increased sensitivity when fluorescence is used. This suggests that miniaturization is advantageous and feasible. For example, if open-tubular capillary columns are used, the volumes involved (sample peak) will be of the order of 10 nL rather than the 100 µL here. The high signal levels of fluorescence should make shot noise still negligible. The concentrations that can be detected should thus be the same. So, detection of 0.01 pg of material is foreseen. This fits nicely with requirements of molecular biology, where small sample sizes and low concentrations are common.

ACKNOWLEDGEMENTS

The author thanks J. C. Kuo, W. G. Tong, D. R. Bobbitt and R. E. Syrovec for parts of this work. Ames Laboratory is operated for the U. S. Department of Energy by Iowa State University under contract No. W-7405-Eng-82. This work was supported by the Director for Energy Research, Office of Basic Energy Sciences.

REFERENCES

1. E. S. Yeung, L. E. Stenhoek, S. D. Woodruff and J. C. Kuo, Anal. Chem. 52, 1399 (1980).
2. C. Wieman and T. W. Hansch, Phys. Rev. Lett. 36, 1170 (1976).
3. D. Heiman, R. W. Hellworth, M. D. Levenson and G. Martin, Phys. Rev. Lett. 36, 189 (1976).
4. M. Ducloy and J. J. Snyder, Proc. SPIE 426, 87 (1983).
5. D. H. Turner, I. Tinoco, Jr. and M. F. Maestre, Biochem. 14, 3794 (1975).

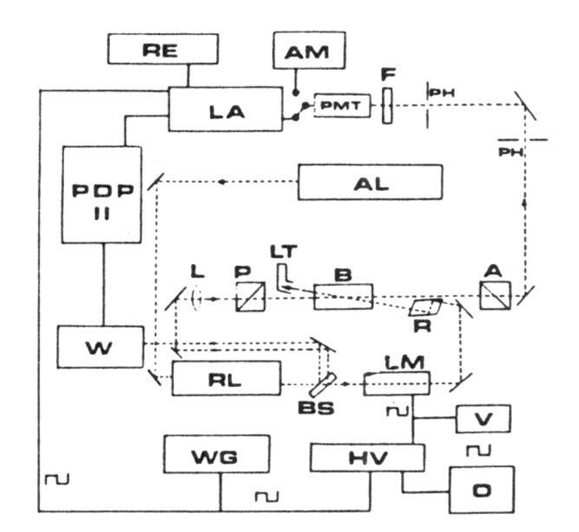

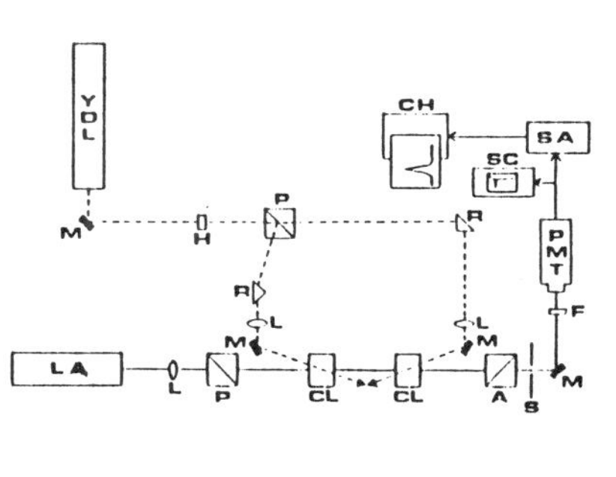

Figure 1. Experimental arrangement for polarization spectroscopy. A, analyzer; AL, argon ion laser; AM, nanoammeter; B, slot burner; BS, beam splitter; F, filter; HV, high-voltage op amp; L, lens; LA, lock-in amplifier; LM, light modulator; LT, light trap; O, oscilloscope; P, polarizer; PDP 11, computer; PH, pin-hole aperture; PMT, photomultiplier tube; R, Fresnel rhomb; RE, chart recorder; RL, ring laser; V, voltmeter; W, wavemeter; WG, waveform generator.

Figure 2. Experimental arrangement for RIKE. YDL, YAG dye laser; M, mirror; H, half-wave plate; LA, probe laser; P, polarizer; A, analyzer; R, right angle prism; L, lens; CL, detection cell; S, spatial filter; F, filter; PMT, photomultiplier tube; SA, boxcar signal averager; SC, fast oscilloscope; and CH, chart recorder.

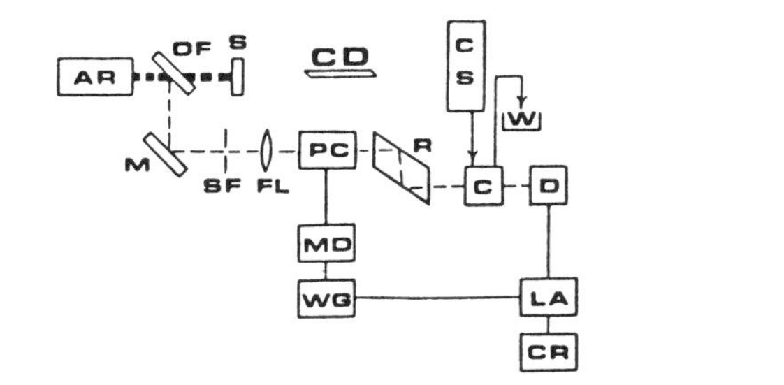

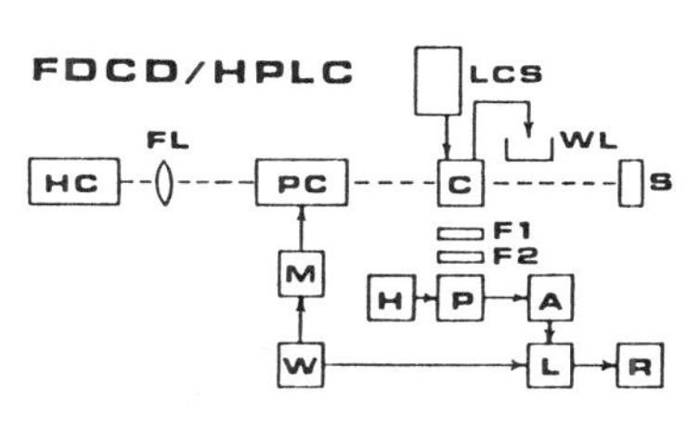

Figure 3. CD experimental configuration for HPLC; AR, argon ion (488 nm) cw laser; OF, optical flat; S, beam stop; M, mirror; SF, spatial filter; FL, 33-cm focal-length lens; PC, Pockels cell; MD, modulation driver; WG, wave form generator; R, rhomb prism; C, detection cell; D, photodetector; LA, lock-in amplifier; CR, chart recorder; CS, chromatography system; W, waste.

Figure 4. FDCD-HPLC system. HC = HeCd laser, 8 mW; FL = 50-cm focal length quartz lens; PC = Pockels cell; M = modulation driver; W = waveform generator; LCS = liquid chromatography system; WL = waste liquid; C = detection cell; S = beam stop; F1 = 4-65 Corning filter; F2 = 0-52 Corning filter; P = photomultiplier tube; H = high voltage power supply; A = AC amplifier; L = lock-in amplifier; R = chart recorder.

the beam axis (squeezing the jet from both sides) causes further narrowing of the vertical concentration profile and and displacement in the (-x)-direction

Horizontal profiles were measured by moving the probe beam in the y-direction. They are flat-topped, and their width appears to correspond with the projection of the desorption spot width onto the observation plane. Concentration profiles with Ar as source gas are about twice as broad as for He and asymmetrical. We thus attribute the narrow vertical profiles in He to a gas-dynamic separation effect caused by the large He/perylene mass ratio (4/252).

We estimated that with our particular arrangement, which is not optimized, about 1.0 % of the entrained vapor can be extracted through a 4 x 1 mm skimmer in the form of a 1 x 4 x 9 mm^3 packet. We therefore conclude that the entrainment technique should have useful sensitivity for identifying molecules adsorbed on surfaces by resonant multiphoton ionization or fluorescence spectroscopy, in cases where the adsorbate lends itself to these methods of detection.

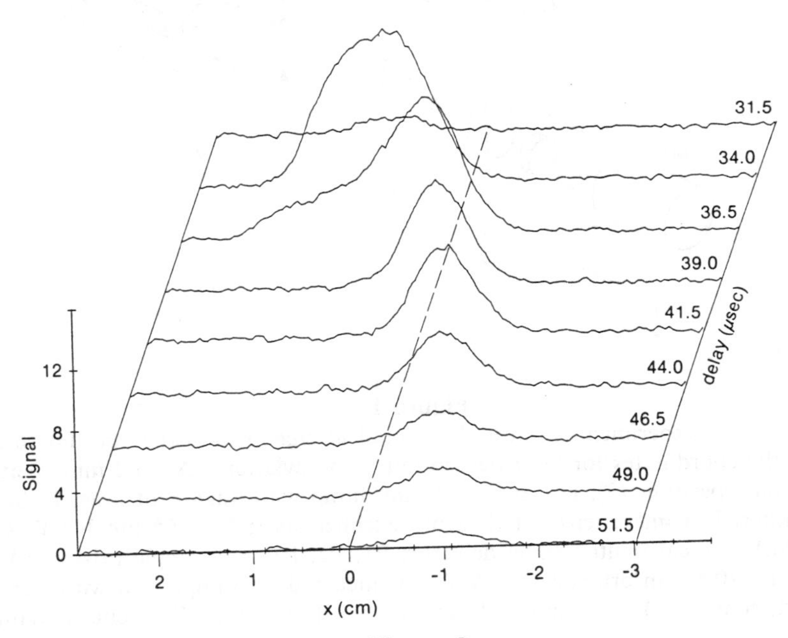

Figure 2

Time evolution of the vertical concentration profile for the inner spot at high desorption pulse energy. x_s = - 0.5 mm, z_s = 0.7 mm.

References

1. K.Domen and T.J.Chuang, Phys.Rev.Lett. 59(13), 1484 (1987).

DIRECT MEASUREMENT OF HCl-AEROSOL STICKING COEFFICIENT

N. A. Abul-Haj, L. R. Martin, and D. M. Brenner
The Aerospace Corporation, Los Angeles, California 90009

ABSTRACT

A technique based on laser induced fluorescence has been developed to measure mass transport of gases into aerosols. A stream of monodisperse, aqueous aerosol (0.07 mm dia.), seeded with quinine, is introduced into an atmosphere containing HCl gas. For a well defined exposure time to the HCl gas, the dependence of the quinine fluorescence spectrum on HCl partial pressure is recorded for individual droplets. The fluorescence spectrum directly determines the extent of titration of the quinine and therefore the mass transport of HCl into the aerosol. This allows the sticking coefficient of HCl on water to be calculated subject to the validity of a simple mass transport model. Preliminary measurements suggest that the sticking coefficient may be less than 0.1.

INTRODUCTION

The objective of this work is to measure the rate at which a soluble gas is partitioned between the gas and liquid phases of an aerosol. Macroscopically, this measurement is the mass transport rate. However, on the molecular scale the fundamental parameter of interest is the sticking coefficient of a gas on a liquid, defined as the fraction of collisions between HCl gas molecules and the liquid surface which result in the HCl molecules being absorbed into the liquid.

Sticking coefficients such as HCl on water are important to theoretical models of acid rain and atmospheric washout. For lack of an experimentally determined value atmospheric scientists have assumed that the value of this sticking coefficient is 1.0. An important application of this modeling for HCl-aerosol interactions is the launch of solid rocket boosters. A major by-product of the combustion of solid rocket propellant is HCl gas. This interacts with ambient and man-made water aerosols in the vicinity of the exhaust duct to produce a toxic environment around and downwind of the launch site.

EXPERIMENTAL

The experiment measures the amount of dissolved HCl in the droplets after a known exposure time to the HCl atmosphere. This is accomplished by seeding the droplets with an acid-base fluorescent indicator (in this case, quinine hydrochloride) and exciting the indicator with a laser. The amount of HCl dissolved in the droplets at the time of laser interrogation is determined from the fluorescence spectrum. The droplets are typically 0.07 mm in diameter and are exposed for about 0.4 ms to the HCl gas at concentrations varying from 0 ppm to 600 ppm at total pressures between 0.1 and 1.0 atmosphere. The experimental runs presented here were done at room temperature and with zero percent relative humidity in the bulk gas surrounding the droplets.

The experimental measurements are fit to a simple mass transport model which sets the rate of uptake of HCl by the droplets proportional to the frequency of collisions between the gas molecules and a droplet. With the droplet diameter, droplet exposure time, HCl gas concentration, and with the droplet's fluorescence spectrum to determine the amount of dissolved HCl, the sticking coefficient of HCl on water can be estimated.

RESULTS AND DISCUSSION

Figure 1 shows how the amount of dissolved HCl, determined as described above, varies with the partial pressure of HCl gas. Each point on the plot represents an experiment performed at atmospheric pressure and room temperature with a different mole fraction of HCl gas in helium. For low partial pressures a straight line can be drawn approximately through the first six points. This straight line indicates an observed sticking coefficient of 0.02 computed on the basis of the simple model described above. (That the points at higher partial pressures fall away from the line suggests either aqueous phase diffusion limitation or saturation of the indicator near the droplet surface.)

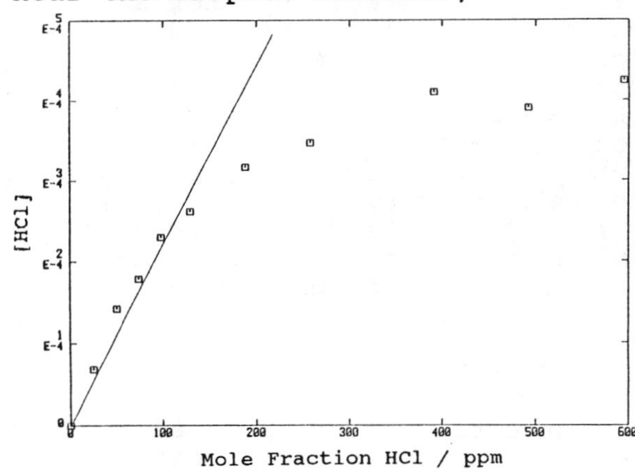

Figure 1. The concentration of dissolved HCl in the droplets versus the mole fraction of HCl in the gas phase. These experiments were performed at constant atmospheric pressure by varying the mixing ratio of HCl and He gases.

If the total pressure in the region around the droplets were reduced, thereby increasing the diffusivity of HCl gas molecules, and if gas phase diffusion of HCl were not the rate limiting step in mass transport, then there should be no change in the observed sticking coefficient. Figure 2 shows the results of such an experiment; the amount of dissolved HCl is plotted against the total pressure around the droplets for a fixed mole fraction of HCl gas. The higher pressure data points do not all fall on the same straight line, implying that the observed sticking coefficient at higher pressures is not constant. This indicates that gas phase diffusion limits the rate of mass transport of HCl to the interior of the droplets. At lower pressures the points do fit approximately on a line, the slope of which suggests a sticking coefficient of 0.03. Thus, 0.03 represents a lower limit to the sticking coefficient of HCl on water.

Another limitation to this measurement, which may lead to an underestimation of the sticking coefficient, arises from the lack of water vapor in the bulk gas around the droplets. The low humidity causes the droplets to evaporate, thus shielding them from incoming HCl and decreasing the local diffusivity of the HCl molecules. Experiments are in progress to resolve this issue.

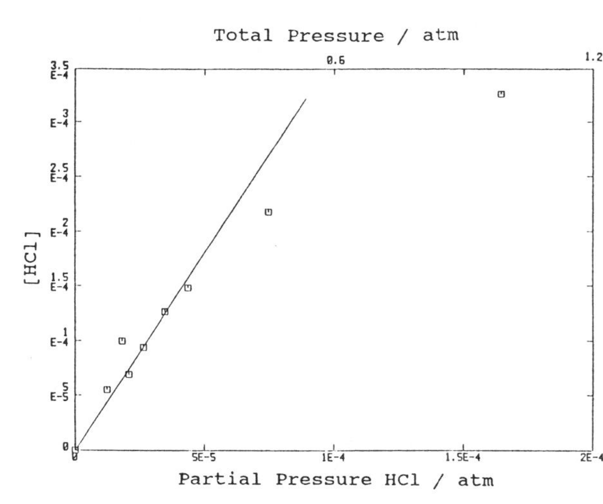

Figure 2. The concentration of dissolved HCl in the droplets versus the partial pressure of HCl gas around the droplets.These data were collected by varying the total pressure (shown on the top axis) for constant mixing ratio of HCl and He. The point at the origin was actually taken at 80 Torr of He with no HCl.

ACKNOWLEDGEMENTS

Support from the U. S. Air Force under contract F-04701-85-C-0086P000-19 is acknowledged. This project is supported jointly by Aerospace Mission Oriented Investigation and Engineering, Space Division Bioastronautics Office and Tyndall Air Force Base.

VERY HIGH RESOLUTION SATURATION SPECTROSCOPY OF LUTETIUM ISOTOPES VIA CW SINGLE-FREQUENCY LASER RESONANCE IONIZATION MASS SPECTROMETRY

B. L. Fearey,[1] D. C. Parent,[2†] R. A. Keller,[2] and C. M. Miller[1]

[1]Isotope and Nuclear Chemistry Group INC-7
[2]Chemical and Laser Sciences Group CLS-2
Los Alamos National Laboratory, Los Alamos, NM 87545

ABSTRACT

In this paper, we discuss the use of Resonance Ionization Mass Spectrometry (RIMS) to perform isotopically selective saturation spectroscopy of lutetium isotopes. Utilizing this technique, it is shown that accurate measurements of the relative frequencies of hyperfine (HF) components for different isotopes easily can be made without the need for an isotopically enriched sample. The precision with which the HF splitting constants can be determined is estimated to be ~5 times greater than in previous work.

INTRODUCTION

Considerable interest in the technique of RIMS has developed over recent years due to: 1) the selective manner in which isobaric interferences are discriminated against,[1,2] and 2) the large dynamic range available for measuring isotopic ratios.[3] Earlier work[3] has demonstrated the ability to measure lutetium isotopic ratios down to the 0.4 ppm level on very small samples (~60 ng). In this previous work, the isotope ratio dynamic range was basically limited by low signal. Recently however, we have demonstrated that using a second, high-power, non-resonant laser for the ionization step can dramatically increase ionization efficiency.[4,5] Here, the mass spectrometer resolution ultimately limits the dynamic range. In order to significantly increase the dynamic range even further, it may become necessary to perform isotopically selective resonance ionization. In the present work, Doppler-free saturation spectroscopy via RIMS is utilized for this goal, i.e., to obtain precise determination of the various HF components for the lutetium isotopes.

EXPERIMENTAL SECTION

Briefly, an ultraviolet Ar^+ laser pumped ring dye laser operating with Stilbene 3 (~65 mW) was tuned to the one-photon transition of lutetium (5d6s6p $^2D^0{}_{3/2} \leftarrow$ 5d6s^2 $^2D_{3/2}$) at ~22125 cm^{-1} (~452 nm). To increase the ionization efficiency,[4,5] a second Ar^+ laser tuned to the 457.9 nm line ionized the excited atom. Laser beams were propagated parallel to and ~2 mm above the sample filament, with typical beam diameters of ~100 μm. A spherical mirror was inserted to retroreflect the laser beams collinearly. A magnetic sector mass spectrometer, equipped with pulse counting electronics, was used for detection of the lutetium ions. Samples were prepared by depositing 1-2 μg of total lutetium onto a zone-refined rhenium filament, along with a similar amount of uranium oxide to provide a diffusive barrier. Spectra were taken at a filament temperature of ~1225° C.

RESULTS AND DISCUSSION

Saturation spectroscopy was discovered soon after the advent of the first gas laser. Bennet[6] and Lamb[7] recognized that the narrow resonances (Lamb dips) that appeared in the center of inhomogeneously broadened gain profiles interacting with counterpropagating laser beams resulted from "holes" burned into the Maxwell-Boltzmann velocity distribution.

† Present address: Naval Research Laboratory, Washington, DC 20375

This phenomenon simultaneously provides a means for determining the center frequency of a transition and for removing the inhomogeneous line-broadening.

Much saturation spectroscopy has been performed in a pump-probe scheme using a relatively high pressure static cell, with inherent difficulty in examining rare isotopes. In contrast, for the present experiments, the pump and probe are the same laser, and a low-pressure mass spectrometer is used as the detector. This removes possible pressure broadening effects and permits simple isotopically selective saturation spectroscopy of unenriched samples. Two features generally are observed: (1) dips which reflect line centers approaching their natural linewidth, and (2) crossover peaks which occur at the mean frequency of two hyperfine lines whose Doppler profiles overlap. The details of the exact processes occurring in these experiments will be discussed more fully in a forthcoming paper.[8]

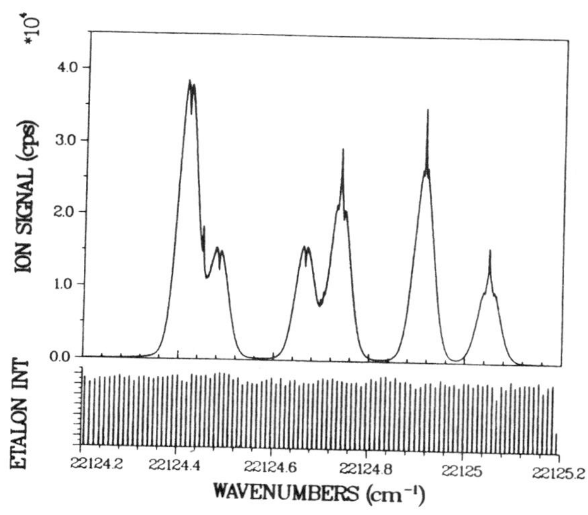

Figure 1. Saturation spectrum of (5d6s6p $^2D^0_{3/2} \leftarrow 5d6s^2\ ^2D_{3/2}$) transition of ^{175}Lu at $\sim$22125 cm^{-1}.

A typical experimental spectrum for ^{175}Lu is shown in Figure 1. The dye and Ar^+ powers were $\sim$40 W/cm^2 and $\sim$8.5 kW/cm^2, respectively. Similar ^{176}Lu spectra were also observed but are not shown. The dips and peaks correspond to the line centers and crossovers for the atomic HF components for lutetium (see reference 8 for details). Excellent signal-to-noise (S/N) was observed for both ^{175}Lu and ^{176}Lu spectra. Figure 2 is an unsmoothed, expanded view of the first two bands in the ^{175}Lu spectrum and illustrates the S/N and resolution attained using this technique.

Included in the spectra, Figures 1 and 2, are the transmission peaks of a $\sim$300 MHz confocal etalon used for frequency calibration. For increased precision, the free spectral range of the etalon was determined to better than 2 ppm using the Los Alamos National Laboratory Fourier Transform Spectrometer (LANL FTS), greatly exceeding data requirements. Because of this precision and the near natural linewidth characterizing the dips and peaks, the hyperfine splitting constants for the ^{175}Lu excited states determined from the dip spectra (see Figures 1 and 2) are expected to be $\sim$5 times more precise than previously determined.[9,10] The exact determination of the HFS constants and their precision is presently underway and will be presented in detail in reference 8.

With a simple experimental modification, i.e., the inclusion of a "vibrating" mirror, removal of the Doppler pedestal is possible. Earlier studies in a standard gas cell[11] have shown that this can be accomplished by modulating the mirror and phase detecting the ionization signal.

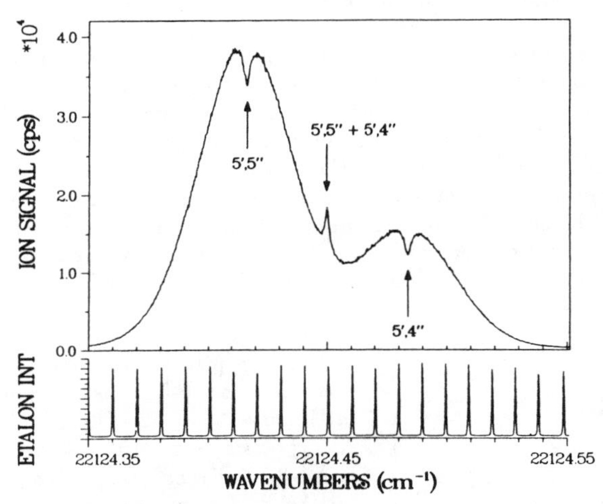

Figure 2. Expanded, unsmoothed spectrum of first two bands of ^{175}Lu shown in Figure 1. The arrows point out the dips and crossover for the indicated hyperfine transitions.

CONCLUSIONS

A new technique utilizing RIMS for obtaining very high resolution atomic spectra with isotopic selectivity was demonstrated. This technique allows the precise determination of HF splitting constants, limited only by the transition's natural linewidth. In addition, it is also feasible with this technique to accurately determine atomic isotope shifts. The exact determination of HF component line positions provides data for isotopically selective ionization which, in turn, will increase RIMS' dynamic range. Future work includes the incorporation of a "vibrating" mirror and the study of rarer isotopes, i.e., ^{174}Lu, ^{173}Lu, ^{172}Lu, ^{171}Lu, and possibly, ^{170}Lu.

LITERATURE CITED

1. C. M. Miller, N. S. Nogar, E. C. Apel, and S. W. Downey, in: Resonance Ionization Spectroscopy 1986, G. S. Hurst, and C. G. Morgan eds. (Inst. of Physics, Bristol, England), 109 (1986).
2. L. J. Moore, J. D. Fassett, and J. C. Travis, Anal. Chem., 56, 2770 (1984).
3. N. S. Nogar, S. W. Downey, and C. M. Miller, in: Resonance Ionization Spectroscopy 1984, G. S. Hurst, and M. G. Payne eds. (Inst. of Physics, Bristol, England), 91 (1984).
4. D. C. Parent, B. L. Fearey, C. M. Miller, and R. A. Keller, in: 35th ASMS Conference Proceedings, 1006 (1987).
5. D. C. Parent, B. L. Fearey, R. A. Keller, and C. M. Miller, J. Opt. Soc. Am. B, (in preparation) (1987).
6. W. R Bennet, Jr., Phys. Rev., 126, 580 (1962).
7. W. E. Lamb, Phys. Rev., 134A, 1429 (1964).
8. B. L. Fearey, D. C. Parent, R. A. Keller, and C. M. Miller, Chem. Phys. Lett., (in preparation) (1987).
9. R. Engleman, Jr., R. A. Keller, and C. M. Miller, J. Opt. Soc. Am. B, 2, 897 (1985).
10. C. M. Miller, R. Engleman, Jr., and R. A. Keller, J. Opt. Soc. Am. B, 2, 1503 (1985).
11. T. P. Duffy, D. Kammen, A. L. Schawlow, S. Svanberg, H. R. Xia, G. G. Xiao, and G.-Y. Yan, Opt. Lett., 10, 597 (1985).

SUPPORT

B.L.F. thanks Los Alamos National Laboratory for postdoctoral fellowship support during the performance of this work. This research was supported by the U.S. Department of Energy under contract W-7405-ENG-36.

Author Index

D

E

F

G

H

I

J

K

L

M

N

O

P

Q

R

S

T

AIP Conference Proceedings

		L.C. Number	ISBN
No. 1	Feedback and Dynamic Control of Plasmas – 1970	70-141596	0-88318-100-2
No. 2	Particles and Fields – 1971 (Rochester)	71-184662	0-88318-101-0
No. 3	Thermal Expansion – 1971 (Corning)	72-76970	0-88318-102-9
No. 4	Superconductivity in *d*- and *f*-Band Metals (Rochester, 1971)	74-18879	0-88318-103-7
No. 5	Magnetism and Magnetic Materials – 1971 (2 parts) (Chicago)	59-2468	0-88318-104-5
No. 6	Particle Physics (Irvine, 1971)	72-81239	0-88318-105-3
No. 7	Exploring the History of Nuclear Physics – 1972	72-81883	0-88318-106-1
No. 8	Experimental Meson Spectroscopy –1972	72-88226	0-88318-107-X
No. 9	Cyclotrons – 1972 (Vancouver)	72-92798	0-88318-108-8
No. 10	Magnetism and Magnetic Materials – 1972	72-623469	0-88318-109-6
No. 11	Transport Phenomena – 1973 (Brown University Conference)	73-80682	0-88318-110-X
No. 12	Experiments on High Energy Particle Collisions – 1973 (Vanderbilt Conference)	73-81705	0-88318-111–8
No. 13	π-π Scattering – 1973 (Tallahassee Conference)	73-81704	0-88318-112-6
No. 14	Particles and Fields – 1973 (APS/DPF Berkeley)	73-91923	0-88318-113-4
No. 15	High Energy Collisions – 1973 (Stony Brook)	73-92324	0-88318-114-2
No. 16	Causality and Physical Theories (Wayne State University, 1973)	73-93420	0-88318-115-0
No. 17	Thermal Expansion – 1973 (Lake of the Ozarks)	73-94415	0-88318-116-9
No. 18	Magnetism and Magnetic Materials – 1973 (2 parts) (Boston)	59-2468	0-88318-117-7
No. 19	Physics and the Energy Problem – 1974 (APS Chicago)	73-94416	0-88318-118-5
No. 20	Tetrahedrally Bonded Amorphous Semiconductors (Yorktown Heights, 1974)	74-80145	0-88318-119-3
No. 21	Experimental Meson Spectroscopy – 1974 (Boston)	74-82628	0-88318-120-7
No. 22	Neutrinos – 1974 (Philadelphia)	74-82413	0-88318-121-5
No. 23	Particles and Fields – 1974 (APS/DPF Williamsburg)	74-27575	0-88318-122-3
No. 24	Magnetism and Magnetic Materials – 1974 (20th Annual Conference, San Francisco)	75-2647	0-88318-123-1
No. 25	Efficient Use of Energy (The APS Studies on the Technical Aspects of the More Efficient Use of Energy)	75-18227	0-88318-124-X

No. 26	High-Energy Physics and Nuclear Structure – 1975 (Santa Fe and Los Alamos)	75-26411	0-88318-125-8
No. 27	Topics in Statistical Mechanics and Biophysics: A Memorial to Julius L. Jackson (Wayne State University, 1975)	75-36309	0-88318-126-6
No. 28	Physics and Our World: A Symposium in Honor of Victor F. Weisskopf (M.I.T., 1974)	76-7207	0-88318-127-4
No. 29	Magnetism and Magnetic Materials – 1975 (21st Annual Conference, Philadelphia)	76-10931	0-88318-128-2
No. 30	Particle Searches and Discoveries – 1976 (Vanderbilt Conference)	76-19949	0-88318-129-0
No. 31	Structure and Excitations of Amorphous Solids (Williamsburg, VA, 1976)	76-22279	0-88318-130-4
No. 32	Materials Technology – 1976 (APS New York Meeting)	76-27967	0-88318-131-2
No. 33	Meson-Nuclear Physics – 1976 (Carnegie-Mellon Conference)	76-26811	0-88318-132-0
No. 34	Magnetism and Magnetic Materials – 1976 (Joint MMM-Intermag Conference, Pittsburgh)	76-47106	0-88318-133-9
No. 35	High Energy Physics with Polarized Beams and Targets (Argonne, 1976)	76-50181	0-88318-134-7
No. 36	Momentum Wave Functions – 1976 (Indiana University)	77-82145	0-88318-135-5
No. 37	Weak Interaction Physics – 1977 (Indiana University)	77-83344	0-88318-136-3
No. 38	Workshop on New Directions in Mossbauer Spectroscopy (Argonne, 1977)	77-90635	0-88318-137-1
No. 39	Physics Careers, Employment and Education (Penn State, 1977)	77-94053	0-88318-138-X
No. 40	Electrical Transport and Optical Properties of Inhomogeneous Media (Ohio State University, 1977)	78-54319	0-88318-139-8
No. 41	Nucleon-Nucleon Interactions – 1977 (Vancouver)	78-54249	0-88318-140-1
No. 42	Higher Energy Polarized Proton Beams (Ann Arbor, 1977)	78-55682	0-88318-141-X
No. 43	Particles and Fields – 1977 (APS/DPF, Argonne)	78-55683	0-88318-142-8
No. 44	Future Trends in Superconductive Electronics (Charlottesville, 1978)	77-9240	0-88318-143-6
No. 45	New Results in High Energy Physics – 1978 (Vanderbilt Conference)	78-67196	0-88318-144-4
No. 46	Topics in Nonlinear Dynamics (La Jolla Institute)	78-57870	0-88318-145-2
No. 47	Clustering Aspects of Nuclear Structure and Nuclear Reactions (Winnepeg, 1978)	78-64942	0-88318-146-0
No. 48	Current Trends in the Theory of Fields (Tallahassee, 1978)	78-72948	0-88318-147-9

No. 49	Cosmic Rays and Particle Physics – 1978 (Bartol Conference)	79-50489	0-88318-148-7
No. 50	Laser-Solid Interactions and Laser Processing – 1978 (Boston)	79-51564	0-88318-149-5
No. 51	High Energy Physics with Polarized Beams and Polarized Targets (Argonne, 1978)	79-64565	0-88318-150-9
No. 52	Long-Distance Neutrino Detection – 1978 (C.L. Cowan Memorial Symposium)	79-52078	0-88318-151-7
No. 53	Modulated Structures – 1979 (Kailua Kona, Hawaii)	79-53846	0-88318-152-5
No. 54	Meson-Nuclear Physics – 1979 (Houston)	79-53978	0-88318-153-3
No. 55	Quantum Chromodynamics (La Jolla, 1978)	79-54969	0-88318-154-1
No. 56	Particle Acceleration Mechanisms in Astrophysics (La Jolla, 1979)	79-55844	0-88318-155-X
No. 57	Nonlinear Dynamics and the Beam-Beam Interaction (Brookhaven, 1979)	79-57341	0-88318-156-8
No. 58	Inhomogeneous Superconductors – 1979 (Berkeley Springs, W.V.)	79-57620	0-88318-157-6
No. 59	Particles and Fields – 1979 (APS/DPF Montreal)	80-66631	0-88318-158-4
No. 60	History of the ZGS (Argonne, 1979)	80-67694	0-88318-159-2
No. 61	Aspects of the Kinetics and Dynamics of Surface Reactions (La Jolla Institute, 1979)	80-68004	0-88318-160-6
No. 62	High Energy e^+e^- Interactions (Vanderbilt, 1980)	80-53377	0-88318-161-4
No. 63	Supernovae Spectra (La Jolla, 1980)	80-70019	0-88318-162-2
No. 64	Laboratory EXAFS Facilities – 1980 (Univ. of Washington)	80-70579	0-88318-163-0
No. 65	Optics in Four Dimensions – 1980 (ICO, Ensenada)	80-70771	0-88318-164-9
No. 66	Physics in the Automotive Industry – 1980 (APS/AAPT Topical Conference)	80-70987	0-88318-165-7
No. 67	Experimental Meson Spectroscopy – 1980 (Sixth International Conference, Brookhaven)	80-71123	0-88318-166-5
No. 68	High Energy Physics – 1980 (XX International Conference, Madison)	81-65032	0-88318-167-3
No. 69	Polarization Phenomena in Nuclear Physics – 1980 (Fifth International Symposium, Santa Fe)	81-65107	0-88318-168-1
No. 70	Chemistry and Physics of Coal Utilization – 1980 (APS, Morgantown)	81-65106	0-88318-169-X
No. 71	Group Theory and its Applications in Physics – 1980 (Latin American School of Physics, Mexico City)	81-66132	0-88318-170-3
No. 72	Weak Interactions as a Probe of Unification (Virginia Polytechnic Institute – 1980)	81-67184	0-88318-171-1
No. 73	Tetrahedrally Bonded Amorphous Semiconductors (Carefree, Arizona, 1981)	81-67419	0-88318-172-X

No. 74	Perturbative Quantum Chromodynamics (Tallahassee, 1981)	81-70372	0-88318-173-8
No. 75	Low Energy X-Ray Diagnostics – 1981 (Monterey)	81-69841	0-88318-174-6
No. 76	Nonlinear Properties of Internal Waves (La Jolla Institute, 1981)	81-71062	0-88318-175-4
No. 77	Gamma Ray Transients and Related Astrophysical Phenomena (La Jolla Institute, 1981)	81-71543	0-88318-176-2
No. 78	Shock Waves in Condensed Matter – 1981 (Menlo Park)	82-70014	0-88318-177-0
No. 79	Pion Production and Absorption in Nuclei – 1981 (Indiana University Cyclotron Facility)	82-70678	0-88318-178-9
No. 80	Polarized Proton Ion Sources (Ann Arbor, 1981)	82-71025	0-88318-179-7
No. 81	Particles and Fields –1981: Testing the Standard Model (APS/DPF, Santa Cruz)	82-71156	0-88318-180-0
No. 82	Interpretation of Climate and Photochemical Models, Ozone and Temperature Measurements (La Jolla Institute, 1981)	82-71345	0-88318-181-9
No. 83	The Galactic Center (Cal. Inst. of Tech., 1982)	82-71635	0-88318-182-7
No. 84	Physics in the Steel Industry (APS/AISI, Lehigh University, 1981)	82-72033	0-88318-183-5
No. 85	Proton-Antiproton Collider Physics –1981 (Madison, Wisconsin)	82-72141	0-88318-184-3
No. 86	Momentum Wave Functions – 1982 (Adelaide, Australia)	82-72375	0-88318-185-1
No. 87	Physics of High Energy Particle Accelerators (Fermilab Summer School, 1981)	82-72421	0-88318-186-X
No. 88	Mathematical Methods in Hydrodynamics and Integrability in Dynamical Systems (La Jolla Institute, 1981)	82-72462	0-88318-187-8
No. 89	Neutron Scattering – 1981 (Argonne National Laboratory)	82-73094	0-88318-188-6
No. 90	Laser Techniques for Extreme Ultraviolt Spectroscopy (Boulder, 1982)	82-73205	0-88318-189-4
No. 91	Laser Acceleration of Particles (Los Alamos, 1982)	82-73361	0-88318-190-8
No. 92	The State of Particle Accelerators and High Energy Physics (Fermilab, 1981)	82-73861	0-88318-191-6
No. 93	Novel Results in Particle Physics (Vanderbilt, 1982)	82-73954	0-88318-192-4
No. 94	X-Ray and Atomic Inner-Shell Physics – 1982 (International Conference, U. of Oregon)	82-74075	0-88318-193-2
No. 95	High Energy Spin Physics – 1982 (Brookhaven National Laboratory)	83-70154	0-88318-194-0
No. 96	Science Underground (Los Alamos, 1982)	83-70377	0-88318-195-9

No. 97	The Interaction Between Medium Energy Nucleons in Nuclei – 1982 (Indiana University)	83-70649	0-88318-196-7
No. 98	Particles and Fields – 1982 (APS/DPF University of Maryland)	83-70807	0-88318-197-5
No. 99	Neutrino Mass and Gauge Structure of Weak Interactions (Telemark, 1982)	83-71072	0-88318-198-3
No. 100	Excimer Lasers – 1983 (OSA, Lake Tahoe, Nevada)	83-71437	0-88318-199-1
No. 101	Positron-Electron Pairs in Astrophysics (Goddard Space Flight Center, 1983)	83-71926	0-88318-200-9
No. 102	Intense Medium Energy Sources of Strangeness (UC-Sant Cruz, 1983)	83-72261	0-88318-201-7
No. 103	Quantum Fluids and Solids – 1983 (Sanibel Island, Florida)	83-72440	0-88318-202-5
No. 104	Physics, Technology and the Nuclear Arms Race (APS Baltimore –1983)	83-72533	0-88318-203-3
No. 105	Physics of High Energy Particle Accelerators (SLAC Summer School, 1982)	83-72986	0-88318-304-8
No. 106	Predictability of Fluid Motions (La Jolla Institute, 1983)	83-73641	0-88318-305-6
No. 107	Physics and Chemistry of Porous Media (Schlumberger-Doll Research, 1983)	83-73640	0-88318-306-4
No. 108	The Time Projection Chamber (TRIUMF, Vancouver, 1983)	83-83445	0-88318-307-2
No. 109	Random Walks and Their Applications in the Physical and Biological Sciences (NBS/La Jolla Institute, 1982)	84-70208	0-88318-308-0
No. 110	Hadron Substructure in Nuclear Physics (Indiana University, 1983)	84-70165	0-88318-309-9
No. 111	Production and Neutralization of Negative Ions and Beams (3rd Int'l Symposium, Brookhaven, 1983)	84-70379	0-88318-310-2
No. 112	Particles and Fields – 1983 (APS/DPF, Blacksburg, VA)	84-70378	0-88318-311-0
No. 113	Experimental Meson Spectroscopy – 1983 (Seventh International Conference, Brookhaven)	84-70910	0-88318-312-9
No. 114	Low Energy Tests of Conservation Laws in Particle Physics (Blacksburg, VA, 1983)	84-71157	0-88318-313-7
No. 115	High Energy Transients in Astrophysics (Santa Cruz, CA, 1983)	84-71205	0-88318-314-5
No. 116	Problems in Unification and Supergravity (La Jolla Institute, 1983)	84-71246	0-88318-315-3
No. 117	Polarized Proton Ion Sources (TRIUMF, Vancouver, 1983)	84-71235	0-88318-316-1

No. 118	Free Electron Generation of Extreme Ultraviolet Coherent Radiation (Brookhaven/OSA, 1983)	84-71539	0-88318-317-X
No. 119	Laser Techniques in the Extreme Ultraviolet (OSA, Boulder, Colorado, 1984)	84-72128	0-88318-318-8
No. 120	Optical Effects in Amorphous Semiconductors (Snowbird, Utah, 1984)	84-72419	0-88318-319-6
No. 121	High Energy e^+e^- Interactions (Vanderbilt, 1984)	84-72632	0-88318-320-X
No. 122	The Physics of VLSI (Xerox, Palo Alto, 1984)	84-72729	0-88318-321-8
No. 123	Intersections Between Particle and Nuclear Physics (Steamboat Springs, 1984)	84-72790	0-88318-322-6
No. 124	Neutron-Nucleus Collisions – A Probe of Nuclear Structure (Burr Oak State Park - 1984)	84-73216	0-88318-323-4
No. 125	Capture Gamma-Ray Spectroscopy and Related Topics – 1984 (Internat. Symposium, Knoxville)	84-73303	0-88318-324-2
No. 126	Solar Neutrinos and Neutrino Astronomy (Homestake, 1984)	84-63143	0-88318-325-0
No. 127	Physics of High Energy Particle Accelerators (BNL/SUNY Summer School, 1983)	85-70057	0-88318-326-9
No. 128	Nuclear Physics with Stored, Cooled Beams (McCormick's Creek State Park, Indiana, 1984)	85-71167	0-88318-327-7
No. 129	Radiofrequency Plasma Heating (Sixth Topical Conference, Callaway Gardens, GA, 1985)	85-48027	0-88318-328-5
No. 130	Laser Acceleration of Particles (Malibu, California, 1985)	85-48028	0-88318-329-3
No. 131	Workshop on Polarized ^{3}He Beams and Targets (Princeton, New Jersey, 1984)	85-48026	0-88318-330-7
No. 132	Hadron Spectroscopy–1985 (International Conference, Univ. of Maryland)	85-72537	0-88318-331-5
No. 133	Hadronic Probes and Nuclear Interactions (Arizona State University, 1985)	85-72638	0-88318-332-3
No. 134	The State of High Energy Physics (BNL/SUNY Summer School, 1983)	85-73170	0-88318-333-1
No. 135	Energy Sources: Conservation and Renewables (APS, Washington, DC, 1985)	85-73019	0-88318-334-X
No. 136	Atomic Theory Workshop on Relativistic and QED Effects in Heavy Atoms	85-73790	0-88318-335-8
No. 137	Polymer-Flow Interaction (La Jolla Institute, 1985)	85-73915	0-88318-336-6
No. 138	Frontiers in Electronic Materials and Processing (Houston, TX, 1985)	86-70108	0-88318-337-4
No. 139	High-Current, High-Brightness, and High-Duty Factor Ion Injectors (La Jolla Institute, 1985)	86-70245	0-88318-338-2

No. 140	Boron-Rich Solids (Albuquerque, NM, 1985)	86-70246	0-88318-339-0
No. 141	Gamma-Ray Bursts (Stanford, CA, 1984)	86-70761	0-88318-340-4
No. 142	Nuclear Structure at High Spin, Excitation, and Momentum Transfer (Indiana University, 1985)	86-70837	0-88318-341-2
No. 143	Mexican School of Particles and Fields (Oaxtepec, México, 1984)	86-81187	0-88318-342-0
No. 144	Magnetospheric Phenomena in Astrophysics (Los Alamos, 1984)	86-71149	0-88318-343-9
No. 145	Polarized Beams at SSC & Polarized Antiprotons (Ann Arbor, MI & Bodega Bay, CA, 1985)	86-71343	0-88318-344-7
No. 146	Advances in Laser Science–I (Dallas, TX, 1985)	86-71536	0-88318-345-5
No. 147	Short Wavelength Coherent Radiation: Generation and Applications (Monterey, CA, 1986)	86-71674	0-88318-346-3
No. 148	Space Colonization: Technology and The Liberal Arts (Geneva, NY, 1985)	86-71675	0-88318-347-1
No. 149	Physics and Chemistry of Protective Coatings (Universal City, CA, 1985)	86-72019	0-88318-348-X
No. 150	Intersections Between Particle and Nuclear Physics (Lake Louise, Canada, 1986)	86-72018	0-88318-349-8
No. 151	Neural Networks for Computing (Snowbird, UT, 1986)	86-72481	0-88318-351-X
No. 152	Heavy Ion Inertial Fusion (Washington, DC, 1986)	86-73185	0-88318-352-8
No. 153	Physics of Particle Accelerators (SLAC Summer School, 1985) (Fermilab Summer School, 1984)	87-70103	0-88318-353-6
No. 154	Physics and Chemistry of Porous Media—II (Ridge Field, CT, 1986)	83-73640	0-88318-354-4
No. 155	The Galactic Center: Proceedings of the Symposium Honoring C. H. Townes (Berkeley, CA, 1986)	86-73186	0-88318-355-2
No. 156	Advanced Accelerator Concepts (Madison, WI, 1986)	87-70635	0-88318-358-0
No. 157	Stability of Amorphous Silicon Alloy Materials and Devices (Palo Alto, CA, 1987)	87-70990	0-88318-359-9
No. 158	Production and Neutralization of Negative Ions and Beams (Brookhaven, NY, 1986)	87-71695	0-88318-358-7

No. 159	Applications of Radio-Frequency Power to Plasma: Seventh Topical Conference (Kissimmee, FL, 1987)	87-71812	0-88318-359-5
No. 160	Advances in Laser Science–II (Seattle, WA, 1986)	87-71962	0-88318-360-9
No. 161	Electron Scattering in Nuclear and Particle Science: In Commemoration of the 35th Anniversary of the Lyman-Hanson-Scott Experiment (Urbana, IL, 1986)	87-72403	0-88318-361-7
No. 162	Few-Body Systems and Multiparticle Dynamics (Crystal City, VA, 1987)	87-72594	0-88318-362-5
No. 163	Pion–Nucleus Physics: Future Directions and New Facilities at LAMPF (Los Alamos, NM, 1987)	87-72961	0-88318-363-3
No. 164	Nuclei Far from Stability: Fifth International Conference (Rosseau Lake, ON, 1987)	87-73214	0-88318-364-1
No. 165	Thin Film Processing and Characterization of High-Temperature Superconductors	87-73420	0-88318-365-X
No. 166	Photovoltaic Safety (Denver, CO, 1988)	88-42854	0-88318-366-8
No. 167	Deposition and Growth: Limits for Microelectronics (Anaheim, CA, 1987)	88-71432	0-88318-367-6
No. 168	Atomic Processes in Plasmas (Santa Fe, NM, 1987)	88-71273	0-88318-368-4
No. 169	Modern Physics in America: A Michelson-Morley Centennial Symposium (Cleveland, OH, 1987)	88-71348	0-88318-369-2
No. 170	Nuclear Spectroscopy of Astrophysical Sources (Washington, D.C., 1987)	88-71625	0-88318-370-6
No. 171	Vacuum Design of Advanced and Compact Synchrotron Light Sources (Upton, NY, 1988)	88-71824	0-88318-371-4